T0140303

Advances in Intelligent Systems and Computing

Volume 817

Series editor

Janusz Kacprzyk, Polish Academy of Sciences, Warsaw, Poland
e-mail: kacprzyk@ibspan.waw.pl

The series "Advances in Intelligent Systems and Computing" contains publications on theory, applications, and design methods of Intelligent Systems and Intelligent Computing. Virtually all disciplines such as engineering, natural sciences, computer and information science, ICT, economics, business, e-commerce, environment, healthcare, life science are covered. The list of topics spans all the areas of modern intelligent systems and computing such as: computational intelligence, soft computing including neural networks, fuzzy systems, evolutionary computing and the fusion of these paradigms, social intelligence, ambient intelligence, computational neuroscience, artificial life, virtual worlds and society, cognitive science and systems, Perception and Vision, DNA and immune based systems, self-organizing and adaptive systems, e-Learning and teaching, human-centered and human-centric computing, recommender systems, intelligent control, robotics and mechatronics including human-machine teaming, knowledge-based paradigms, learning paradigms, machine ethics, intelligent data analysis, knowledge management, intelligent agents, intelligent decision making and support, intelligent network security, trust management, interactive entertainment, Web intelligence and multimedia.

The publications within "Advances in Intelligent Systems and Computing" are primarily proceedings of important conferences, symposia and congresses. They cover significant recent developments in the field, both of a foundational and applicable character. An important characteristic feature of the series is the short publication time and world-wide distribution. This permits a rapid and broad dissemination of research results.

More information about this series at http://www.springer.com/series/11156

Jagdish Chand Bansal · Kedar Nath Das
Atulya Nagar · Kusum Deep
Akshay Kumar Ojha
Editors

Soft Computing for Problem Solving

SocProS 2017, Volume 2

 Springer

Editors
Jagdish Chand Bansal
Department of Mathematics
South Asian University
New Delhi, India

Kedar Nath Das
Department of Mathematics
National Institute of Technology Silchar
Silchar, Assam, India

Atulya Nagar
Department of Mathematics and Computer
  Science, Faculty of Science
Liverpool Hope University
Liverpool, UK

Kusum Deep
Department of Mathematics
Indian Institute of Technology Roorkee
Roorkee, Uttarakhand, India

Akshay Kumar Ojha
School of Basic Sciences
Indian Institute of Technology Bhubaneswar
Bhubaneswar, Odisha, India

ISSN 2194-5357   ISSN 2194-5365 (electronic)
Advances in Intelligent Systems and Computing
ISBN 978-981-13-1594-7   ISBN 978-981-13-1595-4 (eBook)
https://doi.org/10.1007/978-981-13-1595-4

Library of Congress Control Number: 2018947855

This Springer imprint is published by the registered company Springer Nature Singapore Pte Ltd.
The registered company address is: 152 Beach Road, #21-01/04 Gateway East, Singapore 189721, Singapore

Preface

SocProS, which stands for 'Soft Computing for Problem Solving,' is entering its seventh edition as an established and flagship international conference. This particular annual event is a joint collaboration between a group of faculty members from the institutes of repute like South Asian University, New Delhi; NIT Silchar; Liverpool Hope University, UK; IIT Roorkee; and IIT Bhubaneswar.

The first in the series of SocProS started in 2011 and was held from 20th to 22nd December on the IIT Roorkee Campus with Prof. Deep (IITR) and Prof. Nagar (Liverpool Hope University) as the general chairs. JKLU Jaipur hosted the second SocProS from December 28 to 30, 2012. Coinciding with the Golden Jubilee of the IIT Roorkee's Saharanpur Campus, the third edition of this international conference, which has by now become a brand name, took place at the Greater Noida Extension Centre of IIT Roorkee during December 26–28, 2013. Afterward, in 2014, it has been organized at NIT Silchar, Assam, during December 27–29, 2014. The next conference series was held at Saharanpur Campus of IIT Roorkee during December 18–20, 2015. In the last year, Thapar University, Patiala, has hosted the conference during December 23–24, 2016.

Like earlier SocProS conferences, the focus of SocProS 2017 is on soft computing and its applications to real-life problems arising in diverse areas of medical and health care, supply chain management, signal processing and multimedia, industrial optimization, image processing, cryptanalysis, etc. SocProS 2017 attracted a wide spectrum of thought-provoking articles. A total of 164 high-quality research papers have been selected for publication in the form of this two-volume proceeding.

We hope that the papers contained in this proceeding will prove helpful toward improving the understanding of soft computing at teaching as well as research level and will inspire more and more researchers to work in the field of soft computing.

The editors would like to express their sincere gratitude to SocProS 2017 patron, plenary speakers, invited speakers, reviewers, program committee members, international advisory committee, and local organizing committee; without whose support, the quality and standards of the conference could not be maintained. We express special thanks to Springer and its team for this valuable support in the publication of this proceeding.

Over and above, we would like to express our deepest sense of gratitude to the 'Indian Institute of Technology (IIT) Bhubaneswar' to facilitate the hosting of this conference. Our sincere thanks to all the sponsors of SocProS 2017.

SAU New Delhi, India	Jagdish Chand Bansal
NIT Silchar, India	Kedar Nath Das
LHU, Liverpool, UK	Atulya Nagar
IIT Roorkee, India	Kusum Deep
IIT Bhubaneswar, India	Akshay Kumar Ojha

About the Book

The proceedings of SocProS 2017 will serve as an academic bonanza for scientists and researchers working in the field of soft computing. This book contains theoretical as well as practical aspects using fuzzy logic, neural networks, evolutionary algorithms, swarm intelligence algorithms, etc., with many applications under the umbrella of 'soft computing.' The book will be beneficial for young as well as experienced researchers dealing across complex and intricate real-world problems for which finding a solution by traditional methods is a difficult task.

The different application areas covered in the proceedings are image processing, cryptanalysis, industrial optimization, supply chain management, newly proposed nature-inspired algorithms, signal processing, problems related to medical and health care, networking optimization problems, etc.

Contents

About the Editors

Dr. Jagdish Chand Bansal is Assistant Professor at the South Asian University, New Delhi, India, and Visiting Research Fellow at Liverpool Hope University, Liverpool, UK. He has an excellent academic record and is a leading researcher in the field of swarm intelligence, and he has published numerous research papers in respected international and national journals.

Dr. Kedar Nath Das is Assistant Professor in the Department of Mathematics, National Institute of Technology Silchar, Assam, India. Over the past 10 years, he has made substantial contributions to research on 'soft computing.' He has published several research papers in prominent national and international journals. His chief area of interest is evolutionary and bio-inspired algorithms for optimization.

Prof. Atulya Nagar holds the Foundation Chair as Professor of Mathematical Sciences and is Dean of the Faculty of Science, Liverpool Hope University, UK. He is an internationally respected scholar working at the cutting edge of theoretical computer science, applied mathematical analysis, operations research, and systems engineering.

Prof. Kusum Deep is Professor in the Department of Mathematics, Indian Institute of Technology Roorkee, India. Over the past 25 years, her research has made her a central international figure in the area of nature-inspired optimization techniques, genetic algorithms, and particle swarm optimization.

Dr. Akshay Kumar Ojha is Associate Professor at the School of Basic Sciences, Indian Institute of Technology Bhubaneswar, Odisha, India. He completed his B.Sc., M.Sc., and Ph.D. at Utkal University in 1978, 1980, and 1997, respectively. His research interest areas are geometric programming, artificial neural networks, genetic algorithms, particle swarm optimization, fractional programming, nonlinear optimization, data analysis and optimization, and portfolio optimization. He has 34 years of experience and has published over 30 journal articles and 6 books.

Satellite Horizon Effects on Temporal GPS Receiver Position Accuracy over Coastal Area of South India

G. Sasibhushana Rao, B. Lavanya and N. Ashok Kumar

Abstract The spatial distribution of Global Positioning System (GPS) satellites over a particular geographic location is one of the predominant factors that determine the receiver position accuracy referred to as Geometrical Dilution of Precision (GDOP). This parameter is mainly dependent on satellite vehicle (SV) distribution over the horizon ($3°$–$7°$) of a particular geographic location. This paper presents the analysis of temporal variations of GPS receiver position error due to GDOP. Wide variation of GDOP (values are 1.24, 1.76, 3.4, and 8.6 for 11, 9, 7, and 4 visible satellite configurations, respectively) and position errors (40–51.6 m, 76.2–110 m, and 21.8–33.3 m for x-, y-, and z-coordinates, respectively) are observed with respect to the time-varying satellite geometry due to the disappearance of SVs nearby horizon and are different in case of mid-latitude regions of the Indian subcontinent. This problem needs to be addressed while finalizing the deployment of IRNSS SVs over the Indian subcontinent.

Keywords GDOP · Horizon effects · Position accuracy · Least squares

1 Introduction

The GPS is a 31 satellite constellation and is an all-weather satellite-based navigation system [1] developed by Department of Defense (DoD) of USA. It provides a 3D position, velocity, and time by broadcasting two L-band signal frequencies (1575.24 and 1227.6 MHz), 24 hour a day, anywhere on or above the earth. Apart from GPS,

G. Sasibhushana Rao · B. Lavanya (✉) · N. Ashok Kumar
Department of Electronics and Communication Engineering, Andhra University
College of Engineering (A), Visakhapatnam 530003, Andhra Pradesh, India
e-mail: lavanyabagadi@gmail.com

G. Sasibhushana Rao
e-mail: sasigps@gmail.com

N. Ashok Kumar
e-mail: ashok0709@gmail.com

© Springer Nature Singapore Pte Ltd. 2019
J. C. Bansal et al. (eds.), *Soft Computing for Problem Solving*, Advances in Intelligent Systems and Computing 817, https://doi.org/10.1007/978-981-13-1595-4_1

1

there are three more global satellite constellations (GLONASS, Galileo, BeiDou) and few regional satellite constellations like Indian Regional Navigation Satellite System (IRNSS) which is in development stage. Though there are many satellite constellations available in the world, GPS is preferred because of fully functional and highest global coverage, i.e., 8–14 SVs visibility between ±75° Latitude and used for civil aviation and defense sector. GPS position is mainly dependent on the pseudorange measurements [2].

The position accuracy of GPS receiver is affected by geometrical orientation of satellites with respect to receiver position. This paper discusses analysis of GPS position error due to GDOP for the coastal area of Visakhapatnam, South India. GDOP depends on spatial distribution of satellite vehicles by considering the factors like elevation, azimuth angles, and number of satellite vehicles in visibility over a given geographical location. More number of satellites for position estimation will lead to better values of GDOP [3–6]. The spatial distribution of the satellites is plotted for the epochs where maximum and minimum GDOP occurred during a period of 24 hour. For calculation of GDOP factor, it is necessary to know estimated position of receiver which is determined from iterative least squares method. In Sect. 2, position error analysis using iterative least squares is discussed. GDOP calculation is discussed in Sect. 3 followed by results and discussion in Sect. 4. In Sect. 5, conclusions are given.

2   Iterative Least Squares for Position Error Analysis

The iterative least squares method is used for computation of position using pseudorange measurement. To estimate the true receiver position, as there are four unknown parameters (x^s, y^s, z^s, and $\partial\tau$), information from at least four satellites is required for pseudorange calculation as shown in Eq. (1). It is possible to receive GPS signals from more than four satellites at anytime from the surface of earth. When the receiver tracks more than four satellites, the system of equations becomes inconsistent and the method of least squares [7] can be used to find the solution.

$$p = \sqrt{(x^s - x_r)^2 + (y^s - y_r)^2 + (z^s - z_r)^2} - c\partial\tau \qquad (1)$$

where x^s, y^s, z^s are satellite coordinates,

x_r, y_r, z_r are receiver coordinates,

c = velocity of the signal ≈ 3×10^8 m/s, and

$\partial\tau$ is the receiver clock offset error in seconds.

Equation (1) is nonlinear and it can be linearized around the approximate receiver location corresponding to x_r, y_r, z_r using Taylor series expansion up to the first order. Let $b = -c\,\partial\tau$ which is GPS receiver clock error in terms of distance. The general Taylor series expansion for f(x) is given by

$$f(x) = \sum_{i=0}^{n} \frac{f^n(x_0)(x - x_0)^n}{n!} \tag{2}$$

The linearized pseudorange equation can be expressed as:

$$p(x, y, z, b) = p(x_0, y_0, z_0, b_0) + (x - x_0)\frac{\partial P}{\partial x_0} + (y - y_0)\frac{\partial P}{\partial y_0}$$

$$+ (z - z_0)\frac{\partial P}{\partial z_0} + (b - b_0)\frac{\partial P}{\partial b_0} \tag{3}$$

$$p(x, y, z, b) - p(x_0, y_0, z_0, b_0) = \frac{\partial P}{\partial x_0}\delta x + \frac{\partial P}{\partial y_0}\delta y + \frac{\partial P}{\partial z_0}\delta z + \frac{\partial P}{\partial b_0}\delta b \tag{4}$$

Pseudorange is computed for the receiver by assuming a position initially. Then, error in pseudorange δP is calculated, i.e., the difference between observed and predicted pseudoranges. Then, the design matrix A is computed.

$$\delta P = P_{observed} - P_{predicted} = \frac{\partial P}{\partial x_0}\delta x + \frac{\partial P}{\partial y_0}\delta y + \frac{\partial P}{\partial z_0}\delta z + \frac{\partial P}{\partial b_0}\delta b \tag{5}$$

For n satellites, Eq. (5) can be written in matrix form as:

$$\begin{bmatrix} \delta P^1 \\ \delta P^2 \\ \delta P^3 \\ \cdot \\ \cdot \\ \delta P^n \end{bmatrix} = \begin{bmatrix} \frac{\partial P^1}{\partial x_0} & \frac{\partial P^1}{\partial y_0} & \frac{\partial P^1}{\partial z_0} & \frac{\partial P^1}{\partial b_0} \\ \frac{\partial P^2}{\partial x_0} & \frac{\partial P^2}{\partial y_0} & \frac{\partial P^2}{\partial z_0} & \frac{\partial P^2}{\partial b_0} \\ \frac{\partial P^3}{\partial x_0} & \frac{\partial P^3}{\partial y_0} & \frac{\partial P^3}{\partial z_0} & \frac{\partial P^3}{\partial b_0} \\ \cdot & \cdot & \cdot & \cdot \\ \cdot & \cdot & \cdot & \cdot \\ \frac{\partial P^n}{\partial x_0} & \frac{\partial P^n}{\partial y_0} & \frac{\partial P^n}{\partial z_0} & \frac{\partial P^n}{\partial b_0} \end{bmatrix} \begin{bmatrix} \delta x \\ \delta y \\ \delta z \\ \delta b \end{bmatrix} \tag{6}$$

$$\begin{bmatrix} \delta P^1 \\ \delta P^2 \\ \delta P^3 \\ \cdot \\ \cdot \\ \delta P^n \end{bmatrix} = \begin{bmatrix} \frac{x_0-x^1}{P} & \frac{y_0-y^1}{P} & \frac{z_0-z^1}{P} & 1 \\ \frac{x_0-x^2}{P} & \frac{y_0-y^2}{P} & \frac{z_0-z^2}{P} & 1 \\ \frac{x_0-x^3}{P} & \frac{y_0-y^3}{P} & \frac{z_0-z^3}{P} & 1 \\ \cdot & \cdot & \cdot & \cdot \\ \cdot & \cdot & \cdot & \cdot \\ \frac{x_0-x^n}{P} & \frac{y_0-y^n}{P} & \frac{z_0-z^n}{P} & 1 \end{bmatrix} \begin{bmatrix} \delta x \\ \delta y \\ \delta z \\ \delta b \end{bmatrix} \tag{7}$$

$$\delta P = A\delta x \tag{8}$$

Equation (8) expresses a linear relationship between δP and the unknown correction to parameters δx. Thus, the linearized equation can be written as $\delta P - A\delta x = 0$.

Table 1 DOP ratings

DOP value	Ratings
<1	Ideal
1–2	Excellent
2–5	Good
5–10	Moderate
10–20	Fair
>20	Poor

Using least squares solution, $\delta x = (A^T A)^{-1} A^T \delta P$ error is found and then this error is added to initially assumed value of receiver position to update its value. The process is repeated until the error computed is within the tolerance limits.

The receiver position coordinates can be updated from the initial estimate of receiver position x_0, y_0, z_0 to x_r, y_r, z_r iteratively as

$$x_r = x_0 + \delta x \tag{9}$$

$$y_r = y_0 + \delta y \tag{10}$$

$$z_r = z_0 + \delta z \tag{11}$$

The least squares solution can be found by varying the value of x until it satisfies the condition given below

$$\sum (\delta P - A\delta x)^2 < tolerance \tag{12}$$

3 Geometrical Dilution of Precision (GDOP)

The GDOP is calculated by using Eq. (13)

$$GDOP = \sqrt{trace(A^T A)^{-1}} \tag{13}$$

where A is the designed matrix obtained from Eq. (7).

The approach is selecting a group of satellites based on their elevation angle and their disappearance in further calculation. To enhance the GDOP value, the effort has been made in each step by considering the optimal set of satellites [8] which are nearer to horizon and as they disappear in next epoch consideration. With increase in the number of satellites for GDOP calculation, the value of GDOP factor decreases which indicates that selection of all satellites for GDOP results in good accuracy.

The quality of DOP ratings is evaluated according to Table 1 [7].

Table 2 Position of satellites for maximum and minimum GDOP scenario

S. no.	Minimum GDOP = 1.243			Maximum GDOP = 2.28		
	SV PRNs	Elevation angle	Azimuth angle	SV PRNs	Elevation angle	Azimuth angle
1	14	50	234	14	42	290
2	6	23	247	6	15	228
3	18	35	170	18	35	200
4	21	60	139	21	45	172
5	3	16	246	27	27	116
6	31	7	336	9	36	103
7	27	18	136	22	25	234
8	9	30	132	29	62	2
9	22	31	203	15	29	152
10	29	11	16			
11	15	7	136			

4 Results and Discussion

In this paper, GPS receiver position and position error in three dimensions are estimated using least squares method, which is well suited for static receiver position estimation. Real-time data is collected from dual frequency GPS receiver located at Department of Electronics and Communication Engineering, Andhra University, Visakhapatnam (Latitude 17.73° N/Longitude 83.31° E), South India and receiver position error analysis is carried over a period of 24 hour for experimental analysis. The data captured by the receiver is in Receiver Independent Exchange (RINEX) format and it comprises two files: (i) navigation and (ii) observation data files. The navigation data file includes ephemeris data used to calculate satellite position. The pseudoranges are obtained from observation data file that is visible at a particular epoch of time. These two files are processed to estimate the receiver position and position error. Table 2 shows the total availability of visible satellites at only minimum and maximum GDOP values. Available number of visible satellites is 11 at minimum GDOP (06:57 AM) and 9 at maximum GDOP (05:39 AM). The Satellite Vehicle Pseudorandom Noise sequences (SVPRNs) 3 and 31 are missing for maximum GDOP, whereas remaining nine SVPRNs are the same.

From Table 2, it is seen that some of the satellites having low elevation angle are about to disappear and excluding these disappearing satellites one by one (SVPRNs 15, 31, 29, 3, 27, 6, 9, and 22); the position errors are calculated along with the GDOP.

Figures 1 and 2 show sky plots for epochs which brings minimum and maximum GDOP with SVPRNs shown and it is clear that spatial distribution of satellites is wide for minimum GDOP (at 06:57 AM), whereas they are nearby for maximum GDOP (at 05:39AM). In minimum GDOP case, four satellites are widely distributed which forms a tetrahedron with SVPRNs 3, 27, 29, and 31. For maximum GDOP, there

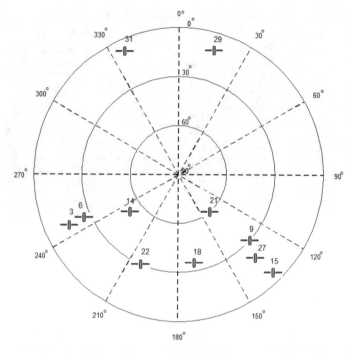

Fig. 1 Sky plot for minimum GDOP

is no satellite in first quadrant (0°–90° azimuthal) to form the tetrahedron. In order to show how the GDOP coefficient changes due to satellite distribution is shown in Figs. 1 and 2. These sky plots represent location of satellites at a particular elevation and azimuthal angles in space with respect to GPS receiver located on earth surface.

The variation of position errors in 3D with respect to number of visible satellites is shown in Fig. 3. It is seen that the position gets affected due to visible satellites at nearby horizon and the plots corresponding to x, y, and z error are plotted in Fig. 3.

Figure 4 shows that with increase in number of satellites GDOP value decreases [9, 10]. But lower the GDOP better is the accuracy. For example, GDOP value is 4.3 for selection of 6 satellites and 1.61 for selection of 10 satellites. Hence, good GDOP occurs if all satellites are included for GDOP calculation.

Figure 5 shows that increase in GDOP leads to increase in position error. For example, for a GDOP of 1.24 the position error is ($X_e = 44.3$ m, $Y_e = 85.1$ m, and $Z_e = 32.6$ m), and for a GDOP of 8.3, the position error is ($X_e = 50.9$ m, $Y_e = 86.5$ m and $Z_e = 23.8$ m).

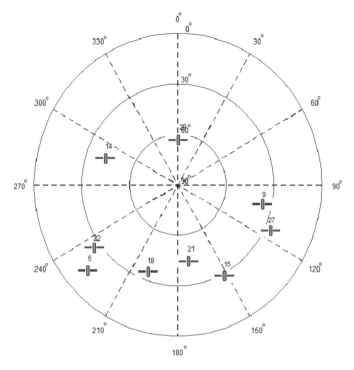

Fig. 2 Sky plot for maximum GDOP

5 Conclusions

The GPS receiver position accuracy gets affected significantly due to number of satellites visible and their spatial distribution over the horizon. It is found that the satellites which are nearby horizon and are about to disappear will have more impact on position dilution. For the GPS receiver located at Andhra University, Visakhapatnam, satellites with low elevation angle have high impact on GDOP and their disappearance lead to rise in GDOP. The GDOP value is low (1.24) when all satellites are considered for calculation and low value of GDOP lead to good accuracy. Better (lower) value of GDOP is obtained for wide distribution of satellites. Wide variations of GDOP (values are 1.24, 1.76, 3.4, and 8.6 for 11, 9, 7, and 4 visible satellite configurations, respectively) and position errors (x ranging from 40 to 51.6 m, y ranging from 76.2 to 110 m, and z ranging from 21.8 to 33.3 m) are observed with respect to the number of SVs visible. These variations are due to the disappearance of SVs at nearby horizon and are different in case of mid-latitude regions. Hence, for better accuracy of the receiver position, the deployment of GPS SVs or the proposed IRNSS SVs in such a way that at all times SVs are available nearby horizon of the Indian subcontinent can be considered.

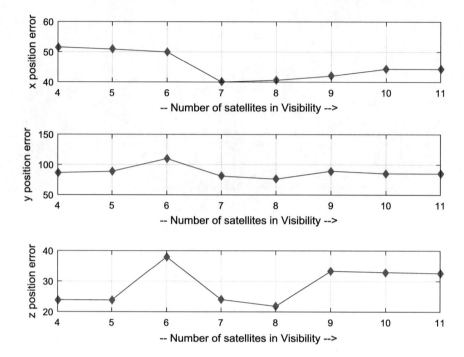

Fig. 3 Variation of position errors with number of satellites in 3D

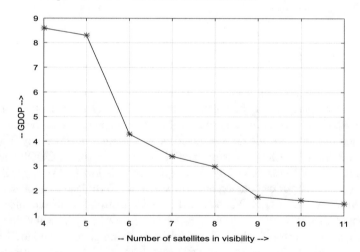

Fig. 4 Variation of GDOP with number of satellites

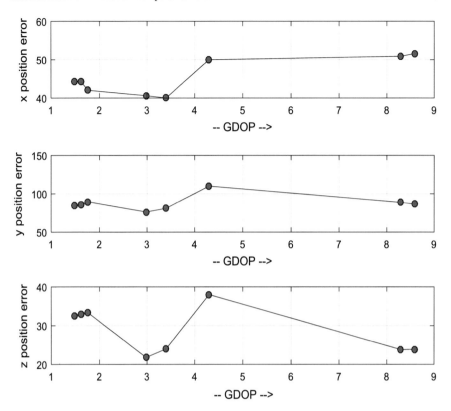

Fig. 5 Variation of position errors with GDOP

References

1. Rao, G.S.: Global Navigation Satellite Systems. McGraw Hill Education Private Limited (2010). ISBN (13): 978-0-07-070029-1
2. Swaszek, P., Hartnett, R., Seals, K.: Lower bounds on DOP. J. Navig. **70**(5), 1041–1061 (2017)
3. Blanco-Delgado, N., Nunes, F.D., Seco-Granados, G.: On the relation between GDOP and the volume described by the user-to-satellite unit vectors for GNSS positioning. GPS Solut. **21**(3), 1139–1147 (2017)
4. Nie, Z., Gao, Y., Wang, Z., Ji, S.: A new method for satellite selection with controllable weighted PDOP threshold. Surv. Rev. **49**(355), 285–293 (2017)
5. Teng, Y., Wang, J.: Some remarks on PDOP and TDOP for multi-GNSS constellations. J. Navig. **69**, 145–155 (2016)
6. Teng, Y., Wang, J.: New characteristics of geometric dilution of precision (GDOP) for multi-GNSS constellations. J. Navig. **67**(6), 1018–1028 (2014)
7. https://en.wikipedia.org/wiki/Dilution_of_precision_(navigation) (2013)
8. Sharp, I., Kegen, Y., Guo, Y.J.: GDOP analysis for positioning system design. IEEE Trans. Veh. Technol. **58**(7), 371–3382 (2009)
9. Zhang, M., Zhang, J.: A fast satellite selection algorithm: beyond four satellites. IEEE J. Sel. Top. Signal Process. **3**(5), 740–747 (2009)
10. Won, D.H., Ahn, J., et al.: Geometric sensitivity index for the GNSS using inner products of line of sight vectors. Int. J. Aeronaut. Space Sci. **16**(3), 437–444 (2015)

GA_NN: An Intelligent Classification System for Diabetes

Dilip Kumar Choubey, Sanchita Paul and Vinay Kumar Dhandhania

Abstract In this modern era, one of the prime most facilities available to this generation is state-of-the-art health care, and still diabetes has emerged as one the leading chronic disease. Diabetes is a condition which implies the glucose level is more than the inquisitive level on a managed premise. The prime motto of this study is to provide a good classification of diabetes. There are existing methods, which are for the classification of diabetes popularly datasets "Pima Indian Diabetes Dataset." Here, the proposed work comprises of four phases: In the first stage, a "Localized Diabetes Dataset" has been compiled and collected from Bombay Medical Hall, Upper Bazar Ranchi, India. In the second stage, neural networks has been used as the classification technique on localized diabetes dataset. In the third stage, GA has been used as a feature selection technique through which six features among twelve features have been obtained. Lastly in the fourth stage, neural networks have been used for classification on suitable attributes produced by GA. In this study, the results have been compared with and without GA for used classification technique. It has been concluded in this work that GA is helpful in removing not only significant attributes, deducing the cost and computation time but also enhancing the ROC and accuracy. The utilized strategy may likewise be executed in other medical issues.

Keywords Localized diabetes dataset · Genetic algorithm · RBF NN
MLP NN · Diagnosis · Feature selection · Classification

D. K. Choubey (✉) · S. Paul
CSE, BIT Mesra, Ranchi, India
e-mail: dilipchoubey_1988@yahoo.in

S. Paul
e-mail: sanchita07@gmail.com

V. K. Dhandhania
Bombay Medical Hall, Upper Bazar Ranchi, India
e-mail: vsudha72@yahoo.com

© Springer Nature Singapore Pte Ltd. 2019
J. C. Bansal et al. (eds.), *Soft Computing for Problem Solving*, Advances in Intelligent
Systems and Computing 817, https://doi.org/10.1007/978-981-13-1595-4_2

11

1 Introduction

Diabetes means glucose level is more than the curious level on a sustained basis. Diabetes happens when a body cannot deliver appropriately to insulin which is expected to keep up the rate of glucose. Diabetes can be controlled with the assistance of insulin infusions, taking oral prescriptions, a controlled eating routine, and doing consistent physical exercise yet no whole cure is yet existed.

In this manuscript, for the analysis of a diagnostic technique, genetic algorithm (GA) has been implemented for an attribute or feature selection and radial basis function neural network (RBF NN) and multilayer perceptron neural network (MLP NN) for classification. By using GA technique, suitable attributes are filtered from the pool of available attributes. The mentioned techniques have been implemented on a dataset collected from Bombay Medical Hall, Mahabir Chowk, Pyada Toli, Upper Bazar Ranchi, Jharkhand, India, which is quite precise, has no missing value, and is basically a noisy-free dataset.

There are several existing methods, which have been used for diagnosis of diabetes, and the popularly used dataset is Pima Indian Diabetes Dataset (PIDD). The PIDD was available in the (UCI Repository of Bioinformatics Databases) [1]. The same dataset has been used [2–26] for classification. Bala et al. [27, 28] have summarized different soft computing and data mining techniques for classification of thunderstorm.

The manuscript distributed is as follows: Problem specifications are given in Sect. 2, proposed methodology is introduced in Sect. 3, results and discussion of proposed methodology are discussed in Sect. 4, conclusion and future directions are present in Sect. 5.

2 Problem Specifications

As we know that previous few years back doctor's determined the disease with experience and by the help of laboratory test reports which are clinical data. The research facility test or laboratory tests' reports may fluctuate contingent upon dinners, work out, affliction, worry, because of little changes in temperature, distinctive gear utilized, and method for test taking care of. So, this type of diagnosing the disease is time-consuming because it is entirely dependent up on the availability and the experience of physician's who deal with not precise and uncertain clinical of the patient's. So, to enhance the decision making with laboratory data and to deduce time consumption, a intelligent classification system is needed, where just the input data (patient's data, i.e., localized diabetes dataset) are analyzed and based on that to design a correct description or model for each class using the features in the dataset by which further easily based on the same concept, classification system may be designed.

3 Proposed Methodology

In the proposed methodology, GA has served as an attribute selection technique and RBF NN, MLP NN for the classification technique on localized diabetes dataset (LDD).

The proposed methodology consists of the following four phases:

1. Compilation or selection of dataset. Here the LDD has been collected from Bombay Medical Hall, Mahabir Chowk, Pyada Toli, Upper Bazar, Ranchi, Jharkhand, India, during September 2015 to December 2015.
2. Perform the classification technique by using RBF NN, MLP NN on LDD.
3. Perform GA as a feature selection technique.
4. Perform the classification technique by using RBF NN, MLP NN on selected features produced by GA in LDD.

3.1 Used Diabetes Disease Dataset

The localized diabetes dataset (LDD) has been collected from Bombay Medical Hall, Mahabir Chowk, Pyada Toli, Upper Bazar, Ranchi, Jharkhand, India. This dataset has been collected during September 2015 to December 2015. All patients in this database belong from age 3 to 80. In this database, there are 1058 instances and 12 attributes or features plus one class variable. All the features of this database are as noted below:

1. Age in years (min 3, max 80)
2. Sex (male/female)
3. Fasting blood sugar measured in milligram per deciliter (mg/dL) (min 64 and max 680)
4. Diastolic blood pressure measured in mille meter per mercury (mmHg) (min 60 and max 110)
5. Systolic blood pressure measured in mille meter per mercury (mmHg) (min 100 and max 220)
6. Body mass index (BMI) measured in kilogram per square meter (kg/m^2) (min 10.94 and max 42.81)
7. Average blood sugar (HbA1C in %) (min 4.4 and max 17.1)
8. Waist circumference measured in centimeter (cm) (min 48 and max 140)
9. Hip circumference measured in centimeter (cm) (min 46 and max 136)
10. Total cholesterol is measured in mille gram per deciliter (mg/dL) (min 149 and max 369)
11. Family history of diabetes (yes/no)
12. 2 h postglucose load or oral glucose tolerance test (OGTT) measured in milligram per deciliter (mg/dL) (min 94 and max 592)
13. Class or output (tested_negative or tested_positive)

Last attribute is class, where two classes are the diabetic and non-diabetic. Here diabetic indicates tested_positive and non-diabetic for tested_negative.

3.2   Used Classification Technique

The following classification techniques have been used RBF NN, MLP NN on localized diabetes dataset. The proposed algorithm for this technique is mentioned below:

3.2.1   Radial Basis Function Neural Network (RBF NN) for Classification

Broomhead and Lowe first proposed the RBF NN in 1988. RBF NN is supervised feedforward neural network with one hidden layer technique for classification. Since the RBF NN is a type of supervised learning, it requires a curious response to be trained. The RBF NN is more briefly explained in [10].

The following steps occur in RBF NN:

1. Initialization (select two radial basis function units).
2. Output calculation (present an input–output pair, calculate network output and error).
3. Weight updation (delta rule).
4. Center adaptation (based on parameters ε_1 and ε_2).
5. Check for convergence (exit if network error is less than threshold; otherwise, add a radial basis function unit, iterate through steps 2 through 5).

Algorithm RBF NN

1. Choose two initial centers
2. Compute network output:

$$Y = \sum_{J=1}^{p} W_j \phi_j(x); \ where \ \phi_j(x) = \exp\left(-\frac{||x - c_j||^2}{2\sigma_j^2}\right) \tag{1}$$

$$Y = \sum_{J=1}^{p} W_j \exp\left(-\frac{||x - c_j||^2}{2\sigma_j^2}\right) \tag{2}$$

3. Calculate error

$$e = D - Y \tag{3}$$

where

D: desired output,
Y: actual output

4. Set learning parameter values ε_1 using the Heuristic table calculate,

$$\varepsilon_2 = \varepsilon_1^2 * \varepsilon_1^2 \tag{4}$$

5. Move Centers
   Find best matching unit (BMU) using

$$||(x - c_j)|| \tag{5}$$

Move BMU:

$$c_{jbmu}(new) = c_{jbmu}(old) + \varepsilon_1 ||(c_j - x)|| \tag{6}$$

Move other centers:

$$c_{jneighbor}(new) = c_{jneighbor}(old) + \varepsilon_2 ||(c_j - x)|| \tag{7}$$

6. Perform weight updation

$$W_{ij}(new) = W_{ij}(old) + n(D - Y)\phi_j(x) \tag{8}$$

7. Insert a new center if error between successive epochs does not fall below the threshold
8. Repeat 2 through 7 until classification is achieved.

3.2.2 Multilayer Perceptron Neural Network for Classification (MLP NN)

Frank Rosenblatt designed the perceptron algorithm in the mid-50s. MLP NN is supervised learning, feedforward technique for classification. Since the MLP NN is supervised learning, where weights are adjusted to reduce error whenever the actual output does not match the desired output, they require a curious response to be trained. The MLP NN is more briefly explained, and the same procedures [21] here are used.

The working of MLP NN is briefly explained below:

1. In input layer, the input data is provided for processing which produces the predicted output.
2. For error value calculation, the predicted output is subtracted from actual or desired output.
3. For weight adjusting, the BP algorithm is used.
4. For weight adjusting, it starts from output layer node and last hidden layer nodes and works backward through networks.

5. The forwarding process starts when BP is finished.
6. The process is iterated until the error between predicted and actual output is minimized.

Backpropagation (BP) Algorithm Network for Adjusting Weight Attributes

Rumelhart, Hinton, and Williams have designed the backpropagation (BP) network in 1986. The highly popular training algorithm for multilayer and feedforward network is BP, which is used to learn the weights. The algorithm is based on the error correcting learning rule. This is done to enhance weights during processing.

The used [25] BP algorithm is mentioned below:

BP Algorithm

Step1: Run the network forward propagation with your input training data to get the network output

Step2: For each output node, compute

$$\delta_k^l = O_k^l(1 - O_k^l)(O_k^l - t_k^l) \tag{9}$$

Step3: For each hidden unit, calculate

$$\delta_h^l = O_h^l(1 - O_h^l) \sum_{k \varepsilon outputs} \delta_k^l W_{hk}^l \tag{10}$$

Step4: Update the weights and biases as follows:

Given

$$\Delta W_{i,j}^l = -\eta \delta_j^l O_i^l \tag{11}$$

$$\Delta \theta_{i,j}^l = -\eta \delta_j^l \tag{12}$$

Apply

$$W_{i,j}^l + \Delta W_{i,j}^l \rightarrow W_{i,j}^l \tag{13}$$

$$\theta_{i,j}^l + \Delta \theta_{i,j}^l \rightarrow \theta_{i,j}^l \tag{14}$$

where δ_k^l: for an output layer node k; δ_h^l: for a hidden layer node h; t_k^l: target output (label) of output unit k; O_i^l: output layer of node i; O_j^l: output layer of node j; O_k^l: output layer of node k; η: is the learning rate; $W_{i,j}^l$ is the weight of node i to j; $\theta_{i,j}^l$ is the threshold of node i to j.

3.3   GA for Attributes Selection

The method and pseudo-code of GA is briefly illustrated in [10, 17, 21, 26].

In the medical world, if any disease has to be diagnosed, then there are certain tests which will be performed for an accurate diagnosis. After obtaining the tests of the reports, the diagnosed may be doing well. For understanding easily to attribute selection or in layman language, consider each and every test as a feature. If a few tests are required/perform among several tests then indeed set of chemicals, equipments, human resource, and time will be save. Basically, here an attribute selection informed that which tests are more relevant or significant for the diagnosis. So it is clear that by reducing the number of tests, cost will be saved. In the same way, GA will provide benefitted as an attribute selection which have selected six among twelve attributes. GA deduces the cost, storage capacity, and computation time by removing insignificant/irrelevant/noisy/redundant attributes.

3.4   RBF NN, MLP NN for Classification on Selected Features

The same briefly explained Sects. 3.2.1 and 3.2.2 methodology has been applied in the current section on the selected attributes produced by GA.

4   Results and Discussion of Proposed Methodology

In experimental work, the dataset has been divided 70–30% (741–317) for training and test of with and without GA for RBF NN, MLP NN technique. As per the table number 3, GA technique provides six features among twelve attributes. It means that decreased the cost to $s(x) = 6/12 = 0.5$ from 1 and have taken place an enhancement on the training and classification by a factor of 2.

It is well known that the classification method is usually measured in terms of accuracy, precision, recall, F-measure, ROC, confusion matrix, kappa statistics (KS), mean absolute error (MAE), root mean squared error (RMSE), relative absolute error (RAE), root relative squared error (RRSE) which are noted in table for performance evaluation are briefly explained in [10, 17, 21, 26].

Table 1 shows the results of the training set evaluation by using RBF NN and MLP NN technique for LDD based on several measures, which are noted below.

In Table 1, it may be seen that every classification technique provides good training results. If in order to find the best classification technique among both in Table 1, then MLP NN is good which produces good results.

Table 2 shows the results of the testing set evaluation by using RBF NN and MLP NN technique for LDD based on several measures, which are noted below.

Table 1 Training set evaluations of RBF NN and MLP NN performance for LDD

Measure	RBF NN	MLP NN
Time taken to build model (s)	1.22	3.72
Precision	0.95	0.99
Recall	0.95	0.99
F-measure	0.95	0.99
Accuracy (%)	95.14	99.06
ROC	0.98	1
KS	0.86	0.98
MAE	0.08	0.02
RMSE	0.2	0.09
RAE (%)	19.08	4.81
RRSE (%)	43.88	21.54

Table 2 Testing set evaluations of RBF NN and MLP NN performance for LDD

Measure	RBF NN	MLP NN
Time taken to build model (s)	1.10	3.68
Precision	0.93	0.91
Recall	0.92	0.91
F-measure	0.93	0.91
Accuracy (%)	92.42	91.17
ROC	0.98	0.98
KS	0.82	0.78
MAE	0.10	0.87
RMSE	0.25	0.26
RAE (%)	25.04	21.32
RRSE (%)	55.04	58.75

In Table 2, it may be seen that RBF NN and MLP NN classification technique provides almost good testing results. If in order to find the best classification technique among both in Table 2, then RBF NN is good in almost all of the measures.

Table 3 shows the feature selection by using GA on LDD, which is noted below.

Table 4 shows the results of the training set evaluation by using RBF NN and MLP NN technique for LDD based on the selected features by using GA of several measures, which are noted below.

In Table 4, it may be seen that GA_RBF NN and GA_MLP NN classification technique provides almost good training results. If in order to find the best classification technique in Table 4, then GA_MLP NN is approximate good in several parameters.

Table 5 shows the results of the testing set evaluation by using RBF NN and MLP NN technique for LDD based on the selected features by using GA of some measure, which is noted below.

Table 3 GA for feature selection

Data set	Number of attributes	Name of attributes	No. of instances	No. of classes
Localized diabetes dataset (without genetic algorithm)	12	1. Age in years 2. Sex 3. Fasting blood sugar measured in milligram per deciliter 4. Diastolic blood pressure measured in mille meter per mercury 5. Systolic blood pressure measured in mille meter per mercury 6. Body mass index (BMI) measured in kilogram per square meter 7. Average blood sugar (HbA1C in %) 8. Waist circumference measured in centimeter 9. Hip circumference measured in centimeter 10. Cholesterol measured in mille gram per deciliter 11. Family history of diabetes (yes/no) 12. 2 h post glucose load or Oral glucose tolerance test (OGTT) measured in milligram per deciliter	1058	2
Localized diabetes dataset (with genetic algorithm)	6	1. Age in years 3. Fasting blood sugar measured in milligram per deciliter 6. Body mass Index (BMI) measured in kilogram per Square meter 7. Average blood sugar (HbA1C in %) 10. Cholesterol measured in mille gram per deciliter 12. 2 h post glucose load or Oral glucose tolerance test (OGTT) measured in milligram per deciliter	1058	2

Table 4 Training set evaluations of RBF NN and MLP NN performance for LDD

Measure	GA_RBF NN	GA_MLP NN
Time taken to build model (s)	1.19	1.78
Precision	0.96	0.97
Recall	0.95	0.97
F-measure	0.96	0.97
Accuracy (%)	95.55	96.62
ROC	0.99	0.99
KS	0.89	0.91
MAE	0.06	0.04
RMSE	0.18	0.16
RAE (%)	15.91	11.62
RRSE (%)	39.81	34.91

Table 5 Testing set evaluations of RBF NN and MLP NN performance for LDD

Measure	GA_RBF NN	GA_MLP NN
Time taken to build model (s)	1.09	1.64
Precision	0.95	0.96
Recall	0.95	0.96
F-measure	0.95	0.96
Accuracy (%)	94.64	95.9
ROC	0.98	0.99
KS	0.87	0.90
MAE	0.075	0.06
RMSE	0.20	0.19
RAE (%)	18.45	15.35
RRSE (%)	46.26	42.53

Table 6 shows the results comparison between implemented classification technique, i.e., RBF NN, GA_RBF NN, MLP NN, GA_MLP NN for LDD by several measures, which are noted below.

In Table 6, it may be seen that with a feature selection technique, the improvement has taken place in the case of RBF NN, MLP NN classification technique. Now consider to RBF NN, MLP NN with GA the enhancement has taken place in every parameter.

With feature selection technique, i.e., here GA on RBF NN and MLP NN classification technique achieving a good result so here it is concluded that achieved the goal of feature selection.

Table 6 Results comparison between RBF NN and MLP NN classification technique with and without GA for LDD

Measure	RBF NN	GA_RBF NN	MLP NN	GA_ MLP NN
Time taken to build model (s)	1.10	1.09	3.68	1.64
Precision	0.93	0.95	0.91	0.96
Recall	0.92	0.95	0.91	0.96
F-measure	0.93	0.95	0.91	0.96
Accuracy (%)	92.42	94.64	91.17	95.9
ROC	0.98	0.98	0.98	0.99
KS	0.82	0.87	0.78	0.90
MAE	0.10	0.075	0.87	0.06
RMSE	0.25	0.20	0.26	0.19
RAE (%)	25.04	18.45	21.32	15.35
RRSE (%)	55.04	46.26	58.75	42.53

5   Conclusions and Future Work

Diabetes implies glucose is over the coveted level on a maintained premise. This is a standout among the most world's far reaching infections and extremely pervasive in the advanced time. Diabetes contributes to blindness, blood pressure, heart disease, kidney disease, and nerve damage, etc., which is hazardous to health. An efficient diabetes detection system is therefore quite essential and requires a proper dataset for fits formulation. The dataset used here has been collected from Bombay Medical Hall, Mahabir Chowk, Pyada Toli, Upper Bazar Ranchi, Jharkhand, India. In this manuscript, firstly the classification has been performed on localized diabetes dataset by using RBF NN, MLP NN and then using GA for features selection, and there by performed classification on the selected features. It can be seen that the proposed methodology provide the good results.

The problem may be that the dataset has been collected in a particular demographic and not of a large number of instances (either patient's or normal).

In this work, with the attribute selection technique, the improvement has been observed in almost every parameter so ultimately achieve the goal of feature selection technique.

Feature selection technique once again facilitates to save the storage capacity, reduce computation time (shorter training time and test time), decrease computation cost, lesser processor requirements, accuracy, and ROC.

For the future direction, it is proposed to design such a classification technique, which provides the same result or even better as the result received in the manuscript, which could fundamentally diminish human services costs by means of early expectation and finding of diabetes infection. The implemented technique can be utilized for any types of medical diseases but not sure that in all the medical diseases either same or more prominent than the current outcomes can be accomplished. Results that are more curious may also occur from the exploration of the datasets also.

Acknowledgements The work done by authors fulfills all the ethical terms and conditions. The data used in the research work were selective and anonymous. Confidentiality of personal and medical data of the patients has been maintained in all aspects. The authors would like to thank firstly all the patients of Bombay Medical Hall, Mahabir Chowk, Pyada Toli, Upper Bazar, Ranchi, Jharkhand, India, who gave us information very patiently and then Dr. Vinay Kumar Dhandhenia, Diabetologist; M/s Sneha Verma Dietitian; Linus ji, and remaining all the staff of Bombay Medical Hall, Ranchi, India, who helped us to collect and compile the dataset of diabetes and non-diabetes patients.

References

1. UCI Repository of Bioinformatics Databases [online]. http://www.ics.uci.edu/~mlearn/MLRe pository.html
2. Dogantekin, E., Dogantekin, A., Avci, D., Avci, L.: An intelligent diagnosis system for diabetes on linear discriminant analysis and adaptive network based fuzzy inference system: LDA-ANFIS. Digit. Signal Process. (Elsevier) **20**, 1248–1255 (2010)
3. Polat, K., Gunes, S.: An expert system approach based on principal component analysis and adaptive neuro-fuzzy inference system to diagnosis of diabetes disease. Digit. Signal Process. (Elsevier) **17**, 702–710 (2007)
4. Seera, M., Lim, C.P.: A hybrid intelligent system for medical data classification. Expert Syst. Appl. (Elsevier) **41**, 2239–2249 (2014)
5. Hasan Orkcu, H., Bal, H.: Comparing performances of backpropagation and genetic algorithms in the data classification. Expert Syst. Appl. (Elsevier) **38**, 3703–3709 (2011)
6. Lukka, P.: Feature selection using fuzzy entropy measures with similarity classifier. Expert Syst. Appl. (Elsevier) **38**, 4600–4607 (2011)
7. Temurtas, H., Yumusak, N., Temurtas, F.: A comparative study on diabetes disease diagnosis using neural networks. Expert Syst. Appl. (Elsevier) **36**, 8610–8615 (2009)
8. Aslam, M.W., Zhu, Z., Nandi, A.K.: Feature generation using genetic programming with comparative partner selection for diabetes classification. Expert Syst. Appl. (Elsevier) **40**, 5402–5412 (2013)
9. Goncalves, L.B., Bernardes, M.M., Vellasco, R.: Inverted hierarchical neuro-fuzzy BSP system: a novel neuro-fuzzy model for pattern classification and rule extraction in databases. IEEE Trans. Syst. Man Cybern. Part C: Appl. Rev. **36**(2), 236–248 (2006)
10. Choubey, D.K., Paul, S.: GA_RBF NN: a classification system for diabetes. Int. J. Biomed. Eng. Technol. (IJBET) Intersci. **23**(1), 71–93 (2017)
11. Choubey, D.K., Paul, S., Kumar, S., Kumar, S.: Classification of Pima Indian diabetes dataset using Naive Bayes with genetic algorithm as an attribute selection. In: Proceedings of the International Conference on Communication and Computing System (ICCCS 2016), pp. 451–455. CRC Press Taylor Francis (2017)
12. Barakat, N.H., Bradley, A.P., Barakat, M.N.H.: Intelligible support vector machines for diagnosis of diabetes mellitus. IEEE Trans. Inf. Technol. Biomed. **14**(4), 1114–1120 (2010)
13. Polat, K., Guneh, S., Arslan, A.: A cascade learning system for classification of diabetes disease: generalized discriminant analysis and least square support vector machine. Expert Syst. Appl. (Elsevier) **34**, 482–487 (2008)
14. Choubey, D.K., Paul, S., Bhattacharjee, J.: Soft computing approaches for diabetes disease diagnosis: a survey. Int. J. Appl. Eng. Res. RIP **9**, 11715–11726 (2014)
15. Ganji, M.F., Abadeh, M.S.: A fuzzy classification system based on ant colony optimization for diabetes disease diagnosis. Expert Syst. Appl. (Elsevier) **38**, 14650–14659 (2011)

16. Choubey, D.K., Paul, S., Dhandhenia, V.K.: Rule based diagnosis system for diabetes. Biomed. Res. (Allied Academies) **28**(12), 5196–5209 (2017)
17. Choubey, D.K., Paul, S.: GA_J48graft DT: a hybrid intelligent system for diabetes disease diagnosis. Int. J. Bio-Sci. Bio-Technol. (IJBSBT) (SERSC) **7**(5), 135–150 (2015). ISSN: 2233-7849
18. Kahramanli, H., Allahverdi, N.: Design of a hybrid system for the diabetes and heart diseases. Expert Syst. Appl. (Elsevier) **35**, 82–89 (2008)
19. Choubey, D.K., Paul, S.: Classification techniques for diagnosis of diabetes disease: a review. Int. J. Biomed. Eng. Technol. (IJBET) Intersci. **21**(1), 15–39 (2016)
20. Ephzibah, E.P.: Cost effective approach on feature selection using genetic algorithms and fuzzy logic for diabetes diagnosis. Int. J. Soft Comput. (IJSC) **2**(1), 1–10 (2011)
21. Choubey, D.K., Paul, S.: GA_MLP NN: a hybrid intelligent system for diabetes disease diagnosis. Int. J. Intell. Syst. Appl. (IJISA) MECS **8**(1), 49–59 (2016)
22. Karegowda, A.G., Manjunath, A.S., Jayaram, M.A.: Application of genetic algorithm optimized neural network connection weights for medical diagnosis of Pima Indians diabetes. Int. J. Soft Comput. (IJSC) **2**(2), 15–23 (2011)
23. Qasem, S.N., Shamsuddin, S.M.: Radial basis function network based on time variant multi-objective particle swarm optimization for medical diseases diagnosis. Appl. Soft Comput. (Elsevier) **11**, 1427–1438 (2011)
24. Jayalakshmi, T., Santhakumaran, A.: A novel classification method for diagnosis of diabetes mellitus using artificial neural networks. In: International Conference on Data Storage and Data Engineering, pp. 159–163. IEEE (2010)
25. Han, J., Kamber, M.: Data Mining: Concepts and Techniques, 2nd edn, pp. 1–702
26. Choubey, D.K., Paul, S.: GA_SVM—a classification system for diagnosis of diabetes. In: Handbook of Research on Nature Inspired Soft Computing and Algorithms, pp. 359–397. IGI Global (2017)
27. Bala, K., Choubey, D.K., Paul, S., Lala, M.G.N.: Classification techniques for thunderstorms and lightning prediction—a survey. In: Soft Computing-Based Nonlinear Control Systems Design, pp. 1–17. IGI Global (2018)
28. Bala, K., Choubey, D.K., Paul, S.: Soft computing and data mining techniques for thunderstorms and lightning prediction: a survey. In: International Conference of Electronics, Communication and Aerospace Technology (ICECA 2017), RVS Technical Campus, Coimbatore, Tamilnadu, India, vol. 1, pp. 42–46. IEEE (2017)

Application of Computer Simulation in Exploring Influence of Alcohol on Aqueous Milieu of a Gut-Brain Octapeptide, Cholecystokinin-8

Apramita Chand, Pragin Chettiyankandy and Snehasis Chowdhuri

Abstract Computer Simulations have been employed by applying techniques of classical molecular dynamics simulations to explore the hydrogen bonding structure and dynamics in aqueous environment of cholecystokinin-8, in the presence of two different concentrations of ethanol (EtOH) and 2,2,2-trifluoroethanol (TFE). Different site–site and centre-of-mass radial distribution functions have been presented here to give an idea of the various microscopic interactions between different species in solution. It is observed that EtOH solution facilitates hydrogen bonding of CCK8 with its aqueous milieu, whereas TFE prefers to envelope the peptide and protects it from water due to its tendency of clustering and encouraging segregation of water. Significant decrease of methionine CH_3-groups hydrophobic solvation is noted with increasing alcohol concentration. The structural relaxation lifetimes of peptide–water hydrogen bonds are observed to be lengthened with EtOH concentration in the solution while highest lifetimes are seen for CCK8–TFE hydrogen bonds. Stronger water–ethanol hydrogen bonds may cause slower translational motion of water molecules in concentrated ethanol than in TFE solution. Conformational clustering analysis shows higher number of similar compact structures of CCK8 in TFE solution relative to aqueous/EtOH solution.

Keywords Molecular dynamics · Cholecystokinin-8 · Radial distribution functions · Self-diffusion coefficients · Structural relaxation lifetimes

A. Chand (✉) · P. Chettiyankandy · S. Chowdhuri
School of Basic Sciences, Indian Institute
of Technology Bhubaneswar, Bhubaneswar, India
e-mail: ac12@iitbbs.ac.in

S. Chowdhuri
e-mail: snehasis@iitbbs.ac.in

© Springer Nature Singapore Pte Ltd. 2019
J. C. Bansal et al. (eds.), *Soft Computing for Problem Solving*, Advances in Intelligent Systems and Computing 817, https://doi.org/10.1007/978-981-13-1595-4_3

1   Introduction

Computer simulations have, in the past several decades, revolutionized the interpretation of structure, dynamics and conformational changes in biomolecular systems [1]. By bridging the abyss between theory and experiment, computer simulations have not only helped in analyzing large chunks of numerical data but also in visualization of real-world biological systems and comprehending how bulk thermodynamic properties may be predicted by examining behaviour of assemblies of particles. Data excavation, pattern detection and grouping from computer simulations results have been facilitated by several soft computing techniques like fuzzy sets, evolutionary algorithms and artificial neural networks which are particularly pertinent for protein structure prediction and analysis [2, 3]. The conundrum faced by experimentalists in explaining peptide and protein-folding pathways, through X-ray crystallography and nuclear magnetic resonance spectroscopy, has been greatly resolved by the use of molecular modelling and molecular dynamics simulations [4]. Analysis of the resulting trajectory from MD simulations in a stipulated time framework can yield microscopic information about interactions amongst particles, translational and rotational dynamics of species in solution and various structural changes occurring in biomolecules like peptides and proteins in the course of a typical simulation. Soft computing techniques like fuzzy clustering strategies have been applied to molecular dynamics trajectories [5, 6] to explore conformations of peptides and proteins like roseotoxin B and human parathyroid hormone.

In this paper, we have applied computer simulations to study the impact of alcohol on aqueous solvation structure and dynamics of an important peptide called Cholecystokinin-8; an octapeptide which is distributed widely in the gastrointestinal as well as central nervous system and yet is understudied in terms of its solvation behaviour. Aqueous alcohol media can perturb or stabilize peptide/protein structure according to the nature and strength of the amphipathic nature of alcohol–water interaction. For this purpose, we have chosen ethanol and trifluoroethanol, both of which exert variety of effects on structure and stability of peptides and proteins in their aqueous solutions [7, 8] but differ in the nature of their concentration-dependent effects on conformational stability and hydrogen bonding interactions [9]. The choice of alcohol as a co-solvent in aqueous CCK8 environment is particularly significant since it has been established from the literature that administration of CCK8 reduces ethanol consumption in rats [10–12] and protects against ethanol-induced gastric wounds in mice [13].

In Sect. 2, a description of the models and methodologies used to carry out this study has been detailed. In Sect. 3.1, we have presented some of the various correlations between peptide and its aqueous environment containing two different concentrations of alcohol (EtOH/TFE). In Sect. 3.2, our calculated hydrogen bond number and lifetime are presented while Sect. 3.3 gives a brief idea of the solvent dynamics in the different solutions.

2 Models and Simulations

2.1 Principle of Molecular Dynamics Simulations

Molecular dynamic simulations typically start with knowledge of the potential energy function for typical biomolecules with respect to initial atomic positions and velocities and this may be used to compute forces on atoms [14, 15].

$$\mathbf{F}_i = -\nabla V(\mathbf{r_i}) \tag{1}$$

The movement of the system of particles (where $\mathbf{r_i}$ are the position vectors and $i = 1, 2, ..., N$ may be calculated through iteratively solving the Newton's equations of motion.

$$m_i \frac{d^2 \mathbf{r_i}}{dt^2} = \mathbf{F_i} \tag{2}$$

The potential energy function contains terms describing the interactions of the system [16]:

$$
\begin{aligned}
V(r) = &\sum_{bonds} k_b(b - b_0)^2 + \sum_{angles} k_\theta(\theta - \theta_0)^2 \\
&+ \sum_{dihedrals} k_\phi(1 + \cos(n\phi - \phi_0)) + \sum_{impropers} k_\psi(\psi - \psi_0)^2 \\
&+ \sum_{non-bondedpairs(i,j)} 4\varepsilon_{ij}\left[\left(\frac{\sigma_{ij}}{r_{ij}}\right)^{12} - \left(\frac{\sigma_{ij}}{r_{ij}}\right)^6\right] + \sum_{non-bondedpairs(i,j)} \frac{q_i q_j}{\varepsilon_D r_{ij}}
\end{aligned} \tag{3}
$$

where the first four terms include expressions for bond stretching, bending, and torsional motions while the last two terms are for non-bonded interactions; a 12-6 Lennard-Jones potential followed by Coulombic interactions.

So, in a typical MD simulation all the atom positions, velocities are initialized and decisions like initial temperature/pressure conditions, number of steps are finalized. Then, forces on atoms are calculated which also include non-bonded inter-atomic terms. Updated positions and velocities are obtained by numerical integration of Newton's equations of motion and the process continues. The three steps of any MD simulation involve initialization, equilibration, and production. In equilibration steps, the system evolves from its initial state and achieves steady fluctuations around average values of observable thermodynamic properties like pressure, temperature, density. The production run is the final step where thermodynamic properties are calculated.

2.2   Models and Methodologies

All molecular dynamic simulations were carried out using GROMACS package [17] (version 4.5.7) using the GROMOS53a6 [18] parameter set with a total simulation time of 300 ns (including equilibration time). The starting structure of CCK8 from the work of Pellegrini and Mierke [19], ethanol coordinates according to Jorgensen et al. [20], TFE model according to Fiorini [9] and SPC/E water model [21] was chosen for the simulations. At $X_{ETH} = 0.167$ (mole fraction of ethanol in ethanol–water binary solution), our calculated density at 300 K was 0.942 g/cm^3 which was in good concordance with experimental value (0.9419–0.9385 g/cm^3) [22]. In case of water–TFE solution, our calculated density (0.1319 g/cm^3) was within range of 5% error from experimental values (0.18–0.19 g/cm^3) [23]. The system was accommodated in a cubic box with distance of 1.4 nm between solute and the box to satisfy minimum image conventions. For preparation of each system, the CCK8 molecule was solvated with appropriate number of alcohol and water molecules to achieve two desired concentrations (5m and 10m, respectively) and overlapping solvent molecules were deleted. All bond lengths were constrained using the LINCS algorithm [24] while specifically water molecules were restrained by SETTLE algorithm [25]. Simulation boxes are energy minimized using steepest descent algorithm and equilibrated to achieve the appropriate volume A 100 ps NVT equilibration was carried out using v-rescale algorithm [26] by coupling every separate component (peptide, water and alcohol) to a temperature bath at 300 K with tau_t $= 0.1$ ps. This was followed by three-step NPT equilibration of 15–20 ns maintaining the pressure at 1 bar with Berendsen barostat [27] for isotropic pressure coupling with 0.5 ps relaxation time and isothermal compressibility of 4.5×10^{-5} bar^{-1}. In the first step of NPT equilibration, the solvent molecules were allowed to relax around the constrained peptide and in the subsequent steps the position restraints were gradually removed from the peptide and the desired pressure was reached. Short- and long-range non-bonded interactions were calculated by applying twin-range cut-offs of 0.9 and 1.4 nm, respectively.

The production MD simulations were run for 40 ns for each system, maintaining target pressure of 1 bar and temperature (300 K). Periodic boundary conditions were applied in all directions. The leap-frog algorithm (time step of 2 fs) was used for integrating the equation of motion [28], and the neighbour list is updated every two steps. Long-range electrostatics is treated using the particle mesh Ewald method [29] with cubic interpolation and a grid spacing of 0.16 nm. The box length varied from 3.159 nm in case of CCK8–water system to 3.295 nm, 3.436 nm, 3.381 nm and 3.549 nm in 5m EtOH, 10m EtOH, 5m TFE and 10m TFE solution, respectively.

3 Results and Discussion

3.1 Solvation Structure

The solvation structure of the CCK8 molecule in alcohol solution can be interpreted with the help of radial distribution functions (RDFs) which give idea about the local correlation density arising between atoms (or groups of atoms) compared to the correlations in the bulk density.

We have investigated the centre-of-mass RDFs between CCK8 and water/alcohol at two different alcohol concentrations ($5m$ and $10m$). From Fig. 1, we can observe that increasing concentration of alcohol reduces probability of CCK8–water interactions in aqueous TFE solutions while EtOH uplifts CCK8–water interactions. It is interesting to note that though probability of CCK8–water is low in TFE solution, the first peak is centred at 0.176 nm which is at a shorter hydrogen bonding distance than that of the first peak of CCK8–water COM–RDF in EtOH solution which is centred at 0.185 nm. This might indicate that access of water molecules to peptide sites are restricted by TFE peptide but wherever there is a possibility of water intrusion, the interactions are stronger owing to rigid framework of TFE molecules confining a few water molecules.

Strong hydrogen bonding between TFE and CCK8 sites may be seen from Fig. 1, where CCK8–TFE interactions are centred at 0.154 nm while peptide–EtOH interaction surface at 0.175 nm with lower probabilities. It is noticed that at higher concentrations of TFE, peptide–TFE correlations are attenuated but deeper minima indicate enhanced stiffness of hydrogen bond network at $10m$ TFE. Since water–peptide and EtOH–peptide interactions occur at nearly the same hydrogen bonding distance and have similar probabilities, the two types of hydrogen bonds are always in competition with each other in solution.

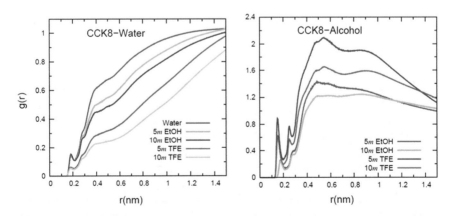

Fig. 1 Radial distribution function between centre-of-mass of CCK8–peptide and water/alcohol in water, $5m$ EtOH/TFE and $10m$ EtOH/TFE

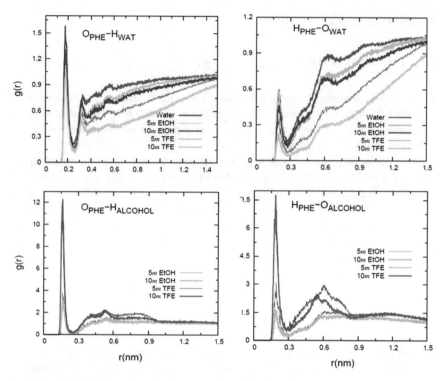

Fig. 2 Radial distribution functions between phenylalanine carbonyl oxygen–water (hydrogen) and PHE-amide hydrogen–oxygen (alcohol)

Each residue has varying affinities towards each type of solvent molecule. For instance, glycine interacts well with both ethanol and TFE sites. Amide hydrogen of tyrosine residue has very low preference to interact with water oxygen in water/aqueous EtOH/aqueous TFE solution. However, the terminal residues, PHE-8 and ASP-1, are exposed for interactions in all solvents. We have probed the trend of site–site correlations in Fig. 2, where hydrogen bonding interactions have been investigated between PHE and solvent molecules.

It is seen from Fig. 2 that O_{PHE}–H_{WAT} correlations occurring at peak position of 0.172 nm are highest in 10m EtOH while the probability reduces in case of increasing TFE concentration. This is explained by high probability of O_{PHE}–H_{TFE} interactions which supercede that of O_{PHE}–$H_{WAT/EtOH}$. The amide hydrogen has lesser affinity for solvent molecules than the carbonyl oxygen as is seen from peak positions occurring at 0.185 nm. H_{PHE}–O_{WAT} interactions are not significantly affected by rise in EtOH concentration while TFE does not encourage the hydrogen bonding interaction and prefers to itself hydrogen bond with PHE-amide hydrogen. The second solvation shell of H_{PHE}...O_{TFE} RDF is well defined at 0.5–0.6 nm as compared to its correlations with ethanol or water oxygens.

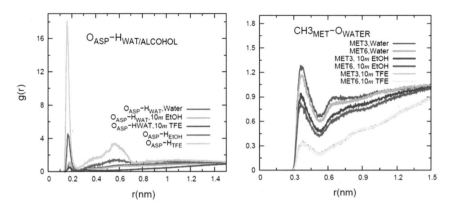

Fig. 3 Radial distribution functions depicting hydrogen bonding between carbonyl oxygen of ASP-1 and hydrogen of solvent (water/alcohol) as well as correlations between methyl groups of methionine residues with water oxygen in different solutions

The same trends are seen for N-terminal residue ASP-1, where we observe from Fig. 3, that though ethanol solution uplifts O_{ASP}-H_{WAT} interactions, very high probability is seen for the correlation of O_{ASP} site with TFE hydrogen which is well-reputed to be an excellent hydrogen bond donor in comparison with ethanol or water hydrogen from experimental studies [30, 31]. The high affinity of TFE hydrogen for carbonyl amide oxygen has been explained by studies of excess enthalpy of NMA by Pikkarainen [32, 33] in ethanol and TFE solution. We have also tried to explore the comparative hydrophobic solvation of methyl groups of the two methionine residues in the different solutions in Fig. 3, and we see that hydrophobic solvation of methyl residues is better in MET-3 residue than MET-6 residue for all solutions and occurs at 0.48 nm. Also, the decrease in solvation of MET–CH$_3$ groups is more apparent at higher TFE concentration than for EtOH solution.

We have observed the nature of aggregation of solvent molecules around the CCK8 peptide through snapshots of the system at 10m alcohol concentration. It is seen from Fig. 4 that clear segregation of water molecules and TFE molecules occurs in TFE solution. TFE has a tendency to envelope the CCK8 peptide while distinct clumps of majority of water molecules are separated out. However, this clustering tendency is not significant in aqueous EtOH mixture where the peptide has no exclusivity of interactions due to segregation. This is in excellent agreement with our results for N-methylacetamide in EtOH/TFE aqueous solution where TFE and peptide molecules act in coordination to segregate out regions of water [34]. Our results are also in concordance with the clustering tendencies seen for ethanol and TFE in the literature [30, 35, 36].

We have also presented the final structural conformation of the CCK8 peptide in alcohol solution which is coloured according to b-factor or temperature factors considering the root-mean-square fluctuations of each residue. The blue regions represent rigid residues while colour gradient towards red is representative of highly

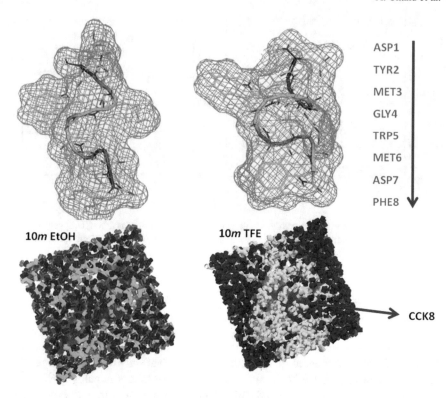

ASP1
TYR2
MET3
GLY4
TRP5
MET6
ASP7
PHE8

10m EtOH   10m TFE

CCK8

Fig. 4 Snapshots of peptide conformation in 10m EtOH/TFE as well as representation of clustering tendency of alcohol molecules in the two solutions

fluctuating residues. It is seen that CCK8 peptide adopts a more coiled conformation in TFE solution than in EtOH and has comparatively less flexible residues. A mesh has been drawn around the peptide to provide a visual representation of pockets of accessible area for solvent molecules. Though not directly comparable, we have observed approximately similar range of our calculated $^3J_{NH}$ coupling constants of CCK8 residues in alcohol solution and the experimental NMR values in DMSO [37]. The calculated values for TRP-5 and MET-6 in 10m TFE (7.86 and 8.07 Hz) are close to the experimental values for these residues (7.60 and 8.14 Hz, respectively). GLY-4 experimental values are higher than our calculated values for all solutions.

3.2 Hydrogen Bonding Properties and Dynamics

Hydrogen bond properties and dynamics for CCK8–water, CCK8–alcohol, water–water, water–alcohol hydrogen bonds have been calculated by defining a set of *geometric criteria* [38, 39], for existence of hydrogen bond between two species subject to agreement with the conditions, i.e., $R^{(OX)} < R_c^{(OX)}$, $R^{(OH)} < R_c^{(OH)}$ (distance criteria) and $\theta < \theta_c$ (angular criteria). In case of water–water and water–ethanol/TFE hydrogen bonds, the distance and angular criteria from previous studies have been taken [34, 38] and the first minima of the respective radial distribution functions can yield indications for the distance cut-off values for peptide–water/alcohol hydrogen bonds. For amide hydrogen of a particular residue of CCK8–oxygen of water hydrogen bond, $R^{(OX)}$ and $R^{(OH)}$ denote the oxygen (water)–nitrogen (CCK8 residue) and oxygen (water)–hydrogen (CCK8 residue) distances, and angle $\theta = \theta(^N_{CCK8}{}^O_{WAT}{}^H_{CCK8})$ is the nitrogen (CCK8)–oxygen (WAT)–hydrogen (CCK8) angle. The cut-off angle $\theta_c = 45°$ is used for the existence of CCK8–water, water–water and CCK8–alcohol hydrogen bonds that may also take into account flexibility due to thermal motion [40].

Hydrogen bond breaking and reforming kinetics can be observed through the intermittent hydrogen bond correlation function defined as $C_{HB}(t)$. It is defined as [41, 42],

$$C_{HB}(t) = \langle h(0)\ h(t) \rangle / \langle h \rangle. \tag{4}$$

The correlation function $C_{HB}(t)$, defined by Luzar and Chandler [42, 43], describes the probability that a hydrogen bond is intact at time t = 0 and t, unconcerned of any rupture in the intermediate period. Here, hydrogen bond population variable $h(t)$ is defined where $h(t)$ = one or zero depending on whether a particular peptide–water/alcohol, water–water pair is hydrogen bonded or not, respectively, at time t according to hydrogen bonding criteria mentioned previously. Thus, the dynamics of $C_{HB}(t)$ describes the hydrogen bond structural relaxation, and the related relaxation times τ_{HB} can be comprehended as the reorganization time scale of peptide–water, peptide–alcohol and water–water bonds. The decay of time correlation function is calculated up to 3000 ps depending on the smooth convergence of $C_{HB}(t)$.

This correlation function may be used to calculate dynamics of hydrogen bond breaking and reformation [44].

$$K(t) = \frac{-dC(t)}{dt} \tag{5}$$

$$K(t) = k_1 C(t) - k_2 n(t) \tag{6}$$

where K(t) is the reactive flux correlation function, whereas k_1 and k_2 are forward and backward rate constants for H-bond breakage and reformation, respectively, and also n(t) is the probability that two hydrogen-bonded groups are within the bonding distance but the H-bond existing at t = 0 is broken. In this way, the inverse of forward rate constant can give us the hydrogen bond lifetime.

$$\tau_{HB} = \frac{1}{k1} \tag{7}$$

Table 1 Average number of peptide–water/alcohol, water–water and water–alcohol hydrogen bonds with corresponding structural relaxation times

	CCK8-WAT	CCK8-ETH/TFE	WAT-WAT	WAT-ETH/TFE
n_{HB}				
Water	34.77	–	2042.73	–
5m ETH	28.24	5.32	1725.97	194.72
5m TFE	11.31	20.5	1759.92	143.14
10m ETH	23.78	7.86	1496.94	315.73
10m TFE	7.62	24.07	1547.12	233.16
τ_{HB} (in ps)				
Water	5.52	–	4.84	–
5m ETH	7.39	21.19	6.47	7.48
5m TFE	5.46	159.98	5.59	7.81
10m ETH	8.36	21.10	7.35	8.59
10m TFE	6.68	100.67	5.90	8.52

The average number of hydrogen bonds between CCK8–water, CCK-–ethanol/TFE, water–water, water–alcohol in the solution and the alcohol concentration dependence results of (forward) structural relaxation times is presented in Table 1.

It is shown in Table 1 that TFE significantly reduces average number of CCK8–water hydrogen bonds with increase in TFE concentration, while EtOH does not disrupt as many peptide–water hydrogen bonds as TFE. Higher number of CCK8–TFE hydrogen bonds (20.5 and 24.07 at $5m$ and $10m$ TFE concentration, respectively) is observed relative to CCK8–ethanol hydrogen bonds (only 7.86 at $10m$ EtOH). This is substantiated by the high lifetimes of CCK8–TFE hydrogen bonds which are more than 5–7 times that of CCK8–EtOH hydrogen bond lifetime signifying that CCK8–EtOH hydrogen bonds may break easily due to other competing interactions like water–ethanol or water–CCK8 hydrogen bonding. It has been proposed by Fioroni et al. [9] that TFE can hydrogen bond to peptide moieties even at the expense of water–peptide contacts but ethanol may not significantly retard the aqueous solvation tendencies. It is interesting to note that doubling the concentration reduces the CCK8–TFE lifetime by nearly 60 ps which indicates possibility of flexibility introduced in the CCK8–TFE hydrogen bond network by water–TFE hydrogen bonds in the vicinity which increases in number at $10m$ (from 143.14 at 5m TFE to 233.16).

3.3 Solvent Translational Dynamics

The long-time limit of the mean square displacement (MSD) can be used for evaluating the translational self-diffusion coefficient as follows:

$$D_i = \lim_{t \to \infty} \frac{\langle |\mathbf{r}(t) - \mathbf{r}(0)|^2 \rangle}{6t}, \tag{8}$$

where $\mathbf{r}(t)$ is the position of a species i at time t, by a least square fit of the long-time region of MSD as obtained from simulation. From our calculated self-diffusion coefficient values of solvent molecules in the solution, it is seen that self-diffusion coefficients of water in pure water (D (10^{-5} cm^2/s) = 2.395) is reduced to a greater extent in 5m EtOH solution (D (10^{-5} cm^2/s) = 1.715) than in 5m TFE aqueous mixture (D (10^{-5} cm^2/s) = 1.820), and as concentration of alcohol in the mixture increases, diffusion values of water decrease further to 1.332 and 1.445, respectively. This may be due to formation of water–alcohol complexes on initial addition of alcohol, particularly ethanol at higher concentrations. Similarly, diffusion values for ethanol (D (10^{-5} cm^2/s) = 0.802) and TFE (D (10^{-5} cm^2/s) = 0.515), decrease at higher concentration (10m) to 0.789 (10^{-5} cm^2/s) and 0.494 (10^{-5} cm^2/s), respectively, suggesting impeded translational motion of the alcohol molecules as concentration increases.

3.4 Clustering Analysis

The simulation data sets often result in large ensembles of statistically weighted configurations and enormous amount of conformational information is generated which leads us to the motive of separation of structural ensembles through various clustering algorithms. In particular, geometric clustering algorithms implemented in GROMACS are efficient in grouping conformationally analogous structures as "clusters". Here, we adopt a "non-fuzzy" or crisp approach [45] where if A is a data set partitioned into small clusters, then:

$$A = \{a_1, a_2, a_3, \ldots, a_l\} \tag{9}$$

$$\text{where } ai \cap aj = 0 \, \forall \, i, j \text{ with } i \neq j \tag{10}$$

Here, each data set in A is separate from the other unlike in fuzzy clustering techniques [5, 6] where data points can belong to more than one cluster and hence the membership of individual points is "fuzzy". Pairwise distances between data points in set A may be utilized to build distance matrix D permuted to adopt block diagonal form, where each block is a cluster and matrix elements in these blocks are smaller than other elements.

$$\mathbf{D'} = \mathbf{PDP}^{-1} \tag{11}$$

The neighbour algorithm by Daura et al. [45] has been found to be robust as well as faster and was used to cluster conformations based on root-mean-square deviation (RMSD). Here, the data point with the highest number of neighbours(within a

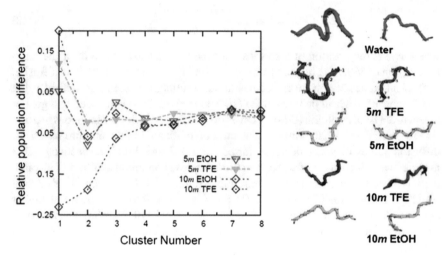

Fig. 5 Difference in fractional population of structures in different clusters (alcohol solution–bulk solution) and the representative or middle structures of the first two dominant clusters

specified cut-off) was considered to be the middle of the first cluster and this data point along with associated neighbours was considered to be the initial cluster and eliminated from further analysis of the structure pool and thereby, the iteration was continued. Various cut-offs for RMSD were tested and RMSD cut-off of 0.175 nm was found appropriate in the analysis of 4000 structures. In water, 25 clusters are seen with the first three clusters (1458, 944, 519 structures, respectively) accounting for 73.02% of the total no. of structures. In Fig. 5, we see that as compared to bulk solution, the first cluster in water–TFE solution is more populated (2266 structures in the first cluster), particularly at higher concentrations which means that higher number of structures are conformationally alike in TFE solution. But in EtOH solution, we find that less number of structures with similar conformations are available and the peptide samples many small-sized clusters and hence is more flexible, which is also seen from the extended conformations in the second cluster. Conformations of the first two dominant clusters in water and in alcohol (5m and 10m) have also been represented and peptide length compaction in TFE solutions is more apparent than in aqueous EtOH. Differences in the fractional population are negligible from the fourth cluster onwards.

Inspection of the trajectories reveals that some structures belonging to different clusters (having lower populations) differ only slightly. In such cases of minor angle variations, similar motifs or in case of transition conformations, fuzzy clustering techniques might be useful rather than crisp clustering analysis since some conformations might belong to more than one cluster. It is pertinent to mention that once a fuzzy membership matrix is constructed it can be converted to crisp membership data and individual conformations can also be extracted [5].

4 Conclusions

We have presented the initial results of our investigations on the impact of alcohol on solvation structure and dynamics in aqueous environment of cholecystokinin-8 by applying computer simulation techniques. We have calculated properties like site–site correlations, solvent diffusion and hydrogen bonding number and lifetimes at two different concentrations each ($5m$ and $10m$) of EtOH and TFE, apart from aqueous solution.

From radial distribution functions, we have observed that TFE preferentially hydrogen bonds to the peptide, shielding it from water molecules whereas ethanol boosts peptide–water interactions. Methyl groups of methionine are demonstrated to be subject to less hydrophobic solvation in more alcohol concentration, particularly for TFE solution. Water oxygens prefer to solvate the MET-3 residue rather than MET-6 in all the solutions. Clustering of TFE molecules is more efficient than EtOH due to cooperative action of three fluorine atoms which enables exclusive TFE contacts to the peptide sites and segregation of water molecules.

Slowdown of translational motion of water as well as alcohol molecules at higher alcohol concentrations is seen which is more conspicuous in case of EtOH due to probable formation of water–EtOH complexes.

We have investigated average number of hydrogen bonds as well as hydrogen bond breaking dynamics of CCK8–water, water–water, water–alcohol and CCK8–alcohol hydrogen bonds. Presence of TFE molecules encourage higher average number of water–water hydrogen bonding but fail to retain their hydrogen bonding affinities towards water at higher concentration. Ethanol prefers to form water–ethanol hydrogen bonds and also facilitates hydration of the peptide, whereas TFE forms highly stable CCK–TFE hydrogen bonds with longer breaking lifetimes while disrupting peptide–water hydrogen bonds. As a result, the CCK8 peptide adopts a coiled conformation in TFE solution while resulting in an extended conformation in EtOH solution which is also seen from RMSD-based clustering analysis.

Since alcohol and CCK8 have profound effects on each other, it is exciting to explore effects of CCK8 administration on chronic alcoholism in humans. It is important to carry out detailed investigations of the conformational states that CCK8 peptide might pass through at different concentrations of alcohols of varying hydrophobicity. In this context, we are presently exploring different clustering analysis methods including fuzzy clustering techniques.

This work would hopefully serve as a prelude to further computer simulations which can pinpoint bioactive conformations of short peptides in which they bind to receptors and whether co-solvent action might change their conformational and solvation preferences.

Acknowledgements Authors are grateful to the Indian Institute of Technology, Bhubaneswar, for infrastructural support and Council of Scientific and Industrial Research (CSIR), Government of India, for research fellowship.

References

1. Sansom, C.E., Smith, C.A.: Computer applications in the biomolecular sciences. Part 1: Molecular modelling. Biochem. Mol. Biol. Educ. **26**(2), 103–110 (1998)
2. Mitra, S., Hayashi, Y.: Bioinformatics with soft computing. IEEE Trans. Syst. Man Cybern. Part C Appl. Rev. **36**(5), 616–635 (2006)
3. Belda, I., Llorà, X., Giralt, E.: Evolutionary algorithms and de novo peptide design. Soft Comput. Fusion Found. Appl. **10**(4), 295–304 (2006)
4. Fersht, A.R.: From the first protein structures to our current knowledge of protein folding: delights and scepticisms. Nat. Rev. Mol. Cell Biol. **9**(8), 650 (2008)
5. Feher, M., Schmidt, J.M.: Fuzzy clustering as a means of selecting representative conformers and molecular alignments. J. Chem. Inf. Comput. Sci. **43**(3), 810–818 (2003)
6. Gordon, H.L., Somorjai, R.L.: Fuzzy cluster analysis of molecular dynamics trajectories. Proteins Struct. Funct. Bioinf. **14**(2), 249–264 (1992)
7. Ghosh, R., Roy, S., Bagchi, B.: Solvent sensitivity of protein unfolding: dynamical study of chicken villin headpiece subdomain in water ethanol binary mixture. J. Phys. Chem. B **117**(49), 15625–15638 (2013)
8. Buck, M.: Trifluoroethanol and colleagues: cosolvents come of age. Recent studies with peptides and proteins. Q. Rev. Biophys. **31**(3), 297–355 (1998)
9. Fioroni, M., Diaz, M., Burger, K., Berger, S.: Solvation phenomena of a tetrapeptide in water/trifluoroethanol and water/ethanol mixtures: a diffusion NMR, intermolecular NOE, and molecular dynamics study. J. Am. Chem. Soc. **124**(26), 7737–7744 (2002)
10. Kulkosky, P.J.: Effect of cholecystokinin octapeptide on ethanol intake in the rat. Alcohol. **1**(2), 125–128 (1984)
11. Kulkosky, P.J., Sanchez, M.R., Glazner, G.W.: Cholecystokinin octapeptide: effect on the ethogram of ethanol consumption and blood ethanol levels in the rat. Physiol. Psychol. **14**(2), 23–30 (1986)
12. Toth, P., Shaw, C., Perlanski, E., Grupp, L.: Cholecystokinin octapeptide reduces ethanol intake in food-and water-sated rats. Pharmacol. Biochem. Behav. **35**(2), 493–495 (1990)
13. Evangelista, S., Maggi, C.: Protection induced by cholecystokinin-8 (CCK-8) in ethanol-induced gastric lesions is mediated via vagal capsaicin-sensitive fibres and CCKA receptors. Br. J. Pharmacol. **102**(1), 119–122 (1991)
14. Rapaport, D.C., Blumberg, R.L., McKay, S.R., Christian, W.: The art of molecular dynamics simulation. Comput. Phys. **10**(5), 456–456 (1996)
15. Frenkel, D., Smith, B.: Understanding Molecular Simulation: From Algorithms to Applications. Academic Press, New York (1996)
16. Nucleic Acids Flexibility (n.d.). http://mmb.irbbarcelona.org/NAFlex/help.php?id=md. Accessed 22 Nov 2017
17. Pronk, S., Páll, S., Schulz, R., Larsson, P., Bjelkmar, P., Apostolov, R., Shirts, M.R., Smith, J.C., Kasson, P.M., van der Spoel, D.: GROMACS 4.5: a high-throughput and highly parallel open source molecular simulation toolkit. Bioinformatics **29**(7), 845–854 (2013)
18. Oostenbrink, C., Villa, A., Mark, A.E., Van Gunsteren, W.F.: A biomolecular force field based on the free enthalpy of hydration and solvation: the GROMOS force-field parameter sets 53A5 and 53A6. J. Comput. Chem. **25**(13), 1656–1676 (2004)
19. Pellegrini, M., Mierke, D.F.: Molecular complex of cholecystokinin-8 and N-terminus of the cholecystokinin A receptor by NMR spectroscopy. Biochemistry (Mosc.) **38**(45), 14775–14783 (1999)

20. Jorgensen, W.L., Maxwell, D.S., Tirado-Rives, J.: Development and testing of the OPLS all-atom force field on conformational energetics and properties of organic liquids. J. Am. Chem. Soc. **118**(45), 11225–11236 (1996)
21. Berendsen, H.J.C., Grigera, J.R., Straatsma, T.P.: The missing term in effective pair potentials. J. Phys. Chem. **91**(24), 6269–6271 (1987)
22. Khattab, I.S., Bandarkar, F., Fakhree, M.A.A., Jouyban, A.: Density, viscosity, and surface tension of water + ethanol mixtures from 293 to 323 K. Korean J. Chem. Eng. **29**(6), 812–817 (2012)
23. Harris, K.R., Newitt, P.J.: Self-diffusion of water at low temperatures and high pressure. J. Chem. Eng. Data **42**(2), 346–348 (1997)
24. Hess, B., Bekker, H., Berendsen, H.J., Fraaije, V.: LINCS: a linear constraint solver for molecular simulations. J. Comput. Chem. **18**(12), 1463–1472 (1997)
25. Miyamoto, S., Kollman, P.A.: SETTLE: an analytical version of the SHAKE and RATTLE algorithm for rigid water models. J. Comput. Chem. **13**(8), 952–962 (1992)
26. Bussi, G., Donadio, D., Parrinello, M.: Canonical sampling through velocity rescaling. J. Chem. Phys. **126**(1), 014101 (2007)
27. Berendsen, H.J., van Postma, J., van Gunsteren, W.F., DiNola, A., Haak, J.R.: Molecular dynamics with coupling to an external bath. J. Chem. Phys. **81**(8), 3684–3690 (1984)
28. Cuendet, M.A., van Gunsteren, W.F.: On the calculation of velocity-dependent properties in molecular dynamics simulations using the leapfrog integration algorithm. J. Chem. Phys. **127**(18), 184102 (2007)
29. Darden, T., York, D., Pedersen, L.: Particle mesh Ewald: an $N \cdot \log(N)$ method for Ewald sums in large systems. J. Chem. Phys. **98**(12), 10089–10092 (1993)
30. Matsugami, M., Yamamoto, R., Kumai, T., Tanaka, M., Umecky, T., Takamuku, T.: Hydrogen bonding in ethanol–water and trifluoroethanol–water mixtures studied by NMR and molecular dynamics simulation. J. Mol. Liq. **217**, 3–11 (2016)
31. Rajan, R., Balaram, P.: A model for the interaction of trifluoroethanol with peptides and proteins. Chem. Biol. Drug Des. **48**(4), 328–336 (1996)
32. Pikkarainen, L.: Excess enthalpies of (2,2,2-trifluoroethanol+2-butanone or N-methylacetamide or N,N-dimethylacetamide). J. Chem. Thermodyn. **20**(4), 481–484 (1988)
33. Pikkarainen, L.: Excess enthalpies of binary solvent mixtures of N-methylacetamide with aliphatic alcohols. J. Solut. Chem. **16**(2), 125–132 (1987)
34. Chand, A., Chowdhuri, S.: Behaviour of aqueous N-methylacetamide solution in presence of ethanol and 2,2,2 tri-fluoroethanol: hydrogen bonding structure and dynamics. J. Mol. Liq. **224**, 1370–1379 (2016)
35. Murto, J.: Hydroxide and alkoxide ions in alkanol-water mixtures. Acta Chem. Scand. **18**(5), 1029–1042 (1964)
36. Rao, C., Goldman, G., Balasubramanian, A.: Structural and solvent effects on the $n \rightarrow \pi*('U \leftarrow' A)$ transitions in aliphatic carbonyl derivatives: evidence for hyperconjugation in the electronically excited states of molecules. Can. J. Chem. **38**(12), 2508–2513 (1960)
37. Loomis, R.E., Lee, P.C., Tseng, C.C.: Conformational analysis of the cholecystokinin C-terminal octapeptide: a nuclear magnetic resonance and computer-simulation approach. Biochim. Biophys. Acta BBA-Protein Struct. Mol. Enzymol. **911**(2), 168–179 (1987)
38. Pattanayak, S.K., Chowdhuri, S.: Effect of water on solvation structure and dynamics of ions in the peptide bond environment: importance of hydrogen bonding and dynamics of the solvents. J. Phys. Chem. B **115**(45), 13241–13252 (2011)
39. Pattanayak, S.K., Chettiyankandy, P., Chowdhuri, S.: Effects of co-solutes on the hydrogen bonding structure and dynamics in aqueous N-methylacetamide solution: a molecular dynamics simulations study. Mol. Phys. **112**(22), 2906–2919 (2014)

40. Chowdhuri, S., Chandra, A.: Dynamics of halide ion-water hydrogen bonds in aqueous solutions: dependence on ion size and temperature. J. Phys. Chem. B. **110**(19), 9674–9680 (2006)
41. Rapaport, D.: Hydrogen bonds in water: network organization and lifetimes. Mol Phys. **50**(5)s, 1151–1162 (1983)
42. Luzar, A., Chandler, D.: Hydrogen bond kinetics in liquid water. Nat. Lond. **379**(6560), 55–57 (1996)
43. Luzar, A.: Resolving the hydrogen bond dynamics conundrum. J. Chem. Phys. **113**(23), 10663–10675 (2000)
44. van der Spoel, D., van Maaren, P.J., Larsson, P., Tîmneanu, N.: Thermodynamics of hydrogen bonding in hydrophilic and hydrophobic media. J. Phys. Chem. B. **110**(9), 4393–4398 (2006)
45. Keller, B., Daura, X., van Gunsteren, W.F.: Comparing geometric and kinetic cluster algorithms for molecular simulation data. J. Chem. Phys. **132**(7), 02B610 (2010)

Long Short-Term Memory Recurrent Neural Network Architectures for Melody Generation

Abhinav Mishra, Kshitij Tripathi, Lakshay Gupta and Krishna Pratap Singh

Abstract Recent work in deep learning has led to more powerful artificial neural network designs, including recurrent neural networks (RNNs) that can process input sequences of arbitrary length. We focus on a special kind of RNN known as a long short-term memory (LSTM) network. LSTM networks have enhanced memory capability which helps them in learning sequences like melodies. This paper focuses on generating melodies using LSTM networks and conducting a survey for verifying quality of melody generated. We used the Nottingham ABC Dataset, which is a database of over 14,000 folk songs in ABC notation and serves as the training input for our RNN model. We have also conducted a Turing Test to give the quality of the music generated by the model. We will also discuss the overall performance, design of the model and adjustments made in order to improve performance.

Keywords Recurrent neural network · Long short-term memory
Sequential learning

A. Mishra (✉) · K. Tripathi · L. Gupta · K. P. Singh
Machine Learning and Optimization Laboratory, IIIT Allahabad, Allahabad, India
e-mail: iim201401@iiita.ac.in; iim2014003@iiita.ac.in

K. Tripathi
e-mail: icm2014007@iiita.ac.in

L. Gupta
e-mail: iit2014044@iiita.ac.in

K. P. Singh
e-mail: kpsingh@iiita.ac.in

© Springer Nature Singapore Pte Ltd. 2019   41
J. C. Bansal et al. (eds.), *Soft Computing for Problem Solving*, Advances in Intelligent
Systems and Computing 817, https://doi.org/10.1007/978-981-13-1595-4_4

1   Introduction

In recent years, neural networks have become widely popular and are often mentioned along with terms such as machine learning, deep learning, data mining and big data. Deep learning methods perform better than traditional machine learning approaches on virtually every single metric. From Google's DeepDream that can learn an artists style, to AlphaGo learning an immensely complicated game as Go, the programs are capable of learning to solve problems in a way our brains can do naturally. To clarify, deep learning, first recognized in the 80s, is one paradigm for performing machine learning. Unlike other machine learning algorithms that rely on hard-coded feature extraction and domain expertise, deep learning models are more powerful because they are capable of automatically discovering representations needed for detection or classification based on the raw data they are fed.

Artificial Neural networks are biologically inspired paradigm which enables a machine to learn from observational data. They are very robust and powerful learning models that have resulted in the best and trendsetting results in variety of supervised and unsupervised problems of machine learning and data science. This success has been possible only due to the ability of these models to learn hierarchical representation. These models are able to extract the underlying patterns that have not been possible with other algorithms [1].

Recurrent neural networks (RNNs) are connectionist models. RNNs have the capability to pass information which are relevant across the sequence steps as they process the sequential data elements one at a time. RNNs are able to model sequential set of input to the corresponding set of outputs that may be dependent on other input element or set of input elements.

Why recurrent neural networks are useful and why it's worth inspecting? It is a question that we will try to answer in the following subsections. We are focused and motivated by the desire to achieve empirical results. RNNs have its root in both cognitive modelling and machine learning, but we will focus on achieving the empirical result. Many foundational papers [2–4] have devalued the biological inspiration in favour of achieving empirical results on various problems. Before diving into the architecture of LSTM networks, we will begin by the architecture of a regular recurrent neural network and its issues, and how LSTMs resolve that issue.

2   Recurrent Neural Networks

Recurrent neural networks (RNNs) are a class of neuron-based models for processing continuous and interdependent sequence. Various architectures of neural network are specialized in solving different types of problems. For example, convolutional neural networks are specialized in processing and finding hidden patterns in images; recurrent neural network on the other hand is suited for other group of tasks. RNNs are best suited for tasks in which we have to derive a relationship from a sequence of values $x^{(1)}, \ldots, x^{(\tau)}$ and the corresponding output $y^{(1)}, \ldots, y^{(\tau)}$. RNNs are same as feedforward neural network, only difference being the edges that span along time steps. These edges are called recurrent edges. These cycles are responsible for the interdependency of the present value of a node on its value at next or upcoming time step and brings a notion of time to the model. RNNs can scale to much longer sequences which is not possible by a model without sequence-based specialization. RNNs are also capable of processing sequences of variable lengths. We usually apply recurrent neural network by operating on mini-batches. The time step index t used in the input vector $x^{(t)}$ generally not refers to the notion of the time and may refer the successive positions in the sequential input. A well known result was proposed by Siegelman and Sontag in 1991 that a recurrent neural network having a finite size with the sigmoid as activation function can simulate universal Turing machine [5].

There are various ways possible for building a RNN. Recurrent neural network has the properties that allows it to model any function which involves a previous time step dependence or recurrence where as feedforward networks cannot model functions where there is a notion of time or recurrence involved. The following equation is widely used by RNNs to define the values of their hidden units.

$$h^{(t)} = f(h^{(t-1)}, x^{(t)}; \theta)$$

There are some additional architectural advancements such as output layers that use state information $h^{(t)}$ to predict the output $\hat{y}^{(t)}$ at the tth time step. The network learns to use $h^{(t)}$ as a kind of compact summary of the task-relevant aspect of the already processed sequence which allows the network to perform task that requires prediction of future from the available information in the past. The network makes a mapping of the processed sequence $(x^{(t)}, x^{(t-1)}, \ldots, x^{(2)}, x^{(1)})$ to a arbitrary lengthed vector $h^{(t)}$. There is a possibility that the relevant information from past stored in $h^{(t)}$ might contain some features of the past with more precision and some it may forget depending on the training criterion. Figure 1 depicts a simple recurrent neural network, and we try to develop the equations for the forward pass. A loss function L represents the negative log-likelihood which gives the measurement of how far each $o^{(t)}$ is from the corresponding $y^{(t)}$. The training of the recurrent neural network is done in two passes similar to feedforward neural network, forward pass and backward pass. Following equations specify all the computational calculations at different time steps during the forward pass.

$$a^{(t)} = b + Wh^{(t-1)} + Ux^{(t)}$$
$$h^{(t)} = \tanh(a^{(t)})$$
$$o^{(t)} = c + Vh^{(t)}$$
$$\hat{y}^{(t)} = \text{softmax}(o^{(t)})$$

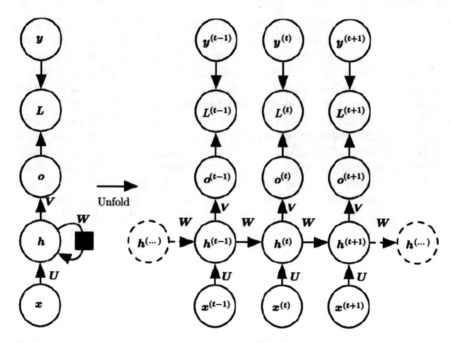

Fig. 1 Recurrent neural network unfolded for the each time step to compute the training loss. Input sequence $x^{(t)}$ are mapped to corresponding sequence of output values $\hat{y}^{(t)}$

Here, the connections between the layers of RNNs are parametrized by weight matrices. The connections between input to hidden layer are parametrized by U, connections between hidden to hidden layer by W and connections between hidden to output layer by V. b and c represent the bias parameters which is necessary for the model to learn the offset.

L represents the overall cost for a continuous sequence of x values along with a continuous sequence of y values which will be the aggregate of all the individual losses over all the time steps. Given $x^{(1)}, \ldots, x^{(t)}$, the negative log-likelihood of $y^{(t)}$ is given by $L^{(t)}$ as

$$L\left(\{x^{(1)}, \ldots, x^{(\tau)}\}, \{y^{(1)}, \ldots, y^{(\tau)}\}\right) = \sum_t L^{(t)}$$

$$= -\sum_t \log p_{model}\left(y^{(t)} | \{x^{(1)}, \ldots, x^{(t)}\}\right),$$

where the value of $p_{model}\left(y^{(t)} | \{x^{(1)}, \ldots, x^{(t)}\}\right)$ is determined by the corresponding entry of $y^{(t)}$ in the output vector of t time step $\hat{y}^{(t)}$. We need to compute the gradient of the loss function with respect to the parameters in order to train the model. The gradient calculation is computationally expensive operation and cannot be reduced by parallel computation due to the inherent sequentiality of the nodes. The network can be trained across many time steps using backpropagation. The expansion of backpropagation across time steps is called *backpropagation through time* (BPTT) [6]. The run-time complexity and the memory complexity of the forward pass is $O(\tau)$.

2.1 Training Recurrent Networks

Training with recurrent networks has always been considered a difficult task. Even for a single-layered feedforward networks, the optimization task is NP-complete [7] and learning the long-range dependencies makes the job of learning with recurrent network even more difficult. For training the RNN, we need to compute the gradient of loss function with respect to parameters. For this, we simply apply the generalized backpropagation algorithm to the unfolded RNN across time steps. Gradient resulted by backpropagation can be employed with different gradient-based training technique such as Gradient Descent, Adagrad. The nodes of our RNN include parameters U, V, W, b and c. It also includes the respective input $x^{(t)}$, hidden state $h^{(t)}$, output preactivation $o^{(t)}$ and the loss function $L^{(t)}$ at the t time step. The gradient computation of each node N, $\nabla_N L$ is required to be done recursively, which depend on the gradient of precomputed nodes. The recursion begins with the nodes preceding the final loss.

$$\frac{\partial L}{\partial L^{(t)}} = 1$$

In the following equations, we will assume that we are using softmax function for calculating the output probability vector $\hat{y}^t$ from output $o^{(t)}$. We also assume that the loss is the cross-entropy loss. The gradient with respect to the outputs $o^{(t)}$ at the time step t is

$$(\nabla_{o^{(t)}} L)_i = \frac{\partial L}{\partial o_i^{(t)}} = \frac{\partial L}{\partial L^{(t)}} \frac{\partial L^{(t)}}{\partial o_i^{(t)}} = \hat{y}_i^{(t)} - 1_{i,y^{(t)}}$$

The gradient calculation begins from the end of the sequence. The gradient calculation for $h^{(t)}$ depends on whether t is the final time step. If $t = \tau$, where τ is the final time step, gradient of loss function with respect to $h^{(\tau)}$ is:

$$\nabla_{h^{(\tau)}} L = V^\top \nabla_{o^{(\tau)}} L.$$

From $t = \tau - 1$, the gradient is then propagated backwards in time to $t = 1$. Since both $o^{(t)}$ and $h^{(t+1)}$ are descendents of $h^{(t)}$, hence its gradient is computed as

$$(\nabla_{h^{(t)}} L) = \frac{\partial L}{\partial h^{(t)}} = \frac{\partial L}{\partial h^{(t+1)}} \frac{\partial h^{(t+1)}}{\partial h^{(t)}} + \frac{\partial L}{\partial o^{(t)}} \frac{\partial o^{(t)}}{\partial h^{(t)}}$$

$$= W^{\top} (\nabla_{h^{(t+1)}} L) diag\left(1 - \left(h^{(t+1)}\right)^2\right) + V^{\top} (\nabla_{o^{(t)}} L)$$

As the gradients of the loss function with respect to the internal nodes are calculated, gradients on the parameter nodes are obtained.

2.2   Challenges of Long-Term Dependencies

The major difficulty that optimization algorithms like backpropagation through time (BPTT) faces when the depth of network increases too much. Depth of the network represents the sequence length or the time sequence duration. When deep computational graphs are constructed by repeatedly applying the same operation at each time step of a long temporal sequence, it gives rise to noticeable difficulties. We are able to see that activation function such as $tanh$ (Fig. 2) and $sigmoid$ tend to approach a flat line when value of the input activation function increases too much, resulting in the derivatives limiting to zero at both ends and the saturation of the corresponding neurons. The neurons have a zero gradient, and it results in driving other gradients in previous layers towards zero. Hence, the multiple matrix multiplications result in exponential shrinking of the gradient values, vanishing after few time steps only. Since gradient contributions of steps which are far become negligible, it does not contribute to learning procedure making the model unable to learn the long-term dependencies. This problem is called vanishing gradient problem [8], and it is not

Fig. 2 The red line represents the gradient of *tanh(x)*. The gradient reaches zero as |x| takes large value

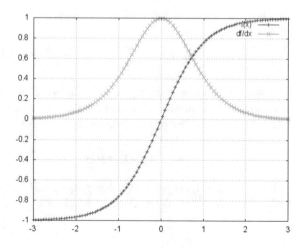

exclusive only to RNNs and may occur in all connectionist models in which depth of layers increases.

It is possible to encounter exploding gradients instead of vanishing gradients hinging on the activation functions and the architecture of the network. There is a twofold reason as to why it is vanishing gradient problem only that is more talked about than exploding gradient problem. First one being that if the gradient becomes two large it will NaN (not a number). The second reason is that gradients can be clipped after a defined threshold. Vanishing gradients are not this easy to tackle, and unlike exploding gradients it is not obvious when this problem may occur.

The vanishing gradient problem can be easily tackled. Regularization and initialization of weight matrices with proper weights are two ways to combat the problem. A more better and commonly preferred solution is the usage of Rectifying Linear Unit (ReLU) activation function instead of using *sigmoid* or *tanh* activation functions. ReLU allows only two possible gradient values 0 and 1, which makes it less susceptible to vanishing gradient problem. Other possible solution for this results in change in basic architecture of recurrent neural network architecture. *Long short-term memory* (LSTM) and *gated recurrent unit* (GRU) [9] architectures are advancements in vanilla RNN architecture which has resulted in better learning and long-range dependencies. LSTMs are one of the most widely used models in Natural Language Processing and were proposed in 1997. Both LSTM and GRU architectures are designed explicitly for dealing with the problem of vanishing gradient and for learning long-range dependencies. Improved architectures and better gradient-following heuristics have made training of RNN possible and feasible. Packages and libraries like *Theano*, *Torch* and *Tensorflow* have made possible the efficient implementation of challenging architectures of multilayered neural networks.

3   LSTM Architecture

The vanilla RNN architecture suffers from the various complications like vanishing gradient problem, convergence problems. There was need of more robust architectures that do not suffer same complications as RNN. In 1997, two research papers published gave the most successful recurrent neural network architecture for sequence modelling. The first paper, *Long Short-Term Memory* (LSTM) model [10], proposes a brilliant idea of introduction of self-loops in order to produce paths where the flow of gradient for long duration is possible. A memory cell is introduced which acts as a unit of computation and replaces the traditional hidden nodes of the network.

3.1   *Long Short-Term Memory (LSTM)*

The LSTM model was introduced by Hochreiter and Schmidhuber in 1997. LSTMs are explicitly designed to tackle the problem of long-term dependencies. LSTM-RNN is a type of gated recurrent neural network. Gated RNNs are able to make

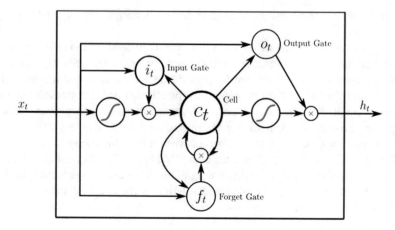

Fig. 3 A LSTM block with input, output and forget gates

paths through time whose derivatives neither explode nor vanish. Rather than being fixed, the weight of the self-loop is conditioned on the context. The structure of RNNs is like a chain of repeated modules. In vanilla RNNs, this repeating module is something as simple as a *tanh* layer, however, in LSTMs [11] this chain-like structure comprises of module having different architecture. In this architecture, there are four neural network layers instead of one, interacting in a very special way. Machine translation, unconstrained handwriting recognition [12], handwriting generation, protein sequence prediction [13], image captioning and parsing, etc., are some of the applications in which LSTM has proved to give groundbreaking results and has been used recurrently by the scientific community.

The LSTM block diagram is illustrated in Fig. 3. LSTMs contain LSTM cells having recurrence relation internally instead of traditional nodes in vanilla RNN that simply apply a element wise nonlinearity of inputs and recurrent units. Each of these cells works as a ordinary recurrent network, and there is a system of gated units which control the flow of information.

The most crucial part of the state unit here is $s_i^{(t)}$ that contains a linear self-loop; however, the controlling of the self-loop weights is done by a **forget gate** unit f_i^t for the ith cell and tth time step. This unit sets the weight of the self-loop to a value between 0 and 1 using activation function.

$$f_i^{(t)} = \sigma\left(b_i^f + \sum_j U_{i,j}^f x_j^{(t)} + \sum_j W_{i,j}^f h_j^{(t-1)}\right),$$

where $x^{(t)}$ represents input vector and $h^{(t)}$ represents hidden layer at t time step. b^f, U^f, W^f represent biases, weights and recurrent weights for the forget gates, respectively. Internal states are updated as follows:

$$s_i^{(t)} = f_i^{(t)} s_i^{(t-1)} + g_i^{(t)} \sigma \left(b_i + \sum_j U_{i,j} x_j^{(t)} + \sum_j W_{i,j} h_j^{(t-1)} \right),$$

Here $\boldsymbol{b}$, $\boldsymbol{U}$, $\boldsymbol{W}$ represent the biases, input weights and recurrent weights in the LSTM cell, respectively. $g_i^{(t)}$ is called **external input gate** unit and is computed the same way as the forget gate. It gives a gating value between 0 and 1 for the current input value and uses its own parameters $\boldsymbol{b}^g$, $\boldsymbol{U}^g$, $\boldsymbol{W}^g$:

$$g_i^{(t)} = \sigma \left(b_i^g + \sum_j U_{i,j}^g x_j^{(t)} + \sum_j W_{i,j}^g h_j^{(t-1)} \right),$$

There is also a **output gate** $q_i^{(t)}$ which controls the output $h_i^{(t)}$ of the LSTM and uses sigmoid unit for gating:

$$q_i^{(t)} = \sigma \left(b_i^o + \sum_j U_{i,j}^o x_j^{(t)} + \sum_j W_{i,j}^o h_j^{(t-1)} \right)$$

$$h_i^{(t)} = \tanh \left(s_i^{(t)} \right) q_i^{(t)}$$

Here $\boldsymbol{b}^o$, $\boldsymbol{U}^o$, $\boldsymbol{W}^o$ denote the biases, input weights and recurrent weights for the output gate of the LSTM cell.

Researches have been able to achieve exceptional results and accuracies in variety of applications using RNNs and fare share of those is achieved using LSTMs only. Several models and variations have been proposed based on long short-term memory model. The Forget gates were not proposed in the original LSTM design, and it was later proposed in 2000. Use of forget gates have been effective and hence are used as a standard design in the implementation. Grid LSTM by Kalchbrenner [14] also seems extremely promising architecture. LSTM was a big step in what we can achieve and accomplish with RNNs and a great possibility of future work. This field of machine learning has seen a tremendous run in the last few years and coming ones promise to only be more so.

4 Overview of Training Process

Learning problems can be grouped into two basic categories: supervised and unsupervised. Supervised learning includes classification, prediction and regression, where the input vectors have a corresponding target (output) vectors. The goal is to predict the output vectors based on the input vectors. In unsupervised learning, such as clustering, there are no target values and the goal is to describe the associations and patterns among a set of input vectors [15]. The LSTM implementation solves a supervised learning problem: given a sequence of inputs, we want to predict the probability of the next output.

Normally to perform machine learning, it is best to break a given dataset into three parts: a training set, a validation set and a test set. The training set is used for learning; the validation set is used to estimate the prediction error for model selection; the test set is used for assessment of the generalization error of the final chosen model. A general rule of thumb is to split the dataset 50% for training and 25% each for validation and testing.

In our case, our goal is to build a model that can predict the next word or that next music note; this is a generative model in which we can generate new text or music by sampling from the output probabilities. Therefore, we will be having training and validation set to fine tune the model, but we will not be having a test set. Instead, to generate an output we can feed in a randomly selected batch of data from the training.

5   Implementation of Melody Generation

5.1   Char-RNN Model

The char-RNN model is a robust model which uses LSTM architecture to learn the structure of the input in the form of text file. It takes a single text file as an input and feeds it into the RNN algorithm that learns to predict the next character in the sequence. After training the RNN, it can generate text character by character that looks stylistically similar to the original dataset. The code is written in Python and uses Numpy library. In this model, we have option to have multiple layers, supporting code for model checkpointing, and using mini-batches to make the learning process efficient.

5.2   Input Dataset: ABC Notation

We used Nottingham Folk Music Dataset as input for the char-RNN model. It is a collection of tunes in ABC format. Nottingham Music Dataset is a combination of 14,000 British and American folk tunes. The ABC notation was developed by Chris Walshow so that music can be represented using ASCII symbols [16]. The basic structure of the abc notation is the header and the notes. The header which contains background information about the song can look something like below:

```
X: 145
T: A and A's Waltz
S: Mick Peat
M: 6/8
K: D
```

```
d4A2c2  |  d4d4  |  A3cd2g2  |  c2A2G4  |
A3cd2g2  |  d2c4A2  |  c3AG2E2  |  D8  |
G2^F2G4  |  G3Ac2d2  |  d4c2d2  |  G4D4  |
A3cd2g2  |  d2c4A2  |  c3AG2E2  |  D8  |
d8  |  c3AG4  |
A3cd2g2  |  d2c4A2  |  c3AG2E2  |  D8
```

Fig. 4 ABC notation

These lines are known as fields, where X is the reference number, T is the title of the song, C is the composer, M is the metre, and K is the key. The notes portion looks something like in Fig. 4.

6 Results and Discussion

6.1 Checkpoints and Minimum Validation Loss

During training, a checkpoint is saved every 1,000 iterations and at every checkpoint. The checkpoint file saves the current values for all the weights in the model. The smaller the loss, the better the checkpoint file works when generating the music. Due to possible overfitting, the minimum cross-entropy validation loss is not necessarily at the end of the training. For example, the Table 1 and Fig. 5 show all the saved checkpoints during a single training.

The minimum validation loss occurs at the 7,000th iteration out of 10,000 iterations. Rerunning the code on the same dataset produced the same loss values.

Table 1 Iteration versus cross-entropy loss

Nth iteration (out of 10000)	Cross-entropy validation loss
1000	381.34
2000	216.56
3000	131.21
4000	80.75
5000	73.20
6000	71.92
7000	69.32
8000	72.64
9000	74.21
10,000	86.52

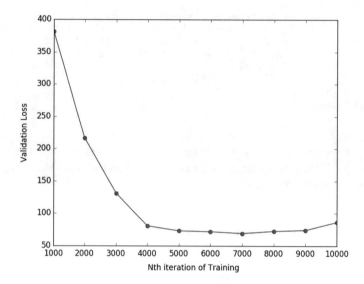

Fig. 5 Plot of validation losses for various checkpoints saved at every 1000th iteration

Each iteration on average takes about 0.33 s, and in total the model takes about 3–4 h to train on CPU.

6.2 Decreasing the Batch Parameter

These default parameters worked well for the dataset. As an experiment, we decreased the batch size from the default size of 100–50. The batch size specifies how many streams of data are processed in parallel at one time. If the input text file has N characters, these get split into chunks of size [batch size] × [sequence length] (length

Table 2 Iteration versus cross-entropy loss (batch size = 50)

Nth iteration (out of 10,000)	Cross-entropy validation loss
1000	411.32
2000	226.65
3000	121.38
4000	85.25
5000	73.61
6000	66.62
7000	69.82
8000	74.41
9000	74.28
10,000	79.52

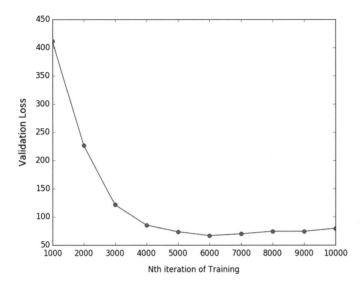

Fig. 6 Plot of validation losses for various checkpoints saved at every 1000th iteration where batch size = 10,000

```
X:11
% Nottingham Music Database
F: http://www.youtube.com/watch?v=6qRO2Tq-I
S:Ralchel
M:6/8
K:D
"Em"e/2d/2e/2d/2 e/2B/2A/2B/2d/2|"A"f/4e/4 e/2G/2c|
|"D"d/2c/2d/2e/2 af/2g/2|g/2f/2e/2d/2 AG|
:"A7"D2B, "Bb"B/2d/2B/2e/2|"B7""b"d2 d3/2B/2|
 [1"C"GF "F#7"f2c|"A"^FE E2|
"D"DA zd|"C"e/2g/2f/2g/2 a/2|B/2d/2 "g#m"b2|\
"D"f/2g/2f/2f/2 c/2B/2A|\
"F#m"A/2B/2A/2c/2 BA|"G"B3/2d/2 dg/2a/2|\
"G"Bd GB/2B/2|
"Am"c/2f/2d/2|\
"C"g/2d/2B/2d/2 "C7"ef|"Eb/g"g/2c/2B/2A/2 FA|"D"dd fA/2c/2|
"Bm"dB/2A/2 B/2A/2G/2B/2|F/2G/2D/2F/2 F/2E/2D/2A/2|
"G/b"d/2z/4f/4 dd
```

Fig. 7 Sample output text

of each step, which is also the length at which the gradients can propagate backwards in time).

The resulting validation loss was less than when trained with batch size of 50, as shown in the Table 2 and Fig. 6.

Figure 7 is a sample output text from the model (default parameters).

Table 3 Confusion matrix

	Machine-generated	Human-generated	
Machine-generated	611	889	40.7%
Human-generated	233	1267	84.4%
	72%	58%	62.6%

6.3 Turing Test

We collected a set of 15 artificially produced melodies by our model and 15 human-generated melodies and took a survey to classify the music as human-generated or machine-generated on a set of 100 people. The results of the survey are represented as confusion matrix as follows.

As per the Turing Test, it was found that people were not able to distinguish between the human-generated and machine-generated melodies. After conducting the test, we found that over 62% of the test subjects were convinced that the melodies were at par with human music.

The challenge of getting a successful result from this char-RNN model was the limitation of the small dataset. There are not many songs that are in abc format, and the songs themselves are short (only about 10–20 measures per song) and very simple tunes. The first time we ran the char-RNN it was on a dataset of 277 KB, which is far less than the minimum required size of 1 MB for the char-RNN to produce tangible results. After finding more music from another source, the dataset increased to 451 KB, and the results were significantly better. Overall, music generated from the model with the larger dataset sounded better (Table 3).

7 Conclusion

The LSTM-RNN is able to learn from the songs and generate melodies on its own which sounds similar to human made music. However, there are some outputs where occasionally there would be a note or two that does not sound cohesive with the rest of the music, which may be because the note is not part of the music's scale. LSTM models have proven to be robust in generating music that is very similar to the given input dataset.

There are many optimization techniques involved in neural networks that people may explore further. Overall we are impressed with the progress and results deep learning techniques have accomplished in the recent years. It is fascinating to see deep neural networks' capability to create something that is aligned with the style of the given input, and yet have its own unique taste that can be sometimes bizarre but sometimes incredible. Researchers may collaborate with people with background in music theory in order to improve the accuracy of the model.

References

1. Karpathy, A.: The Unreasonable Effectiveness of Recurrent Neural Networks, Github, 21 May 2015, Web 04 May 2016
2. Schuster, M., Paliwal, K.K.: Bidirectional recurrent neural networks. IEEE Trans. Signal Process. **45**(11), 2673–2681 (1997)
3. Socher, R., Karpathy, A., Le, Q.V., Manning, C.D., Andrew, Y.Ng.: Grounded compositional semantics for finding and describing images with sentences. Trans. Assoc. Computat. Linguist. **2**, 207–218 (2014)
4. Karpathy, A., Fei-Fei, L.: Deep Visual-Semantic Alignments for Generating Image Descriptions (2014). arXiv:1412.2306
5. Siegelmann, H.T., Sontag, E.D.: Turing computability with neural nets. Appl. Math. Lett. **4**(6), 77–80 (1991)
6. Werbos, P.J.: Backpropagation through time: what it does and how to do it. Proc. IEEE **78**(10), 1550–1560 (1990)
7. Blum, A.L., Rivest, R.L.: Training a 3-node neural network is NPcomplete. In: Machine Learning: From Theory to Applications, pp. 9–28. Springer (1993)
8. Hochreiter, S.: The vanishing gradient problem during learning recurrent neural nets and problem solutions. Int. J. Uncertain. Fuzziness Knowl. Based Syst. (1995)
9. Graves, A., Schmidhuber, J.: Framewise phoneme classification with bidirectional LSTM and other neural network architectures. Neural Netw. **18**(5), 602–610 (2005)
10. Hochreiter, S., Schmidhuber, J.: Long short-term memory. Neural Comput. **9**(8), 1735–1780 (1997)
11. Gers, F.A.: Long short-term memory in recurrent neural networks. Unpublished Ph.D. dissertation, Ecole Polytechnique Federale de Lausanne, Lau sanne, Switzerland, 2001
12. Graves, A., Liwicki, M., Bunke, H., Schmidhuber, J., FernÃandez, S.: Unconstrained on-line handwriting recognition with recurrent neural networks. In: Platt, J., Koller, D., Singer, Y., Roweis, S. (eds.) NIPS'2007, pp. 577–584 (2008)
13. Baldi, P., Brunak, S., Frasconi, P., Soda, G., Pollastri, G.: Exploiting the past and the future in protein secondary structure prediction. Bioinformatics **15**(11), 937–946 (1999)
14. Kalchbrenner, N., Danihelka, I., Graves, A.: Grid Long Short-Term Memory (2015). arXiv:1507.01526
15. Graves, A.: Supervised Sequence Labelling with Recurrent Neural Networks, vol. 385. Springer (2012)
16. Walshaw, C.: How to Understand Abc (the Basics). Web log post. ABC Notation Blog. WordPress, 23 Dec 2009, Web 18 Dec 2015

A Survey on Recurrent Neural Network Architectures for Sequential Learning

B. Shiva Prakash, K. V. Sanjeev, Ramesh Prakash
and K. Chandrasekaran

Abstract The expanding textual information and significance of examining the substance has started a colossal research in the field of synopsis. Text summarization is the process of conveying the gist of a text with a minimized representation. The requirement for automation of the procedure is at its apex with exponential burst of information because of digitization. Text captioning comes under the branch of abstractive summarization which captures the gist of the article in a few words. In this paper, we present an approach to text captioning using recurrent neural networks which comprise of an encoder–decoder model. The key challenges dealt here was to figure out the ideal input required to produce the desired output. The model performs better when the input is fed with the summary as compared to the original article itself. The recurrent neural network model with LSTM results has been effective in transcribing a caption for the textual data.

1 Introduction

Topic modeling is a popular tool for extracting the different topics in a given document and is useful in mining the concealed semantic knowledge. Latent Dirichlet allocation is a popular topic modeling tool which can be classified under

B. Shiva Prakash (✉) · K. V. Sanjeev · R. Prakash · K. Chandrasekaran
Department of Computer Science and Engineering, National Institute
of Technology Karnataka, Surathkal 575025, Karnataka, India
e-mail: shiva96b@gmail.com

K. V. Sanjeev
e-mail: sanjeev.vadiraj@gmail.com

R. Prakash
e-mail: rameshprakash6196@gmail.com

K. Chandrasekaran
e-mail: kchnitk@ieee.org

© Springer Nature Singapore Pte Ltd. 2019
J. C. Bansal et al. (eds.), *Soft Computing for Problem Solving*, Advances in Intelligent
Systems and Computing 817, https://doi.org/10.1007/978-981-13-1595-4_5

statistical and probabilistic models. Classification of information is very demanding today which implies the difficulty in accessing relevant data in the desired form. Topic modeling is a robust method to help obtain the required data based on the query. This text mining algorithm determines the point of conversation in the document. Information retrieval from unstructured textual data and document organization are some of the key applications which benefit from topic modeling.

Automatic refers to doing any task with very less or no human involvement. Content is any arrangement of characters which is discernable by a human. By and large, these groupings of characters consolidate to pass on some message which has a significance. Abridge intends to incorporate just the vital parts of the subject without going into particular little points of interest. Automatic text summarization is an automation of the process of creating a rundown of the given content, that will spare sizable human time.

Recurrent neural networks have as of late been observed to be extremely compelling for many errands—that is changing and transforming content representation. Cases of such applications incorporate machine interpretation and machine intelligence. Recurrent neural networks have additionally been connected as of late to perusing comprehension. There are models which are trained to perceive and comprehend passages and generate their interpretation of the same.

This paper presents an approach to process the text and generate a caption or heading with the help of recurrent neural networks. In the initial phase of the algorithm, important sentences of the given text are extracted through traditional extractive methods and fed it to the neural network after embeddings and vectorization. The recurrent network consists of long short-term memory (LSTM) nodes which along with the attention weights help caption or summarize the text in a few words. An encoder–decoder model is generated using recurrent neural network to generate the headline for a given document. The applications of this algorithm are spread over all domains where topic modeling or headline generation is prescribed. Annotating documents is one other app.

The rest of the paper is organized as follows. The next section introduces important concepts associated with the approach. In Sect. 3, we delve into particulars of the algorithm. Section 4 presents the implementation of the approach followed. This is followed by the result and analysis of the model. In Sect. 5, we discuss the limitations and possible improvements for our model and conclude the paper. Section 6 lists all the papers and works which have impacted this paper.

2 Key Concepts

2.1 Recurrent Neural Network

A recurrent neural network (RNN) belongs to a set of neural networks where the neurons form a directed cycle, resulting in a feedback to some part of the network.

Fig. 1 Long short-term
memory

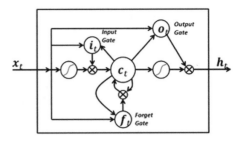

This helps in RNN to adapt itself to dynamic changes in the pattern of the data. RNNs perform extremely well over other algorithms in handwriting recognition and speech recognition. RNN is also the most suited NN for the problem statements involving sequence-in architecture and sequence-out architecture. For NLP, these networks are best suited.

2.2 LSTM

Long short-term memory (LSTM) is a type of RNN which has the unique ability to store the states of the internal neurons. The main advantage of LSTM in many applications over other RNNs and HMMs is it is relative insensitivity to gap length. Each unit of LSTM has three gates, namely read gate, write gate, and forget gate. These gates help to manipulate, store, and read the values in the unit as is necessary. The rate at which information should be dropped or learnt from other units can also be decided for these units (Fig. 1).

2.3 Glove

Glove is an alternate approach to word2vec which is used to get the word embedding. It helps to convert a word into a vector representation. For all the words, based on the words occurring in the context of the given word, a high-dimensional matrix of co-occurrence is constructed. While word2vec is a predictive model, Glove is a count-based model. The dimensionality reduction of the matrix results in a vector representation of each word. As Glove is more easily parallelizable than word2vec, it trains more faster and is used in this paper.

2.4 Sequence-to-Sequence Model

This model, unlike most other machine learning models, maps an input sequence to an output sequence. The advantage of this model is from the fact that it can accommodate a sequence output than a single output term as in classification and regression. Recurrent neural nets are used as the encoders and decoders. Feedback from the output to attention mechanism makes it dynamic and the process of output generation iterative.

3 Proposed Methodology

The entire approach consists of extracting the important sentences in the input text followed by feeding those sentences into a RNN model to get the headline.

Recursive neural network (RNN) is used as encoder and decoder to generate the headline from the given extracted summary. One-million news articles and their headlines are used to train the model. News articles are initially subjected to extraction, and these extracted sentences are fed into the tokenizer. Tokenized words are replaced by their word embedding, which is the vector representation of the word under consideration. This word embedding forms the first layer of the encoder neural net. Similarly, each word of the input text is fed into the encoder in series to obtain the intermediate representation of the input text at the output of the encoder. The last symbol of the input marks the final intermediate representation. This intermediate representation is fed into the decoder RNN. The decoder will have the same architecture as the encoder along with the feedback from the output of itself. Based on the intermediate representation which is fed in from the encoder, decoder outputs the first word (word embedding) of the headline. Comparing this with the first word of the actual headline, the error is back propagated and the weights are corrected to minimize the error.

Algorithm 1 Extraction and Abstraction

---

procedure TESTING
    Data ← Cleaning (Input Article)
    Features ← Feature Extraction (Data)
    Clusters ← K-Means Clustering (Features)
    Extracted Summary ← Selection (Clusters)
    Input Vectors ← Tokenization, Glove Vectorization (Summary)
    Output Vector ← Recursive Neural Network (Vectors)
    Caption ← Glove Transform (Output Vector)
end procedure

The decoder RNN, along with the first word, also gives the weights of each input word over that output word and helps in determining the importance of every input

Fig. 2 Methodology

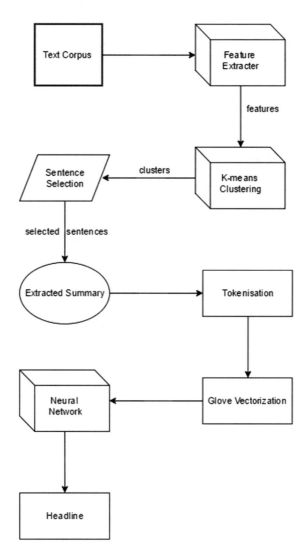

word. The above weights are used to calculate the weighted average of the intermediate representation of the encoder. The intermediate representations are stored in the LSTMs. The process from here changes for training and testing. In the training phase, this weighted average and the Nth word in the actual headline is used to predict the Nth word of the output, while in the testing phase, $(N - 1)$th word and the weighted average is used to predict the Nth word of the output headline. To overcome this imbalance in training and testing mechanism, in 15% of the training data, we use the feedback from the decoder instead of the actual word from the headline (Fig. 2).

The initial step of both training and testing phase is extraction. The text whose summary is to be generated is initially fed into the feature extractor. The feature extractor module extracts all the necessary features which compromise of the sentence length, tf-idf of each word in each sentence, frequency of each word in each sentence, and position of the sentence in the paragraph. All the features extracted for each sentence are next fed into the k-means clustering algorithm. For the k-means algorithm, each sentence is a new data instance, with each instance having its own values for the features extracted in the feature extractor. Based on the Euclidean distance, similar sentences are grouped under the same cluster. The number of clusters has to be predecided in the k-means clustering algorithm. This problem is overcome by fixing the number of clusters to be based on the tf-idf scores extracted from the summary. An alternate approach would be to set the number of clusters to be same as the number of sentences required for the engendered summary. From each of the clusters, one sentence which is a representative of the entire cluster of sentences is selected based on the attributes extracted. Hence, this sentence forms a base sentence around which the other sentences belonging to the same cluster are clubbed to form the summary.

Fig. 3 RNN
encoder–decoder mechanism

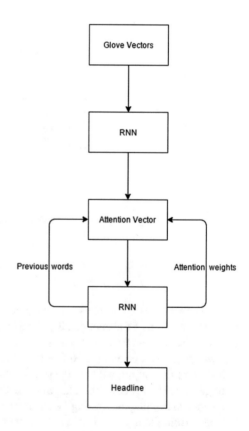

4 Data and Implementation

The signal media one-million news articles' dataset was used for the training and testing of the model. The dataset is mainly used for information retrieval. The dataset provided the required diversity, as it was compiled over a period of one month and consisted of news articles from various news agencies and sources. The data are present in a json format which is easily accessible and parsable. Each json had an unique id, title, content, and source along with the date of publication and article class which was not used in this methodology. The news articles are from approximately 93,000 different sources containing around 250,000 blog articles and 700,000 news articles.

As far as the preprocessing of data, the anomalies in the dataset where the title exceeds 10 tokens were summarized to less than 10 words. In cases, where the text title is missing, the data fragment was ignored. The text was cleaned, tokenized, and lowercased. The processed data were divided into a 75–25 ratio (training : testing). As the initial phase of the algorithm was extraction-based summarization, it required less processing power compared to the training of the neural net. The neural net was trained on a Nvidia GTX 1060 GPU to reduce the training time. The complete training of the dataset required 2 days (Fig. 3).

5 Experimental Results and Evaluation

Bilingual evaluation understudy (BLEU) is an algorithm for evaluating the quality of text which has been machine-translated from one natural language to another. BLEU is designed to approximate human judgement at a corpus level and performs badly if used to evaluate the quality of individual sentences. In statistical analysis, the F-score is a measure of a test's accuracy. It considers the harmonic mean of the precision and the recall of the test to compute the score (Table 1 and Fig. 4).

The BLEU score is a standard evaluation measure, but here another evaluation technique is used. The similarity between the captioned text and the original title is calculated with the help of doc2vec. Doc2Vec calculates the cosine distance between the extracted topic and original topic, which is a standardized value between 0 and 1. The approach is similar to word2vec where each word is represented as a vector after which similar or dissimilar words can be derived (Table 2 and Fig. 5).

Table 1 BLEU scores after each epoch

Epochs	BLEU score
1	0.03
2	0.04
3	0.06
4	0.07

Fig. 4 Results

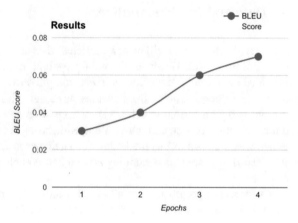

Table 2 Evaluation based on Doc2Vec similarity

Length of headline	Average cosine similarity
3	0.55
4	0.56
5	0.68
6	0.61
7	0.54
8	0.46
9	0.48

Fig. 5 Evaluation

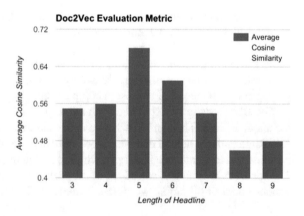

From the results obtained in the similarity of topics, the best results can be observed when the length of the heading is at 5 words. The results for headlines where the length of the topic is greater than 5 is always lesser than the similarity achieved at the 5-word topic. This might be because the predicted topic after five words does not hold the same significance as the initial words. Going by the charts, the ideal topic length is 5 which summarizes the article in an optimal way.

6 Conclusion and Future Work

This model hits an average BLEU score of 0.07 after four epochs and is seen to be a non-decreasing function with the mentioned parameters. Better results are seen when the topics are evaluated by the similarity using doc2vec. The values are promising and provide substantial support to the existing methods of topic modeling. Captioning and summarization models need to comprehensively interpret the data in order to engender coherent summaries. A similar approach to generate the whole summary is computation expensive. Combining extraction with the neural nets makes the headline generation more directed and informative and also decreases the computation required. There is a tremendous scope for models that automatically deliver accurate summaries of longer text which can be useful for digesting large amounts of information.

To further extend the work, the recurrent neural network can be improved by increasing the decision parameters while updating the neurons. The word predicted by the neural net is based on the neural weights and the previous word only. A better mechanism would be to increase the significance of other contributing words as well. Also for the model, there is scope in improving the grammar. The neural net could better interpret the linguistic and semantic structure of the language.

References

1. Das, D., Martins, A.F.T.: A survey on automatic text summarization. Lit. Surv. Lang. Stat. II Course CMU **4**, 192–195 (2007)
2. Torres-Moreno, J.-M.: Automatic Text Summarization. Wiley (2014)
3. Barzilay, R., Elhadad, M.: Using lexical chains for text summarization. Adv. Autom. Text Summ. 111–121 (1999)
4. Luhn, H.P.: The automatic creation of literature abstracts. IBM J. Res. Dev. **2**(2), 159–165 (1958)
5. Radev, D.R., Jing, H., Budzikowska, M.: Centroid-based summarization of multiple documents: sentence extraction, utility-based evaluation, and user studies. In: Proceedings of the 2000 NAACL-ANLP Workshop on Automatic Summarization. Association for Computational Linguistics (2000)
6. Page, L., et al.: The PageRank Citation Ranking: Bringing Order to the Web (1999)
7. Litvak, M., Vanetik, N.: Multi-document summarization using tensor decomposition. Computacin y Sistemas **18**(3), 581–589 (2014)
8. https://www.lexalytics.com/lexical-chaining
9. Rocktschel, T., et al.: Reasoning About Entailment with Neural Attention (2015). arXiv:1509.06664
10. Banko, M., Mittal, V.O., Witbrock, M.J.: Headline generation based on statistical translation. In: Proceedings of the 38th Annual Meeting on Association for Computational Linguistics. Association for Computational Linguistics (2000)
11. Dorr, B., Zajic, D., Schwartz, R.: Hedge trimmer: a parse-and-trim approach to headline generation. In: Proceedings of the HLT-NAACL 03 on Text Summarization Workshop, vol. 5. Association for Computational Linguistics (2003)

12. Lopyrev, K.: Generating News Headlines with Recurrent Neural Networks (2015). arXiv:1512.01712
13. Zhang, X., LeCun, Y.: Text Understanding from Scratch (2015). arXiv:1502.01710
14. Shen, S., Liu, Z., Sun, M.: Neural Headline Generation with Minimum Risk Training (2016). arXiv:1604.01904

Test Case Generation and Optimization for Critical Path Testing Using Genetic Algorithm

Deepti Bala Mishra, Rajashree Mishra, Kedar Nath Das and Arup Abhinna Acharya

Abstract This paper presents a method for path testing by generating the test data automatically and optimizing the test data to test the critical paths for a software under test (SUT), using real-coded genetic algorithm. Real encoding is used for automatic test data generation, and a representative test suite, which achieves 100% path coverage, is found as an optimum result. In this paper, the proposed real-coded genetic algorithm for path coverage (RCGAPC) generates a set of inputs for testing a specific software and outperforms by giving effective and efficient results in terms of less number of test data generation counts. In the proposed approach, one-to-one injective mapping scheme is used for mapping the test data to the corresponding path and the most critical path is covered during path testing of a specific software. It seems to be faster than the traditional GA in covering critical path. The proposed method can reduce the number of test data generation required for path testing of a SUT and give an optimized Test suite that covers 100% path for specific software.

Keywords Software test case · Real-coded genetic algorithm
Critical path testing · Test suite optimization

D. B. Mishra · A. A. Acharya
School of Computer Engineering, KIIT University, Bhubaneswar 751024, India
e-mail: dbm2980@gmail.com

A. A. Acharya
e-mail: aacharyafcs@kiit.ac.in

R. Mishra (✉)
School of Applied Sciences, KIIT University, Bhubaneswar 751024, India
e-mail: rajashreemishra011@gmail.com

K. N. Das
Department of Mathematics, National Institute
of Technology, Silchar, Silchar 788010 , Assam, India
e-mail: kedar.iitr@gmail.com

© Springer Nature Singapore Pte Ltd. 2019
J. C. Bansal et al. (eds.), *Soft Computing for Problem Solving*, Advances in Intelligent Systems and Computing 817, https://doi.org/10.1007/978-981-13-1595-4_6

1 Introduction

In software testing, the runtime standard of the software is tested to maximum limits for a qualitative software. Software industry suffers from a heavy loss of $500 billion per year due to decrease in software quality. For a high-quality software that satisfies the user specifications and requirements, testing is required [1, 2]. Manual testing is a kind of software testing that was used in the early stages of the software industry where test cases are executed manually by the testers, and there is no need for automation tools. It was the sole process to find bugs and erase them from software system. Software failure causes by different faults and those faults can be detected by software testing [2]. Software testing is a very time consuming, labor-intensive, and complex process. The time and cost of manual testing can be reduced by automation process. We can improve the quality of software by performing automated testing and deliver the fully tested software, which can meet the user's specifications and requirements [3]. Manual testing suffers from the drawbacks such as operation speed, high investment of cost and time, limited availability of resources, redundancy of test cases, inefficient and inaccurate test checking. These drawbacks can be overcome by automated testing [4]. In manual testing, it is found that half of the resources of the software development are consumed for successful testing. So the development cost and time can be decreased by automatic testing [2].

Structural testing mainly involves testing process of a unit or modules and is very important for software developer. In structural testing phase, the developer tests each module of the software under test by going through the source code of the software. Coverage-based test data such as statement, condition, multiple conditions; path coverage play a vital role during structural testing and among all type of testing path coverage-based testing can detect about 65% of defects in an SUT [5].

In software testing, many optimization problems related to total cost needed for a specific software, time consumption, and huge number of test cases required for testing are arises [2]. In order to solve those problems, GA is frequently used by many researchers. The test cases are automatically generated in this process for structural testing, as parallelism and search space operations are the important characteristics [6]. GA is efficiently used to solve many optimization problems to provide near global optimum solution [7].

This paper presents an approach to find total path coverage test data generation for software testing. The rest of the paper is organized as: Sect. 2 describes the operators used in GA, Sect. 3 discusses related work on path coverage-based testing using GA, Sect. 4 describes path coverage in structural testing and gives a brief description about our proposed method using real-coded genetic algorithm for path coverage (RCGAPC), Sect. 5 describes the implementation of the proposed approach with a case study. In Sect. 6, the efficiency of the proposed technique is justified through a fair comparison over other related works. In Sect. 7, the proposed method is implemented in another case study for performance evaluation, and the conclusion of the paper is drawn in Sect. 8, which also includes few highlights on the scopes for future work.

2 Genetic Algorithm

Genetic algorithm is an evolutionary algorithm, which is developed by John Holland in 1975. It can be used to solve many complex and real-life problems by producing high-quality test data automatically [7]. GA has emerged as a practical, robust optimization technique, and search method, and it is inspired by the way nature evolves species using natural selection of the fittest individuals [8]. Basically, GA uses four different types of operators such as selection, crossover, mutation, and elitism (optional) [7].

In this paper, real-coded GA is used for the proposed method, which takes real values instead of binary values. It is because the binary-coded GA takes more time than real-coded GA due to the swapping of bits and repeated evaluation of parameters to fit them in their respective specified range. Hence, the real-coded GA is employed and is used in this present work.

2.1 Real-Coded Genetic Algorithm

In real-coded GA, the encoding is purely real type, which is very simple and straight-forward [9]. In binary encoding, parents are encoded by binary digits only, i.e., in the form of 0 and 1, as shown in Fig. 1 and the corresponding real coding value of Fig. 1 is shown in Fig. 2.

2.2 Average Crossover

Average crossover takes two parents to perform crossover and creates only one offspring by finding the average of two parents [10]. New offspring can be found by averaging the genes of both the parents of Fig. 2, and the resultant offspring is shown in Fig. 3

Fig. 1 Binary numbers   **Parent 1: 1001, 1011, 1000, 1010**

Parent 2: 1100, 1110, 0000, 0001

Fig. 2 Real numbers   **Parent 1: 9, 11, 8, 10**

Parent 2: 12, 14, 0, 1

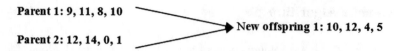

Fig. 3 Average crossover

2.3 Gaussian Mutation

The Gaussian mutation operator [11] changes one real variable x to a new value by using Eq. (1).

$$G(x) = Min(Max(N(x, \sigma)), a), b)$$ (1)

here, a and b are the bounds of the Chromosomes. σ depends on the length or time in GA, where length can be computed in terms of bounds and time can be computed in terms of generations. The mutation operation is performed with a probability factor of very less. In our approach, σ depends on the length of the Chromosome.

2.4 Fitness Function

The fitness function evaluates individual's performance. Based on the fitness value, the individuals with higher are selected to the next generation for better optimum solution [8]. Our fitness function is defined in Eq. (2) of Sect. 6 based on the number of path covered by a particular chromosome.

3 Literature Survey

Lin and Yeh [12] started the path coverage-based software testing using GA by extending Jones et al. work [13] of branch coverage-based testing. They have used extended hamming distance (EHD), where $n > 1$ to measure the path distance. Their fitness function is named as SIMILARITY, which calculates the distance between current executed path and the target path and they found their approach performed very well to generate test data which are normally very difficult to find out. Hermadi and Ahmed [14] developed an evolutionary test data generator for path coverage-based testing. Their fitness function uses path traversal, neighborhood influence, and normalization techniques to achieve path coverage and found their proposed method improves the GA performance in terms of space exploitation and exploration with a faster convergence. Again, in 2007, authors have used a modified fitness function with the addition of a new rewarding scheme. They have taken six numbers of benchmark problems for their experiment and found the new GA-based multiple path test data generator gives better results than other previous work [15]. Chen and Zhong [16] presented a multipopulation GA (MPGA) to generate automatically path based test data. They have taken the triangle classifier problem for their experiment and found the time of test data generation is very less than using simple GA but the number of data generation is still high to cover total number of paths. Srivastava and Kim [4] proposed a GA-based method to optimize test cases for path testing by identifying the most difficult path of a program under test. They have assigned different weights to the edges of the CFG of their experiment by applying 80–20 rule. They assigned 80 to all type of edges containing conditions and loops and 20 to each of the sequential nodes. Finally, they calculate the total weight value and according to the total value they identified the criticality of the path.

Boopathi et al. [17] presented an approach to estimate the reliability of a software by using *Markov Chain method*. Their method can generate optimized test cases to achieve total path coverage in less cost and time. They used *Markov transition probability matrix* for assigning the weights to the edges of their dd-graph. Khan et al. [18] used GA to generate test cases automatically for path testing. Their proposed method takes a set of inputs for total path coverage and gives optimized results. They designed their fitness by taking all du-path adequacy criteria. Thi et al. [5] proposed an improved GA to find total path coverage test data by identifying the critical paths, which are not covered easily for a specific program under test. They have used static program analysis to achieve full path coverage and found their approach can generate test data to cover the most difficult paths, which are not covered by normal GA. Shimin and Zhangang [19] proposed an approach for generating test data automatically and from experimental result they found the path oriented test data can be achieved using GA.

4 Proposed Algorithm for Total Path Coverage

In path testing, test cases are designed in such a way that all linearly independent paths of a particular program under test should be executed at least once. A linearly independent path can be obtained by the control flow graph (CFG) of a program which shows the flow of sequence in a program [20]. The path testing is considered as structural testing or white box testing method that helps in designing test cases. With the help of path testing, the test cases are created and executed for all possible paths results in 100% statement coverage and 100% branch coverage [18].

The proposed algorithm aims to find a representative test suite which achieves 100% path coverage and the algorithm takes real integer values as inputs and gives one set of values to achieve full path coverage. We take the initial population 50 in the mating pool, representing each as a test suite. The algorithm is named as real-coded GA for path coverage (RCGAPC) as it creates the real chromosomes automatically indicating the initial population. It employees the operators such as

72 D. B. Mishra et al.

Fig. 4 Flow of RCGAPC

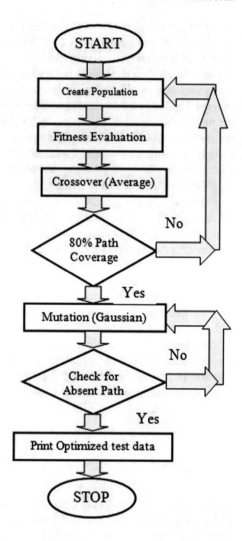

average crossover (AX) [10], Gaussian mutation $M_G(x)$ [11] along with random selection based on higher fitness. RCGAPC uses the one-to-one injective function for mapping the test data to its corresponding path [21]. To implement our algorithm, the triangle classifier problem is taken from [6, 11, 13, 15], which displays the type of a triangle as equilateral, isosceles, scalene, or not a Triangle, according to the input value for three sides of the triangle. The pseudocode for the proposed RCGAPC is outlined as follows, and Fig. 4 represents the flow diagram of the proposed algorithm.

Fig. 5 Initial population

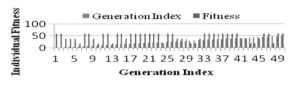

Fig. 6 Path generated for
initial population

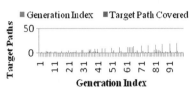

Pseudocode for RCGAPC:

Step 1. Create Initial Population
Step 2. Initialize the Population as Test Suites
        (Test Suite = Number of Chromosomes)
Step 3. Calculate Fitness
Step 4. Select Best Two Chromosomes Based on Fitness
Step 5. (Apply Crossover Operator)
    Do Average Crossover for Selected Chromosomes
        While (Fitness >=80%)
        End While
Step 6. (Apply Gaussian Mutation)
    Do Replace the New Chromosome with the Duplicate one
        While (Absent path is not found)
        End While
Step 7. Check for the Feasible Solution
        If (Total Path Covered=100%)
        Print the Solution
        Else go to **Step 1**
    End

Basically, RCGAPC creates initial population randomly and evaluate the fitness for each chromosome. The selection procedure is based on the fitness of the chromosomes, which are taken for the mating pool. The proposed algorithm selects two best chromosomes for next generation and applies the average crossover between the chromosomes, and finally Gaussian mutation with a probability 0.07 is applied to reproduce a new trait, which covers the most critical path, i.e., the path that cannot be covered by normal GA [5, 12, 15, 16, 19].

5 Implementation of RCGAPC

To evaluate the effectiveness and efficiency of the proposed RCGAPC, two most popular search-based programs have been taken from the literature as triangle classifier program [6, 11, 13, 15] and finding Greatest Common Divisor (GCD) of two numbers [4]. Further, their statistical results have been compared with other existing works.

5.1 *Case Study 1: Triangle Classifier Problem*

The proposed algorithm is first implemented with the triangle classifier problem, in which the user has to input three sides of a triangle and check the types of the triangle, i.e., equilateral, isosceles, scalene, or not a Triangle. The corresponding paths are taken based on the types of triangle as shown below, and the most critical path is path 5, i.e., the path for equilateral triangle as it needed three equal and positive values for three sides so that the corresponding path will be executed. From the literature, it is seen that this specific path is very difficult to cover by random testing and the path can only be covered after a huge number of test data generation. It is also seen that normal GA could not cover this critical path. So the main aim of RCGAPC is to cover all paths including the most critical path in less time, i.e., in terms of less data generation.

Different paths for Triangle Classifier:

PATH 1: (Invalid input)
PATH 2: (Not a triangle)
PATH 3: (Scalene)
PATH 4: (Isosceles)
PATH 5: (Equilateral)

6 Experimental Setup

The triangle classifier problem is first taken for experiment, and it is aimed at searching one candidate test suite which is able to cover all the feasible target paths for the case study, especially to cover the most critical path. The fitness defined for the above problem represents the percentage of path coverage by a test suite. The fitness function is designed by considering the total number of path covered by a chromosome and it is defined in Eq. (2). The initial data taken for our experiment is shown in Table 1.

$$f(x) = \left(\frac{Number\ of\ path\ Covered\ by\ a\ Chromose}{Total\ Number\ of\ path} \right) \times 100 \qquad (2)$$

Table 1 Parameters used in RCGAPC for case study 1

S. no.	Parameters	Value
1	Population size	50
2	Input range	−5 to 5
3	Encoding type	Real encoding
4	Crossover type and rate	Average crossover (AX), 0.8
5	Mutation type and rate	Gaussian mutation $M_G(x)$, 0.02, 0.03, 0.07
6	Generation	100
7	Run	10

Table 2 Initial population

Pop no.	Target path	Fitness
1	1, 4, 1, 2, 1	0.6
2	1, 1, 1, 1, 1	0.2
3	1, 4, 1, 4, 4	0.4

Table 3 Best population

Pop no.	Target path	Fitness
1	1, 4, 1, 2, 1	0.6
3	1, 4, 1, 4, 4	0.4

Table 4 New trait (after crossover)

Pop no.	Target path	Fitness
1	1, 4, 1, 3, 2	0.8

6.1 Procedure in RCGAPC

The various steps of RCGAPC are shown in Tables 2, 3 and 4.

Initial Population: Suppose the initial population contains three numbers of chromosomes as:

1. {C1, C2, C3, C4, C5} = {(0, 2, 4), (2, 4, 2), (4, −2, 3), (1, 2, 3), (−2, 4, 2)}
2. {C1, C2, C3, C4, C5} = {(−1, 0, −4), (4, 1, 3), (0, 4, 3), (1, 0, 4), (−2, −1, 2)}
3. {C1, C2, C3, C4, C5} = {(0, −1, 3), (4, 2, 4), (−1, 2, 3), (2, 1, 2), (2, 1, 2)}

Now the fitness of current population can be obtained by using Eq. (2), after mapping the chromosomes into their corresponding path shown in Table 2.

Selection: The proposed method randomly selects two best chromosomes based on the fitness shown in Table 3.

Crossover and Mutation: Average crossover is applied on the selected chromosomes and Gaussian mutation operator is applied after getting the desired chromosome having 80% path coverage shown in Table 4.

Fig. 7 Result after
successive runs

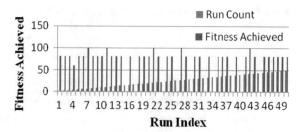

6.2 Result Analysis

The proposed algorithm RCGAPC has been performed and ran in Dev C++ IDE-5.11 for several times with same population and after a series of experiments the maximum fitness is achieved representing the test suite to cover all target paths and is shown in Fig. 7. The initial population with their fitness is shown in Fig. 5, and the generated paths are shown in Fig. 6.

From Fig. 6, it is clearly seen that all paths are not covered at the same time by different populations and in many of the cases the path for equilateral cannot be covered as it is the most difficult path [5, 12, 15, 16, 19]. During the experiment, it is observed that the most critical path for equilateral is covered by some test data, so in that case the path which is not covered is taken as the critical path. Therefore, after crossover when a chromosome, with 80% fitness is found, immediately the Gaussian Mutation operator is applied on that particular chromosome to find the absent path, which is not covered previously. After a number of generations when the algorithm finds that absent path, then it will check with the original set of data and replace the new data with the duplicate one. The efficiency of the proposed algorithm is represented in terms of average number of test data generated to cover all paths of the SUT. Further, the statistical results are compared with other related work which shows the proposed algorithm performs better in comparison to previous related work.

6.3 Comparison with Related Work

The statistical result obtained by the proposed RCGAPC is compared with the previous related work on the same program under test, and the performance result is shown in Tables 5 and 6 in terms of test data generation count and run count, respectively, to cover all paths including the most critical path for triangle classifier.

From Tables 5 and 6, it is observed that the proposed method RCGAPC is more effective in terms of test data generation count and run count to find the total path coverage of a specific program under test. The method can generate test data automatically and achieve 100% path coverage in a less number of test data generations

Table 5 Performance comparison result (in terms of test data generation count)

Problem type	Authors	Method used	Number of path taken	Test data generation count for 100% path coverage
Triangle type	Lin and Yeh [12], 2000	GA	4	10,000
Triangle type	Chen and Zhong [16], 2008	MPGA	4	21,073
Triangle type	Thi et al. [5], 2016	Improved GA	4	3198
Triangle type	Mishra et al. (authors of the proposed method)	**RCGAPC**	**5**	**1737**

Table 6 Performance comparison result (in terms of run count)

Problem type	Authors	Method used	Number of path taken	Run count
Triangle type	Ahmed and Hermadi [15]	GA	4	10
Triangle type	Shimin and Zhangang [19]	GA	5	65
Triangle type	Mishra et al. (authors of the proposed method)	**RCGAPC**	**5**	**7**

(1737 test data) as compared to [5, 12, 16] shown in Table 5. Again the proposed method is compared with [15, 19] in terms of run count to find a representative test suite that achieves 100% path coverage and the result is shown in Table 6.

7 Evaluation and Validation of RCGAPC

From the statistical results, it is observed that the proposed algorithm performs very well for path coverage-based testing, i.e., testing in unit level and the algorithm successfully covers the most critical path in a very less number of test data generation which leads to saving the testing time. For further evaluation and validation, the proposed algorithm is again implemented in the second case study as finding GCD of two inputted numbers [4].

Table 7 Parameters used in RCGAPC for case study 2

S. n.	Parameters	Value
1	Population size	50
2	Input range	1–50
3	Encoding type	Real encoding
4	Crossover type and rate	Average crossover (AX), 0.8
5	Mutation type and rate	Gaussian mutation $M_G(x)$, 0.02, 0.03, 0.07
6	Generation	5

Table 8 Optimized test data (after second generation)

Test data	Target path	Fitness
(3, 3), (9, 28), (24, 7), (27, 9), (8, 24)	1, 2, 3, 4, 5	100%

7.1 Case Study 2: Finding GCD of Two Numbers

The proposed algorithm RCGAPC is again implemented in the GCD (int m, int n) problem which takes two positive integers m and n as inputs. The function computes the reminder r and returns the greatest common divisor as its output. The corresponding paths can be found from CFG [20], and it is shown below as:

PATH 1: if (m==n)
PATH 2: if (n>m and r!=0)
PATH 3: if (m>n and r!=0)
PATH 4: if (m>n and r=0)
PATH 5: if (n>m and r=0)

The parameters taken for GCD problem are shown in Table 7, and the optimized test suite is shown in Table 8 with 100% path coverage.

7.2 Comparison with Related Work

The optimized result of the case study 2 is partially compared with one previous related work [4], but it is not completely compared as authors have not mentioned the critical path in their study. Table 9 shows the performance result of RCGAPC in terms of less number of generation counts than other related work.

Table 9 Performance comparison result (in terms of generation)

Problem type	Authors	Method used	Total path	Parameters	Generation count
GCD of two numbers	Srivastava and Kim [4]	GA	4	Gen-3, Pop-4 Em-Binary, Sm-Random, Cm-Pair Wise, Cr-0.8, Mm-Flip, Mr-0.3	3
GCD of two numbers	Mishra et al. (authors of the proposed method)	**RCGA PC**	**5**	Gen-5, Pop-10 Em-Real, Sm-Random, Cm-Average, Cr-0.8, Mm-Gaussian, Mr-0.02, 0.03, 0.07	**2**

Pop population Size, *Gen* number of generation, *Sm* selection method, *Cr* crossover rate, *Cm* crossover method, *Mr* mutation rate, *Mm* mutation method, *Em* encoding method

8 Conclusion and Future Work

Automatic test data generation has a key role in software testing phase. Test cases with ability to cover all the basic paths of a program under test are more effective for unit-level testing. From the experimental results, it is found that the proposed method can generate test data for most difficult paths, which are normally, cannot be generated in traditional GA and gives a test suite that covers multiple target paths at a time with less number of test data generations. The proposed algorithm RCGAPC shows very good performance in comparison with previously discussed related works to find total path coverage-based test data for a specific SUT. In future, it is planned to develop and apply a new hybridized algorithm to generate test cases automatically and find the test data, which covers the most critical path for a specific SUT more quickly. It is also planned to carry out more complex real-life software to measure the efficiency of the proposed method in generating test data automatically.

References

1. Mathur, A.P.: Foundations of Software Testing, 2 edn. Pearson Education, India (2013)
2. Mishra, D.B., Bilgaiyan, S., Mishra, R., Acharya, A.A., Mishra, S.: A review of random test case generation using genetic algorithm. Indian J. Sci. Technol. **10**(30) (2017)
3. Chauhan, N.: Software Testing: Principles and Practices. Oxford University Press (2010)
4. Srivastava, P.R., Kim, T.H.: Application of genetic algorithm in software testing. Int. J. Softw. Eng. Appl. **3**(4), 87–96 (2009)

5. Thi, D.N., Hieu, V.D., Ha, N.V.: November. A technique for generating test data using genetic algorithm. In: 2016 International Conference on Advanced Computing and Applications (ACOMP), pp. 67–73. IEEE (2016)
6. Sharma, A., Rishon, P., Aggarwal, A.: Software testing using genetic algorithms. Int. J. Comput. Sci. Eng. Surv. (IJCSES) **7**(2), 21–33 (2016)
7. Deb, K.: Optimization for Engineering Design: Algorithms and Examples. PHI Learning Pvt. Ltd. (2012)
8. Mishra, D.B., Mishra, R., Das, K.N., Acharya, A.A.: A systematic review of software testing using evolutionary techniques. In: Proceedings of Sixth International Conference on Soft Computing for Problem Solving, pp. 174–184. Springer, Singapore (2017)
9. Goldberg, D.E.: Real-coded genetic algorithms, virtual alphabets, and blocking. Complex Syst. **5**(2), 139–167 (1991)
10. Umbarkar, A.J., Sheth, P.D.: Crossover operators in genetic algorithms: a review. ICTACT J. Soft Comput. **6**(1) (2015)
11. Heitzinger, C.: Simulation and inverse modeling of semiconductor manufacturing processes (2002)
12. Lin, J.C., Yeh, P.L.: Using genetic algorithms for test case generation in path testing. In: Proceedings of the Ninth Asian Test Symposium, 2000 (ATS 2000), pp. 241–246. IEEE (2000)
13. Jones, B.F., Sthamer, H.H., Eyres, D.E.: Automatic structural testing using genetic algorithms. Softw. Eng. J. **11**(5), 299–306 (1996)
14. Hermadi, I., Ahmed, M.A.: Genetic algorithm based test data generator. In: The 2003 Congress on Evolutionary Computation, 2003, CEC'03, vol. 1, pp. 85–91. IEEE (2003)
15. Ahmed, M.A., Hermadi, I.: GA-based multiple paths test data generator. Comput. Oper. Res. **35**(10), 3107–3124 (2008)
16. Chen, Y., Zhong, Y.: Automatic path-oriented test data generation using a multi-population genetic algorithm. In: Fourth International Conference on Natural Computation, 2008, ICNC'08, vol. 1, pp. 566–570. IEEE (2008)
17. Boopathi, M., Sujatha, R., Kumar, C.S., Narasimman, S.: The mathematics of software testing using genetic algorithm. In: 2014 3rd International Conference on Reliability, Infocom Technologies and Optimization (ICRITO). Trends and Future Directions, pp. 1–6. IEEE (2014)
18. Khan, R., Amjad, M., Srivastava, A.K.: Optimization of Automatic Generated Test Cases for Path Testing Using Genetic Algorithm. In: 2016 Second International Conference on Computational Intelligence & Communication Technology (CICT), Ghaziabad, 2016, pp. 32–36 (2016)
19. Shimin, L., Zhangang, W.: Genetic algorithm and its application in the path-oriented test data automatic generation. Procedia Eng. **15**, 1186–1190 (2011)
20. Ghiduk, A.S.: Automatic generation of basis test paths using variable length genetic algorithm. Inf. Process. Lett. **114**(6), 304–316 (2014)
21. Bartle, R.G., Bartle, R.G.: The Elements of Real Analysis, vol. 2. Wiley, New York (1964)

A Novel PID Controller Designed via Polynomial Approach for Stable/Unstable Second-Order Process with Time Delay

M. Praveen Kumar, M. Manimozhi, P. Ponnambalam and G. Gokulakrishnan

Abstract The present work proposes a new proportional–integral-derivative (PID) controller for time-delayed stable and unstable second-order processes. The performance of PID controller is augmented by associating a second-order filter. The parameters of the controller as well as the filter are derived analytically using polynomial approach. The Lagrange shift approximation of time delay is employed in the process of deriving controller parameters. The tuning parameter is selected based on the value of maximum sensitivity (MS). A set point filter is employed in order to reduce the overshoot in the servo response. The proposed work is verified against recently published works with the help of various benchmarking examples.

Keywords PID · Unstable processes · Time delay · Maximum sensitivity
Set point filter

1 Introduction

PID control is being the widely used controller among the industries. The evaluation of PID controller deserves a discussion here. Pneumatic/hydraulic PID controllers are replaced by electronic PID controllers which offer many advantages over the former controllers. Thereafter, electronic PID controllers have been replaced by computer (i.e., computer program). Nowadays, PID controller is a software program rather than a controller with physical components. This enabled the researchers to design various control techniques involving PID controller with filters, multiloop control techniques involving more than one controller, optimization-based controllers, online tuning of controllers, etc.

Dealing with the integrating and unstable processes is always critical when compared to that of stable processes. Unfortunately, many chemical processes are integrating or unstable in nature. Typical examples are bottom level control of distillation

M. Praveen Kumar (✉) · M. Manimozhi · P. Ponnambalam · G. Gokulakrishnan
VIT University, Vellore 632014, Tamil Nadu, India
e-mail: praveen.m@vit.ac.in

© Springer Nature Singapore Pte Ltd. 2019
J. C. Bansal et al. (eds.), *Soft Computing for Problem Solving*, Advances in Intelligent
Systems and Computing 817, https://doi.org/10.1007/978-981-13-1595-4_7

M. Praveen Kumar et al.

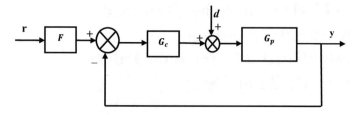

Fig. 1 Proposed control structure

column, chemical reactors, bioreactors, etc. The complexity in control gets enhanced if the processes are associated with time delay. Time delay may be due to process lag or measurement lag. Several researchers [1–17] have proposed various control schemes for unstable and/or integrating processes which include internal model control (IMC)-based control structures [4, 10, 13, 16], direct synthesis-based controller design [1, 3, 5], modified smith predictor-based control structures [8, 12, 17].

The control schemes with multiloops and controllers are complex in nature and need cumbersome tuning process because of multiple tuning parameters. The aim of the present work is to design a simple control strategy for stable/unstable second-order process with time delay. A polynomial approach-based design is carried out in the process of deriving PID controller cascaded with a second-order filter. The proposed control strategy is compared with the existing works in terms of various performance indices.

The paper is organized as follows: Sect. 2 deals with the design of controller, set point filter is elaborated in Sect. 3, simulation analysis is carried out in Sect. 4 followed by conclusion and future scope in Sect. 5.

2 Design of Controller

The proposed control structure is shown in Fig. 1 where: G_c: controller, G_p: stable/unstable second-order process, F: set point filter, r: set point, d: disturbance and y: output. The control structure is pretty similar to conventional control structure. The proposed method employs a PID controller cascaded with a second-order filter. The set point filter is employed to reduce the overshoot in the servo response which is due to the controller induced zeroes.

Servo and regulatory responses are derived from Fig. 1 as

$$\frac{y}{r} = \frac{G_c G_p F}{1 + G_c G_p} \tag{1}$$

$$\frac{y}{d} = \frac{G_p}{1 + G_c G_p} \tag{2}$$

The proposed PID controller which is cascaded with second-order filter is presented in Eqs. (3a)–(3c) and has been represented as ratio of two polynomials.

$$G_c(s) = \frac{q}{p} = \left(k_p + \frac{k_i}{s} + k_d s \right) \left(\frac{(a_1 s + 1)^2}{b_2 s^2 + b_1 s + 1} \right) \tag{3a}$$

where

$$q = (k_d s^2 + k_p s + k_i)(a_1 s + 1)^2 \tag{3b}$$

$$p = s(b_2 s^2 + b_1 s + 1) \tag{3c}$$

2.1 Design of G_c

A second order is process is assumed as ratio of two polynomials as shown in Eqs. (4a)–(4c).

$$G_p(s) = \frac{k}{(\tau_2 s + 1)(\tau_1 s + 1)} e^{-s\theta} = \frac{b}{a} e^{-s\theta} \tag{4a}$$

where

$$b = k \tag{4b}$$

$$a = (\tau_2 s + 1)(\tau_1 s + 1) \tag{4c}$$

By substituting Eqs. (3a)–(3c) and (4a)–(4c) in Eqs. (1) and (2),

$$\frac{y}{r} = \frac{Fbqe^{-s\theta}}{ap + bqe^{-s\theta}} = \frac{Fk(k_d s^2 + k_p s + k_i)(a_1 s + 1)^2 e^{-s\theta}}{s(\tau_2 s + 1)(\tau_1 s + 1)(b_2 s^2 + b_1 s + 1) + k(k_d s^2 + k_p s + k_i)(a_1 s + 1)^2 e^{-s\theta}} \tag{5}$$

$$\frac{y}{d} = \frac{bpe^{-s\theta}}{ap + bqe^{-s\theta}} = \frac{ks(b_2 s^2 + b_1 s + 1)e^{-s\theta}}{s(\tau_2 s + 1)(\tau_1 s + 1)(b_2 s^2 + b_1 s + 1) + k(k_d s^2 + k_p s + k_i)(a_1 s + 1)^2 e^{-s\theta}} \tag{6}$$

The characteristic equation (CE) of the servo and regulatory responses is crucial as it is the governing equation for the stability of closed-loop control system. CE can be written as

$$s(\tau_2 s + 1)(\tau_1 s + 1)(b_2 s^2 + b_1 s + 1) + k(k_d s^2 + k_p s + k_i)(a_1 s + 1)^2 e^{-s\theta} = 0 \tag{7}$$

The time delay in the process is approximated using second-order Laguerre shift as shown in Eq. (8).

$$e^{-s\theta} = \frac{\left(1 - \frac{s\theta}{4}\right)^2}{\left(1 + \frac{s\theta}{4}\right)^2} \tag{8}$$

Considering $a_1 = 0.25\theta$ in Eq. (7) and substituting Eq. (8) in Eq. (7),

$$s(\tau_2 s + 1)(\tau_1 s + 1)\left(b_2 s^2 + b_1 s + 1\right) + k\left(k_d s^2 + k_p s + k_i\right)\left(1 - \frac{s\theta}{4}\right)^2 = 0 \tag{9}$$

On simplifying Eq. (9),

$$kk_i\left(c_5 s^5 + c_4 s^4 + c_3 s^3 + c_2 s^2 + c_1 + 1\right) = 0 \tag{10a}$$

where

$$c_5 = \frac{b_2 \tau_1 \tau_2}{kk_i} \tag{10b}$$

$$c_4 = \frac{kk_d \theta^2 + 16b_2(\tau_1 + \tau_2) + 16b_1 \tau_1 \tau_2}{16kk_i} \tag{10c}$$

$$c_3 = \frac{kk_p \theta^2 - 8kk_d \theta + 16b_2 + 16b_1(\tau_1 + \tau_2) + 16\tau_1 \tau_2}{16kk_i} \tag{10d}$$

$$c_2 = \frac{kk_i \theta^2 - 8kk_p \theta + 16(b_1 + \tau_1 + \tau_2) + 16kk_d}{16kk_i} \tag{10e}$$

$$c_1 = \frac{2kk_p - kk_i \theta + 2}{2kk_i} \tag{10f}$$

The CE must be solved with a desired or target CE which ensures the pole location on the left half of s plane.

$$\left(c_5 s^5 + c_4 s^4 + c_3 s^3 + c_2 s^2 + c_1 + 1\right) = (\lambda s + 1)^3 (0.25\theta s + 1)^2 \tag{11}$$

In the equation of desired CE (Eq. (11)), λ is tuning parameter which should be positive, thus ensuring the poles on the left half of s plane at $s = -1/\lambda$. It is clear from the servo response that the controller is causing zeroes at $s = -1/a_1$, i.e., $s = -1/0.25\theta$. In order to cancel these zeroes, two poles are placed at $s = -1/0.25\theta$ in the desired CE.

2.2 Selection of λ

Selection of λ is highly crucial as it is solely responsible for the derived controller parameters. The proposed method employed MS-based selection for λ. MS is a well-known measure for analyzing the relative/robust stability of a control structure. MS is defined as the maximum possible sensitivity(S) of a control structure.

$$MS = max\left(\left|\frac{1}{1+L}\right|\right) \tag{12}$$

Here, L is loop transfer function of closed loop control system. In other words, MS is the shortest possible distance of the Nyquist curve of loop transfer function to the critical point. The lower the value of MS, the higher will be the robust stability of the system. However, lower value of MS also corresponds to slower responses. Higher value of MS corresponds to faster responses. But high MS is also an indication of less robust stability. As a compromise, the MS is selected between 1 and 2 for stable systems. However, many a time, researchers [1, 3, 10, 13, 14] have used MS values greater than 2 for integrating and unstable systems as it is not possible to achieve lower values of MS.

3 Design of Set Point Filter (F)

It is clear from the servo response (see Eq. (5)) that the controller is introducing zeroes in the transfer function of servo response. These zeroes may cause overshoot if they are close to imaginary axis. However, two of these zeroes are canceled by assuming a suitable desired CE (Eq. (11)). To further cancel more zeroes, the following set point filter (F) is implemented.

$$F = \frac{1}{\frac{k_d}{k_i}s^2 + \frac{k_p}{k_i}s + 1} \tag{13}$$

The filter parameters are assumed so as to cancel the two more zeroes introduced by the controller in servo response (Eq. (5)).

4 Simulation Analysis

In this section, various benchmarking examples are considered to verify the proposed method over the existing methods. The evaluation is carried out in terms of various performance indices. Widely used performance indices are as follows:

$$\text{Integral absolute error (IAE)} = \int_0^\infty |e|dt \tag{14}$$

$$\text{Integral square error (ISE)} = \int_0^\infty e^2 dt \tag{15}$$

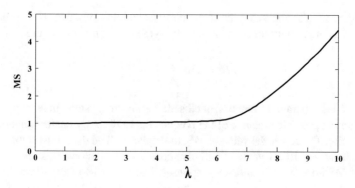

Fig. 2 Variation of MS with λ for Example 1

$$\text{Integral time absolute error (ITAE)} = \int_0^\infty t|e|dt \tag{16}$$

Total variation (TV) is a measure of smoothness of control signal which is defined as

$$\text{TV} = \sum_{i=0}^\infty |u_{i+1} - u_i| \tag{17}$$

where u_i is process input at ith instant of time. Higher TV corresponds to aggressive variations in control signal whereas smaller TV corresponds to smoother control signal. A sample period of 0.1 s is used in the simulations. Settling time (t_s) is another factor which is specified as the time taken by the response to be settled in a prescribed error band.

Example 1 A stable second-order process is considered in this example as shown in Eq. (18). Comparing the process with Eqs. (4a)–(4c): $k = 1, \tau_1 = 1, \tau_2 = 0.7, \theta = 2$.

$$G_p(s) = \frac{1}{(s+1)(0.7s+1)}e^{-2s} \tag{18}$$

The controller parameters for this process are derived at various values of λ by solving Eq. (11) using MATLAB. The variation of MS value is shown Fig. 2.

The λ is selected as 1 where the MS value is 1.03. The proposed method is compared with the method proposed by [10]. The method of [10] is an IMC-based PID controller for various classes of systems. Set point weighting (ε) is employed by [10] in order to reduce the overshoot in servo response. The method of [10] is proved to be offering superior response when compared to the existing methods proposed by [6, 15]. For the present example, [10] proposed the controller at an MS value 1.64. The proposed controller as well as the proposed controller by [10] is presented in Table 1.

Table 1 Details of controller structure and parameters

Process	Method	Controller
$\frac{e^{-2s}}{(s+1)(0.7s+1)}$	Proposed[a] Wang et al.[b]	$\left(0.2108s + 0.5135 + \frac{0.3027}{s}\right)$ $\left(\frac{(0.5s+1)^2}{0.1081s^2+0.419s+1}\right)$ $\left(0.174s + 0.435 + \frac{0.263}{s}\right)$
$\frac{e^{-0.5s}}{(2s+1)(0.5s+1)(5s-1)}$	Proposed[c] Wang et al.[d]	$\left(11.4676s + 9.722 + \frac{2.2143}{s}\right)$ $\left(\frac{(0.2347s+1)^2}{0.0118s^2+0.0714s+1}\right)$ $\left(8.79s + 6.63 + \frac{1.151}{s}\right)\left(\frac{0.4695s+1}{0.01s+1}\right)$

[a] $F = \frac{1}{0.696s^2+1.696s+1}$
[b] $\varepsilon = 0.4$
[c] $F = \frac{1}{5.179s^2+4.391s+1}$
[d] $\varepsilon = 0.2$

Table 2 Performance analysis under nominal conditions

Process	Method	Set point tracking			Disturbance rejection		
		$t_s(s)$	IAE	TV	$t_s(s)$	IAE	TV
$\frac{e^{-2s}}{(s+1)(0.7s+1)}$	Proposed	9.612	3.304	1.005	11.967	0.661	0.200
	Wang et al.	13.42	5.23	1.156	16.45	0.838	0.228
$\frac{e^{-0.5s}}{(2s+1)(0.5s+1)(5s-1)}$	Proposed	8.953	3.94	6.76	11.104	0.9038	5.62
	Wang et al.	7.97	3.74	132	12.472	1.738	12.11

A unit step change in set point at $t = 0$ s and a negative input disturbance to with a magnitude of 2 are considered at $t = 50$ s. The nominal performance is presented in Fig. 3. Corresponding performance indices are presented in Table 2. It is clear that the proposed method offers a better response in terms of all the performance indices. However, the actual responsibility of the controller is to sustain the performance when the process parameters are changed. In practical, the estimated model may not always represent actual model because of the nonlinearities in the process, modeling errors, change of operating conditions, etc. So an efficient controller must be able to furnish good performance under robust conditions too. To verify this, a +20% change in process gain (k) and time delay (θ) is assumed. The resulting analysis is presented in Fig. 4 and Table 3. It is observed that the proposed method is offering superior robust performance. The notable point is that the proposed method is able to offer faster response with lower TV when compared to the other method [10]. In addition, the faster response is obtained at a lower MS (high robustness) value when compared to other method proposed by [10].

Example 2 In practical, the real-world processes are of higher order or nonlinear in nature. To implement a control strategy, they are approximated to lower-order

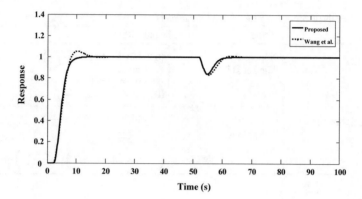

Fig. 3 Nominal response of Example 1

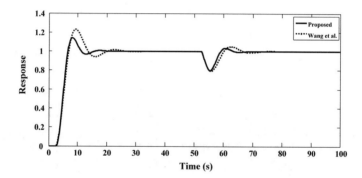

Fig. 4 Perturbed response of Example 1

Table 3 Performance analysis under perturbed conditions

Perturbed process	Method	Set point tracking			Disturbance rejection		
		$t_s(s)$	IAE	TV	$t_s(s)$	IAE	TV
$\frac{1.2e^{-2.4s}}{(s+1)(0.7s+1)}$	Proposed	14.25	3.582	1.325	16.89	0.859	0.318
	Wang et al.	18.87	6.147	1.485	22.13	1.201	0.344
$\frac{e^{-0.6s}}{(1.8s+1)(0.45s+1)(4.5s-1)}$	Proposed	9.699	4.025	9.67	11.20	0.907	10.90
	Wang et al.	10.51	3.89	134	14.726	1.739	15.464

processes. A higher-order process with one unstable pole as shown in Eq. (19) is discussed in this example.

$$G_p(s) = \frac{1}{(2s+1)(0.5s+1)(5s-1)}e^{-0.5s} \tag{19}$$

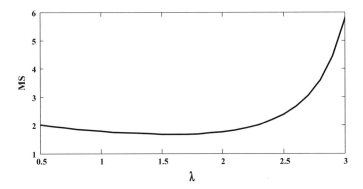

Fig. 5 Variation of MS with λ for Example 2

The higher-order process is approximated to a second-order process [13] as shown in Eq. (20).

$$G_p(s) = \frac{1}{(2.07s + 1)(5s - 1)} e^{-0.939s} \qquad (20)$$

Comparing Eq. (20) with Eqs. (4a)–(4c), $k = -1, \tau_1 = 2.07, \tau_2 = -5, \theta = 0.939$. Using this approximated model of the process, the controller parameters are derived for $\lambda = 1$ at which an MS value of 1.7887 is obtained. The variation of MS is presented in Fig. 5. The derived controller parameters are presented in Table 1. The method proposed by [10] has used an MS value of 2.24. This method is proved to be superior to the methods proposed by [13, 14].

Change in set point and disturbance magnitudes are considered similar to Example 1. The nominal response is shown in Fig. 6, and the analysis is presented in Table 2. For robustness analysis, +20% change in k and θ, -10% change in τ_1, τ_2 are assumed. The perturbed response is presented in Fig. 7, and the corresponding analysis is presented in Table 3. From the performance indices presented in Tables 2 and 3, it is clear that the proposed method is far superior to the method proposed by [10] in terms of disturbance rejection. However, in case of servo response, both the methods have performed almost equally well in terms of IAE and settling time. But the method of [10] registered substantially large TV when compared to the proposed method. The authors observed that the control signal is suffering from high-frequency fluctuations with the method of [10]. This is shown in Fig. 8.

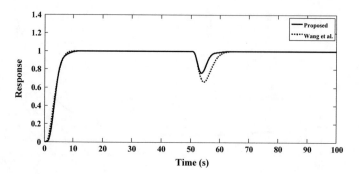

Fig. 6 Nominal response of Example 2

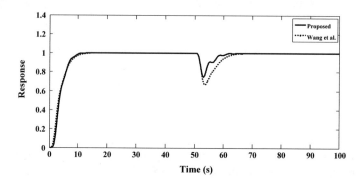

Fig. 7 Perturbed response of Example 2

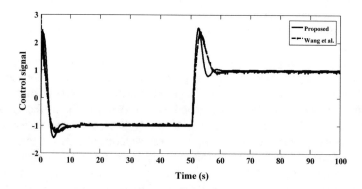

Fig. 8 Control signal for nominal response of Example 2

5 Conclusion and Future Scope

A novel PID controller for stable/unstable second-order process is proposed. The proposed method is simple in terms of control structure. An analytically derived second-order filter is cascaded to the PID controller. A set point filter is employed to reduce the overshoot in servo response. The proposed method is able to produce improved performance over existing methods. However, the performance is degraded with the deviation in the process model parameters. In future, the authors would like to address this by incorporating intelligence to the controller. The intelligence could be a mechanism which identifies the process model online and updates the tuning parameter accordingly.

References

1. Ajmeri, M., Ali, A.: Two degree of freedom control scheme for unstable processes with small time delay. ISA Trans. **56**, 308–326 (2015). https://doi.org/10.1016/j.isatra.2014.12.007
2. Ali, A., Majhi, S.: PID controller tuning for integrating processes. ISA Trans. **49**(1), 70–78 (2010). https://doi.org/10.1016/j.isatra.2009.09.001
3. Anil, C., Sree, R.P.: Tuning of PID controllers for integrating systems using direct synthesis method. ISA Trans. **57**, 211–219 (2015). https://doi.org/10.1016/j.isatra.2015.03.002
4. Begum, K.G., Rao, A.S., Radhakrishnan, T.K.: Enhanced IMC based PID controller design for non-minimum phase (NMP) integrating processes with time delays. ISA Trans. **68**, 223–234 (2017). https://doi.org/10.1016/j.isatra.2017.03.005
5. Chen, D., Seborg, D.E.: PI/PID controller design based on direct synthesis and disturbance rejection. Ind. Eng. Chem. Res. **41**(19), 4807–4822 (2002). https://doi.org/10.1021/ie010756m
6. Lee, J., Cho, W., Edgar, T.F.: Simple analytic PID controller tuning rules revisited. Ind. Eng. Chem. Res. **53**(13), 5038–5047 (2013). https://doi.org/10.1021/ie4009919
7. Oliveira, V.A., Cossi, L.V., Teixeira, M.C., Silva, A.M.: Synthesis of PID controllers for a class of time delay systems. Automatica **45**(7), 1778–1782 (2009). https://doi.org/10.1016/j.automatica.2009.03.018
8. Padhan, D.G., Majhi, S.: Modified Smith predictor based cascade control of unstable time delay processes. ISA Trans. **51**(1), 95–104 (2012). https://doi.org/10.1016/j.isatra.2011.08.002
9. Panda, R.C.: Synthesis of PID controller for unstable and integrating processes. Chem. Eng. Sci. **64**(12), 2807–2816 (2009). https://doi.org/10.1016/j.ces.2009.02.051
10. Wang, Q., Lu, C., Pan, W.: IMC PID controller tuning for stable and unstable processes with time delay. Chem. Eng. Res. Des. **105**, 120–129 (2016). https://doi.org/10.1016/j.cherd.2015.11.011
11. Raja, G.L., Ali, A.: Modified parallel cascade control strategy for stable, unstable and integrating processes. ISA Trans. **65**, 394–406 (2016). https://doi.org/10.1016/j.isatra.2016.07.008
12. Rao, A.S., Chidambaram, M.: Enhanced Smith predictor for unstable processes with time delay. Ind. Eng. Chem. Res. **44**(22), 8291–8299 (2005). https://doi.org/10.1021/ie0503161
13. Shamsuzzoha, M., Lee, M.: IMC−PID controller design for improved disturbance rejection of time-delayed processes. Ind. Eng. Chem. Res. **46**(7), 2077–2091 (2007). https://doi.org/10.1021/ie0612360
14. Shamsuzzoha, M., Lee, M.: Design of advanced PID controller for enhanced disturbance rejection of second-order processes with time delay. AIChE J. **54**(6), 1526–1536 (2008). https://doi.org/10.1002/aic.11483
15. Skogestad, S.: Simple analytic rules for model reduction and PID controller tuning. J. Process Control **13**(4), 291–309 (2003)

16. Yang, X.P., Wang, Q.G., Hang, C.C., Lin, C.: IMC-based control system design for unstable processes. Ind. Eng. Chem. Res. **41**(17), 4288–4294 (2002). https://doi.org/10.1021/ie010812j
17. Zhang, W., Gu, D., Wang, W., Xu, X.: Quantitative performance design of a modified Smith predictor for unstable processes with time delay. Ind. Eng. Chem. Res. **43**(1), 56–62 (2004). https://doi.org/10.1021/ie020732v

Identification of Bad Data from Phasor Measurement Units Using Evolutionary Algorithms

Polly Thomas, Emil Ninan Skariah, Sheena Thomas, Sandy J. Thomson and Shanmugam Prabhakar Karthikeyan

Abstract Phasor measurement units (PMUs) are power system devices placed at various locations in electrical power network that provide the measurement of phasors of voltages and currents from the respective meters or instruments placed at locations. Data provided by the PMUs are used for system analysis and also help to analyze the sequence of events that may have contributed to the failure of the power system. The presence of bad data in PMUs is increasing in today's power system due to the system complexity and associated issues. This will lead to inaccurate measurement of voltage magnitude and phase angle. Hence, it is important to detect the presence of bad data and identify the PMU which contains the error. In order to identify the bad data, state estimation is carried out to obtain the true state variables by the application of Load Flow Analysis (LFA). The application of Engineering Optimization is called into this work for improving the detection of bad data and its correction. Firefly optimization and particle swarm optimization techniques are developed for the above problem, and later results are compared with respect to the accuracy of results. The system is modeled in MATLAB platform.

Keywords Power system · State estimation · Phasor measurement units
Bad data · Evolutionary algorithms

P. Thomas (✉) · E. N. Skariah · S. Thomas · S. J. Thomson
SAINTGITS College of Engineering, Pathamuttam, Kerala, India
e-mail: polly.thomas@saintgits.org

E. N. Skariah
e-mail: emil.ninan@saintgits.org

S. Thomas
e-mail: Sheena.thomas@saintgits.org

S. Prabhakar Karthikeyan
VIT University, Vellore, Tamil Nadu, India
e-mail: sprabhakarkarthikeya@vit.ac.in

© Springer Nature Singapore Pte Ltd. 2019
J. C. Bansal et al. (eds.), *Soft Computing for Problem Solving*, Advances in Intelligent Systems and Computing 817, https://doi.org/10.1007/978-981-13-1595-4_8

1 Introduction

A power system is a complex system that connects power generators through transmission and distribution networks across a large area. Power systems have become increasingly large and complicated over the time. Thus, it has become difficult and challenging to control the power system. Thus, reliability of power system becomes a major issue that has to be dealt with nowadays in power system. The reliability of the system can be maintained by enabling the system to be in a normal and secure state. To accomplish this goal, real-time monitoring of the system state must be provided [1].

State estimation is a process where data from network measuring points can be formed into a set of reliable data for real-time monitoring of the system. The basic idea of SE is to obtain the estimated states of a power system [2]. State estimation methods can be mainly classified into two groups. The first group is categorized as mathematical methods [3, 4], and the second group is categorized as intelligent methods. Weighted Linear Least (WLS), Weighted Least Absolute Value (WLAV), and Estimation with Non-Fixed Error (M-Estimator) are the mathematical methods, whereas Fuzzy Inference System (FIS), state estimation based on Neural Network (NT), and Adaptive Neuron Fuzzy Inference System (ANFIS) are the intelligent methods [5, 6]. Intelligent methods require less time, but they are not sufficiently accurate when compared with mathematical methods. The accuracy and speed of WLS and WLAV are better than the other mathematical methods [7].

Bad data detection is considered as the best solution to detect malicious data which is applied to reduce the observation errors and detect the false data. There are various bad data detection techniques designed. In [8] compares various methods of bad data identification. The methods are identification by elimination, identification by HT (Hypothesis Testing), and identification by NQC (Non-Quadratic Criteria). The result of comparison is that the Hypothesis Testing Identification method combines the effectiveness, reliability, and compatibility with online implementation requirements. In [9], bad data is identified and corrected using calibration factor method. The calibration factor is calculated by using nonlinear optimal estimation theory with a traditional model of a non transposed transmission line with unbalanced load. But in most of the methods mentioned above, the bad data detection in the system is limited to detect its presence only and no provision for further information regarding the error. The elimination of error is not ensured so that it may lead to system contingency.

Phasor measurement units (PMUs) are units which provide phasors of the voltages, currents, and power signals throughout the power system. With PMU, the wide area snapshot of a power system can be obtained [10]. Phasor information is useful in both normal and abnormal conditions. During abnormal condition, it is used for control and protection. During normal operating conditions, they help in monitoring the system state. Data obtained from the PMUs are accurate, and it helps system analysts to determine the sequence of events which may have contributed to the system contingencies like blackout and also helps to determine the events that may have contributed to the failure of the power system. With the usage of PMUs, there are various applications [11, 12] such as improved real-time monitoring and control of power system, power congestion management, state estimation of the power system, post disturbance analysis of the power system, overload monitoring and dynamic rating, restoration of power system, protection and control application of distributed generation.

Measurements may contain errors due to the various reasons such as internal meter errors, wrong/reverse connection, and incorrect topology information. Thus, any abnormality in the phasor information greatly affects the operation of power system. Thus, it is essential to identify the presence of bad data in PMU and to correct it.

The rest of the paper is structured as following. Section 2 describes the system used for the proposed method. Section 3 presents the process of state estimation. And Sect. 4 outlines the proposed methods. Finally, conclusions are drawn in Sect. 5 by taking IEEE 14 bus as the test system.

2 System Description

Consider a 14-bus system modeled as shown in Fig. 1 which consists of five generation units and ten loads that operate under normal condition. There are 19 transmission lines which indicate the power flow. Fourteen PMUs (each at a bus) indicate the voltage magnitude and the phase angle of the system. These readings are sent to the Phasor Data Collector which is further used for the control and protection purposes. The system shown above is taken as the test case. Table 1 shows the per unit values of the reactance obtained after modeling the test system and simulating using ETAP.

3 State Estimation

State estimation is the process of assigning a value to an unknown system state variable based on measurements from that system according to some criteria. In a power system, the state variables are the voltage magnitudes and relative phase angles at the system nodes. The estimator is designed to produce the "best estimate" of the system voltage and phase angles, recognizing that there are errors in measured quantities and that there may be redundant measurements [13].

State estimation problem can be described as

$$\min_{x} J(x) = \sum_{i=1}^{Nm} \frac{\left[z_i - f_i(x)\right]^2}{\sigma_i^2} \qquad (1)$$

We first form the gradient of J(x) as

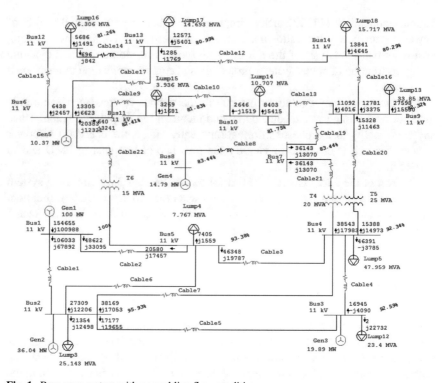

Fig. 1 Base case system with normal line flow condition

$$\nabla_x J(x) = \begin{bmatrix} \dfrac{\partial J(x)}{\partial x_1} \\ \dfrac{\partial J(x)}{\partial x_2} \end{bmatrix} \tag{2}$$

$$= -2 \begin{bmatrix} \dfrac{\partial f_1}{\partial x_1} & \dfrac{\partial f_2}{\partial x_1} & \dfrac{\partial f_3}{\partial x_1} & \cdots \\ \dfrac{\partial f_1}{\partial x_2} & \dfrac{\partial f_2}{\partial x_2} & \dfrac{\partial f_3}{\partial x_2} & \cdots \\ \vdots & \vdots & \vdots & \end{bmatrix} \begin{bmatrix} \dfrac{1}{\sigma_1^2} & & \\ & \dfrac{1}{\sigma_2^2} & \\ & & \ddots \end{bmatrix} \begin{bmatrix} [z_i - f_i(x)] \\ [z_i - f_i(x)] \\ \vdots \end{bmatrix} \tag{3}$$

If we put the $f_i(x)$ functions in a vector form $f(x)$ and calculate the Jacobian of $f(x)$, we would obtain

$$\frac{\partial f(x)}{\partial x} = \begin{bmatrix} \dfrac{\partial f_1}{\partial x_1} & \dfrac{\partial f_2}{\partial x_2} & \dfrac{\partial f_3}{\partial x_3} & \cdots \\ \dfrac{\partial f_1}{\partial x_1} & \dfrac{\partial f_2}{\partial x_1} & \dfrac{\partial f_3}{\partial x_1} & \cdots \\ \vdots & \vdots & \vdots & \end{bmatrix} \tag{4}$$

Table 1 System line data

Sr. no.	From bus	To bus	R(pu)	X(pu)	B/2(pu)
1	1	2	0.01938	0.05917	0.0264
2	1	5	0.05403	0.22304	0.0246
3	2	3	0.04699	0.19797	0.0219
4	2	4	0.05811	0.17632	0.0170
5	2	5	0.05695	0.17388	0.0173
6	3	4	0.06701	0.17103	0.0064
7	4	5	0.01335	0.04211	0.0000
8	4	7	0.0000	0.20912	0.0000
9	4	9	0.0000	0.55618	0.0000
10	5	6	0.0000	0.25202	0.0000
11	6	11	0.09498	0.19890	0.0000
12	12	6	0.12291	0.25581	0.0000
13	13	6	0.06615	0.13027	0.0000
14	14	7	0.0000	0.17615	0.0000
15	15	7	0.0000	0.11001	0.0000
16	16	9	0.03181	0.08450	0.0000
17	17	9	0.12711	0.27038	0.0000
18	18	10	0.08205	0.19207	0.0000
19	19	12	0.22092	0.19988	0.0000
20	20	13	0.17093	0.34802	0.0000

This matrix is expressed as [H]. Then

$$[H] = \begin{bmatrix} \frac{\partial f_1}{\partial x_1} & \frac{\partial f_2}{\partial x_2} & \frac{\partial f_3}{\partial x_3} & \cdots \\ \frac{\partial f_1}{\partial x_1} & \frac{\partial f_2}{\partial x_2} & \frac{\partial f_3}{\partial x_3} & \cdots \\ \vdots & \vdots & \vdots & \end{bmatrix} \tag{5}$$

And its transpose is

$$[H]^T = \begin{bmatrix} \frac{\partial f_1}{\partial x_1} & \frac{\partial f_2}{\partial x_1} & \frac{\partial f_3}{\partial x_1} & \cdots \\ \frac{\partial f_1}{\partial x_2} & \frac{\partial f_2}{\partial x_2} & \frac{\partial f_3}{\partial x_2} & \cdots \\ \vdots & \vdots & \vdots & \end{bmatrix} \tag{6}$$

To make $\nabla_x J(x)$ equal zero, we apply Newton's method, then

$$\Delta_x = \left[\frac{\partial \nabla_x J(x)}{\partial x} \right]^{-1} [-\nabla_x J(x)] \tag{7}$$

The Jacobian $\nabla_x J(x)$ is calculated by treating $[H]$ as a constant matrix

$$\Delta_x = \left[[H]^T [R]^{-1}[H]\right]^{-1} [H]^T [R]^{-1} \begin{bmatrix} [z_i - f_i(x)] \\ [z_i - f_i(x)] \\ \vdots \end{bmatrix} \tag{8}$$

4 Optimization Algorithm

Optimization means maximization or minimization of one or more functions with any possible constraints. Optimization can be defined as a process of minimizing undesirable things such as cost, error, losses or maximizing desirable things such as profit, efficiency. It is the process of finding the variables that can minimize or maximize an objective function.

The choice of solution to an optimization problem mainly depends on the type of optimization problem. To solve such problems, various classical methods are present. The development of fast digital computers led to the advancement in optimization. In recent days, many advanced optimization techniques are available to solve complex problems.

Nowadays, the optimization problems have many intricacies such as discrete, continuous, or mixed variables, multiple conflicting objectives, nonlinearity, discontinuity. Main disadvantages of such problems are that the search space may be so large that the optimum value cannot be found in appropriate amount of time. Such problems cannot be solved by the classical methods.

Evolutionary algorithms (EAs) are mainly implemented to achieve the almost optimum solutions for complex optimization problems. EAs mainly deal with the problems that have numerous decision variables and nonlinear objective functions. EAs are inspired by the process of natural biological evolution or social behavior, for example, during migration, how birds find their destination and how ants find the shortest way to a source of food. The evolutionary algorithms are developed according to population-based search procedures which incorporate random variation and selection. The genetic algorithm (GA)—first optimization technique developed—was evolutionary based. There are so many algorithms like ant colony optimization (ACO), particle swarm optimization (PSO), firefly optimization, and estimation of distribution algorithm (EDA) that comes under EA.

4.1 Particle Swarm Optimization

The PSO algorithm [14] is developed by maintaining several candidate solutions within a given search space. The algorithm is developed on the assumption that each candidate solution is a particle which is flying through the search space so that the maximum or minimum of the objective function can be found out. The candidate solutions that composed of all the possible solutions are randomly chosen from search space initially. The algorithm has no idea about the objective function and thus has no information about the candidate solutions that is whether they are near to or far away from the local or global maximum [15].

The PSO algorithm consists of just three steps, which are repeated until maximum iteration (stopping criteria) is reached.

1. Evaluate the fitness of each particle.
2. Update individual and global best finesses and positions.
3. Update velocity and position of each particle.

Velocity of a particle defined as

$$v = v + c_1 * rand * (pBest - p) + c_2 * rand * (gBest - p) \qquad (9)$$

where

p	particle's position
v	path direction
c_1	weight of local information
c_2	weight of global information
$pBest$	best position of the particle
$gBest$	best position of the swarm
$rand$	random variable

The terms in the velocity update equation have three different roles in the algorithm. The first term is known as the inertia component. It is responsible of keeping the particle moving in the same direction it was actually moving to. The inertia coefficient's value lies between 0.8 and 1.2. The second term is known as the cognitive component. It acts as the particle's memory, which helps the particle to return to the regions of the search space in which it has best individual objective function. The value of cognitive coefficient c_1 lies close to 2. The size of the step that the particle takes toward its individual best solution depends on the cognitive coefficient. The third term is termed as the social component. It causes the particle to move to the best location that has been found so far. The value of social coefficient c_2 lies close to 2. The size of the step that the particle takes toward the global best solution $g(x)$ that has been found up until that point depends on the social component.

4.2 Firefly Optimization

The firefly algorithm [16] which is inspired by the mating or flashing behavior of fireflies is called into the reference system to analyze the behavior response of the system. There are mainly two points that must be considered when firefly algorithm is taken. First one is the light intensity, and second one is the attractiveness. The brightness of the firefly is determined by the objective function. Secondly, the variation of light intensity is defined in order to formulate the change of the attractiveness. The light intensity decreases with the distance from its source and the media will absorb the light, so in our simulation we suppose the light intensity I varies with the distance r and light absorption parameter γ exponentially and monotonically.

The amount of movement of firefly i to another more attractive (brighter) firefly j is determined by

$$x_i = x_i + \beta_0 e^{\wedge}\left(-\gamma r^2\right)\left(x_j - x_i\right) + \alpha\,\varepsilon_i \tag{10}$$

where the first term is the current location of firefly i, the second term is due to the attraction, while the third term is randomization with α being the randomization parameter and ε being the vector of random numbers drawn from a Gaussian distribution.

5 Results and Discussion

The bad data identification and correction using PSO and firefly optimization has been performed on the IEEE 14-bus. The implementing program is written in MATLAB programming language. State estimation is done with analog measurements such as real and reactive power injections, and flows and the estimated values are obtained. Error is injected in the analog measurements, and the values are obtained which are taken as the measured readings. The error is calculated by the difference of estimated and measured readings. The main aim of the optimization is to find the optimal values of measurements such that minimum error is obtained. The objective function is taken as the sum of error developed in voltage magnitude and phase angle when error is injected in the analog measurements.

5.1 Error Correction Using Particle Swarm Optimization

Here, the population size is chosen as 25. The maximum number of iterations is chosen as 1000 which is the stopping criterion, and the other parameters used are was 0.99, c_1 as 1.5 and c_2 as 2.

Figure 2a shows the case when error is injected in measurements of real power injections. The difference between the estimated reading and the actual reading could be clearly identified, and hence, the deviation at that point is specifically shown in the figure. The case depicts only the values of voltage magnitude. It is shown as two sections, i.e., Fig. A shows the readings before correction of bad data, whereas Fig. 2b shows the readings after the correction.

Deviation zero means that the particular meter contains no error, whereas nonzero deviations mean that the particular meter contains error and it is displayed. In Fig. 2b,

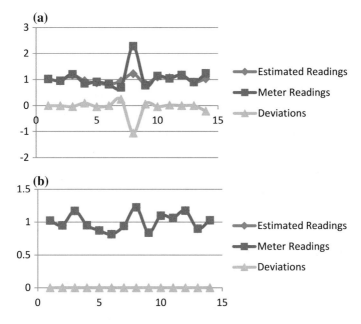

Fig. 2 Case when error occurs in real power injections (voltage magnitude) **a** before correction and **b** after correction

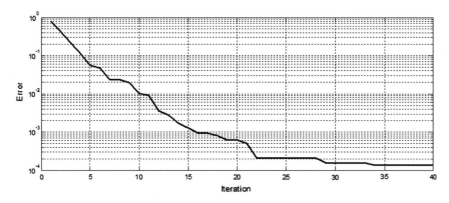

Fig. 3 Convergence graph

the overlapping of the meter reading depicts the similarity in the readings which in turn mean that the error is nil. If the meter readings show zero, deviations which confirm that all bad data in meters are corrected.

Figure 3 shows the error convergence obtained at the end of each iteration. The readings after the correction are obtained which is similar to the estimated readings that confirm the correction of error.

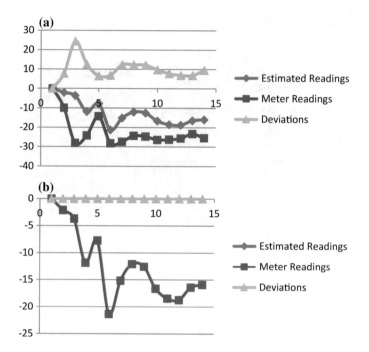

Fig. 4 Case when error occurs in real power flow (phase angle) **a** before correction and **b** after correction

5.2 Error Correction Using Firefly Optimization

Prevailing the earlier conditions, the other parameters used are γ as 1, β_0 as 2 and α as 0.2. Figure 4a shows the case when error is injected in measurements of real power flow. The case depicts only the values of phase angle. It is shown as two sections, i.e., Fig. A shows the readings before correction of bad data, whereas Fig. 4b shows the readings after the correction. Deviation zero means that the particular meter contains no error, whereas nonzero deviations mean that the particular meter contains error and it is displayed. The entire meter contains zero deviations which confirm that all bad data in meters are corrected.

The error obtained at the end of each iteration is shown in the convergence graph below. The readings after the correction are obtained which is similar to the estimated readings that confirm the correction of error. The convergence graph of the case is shown in Fig. 5.

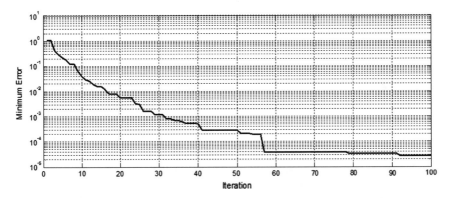

Fig. 5 Convergence graph

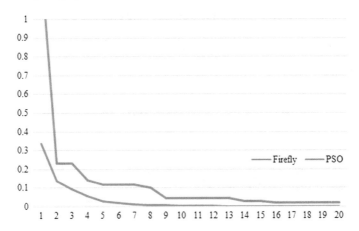

Fig. 6 Comparison graph of firefly and PSO optimization

5.3   *Comparison Between Particle Swarm and Firefly Optimization*

The comparison of the optimization technique is done by taking 20 iterations as the sample case. PSO requires more number of iterations than firefly to obtain the minimum error or zero error. The graph and tables shown below confirm that firefly optimization technique is better than PSO. From the convergence graph of both the optimization techniques, we can understand that the rate of decrease of error is more in case of firefly than PSO optimization technique. From Fig. 6, we can understand that magnitude of error decreased in case of firefly is more than the PSO optimization technique, since the settling of firefly is faster when compared to the PSO. Also, we can conclude that firefly optimization technique is better than PSO optimization technique in terms of its error correction.

6   Conclusion

This paper introduces a method to detect and correct the presence of bad data in PMU. Optimization techniques were developed for improving the detection of such errors and its correction. Particle swarm and firefly optimization methods are implemented for the same. The two methods were tested on IEEE 14-bus system, to study the bad data detection as well as its correction. The system efficiently detects the bad data and performs the error correction operation by means of optimization. The results were compared and concluded that the firefly optimization technique is better than the particle swarm optimization technique in terms of error correction and convergence. The convergence characteristics as well as the time taken are convincingly better in the case of firefly algorithm than the long-standing particle swarm optimization for the reason that the points are concentrated around one single point in the case of firefly where it is scattered in the entire region in the case of PSO.

References

1. Kundur, P., Balu, N.J., Lauby, M.G.: Power System Stability and Control, vol. 7. McGraw-Hill, New York (1994)
2. Abur, A., Exposito, A.G.: Power System State Estimation: Theory and Implementation. CRC Press (2004)
3. Naka, S., Genji, T., Yura, T., Fukuyama, Y.: A hybrid particle swarm optimization for distribution state estimation. IEEE Power Eng. Rev. 22(11), 57–57 (2002)
4. Khwanram, J., Damrongkulkamjorn, P.: Multiple bad data identification in power system state estimation using particle swarm optimization. In: 6th International Conference on Electrical Engineering/Electronics, Computer, Telecommunications and Information Technology, 2009, ECTI-CON 2009, vol. 1. IEEE (2009)
5. Kumar, D.V., Srivastava, S., Shah, S., Mathur, S.: Topology processing and static state estimation using artificial neural networks. IEEE Proc. Gener. Transm. Distrib. 143(1), 99–105 (1996)
6. Singh, D., Pandey, J., Chauhan, D.: Topology identification, bad data processing, and state estimation using fuzzy pattern matching. IEEE Trans. Power Syst. 20(3), 1570–1579 (2005)
7. Mahaei, S.M., Navayi, M.R.: Power system state estimation with weighted linear least square. Int. J. Electr. Comput. Eng. 4(2), 169 (2014)
8. Hagh, M.T., Mahaei, S.M., Zare, K.: Improving bad data detection in state estimation of power systems. Int. J. Electr. Comput. Eng. 1(2), 85 (2011)
9. Shi, D., Tylavsky, D.J., Logic, N.: An adaptive method for detection and correction of errors in PMU measurements. IEEE Trans. Smart Grid 3(4), 1575–1583 (2012)
10. Wilson, Robert E.: PMUs [phasor measurement unit]. IEEE Potentials 13(2), 26–28 (1994)
11. Phadke, A.G., Thorp, J.S.: History and applications of phasor measurements. In: Power Systems Conference and Exposition, 2006, PSCE'06. IEEE PES. IEEE (2006)
12. Penshanwar, M.K., Gavande, M., Satarkar, M.R.: Phasor measurement unit technology and its applications—a review. In: International Conference on Energy Systems and Applications, 2015. IEEE (2015)
13. Wood, A.J., Wollenberg, B.F.: Power Generation, Operation, and Control. Wiley (2012)
14. Du, K.L., Swamy, M.N.S.: Particle swarm optimization. In: Search and Optimization by Metaheuristics. Birkhäuser, Cham (2016)

15. Blondin, J.: Particle swarm optimization: a tutorial (2009). http://cs.armstrong.edu/saad/csci8 100/psotutorial.pdf
16. Osaba, E., Carballedo, R., Yang, X.S., Diaz, F.: An evolutionary discrete firefly algorithm with novel operators for solving the vehicle routing problem with time windows. In: Yang, X.S. (ed.) Nature-Inspired Computation in Engineering. Studies in Computational Intelligence, vol. 637. Springer, Cham (2016)
17. Yang, X.-S.: Firefly algorithm, Lévy flights and global optimization. In: Bramer, M., Ellis, R., Petridis, M. (eds.) Research and Development in Intelligent Systems, vol. XXVI, pp. 209–218. Springer, London (2010)

Image Compression Using Neural Network for Biomedical Applications

G. Sasibhushana Rao, G. Vimala Kumari and B. Prabhakara Rao

Abstract As images are of large size and require huge bandwidth and large storage space, an effective compression algorithm is essential. Hence in this paper, feedforward backpropagation neural network with the multilayer perception using resilient backpropagation (RP) algorithm is used with the objective to develop an image compression in the field of biomedical sciences. With the concept of neural network, data compression can be achieved by producing an internal data representation. This network is an application of backpropagation that takes huge content of data as input, compresses it while storing or transmitting, and decompresses the compressed data whenever required. The training algorithm and development architecture give less distortion and considerable compression ratio and also keep up the capability of hypothesizing and are becoming important. The efficiency of the RP is evaluated on x-ray image of rib cage and has given better results of the various performance metrics when compared to the other algorithms.

Keywords Artificial neural network · Backpropagation neural network
Gradient descent algorithm (GD) · Image compression
Resilient backpropagation algorithm (RP)

G. Sasibhushana Rao
Department of Electronics and Communication Engineering, Andhra University
College of Engineering (A), Visakhapatnam 530003, Andhra Pradesh, India
e-mail: sasigps@gmail.com

G. Vimala Kumari (✉)
Department of Electronics and Communication Engineering, M.V.G.R. College
of Engineering, Vizianagaram 535002, India
e-mail: Vimalakumari7@gmail.com

B. Prabhakara Rao
Department of Electronics and Communication Engineering, JNTUK,
Kakinda 533003, Andhra Pradesh, India
e-mail: drbprjntuk@gmail.com

© Springer Nature Singapore Pte Ltd. 2019
J. C. Bansal et al. (eds.), *Soft Computing for Problem Solving*, Advances in Intelligent
Systems and Computing 817, https://doi.org/10.1007/978-981-13-1595-4_9

1   Introduction

In this modern world, trillions and trillions of bits of information are getting transmitted and received every second. This data includes images, videos, and text as well.
As videos and images take maximum amounts of size and if they are transmitted
the same way, the rate of data transfer gets slow, costs lot, and takes more time for
transmitting data. To avoid this major problem, the concept of image compression
comes in handy. The biggest challenge in image compression is error percentage or
the loss of data. Recently, artificial neural networks (ANNs) are proved to be efficient
for providing reduced image compression loss. Artificial neural networks (ANNs)
are archetypes of the biological neuron system and thus have been drawn from the
abilities of a human brain. The architecture of ANN being drawn from the concept of
brain functioning, a neural network is a hugely reticulated network of a huge number
of neurons which are processing elements [1]. ANNs are employed to summarize
and prototype some of the functional aspects of the human brain system in an effort
so as to acquire some of its computational strengths [2]. A neural network (NN) consists of eight components: neurons, signal function, activation state vector, activity
aggregation rule, pattern of connectivity, learning rule, activation rule, and environment [3]. Recently, ANNs are applied in areas in which high rates of computation
are essential and considered as probable solutions to problems of image compression [4]. Generally, two different categories have been put forward for enhancing the
performance of compression methods. Firstly, a method for compression by using
ANN technology has to be developed to improve the design. Secondly, NN has to
be applied to develop compression methods [5–7]. Backpropagation algorithm is
extensively used learning algorithms in ANNs. With generalization ability and high
accuracy, the feedforward neural network architecture is capable of approximating
most problems. This architecture is based on the learning rule of error correction
[8]. Error propagation comprises of two passes, a forward pass and a backward pass
through different layers of network. The effect, of input vector's application to the
sensory nodes of the network, transmits through the network layer-by-layer in the
forward pass. In the end, a set of outputs are produced as an actual response of this
process. All the synaptic weights of the networks are fixed during the forward pass
only and adjusted according to the need of error correction during the back pass. The
error signal is produced when the actual output of the network is subtracted from the
expected output. This error signal is then propagated backward against the direction
of synaptic conditions through the network. Until the actual output of the network
so produced is nearer to the expected output, the synaptic weights are adjusted. To
produce a complex output, the backpropagation neural network is essentially made
of a network of simple processing elements working together [9–12]. From the above
knowledge of backpropagation neural networks, image compression and decompression can be achieved.

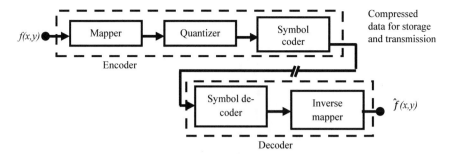

Fig. 1 Basic image compression block diagram

2 Image Compression

The main aim of image compression is to reduce the redundancy in image details for storage or transmission of data in a competent form. There are two types of image compression—lossy and lossless. Compression is obtained by removing one or more than three basic data redundancies for which optimal codes are used. Image compression techniques are used to reduce number of bits required to represent an image. An inverse process called decompression is applied to reconstruct the image. Image compression composes of two different processes which can be expressed in blocks as encoder and decoder. In encoder, the input image is fed to the system and image is mapped. This mapped image is processed to quantizer, and from the quantizer, the data is fed to symbol encoder. The encoder is responsible for reducing the bit size. In decoder, the output of the encoder is given as input. At first stage, symbol decoder processes the data and gives the processed information to inverse mapper to retrieve the original image. Figure 1 shows the block diagram of image compression.

3 Neural Network for Image Compression

ANN is a computer-based algorithm in which the required process works as a neuron cell as present in the human body. The basic functionality of the NN is derived from the functioning of the artificial neuron. Artificial neuron has a self-analyzing mechanism which is similar to the behavior of a biological nerve cell. To achieve image compression process, three different layers of operation are there in NN: input layer, output layer, and hidden layer. The number of nodes in the input layer and output layer are equal and are greater than the nodes of hidden layer. Hidden layer is coupled to both input and output layers. Compression of image is possible when assigning the value of number of neurons at the hidden layers which is less than the neurons at input and output layers. The architecture of the neural network is shown in Fig. 2. The input image is segmented into blocks of $N \times N$ square blocks, and

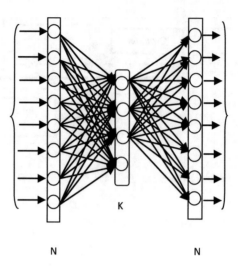

Fig. 2 Neural network architecture

each block is identified with a value, and this value gets added by another value called weight. These weights are calculated using a formula or by a mathematical function. The final value from the nodes is collectively added and saved to a node in the hidden layer. These are retrieved at the output layer with the reference from the input layer so as to extract the original image from the compressed image. An interesting mechanism takes place after retrieving the image. In NNs, loss of data is calculated from the comparison between the original image and decompressed image. The weights used in compression are adjusted such that the loss of data is minimized.

This technique is also called backpropagation neural network. This is a commonly used mechanism to enhance the loss reduction factor of NN. The factors considered in the image compression are peak signal-to-noise ratio (PSNR) and mean square error (MSE).

4 Learning

Neural networks being similar to brain are occasionally called machine learning algorithms due to their acquisition of knowledge from experience. The network learns to solve a problem by varying its connection weights (training). The strength of nodes between the neurons is assigned as a synaptic weight value for the particular node. The ANN acquires novel information by changing these weights of nodes. The learning skill of ANN is influenced by its architecture and the algorithmic method selected for the purpose of training. The learning method will be one of these three patterns: reinforcement learning, unsupervised learning, and supervised learning.

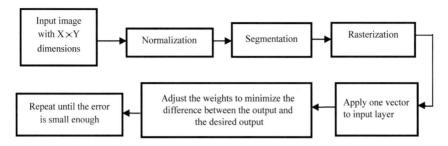

Fig. 3 Block diagram for training process

Supervised learning is essentially used for image compression purposes. The block diagram for training process is shown in Fig. 3.

5　Resilient Backpropagation Neural Network Algorithm

The first step in realizing the given task is to define the network architecture as shown in Fig. 2. The number of input neurons depends upon the type of problem and its requirement. The goal of the network is to compress the image which must be the defining parameter of the network. The constructed network will compress the image which is represented in matrix which has elements of pixels from the interval [0, 255]. Blocks of size 8×8 pixels are separated from the image and fed to a typical neural network with bipolar sigmoid activation function after normalizing the gray values from interval [0, 255] to real numbers from interval $[-1, 1]$. After rasterizing of each block by transformation of two-dimensional blocks into one-dimensional vectors, which means that values of 64 pixels used in network training are presented to the network input layer and expected output vectors, the synaptic weights are modified in order to get a minimal error. To adjust weights, resilient propagation (RP) learning algorithm was considered. The conventional backpropagation (BP) learning algorithm uses the partial derivatives of the error function (i.e., the gradient) to minimize the global error of the neural network by performing a gradient descent as given in Eq. (1).

$$H_{nk}^{(t+1)} = H_{nk}^{t} - \eta \frac{\partial Z^{(t)}}{\partial H_{nk}} \tag{1}$$

The choice of the learning rate parameter η, which scales the derivative, has an important effect on the time needed until convergence. Resilient backpropagation (RP) algorithm is a local adaptive learning method, much faster and stable than other usual variations of the backpropagation algorithms. The elimination of negative and unexpected influence of the size of the partial derivative on the weight step is its basic principle. So, the direction of the weight modification is indicated by only the sign of the derivative. The size of the update is given in Eq. (2) by a weight-specific update value Δ_{nk}, and the update values are determined from Eq. (3)

$$\Delta H_{nk}^t = \begin{cases} -\Delta_{nk}^t, & \text{if } \frac{\partial Z^t}{\partial H_{nk}} > 0 \\ +\Delta_{nk}^t, & \text{if } \frac{\partial Z^t}{\partial H_{nk}} < 0, \\ 0, & \text{else} \end{cases} \tag{2}$$

$$\Delta_{nk}^t = \begin{cases} \eta^+ * \Delta_{nk}^{(t-1)}, & \text{if } \frac{\partial Z^{(t-1)}}{\partial H_{nk}} * \frac{\partial Z^t}{\partial H_{nk}} > 0 \\ \eta^- * \Delta_{nk}^{(t-1)}, & \text{if } \frac{\partial Z^{(t-1)}}{\partial H_{nk}} * \frac{\partial Z^t}{\partial H_{nk}} < 0 \end{cases} \tag{3}$$

$$\text{where } 0 < \eta^- < 1, \eta^+ > 1$$

The modified values are determined from Eq. (4)

$$H_{nk}^{t+1} = H_{nk}^t + \Delta H_{nk}^t \tag{4}$$

The RP learning algorithm is based on learning by epoch; that is, after each training pattern has been obtained and the gradient of the sum of pattern errors is known, the weight update is performed only after the gradient information is completely available. Some authors have noticed that while using the standard backpropagation method, the weights in the hidden layer are updated with much smaller amounts than the weights in output layer and hence modify much slower. All of the weights growing uniformly are another advantage of RP.

6   Results and Discussion

For the evaluation of the experiments with RP backpropagation algorithm, x-ray image is taken. Experiments are conducted on image of size 256×256, namely rib cage with 8-bit pixel amplitude resolution. In this paper, RP algorithm is compared with various training algorithms of feedforward backpropagation neural network using two error metrics, PSNR and MSE. These algorithms are gradient descent (GD), gradient descent with momentum (GDM), gradient descent with variable learning rate (GDX), and Broyden–Fletcher–Goldfarb–Shanno (BFGS). The cumulative squared error between the original and decompressed image defined as MSE is given in Eq. (5).

$$MSE = \frac{\Sigma_{i,j}\left(f(x, y) - \hat{f}(x, y)\right)^2}{M \times N} \tag{5}$$

where $M \times N$ is size of the image, M is number of rows, N is number of columns, f represents the original image, and $\hat{f}$ represents the decompressed image. The quality of image coding is typically evaluated by the PSNR as defined in Eq. (6).

$$PSNR = 10 \log_{10} \left(\frac{255^2}{MSE} \right)$$ (6)

Compression ratio (CR) is the ratio of number of input neurons to the number of hidden neurons and is given in Eq. (7).

$$CR = \frac{n_i}{n_h}$$ (7)

where n_i is number of input neurons and n_h is number of hidden neurons. The network for every training algorithm is made to train on same image using 64 inputs with 4, 9, and 16 hidden neurons with compression rates of 16, 7.12, and 4, respectively. After the training is completed, the performance of algorithms is verified. The code has been run for different epochs using MATLAB R2015a. The results are recorded for all the training algorithms with sigmoid activation function in the hidden and output layers. Table 1 shows the average PSNR values for 8×8 block size with nine hidden neurons of test image of rib cage. These results demonstrated that the average PSNR value obtained by RP algorithm is better than those obtained by GD, GDM, GDX, and BFGS algorithms and observed that the higher value of epoch gives highest PSNR for all training algorithms. MSE values of five algorithms are shown in Table 2. The results of these experiments proved that the MSE of RP algorithm is lower than other four algorithms. Table 3 gives the comparison of the performance metrics of five algorithms when code has been run for 1000 epochs, from which it can be observed that for RP algorithm the PSNR is more and MSE is less when compared to that of other algorithms and found that by using the RP algorithm the simulation and encoding times are less when compared to that of other algorithms. As it can be seen from the table, the training time for RP algorithm is almost similar when compared to GD, GDM, GDX, but it is high for BFGS algorithm. The change in PSNR values with respect to the number of epochs for 8×8 block size with nine hidden neurons for test image of rib cage is shown in graph in Fig. 4. From graph, it can be observed that the PSNR values for RP algorithm are greater than the other four algorithms.

The change in MSE values with respect to the number of epochs for 8 × 8 block size with nine hidden neurons for test image of rib cage is shown in graph in Fig. 5. From graph, it can be observed that the MSE values for RP algorithm are lower than the other four algorithms. Figure 6a–f shows the original image and decompressed images of five training algorithms GD, GDM, GDX, BFGS, and RP, respectively. From the obtained decompressed images, it is observed that image quality is more for RP algorithm when compared to the other algorithms.

The performance plots of the change in MSE have been plotted with respect to 1000 epochs, and the regression plot with target vs output with nine hidden neurons of 8 × 8 image block size for various training algorithms GD, GDM, GDX, BFGS, and RP for x-ray image of rib cage is shown in Fig. 7a, b. From plots, it can be observed that regression plot is more linear for RP algorithm and almost reaches the ideal fitness graph when compared to the other algorithms. Also in the performance plots, the mean square error is less for RP algorithm.

Table 1 Comparison of PSNR values of decompressed images for number of epochs for five algorithms for rib cage image

Epochs	GD	GDM	GDX	BFGS	RP
100	24.9	24.95	24.99	25	25.84
200	24.92	24.98	25.02	25.15	26.34
300	24.94	24.99	25.18	25.43	25.66
400	24.95	25.02	26.4	26.82	26.93
500	24.98	25.28	25.35	26.57	29.8
600	24.99	25.33	28.2	28.73	29.87
700	24.99	25.39	25.71	25.82	26.9
800	25.00	25.11	26.4	28.04	28.81
900	25.02	25.24	27.29	28.65	29.01
1000	25.06	25.79	27.39	29.35	29.86

Table 2 Comparison of MSE values of decompressed images for number of epochs for five algorithms for rib cage image

Epochs	GD	GDM	GDX	BFGS	RP
100	209.6	207.9	207.2	201.2	169.2
200	209.2	206.1	205.6	201.5	150
300	209.0	205.9	197.2	196.25	176.3
400	206.5	205.6	148.9	135	129.5
500	206.2	205.2	189.5	143.1	102.9
600	205.6	204.1	198.22	160	133.4
700	193.5	190.3	174.4	169	131.3
800	190.1	187.1	118.2	170.2	102.2
900	187.4	180.1	152.4	129.5	70.3
1000	172.9	170.7	120.6	108.1	58.8

Table 3 Comparison of performance metrics of five algorithms for rib cage image

Block size 8 × 8 for 9 hidden neurons

Training algorithm	Epochs	PSNR (dB)	MSE	Training time (s)	Simulation time (s)	Encoding time (s)
GD	1000	25.06	172.9	5.746	0.0186	0.0965
GDM	1000	25.79	170.7	5.8411	0.0161	0.1
GDX	1000	27.39	120.6	5.797	0.0152	0.0857
BFGS	1000	29.35	108.1	126.49	0.0155	0.0947
RP	1000	29.86	58.8	5.839	0.0146	0.0836

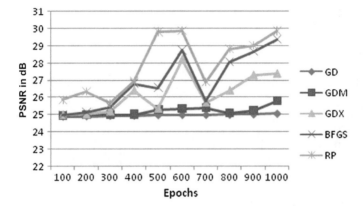

Fig. 4 Comparison of PSNR values of decompressed results for rib cage image

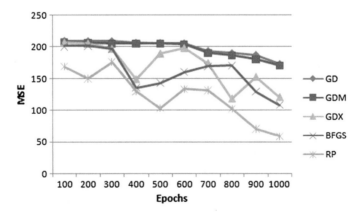

Fig. 5 Comparison of MSE values of decompressed results for rib cage image

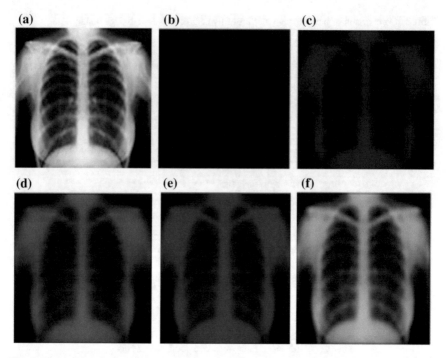

Fig. 6 X-ray image of rib cage: **a** original and **b–f** decompressed images of various training algorithms GD, GDM, GDX, BFG and RP, respectively

7   Conclusions

In this image compression system using artificial neural network, the backpropagation neural network with the multilayer perception is used along with quantizer and Huffman encoder. The NN is trained with a small 8×8 blocks of image and tested. The performance of the five various algorithms for image compression have been studied and compared. The algorithms are tested on x-ray medical image of rib cage. The algorithms have been compared for performance metrics, performance plots, and regression plots. Out of all these five algorithms, it is found that resilient backpropagation algorithm has shown more accurate results with less mean square error and high PSNR when compared to the other algorithms, so it would be very helpful in biomedical applications.

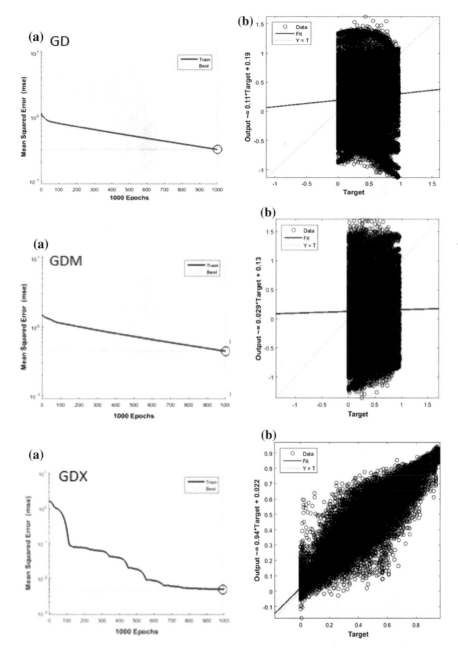

Fig. 7 Performance and regression plots with nine hidden neurons of 8 × 8 image block size for various training algorithms GD, GDM, GDX, BFGS, and RP for x-ray image of rib cage (**a** and **b**), respectively

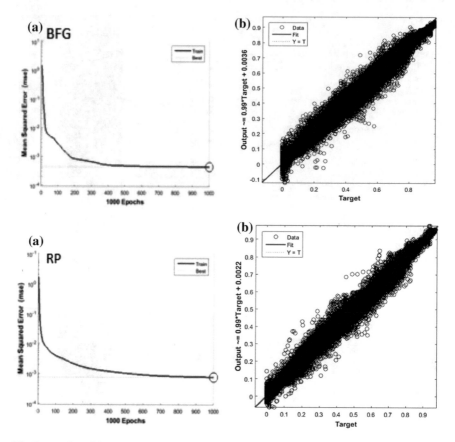

Fig. 7 (continued)

References

1. Robert Dony, D., Haykin, S.: Neural network approaches to image compression. Proc. IEEE **83**(2) (1995)
2. Xu, W., Nandi, A.K., Zhang, J.: Novel fuzzy reinforced learning vector quantisation algorithm and its application in image compression. IEEE Proc. Image Signal Process. **150**(5) (2003)
3. Nagi Reddy, S.K., Vikram, B.R., Rao, K.L., Reddy, S.B.: Image compression and reconstruction using a new approach by artificial neural network. Int. J. Image Process. (IJIP) **6**(2) (2012)
4. Patel, B.K., Agrawal, S.: Image compression techniques using artificial neural network. Int. J. Adv. Res. Comput. Eng. Technol. (IJARCET) **2**(10) (2013)
5. Khalil, A.R., Younis, C.M.: Image compression technique using a hierarchical neural network. J. Comput. Math. **3**(2) (2006)
6. Anusha, K., Madhura, G., Lakshmikantha, S.: Modeling of neural image compression using gradient decent technology. Int. J. Eng. Sci. (IJES) **3**(12) (2014)
7. Wilamowski, B.M., Chen, Y, Malinowski, A.: Efficient algorithm for training neural networks with one hidden layer. In: Proceedings of the IJCNN, vol. 3, pp. 1725–728 (1999)
8. Shankar, T.N., Maheswara Rao, M., Sudha, D., Sahoo, G.: Image compression by using back-propagation. Int. J. Comput. Sci. Appl. **3**(2) (2010)

9. Liu, L.: The progress and analysis of image compression based on BPANN. Micro Comput. Inf. **23**(6), 312–314 (2007)
10. Ahamed, S.A., Chandrashekarappa, K.: ANN implementation for image compression and decompression using backpropagation technique. Int. J. Sci. Res. (IJSR) **3**(6) (2014)
11. Srivastava, R., Singh, O.P.: Lossless image compression using neural network. Int. J. Remote Sens. Geosci. (IJRSG) **4**(3) (2015)
12. Nait, C.H., Salam, F.M.: Neural networks-based image compression system. In: Proceedings of the 43rd IEEE Midwest Symposium on Circuits and Systems (2000)

Symmetrical Cascaded Switched-Diode Multilevel Inverter with Fuzzy Controller

Y. Viswanath, K. Muralikumar, P. Ponnambalam and M. Praveen Kumar

Abstract This chapter presents the experiment of symmetrical cascaded switched-diode multilevel inverter (SCSD MLI) with fuzzy controller for a different number of levels. The objective of this topology is to reduce the number of power semiconductor switches along with its gate driver circuits as the number of level increases; therefore, the complexity of the circuit and installation cost of the converter are reduced when compared with the conventional cascaded multilevel inverter and cascaded half-bridge multilevel inverter. In this document, seven-, nine-, eleven-level SCSD MLIs are analyzed with simulation results which describe the total harmonic distortion reduction with the increment in number of levels. For this circuit topology, phase disposition pulse-width modulation technique is developed to regulate the RMS output voltage of inverter. In order to maintain the RMS output voltage, appropriate fuzzy controllers are constructed. MATLAB/SIMULINK simulation results of seven-level, nine-level, and eleven-level are presented to justify the performance of the suggested topology.

Keywords Multilevel inverter · PWM technique · Fuzzy logic controller

1 Introduction

Among the various popular inverter topologies, multilevel inverters have evolved into an impressive and potential inverter in power electronic circuits for numerous applications. Multilevel inverters are power electronic-based converters which comprise

Y. Viswanath (✉) · K. Muralikumar
School of Electrical Engineering, VIT University, Vellore 632014,
Tamil Nadu, India
e-mail: viswanath5859@gmail.com

P. Ponnambalam · M. Praveen Kumar
Faculty of School of Electrical Engineering, VIT University,
Vellore 632014, Tamil Nadu, India
e-mail: praveen.m@vit.ac.in

© Springer Nature Singapore Pte Ltd. 2019   121
J. C. Bansal et al. (eds.), *Soft Computing for Problem Solving*, Advances in Intelligent
Systems and Computing 817, https://doi.org/10.1007/978-981-13-1595-4_10

of many input DC power sources and elements such as power switches, diodes, gate driver circuits which give the desired output voltage by a relevant switching pattern compared with the conventional three-level inverter. Multilevel inverters have inherent features that can be characterized as lower values of voltage stresses on switches, a tolerable amount of harmonic distortions, power quality, electromagnetic interference, no usage of bulky transformers or high-rating filters, the capability to reduce the installation, and maintenance operating costs in contradiction with the standard inverter. Generally, multilevel inverters are capable of generating peculiar number of levels across the output as a staircase-like waveform by the addition of employing various counts of potential switches, diodes, and DC power sources which can be obtained from sustainable energy integration systems [1] like photovoltaic system, tidal energy, wind energy, and so on. The output waveform quality depends on number of levels of the inverter. As the number of levels increases, the output waveform becomes more and nearer to sinusoidal; as a result, total harmonic distortion content decreases [2].

Basically, the standard topologies of multilevel inverters are classified into three types, namely the cascaded H-bridge multilevel inverter (CHB MLI) [3–5], the diode-clamped or neutral-point-clamped (NPC MLI) multilevel inverter [6–8], and the flying capacitor multilevel inverter (FC MLI) [9, 10]. Among these basic topologies, cascaded H-bridge multilevel inverter has fascinated most consideration due to its uncomplicated structure. The overall operation of cascaded multilevel inverter depends on the number of isolated DC sources used. It has tremendous prospective to be engaged with sustainable energy integration systems. The drawbacks of the flying capacitor and diode-clamped multilevel inverters are troubled with demanding a large number of power semiconductor devices and scheming a charge stabilizing control path for smoothening the unbalanced voltage issues due to the series connection of capacitors. On the other side, researchers have been demonstrated new multilevel inverter topologies which require the less number of switch devices, gate driver circuits, and DC power sources. Of those topologies, the topology "Cascaded Half-Bridge Multilevel Inverter" proposed by Hosseini [11] is adequately reduced half of the switches required when compared with the conventional cascaded H-bridge topology. Then, Alishah established a new topology "Cascaded Switched Diode" [12] which in turn reduces the switch count compared to cascaded half-bridge topology and also generates more levels with minimum number of switches. This paper presents a symmetrical cascaded switched-diode multilevel inverter (SCSD MLI) for obtaining higher number of levels with less switches and also nullifies the voltage spike occurring at the output voltage.

In general to limit the inverter RMS voltage at output, different modulation techniques are developed. The process of converting the switching state of the device from one state to other state is known as modulation. All modulations aims to generate stepped waveform with variable frequency, phase component, and amplitude, which is sinusoidal in steady state, and each of it has different switching configurations to obtain the desired output. Modulation techniques are classified into fundamental switching frequency (FSF) and high switching frequency (HSF). Fundamental switching frequency is divided into space vector control (SVC) and selective har-

Fig. 1 Proposed elementary unit for multilevel inverter

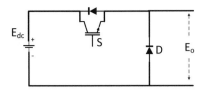

Table 1 Values of E_o for states of switches S1 and D1

State	Switches' states		E_g
	S_1	D_1	
1	1	0	E_{dc}
2	0	1	0

monic elimination pulse-width modulation (SHE PWM), and high switching frequency is divided into sinusoidal pulse-width modulation (SPWM) and space vector PWM. The frequently employed modulation techniques for cascaded inverters are sinusoidal pulse-width modulation and space vector pulse-width modulation. In this paper, a phase disposition pulse-width modulation (PD-PWM) method [13] is implemented to control the proposed SCSD MLI topology. A fuzzy controller is used to get the closed-loop operation for the proposed topology.

2 Topology of the Proposed SCSD MLI

2.1 Seven-Level SCSD MLI

Figure 1 shows the elementary unit of the proposed sub-multilevel inverter. There exists a capacitor or power source equal to E_{dc}, a switch (S), and a diode (D). Table 1 indicates the values of E_o for the states of switch and diode. We can observe that both switch and diode are complementary to each other and cannot be on at a time because a short circuit occurs across the voltage E_{dc}. When the switch S is ON, the output will be E_{dc}, and when the switch is OFF, the output will be zero.

SCSD MLI is a two-stage topology. In Fig. 1, the basic unit of first stage is shown which can be cascaded along with the spike removal (S_R) switch which is connected between the second unit and the Nth unit (where number of cascaded units are represented as N) provides a way for the load reversal current, and stage two contains the full-bridge inverter. For seven-level, whenever the switch S_1 is ON, and S_2 and S_3 are OFF, S_R is turned on, whereas in the other cases S_R is in OFF state. In this chapter, seven-level circuit topology is presented and is shown in Fig. 2. The switching pattern for this topology is shown in Table 2.

The output voltage developed across the stage one as shown in Fig. 2 is given in Eq. 1

Fig. 2 Proposed seven-level
SCSD MLI topology

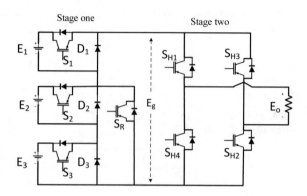

Table 2 Switching states of stage one

State	Switches' states						E_g
	S_1	S_2	S_3	D_1	D_2	D_3	
1	0	0	0	1	1	1	0
2	1	0	0	0	1	1	E_1
3	0	1	0	1	0	1	E_2
4	0	0	1	1	1	0	E_3
5	1	1	0	0	0	1	$E_1 + E_2$
6	0	1	1	1	0	0	$E_2 + E_3$
7	1	0	1	0	1	0	$E_3 + E_1$
8	1	1	1	0	0	0	$E_1 + E_2 + E_3$

$$E_g = E_1 + E_2 + E_3 \tag{1}$$

The stage one can generate output voltage waveform of positive staircase, and the second stage generates both positive cycle and negative cycle output voltages when a full-bridge inverter is added to the stage one. The output waveforms of stage one and stage two are shown in Fig. 3. Table 3 shows the switching states of stage two which is a full-bridge inverter. It is clear that both switches (S_{H1} and S_{H2}) or (S_{H3} and S_{H4}) cannot be on simultaneously because a short circuit occurs across the output voltage E_g.

Sources used in multilevel inverters are of two types, namely symmetrical and asymmetrical. If all the DC sources are equal in magnitude, it is called as symmetrical multilevel inverter, whereas if the magnitudes are different in nature, then it is the asymmetrical multilevel inverter. In this topology, symmetrical concept is implemented. Under the symmetrical method, all DC power sources are equal to E_{dc}. Then, the number of output voltage levels will be

$$M_{level} = 2N + 1 \tag{2}$$

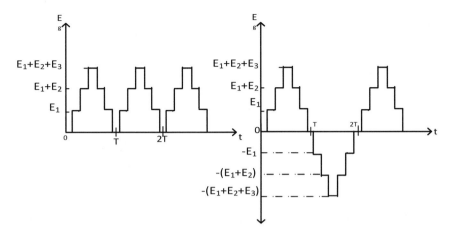

Fig. 3 Output waveforms of stage one and stage two

Table 3 Switching states of stage two

State	Switches' states				E_0
	S_{H1}	S_{H2}	S_{H3}	S_{H4}	
1	1	1	0	0	$+E_0$
2	0	0	1	1	$-E_0$
3	1	0	1	0	0
4	0	1	0	1	0

where N represents the number of cascaded units used in the topology, and M_{level} represents the number of output voltage levels obtained. The overall switches required are

$$N_{IGBT} = \frac{M_{level} + 9}{2} \tag{3}$$

According to the topology shown in Fig. 2, three units (N = 3) are employed; then, the levels obtained are $M_{level} = 7$ level and corresponding switches required are eight switches. In the same manner, we can calculate for M number of levels. In Fig. 4, seven-level SCSD MLI output voltage is shown. The output is grabbed by implementing phase disposition pulse-width modulation (PD-PWM) technique. This scheme gives a better performance when it is applied in closed-loop operation. Hence, the equivalent modulation scheme is implemented for higher levels of SCSD MLI which are commenced in the paper that follows. In Fig. 5, the FFT analysis for seven-level SCSD MLI is shown, and it might be concluded that the THD for seven-level SCSD MLI is 16.24%. The THD value can be decreased by designing suitable filter circuit.

Fig. 4 Seven-level output voltage across the load

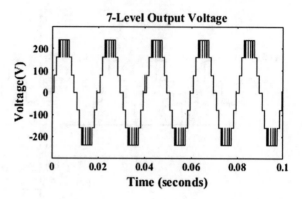

Fig. 5 Seven-levels FFT analysis

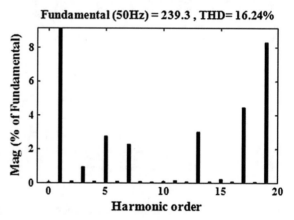

2.2 Nine-Level SCSD MLI

The nine-level SCSD MLI can be achieved by cascading one more switched-diode unit to the existing seven-level topology. In Fig. 6, simulated output for nine-level SCSD MLI is shown. The switching pattern is invented by prolonging switching scheme of seven-level as mentioned in Table 2. In Fig. 7, FFT analysis for nine-level topology is shown by which it might be concluded that THD value is 10.32%.

2.3 Eleven-Level SCSD MLI

The eleven-level SCSD MLI can be obtained by cascading one more switched-diode unit to the nine-level topology. The output voltage across the load for this configuration is shown in Fig. 8. The switching pattern could be achieved by prolonging the switching scheme designed for nine-level topology. The FFT analysis for eleven-

Fig. 6 Nine-level output voltage across the load

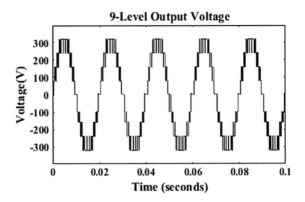

Fig. 7 Nine-level FFT analysis

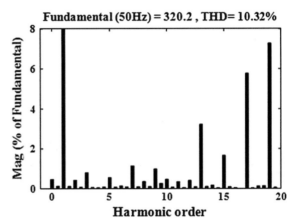

Fig. 8 Eleven-level output voltage across the load

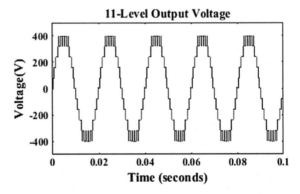

level is shown in Fig. 9 which can be concluded that the THD obtained for this topology is 6.48%.

The THD analysis for seven-level, nine-level, and eleven-level symmetrical cascaded switched-diode multilevel inverters (SCSD MLIs) as shown in Table 4. And it

Fig. 9 Eleven-level FFT analysis

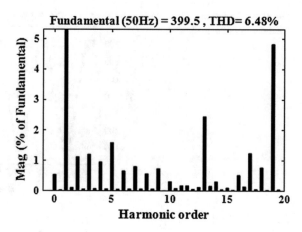

Fundamental (50Hz) = 399.5 , THD= 6.48%

Table 4 THD analysis of SCSD MLI for different levels

S. no.	Number of levels	%THD
1	Seven levels	16.24
2	Nine levels	10.32
3	Eleven levels	6.48

can be concluded from the table that the THD is high for seven-level, i.e., 16.24%, and low for eleven-level, i.e., 6.48%. So, by increasing the number of levels, the THD will be reduced.

3   Fuzzy Controller for SCSD Multilevel Inverter

A new addition to controlling theory is fuzzy logic control. In controller design, by accommodating expert knowledge from all the previous methods, its design philosophy deviates. Fuzzy logic control is derived from fuzzy set theory [14–16] introduced by Zadeh in 1965. The transition between one membership and other membership can be gradual in fuzzy set theory. To make useful for corresponding systems, boundaries of fuzzy sets can be vague and ambiguous. The FLCs can be an attractive option when the mathematical formula determinations are not workable. The other advantages over conventional controllers are: (1) Compared to the conventional nonlinear controllers, FLC can operate with less data storage in the form of rules and membership functions; (2) for imprecise inputs, FLC can work easily; (3) FLC can operate nonlinearly; and (4) by comparing with any other nonlinear controller, it is more robust.

Figure 10 shows the closed-loop fuzzy controller's schematic diagram. To produce an error, the regulated output is compared with the reference input. The error is feed to FLC to generate the output. This process is known as the fuzzy inference process which requires three ideal steps as mentioned in the later section. Figure 11 shows

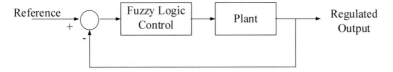

Fig. 10 Closed-loop control using FLC

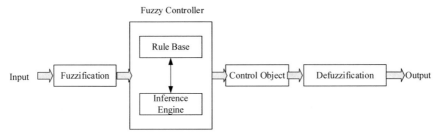

Fig. 11 Fuzzy logic controller structure with control object

the schematic diagram of a closed-loop FLC for the converter. FLC has three main functional blocks. They are (1) fuzzifier; (2) rule evaluator; and (3) defuzzifier. The two databases are, namely databased and rule base. The following sections have described the functions of these five components.

Fuzzification

Fuzzy logic uses linguistic variables instead of numerical variables. The error between reference and output in a closed-loop system is mentioned as decrement or negative small (DS), zero (Z), increment or positive small (IS), etc. Measurable quantities are crisp numbers (real) in the world. The method of converting a real variable (numerical number) into a fuzzy number (linguistic variable) is called fuzzification method. The membership function (triangular) used in fuzzification is shown in Fig. 12. Fuzzifier finds the degree of membership in every linguistic variable, except two which will have zero membership since there is only two overlapping memberships.

Rule Evaluator

In conventional controller, we control gain by a combination of values which are of numerical type that are control laws. The corresponding terms and rules are linguistic in nature of FLC. A standard rule might be written as follows.

R_k: The output is C_i if A_i as E and B_i as CE

where output, error (E), and change of error (CE) are the labels of linguistic variables A_i, B_i, and C_i, respectively. The degree of membership represented in terms of E, CE, and output. Fuzzy set theory is used to execute the above rule.

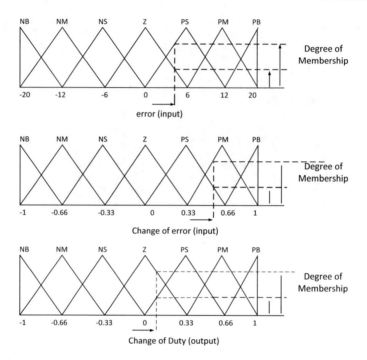

Fig. 12 Membership functions used for fuzzification and defuzzification

Fuzzy Set

Let Z be a group of articles denoted ideally by {Z}, which might be discrete or continuous, and be called as the universe. In the universe, if an element will, let us say z or say x, be a member of fuzzy set A, then it can be mapped as follows

$$\mu(z) \in [0, 1]$$

$$A = [z, \mu(z)|z \in Z]$$

The fuzzy set operations of basic type used for fuzzy rules evaluation are NOT (⁻), OR (U), AND (∩).

(1) NOT → Complement: $\mu_A = 1 - \mu_A(z)$
(2) OR → Union: $\mu\ A \cup B = \max[\mu_b(z), \mu_B(z)]$
(3) AND → Intersection: $\mu\ A \cap B = \min[\mu_A(z), \mu_B(z)]$

Based on AND evaluation rule of definition, R_k results in a minimum of $\mu_{Ai}(z)$, $\mu_{Bi}(z)$ allocated to $\mu_{Ci}(z)$.

Table 5 Rule table

E/CE	DB	DM	NS	Z	IS	IM	IB
DB	IB	IB	IB	IB	IM	IS	Z
DM	IB	IB	IB	IM	IS	Z	DS
DS	IB	IB	IM	IS	Z	DS	DM
Z	IB	IM	IS	Z	DS	DM	DB
IS	IM	IS	Z	DS	DM	DB	DB
IM	IS	Z	DS	DM	DB	DB	DB
IB	Z	DS	DM	DB	DB	DB	DB

Defuzzification

The reverse process of fuzzification is called defuzzification. In linguistic variables (fuzzy numbers), the rules of FLC produce the required output. Linguistic variables have to be transformed to crisp output (real number) according to real-world requirements. Numerous choices are available for defuzzification. The choice of strategy is a compromise between accuracy and computational intensity. Definitions of output membership are used to calculate the particular area for the linguistic output variables from the rule evaluator. Finally, the crisp output can be obtained by

$$Output = \sum A_i * X_i / \sum A_i$$

Knowledgebase

Database:

The database contains membership function definition needed by fuzzifier and defuzzifier. Storage format is a relation between MIPS and available memory of the digital controller chip.

Rule base:

The rule base contains the required rule evaluator's linguistic control rules (decision making logic). Table 5 shows the rule table used in this paper.

3.1 Fuzzy Controller for Seven-Level SCSD Multilevel Inverter

The seven-level symmetrical cascaded switched-diode (SCSD) multilevel inverter with the fuzzy controller is implemented, and analysis of closed loop is executed to develop the RMS voltage value across the circuit output. The output RMS voltage value (V_{out}) is compared with voltage reference value (V_{ref}). The differences among the RMS voltage and the reference voltage values are fed to error signal into fuzzy controller. The seven-level SCSD multilevel inverter topology is cross-checked with the fuzzy controller, and the output voltage across the load is shown in Fig. 13.

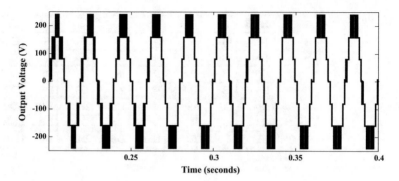

Fig. 13 Fuzzy-controlled output voltage for seven-level

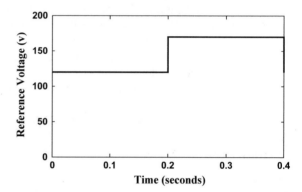

Fig. 14 Seven-level
fuzzy-controlled reference
voltage

From the analysis of fuzzy controller above, the reference and RMS voltage values which are shown in Figs. 14 and 15, can be terminated as until the time reaches 0.2 s the reference voltage is maintained constant at the rate of 120 V. And at time 0.2 s, voltage reference is raised to 170 V and also RMS voltage value is raised to 170 V. From this information, we stated that RMS value changes (increases) with the changes of reference voltage values.

3.2   Fuzzy Controller for Nine-Level SCSD Multilevel Inverter

The nine-level fuzzy controller analysis of SCSD multilevel inverter and its output voltage is shown in Fig. 16. The nine-level SCSD multilevel inverter input and output membership functions are same as the seven-level membership functions. The only variation is in seven-level configuration the input membership function range is [−245 245]. In nine-level topology, the membership function input range is [−325 325]. In Figs. 17 and 18, the reference voltage and RMS output voltages are shown.

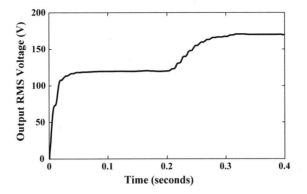

Fig. 15 Seven-level fuzzy-controlled RMS output voltage

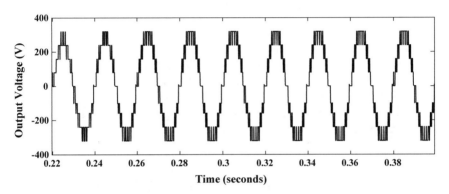

Fig. 16 Fuzzy-controlled nine-level output voltage

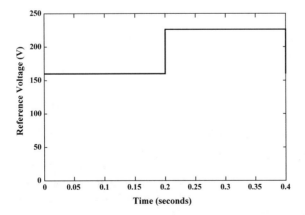

Fig. 17 Nine-level fuzzy-controlled reference voltage

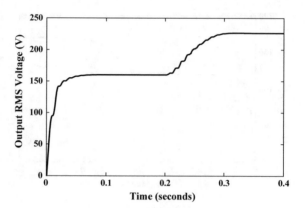

Fig. 18 Nine-level fuzzy-controlled RMS output voltage

From the above analysis of fuzzy controller, the reference and RMS voltage values which are shown in Figs. 17 and 18, can be incidental as until the time reaches 0.2 s the reference voltage is maintained constant at the rate of 160 V. And at time 0.2 s, reference voltage is raised to 230 V and also RMS voltage value is raised to 230 V. From this information, we stated that RMS value changes (increases) with the changes of reference voltage values.

3.3 Fuzzy Controller for Eleven-Level SCSD Multilevel Inverter

The simulation output voltage of closed-loop fuzzy controller for eleven-level SCSD multilevel inverter topology is shown in Fig. 19. The membership functions of input and output are same as the nine-level SCSD multilevel inverter circuit configurations. But the only change is that the input membership function range for nine-level is [−325 325]. In eleven-level SCSD multilevel inverter topology, the input membership function range is [−425 425]. The reference voltage and RMS output voltages are shown in Figs. 20 and 21.

From the above analysis of fuzzy controller, the reference and RMS voltage values which are shown in Figs. 17 and 18, can be incidental as until the time reaches 0.2 s the reference voltage is maintained constant at the rate of 210 V. And at time 0.2 s, reference voltage is increased to 300 V and also RMS voltage value is raised to 300 V. From this information, we stated that RMS value changes (increases) with the changes of reference voltage values.

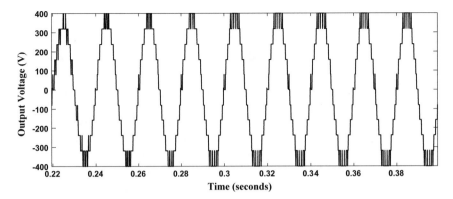

Fig. 19 Output voltage for eleven-level fuzzy control

Fig. 20 Eleven-level fuzzy-controlled reference voltage

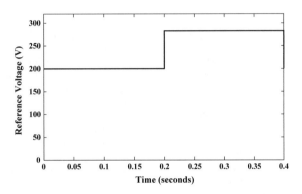

Fig. 21 Eleven-level fuzzy-controlled RMS output voltage

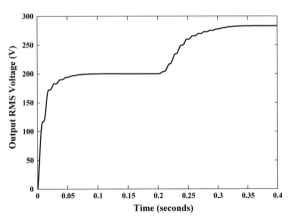

4   Conclusion

The symmetrical cascaded switched-diode multilevel inverter is evaluated for various levels, and corresponding outputs are related to each other by considering total harmonic distortion (FFT analysis) which terminates the interpretation that THD reduces with increase in number of levels. The THDs obtained are 16.24%, 10.32%, and 6.48% for seven-level, nine-level, and eleven-level SCSD MLIs, respectively. Relevant fuzzy logic controller was developed for these SCSD MLIs, and their responses have been verified. Co-relation was built between the outputs of fuzzy-controlled SCSD MLIs by achieving fickle responses of corresponding configurations.

References

1. Wang, L., Wu, Q.H., Tang, W.: Novel cascaded switched-diode multilevel inverter for renewable energy integration. IEEE Trans. Energy Convers. (2017)
2. Saeedifard, M., Barbosa, P.M., Steimer, P.K.: Operation and control of a hybrid seven-level converter. IEEE Trans. Power Electron. 27(2), 652–660 (2012)
3. Villanueva, E., Correa, P., Rodríguez, J., Pacas, M.: Control of a single-phase cascaded H-bridge multilevel inverter for grid-connected photovoltaic systems. IEEE Trans. Ind. Electron. 56(11), 4399–4406 (2009)
4. Malinowski, M., Gopakumar, K., Rodriguez, J., Perez, M.A.: A survey on cascaded multilevel inverters. IEEE Trans. Ind. Electron. 57(7), 2197–2206 (2010)
5. Alishah, R.S., Nazarpour, D., Hosseini, S.H., Sabahi, M.: Novel multilevel inverter topologies for medium and high-voltage applications with lower values of blocked voltage by switches. IET Power Electron. 7(12), 3062–3071 (2014)
6. Nabae, A., Takahashi, I., Akagi, H.: A new neutral-point-clamped PWM inverter. IEEE Trans. Ind. Appl. 5, 518–523 (1981)
7. Khajehoddin, S.A., Bakhshai, A., Jain, P.K.: A simple voltage balancing scheme for m-level diode-clamped multilevel converters based on a generalized current flow model. IEEE Trans. Power Electron. 23(5), 2248–2259 (2008)
8. Lopez, I., Ceballos, S., Pou, J., Zaragoza, J., Andreu, J., Kortabarria, I., Agelidis, V.G.: Modulation strategy for multiphase neutral-point-clamped converters. IEEE Trans. Power Electron. 31(2), 928–941 (2016)
9. Khazraei, M., Sepahvand, H., Corzine, K.A., Ferdowsi, M.: Active capacitor voltage balancing in single-phase flying-capacitor multilevel power converters. IEEE Trans. Ind. Electron. 59(2), 769–778 (2012)
10. Dargahi, V., Sadigh, A.K., Abarzadeh, M., Eskandari, S., Corzine, K.A.: A new family of modular multilevel converter based on modified flying-capacitor multicell converters. IEEE Trans. Power Electron. 30(1), 138–147 (2015)
11. Babaei, E., Hosseini, S.H.: New cascaded multilevel inverter topology with minimum number of switches. Energy Convers. Manag. 50(11), 2761–2767 (2009)
12. Alishah, R.S., Nazarpour, D., Hosseini, S.H., Sabahi, M.: Novel topologies for symmetric, asymmetric, and cascade switched-diode multilevel converter with minimum number of power electronic components. IEEE Trans. Ind. Electron. 61(10) (2014)
13. Palanivel, P., Dash, S.S.: Analysis of THD and output voltage performance for cascaded multilevel inverter using carrier pulse width modulation techniques. IET Power Electron. 4(8), 951–958 (2011)

14. Nguyen, H.T., Sugeno, M. (eds.): Fuzzy Systems: Modeling and Control. Springer Science & Business Media (2012)
15. Driankov, D., Hellendoorn, H., Reinfrank, M.: Introduction. In: An Introduction to Fuzzy Control, pp. 1–36. Springer, Berlin, Heidelberg (1996)
16. Ponnambalam, P., Aroul, K., Reddy, P.P., Muralikumar, K.: Analysis of fuzzy controller for H-bridge flying capacitor multilevel converter. In: Proceedings of Sixth International Conference on Soft Computing for Problem Solving, pp. 307–317. Springer, Singapore (2017)

An Optimized Path Planning for the Mobile Robot Using Potential Field Method and PSO Algorithm

Ravi Kumar Mandava, Sukesh Bondada and Pandu R. Vundavilli

Abstract The main aim of any path/motion planning algorithm in the context of a mobile robot is to produce a collision-/crash-free path among the goal and start points in an environment in which it is present. The past few decades have seen the development of various methodologies to design an optimal path. The present research focuses on the development of an optimized path planning algorithm for the robot using a hybrid method after combining particle swarm optimization (PSO) algorithm with potential field method for static obstacles and potential field method (PFM) prediction for dynamic obstacles. While implementing, PSO-based potential field method, the total potential, that is the sum of repulsive and attractive potentials, is considered as the fitness function which is optimized using PSO algorithm. Further, a 3-point method has been used for smoothing the obtained path. Once the image of the scene is obtained, a clustering method is employed to find the center of obstacle and the location of the robot has been determined by calculating the repulsive potential in each iteration. Finally, the developed algorithms are tested on both the static and dynamic environments in computer simulations and found satisfactory.

Keywords Path planning · Potential field method · PSO algorithm
3-point smoothing · Static and dynamic environments

R. K. Mandava (✉) · S. Bondada · P. R. Vundavilli
School of Mechanical Sciences, IIT Bhubaneswar, Bhubaneswar 752050, Odisha, India
e-mail: rm19@iitbbs.ac.in

S. Bondada
e-mail: sb24@iitbbs.ac.in

P. R. Vundavilli
e-mail: pandu@iitbbs.ac.in

R. K. Mandava
Mechanical Engineering Department, Vignan's Foundation for Science, Technology and Research, Guntur 522213, Andhra Pradesh, India

© Springer Nature Singapore Pte Ltd. 2019   139
J. C. Bansal et al. (eds.), *Soft Computing for Problem Solving*, Advances in Intelligent Systems and Computing 817, https://doi.org/10.1007/978-981-13-1595-4_11

1 Introduction

Deciding the path/motion of the mobile robot is a significant task in the field of autonomous robots, which allows the robot to find a path between the start and goal points, in a given environment. This could be the shortest path between the two points or could be an optimal path that minimizes the time of travel or length of path or some other aspects that the application requires. For the past few decades, many researchers are working on the development of various algorithms/methodologies to decide the shortest/optimal route that are not only important in the context of robotics but also in the fields of network routing, sequencing of genes, and video games. Path planning necessitates a robot that is conscious of its position and the map that represents the environment. This map can either be created by the robot itself while it is moving or can be provided by other external sources like an overhead camera. There are several ways to represent this map of environment and to localize the robot in the map. In order to plan the path, researchers had developed two systematic and easily understandable complimentary approaches. The first one deals with the discrete approximation or estimation, and the second one is focused on continuous approximation. In discrete estimation, the map is separated into pieces of unequal or equal sizes. In the second one, the maps are also called as topological maps. Every piece of map/plot represents a node or vertex that is linked with edges, and the robot/vehicle can be traveled from one node to another. Computationally, the graph can be saved as an incidence or adjacency matrix/list. Moreover, the continuous estimation technique necessitates the description of outer boundaries and inner (obstacles) in the shape of a polygon, whereas the paths/routes could be programmed as a sequence of real numbers. Further, the discrete maps/plots are having leading demonstration in the field of robotics since images or videos from camera generally come in the form of pixels.

2 Related Work

Dijkstra [1] conceived an algorithm/methodology for discovering the shortest/optimal routes among the nodes/vertex in a plot. Many variants are available in the said methodology. However, the original variant of Dijkstra's method was used to establish the shortest/least path among the two vertices. After nine years, Hart et al. [2] made an extension to this algorithm, called A* search algorithm. This A* used heuristics to guide the search and to attain improved performance. Anthony [3] developed a variant of D* algorithm, which combined the ideas of A* and the original D*. Another variant, D* lite, introduced by Koenig et al. [4] was an incremental heuristic search algorithm. Due to the high computational complexity involved with these algorithms, it became difficult to search convex and higher-dimensional spaces. To solve this problem, LaValle [5] proposed rapidly exploring random trees (RRTs), which could solve the problems with obstacles and differential constraints, and widely used to solve the motion planning problems of autonomous robots. Further, Nasrollahy and Javadi [6] developed an approach to identify an optimal path in the unstructured/dynamic environment using PSO algorithm. In [7], the authors proposed an artificial potential field (APF) method to prevent the standstill and non-reachability glitches in the mobility of the mobile robot. The actual front face obstacle data related with the direction of its velocity was used to change the traditional APF (T-APF) methodology. The modified version of the method helped in avoiding the deadlock problem. Moreover, Montiel et al. [8] used a population-based algorithm called bacterial potential field (BPF) algorithm to plan the motion for the mobile robots and to avoid both the static and dynamic obstacles. It was observed that the developed algorithm ensured a reasonable safe and optimal path/route. Further, the outcomes of simulations were compared with the genetic potential field (GPF), artificial potential field (APF) and pseudo-bacterial potential field (PBPF) and method in terms of computational time to determine the optimal route. Han and Seo [9] proposed a new methodology, that is, surrounding point set (SPS) method, for resolving the motion planning glitches. In that work, the obstacles were assumed to be limited by their locations. The early feasible route was generated based on SPS, and they also applied a route development method based on later and former points (PI-FLP), which uses every node in the route to reposition itself based on the two nodes selected on both sides of it. The SPS was suitable to recognize the essential nodes for resolving motion planning glitches, PI-FLP could decrease the total distance of the route, and also the smoothness of the path was improved. Kovács et al. [10] developed an innovative potential field for motion planning of the robots after familiarizing the attributes of animal motion. This algorithm was helpful to develop real-time local motion planning and to support the human robot collaboration. The above method was applied in embedded structure and estimated on MOGI-ETHON, a holonomic drive robot. In [11], Bakdi et al. used fuzzy logic control algorithm and genetic algorithms to obtain the optimal motion planning scheme for the mobile robots. A vision system with image processing methodology was used to extract the depth information. Along with the above techniques, a genetic algorithm was employed to create the optimal route between the start and goal points. The smooth path was obtained by adopting piecewise cubic Hermite interpolating polynomial. An adaptive fuzzy logic controller was designed for tracking the mobile robot on the desired smooth path. A sensor fusion was used to evaluate the current location, and a Kalman filter was used to determine the orientation of the robot. Moreover, Pandey and Parhi [12] introduced a type-1 fuzzy logic system (T1-SFLS) and fuzzy wind-driven optimization hybrid controller for the navigation and collision-free evasion in unknown dynamic and static environments. Further, to optimize the input and output parameters of the membership functions, a wind-driven optimization (WDO) algorithm was used.

In the present manuscript, an attempt is made to find the collision-free time-optimal motion planning algorithm, after utilizing the PSO-based potential field method. The net potential is considered as the fitness function for the proposed PSO-PFM. The developed algorithms are tested for static and dynamic obstacle cases in computer simulations.

3 Problem Formulation

The motion planning problem of the mobile robot is solved in two parts, such as motion planning among static obstacles and dynamic obstacles.

3.1 Motion Planning Among Static Obstacles

The goal of the present research is to find the optimal route through the obstacles while avoiding touching any of them and minimizing either the path length, number of turns made by the robot, time of travel, or a combination of them. In the present problem, depending on the shape, the obstacles are considered as circles, rectangles, and their combinations. The location of the obstacles is determined with the help of image processing technique. Once the location of the obstacles is determined, the exact location of the robot will be determined by the PFM. To solve the motion planning problem using PSO-PFM, the total potential field is used as fitness function which is then optimized by using particle swarm optimization. This hybridization helps the PFM to overcome its limitation of struck at the local minima. Since PSO is a heuristic algorithm, the obtained path is to be further optimized using a 3-point smoothing method.

3.1.1 Potential Field Method

Potential field method is a most popularly used conventional motion planning method for mobile robots. The movement of the robot in the terrain is possible only due to the combined action of repulsive and attractive potentials developed between the robot-obstacle and robot-goal, respectively. The attractive potential inclines to pull the robot near, whereas the repulsive potential pushes the mobile robot away from the obstacle. It is important to mention that the effectiveness of the PFM depends on the type of potential field function utilized in the study. In the present work, parabolic, well-attractive potential field function has been used and can be defined as follows:

$$Attractive\ potential\ at\ node(p) = att \times dist^2(p, goal)$$

where 'att' is an attractive scaling factor, p is the node under observation, and $dist(p, goal)$ is the Euclidean distance between the goal and robot. The repulsive potential field between the obstacle and robot takes the following form.

$$Repulsive\ potential\ at\ node\ p = rep \times \left[\frac{1}{dist(p, obstacle)} \right]^2$$

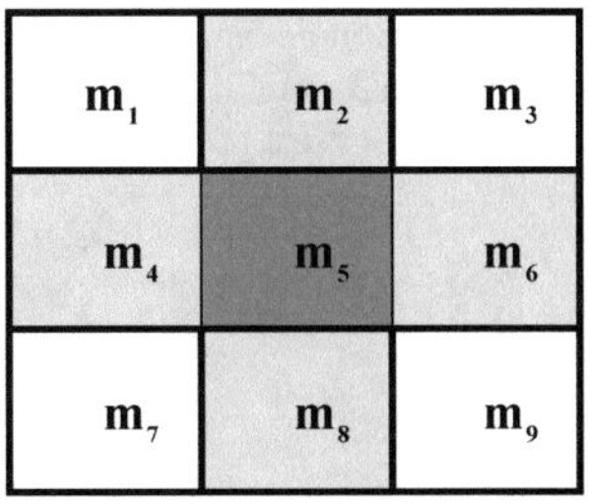

Fig. 1 Schematic diagram
showing the nodes of an
occupancy grip map

where '*rep*' is a positive scaling factor and *dist(p, obstacle)* is the distance between robot *p* and obstacle. Once the attractive and repulsive potentials are calculated, then the robot is allowed to move based on the value of total potential.

3.1.2 Bushfire Algorithm

The bushfire algorithm is used to minimize the distance between the obstacle and the robot. The algorithm consists of eight nodes that surround the original node (ref. to Fig. 1). The minimum distance from the node to the obstacle will be one more than the distance of surrounding node and is given below.

$$Minimum\ distance = dist(m_5, obs) = 1 + min(dist(m_i, obs))$$

where $m_i = i$th node and dist(x, y) means the distance between x and y.

3.1.3 Particle Swarm Optimization (PSO)

PSO is a nature-inspired heuristic algorithm. Its phenomenon is motivated by observing the social behavior of fishes and birds [13]. The birds move in the form of groups during bird flocking. During this process, every bird in the flock is designated with a particular position and velocity associated with it. This value represents the best personal position achieved by each bird. This best value is gauged by the fitness or objective function value of the entire flock. The fitness function in bird flocking phenomenon searches for the food that is available nearby its shelter. Therefore, every bird in the flock is assigned with personal best position based on the objection function value. The bird with best objective or fitness function value is deliberated as the global best position. As the birds in the flock share the information related to the fitness value among themselves, all birds will incline to shift near the global best solution. The velocity and position of every bird are updated by the given equation.

$$V_i(k + 1) = w * V_i(k) + C_1 * rand * \left(Y_{pbest} - Y_i\right) + C_2 * rand * \left(Y_{gbest} - Y_i\right)$$
$$Y_i(k + 1) = Y_i(k) + V_i(k + 1)$$

Fig. 2 Schematic diagram
showing the 3-point
smoothing of three points

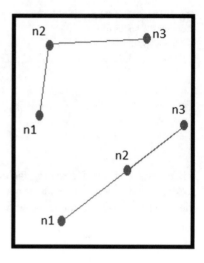

where i, k, and w represent particle, iteration counter, and inertia weight, respectively.
C_1 and C_2 denote the cognitive and social parameters, and Y_{pbest} and Y_{gbest} indicate
the personal and global best positions of the birds, respectively.

3.1.4   3-Point Smoothing Method

The path obtained using the PSO-PFM approach is a heuristic-based path and guarantees the path between the robot and the goal. This path needs to be further optimized
to eliminate sharp turns (that is, smoothing), which causes problems when the real
robot moves on this path. Here, an attempt is made to smooth the path with the help of
a 3-point smoothing method (ref. to Fig. 2). Once all the points on the path obtained
by PSO-PFM approach are available, smoothing is done by taking any three points
and replacing the second point by the average value of the first and third, if the new
position does not lie on the obstacle.

3.2   Motion Planning Among Dynamic Obstacles

The solution for dynamic obstacle path planning is similar to that of path planning
for static obstacles. The only difference is that the obstacles are allowed to move
with time. As the location of the obstacles is continuously changing, the image of
the terrain needs to be continuously sampled and processed to get the information
related to the obstacles and robot. The center locations of the obstacles and robot
are determined with the help of K-means clustering algorithm. Once the locations
are determined, PFM needs to be applied to get the values of attractive and repulsive

potentials to determine the path to be followed by the robot. With the normal potential method, it is very difficult to identify that where the robot will be moving, and there is a chance that the robot and obstacles colloid with each other. To prevent this situation, in the present research, a phenomenon called predicted potential value is implemented. The predicted potential value is obtained from the linear extrapolation and is given by the following equation.

$$Pot_{i+1}(n) = pot_i(n) + \beta \times (pot_i(n) - pot_{i-1}(n))$$

where i is the iteration number, $Pot_i(n)$ represents the potential of nth node in ith iteration, and β indicates the degree of extrapolation. The flowchart showing the steps involved in dynamic obstacle motion planning is shown in Fig. 3.

4 Results and Discussions

Once the algorithms are developed, the effectiveness of the proposed approaches is tested in computer simulations for both static and dynamic obstacles on different maps. It is important to note that the size of the maps related to the static and dynamic obstacles cases is equal in all scenarios.

4.1 Static Obstacles

Figure 4 shows various maps in which the PSO-PFM approach has obtained the paths without and with smoothing effect. The green color indicates the path which is not smoothened. Alongside, the blue circle denotes the start position of the robot and the red circle represents the goal position of the robot. It has been observed that in all the maps the robot has found the path from start position to goal position without hitting any obstacles, that is, a collision-free path, has been developed. Further, the path obtained after applying the 3-point smoothing concept was represented with the help of blue line. From these graphs, it can be observed that the length of the path is shortened with the 3-point smoothing when compared with the unsmoothed paths.

The length of the paths obtained for both the algorithms in different maps is given in Table 1. It can be observed that the length of the path has been decreased after adding 3-point smoothing concept. This may help to travel the robot from start point to goal point with less time and in a smooth manner. From this, it is evident that the smoothing of the path obtained from PSO-PFM yielded shortest path when compared with the unsmoothed path.

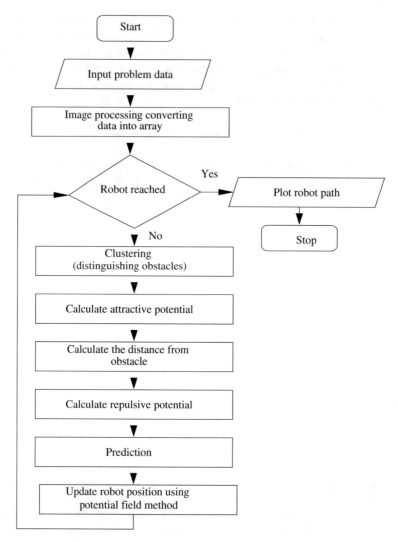

Fig. 3 Flowchart showing the step-by-step procedure for dynamic obstacles

4.2 Dynamic Obstacles

Figures 5, 6, and 7 show the results related to the simulation of dynamic obstacles on different maps. Figures 5a, 6a, and 7a represent the path planning of the robot in dynamic environment without prediction, whereas Figs. 5b, 6b, and 7b show the path planning of the robot after applying the concept of prediction. It can be observed that the length of the path has been decreased after applying the concept of prediction. It may be due to the reason that after prediction the chances of robot colliding with the

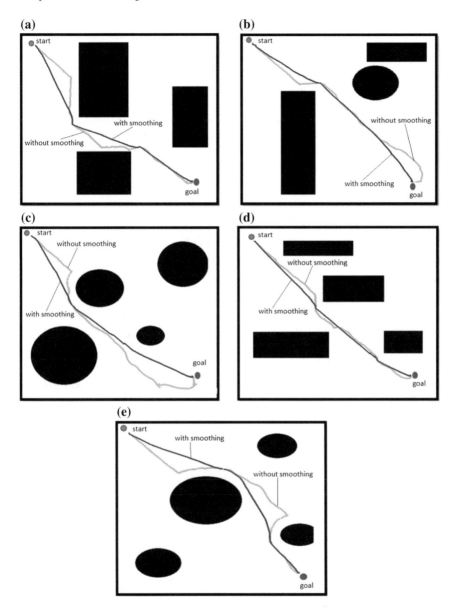

Fig. 4 Schematic diagram showing the path planning for static obstacles in different maps **a** map-1, **b** map-2, **c** map-3, **d** map-4, and **e** map-5

obstacle are reduced and the robot knows its position before moving to the new/next location. Due to the prediction of potential value, the robot has not moved to the already moved position. In the below maps, the start and goal points of the robot are

Table 1 Experimental results for static obstacles

Map	Path length (mm)	
	APF + PSO	APF + PSO + 3 point smoothing
1	819	753
2	714	678
3	806	711
4	687	642
5	822	715

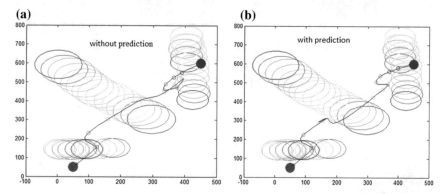

Fig. 5 Schematic diagram showing the path planning for dynamic obstacles **a** map-1 = 752 mm and **b** map-2 = 718 mm

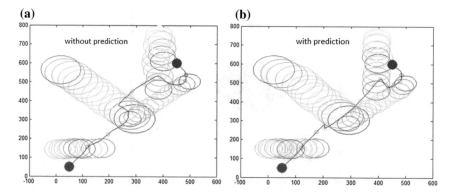

Fig. 6 Schematic diagram showing the path planning for dynamic obstacles **a** map-1 = 761 mm and **b** map-2 = 715 mm

represented by red and blue color dots. The obstacles' starting position was indicated by red color, and while moving, its color changed to blue color.

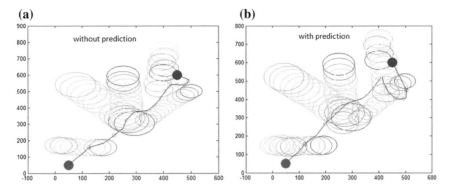

Fig. 7 Schematic diagram showing the path planning for dynamic obstacles **a** map-1 = 772 mm and **b** map-2 = 723 mm

5 Conclusions

In the present study, a novel hierarchical path planning approach for the mobile robot moving in the cluttered environment has been developed. In this approach, a combination of artificial potential field combined with particle swarm optimization and a 3-point smoothing method has been used for static obstacles and a prediction technique is used for finding the robot's previous position during dynamic obstacles. It can be observed that in both static and dynamic obstacles, the length of the path is less after applying the 3-point smoothing method and prediction technique, respectively. The computational time is less for solving the problem using 3-point smoothing and prediction technique for static and dynamic obstacles.

6 Future Scope

In the present manuscript, the authors have implemented the path planning algorithm in static and dynamic environments for regular shapes only. Further, the authors are planning to implement the same algorithm for irregular shapes and also planning to implement in real-time environment. At present, the authors are working on these problems.

References

1. Dijkstra, E.W.: A note on two problems in connexion with graphs. Numer. Math. **1**, 269–271 (1959)
2. Hart, P.E., Nilsson, N.J., Raphael, B.: A formal basis for the heuristic determination of minimum cost paths. IEEE Trans. Syst. Sci. Cybern. SSC4 **4**(2), 100–107 (1968)
3. Anthony, S.: Optimal and efficient path planning for partially-known environments. In: Proceedings of the International Conference on Robotics and Automation, pp. 3310–3317 (1994)
4. Koenig, S., Likhachev, M., Furcy, D.: Lifelong planning A*. Artif. Intell. J. **155**(1–2), 93–146 (2004)
5. LaValle, S.M.: Rapidly-exploring random trees: a new tool for path planning. Technical Report, Computer Science Department, Iowa State University, 98-11 (1998)
6. Nasrollahy, A.Z., Javadi, H.H.S.: Using particle swarm optimization for robot path planning in dynamic environments with moving obstacles and target. In: 2009 Third UKSim European Symposium on Computer Modeling and Simulation (2009)
7. Weerakoon, T., Ishii, K., Nassiraei, A.A.F.: An artificial potential field based mobile robot navigation method to prevent from deadlock. JAISCR **5**(3), 189–203 (2015)
8. Montiel, O., Orozco-Rosas, U., Sepúlveda, R.: Path planning for mobile robots using Bacterial Potential Field for avoiding static and dynamic obstacles. Expert Syst. Appl. **42**, 5177–5191 (2015)
9. Han, J., Seo, Y.: Mobile robot path planning with surrounding point set and path improvement. Appl. Soft Comput. **57**, 35–47 (2017)
10. Kovács, B., Szayer, G., Tajti, F., Burdelis, M., Korondi, P.: A novel potential field method for path planning of mobile robots by adapting animal motion attributes. Robot. Auton. Syst. **82**, 24–34 (2016)
11. Bakdi, A., Hentout, A., Boutami, H., Maoudj, A., Hachour, O., Bouzouiaa, B.: Optimal path planning and execution for mobile robots using genetic algorithm and adaptive fuzzy-logic control. Robot. Auton. Syst. **89**, 95–109 (2017)
12. Pandey, A., Parhi, D.R.: Optimum path planning of mobile robot in unknown static and dynamic environments using fuzzy-wind driven optimization algorithm. Def. Technol. **13**, 47–58 (2017)
13. Kennedy, J., Eberhart, R.: Particle swarm optimization. In: Proceedings of IEEE International Conference on Neural Networks, pp. 1942–1948 (1995)

Recommendation Systems: Techniques, Challenges, Application, and Evaluation

Sandeep K. Raghuwanshi and R. K. Pateriya

Abstract With this tremendous growth of the Internet, mobile devices, and e-business, information load is increasing day by day. That leads to the development of the system, which can filter and prioritize the relevant information for users. Recommendation system solves this issue by enabling users to get knowledge, products, and services of personalized basis. Since the inception of recommender system, researcher has paid much attention and developed various filtering techniques to make these systems effective and efficient in terms of users and system experience. This paper presents a preliminary survey of different recommendation system based on filtering techniques, challenges applications, and evaluation metrics. The motive of work is to introduce researchers and practitioner with the different characteristics and possible filtering techniques of recommendation systems.

Keywords Collaborative filtering · Content-based filtering · Recommendation system

1 Introduction

Multiple choices lead confusion to a human being about what item is right for them or fulfill their requirements. This causes inception to develop a system which could help human being for the selection criteria and eliminate the dilemma. In present or past, human always relies on the suggestions from the outside for one purpose or other. Based on this, recommender system becomes the tools that shrink our options and present most suitable suggestions as per the requirements and our taste. The huge volume of information and user preferences increase the demand for new and

S. K. Raghuwanshi (✉) · R. K. Pateriya
Computer Science & Engineering, Maulana Azad National Institute
of Technology, Bhopal 462003, Madhya Pradesh, India
e-mail: sraghuwanshi@gmail.com

R. K. Pateriya ·
e-mail: pateriyark@gmail.com

© Springer Nature Singapore Pte Ltd. 2019
J. C. Bansal et al. (eds.), *Soft Computing for Problem Solving*, Advances in Intelligent
Systems and Computing 817, https://doi.org/10.1007/978-981-13-1595-4_12

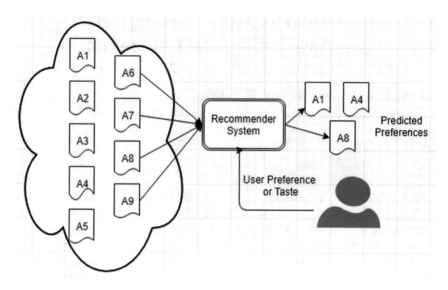

Fig. 1 Recommendation system

effective recommender system in the current age. Isinkaye et al. [1] with this RS must act as information filtering system that urges to predict preferences that user might have for an item over other and predict whether a particular item would prefer or not by him. A precise definition of a recommender system is given as (Fig. 1): A recommender system or a recommendation system (sometimes replacing the system with a synonym such as a platform or an engine) is a subclass of information filtering system that seeks to predict the rating or preference that a user would give to an item [2].

The existence of recommender system had been identified in the late 1970s, ever since many researchers have proposed various approaches to develop efficient recommender system. The first computer-based recommendation system was developed in 1992 by Goldberg et al. [3]. It was called Tapestry, a mail recommender system which was developed at the Xerox Palo Alto Research Centre [3]. Tapestry is an experimental information filtering system that manages the huge incoming stream of documents such as e-mail, news stories, and articles. It predicated documents on the belief that information filtering can be more effective when humans are involved in the filtering process. The primary motivation behind the development of RS is to reduce information load and processing cost by working on personalized information and data through analyzing the interest and behavior of the user to guess his/her preferences over the item. It is beneficiary for both users and service providers [4]. Presently, many organizations such as Google, Twitter, LinkedIn, Netflix, Amazon use recommendation system as a decision maker to either maximize its profits and minimize the risk possibility [5, 6]. Most popular recommender systems of today are Group Lens recommender system, Amazon.com recommender system, Netflix movie recommender system, Google News personalization system, Facebook friend recommendations, link prediction recommender system.

2 Filtering Techniques

Various methods are being proposed to develop an effective recommendation system out of them, two form the basis for the development of other approaches. These methods are content filtering, collaborative filtering [1]. Further, these techniques are extended and in present RS techniques are classified as (Fig. 2):

2.1 Content Filtering Recommendation System

Content-based (CB) techniques use feature list of the item and compare it with items preferred by a specific user previously. The items which match in similarity are recommended to the user. The essential function of content-based filtering works in two steps: It stores a user profile based on item features which are most commonly preferred by the user. These features are used to map the similarity of one item with other by similarity equation. After that, it compares each item's features with the user profile and recommends those who have a high degree of similarity [7]. For content-based system, one has to construct item profile, which is a record of essential characteristics of that item. These characteristics are discovered easily like in a movie the record may contain a list of actors, director, year of release, and genre (Fig. 3).

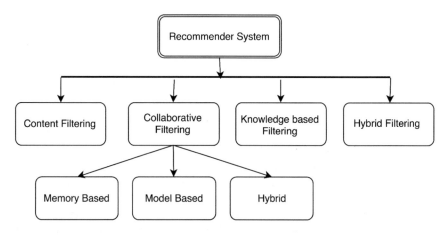

Fig. 2 Classification of recommendation systems

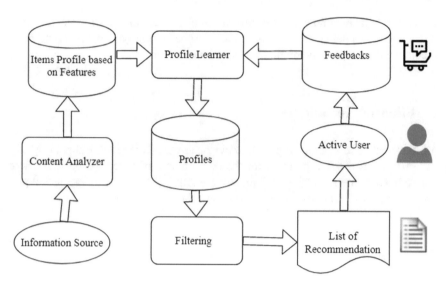

Fig. 3 Content-based filtering

CB filtering is most simple and natural approach to adopt as a recommendation it does not require any feedback from the user. Sometimes a single preference is enough to recommend many items to the user. This approach also extends naturally to the cases where item information is well organized and available such as movies, songs, products, and books. But at the same time, it also comes as a limitation of content-based filtering as item description is not always present in all cases that create difficulty in measuring the similarity between items. These recommendation systems have limitations to produce similar results and are static over the time.

2.2 Collaborative Filtering (CF)

CF is the most popular and used recommendation technique. The basis for collaborative filtering is that users with similar interest are inclined to give same preference for the new and future items. This technique works on two points. First, it serves as a criterion to select a group of similar people whose opinions will be accumulated as a basis for a recommendation (nearest neighbors). Second, it also uses these opinions to form a bigger group and have a greater impact on the recommendation [8]. Collaborative filtering techniques involved very large data sets and applied in diverse application areas like finance, weather forecasting, environmental sensing, e-commerce.

Collaborative filtering techniques make use of a data set of preferences/ratings given by the users for items to predict additional items that an active user might like. The model can be expressed as a preferences/rating matrix of order m × n,

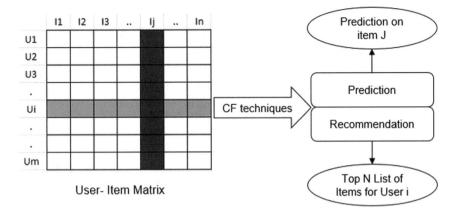

Fig. 4 Collaborative filtering [1]

where m is the number of users $(U_1, U_2, U_3, \ldots, U_m)$ and n is the number of items $(I_1, I_2, I_3, \ldots, I_n)$, rated by the users. The cell value ri, j of the matrix is the rating of item j was given by user i. These ratings can be either implicit (such as purchase for an item) or explicit (feedback by the user on a scale of k). The outcome of collaborative techniques can be of two types: First is the prediction of r_{ij}, a numerical value shows the preference of user I over item j, and second is the recommendation list of top N items that user might like the most [1, 9] and (Fig. 4)

The primary function of recommendation is to predict the utility of an item for a user. The recommendation system is characterizing as a user u is interested in item i with some degree of preference or rating $r(u, i)$. Each user has user profile of his taste, likes, dislikes, or sometimes a rating or feedback is given to a particular item. Each item is characterizing by its feature set such as for a movie the feature set may include movie Id, actors, director, release date, genre. Predicted preference of an active user a, for an item j, is calculated as:

$$P_{a,j} = \overline{r_a} + k \sum_{i=1}^{n} S_{(a,i)} \times r_{i,j} - \overline{r_i} \tag{1}$$

where r_a: the mean rating of user $a \cdot n$: the number of users in the database with nonzero $r_{i,j} \cdot S_{(a,i)}$: Similarity between the active user and each user $i \cdot k$: a normalizing factor such that the absolute values of the weights sum to unity.

There are many techniques used to compute the similarity between the users. Each one has its pros and cons in their areas some of them are:

i. Pearson Correlation Similarity: Pearson correlation defines the linear correlation between two vectors and has a value between -1 and 1. The similarity between the two vectors u and v is defined as:

$$S_{Cosine}(u, v) = \frac{\sum_{i=1}^{n}(r_{u,i} - \overline{r_u}) \times (r_{v,i} - \overline{r_v})}{\sqrt{\sum_{i=1}^{n}(r_{u,i} - \overline{r_u})^2 \times (r_{v,i} - \overline{r_v})^2}} \qquad (2)$$

ii. Cosine Similarity: Cosine is one of the most popular methods of statistics to find similarity between two nonzero real values' vectors. It looked for an angle between two vectors in n-dimensional space and defined as:

$$S_{Pearson}(u, v) = \frac{\sum_{i=1}^{n}(r_{u,i}) \cdot (r_{v,i})}{\sqrt{\sum_{i=1}^{n}(r_{u,i})^2 \times (r_{v,i})^2}} \qquad (3)$$

Collaborative techniques are grouped into memory-based techniques, model-based techniques, and hybrid techniques.

2.2.1 Memory-Based Filtering

Memory-based methods are straightforward and easy to implement. The best-known technique used is memory-based neighborhood-based filtering, which predicts preferences by referring to users who are similar to queried user or to items that are similar to queried item. The accuracy and efficiency of neighborhood technique are profoundly affected by how the similarity between users or items is calculated. This technique can be further extended with default votes, inverse user frequency, and case amplification [10, 11]. These techniques are further classified as user-based filtering techniques and item-based filtering technique.

User-Based Filtering technique computes the similarity between users by comparing their preference over the same item and calculates the predicted preference for items for the active user [1].

Item-Based Filtering techniques calculate predictions using similarity between items. The technique works by retrieving all the items rated by an active user and determines similarities of retrieved items with target item. It then selects top N most similar items to predict the preference of the active user for the target item [1].

2.2.2 Model-Based Filtering

Model-based techniques make use of data mining and machine learning approaches to predict the preference of a user to an item. These techniques include association rule [12], clustering [13], decision tree [14], artificial neural network [15], Bayesian classifier [16], regression [17], link analysis [18], and latent factor models. Among these latent factor models are the most studied and used model-based techniques. These techniques perform dimensionality reduction over user–item preference matrix and learn latent variables to predict preference of the user to an item in the recommendation process. These methods include matrix factorization [19], singular value decomposition [20], probabilistic matrix factorization [21, 22], Bayesian probabilistic matrix factorization [22], low-rank factorization [23], nonnegative matrix factorization [24], and latent Dirichlet allocation [25].

2.2.3 Hybrid Filtering

Some applications combine the advantages of memory-based and model-based approaches to form a hybrid filtering system. It results in better prediction and efficiency. A proper combination can overcome the limitation of collaborative filterings such as sparsity and diversity [9].

2.3 Knowledge-Based Filtering

Knowledge-based filtering uses back-end knowledge or information of users, items, and their relationship. These systems describe how a particular item meets the requirement of the user. The technique requires domain-specific knowledge of users and items. The most traditional knowledge-based system is the case-based system [26].

2.4 Hybrid Filtering

Hybrid filtering techniques is one which combines the advantages of two or more filtering techniques and overcomes their limitations. These techniques provide more effective and enhance results of recommendation [27]. Hybrid techniques can adopt one of the following strategies to develop a hybrid filtering method:

1. Use content-based and collaborative-based filterings to produce separate recommendation, and then use a linear combination of this two recommendations to provide a single recommendation [28].
2. Collaborative filtering can be used with content-based characteristics to calculate the similarity between users and find out neighbors to predict the recommendation [29].
3. Content-based techniques can be added to collaborative filtering characteristics, such as latent factor model with the content-based approach [30].
4. A conventional probabilistic method for combining collaborative and content-based technique to predict recommendation [31, 32].

Burke [33] performed over hybrid recommender systems and grouped them into seven classes as weighted hybridization, switching hybridization, mixed hybridization, feature-combining hybridization, cascade hybridization, feature-augmenting hybridization, and meta-level hybridization.

3 Challenges

3.1 Data Sparsity

Many e-commerce and shopping Web sites use recommender system and evaluate a very large item sets. With large item sets, the user–item metric becomes sparse and results as a limitation to many recommender systems. Few values of ratings/preferences in user–item metric lead to poor predictions. New items cannot be recommended until some users rate them, and similarly new users are also not getting good recommendations due to lack of their preference history. To deal with data sparsity problem, many techniques have proposed out of them dimensionality reduction like singular value decomposition [20], probabilistic matrix factorization [21, 22], and hybrid techniques such as content boosted are popular and mostly used (Table 1).

3.2 Scalability

Scalability has always been the challenge for recommendation systems. The performance of mostly traditional CF algorithms started to suffer from the increase of size in users and items. The tremendous increase in a database leads to a poor performance of algorithm as computational capabilities went beyond the practical limits [34].

3.3 Cold Start

Business adhered with recommendation systems has a cold start. Initially, for a new user case, do not have sufficient information. A considerable amount of time is required to lure a user and getting them know. However, many networks promote users to fill information to provide them more options. Items also have cold start when they are not rated [35].

3.4 Gray Sheep

User whose opinions do not consistent with any group of people is known as gray sheep. These users do not support the smooth functioning of collaborative filtering [28]. On the other, a special class of users known as Black sheep whose idiosyncratic behavior makes recommendations nearly impossible. With an optimal combination of content-based and collaborative filtering (hybrid techniques) is helping to solve gray sheep problem [36].

Table 1 Comparison of different recommendation systems

Filtering technique comparison

Filtering technique		Method	Advantages	Limitations
Content-based filtering		Use implicit and explicit feedback of users	• User independence • Transparent • Easy to recommend new items	• Hard to learn user preference • Limited degree of novelty • Static
Collaborative filtering	Memory-based	Neighbor-based approach	• Easy implementation • Does not need user profile and item features • Scalable with co-rated items • New data can be added easily	• User preference is needed • Performance decreases with sparsity • New user problem
	Model-based	Data mining, machine learning, dimensionality reduction	• Work well with sparse data • Scalable • Better prediction performance	• Loss of information due to dimensionality reduction • Trade-off between prediction performance and scalability
	Hybrid	Combine memory and model-based	• Improved prediction performance • Overcome problems such as sparsity and gray sheep	• Complex • Expensive implementation • Sometimes need explicit information
Knowledge-based filtering		Case-based, constraint-based, ontologies	• Improved personalized prediction • Handle new user and cold start problem well	• Expensive and complex • Need external domain-specific knowledge
Hybrid filtering		Combine two or more filtering techniques	• More accurate and effective recommendation • Suppress the limitation of individual techniques	• Expensive and complex

3.5 Synonymy

Synonymy refers to the tendency of a number of the same or very similar items to have different names or entries. Most recommender systems are unable to discover this latent association and thus treat these products differently [37, 38].

3.6 Privacy Breach

Recommendation system anyways leaks the information to users. The best example of this is the people you may know feature of Facebook. The issue of trust arises when evaluating a customer [39].

3.7 Shilling Attack

The recommendation is a public activity, so peoples get biased for their feedbacks and give millions of positive reviews for their products or items and sometimes negative views for their competitors. So, it becomes necessary for the system to incorporate a kind of mechanism to discourage this sort of phenomenon [40].

4 Applications

Recommender system applications are now ranging in personal, social, business services. All these areas have their practical applications in human life and have a great impact too. Researchers paid more emphasis over business applications and improved them a lot in recent past. The primary objective of research involves a practical aspect of implementation and effectiveness of recommendations. Mainly recommendation system applications are classified as:

E-commerce/E-shopping: The system was developed to provide guidelines for online customers. It is most popular and specialized field and employed through ratings/preference, which subsequently use to make recommendations. Tagging and reviews are other ways to connect user–item relationship. iTunes, Amazon, and eBay are some of the popular recommendation systems of the e-commerce world.

Entertainment: With the extensive growth of movies, videos, and music, users get frustrated while searching for the right content of their taste. This leads to the development of more effective and personalized recommendation system. Collaborative filtering has mostly used the technique in these systems. For videos content such as TV(Netflix) and YouTube, social and context-aware techniques play an effective role in traditional content-based and collaborative methods.

Contents: In recent years, recommender system has become the key of the e-content system to locate information and knowledge in the digital library. It covers personalized Web pages, a new article, e-mail filtering, etc.

Service Oriented: The Internet and mobile devices open a great opportunity to access various types of information. That also gives essence for the development of many service-based recommendation systems such as tourist recommendation, travel services, matchmaking services, consultation services.

5 Evaluation

The quality of recommendation system is measured through various types of evaluation metric based on the accuracy of prediction and coverage. The selection of metric depends on filtering technique, features of data set, and the task of recommendation system. According to Herlocker [41], evaluation metrics are categorized as prediction accuracy metrics (MAE, RMSE) and classification accuracy metrics (precision, recall, F-measures).

MAE: Mean Absolute Error is the average of the absolute difference between the predictions and actual values.

$$MAE = \frac{1}{N} \sum_{i=1}^{m} \sum_{j=1}^{n} |r_{(i,j)} - \widehat{r_{i,j}}| \qquad (4)$$

RMSE: Root Mean Square Error is computed by the square root of the average of the difference between predictions and actual values. Lower the RMSE is better the recommendation.

$$RMSE = \sqrt{\frac{1}{N} \sum_{i=1}^{m} \sum_{j=1}^{n} |r_{(i,j)} - \widehat{r_{i,j}}|^2} \qquad (5)$$

Classification accuracy metrics are used to measure the performance of recommendation system based on classification techniques. These metrics are computed on confusion metric of predicted and actual values of classification (Table 2).

Table 2 Confusion metric

		Predicted values	
Actual values		Positive	Negative
	Positive	TP	FN
	Negative	FP	TN

Precision: A measure of exactness determines the fraction of relevant items retrieved out of all items retrieved.

$$Precision = \frac{TP}{TP + FP} \tag{6}$$

Recall: A measure of completeness determines the fraction of relevant items retrieved out of all relevant items.

$$Recall = \frac{TP}{TP + FN} \tag{7}$$

F-measure: Harmonic mean of precision and recall to get a single value for comparison purpose.

$$F - measure = \frac{2(Precision * Recall)}{Precision + Recall} \tag{8}$$

6 Conclusion

Recommender systems are the part of everyone's daily life. With the tremendous growth of information and knowledge over the Internet, it is become necessary to have more and more effective and efficient recommendation systems. These systems enable their users to access services and products of their taste, which are not readily available. This paper discusses and highlights various recommendation system with their techniques, challenges, applications, and their evaluation metrics. Presently, different hybridization techniques are used to develop recommendation systems required on task and user personalized basis. The paper helps the researcher to understand and improve the state of current recommendation system.

References

1. Isinkaye, F.O., Folajimi, Y.O., Ojokoh, B.A.: Recommendation systems: principles, methods, and evaluation. Egypt. Inf. J. 261–273 (2015)
2. Recommender System Definition: available at, https://en.wikipedia.org/wiki/Recommender_system

3. Goldberg, D., Nichols, D., Oki, B.M., Terry, D: Using collaborative filtering to weave an information tapestry. Commu. ACM **35**(12), 61–70 (1992)
4. Pu, P., Chen, L., Hu, R.: A user-centric evaluation framework for recommender systems. In: Proceedings of the fifth ACM conference on Recommender Systems (RecSys'11), pp. 57–164. ACM, New York, NY, USA (2011)
5. Bouneffouf, D., Bouzeghoub, A., Ganarski, A.L.: Risk-aware recommender systems. In: Neural Information Processing, pp. 57–65. Springer, Berlin, Heidelberg (2013)
6. Chen, L.S., Hsu, F.H., Chen, M.C., Hsu, Y.C.: Developing recommender systems with the consideration of product profitability for sellers. Inf. Sci. **178**(4), 1032–1048 (2008)
7. Pazzani, M., Billsus, D.: Content-based recommendation systems. In: The Adaptive Web, pp. 325–341. Springer, Berlin, Heidelberg (2007)
8. Guo, G., Zhang, J., Yorke-Smith, N.: A novel evidence based bayesian similarity measure for recommendation systems. J. ACM Trans Web **10**(2), 8.1–8.30 (2016)
9. Su, X., Khoshgoftaar, T.M.: A survey of collaborative filtering techniques. Adv. Artif. Intell. 1–20 (2009)
10. Breese, J.S., Heckerman, D., Kadie C.: Empirical analysis of predictive algorithms for collaborative filtering. In: Proceedings of the Fourteenth Annual Conference on Uncertainty in Artificial Intelligence, pp. 43–52. July 1998
11. Joonseok, L., Sun, M., Lebanon, G.: A Comparative Study of Collaborative Filtering Algorithms (2012)
12. Mobasher, B., Jin, X., Zhou, Y.: Semantically enhanced collaborative filtering on the web. In: Web Mining: from web to semantic web, pp. 57–76. Berlin, Heidelberg, Springer 2004
13. Ku Zalewski U.: Advantages of information granulation in clustering algorithms. In: Agents and artificial intelligence, pp. 131–145. NY, Springer (2013)
14. Michael, J.A., Berry, A., Gordon, S., Linoff, L.: Data mining techniques, 2nd ed. Wiley Publishing Inc., (2004)
15. Larose, T.D.: Discovering knowledge in data. Wiley, Hoboken, (New Jersey) (2005)
16. Friedman, N., Geiger, D., Goldszmidt, M.: Bayesian network classifiers. Mach. Learn. **29**(2–3), 131–63 (1997)
17. Vucetic, S., Obradovic, Z.: Collaborative filtering using a regression based approach. Knowl. Inf. Syst. 1–22 (2005)
18. Berry, M.J.A., Linoff, G.: Data mining techniques: for marketing, sales, and customer support. Wiley Computer Publishing, New York (1997)
19. Bell, R., Koren, Y., Volinsky, C.: Matrix factorization techniques for recommender systems. Computer **42**(8), 30–37 (2009)
20. Sali, S.: Movie rating prediction using singular value decomposition. In: Machine Learning Project Report by University of California, Santa Cruz (2008)
21. Hofmann, T.: Probabilistic latent semantic analysis. In: Proceedings of the 15th Conference on Uncertainty in AI, pp. 289–296. San Fransisco, California (1999)
22. Salakhutdinov, R., Mnih, A.: Probabilistic matrix factorization. In: Proceedings of the 20th International Conference on Neural Information Processing Systems (NIPS'07) (2007)
23. Lu, yuan, Yang Jie, Notes on "Low-Rank Matrix Factorization", e-print (2015) arXiv:1507.00333
24. Patrik Hoyer, O.: Non-negative matrix factorization with sparseness constraints. J. Mach. Learn. Res. **5**, 1457–1469 (2004)
25. David Blei, M., Andrew Ng, Y., Jordan, M.I.: Latent Dirichlet Allocation. J. Mach. Learn. Res. **3**, 993–1022 (2003)
26. Bridge, D., Mehmet Gker, H., McGinty, L., Smyth, B.: Case-based recommender systems. Knowl. Eng. Rev. **20**(3), 315–320 (2005)
27. Adomavicius, G., Zhang, J.: Impact of data characteristics on recommender systems performance. ACM Trans. Manage Inf. Syst. **3**(1), 3.1–3.17 (2012)
28. Claypool, M., Gokhale, A., Miranda, T., Murnikov, P., Netes, D., Sartin, M.: Combining content-based and collaborative filters in an online newspaper. In: Proceedings of ACM SIGIR Workshop on Recommender Systems: algorithms and evaluation. Berkeley, California (1999)

29. Billsus, D., Pazzani, M.J.: A hybrid user model for news story classification. In: Kay, J. (ed.) Proceedings of the seventh International Conference on user Modelling, pp. 99–108. Banff, Canada, Springer, Newyork (1999)
30. Soboroff, I., Nicholas, K.C., Pazzani, M.J.: Workshop on recommender systems: algorithms and evaluation. In: Conference Proceedings SIGIR Forum, vol. 33, no. 1, pp. 36–43 (1999)
31. Shein, I., Popescul, A., Ungar, L.H., Pennock, D.M.: Methods and metrics for cold-start recommendations. In: Proceedings of the 25th International ACM SIGIR Conference on Research and Development in Information Retrieval SIGIR'02, pp. 253–260. ACM, New York, NY, USA (2002)
32. Popescul, A., Ungar, L.H., Pennock, D.M., Lawrence, S.: Probabilistic models for unified collaborative and content-based recommendation in sparse data environments. In: Proceedings of the 17th Conference on Uncertainty in Artificial Intelligence, UAI'01, pp. 437–444 (2001)
33. Burke, R.: Hybrid recommender systems: survey and experiments. User Model. User-Adapt. Interact. **12**(4), 331–370 (2002)
34. Linden, G., Smith, B., York, J.: Recommendations: item-to-item collaborative filtering. IEEE Internet Comput. **7**(1), 76–80 (2003). www.Amazon.com
35. Rana, M.C.: Survey paper on recommendation system. Int. J. Comput. Sci. Inf. Technol. **3**(2), 3460–3462 (2012)
36. Mahony, M.O., Hurley, N., Kushmerick, N., Silvestre, G.: Collaborative recommendation: a robustness analysis. ACM Trans. Internet Technol. **4**(4), 344–377 (2004)
37. Jones, S.K.: A statistical interpretation of term specificity and its applications in retrieval. J. Doc. **28**(1) 11–21)(1972)
38. Gong, M., Xu, Z., Xu, L., Li, Y., Chen, L.: Recommending web service based on user relationships and preferences. In: 20th International conference on web services. IEEE (2013)
39. Canny, J.: Collaborative filtering with privacy via factor analysis. In: Proceedings of the 25th Annual International ACM SIGIR Conference on Research and Development in Information Retrieval, pp. 238–245 (2002)
40. Resnick, P., Varian, H.R.: Recommender systems. Commun. ACM **40**(3), 56–58 (1997)
41. Herlocker, J.L., Konstan, J.A., Terveen, L.G., Reidll, J.T.: Evaluating recommendation systems. ACM Trans. Inf. Syst. **22**(1), 5–53 (2004)

Parametric Optimization of Turning Process Using Evolutionary Optimization Techniques—A Review (2000–2016)

Parthiv B. Rana, Jigar L. Patel and D. I. Lalwani

Abstract Manufacturing is the foundation of any industrialized country that involves making products from raw materials using various processes. Usually, casting and metal forming processes are used to produce most of the parts, and afterword, the parts are machined to obtain the desired size, shape and surface finish. The traditional machining processes, i.e., turning, milling, grinding, drilling, are widely used to obtain the desired product. The proper selection of process parameters is required to produce products at low cost, high quality and within time bound. Therefore, in past, researchers had optimized the process parameters to obtain the desired product. In the present study, the overview of turning process and the review of research related to optimization of turning process using evolutionary optimization techniques are carried out. The review period is selected from the year 2000 to 2016. The present review work is well-organized information in terms of objectives, process parameters, constraints and optimization techniques for optimization of turning process that can be beneficial for succeeding researchers to ascertain the gap in the research.

Keywords Turning process · Parameter optimization · Evolutionary optimization techniques

1 Introduction

The industries are growing very fast, and the technological improvements become essential to meet the demand in various domains, such as manufacturing engineering, nuclear engineering, aerospace engineering, marine engineering, product improve-

P. B. Rana (✉) · J. L. Patel · D. I. Lalwani
Mechanical Engineering Department, Sardar Vallabhbhai National Institute
of Technology, Surat, Surat, Gujarat, India
e-mail: ranaparthiv@gmail.com

J. L. Patel
e-mail: jigu.mech1515@gmail.com

D. I. Lalwani
e-mail: dil@med.svnit.ac.in

© Springer Nature Singapore Pte Ltd. 2019
J. C. Bansal et al. (eds.), *Soft Computing for Problem Solving*, Advances in Intelligent
Systems and Computing 817, https://doi.org/10.1007/978-981-13-1595-4_13

ment (or design). The production of a product is an area where the number of manufacturing processes is involved, but traditional machining processes are widely used in industries to produce finish product from the raw material. The requirements of a product with high quality, low cost and within time bound increase day by day. Therefore, industries become more attentive to fulfil such requirements. As the machining processes are influenced by a number process parameters (decision variables) and practical constraints, the appropriate selection of process parameters is required to get better machining performance (i.e., production rate, machining time, surface roughness) [1]. The parameter selection becomes more complex when number of objectives, process parameters and constraints increase in the machining process.

To select the process parameters for the favourable outcome, parameter optimization is carried out using conventional or evolutionary optimization techniques. The conventional optimization techniques (COTs) are not preferred as they have some limitations; that is, they are trapped at local optimum for multimodal problems, they are not perfect for multi-objective, non-differentiable and continuous problems, and they are not suitable when the number of constraints increases in the problem [2]. Therefore, evolutionary optimization techniques (EOTs) draw a great attention of researchers because of their capability to overcome limitations raised in COTs. In the present study, efforts have been applied to conduct a review of real-life optimization problems on turning processes. The next section illustrates the review of process parameter optimization of turning process.

2   Process Parameter Optimization of Turning Process

Turning process is one of the primary and widely used machining processes where the material is removed from the rotating workpiece, and tool performs radial, axial or both motions simultaneously to obtain the required shape. The turning process can be carried out in single cut (pass) or multiple cuts (passes). To produce the product with low cost and high quality within time bound, the proper selection of process parameters is essential. There are various process parameters in turning processes, such as cutting speed, feed rate, depth of cut, tool material and tool geometry, number of passes, workpiece material and its properties, power consumption, properties and characteristics of cutting fluid, cutting forces and cutting temperature. The turning process parameters are optimized to increase production rate, to increase tool life, to improve geometrical accuracy, to increase surface finish and to reduce production time as well as production cost [3]. Table 1 shows the summary of process parameters optimization of turning process in the form of objectives, process parameters and constraints considered by various researchers to optimize turning process using various EOTs.

In Table 1, review of the researchers is presented in chronological year of publication. Many researchers optimized the model that is proposed by Chen and Tsai [4] as it is and some of them have modified or relaxed some constraints and process parameters using various EOTs [5, 8–10, 12, 14, 18, 20, 23, 26, 29, 30, 36, 39–41, 44,

Table 1 Review on process parameter optimization of turning process

Author(s)/Year	Optimization technique(s) considered	Objective(s)	Process parameters	Constraints	Remarks
Chen and Tsai [4]	• Simulated annealing (SA) • Pattern search (PS)	Production cost	For rough and finish pass • Cutting speed • Feed rate • Depth of cut	For rough and finish pass: • Parameter bounds • Tool life • Cutting force • Cutting power • Stable cutting region • Chip tool interface temperature For finish pass: • Surface finish • Rough and finish machining process parameter relations	• They developed a model for production cost, and process parameters are optimized using SA and PS • They considered cutting force constant as 5 kgf and that is infeasible
Onwubolu and Kumalo [5]	Genetic algorithm (GA)	Production cost	Same as Chen and Tsai [4]	Same as Chen and Tsai [4]	• They considered cutting force constant as 5 kgf, and it is infeasible
Wang et al. [6]	Genetic algorithm (GA)	• Surface roughness • Cutting force • Material removal rate • Chip breakability • Tool life	• Cutting speed • Feed rate • Depth of cut	• Parameter bounds • Surface roughness • Material removal rate • Cutting force • Tool life • Chip breakability • Total depth of cut	• They considered tool wear effect to predict objectives
Wang et al. [7]	Genetic algorithm (GA)	• Surface roughness • Cutting force • Material removal rate • Chip breakability • Tool life	• Cutting speed • Feed rate • Depth of cut	• Parameter bounds' • Surface roughness • Material removal rate • Cutting force • Tool life • Chip breakability • Total depth of cut	• They controlled the combined objective function using weight factors
Chen and Chen [8]	Float encoding genetic algorithm (FEGA)	Production cost	Same as Chen and Tsai [4]	Same as Chen and Tsai [4]	• They reported that FEGA performed equally to SA in terms of objective function • They have not reported the value of optimum process parameters
Vijaykumar et al. [9]	Ant colony optimization (ACO)	Production cost	Same as Chen and Tsai [4]	Same as Chen and Tsai [4]	• They reported that number of passes for rough pass is less than one, and it is not possible
Chen [10]	Scatter search (SC)	Production cost	Same as Chen and Tsai [4]	Same as Chen and Tsai [4]	• They concluded that SS technique is outperformed to GA and SA
Sardinas et al. [11]	Genetic algorithm (GA)	• Production rate • Used tool life	• Cutting speed • Feed rate • Depth of cut	• Parameter bounds • Cutting force • Cutting power • Surface roughness	• They remarked that the multi-objective optimization approach was desirable over the single-objective one
Wang [12]	Ant colony optimization (ACO)	Production cost	Same as Chen and Tsai [4]	Same as Chen and Tsai [4]	• He proved that problem solved by Vijaykumar et al. [9] is invalid

(continued)

Table 1 (continued)

Author(s)/Year	Optimization technique(s) considered	Objective(s)	Process parameters	Constraints	Remarks
Sarvanan and Janakiraman [13]	Genetic algorithm (GA)	Machining time	• Cutting speed • Feed rate	• Cutting speed • Feed rate • Tool life • Cutting force • Surface finish • Maximum power	• They optimized the process parameters using GA
Siva-Sankar et al. [14]	Modified genetic algorithm (MGA)	Production cost	Same as Chen and Tsai [4]	Same as Chen and Tsai [4]	• They modified the crossover and mutation operator to improve the search pattern for better performance of MGA
Prasad et al. [15]	Genetic algorithm (GA)	Tool wear	• Cutting speed • Feed rate • Depth of cut	• Parameter bounds • Cutting force • Cutting temperature	• They compared their results with dynamic programming and found that GA is performed better
Singh and Rao [16]	Genetic algorithm (GA)	Surface roughness	• Cutting speed • Feed rate • Effective rack angle • Nose radius	• Parameter bounds	• They obtained surface roughness model using experiments, and the process parameters of the model were optimized using GA
Abburi and Dixit [17]	• Real-parameter genetic algorithm (RGA) • Sequential quadratic programming	• Production time • Production cost	Same as Chen and Tsai [4]	• Parameter bounds • Surface finish • Cutting force • Machine power • Geometric relation	• They used RGA to optimize the process parameters. Further, SQP was used to optimize process to improve the solution
Yildiz [18]	Hybrid Taguchi-Harmony search algorithm (HTHSA)	Production cost	Same as Chen and Tsai [4]	Same as Chen and Tsai [4]	• He concluded that HTHSA gives better results than other techniques (used by other researchers) • They have not reported value of optimum process parameters
Tang et al. [19]	Particle swarm optimization (PSO)	Machining time	• Cutting speed • Feed rate • Depth of cut	• Parameter bounds • Cutting power • Cutting force • Tool life	• They optimized multi-tool parallel turning process with n number of features using PSO
Wu and Yao [20]	Modified continuous ant colony optimization (MCACO)	Production cost	Same as Chen and Tsai [4]	Same as Chen and Tsai [4]	• They suggested that MCACO technique is better for rapid selection of process parameters

(continued)

Table 1 (continued)

Author(s)/Year	Optimization technique(s) considered	Objective(s)	Process parameters	Constraints	Remarks
Kim et al. [21]	Real-coded genetic algorithm (RCGA)	Production cost for multi-pass (mild steel workpiece and carbide tool)	• Cutting speed • Feed rate	• Cutting force • Cutting power • Tool life • Temperature	• They optimized five models (objectives) of turning process using RCGA and proved that RCGA is reliable and accurate technique to solve nonlinear optimization models
		Production cost for single pass	• Cutting speed • Feed rate	• Surface finish • Feed rate • Cutting power	
		Production cost for single pass (medium carbon steel and carbide tool)	• Cutting speed • Feed rate	• Surface finish • Cutting power	
		Production cost for single pass	• Cutting speed • Feed rate	• Surface finish • Feed rate • Cutting power	
		Production cost for multi-pass (medium carbon tool)	• Number of passes • Cutting speed • Feed rate • Depth of Cut	• Parameter bound • Cutting force • Stable cutting region related to cutting surface • Surface roughness • Cutting power	
Kolahan and Abachizadeh [22]	Simulated annealing (SA)	Production cost	• Cutting speed • Feed rate • Depth of cut	• Parameter bounds • Tool life • Cutting force • Cutting power	• They modified original Chen and Tsai [4] model and optimized it using SA
Yildiz [23]	Particle swarm and receptor editing (PSRE)	Production cost	Same as Chen and Tsai [4]	Same as Chen and Tsai [4]	• He developed an algorithm using immune system and reported that PSRE provided better results than other techniques (used by other researchers) • They have not reported the value of optimum process parameters
Cus et al. [24]	Ant colony optimization (ACO)	• Production cost • Production rate • Cutting roughness	• Cutting speed • Feed rate • Depth of cut	• Parameter bounds • Tool life • Cutting power	• They reported that ACO technique is superior to GA and SA
Srinivas et al. [25]	Particle swarm optimization (PSO)	• Production cost • Machining time	• Cutting speed • Feed rate • Depth of Cut	Same as Chen and Tsai [4]	• They compared the production cost and constraint violation to obtain best solution

(continued)

Table 1 (continued)

Author(s)/Year	Optimization technique(s) considered	Objective(s)	Process parameters	Constraints	Remarks
Zheng and Ponnambalam [26]	Genetic algorithm-artificial immune system (GA-AIS)	Production cost	Same as Chen and Tsai [4]	Same as Chen and Tsai [4]	• They considered cutting force constant as 5 kgf, and it is infeasible
Raj and Namboothiri [27]	Improved Genetic Algorithm (IGA)	Surface roughness	• Cutting speed • Feed rate • Depth of cut • Nose radius	• Parameter bounds	• They reported that IGA worked better than the Conventional genetic algorithm (CGA)
Raja and Baskar [28]	• Genetic algorithm (GA) • Simulated annealing (SA) • Particle swarm optimization (PSO)	1. Production cost and Production time for single pass 2. Production cost for multi-pass	Objective 1 • Cutting speed • Feed rate • Depth of cut Objective 2 Same as Chen and Tsai [4]	Objective 1 • Parameter bounds • Cutting power • Surface finish • Temperature • Cutting force Objective 2 For rough and finish pass • Parameter bounds • Surface finish • Cutting force • Cutting power	• They reported that PSO gives the best results
Xie and Pan [29]	Genetic algorithm (GA)	Production cost	Same as Chen and Tsai [4]	Same as Chen and Tsai [4]	• They have not reported value of optimum process parameters
Zheng and Ponnambalam [30]	Particle swarm optimization (PSO)	Production cost	Same as Chen and Tsai [4]	Same as Chen and Tsai [4]	• They considered cutting force constant as 5 kgf, and it is infeasible
Raja and Baskar [31]	• Simulated annealing (SA) • Genetic algorithm (GA) • Particle swarm optimization (PSO) • Memetic algorithm (MA) • Hybrid algorithm (HA)	• Production cost • Production time	• Cutting speed • Feed rate • Depth of cut	• Parameter bounds • Cutting power • Surface finish • Temperature • Cutting force	• They used various optimization techniques to optimize the turning process model developed by Agapiou [32], and the obtained results were compared • They reported that PSO technique gave better results than other techniques

(continued)

45, 48, 51–57, 62, 65, 67]; the rest of the researchers proposed a model by conducting experiments and later process parameters of that model were optimized using EOT.

3 Conclusions

The literature review related to process parameters optimization of turning process is carried out in-depth. The detail literature review on parameters optimization of turning process from 2000 to 2016 is made and summarized in tabular form. The

Table 1 (continued)

Author(s)/Year	Optimization technique(s) considered	Objective(s)	Process parameters	Constraints	Remarks
Yang and Natarajan [33]	• Multi-objective differential evolution (MODE) • Non-dominated sorting genetic algorithm—II (NSGA II)	• Tool wear • Material removal rate	• Cutting speed • Feed rate • Depth of cut	• Parameter bounds • Surface finish • Temperature	• They used the MODE and NSGA II techniques to optimize the process parameters and reported that the results obtained using MODE technique is better than NSGA II
Raja and Baskar [34]	Particle swarm optimization (PSO)	Production time	• Cutting speed • Feed rate • Depth of cut	• Parameter bounds • Surface finish	• They considered cutting force as 5 kgf and that is infeasible
An [35]	Genetic algorithm	Production cost	• Cutting speed • Feed rate • Depth of cut	• Parameter bound • Tool life • Surface finish • Cutting force • Cutting power	• He minimized the production cost using genetic algorithm for various depth of cut
Costa et al. [36]	Hybrid particle swarm optimization (HPSO)	Production cost	Same as Chen and Tsai [4]	Same as Chen and Tsai [4]	• They reported that HPSO is performed superior to other techniques (used by other researchers)
Aungkulanon et al. [37]	• Particle swarm optimization (PSO) • Fire fly algorithm (FFA)	Production cost	• Cutting speed • Feed rate • Depth of cut	• Parameter bounds • Cutting force • Cutting power • Tool life • Temperature • Depth of cut for n passes • Surface finish • Stable cutting region related to cutting surface	• They considered four turning process models to optimize process parameters, and the obtained results were compared in terms of computational time, convergence speed, and quality of the results • They reported that FFA is more reliable and converges consistently
Hippalgaonkar and Shin [38]	Dynamic-objective particle swarm optimization (DOPSO)	Production cost	• Cutting speed • Feed rate • Depth of cut	• Parameter bound • Cutting force • Cutting power • Rough and finishing parameter relations • Surface finish • Total depth of cut for rough and finishing passes	• They minimized the production cost by an appropriate probabilistic tool life model that is incorporated by stochastic tool failure to make the problem more realistic
Singh and Chauhan [39]	Differential evolution (DE)	Production cost	Same as Chen and Tsai [4]	Same as Chen and Tsai [4]	• The selected tool life model is not as per the model considered by Chen and Tsai [4]

(continued)

Table 1 (continued)

Author(s)/Year	Optimization technique(s) considered	Objective(s)	Process parameters	Constraints	Remarks
Yildiz [40]	Differential evolution algorithm and receptor editing (DERE)	Production cost	Same as Chen and Tsai [4]	Same as Chen and Tsai [4]	• He compared the results obtained using DERE with other techniques and reported that DERE is superior to other techniques (used by other researchers) • They have not reported the value of optimum process parameters
Aryanfar and Solimanpur [41]	Genetic algorithms (GA)	Production cost	Same as Chen and Tsai [4]	Same as Chen and Tsai [4]	• They reported that the production cost obtained using GA was 6.26% and 7.23% lower than PSO and SA/PS, respectively
Khan et al. [42]	• Genetic algorithm-based neural network (GA-NN) • Particle swarm optimization-based neural network (PSO-NN)	Production cost	• Cutting speed • Feed rate • Depth of cut	• Parameter bounds • Tool life • Actual machining time • Surface finish • Cutting power	• They reported that PSO-NN technique gives better results than GA-NN
Rao and Kalyankar [43]	Teaching–learning-based optimization (TLBO)	• Production time • Production cost • Surface roughness	• Cutting speed • Feed rate • Depth of cut	• Parameter bounds • Cutting power • Cutting force	• They remarked that TLBO gives better results in less computational time than other techniques
Lee and Ponnambalam [44]	• Particle swarm optimization (PSO) • Genetic algorithm and artificial immune system (GA-AIS)	Production cost	• Cutting speed • Feed rate • Depth of cut	Same as Chen and Tsai [4]	• They concluded that GA-AIS performed superior to PSO
Xie and Guo [45]	Ant colony optimization (ACO)	Production cost	• Cutting speed • Feed rate • Depth of Cut	Same as Chen and Tsai [4]	• They reported that ACO gave better results than other techniques
Lu et al. [46]	• Hybrid solver (genetic algorithm (GA) + sequential quadratic programming (SQP)) • Dynamic programming	Production time	• Cutting speed • Feed rate • Depth of cut • No. of cutting passes • Sequence of cutting passes	• Parameter bounds • Tolerance • Chatter free • Tool life constraint • Cutting power • Geometric relation • Surface roughness	• They reported that the given partition of depth of cut can be used to minimize production time

(continued)

Table 1 (continued)

Author(s)/Year	Optimization technique(s) considered	Objective(s)	Process parameters	Constraints	Remarks
Ahilan et al. [47]	• Neural network using genetic algorithm (NNGA) • Neural network using particle swarm optimization (NNPSO)	• Power consumption • Surface roughness	• Cutting speed • Feed rate • Depth of cut • Nose radius	• Parameter bounds	• They performed experiments using Taguchi L_{27} orthogonal array to obtain power consumption and surface roughness models • They optimized the process parameters of the models using NNGA and NNPSO, and reported that NNPSO gave better results than NNGA
Belloufi et al. [48]	Hybrid genetic algorithm (genetic algorithm + sequential quadratic programming)	Production cost	For rough and finish passes • Cutting speed • Feed rate • Depth of cut	Same as Chen and Tsai [4]	• They remarked that the hybrid GA was significantly reduced the production cost
Jabri et al. [49]	Genetic algorithm (GA)	• Cutting cost • Used tool life	Same as Chen and Tsai [4]	For rough and finish pass • Parameter bound • Cutting force • Cutting power • Chip tool interface temperature For finish pass • Surface finish	• They used the Pareto frontier graphic to select optimal process parameters of multi-objective turning process
Rao and Kalyankar [50]	Teaching–learning-based optimization (TLBO)	1. Multi-objective: • Production time • Tool life 2. Single-objective: • Production cost	Objective 1 • Cutting speed • Feed rate • Depth of cut Objective 2 Same as Chen and Tsai [4]	Objective 1 • Parameter bounds • Cutting force • Cutting power • Surface finish Objective 2 Same as Chen and Tsai [4]	• They reported that the TLBO gives better results than other techniques
Yildiz [51]	Hybrid robust differential evolution (HRDE)	Production cost	Same as Chen and Tsai [4]	Same as Chen and Tsai [4]	• He reported that HRDE provided better results than other techniques (used by other researchers) • They have not reported the value of optimum process parameters
Yildiz [52]	Hybrid artificial bee colony (HABC)	Production cost	Same as Chen and Tsai [4]	Same as Chen and Tsai [4]	• He remarked that the results obtained using HABC were better than other techniques (used by other researchers) • They have not reported the value of optimum process parameters

(continued)

Table 1 (continued)

Author(s)/Year	Optimization technique(s) considered	Objective(s)	Process parameters	Constraints	Remarks
Xie and Pan [53]	Estimation of distribution algorithm (EDA)	Production cost	For rough and finish passes • Cutting speed • Feed rate • Depth of cut	• Parameter bounds • Tool life • Surface finish • Stable cutting region • Chip tool interface temperature • Rough and finishing parameter relations	• They incorporated parallel turning of rough and finish passes in the model developed by Chen and Tsai [4], and the process parameters of the model were optimized using EDA
Yildiz [54]	Hybrid teaching–learning-based optimization (HRTLBO)	Production cost	For rough pass • Cutting speed • Feed rate Finish pass • Cutting speed • Feed rate • Depth of cut	Same as Chen and Tsai [4]	• He remarked that HRTLBO was superior to other techniques
Singh and Chauhan [55]	Laplace crossover power mutation genetic algorithm (LXPM-GA)	Production cost	Same as Chen and Tsai [4]	Same as Chen and Tsai [4]	• They concluded that the results obtained by LXPM-GA is better than other techniques (used by other researchers), and it requires less number of iterations and computational time
Chauhan et al. [56]	Totally distributed particle swarm optimization (TDPSO)	Production cost	Same as Chen and Tsai [4]	Same as Chen and Tsai [4]	• They reported that TDPSO outperformed to GA, SA and PSO
Belloufi et al. [57]	Fire fly algorithm (FFA)	Production cost	For rough and finish passes • Cutting speed • Feed rate • Depth of cut	Same as Chen and Tsai [4]	• They concluded that FFA gives near optimum solution
Hreja et al. [58]	• Gravitational search algorithm (GSA) • Particle swarm optimization (PSO)	• Cutting force • Surface roughness • Tool life	For rough and finish passes • Cutting speed • Feed rate • Depth of cut	Parameter bounds	• They performed experiments to obtain initial data, and the same data were used to optimize the model using GSA and PSO • They reported that GSA and PSO techniques provided less deviation compared to expected values. However, GSA required less optimization time than PSO

(continued)

Table 1 (continued)

Author(s)/Year	Optimization technique(s) considered	Objective(s)	Process parameters	Constraints	Remarks
Hrelja et al. [59]	Particle swarm optimization (PSO)	• Cutting force • Surface roughness • Tool life	For rough and finish passes • Cutting speed • Feed rate • Depth of cut	Parameter bounds	• They performed the experiments and obtained the model (coefficients of model) using PSO. Further, process parameters of the model are optimized
Acayaba and Escalona [60]	Simulated annealing (SA)	Surface roughness	• Cutting speed • Feed rate • Depth of cut • Nose radius	Parameter bounds	• They performed experiments to develop a prediction model using artificial neural network and multiple linear regression • Further, they integrated developed model with SA algorithm to obtain optimum process parameters that provided low surface roughness
Gayatri and Baskar [61]	Hybrid genetic simulated swarm (HGSS)	Production cost	For rough and finish passes • Cutting speed • Feed rate • Depth of cut	• Parameter bounds • Tool life • Cutting force • Cutting power • Stable cutting region • Chip tool interface • Surface finish	• They remarked that the results obtained using HGSS were better than GA, SA and PSO
Gayatri and Baskar [62]	• Genetic algorithm (GA) • Simulated annealing (SA) • Particle swarm optimization (PSO)	Production cost	For rough and finish passes • Cutting speed • Feed rate • Depth of cut	Same as Chen and Tsai [4]	• They used the model proposed by Chen and Tsai [4] to analyse algorithmic parameters • They carried out sensitivity analysis and performance analysis by varying algorithmic parameters of GA, SA and PSO
Kubler et al. [63]	Genetic algorithm (GA)	• Manufacturing process time • Tool wear • Process energy	• Cutting speed • Feed rate • Depth of cut	• Parameter bounds • Cutting power • Surface roughness	• They performed the experiments using central composite design (CCD) to develop multi-objective model • They used Pareto front to select optimum process parameters for multi-objective problem
Lin et al. [64]	Multi-objective teaching–learning-based optimization (MOTLBO)	• Carbon emission • Operation time	• Cutting speed • Feed rate • Depth of cut • Rotation speed	• Parameter bounds	• They developed a method to measure the amount of carbon emission during dry turning process • They proposed MOTLBO to optimize the process parameters of multi-objective problem

(continued)

Table 1 (continued)

Author(s)/Year	Optimization technique(s) considered	Objective(s)	Process parameters	Constraints	Remarks
Chauhan et al. [65]	Totally distributed particle swarm optimization (TDPSO)	Production cost	Model 1 For rough and finish passes • Cutting speed • Feed rate • Depth of cut Model 2 Same as Model 1	Model 1 Rough and finish passes • Parameter bounds • Tool life • Cutting force • Cutting power • Surface finish For rough pass • Total depth of cut Model 2 Same as Chen and Tsai [4]	• They reported that TDPSO gives better results than other techniques (used by other researchers) for various depth of cut
Jiang et al. [66]	Hybrid genetic algorithm (Hybrid GA)	• Process cost • Cutting fluid consumption	• Cutting speed • Feed rate • Depth of cut • Cutting fluid flow rate	• Cutting force • Cutting power	• They proposed a hybrid GA to optimize the process parameters of multi-objective problem
Mellal and Williams [67]	Cuckoo optimization algorithm (COA)	Production cost	Same as Chen and Tsai [4]	Same as Chen and Tsai [4]	• They concluded that results obtained using COA were better than other techniques and highly competitive to HPSO
Lu et al. [68]	Multi-objective backtracking search algorithm (MOBSA)	• Energy consumption • Machining precision	Same as Chen and Tsai [4]	Same as Chen and Tsai [4]	• They proposed MOBSA to optimize the process parameters of multi-objective, multi-pass problem
Park et al. [69]	• Non-dominated sorting genetic algorithm (NSGA II)	• Specific cutting energy • Energy efficiency	• Cutting speed • Feed rate • Nose radius • Edge radius • Rake angle • Relief angle	• Parameter bound	• They performed experiments on hardened steel (AISI 4240) using Box–Behnken design (BBD) to develop multi-objective model • They optimized the process parameters of model using NSGA II

work carried out by various researchers is summarized as objectives, process parameters, constraints and optimization techniques considered for optimization of turning process. The following observations are made based on the present review:

- Lots of research work was conducted on parameters optimization of turning process using evolutionary optimization techniques (EOTs).
- The model developed by Chen and Tsai [4] is widely used by many researchers to minimize the production cost using various EOTs developed by various researchers.
- A few researchers modified the previous turning process model by incorporating either the energy consumption, energy efficiency and cutting fluid consumption or any combination as objective.

4 Future Scope

A few researchers have updated some old models by incorporating constraints. There is still scope to update the existing turning models by incorporating real-life working conditions, such as combination of tool and workpiece material, machine conditions, quality requirements, as constraints. The updated models of turning process can be optimized using various EOTs to check whether further improvement is possible or not.

References

1. Benardos, P.G., Vosnaikos, G.C.: Predicting surface roughness in machining: a review. Int. J. Mach. Tools Manuf **43**, 833–844 (2003)
2. Savsani, Poonam: Jhalab, R. L., Savsani, Vimal: Effect of hybridizing Biogeography-based Optimization (BBO) technique with Artificial Immune Algorithm (AIA) and Ant Colony Optimization (ACO). Appl. Soft Comput. **21**, 542–553 (2014)
3. Rana, P.B., Lalwani, D.I.: Optimization of turning process using amended differential evolution algorithm. Eng. Sci. Technol. Int. J. **20**, 1285–1301 (2017)
4. Chen, M.C., Tsai, D.M.: A simulated annealing approach for optimization of multi-pass turning operations. Int. J. Prod. Res. **34**(10), 2803–2825 (1996)
5. Onwubolu, G.C., Kumalo, T.: Optimization of multi-pass turning operations with genetic algorithm. Int. J. Prod. Res. **39**, 3727–3745 (2001)
6. Wang, X., Jawahir, I.S.: IFSA World Congress and 20th NAFIPS International Conference, 2001, Vancouver (2001)
7. Wang, X., Da, Z.J., Balaji, A. K., Jawahir. I.S.: Performance-based optimal selection of cutting conditions and cutting tools in multipass turning operations using genetic algorithms. Int. J. Prod. Res. **40**(9), 3727–3745 (2002)
8. Chen, M.C, Chen K.Y.: Optimization of multi-pass turning operations with genetic algorithms: a note. Int. J. Prod. Res. **41**, 3385–3388 (2003)
9. Vijayakumar, K., Prabhaharan, G., Asokan, P., Saravanan, R.: Optimization of multi-pass turning operations using ant colony system. Int. J. Mach. Tools Manuf. 43, 1633–1639 (2003)
10. Chen, M.C.: Optimizing machining economics models of turning operations using the scatter search approach. Int. J. Prod. Res. **42**, 2611–2625 (2004)
11. Sardinas, R.Q., Santana M.R., Brindis E.A.: Genetic algorithm-based multi-objective optimization of cutting parameters in turning processes. Eng. Appl. Artif. Intell. **19**, 127–133 (2006)
12. Wang, Y.C.: A note on 'optimization of multi-pass turning operations using ant colony system'. Int. J. Mach. Tools Manuf. **47**, 2057–2059 (2007)
13. Sarvanan, R., Janakiraman, V.: study on reduction of machining time in cnc turning centre by genetic algorithm. In: International Conference on Computational Intelligence and Multimedia Applications, pp. 481–486. Sivakashi, Tamil Nadu (2007)
14. Siva-Sankar, R., Asokan, P., Sarvanan, R., Kumanan, S., Prabhaharan, G.: Selection of machining parameters for constrained machining problem using evolutionary computation. Int. J. Adv. Manuf. Technol. **32**, 892–901 (2007)
15. Prasad, C.F., Jaybal, S., Natrajan, U.: Optimization of tool wear in turning using genetic algorithm. Indian J. Eng. Mater. Sci. **14**, 403–407 (2007)
16. Singh, D., Rao, P.V.: Optimization of tool geometry and cutting parameters for hard turning. Mater. Manuf. Process. **22**, 15–21 (2007)
17. Abburi, N.R., Dixit, U.S.: Multi-objective optimization of multipass turning processes. Int. J. Adv. Manuf. Technol. **32**, 902–910 (2007)

18. Yildiz, A.R.: Hybrid Taguchi-harmony search algorithm for solving engineering optimization problems. Int. J. Ind. Eng. **15**, 286–293 (2008)
19. Tang, L., Landers, R.G., Balakrishnan, S.N.: Parallel turning process parameter optimization based on a novel heuristic approach. J. Manuf. Sci. Eng. 130(031002), 1–12, (2008)
20. Wu, J., Yao, Y.: A modified ant colony system for the selection of machining parameters. In: International Conference on Grid and Cooperative Computing. China (2008)
21. Kim, S.S., Kim, I.H., Mani, V., Kim, H.J.: Real-coded genetic algorithm for machining condition optimization. Int. J. Adv. Manuf Technol. **38**, 884–895 (2008)
22. Kolahan, F., Abachizadeh, M.: Optimizing turning parameters for cylindrical parts using simulated annealing. In: Proceedings of World Academy of Science, Engineering and Technology vol. 36. Dec 2008. ISSN 2070-3740
23. Yildiz, A.R.: A novel particle swarm optimization approach for product design and manufacturing. Int. J. Adv. Manuf. Technol. 40, 617–628 (2009)
24. Cus, F., Balic, J., Zuperl, U.: Hybrid ANFIS-ants system based optimisation of turning parameters. J. Achiev. Mater. Manuf. Eng. **36**(1) 79–86 (2009)
25. Srinivas, J., Giri, R., Yang, S.: Optimization of multi-pass turning using particle swarm intelligence. Int. J. Adv. Manuf. Technol. **40**, 56–66 (2009)
26. Zheng, L.Y., Ponnambalam, SG.: A Hybrid GA-AIS Heuristic for Optimization of Multipass Turning Operations. In: Third International Conference ICIRA 2010, Part II, LNAI 6425, pp. 599–611. Shanghai, China (2010)
27. Raj, T.G.A., Namboothiri, V.N.N.: An Improved genetic algorithm for the prediction of surface finish in dry turning of SS 420 materials. Int. J. Adv. Manuf. Technol. **47**, 313–324 (2010)
28. Raja, S.B., Baskar, N.: Optimization techniques for machining operations: a retrospective research based on various mathematical models. Int. J. Adv. Manuf. Technol. **48**, 1075–1090 (2010)
29. Xie, S., Pan, L.: Selection of machining parameters using genetic algorithms. In: The 5th International Conference on Computer Science & Education, pp. 1147–1150. Hefei, China, 24–27 Aug 2010
30. Zheng, L.Y., Ponnambalam, S.G.: Optimization of multipass turning operations using particle swarm optimization. In: Proceeding of the 7th International Symposium on Mechatronics and its Applications (ISMA10), pp. 1–6. Sharjah, UAE, 20–22 April 2010
31. Raja, S.B.: Investigation of optimal machining parameters for turning operation using intelligent techniques. Int. J. Mach. Mach. Mater. **8**(1/2), 146–166 (2010)
32. Agapiou, J.S.: Optimization of multistage machining systems, Part 1: Mathematical Solution. J. Eng Ind. **114**(4). Trans. ASME J. Eng. Ind. **114** 524–531 (1992)
33. Yang, S.H., Natarajan, U.: Multi-objective optimization of cutting parameters in turning process using differential evolution and non-dominated sorting genetic algorithm-II approaches. Int. J. Adv. Manuf. Technol. **49**, 773–784 (2010)
34. Raja, S.B., Baskar, N.: Particle swarm optimization technique for determining optimal machining parameters of different work piece materials in turning operation. Int. J. Adv. Manuf. Technol. **54**, 445–463 (2011)
35. An, L.: Optimal selection of machining parameters for multi-pass turning operations. Adv. Mater. Research Vol 156–157, pp 956-960 (2011)
36. Costa, A., Celano, G., Fichera, S.: Optimization of multi-pass turning economies through a hybrid partile swarm optimization technique. Int. J. Adv. Manuf. Technol. 53, 421–433 (2011)
37. Aungkulanon, P., Chai-ead, N., Luangpaiboon, P.: Simulated manufacturing process improvement via particle swarm optimisation and firefly algorithms. In: Proceeding of the International Multi conference of Engineering and Computer Scientists, IMECS, vol. 2. Honkong, 16–18 March 2011
38. Hippalgaonkar, R.R., Shin, Y.C.: Robust optimisation of machining conditions with tool life and surface roughness uncertainties. Int. J. Prod. Res. **49**(13), 3963–3978
39. Singh, Y., Chauhan, P.: Analysing constrained machining conditions in turning operations by differential evolution. Adv. Mech. Eng. Appl. (AMEA) **2**(3), (2012). ISSN 2167-6380

40. Yildiz, A.R.: A comparative study of population-based optimization algorithms for turning operations. Information Sciences **210**, 81–88 (2012)
41. Aryanfar, A., Solimanpur, M.: Optimization of multi-pass turning operations using genetic algorithms. In: Proceedings of the 2012 International Conference on Industrial Engineering and Operations Management, pp. 1560–1568. Istanbul, Turkey, 3–6 July 2012
42. Khan, M.A., Kumar, A.S., Poomari, A.: A hybrid algorithm to optimize cutting parameter for machining GFRP composite using alumina cutting tools. Int. J. Adv. Manuf. Technol. **59**, 1047–1056 (2012)
43. Rao, R.V., Kalyankar, V.D.: Parameter optimization of machining processes using a new optimization. Algorithm. Mater. Manuf. Process. **27**, 978–985 (2012)
44. Lee, Y.Z., Ponnambalam, S.G.: Optimisation of multipass turning operations using PSO and GA-AIS algorithms. Int. J. Prod. Res. **50**(22) 6499–6518 (2012)
45. Xie, S., Guo, Y.: Optimisation of machining parameters in multi-pass turnings using ant colony optimisations. Int. J. Mach. Machinability Mater. **11**(2), 204–220 (2012)
46. Lu, K., Jing, M., Zhang, X., Dong, G., Liu, H.: An effective optimization algorithm for multipass turning of flexible workpieces. J. Intell. Manuf. **26**(4), 831–840 (2013)
47. Ahilan, C., Kumanan, S., Sivakumaran, N., Dhas, J.E.R.: Modeling and prediction of machining quality in CNC turning process using intelligent hybrid decision-making tools. Appl. Soft Comput. **13**, 1543–1551 (2013)
48. Belloufi, A., Assas, M., Rezgui, I.: Optimization of turning operations by using a hybrid genetic algorithm with sequential quadratic programming. J. Appl. Res. Technol **11** 88–94 (2013)
49. Jabri, A., Barkany, A. E., Khalifi, A.E.: Multi-objective optimization using genetic algorithms of multi-pass turning process. Engineering **5**, 601–610 (2013)
50. Rao, R.V., Kalyankar, V.D.: Multi-pass turning process parameter optimization using teaching–learning-based optimization algorithm. Scientia Iranica E, **20**(3), 967–974 (2013)
51. Yildiz, A.R.: Hybrid Taguchi-differential evolution algorithm for optimization of multi-pass turning operations. Appl. Soft Comput. **13**, 1433–1439 (2013)
52. Yildiz, A.R.: Optimization of cutting parameters in multi-pass turning using artificial bee colony-based approach. Inf. Sci. **220**, 399–407 (2013)
53. Xie, S., Pan, L.: Optimization of machining parameters for parallel turnings using estimation of distribution algorithms. Adv. Mater. Res. **753–755**, 1192--1195 (2013)
54. Yildiz, A.R.: Optimization of multi-pass turning operations using hybrid teaching learning-based approach. Int. J. Adv. Manuf. Technol. **66**, 1319–1326 (2013)
55. Singh, Y., Chauhan, P.: Selection of optimal machining conditions in multi-pass turning operations using real coded genetic algorithm. Int. J. of Appl. Math. Mech. **10**(4), 73–83 (2014)
56. Chauhan, P., Pant, M., Deep, K.: Parameter optimization of multi-pass turning using chaotic PSO. Int. J. Mach. Learn. Cybernet. **6**(3) 385–397 (2014)
57. Belloufi, A., Assas, M., Rezgui, I.: Intelligent selection of machining parameters in multipass turnings using firefly algorithm. Model. Simul. Eng. **2014**, 6, Article ID 592627 (2014)
58. Hrelja, M., Klancnik, S., Balic, J., Brezocnik, M.: Modelling of a turning process using the gravitational search algorithm. Int. J. Simul. Model. **13**(1), 30–41 (2014)
59. Hrelja, M., Klancnik, S., Irgolic, T., Paulic, M., Balic, J., Brezocnik, M.: Turning parameters optimization using particle swarm optimization. Procedia Eng. **69**, 670--677 (2014)
60. Acayaba, G.M.A., Escalona, P.M.: Prediction of surface roughness in low speed turning of AISI316 austenitic stainless steel. CIRP J. Manuf. Sci. Technol. **11**, 62–67 (2015)
61. Gayatri, R., Baskar, N.: Evaluating process parameters of multi-pass turning process using hybrid genetic simulated swarm algorithm. J. Adv. Manuf. Syst. **14**(4), 215–233 (2015)
62. Gayatri, R., Baskar, N.: Performance analysis of non-traditional algorithmic parameters in machining operation. Int. J. Adv. Manuf. Technol. **77**, 443–460 (2015)
63. Kubler, F., Bohner, J., Steinhilper, R.: Resource efficiency optimization of manufacturing processes using evolutionary computation: a turning case. Procedia CIRP **29**, 822–827 (2015)
64. Lin, W., Yu, D.Y., Wang, S., Chaoyong Z., Sanqiang Z., Huiyu T., Min L., Shengqiang L.: Multi-objective teaching–learning-based optimization algorithm for reducing carbon emissions and operation time in turning operations. Eng. Optim. **47**(7) 994–1007 (2015)

65. Chauhan, P., Pant, M., Deep, K.: Parameter optimization of multi-pass turning using chaotic. Int. J. Mach. Learn. Cyber. **6**, 319–337 (2015)
66. Zhigang J., Fan Z., Hua Z., Yan W., John W. Sutherland: optimization of machining parameters considering minimum cutting fluid consumption. J. Clean. Prod. **108**, 183–191 (2015)
67. Mellal, M.A., Williams, E.J.: Cuckoo optimization algorithm for unit production cost in multi-pass turning operations. Int. J. Adv. Manuf. Technol. **76**(1), 647–656 (2016)
68. Lu, C., Gao, L., Li, X., Chen, P.: Energy-efficient multi-pass turning operation using multi-objective backtracking search algorithm. J. Clean. Prod. **137**, 1516–1531 (2016)
69. Park, H.-S., Nguyen, T.-T., Dang, X.-P.: Multi-objective optimization of turning process of hardened material for energy efficiency. Int. J. Precision Eng. Manuf. **17**(12) 1623–1631 (2016)

Image Captioning-Based Image Search Engine: An Alternative to Retrieval by Metadata

Sethurathienam Iyer, Shubham Chaturvedi and Tirtharaj Dash

Abstract Image retrieval is an integral part of many different search engines. Search based on metadata of the image has been a primary approach in the process of image retrieval. In this work, we implement a search engine for better quality image retrieval using query image. Our implementation uses elastic search for indexing of the available images in the server and intermediate captioning mechanism for both search and retrieval process. The image captioning has been carried out using VGG16 Convolutional Neural Network. The implemented engine has been implemented and tested using the popular benchmark dataset called Flickr-8k dataset. The retrieved image quality demonstrated promising performance and suggests that an intermediate captioning-based image search could be an alternative to metadata-based search engines.

Keywords Image captioning · Image search · Elastic search · Tensorflow
Convolutional neural network

1 Introduction

Image retrieval in a search engine can be carried out using (a) metadata, (b) similar example, or (c) a hybrid of the two approaches. Search by metadata relies on manual annotations or text that embed the image on a Web document. It does not examine the content of the image. Search by example relies solely on the contents of the image. The image is analyzed, quantified, and stored in a database to retrieve similar images. The hybrid approach takes into account both search by metadata as well as the contents of the images. The most popular example of this would be Google

S. Iyer · S. Chaturvedi · T. Dash (✉)
Department of Computer Science and Information Systems,
Birla Institute of Technology and Science, Pilani, K. K. Birla Goa Campus,
Sancoale 403726, Goa, India
e-mail: tirtharaj@goa.bits-pilani.ac.in

© Springer Nature Singapore Pte Ltd. 2019
J. C. Bansal et al. (eds.), *Soft Computing for Problem Solving*, Advances in Intelligent
Systems and Computing 817, https://doi.org/10.1007/978-981-13-1595-4_14

Image Search.[1] A majority of image retrieval process works by direct or indirect use of metadata of the image [1]. Therefore, the ability to automatically describe the contents of the image can have significant practical implications because there are over billion images on the Web and it could significantly improve the efficiency of accessing the information they contain [2]. The problem of automatically associating images with complete sentences—the description of the image—has got a lot of attention and most of the researchers who worked on this problem used deep learning [3]. Deep neural networks typically scale well with high-dimensional data and use such a system as an intermediate tool during the image retrieval process would make the system accurate and efficient. In this work, we aim to build such an image search engine by using available highly scalable libraries.

The principal process pipeline of the image retrieval involves the following sequential steps: (a) image description, typically an algorithm to define the image, (b) extracting features and indexing, (c) defining the similarity metric, (d) performing the search using the defined similarity metric. In our present work, we focus on the use of image captions as the image description for the image retrieval process. The underlying assumption is that the images with similar captions are relevant to each other. However, it should be noted that the quality of the retrieval depends on the quality of the captioning engine.

The rest of the paper is organized into adequate sections as follows. Section 2 describes the methodology used in this work, and Sect. 3 briefly explains the system setup for implementation and evaluation. The results are presented in Sect. 4. Relevant works which are directly or indirectly related to this work are listed in Sect. 5. The paper has been concluded in Sect. 6.

2 Methodology: Image Retrieval Process Workflow

A multimodal recurrent neural network (RNN) has been used that describes images with sentences (caption). It is trained using Joint Image-Text Feature Learning mechanism. We use convolutional neural network (CNN) as image encoder by first training it for image classification task and using the last hidden layer as the input to the RNN decoder which generates the caption [4]. In order to reduce the additional training time, the pre-trained versions of CNN (VGG16 [5]) and RNN models have been used which are readily available in Keras [6]. The implemented image retrieval system in this work uses an open-source package called the elastic search for indexing images and captions. The overall process flow diagram has been presented in Fig. 1.

The fundamental motivation behind the use of elastic search in this work remains with the fact that it provides a lot of merits over other conventional search techniques. Elastic search has been tested to be very efficient with regard to database retrieval time and is useful especially for searching a large number of targeted objects (unstructured text) with the available phrase. The first and crucial step in elastic search involves

---

[1]https://images.google.com.

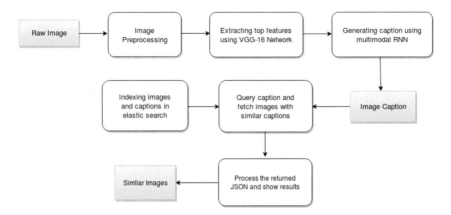

Fig. 1 Process flow diagram of our implemented image retrieval system (User interpretable steps are shown in solid color.). We refer the steps in the diagrams as processes; Process 0: Indexing images and captions in elastic search, Process 1: Image processing, Process 2: Feature extraction using VGG16, Process 3: Caption generation using RNN, Process 4: Query for relevant caption, Process 5: Showing results in user interpretable format

indexing of the documents in the server in which the document as an object is transmitted as a simple JSON object. One could visit elastic search site[2] for further information regarding elastic search and its architecture. For readers' understanding, we briefly explain the relevance scoring mechanism in elastic search in the following subsection.

2.1 Relevance Scoring in Elastic Search

In our case, the feature would be the caption of the image which is an English sentence. Defining similarity metric among sentences or between a sentence and a document is an information retrieval task. For the latter kind of retrieval, elastic search uses vector space model that provides a way of comparing a multi-term query against a document. The output is a single score that represents how well the document matches with the query. In order to do this, the model represents both the document and the query as vectors. In our implementation, we convert the multi-term query to an array of numbers by using term frequency-inverse document frequency (tf-idf), which quantifies the importance of a term in the corpus [7]. The tf-idf measure takes the following factors into account: the term frequency, the inverse document frequency, and the field-length Norm [8]. Term frequency (tf) is dependent on term (t) and the document (d) and is calculated as follows

---

[2]https://www.elastic.co/products/elasticsearch.

$$tf\,(t, d) = \sqrt{frequency} \tag{1}$$

Inverse document frequency (*idf*) is dependent on total number of documents (*numDocs*) in the corpus and number of documents which contains the desired term (*docFreq*) and is calculated as shown in (2).

$$idf\,(t) = 1 + \log\left(\frac{numDocs}{docFreq + 1}\right) \tag{2}$$

Field length norm is the inverse square root of number of terms in a field (*fieldFreq*) as,

$$norm(d) = \frac{1}{\sqrt{fieldFreq}} \tag{3}$$

Elastic search combines the above three measures to compute the relevance score and calculates the weight of a single term in a particular document.

2.2 Steps

2.2.1 Process 0: Indexing of Images and Captions

Indexing images in elastic search server are done by sending HTTP POST request. We encode the data path in our local system and its caption in JSON format and send them to the elastic search server so that it indexes it and becomes available for search.

2.2.2 Process 1, 2, and 3 (Image Caption Generation)

The task of generating captions from images has been studied in detail over these past few years [4]. We used VGG16 [5] to extract the features from the image which is passed as input to the RNN—an LSTM generator—that uses 8197 unique words. Finally, the caption for the image is generated using these words. The process of RNN training could be briefly explained as follows. The multimodal[3] RNN, in this work, takes the image pixels I and a sequence of input vectors $(x_1, x_2, \ldots, x_t)$ and produces a sequence of hidden vectors $(h_1, h_2, \ldots, h_t)$ and output vectors $(y_1, y_2, \ldots, y_t)$ by iterating the following recurrence relation for $t = 1, \ldots, T$, where T is the amount of time taken to generate the caption.

$$b_v = W_{hi}[CNN_{\theta_c}(I)] \tag{4}$$

---

[3]The input is image and output is a set of words.

$$h_t = f(W_{hx}x_t + W_{hh}h_{t-1} + b_h + \mathbb{1}(t = 1) \odot b_v) \tag{5}$$

$$y_t = Softmax(W_{oh}h_t + b_t) \tag{6}$$

In the equations above, W_{hi}, W_{hx}, W_{hh}, W_{oh}, x_i, b_o, b_h are learnable parameters and $CNN_{\theta_c}(I)$ is the last hidden layer's output of the VGG16 CNN.

Initially, we set $h_0 = \mathbf{0}$, x_1 to be a special START vector and the desired label y_1 to be the first word in sequence. Analogously, we set x_2 to be the word vector of x_1 and we expect the network to predict the second word until it encounters the last word; the target label is set to a special END token. The objective is to maximize the log probability associated with the target labels using softmax transfer function with cross entropy. To predict a sentence, we first compute the image representation b_v, and set $h_0 = 0$, x_1 to be the special START vector, and then we compute distribution over first word y_1. We pick the word with maximum log probability in y_1 and set its embedding vector as x_2 and repeat this process till END token is generated.

2.2.3 Process 4, 5

The captions generated in after the process 3 are sent to elastic search server to query for the image path, its captions, and the retrieval score.

3 Implementation and Evaluation

The implementation is carried out using Python installed on a system with 16 GB memory and i7 processor and 2 GB NVIDIA GTX GeForce 965 m. We use Keras [6] that contains inbuilt VGG16 implementation for the image feature extraction. The py-elasticsearch—a low-level Python package—has been used to interact with the elastic search server. In this work, for the preliminary implementation and testing, we used the popular Flickr-8k dataset [9].

It has been mentioned that the quality of captioning directly influence the quality of retrieval. For measuring the quality of captioning, we used BLEU score [10] that results in a score in the range [0, 1], where 1 represents the best quality and 0 represents the lowest quality of captioning. Similarly, retrieval score has been calculated by using the measures described in the Sect. 2.1. Further, we compute the precision, recall, and F-measure to measure the quality of retrieval of the image search engine. These performance metrics are being discussed below. In the following equations, *TP* denotes true-positive image, *FP* denotes false-positive image, *TN* denotes true-negative image, and *FN* denotes false-negative image. A positive image refers to an image that contains the intended subject of the interest. A negative image is an image that does not contain the intended subject of interest. F-measure is a trade-off between precision and recall. It should be noted that higher values of precision, recall, and F-measure are preferred.

$$Precision = \frac{TP}{TP + FP} \qquad (7)$$

$$Recall = \frac{TP}{TP + FN} \qquad (8)$$

$$\text{F-measure} = \frac{(1 + \beta^2) \times Recall \times Precision}{\beta^2 \times Recall + Precision} \qquad (9)$$

where β dictates the importance of precision versus recall. We will be using balanced F-measure with $\beta = 1$ also called the $F1$ score.

4 Results

We evaluated the trained RNN on the Flickr-8k dataset which has four different manual captions as ground truth. We have calculated the BLEU scores against all the available captions and chose the best caption among the resulting captions using the highest BLEU score. The best caption for the image was indexed in the server using the elastic search and the retrieval scores were calculated. The higher the score, the higher is the similarity between two captions and the images. Few images with higher scores were retrieved for the query image. The number of images to be retrieved for the query image could be either specified a priori or can be made dynamic. In the subsections, we show the performance of the image search engine in two different ways that are based on the following facts. First, the query image and the images to be retrieved have some degree of similarity among them; second, activity of the subject presents in the query image.

4.1 Similarity of Query Image and Images to Be Retrieved

In our test simulations, we show the best two retrieved images for the tested images. We present the results of the three query images (a) a surfing image, (b) kid playing football, and (c) a person playing tennis. Table 1 shows the query image along with the retrieved images using our implementation. It could be seen that the retrieved images are quite similar to the query image, thus suggesting a good quality of image retrieval.

Table 2 depicts the retrieval scores for each of the query images shown in Table 1. The second best-retrieved result for the third query image in which a person is playing tennis has a very smaller score as compared to the first retrieved image probably because of the fact that the relative size of the person playing tennis in the second image is very small with regard to the whole image. Moreover, the person standing

Table 1 Query image with two retrieved images; 1. Surfing, 2. Kid with football, 3. A man playing tennis (These images are available in Flickr-8k datasets [9] which has been used in our present work.)

Query image	Retrieved images (best 2)

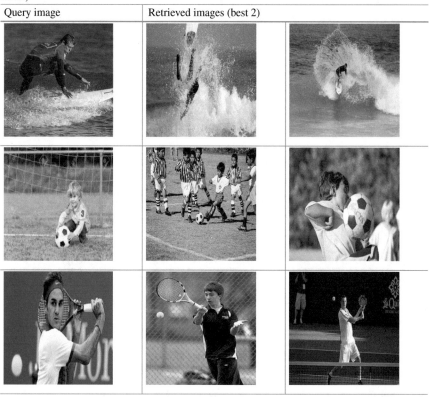

Table 2 Similarity score for the images retrieved for the query image in Table 1

Query image description	Retrieval scores	
	Image 1	Image 2
A surfing image	16.0818	14.2265
Kid playing football	13.6676	11.9925
A person playing tennis	26.3678	18.0666

behind the player in the second retrieved image also does add as a noisy term during the captioning process.

Table 3 Average quality of the image retrieval based on activities when the number of images to be retrieved is fixed at 20 for each query in 10 test images. Specific activities like running, jumping, drinking, and eating has high accuracy while vague activities like playing, looking has low precision

Activity	Precision	Recall	F-measure
Running	0.90	0.03	0.06
Eating	0.85	0.11	0.19
Playing	0.65	0.02	0.04
Jumping	0.80	0.03	0.06
Drinking	0.80	0.31	0.45
Looking	0.65	0.04	0.08

Table 4 Average quality of the image retrieval based on activities when number of images to be retrieved is made dynamic for each query in 10 test images

Activity	Retrieval count	Precision	Recall	F-measure
Running	417	0.93	0.54	0.68
Eating	21	0.89	0.13	0.23
Playing	192	0.85	0.23	0.36
Jumping	147	0.94	0.25	0.06
Drinking	17	0.84	0.32	0.45
Looking	62	0.83	0.20	0.08

4.2 Activity of the Subject Present in the Image

Main components of an image are the subjects present in the image, and the activity being done by the subject. In this section, the performance of the image captioning engine in detecting top six activities and subjects is discussed.

The top six activities present in Flickr-8k datasets are running, eating, playing, jumping, drinking, and looking. A well-known Python library called NLTK library [11] was used to parse the activity from the image descriptions. The performance of the image captioning engine was checked for each of these activities with 10 test images. First, the performance is checked when 20 search results are returned every time. The results are described in Table 3.

The probable reason for the recall being low is that the number of retrieved images is fixed at 20. In our second case for the activity-based image retrieval, the images are retrieved as long as the search score is at least the half of the first image's search score (kindly refer Table 2). The elastic search returns the results in sorted order, so the first image has the highest search score. We can call the retrieval as dynamic because the number of the images to be retrieved is not fixed. Table 4 provides the achieved performance of our search engine.

Furthermore, we produce the results obtained by our image search engine based on subject present in the image. The top six subjects present in the considered dataset

Table 5 Average quality of the image retrieval based on subjects when number of images to be retrieved is fixed at 20 for each query in 10 test images

Subject	Precision	Recall	F-measure
Dog	0.95	0.03	0.06
Boy	0.80	0.01	0.02
Girl	0.70	0.02	0.04
Man	0.90	0.01	0.02
Woman	0.90	0.04	0.08
Shirt	0.95	0.04	0.08

Table 6 Average quality of the image retrieval based on subjects when number of images to be retrieved is made dynamic for each query in 10 test images

Subject	Retrieval count	Precision	Recall	F-measure
Dog	624	0.89	0.33	0.49
Boy	129	0.82	0.12	0.21
Girl	111	0.83	0.12	0.21
Man	178	0.89	0.08	0.15
Woman	79	0.92	0.16	0.27
Shirt	93	0.76	0.19	0.3

are dog, boy, girl, man, woman, and shirt. Again, NLTK library [11] was used to parse the main subject present in the image. Similar to activity-based image retrieval, we retrieve the images by first fixing the retrieval count to 20 and then making it dynamic. Tables 5 and 6 report these two cases of image retrieval by using our search engine.

5 Related Work

Image retrieval has been an interesting topic in the field of multimedia information retrieval and machine learning. It started attracting attention during the 1999–2000 (refer the work of Rui et al. [12] and Smeulders et al. [13]). Since then, many different image retrieval schemes have been proposed which are based on the query text, image properties such as texture, color etc. Few examples include [14, 15]. Many different search engines have been built with different techniques for best quality retrieval such as [16, 17].

6 Conclusion and Future Directions

In this work, we implement an alternative search engine for retrieval of images from the server using an example image. Our implementation uses image captioning mechanism as an intermediate process. The retrieval is carried out by using the captions generated for the query image and the indexed captions for the stored images on the server. The resulting implementation has been tested and simulated with Flickr-8k dataset. Our implementation uses the emerging elastic search mechanism for the indexing of the available images. Although the implementation has not been tested for larger datasets, the present retrieval quality with smaller dataset like Flickr-8k still demonstrates promising results. One could look at the following research directions:

- Elastic search provides higher scalability. Therefore, integrating elastic search with Hadoop would scale the system well for massive datasets. The implementation of our present search engine is scalable as both Tensorflow and elastic search integrate well with big data ecosystems.
- Improvements could be done, by considering capsule networks [18] instead of convolutional neural network or by using attention mechanisms for image feature extraction. This would allow the RNN to discover more latent features of the image and could enhance the quality of the image caption engine.

References

1. Yee, K.P., Swearingen, K., Li, K., Hearst, M.: Faceted metadata for image search and browsing. In: Proceedings of the SIGCHI conference on Human factors in computing systems. pp. 401–408. ACM (2003)
2. Kherfi, M.L., Ziou, D., Bernardi, A.: Image retrieval from the world wide web: issues, techniques, and systems. ACM Comput. Surv. (CSUR) **36**(1), 35–67 (2004)
3. Xu, K., Ba, J., Kiros, R., Cho, K., Courville, A., Salakhudinov, R., Zemel, R., Bengio, Y.: Show, attend and tell: neural image caption generation with visual attention. In: International Conference on Machine Learning. pp. 2048–2057 (2015)
4. Karpathy, A., Fei-Fei, L.: Deep visual-semantic alignments for generating image descriptions. In: Proceedings of the IEEE Conference on Computer Vision and Pattern Recognition. pp. 3128–3137 (2015)
5. Simonyan, K., Zisserman, A.: Very deep convolutional networks for large-scale image recognition (2014). arXiv:1409.1556
6. Chollet, F.: Keras (2015)
7. Ramos, J., et al.: Using TF-IDF to determine word relevance in document queries. In: Proceedings of the first instructional conference on machine learning (2003)
8. Manning, C.D., Raghavan, P., Schütze, H.: Scoring, term weighting and the vector space model. Introd. Inf. Retr. **100**, 2–4 (2008)
9. Rashtchian, C., Young, P., Hodosh, M., Hockenmaier, J.: Collecting image annotations using amazon's mechanical turk. In: Proceedings of the NAACL HLT 2010 Workshop on Creating Speech and Language Data with Amazon's Mechanical Turk. pp. 139–147. Association for Computational Linguistics (2010)
10. Papineni, K., Roukos, S., Ward, T., Zhu, W.J.: Bleu: a method for automatic evaluation of machine translation. In: Proceedings of the 40th annual meeting on association for computational linguistics. pp. 311–318. Association for Computational Linguistics (2002)

11. Bird, S., Klein, E., Loper, E.: Natural language processing with Python: analyzing text with the natural language toolkit. O'Reilly Media, Inc., (2009)
12. Rui, Y., Huang, T.S., Chang, S.F.: Image retrieval: Current techniques, promising directions, and open issues. J. Vis. Commun. Image Represent. **10**(1), 39–62 (1999)
13. Smeulders, A.W., Worring, M., Santini, S., Gupta, A., Jain, R.: Content-based image retrieval at the end of the early years. IEEE Trans. Pattern Anal. Mach. Intell. **22**(12), 1349–1380 (2000)
14. Vipparthi, S.K., Nagar, S.: Expert image retrieval system using directional local motif XoR patterns. Expert Syst. Appl. **41**(17), 8016–8026 (2014)
15. Guo, J.M., Prasetyo, H., Wang, N.J.: Effective image retrieval system using dot-diffused block truncation coding features. IEEE Trans. Multimed. **17**(9), 1576–1590 (2015)
16. Markonis, D., Schaer, R., de Herrera, A.G.S., Müller, H.: The parallel distributed image search engine (paradise) (2017). arXiv:1701.05596
17. Gordo, A., Almazán, J., Revaud, J., Larlus, D.: Deep image retrieval: learning global representations for image search. In: European Conference on Computer Vision. pp. 241–257. Springer (2016)
18. Sabour, S., Frosst, N., Hinton, G.E.: Dynamic routing between capsules (2017). arXiv:1710.09829

Minimization of Makespan for Parallel Machines Using PSO to Enhance Caching of MSA-Based Multi-query Processes

Soniya Lalwani, Harish Sharma, Abhay Verma and Kusum Deep

Abstract This paper proposes two-fold approach that works on minimizing makespan for parallel machines and performs caching operation to reduce the load on data servers during multi-query process. Makespan criterion is an NP-hard problem, which assigns jobs to machines so as to minimize the time by which the last job gets finished. Discrete version of particle swarm optimization (PSO) is implemented to schedule the processing of parallel machines, connected through message passing interface (MPI). Data caching on B-tree is performed on multiple processors for handling subsequent and concurrent queries. The multi-query process addressed here is multiple sequence alignment (MSA), which is one of the most challenging and prominent areas of bioinformatics with high computational cost. State-of-the-art alignment algorithm ClustalW is used for obtaining MSA of highly complex sequence sets. The results show the improvements in process time.

Keywords B-tree · Data caching · Multiple sequence alignment · Makespan
ClustalW · Message passing interface · Particle swarm optimization

1 Introduction

The formulation of makespan problem for the identical parallel machines scheduling includes n independent jobs that require being processed on m identical machines. Each of the jobs can be processed on any of the parallel machines, where the required

S. Lalwani (✉) · H. Sharma · A. Verma
Department of Computer Science & Engineering, Rajasthan Technical University,
Kota, Rajasthan, India
e-mail: slalwani.math@gmail.com

H. Sharma
e-mail: hsharma@rtu.ac.in

K. Deep
Department of Mathematics, Indian Institute of Technology, Roorkee,
Roorkee 247667, Uttarakhand, India
e-mail: kusumdeep@gmail.com

© Springer Nature Singapore Pte Ltd. 2019
J. C. Bansal et al. (eds.), *Soft Computing for Problem Solving*, Advances in Intelligent
Systems and Computing 817, https://doi.org/10.1007/978-981-13-1595-4_15

processing time for job i at any machine is given by t_i. Only one job can be processed at a time on each machine formulated as:

$$\sum_{j=1}^{m} p_{ij} = 1 \quad \text{for } i = 1, 2, \ldots, n \tag{1}$$

here, $p_{ij}=1$, if the ith job is assigned to the jth machine and 0 otherwise. The objective is to minimize the makespan by optimal assignments of jobs on the machines, formulated as:

$$min\{M\}$$

$$M - \sum_{i=1}^{n} t_i \, p_{ij} > 0, \text{for } j = 1, 2, \ldots, m \tag{2}$$

$$\text{where, } M = Max\{C_j\}, \text{for } j = 1, 2, \ldots, m$$

here, M denotes the makespan and C_j is the completion time of the last job on the machine j. Maximum of the completion times of the last jobs at all the machines is M. Due to the NP-hard nature of the makespan problem [1], discrete version of particle swarm optimization (PSO) is implemented in order to obtain optimal schedule. The scheduling is done so as to perform caching operation of multi-query process of multiple sequence alignment (MSA).

Due to large-scale genomic research and next-generation sequencing, massive amount of sequence data are generated. For analysis and processing of these data, MSA is an essential tool. MSA plays crucial role in structure prediction, determining conserved domains, pattern identification, phylogenetic analysis, detecting DNA regularity elements, function prediction, and several more [2]. The wide range of applications of MSA is emerging into the advancement of various tools; hence, several queries get generated over data servers to access datasets. During this process, a few queries regarding MSA will contain common sequences, submitted by different users. If the alignment of these common sequences can be saved in either way, it will result in reduced load on data server with improved performance. For this purpose, data caching could serve itself as an efficient tool [3]. Here, data caching is implemented over B-tree, a balanced multi-way tree, which is suitable for managing large set of strings.

In this paper, discrete PSO is employed to optimize the job scheduling on parallel machines connected via message passing interface (MPI), so as to minimize the makespan. In a multi-query process, several queries are generated by different clients for accessing the database of sequences and/or to compare one or multiple known/new sequences. B-tree is employed for caching the computed alignment results of ClustalW [4].

The paper is classified as follows: Section 2 carries the details of discrete PSO, data caching, ClustalW, and MPI-based parallelization. Section 3 presents the MPI-based parallel implementation of B-tree for MSA with multi-query processes. Section 4 presents the experimental setup. Section 5 explains the results and discussions, followed by the conclusion in Sect. 6.

2 Problem Description

This section carries introduction of discrete PSO, ClustalW, and its alignment strategy, followed by basics of data caching with B-tree. ClustalW details include the elaboration of its progressive alignment strategy with all the three steps followed to form alignment. This section is further elaborated with the description of MPI-based parallelization strategy.

2.1 Discrete Particle Swarm Optimization

PSO algorithm was derived by Kennedy and Eberhart in 1995 by simulating the behavior of a bird flock or fish school [5]. They move in different directions while communicating with each other, updating their positions and velocities for the better position that may provide optimal solution. The objective function to be minimized (maximized) is formulated as follows:

$$\min f(x) \quad \text{s.t.} \quad x \in S \subseteq R^D \tag{3}$$

where x is decision variable matrix, composed of m vectors defined as with dimension D, S is the feasible solution space of the problem. At tth iteration, the previous velocity $v_i(t)$ and position $x_i(t)$ are updated by

$$v^i(t+1) = wv^i(t) + c_1 r_1 (pbest^i(t) - x^i(t)) + c_2 r_2 (gbest(t) - x^i(t)) \tag{4}$$

$$\begin{aligned} x^i(t+1) &= x^i(t) + v^i(t+1) \\ \text{with } x^i(0) &\sim U(x_{\min}, x_{\max}) \end{aligned} \tag{5}$$

The velocity v_i lies between lower and upper bound, i.e., $[v_{min}, v_{max}]$, where $v_{min} = -v_{max}$; w is inertia weight lying between 0 and 1, the scaling factor over the previous velocity; c_1 and c_2 are cognitive and social acceleration coefficients, respectively; r_1 and r_2 are uniform random numbers in range [0 1]. Particles personal best $pbest$ at iteration $(t+1)$ and best of the positions, i.e., $gbest(t)$, are updated as follows:

$$pbest^i(t+1) = \begin{cases} pbest^i(t) \text{ if } f(x^i(t+1)) \geq f(pbest^i(t)) \\ x^i(t+1) \text{ if } f(x^i(t+1)) < f(pbest^i(t)) \end{cases} \tag{6}$$

The best of the positions, i.e., $gbest$ is updated as follows:

$$gbest(t) = x_k \in \left\{ pbest^1(t), pbest^2(t), \ldots, pbest^n(t) \right\} \tag{7}$$

where $f(x_k) = \min \left\{ f(pbest^1(t)), f(pbest^2(t)), \ldots, f(pbest^n(t)) \right\}$

The discrete version of PSO basically implements Eqs. (4) and (5) in a different way, i.e.,

$$v^i(t+1) = v^i(t) \oplus [r_1 \otimes (pbest^i(t) \ominus x^i(t)) \oplus r_2 \otimes (gbest(t) \ominus x^i(t))] \quad (8)$$

$$x^i(t+1) = x^i(t) \oplus v^i(t+1)$$
$$\text{with } x^i(0) \sim U(x_{\min}, x_{\max}) \tag{9}$$

In Eqs. (8) and (9), the mathematical operations for discrete variables are represented by different symbols, i.e., addition by $\oplus$, subtraction by $\ominus$, and multiplication by $\otimes$ inspired by [6]. Also, the inertia weight, cognitive, and social acceleration factors are not present in equations.

2.2 ClustalW

Multiple sequence alignment can be performed in several ways. The major classification of alignment algorithms is, namely progressive algorithms; exact algorithms; iterative algorithms and consistency-based algorithms. Progressive alignment algorithms construct MSA starting from two sequences. Then, the consensus of alignments is aligned with another sequence for expanded MSA. This process is repeated until all the sequences get aligned. The exact alignments are based on the Needleman–Wunsch algorithm, a dynamic backtracking algorithm. Although, these algorithms always produce optimal alignment, the time and space requirement gets grown exponentially with the increasing number of sequences. In all iterative algorithms, the process starts from a non-optimal alignment and optimization is employed to obtain optimal MSA. The iterations continue until they arrive at some stopping criteria. Consistency-based alignment algorithms first create all the possible pairwise alignments, so as to produce a library of alignments for each sequence pair. Hence, obtained library is then employed to assure consistency in the final MSA by creating position-specific substitution matrix [7].

ClustalW alignment strategy is based on progressive algorithm. The alignment process depicted by Fig. 1, explains the three-step process as follows:

Step 1: Dynamic programming-based pairwise alignment for creation of distance matrix: As shown in Fig. 1(i), the first step is aligning all the pairs of sequences separately by following dynamic programming [8, 9] using gap opening and extension penalties. Hence, the obtained alignment score is employed to calculate percent identities scores and then it is converted to distance score. This score is used to create the distance matrix between pairs of sequences.

Step 2: Construction of guide tree: The distance matrix score obtained from *step* 1 is used to compute the guide tree by using neighbor-joining (NJ) method [10] in *step* 2,

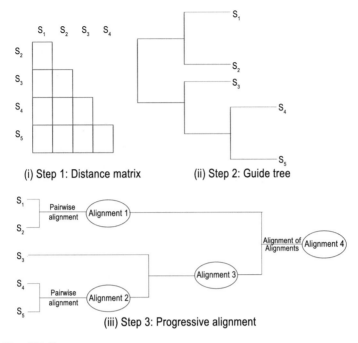

Fig. 1 ClustalW alignment steps

depicted by Fig. 1(ii). NJ method minimizes the sum of branch lengths. The branch length is proportional to the estimated divergence along each branch. The larger the divergence is, the lesser the similarity between sequences is that results in longer branch length.

Step 3: Progressive alignment in accordance with the branching order of guide tree: The process of progressive alignment starts from leaf to the root as shown in Fig. 1(iii). At each stage, the dynamic programming is implemented with score matrix and penalties, so as to align two alignments or sequences. Gaps from previous stage remain at their positions, and new gaps are inserted to align the alignments or sequences.

2.3 Data Caching

Caching is a technique of storing often used information or data in memory, so that whenever it is required next time, it is directly retrieved from the quickly accessed media rather than being generated by the application. B-tree is a dynamic high-performance data structure to arrange and manage massive datasets that are stored on pseudo-random access devices. The information is stored in the form of key-value pairs (k, v). Here, for the key k, v is the related information.

Data retrieval becomes very fast and hardly dependent on the size of the dataset, when assigned on B-trees because the index as well as the retrieval time grows logarithmically with reference to the size of the dataset. B-trees are implemented in many areas, like Web or library queries, banking, e-shopping, data warehouses, high-dimensional databases, geographic databases, spatial databases, multimedia dastabases, text retrieval systems, and the areas wherever highly parallel processing within the databases is needed [11, 12]. The properties due to which B-tree is desirable for several applications are as follows:

B-trees have high branching factor due to which, accessing the number of nodes remains low. Hence, even for large volumes, the height of B-trees remains less; load factor of all leaf nodes is 50%, except for the root node. Even for random keys, the average utilization is 69%; bounds on access costs are an essential requirement for applications such as database query engines; the root are required to be visited by every search and update process, but for most updates, it needs to be read-only. Thus, the root and maybe some of its children are read hot spots, and hence, they are always cached in main memory by the standard caching techniques. Therefore, an extremely fast and highly parallel processing is performed; B-trees offer guaranteed access cost in worst case also [13].

2.4 MPI-Based Parallelization

Message passing is a form of interprocess communication in which a physical copy of an object is sent to another process. The object contains some data, which can be of any type including commands, data, or signals. Message passing is in use since networking was born, the trend of parallel computing has kept it alive. Message passing is not a language or a library, but a model of parallel computing. There are various libraries to support this method of interprocess communication or can be implemented by the programmer quite easily otherwise. In this approach, data are considered as the driving unit of the whole system. MPI enables the system to transfer the data between the different devices thus progressing the processing. The difference between usual procedure calls and message passing approach is that for procedure calls the processes must be on the same machine, however in the case of message passing that does not need to be. MPI was earlier used in high-performance software and distributed servers, but now it is used in general-purpose computing as well. Its effectiveness comes from the fact that more than one process runs on machine asynchronously and use of dedicated machines for the respective operations. Its heterogeneous capability allows an application to communicate over wide ranges of networks and machines [14, 15].

3   Proposed Methodology

The output obtained after performing MSA from ClustalW has three kinds of inter-mediate outcomes, i.e., sets of pairwise aligned sequences and distance matrix; guide tree; and groups of aligned sequences. When recording the time taken by all the three steps, it was observed that pairwise alignment takes most of the time (discussed in Sect. 2.2). So the pairwise alignment results are cached in order to perform the query execution of new similar sequences faster.

Implementation of several variants of PSO for MSA has been a very popular approach [16, 17], but not been implemented in proposed way so far. Here, discrete version of PSO presented by Sect. 2.1 is implemented with objective of minimizing makespan presented by Eq. (2). The operations addition ($\oplus$) stand for crossover oper-ation between two particles, so as to interchange jobs between machines. The number of cut points is taken two here. Subtraction ($\ominus$) operation performs computation of the hamming distance between *pbest* (or *gbest*) and current position. Whenever the results get *pbest* (or *gbest*) and current positions equal, output as the job zero gets its job assigned by longest processing time (LPT) rule. The multiplication operation ($\otimes$) is basically based upon Hadamard product following the current assignment on the machine.

Figure 2 presents the MPI process, in which slaves send a query on the master computer before every pairwise alignment, so as to determine whether to store the alignment or fetch the cache results. For a fetch operation, in which the data are encapsulated in the byte stream, the master server creates a new thread for it, extracts

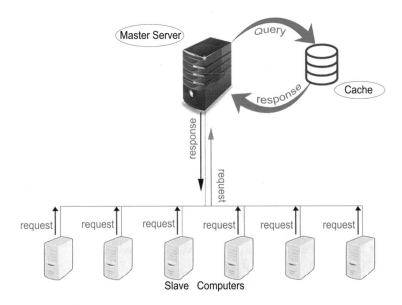

Fig. 2 MPI process in proposed work

the given sequence, then searches for it in the cache, if there is a match, then the score is returned, else -1 is returned.

Each sequence in the dataset is assigned a unique 32-bit number, because the sequence names in dataset may not be unique. To avoid hash collision, the mod of sequence length is also concatenated with the unique 32-bit identifier of that sequence. If score is returned from the server, then no calculation is done on the slave for that pair; otherwise, if -1 is the returned value, it performs the usual calculations and then sends the result to the server for future reference. This ensures that every pair is calculated not more than once on the network. The message passing approach works by sending a message in form of a byte stream, which can have two types of structure based on the type of query; the structure is identified by its first byte:

- if byte $1 = 1$ it is a FETCH query
  Structure: $1 \mid$ <hash(seq1) + hash(seq2) + seq1 $\cdot$ len + seq2 $\cdot$ len>
- if byte $1 = 2$, it is a STORE query
  Structure: $2 \mid score \mid$ <hash(seq1) + hash(seq2) + seq1 $\cdot$ len + seq2 $\cdot$ len>

The first bit is read, and then, the proper action is chosen. For a FETCH query, the sequence is searched through the B-tree, and if a match is found, then the score is sent, else -1 is sent. For the STORE query, the given score is saved.

Initially, time for each process is pre-estimated for each sequence set assigned to each machine. Then, PSO is run for optimal scheduling of parallel machines followed by performing MSA and its caching. Caching is performed over B-tree, one of the most efficient data structure for maintaining sorted data on disk with a balanced tree structure.

4 Experimental Setup

The experiments are carried out on nine general-purpose computers equipped with Intel(R) Core(TM) i7-4790 CPU @ 3.60 GHz running Ubuntu 16.04 on Linux cse37 4.8.0-56-generic with main memory of 4 GB (DIMM DDR3 Synchronous 1600 MHz) on each of them. For the first experiment, to know the role of cache, a single server is occupied. For the other entire experiments, we created our own MPI server with master–slave strategy, one master server with eight general-purpose computers as slaves, added in MPI-based parallelization.

A few changes in the ClustalW codes are done, so as to support data caching using B-tree. Changes in main.cpp, pairwise.cpp, and align function of ClustalW codes are made, for making them suitable to implement in MPI environment. The C++ codes of B-tree are taken from GitHub Repository.

218 protein sequence sets from BAliBASE 4 datasets [18] are used for the experiment. BAliBASE reference set 10 is tested that contains complex protein families. The size of the smallest set is 4 sequences, and of the largest set is 807 sequences.

The length of protein sequence varies from 19 amino acids to 4895 amino acids. The average sequence length is 460 amino acids.

5   Results and Discussions

This section presents the details of obtained results and their analysis supported by Figs. 3 and 4 and Table 1. Table 1 carries the details of the randomly selected sequences for testing, shown in Fig. 3. For each sequence set, L_{MIN} stands for length of the smallest sequence, L_{MAX} for length of the longest sequence, and L_{AVG} for average sequence length. Table 2 carries the details of the tested combinations of machines, jobs, and processing time that results in the mean performance and time consumed for each combination.

Figure 3 presents the requirement of caching. It can be observed by the non-cache results that the most time of the alignment is consumed in performing pairwise alignment. Hence, if the pairwise alignment could be cached, the execution of queries will be speeded up. Proposed work has taken this conclusion into consideration and implemented B-tree for caching pairwise alignment results of ClustalW. The comparison of B-tree for pairwise alignment and multiple sequence alignment with non-caching approach shows that B-tree has made major difference in the time of performance for pairwise alignment; e.g., time has decreased to 1.70693s in B-tree for seq 9 from 96.6123s of non-cache. The time taken (s) by each approach is shown by data labels in the figure. Also, the implementation of discrete PSO for minimization of makespan has remarkably improved the process time. The time taken by B-tree

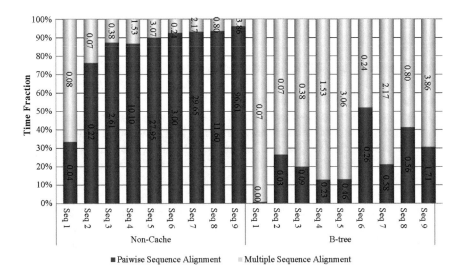

Fig. 3 Comparison between B-tree and Non-Cache approach

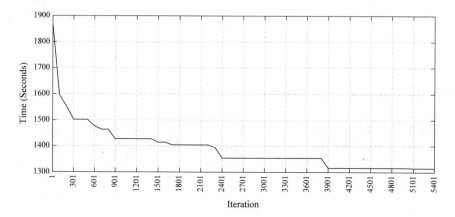

Fig. 4 Iterative improvement in makespan

Table 1 Details of randomly selected sequences related to Fig. 3

Sequence set	No. of Seqs	L_{MIN}	L_{MAX}	L_{AVG}
BBA0192	4	589	799	722.5
BBA0201	29	145	241	208.14
BBA0087	50	336	498	411.32
BBA0187	76	132	1482	563.65
BBA0125	97	181	1112	760.67
BBA0092	105	158	348	200.16
BBA0063	129	121	1001	424.30
BBA0058	143	49	496	264.51
BBA0078	199	170	819	631.47

without makespan was 1964s that has improved to 1320s by the implementation of discrete PSO-based optimization. Figure 4 shows the iterative improvement in makespan with the implementation of discrete PSO.

Table 2 presents the experimental frameworks with the variables responsible for performance, i.e., number of machines; number of jobs; and the distribution of generated job-processing time per job. For each combination, the time taken and the mean performance are also presented. Four combinations of the number of machines are tested, i.e., 3, 4, 5, 6; similarly, three combinations of the number of jobs are tested, i.e., 60, 90, 150 with two uniformly distributed processing time, i.e., 100–200 and 200–400 s/job. Hence, a total of 24 experiment runs, with 20 replications (on randomly generated processing time), resulted a total of 480 runs. The mean performance is the average of ratios obtained (for a total of 20 runs), where each ratio is the division of the experimental value of makespan on the amount of lower bound. Lower bound is calculated by the following equation:

Table 2 Jobs, Machines, and Processing time combinations for time and mean performance

Number of machines	Number of jobs	Processing time (s)	Time (s)	Mean performance
3	60	100–200	0.67	1.0006
	90		0.94	1.0004
	150		1.556	1.0003
	60	200–400	0.641	1.0003
	90		0.992	1.0002
	150		1.531	1.00033
4	60	100–200	0.709	1.0037
	90		1.023	1.0030
	150		1.659	1.0024
	60	200–400	0.702	1.0029
	90		1.009	1.0022
	150		1.639	1.00247
5	60	100–200	0.739	1.0169
	90		1.059	1.0087
	150		1.746	1.0068
	60	200–400	0.776	1.0113
	90		1.093	1.0094
	150		1.765	1.0056
8	60	100–200	0.907	1.0496
	90		1.275	1.0380
	150		2.047	1.0306
	60	200–400	0.822	1.0507
	90		1.301	1.0393
	150		2.049	1.0223

$$LB = \max\left[\frac{\sum_{i=1}^{n} t_i}{n} \;\middle|\; \max_{i=1,2,\ldots,n} t_i\right] \tag{10}$$

The average LB of the scheduling is around 1300. The following inferences can be drawn from the experiment conducted on different sets of machines and the number of jobs:

- The scheduling time optimized by PSO increases as soon as the number of jobs and the number of machines increase;
- For any number of machines, performance improves as the jobs (to be scheduled) increase;
- Hence, grouping more jobs together will give better results, but, then PSO will require more time to optimize.

Here, the question arises "Is the PSO's time scheduling worth taken into consideration?" The answer is explained with an example of eight machines with three sets of t_i, where processing time generated lies between 200–400s.

Case a: For 60 jobs: performance of 1.05073 (p_a) in time 0.822s (t_a)
Case b: For 90 jobs: performance of 1.03798 (p_b) in time 1.301s (t_b)
Case c: For 150 jobs: performance of 1.02227 (p_c) in time 2.049s (t_c)

Here, performance stands for the ratio of the makespan time of the specific case on the LB. Since, scheduling can be parallelized, time taken to schedule 150 jobs in groups of 60 and 90 is $t_d = \max(t_a, t_b)$, and $t_d = 1.301$ (case d). Now, the performance of case d will be $p_d = \max(p_a, p_b) = 1.05073$.

Therefore, makespan of case d = $1.05073 \times 1300 = 1365.949$; makespan for case c = $1.02227 \times 1300 = 1328.951$; improvement of makespan in case c is 36.998; time invested in scheduling 150 jobs by PSO (case c) is $t_c - t_d = 2.049 - 1.301 = 0.748$s. Thus, at the cost of 0.748s, the time saved is 36.998s. Hence, PSO's time scheduling is remarkably worth to be taken into the consideration.

6 Conclusion

Proposed work employs discrete PSO for minimizing the makespan, in order to optimize the job scheduling. The experiments are performed on parallel machines connected through MPI, so as to perform caching operations to reduce the load on data servers during multi-query process. MSA is one of the most popular and challenging areas of bioinformatics. MSA servers get multiple queries generated for accessing database of the sequences for several applications. B-tree is employed for caching the computed pairwise alignment results of MSA performing software ClustalW. The computation results show remarkable improvement in processing time (with reference to the number of machines, jobs, and generated process time) and also the impact of caching.

Efficiency of an algorithm is determined by its response time and the performance. Hence, implementation of a better version of B-tree as well as a modified variant of discrete PSO with multi-objective optimization (taking minimization of machine idle time as the second objective) will be able to improve the performance more. Also, testing on larger sequences with other parallelization strategies for different alignment software can be done as further investigation work. Moreover, implementation of other nature-inspired algorithms can provide a better comparison between performances.

Acknowledgements The first author (S.L.) gratefully acknowledges Science & Engineering Research Board, DST, Government of India for the fellowship (PDF/2016/000008). We are thankful to Dr. Krishna Mohan from BISR, Jaipur, India, for his valuable suggestions throughout the work.

References

1. Robert, C.: Edgar and serafim batzoglou, multiple sequence alignment. Curr. Opin. Struct. Biol. **16**(3), 368–373 (2006)
2. Xiao Lin L., Yu Peng L., Yu W.: Minimising makespan on a batch processing machine using heuristics improved by an enumeration scheme. Int. J. Prod. Res. **55**(1), 176–186 (2017)
3. Schatz, M.C., Trapnell, C., Delcher, A.L., Varshney, A.: High-throughput sequence alignment using graphics processing units. BMC Bioinform. **8**(1), 474 (2007)
4. Julie, D.T., Desmond, G.H., Toby, J.G.: CLUSTAL W: improving the sensitivity of progressive multiple sequence alignment through sequence weighting, position-specific gap penalties and weight matrix choice. Nucleic Acids Res. **22**, 4673–4680 (1994)
5. Kennedy, J.F., Eberhart. R.C.: Particle swarm optimization. In: Proceedings of IEEE International Conference on Neural Networks, pp. 1942–1948. Piscataway, NJ (1995)
6. Husseinzadeh, A.K., Karimi, B.: A discrete particle swarm optimization algorithm for scheduling parallel machines. Comput. Ind. Eng. **56**, 216–223 (2009)
7. Notredame, C.: Recent progress in multiple sequence alignment: a survey. Pharmacogenomics **3**(1), 131–144 (2002)
8. Myers, E.W., Miller, W.: Optimal alignments in linear space. Bioinformatics **4**(1), 11–17 (1988)
9. Smith, T.F., Waterman, M.S., Walter, M.F.: Comparative biosequence metrics. J. Mol. Evol. **18**(1), 38–46 (1981)
10. Saitou, N., Nei, M.: The neighbor-joining method: a new method for reconstructing phylogenetic trees. Mol. Biol. Evol. **4**, 406–425 (1987)
11. Askitis, N., Zobel, J.: B-tries for disk-based string management. VLDB J. Int. J. Very Larg. Data Bases **18**(1), 157–179 (2009)
12. Catalyurek, U., Ferreira, R., Kurc, T., Saltz, J., Stahlberg, E.: Improving performance of multiple sequence alignment analysis in multi-client environments. In: Proceedings of the 16th International Parallel and Distributed Processing Symposium. Washigton, DC, USA (2001)
13. Weikum, G., Vossen, G.: Transactional Information Systems. Morgan Kaufmann Publishers (2002)
14. Gropp, W., Lusk, E., Skjellum, A.: Using MPI: portable parallel programming with the message-passing interface, vol. 1. MIT press (1999)
15. Adamo, J.-M.: Multi-threaded object-oriented MPI-based message passing interface: the ARCH library, vol. 446. Springer Science & Business Media (2012)
16. Lalwani, S., Kumar, R., Gupta, N.: A review on particle swarm optimization variants and their applications to multiple sequence alignment. J. Appl. Math. Bioinformatics. **3**(2), 87–124 (2013)
17. Lalwani, S., Kumar, R., Deep, K.: Multi-objective two-level swarm intelligence approach for multiple RNA sequence-structure alignment. Swarm Evolut. Comput. **34**, 130–144 (2017)
18. Bahr, A., Thompson, J.D., Thierry, J.-C., Poch, O.: BAliBASE (benchmark alignment database): enhancements for repeats, transmembrane sequences and circular permutations. Nucleic Acids Res. **29**(1), 323–326 (2001)

Efficient License Plate Recognition System with Smarter Interpretation Through IoT

K. Tejas, K. Ashok Reddy, D. Pradeep Reddy, K. P. Bharath, R. Karthik and M. Rajesh Kumar

Abstract Vehicles play a vital role in modern-day transportation systems. Number plate provides a standard means of identification for any vehicle. To serve this purpose, automatic license plate recognition system was developed. This consisted of four major steps: preprocessing of obtained image, extraction of license plate region, segmentation, and character recognition. In earlier research, direct application of Sobel edge detection algorithm or applying threshold was used as key steps to extract the license plate region, which do not produce efficient results when captured image is subjected to high intensity of light. The use of morphological operations causes deformity in the characters during segmentation. We propose a novel algorithm to tackle the mentioned issues through a unique edge detection algorithm. It is also a tedious task to create and update the database of required vehicles frequently. This problem is solved by the use of 'Internet of things' where an online database can be created and updated from any module instantly. Also, through IoT, we connect all the cameras in a geographical area to one server to create a 'universal eye' which drastically increases the probability of tracing a vehicle over having manual database attached to each camera for identification purpose.

K. Tejas · K. Ashok Reddy · D. Pradeep Reddy · K. P. Bharath · R. Karthik
M. Rajesh Kumar (✉)
School of Electronics Engineering, VIT University, Vellore 632014, India
e-mail: mrajeshkumar@vit.ac.in

K. Tejas
e-mail: tejastk.reddy@gmail.com

K. Ashok Reddy
e-mail: kadariashokreddy@gmail.com

D. Pradeep Reddy
e-mail: pradeepreddy0003@gmail.com

K. P. Bharath
e-mail: bharathkp25@gmail.com

R. Karthik
e-mail: tkgravikarthik@gmail.com

© Springer Nature Singapore Pte Ltd. 2019
J. C. Bansal et al. (eds.), *Soft Computing for Problem Solving*, Advances in Intelligent Systems and Computing 817, https://doi.org/10.1007/978-981-13-1595-4_16

Keywords License plate extraction · Character segmentation · Edge detection Recognition · Internet of things (IoT)

1 Introduction

The exponential growth in the number of vehicles used in cities and towns is leading to increase in various crimes such as theft of vehicles, violating traffic rules, hit and run accidents, and trespassing into unauthorized areas, etc. Therefore adapting automatic license plate recognition system is much necessary. Digital image processing techniques were used to interpret digitized images in order to extract meaningful knowledge from it. These techniques were broadly classified into four stages: pre-processing of the captured vehicle license plate image, detection and extraction of license plate region, segmentation of characters from the extracted plate using morphological operations and character recognition. In earlier research, the extraction of license plate region was either done through direct application of Sobel edge detector followed by linear dilation and filling connected components with holes or the image was converted into a binary image by applying threshold at a certain value. These methods do not produce efficient results when captured license plate image is subjected to high intensity of light or when the vehicle and license plate are of similar colors. This problem also exists due to variable factors such as distance, weather, and camera quality. In this paper, we discuss a novel algorithm to solve these issues to an extent. In the segmentation process, morphological operations such as dilation and erosion are used which causes deformity in the characters. The original font style and size are manipulated, which makes it hard for character recognition. A unique algorithm is developed for the segmentation of characters without morphological operations for efficient results in this paper. Internet of things (IoT) is a technology where an embedded system is connected to the cloud for easier and smarter analysis. The hardware part of embedded system collects the data, and software part processes the data to produce use full information. This information is passed on to the cloud where it can be interpreted. The use of IoT technology has been increasing exponentially in the recent years. Its drastic growth is due to its ability to make things smart and make it easier to collect and interpret large sets of data through cloud computing. The combination of automatic license plate recognition system with IoT can produce much efficient results. Here the camera acts as the embedded hardware which collects the data through digitized vehicle images, and MATLAB 2016 software is used to process the image to obtain the license plate number, which is then transferred to cloud through Internet for further applications.

All surveillance cameras as well as traffic control cameras can be connected to a cloud which consists of database of required vehicles. By doing this, all the cameras in an area act as a single system where each camera acts as an eye searching for the vehicles listed in the database uploaded by the administrator. In this way, a universal eye could be created making it hard for the accused to hide the vehicle and also saves time on searching for these vehicles.

2 Literature Review

The automatic number plate recognition (ANPR) was invented in 1976 at the Police Scientific Development Branch in the UK. Kim et al. [1] proposed a method using a distributed genetic algorithm to overcome the difficulties dealing with degraded number plates. Arth et al. [2] presented a real-time license plate detection and recognition system where image fragments are taken from real-time video and processed. Ozbay and Ercelebi [3] proposed a method to add a simple step before character segmentation which is more efficient to remove noise and unwanted spots. Sulehria et al. [4] proposed an algorithm that allows the recognition of vehicles number plates using hybrid morphological techniques including hat transformations with morphological gradients and neural networks. Qadri and Asif [5] proposed a method for automatic number plate recognition which uses yellow search algorithm to extract the likelihood ROI in an image, especially for yellow license plate recognition. Roy and Ghoshal [6] proposed a method to overcome the difficult to correctly identify the non-standard number plate characters, by using a pixel-based segmentation algorithm of the alphanumeric characters in the license plate.

Arulmozhi et al. [7] applied skew correction technique for accurate character segmentation, followed by character recognition through centroid-based Hough transform technique. Kaur et al. [8] performed preprocessing and number plate localization by using Ostus methods and feature-based localization methods. Roy et al. [9] proposed a method where image filled with holes is used for license plate segmentation, removing all the connecting edges and applied a threshold of 1000 pixels. Du et al. [10] presented all the methods which are being used till then and have also listed pros and cons of each method in plate segmentation. Hsu et al. [11] used edge clustering formula for solving plate detection for the first time. It is also a novel application of the maximally stable extreme region (MSER) detector to character segmentation. Sulaiman et al. [12] proposed a combination of image processing techniques and OCR to obtain the accurate vehicle plate recognition for vehicles in Malaysia. Bhat et al. [13] have also used license plate segmentation through filling the binary image with holes, but this system had limitations on size of vehicle.

Kaur et al. [14] applied bilateral filter during preprocessing for the image, and remaining method is same as [9, 13]. Chuang et al. [15] proposed an approach to overcome the problems like low-resolution, long-distance image, and blurred image with super resolution, by LBP with the concept of fuzzy. Sharma et al. [16] concentrated on the improvement of the recognition rate and recognition time for recognition of the number and the characters of the vehicle license plate. Rabee et al. [17] proposed an algorithm where SVM was applied at last to construct a classifier to categorize the inputs. Prabhakar et al. [18] proposed an algorithm which uses Hough transform for licence plate segmentation. Raskar et al. [19] used median filter to remove the noise and later followed sobel edge detection algorithm. García-Sánchez et al. [20] proposed a method for identifying the position, length and width of the licence plate that contains the characters in an image, when the plate's location has not been accurate. Dewan et al. [21] proposed an ant colony based number plate extraction method which serves better results in edge detection while applying image segmentation. Kapoor et al. [22] presented a method were IoT and image processing are combined in the field of agriculture the input data is tested comparing with the pre-defined data base of the images which are already fed, this gives the information of character of input.

3 Proposed Method

3.1 Digital Image Processing Techniques

Image processing techniques are used to convert the raw data obtained by the cameras into useful information. This algorithm can be broadly classified into four stages which can be discussed as follows.

3.1.1 Preprocessing

The video or image of the vehicle is obtained from the camera, whose vehicle number is to be identified as illustrated in Fig. 1. If it was a video, the video is converted into frames and these frames are further used for vehicle number identification. This image or frame is then converted from RGB scale to gray scale. A median filter is applied on the obtained grayscale image in order to remove different kinds of noises present in the image. The median filter also concentrates on the high-frequency areas of the image as shown in Fig. 1, which helps to obtain better edge detection results in the later part of the algorithm.

3.1.2 Extraction of License Plate Region

Here our aim is to localize the position of the license plate region and extract a sub-image plot with only the number plate for further analysis as described in Fig. 2. The image obtained after processing through a median filter is now processed through an averaging filter. When this blurred image is subtracted with original grayscale image, the extra acquired values remain and others are deleted resulting in an image as shown in Fig. 3a. Threshold this image at a very low value, to create a binary image. For efficient results threshold at a value of 0.03, where all pixels above this

Fig. 1 Image after applying median filter

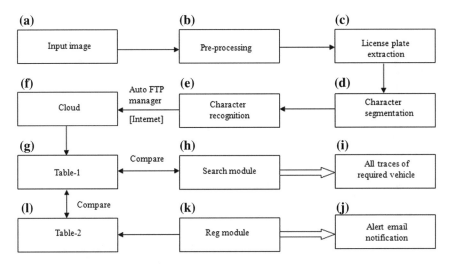

Fig. 2 Flowchart of the proposed algorithm

Fig. 3 **a** Subtracted image, **b** Sobel edge detected image, **c** License plate extracted image

value tend to become unity and others tend to become zero. All the components touching the boundary of the obtained binary image are then deleted.

Sobel edge filter is applied on the obtained binary image in order to obtain accurate boundaries of binary objects in the figure. Scan for all the connected components in the figure and fill them with holes as shown in Fig. 3b. Search for a rectangular area filled with holes in the image which is probably in the size of the license plate. This rectangular region of the license plate is now extracted into a sub-image figure. Multiply the extracted license plate region with the initial grayscale image, in order to isolate the license plate from the original image of the vehicle as shown in Fig. 3c structuring element of size (20 × 20), which returns the center part of the correlation value without zero padding at the edges as a result of which a blurred image is obtained. Now, the blurred image is subtracted from the original grayscale image to obtain intensity difference image. This is because, when an image is blurred by using an average filter, the high-frequency pixels in the image tend to equalize their pixel value with their surroundings. Thus surrounding pixels acquire a higher value.

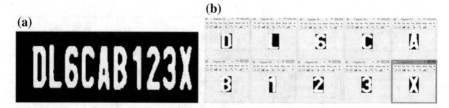

Fig. 4 **a** Image with characters, unnecessary regions removed. **b** Individually segmented characters

3.1.3   Character Segmentation

Threshold this obtained license plate at a low value such as 0.01 to convert the grayscale license plate image into a binary image. The filled regions in the image are now labeled with distinct indices, and the bounding box of four sides is applied on the image to mark all the filled regions of the image. Initially, to remove the extra zero padded regions in the image, bounding box algorithm is applied to obtain the rectangle of maximum size. The license plate is extracted in this step which is re-sized to a fixed size for easier character recognition in the further part. Consider the fixed size to be (175 × 730). The obtained license plate is now complemented, and the boundaries are cleared to erase the license plate border or unnecessary regions leaving only the characters in the image. The regions with less than 1000 pixels and more than 8000 pixels are deleted in order to remove the rusted part, screw holes, or broken regions of the number plate (if any) and results in an image with characters as shown in Fig. 4a. The remaining characters in the image are now individually extracted by applying bounding box algorithm again as shown in Fig. 4b.

3.1.4   Character Recognition

The next step is identifying the information in the image and decoding it for further interpretation as seen in Fig. 2e. Each character from the license plate is extracted by applying bounding box algorithm, and these characters are compared against the alphanumeric database which uses template matching. The extracted character is compared with the template images in all possible positions. A large database of alphanumeric characters can be taken with each character having several font styles and sizes. Neural logics can be applied to train the database in order to identify different font styles. New font styles can be regularly added during the process of character segmentation. In this way, the algorithm becomes smarter making it artificially intelligent. The final output is written in a notepad file as shown in Fig. 5.

3.2   Internet of Things Technology

Raw data are converted into information by the image processing techniques. This information is interpreted using IoT technology. The information produced in the

Fig. 5 Character
recognition output

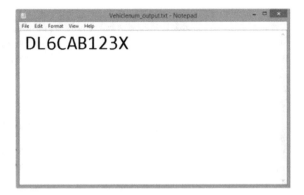

notepad file is then transferred to the cloud for further use as illustrated in Fig. 2f. To interpret the uploaded information, we need an Internet application, and hence, we created a tentative Web site www.searchyourcar.ml.

This Web site was hosted from www.hostinger.in. The database for this Web application is phpMyAdmin, and the online transfers take place using MySQL queries. The front-end user interface part of the Web application in the Web site is coded using hypertext markup language (HTML) and cascading style sheets (CSS). The back-end database for the Web application is phpMyAdmin and is maintained using MySQL queries. The connection between front end and the back end of the Web application is coded using PHP.

The output from the MATLAB software is written and stored in a notepad file. The information in the notepad file is later read and transferred to the cloud using Auto File Transfer Protocol (FTP) Manager once in every 10 s which is later used for interpretation. This information can be accessed and interpreted using two different modules as discussed further.

3.2.1 Search Module

The main aim of search module is to know the vehicle number in interest of the administrator and input the number from user interface of the Web site. This vehicle number is then searched in the database, and all its traces in different locations and time can be accessed by the authorized administrator. The user interface of search module is shown in Fig. 6.

In the server side, we have two different tables for managing the data taken from the front end of the database. Table 1 contains the attributes such as vehicle number, location (latitude and longitude), time, and date. Table 2 is used to manage the register module as shown in Fig. 2h, k. The search module algorithm is discussed in detail in Algorithm 1.

Fig. 6 Search module in the Web application

Table 1 Extracting license plate region by converting input image into binary image

Recent research	Input image	Existing method	Proposed method
(a) Kaur [14]			
(b) Khan [23] and Raskar [19]			

Table 2 Extracting license plate region by applying sobel edge mask on input image

Recent research	Input image	Existing method	Proposed method
(a) Roy et al. [9] and Bhat and mehandia [13]			
(b) Babu et al. [24]			

Fig. 7 Register module in
the Web application

Algorithm 1:

1. Establishing the connection with server by matching username, password, and database name by using connection object in PHP using MySQL queries.
2. Input is taken from the interface and compared with the data in the table.
3. SELECT number, location, time FROM Table 1 WHERE number = input from user.
4. Data from user are transferred to server using Auto FTP Manager in Table 1.
5. Using Jason decoding algorithm, location is retrieved by using IP address from the source http://freegeoip.net.
6. Local time zone is set according to locality in PHP time zone. This time is retrieved to store the vehicle traced time in Table 1.
7. Using Insert MySQL query data is inserted to Table 1.

3.2.2 Register Module

In case we could not trace a vehicle through search module immediately, then we could use register module. Register module obtains data about the required vehicles through attributes such as vehicle number, user email id, mobile number, and the details of the required vehicle. Through this module, the required vehicle number is compared with every input number uploaded from the embedded system. If a match is found, an alert email is sent to the registered email id. Register module algorithm can be discussed in detail as Algorithm 2. The front-end user interface of the register module is as shown in Fig. 7.

Algorithm 2:

1. The data such as vehicle number, email id, mobile number, and details of the required vehicle are collected from the admin using HTML form in the front end of the Web application.
2. Using PHP, the connection is established between the front end and the back end of the website. Here, the Table 2 acts as the back end of the database.

3. The attributes collected in step 1 are stored in Table 2 by using INSERT mySQL query.
4. The vehicle numbers transferred through FTP manager to Table 1 are automatically compared with the vehicle in Table 2. If a match is found then the email id is retrieved from Table 2.
5. Using PHP mailer, an alert mail is immediately sent to the email id retrieved from Table 2.

4 Experimental Results

4.1 Digital image processing

Unique edge detecting algorithm was used to improve the efficiency of extracting the license plate region. A distinct algorithm for segmenting the characters in the number plate was proposed in the paper which avoided producing deformity in the characters. In Table 1, it is observed that direct conversion of the input grayscale image to a binary image by applying threshold at 127-pixel values in order to detect the edges of the license plate does not produce efficient results when the light intensity is not uniform throughout the picture. We can see that in Table 1a, the intensity of light is more toward right side of the picture as compared to the left, and in Table 1b, the intensity of light is non-uniform throughout the picture. The proposed algorithm improves the efficiency in detecting and extracting license plate region for images that are subjected to non-uniform intensity of light. In Table 2, we can see that, applying Sobel edge mask directly on input grayscale image in order to obtain the license plate region does not produce efficient results when the input image is subjected to high intensity of light, when the shadow of the vehicle is on the license plate or when the license plate tends to reflect light.

In Table 2a, we can see that the proposed algorithm could extract the license plate region more efficiently as compared to the existing method. In Table 2b, we can see that the license plate is reflecting light, which hinders efficient license plate detection and extraction. In Table 3, it can be observed that the use of morphological operations such as dilation and erosion produces deformity in the characters. This deformity further hinders the process of character recognition, where characters such as O, D, 0, and B are falsely recognized among each other when filled with holes due to irregular dilation. This deformity can be eliminated by the use of distinct algorithm proposed in this paper where bounding box method is used twice in order to segment the characters.

Table 3 Character Segmentation

Recent research	Input image	Existing method	Proposed method
(a) Dewan [21]		MH12DE1433	MH12DE1433
(b) Agarawal [25]		TN 23 AL 0322	TN 23 AL 0322

4.2 Smarter Recognition Using IoT Technology

After processing the input image through MATLAB, the vehicle number is written and saved in Notepad file. This information written in the file is constantly cleared, and new information is uploaded to the cloud from the notepad with the help of Auto FTP Manager. This uploaded information is stored in Table 1 as shown in Fig. 8a. Now in order to trace a vehicle, we can either use search module to check if the trace of the vehicle already exists in recorded database or we can use register module to register a required vehicle number with an email id so that an alert email notification is sent as soon as the cloud traces the vehicle number. If a trace is found through search module, the administrator can obtain the location, date, and time where the

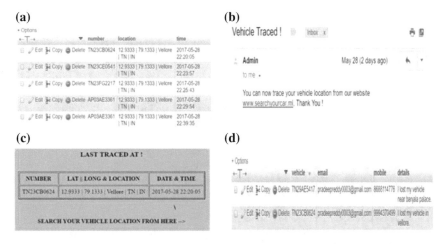

Fig. 8 Results of proposed algorithm where **a** Table 1, **b** Register module results, **c** Search module results and **d** Table 2

Table 4 Results

Units	Number of accuracy	Percentage of accuracy (%)
Extraction	93/95	97.89
Segmentation	94/95	98.94
Recognition	92/95	96.84

vehicle was traced as shown in Fig. 8c. The details of registered numbers are stored in Table 2 as shown in Fig. 8d. The values in Table 2 are constantly compared with vehicle numbers in Table 1, and if the match is found, an alert email is sent to the registered email id as shown in Fig. 7b from which the administrator can obtain the location where the vehicle was traced along with time and date.

A total of 95 images were processed through MATLAB 2016 software. These images were taken in different illumination conditions. Images were obtained from various cameras from 5 to 13 MP clarity. Distance varied from 1 to 20 m. Images of vehicles in various weather conditions were processed. The results after processing are presented in Table 4.

5 Conclusion and Future Work

In this paper, we combined digital image processing techniques with IoT technology to create a much efficient automatic license plate detection and recognition system. In digital image processing techniques, a unique edge detection algorithm was proposed for better license plate extraction. Bounding box was used to extract the characters from the license plate which does not produce any distortion in the segmented characters. This processed information was later sent to cloud for further interpretation. This IoT-based license plate recognition makes it easier to update the required database and also makes it easier to trace the required vehicles if all the cameras in a geographical area are connected to one server. As a part of future work, a genetic algorithm can be introduced which can select the best frame in a video in order to make the process faster. The efficiency of the Web application can be integrated as per the administrators requirement.

Acknowledgements I am thankful to my university, Vellore Institute of Technology, for providing a wonderful platform to learn and work out our ideas into pragmatic projects. I am also thankful to Dr. Rajesh Kumar Muthu, for his valuable guidance, encouragement, and cooperation during the course of the project. The images tested during the experimentation were from various online sources and personally created database to check the efficiency of the proposed algorithm, but the images displayed in the paper are all from online sources (Google) and publicly available images. I, the first author (Tejas K), would like to state that I take the sole responsibility in the future for the images published in this paper.

References

1. Kim, S.K., Kim, D.W, Kim, H.J.: A recognition of vehicle license plate using a genetic algorithm based segmentation. In: Image Processing Proceedings, vol. 2, pp. 661–664. IEEE (1996)
2. Arth, C., Limberger, F., Bischof, H.: Real-time license plate recognition on an embedded DSP-platform. In: IEEE Conference on Computer Vision and Pattern Recognition, CVPR'07, pp. 1–8, June 2007
3. Ozbay, S., Ercelebi, E.: Automatic vehicle identification by plate recognition. World Acad. Sci. Eng. Technol. **1**, 1410–1413 (2007)
4. Sulehria, H.K., Ye Zhang, D.I., Sulehria, A.K.: Vehicle number plate recognition using mathematical morphology and neural networks. WSEAS Trans. Comput. **7**, 781–790 (2008)
5. Qadri, M.T., Asif, M.: Automatic number plate recognition system for vehicle identification using optical character recognition. In: ICETC'09, pp. 335–338. IEEE (2009)
6. Roy, A., Ghoshal, D.P.: Number Plate Recognition for use in different countries using an improved segmentation. In: NCETACS, pp. 1–5. IEEE, Mar 2011
7. Arulmozhi, K., Perumal, S.A., Priyadarsini, C.T., Nallaperumal, K.: Image refinement using skew angle detection and correction for Indian license plates. In: ICCIC, pp. 1–4. IEEE, Dec 2012
8. Kaur, A., Jindal, S., Jindal, R.: License plate recognition using support vector machine (SVM). Int. J. Adv. Res. Comput. Sci. Softw. Eng. 2–7 (2012)
9. Roy, S., Choudhury, A., Mukherjee, J.: An approach towards detection of Indian number plate from vehicle. Int. J. Innov. Technol. Explor. Eng. **2**, 2–4 (2013)
10. Du, S., Ibrahim, M., Shehata, M., Badawy, W.: Automatic license plate recognition (ALPR): a state-of-the-art review. IEEE Trans. Circuits Syst. Video Technol. **23**, 311–325 (2013)
11. Hsu, G.S., Chen, J.C., Chung, Y.Z.: Application-oriented license plate recognition. IEEE Trans. Veh. Technol. **2**, 552–561 (2013)
12. Sulaiman, N, Jalani, S.N.H.M., Mustafa, M., Hawari, K.: Development of automatic vehicle plate detection system. In: IEEE 3rd International Conference on System Engineering and Technology, pp. 130–135. Aug 2013
13. Bhat, R., Mehandia, B.: Recognition of vehicle number plate using matlab. Int. J. Innov. Res. Electr. Electron. Instrum. **2**, 2–8 (2014)
14. Kaur, S.: An efficient approach for number plate extraction from vehicles. Int. J. Comput. Sci. Inf. Technol. **5**, 2954–2959 (2014)
15. Chuang, C.H., Tsai, L.W., Deng, M.S., Hsieh, J.W., Fan, K.C.: Vehicle licence plate recognition using super-resolution technique. In: 11th IEEE International Conference on Advanced Video and Signal Based Surveillance (AVSS), pp. 411–416. IEEE (2014)
16. Sharma, J., Mishra, A., Saxena, K., Kumar, S.: A hybrid technique for license plate recognition based on feature selection of wavelet transform and artificial neural network. In: Optimization, Reliabilty and Information Technology (ICROIT), pp. 347–352 (2014)
17. Rabee, A., Barhumi, I.: License plate detection and recognition in complex scenes using mathematical morphology and support vector machines. In: Systems, Signals and Image Processing (IWSSIP), pp. 59–62. IEEE (2014)
18. Prabhakar, P., Anupama, P., Resmi, S.R.: Automatic vehicle number plate detection and recognition. In: Control, Instrumentation, Communication and Computational Technologies (ICCI-CCT), 2014 International Conference on (pp. 185–190). IEEE (2014, July)
19. Raskar, R.R., Dabhade, R.G.: Automatic number plate recognition (ANPR). Int. J. Emerg. Technol. Adv. Eng. **5**, (2015)
20. García-Sánchez, S., Aubert, S., Iraqui, I., Janbon, G., Ghigo, J.M., d'Enfert, C.: Candida albicans biofilms: a developmental state associated with specific and stable gene expression patterns. Eukaryot. Cell **3**(2), 536–545 (2004)
21. Dewan, S., Bajaj, S., Prakash, S.: Using Ant's Colony Algorithm for improved segmentation for number plate recognition. In: Computer and Information Science (ICIS), 2015 IEEE/ACIS 14th International Conference on (pp. 313–318). IEEE. 2 (2015, June)

22. Kapoor, A., Bhat, S.I., Shidnal, S., Mehra, A.: Implementation of IoT (Internet of Things) and Image processing in smart agriculture. In: Computation System and Information Technology for Sustainable Solutions (CSITSS), International Conference on (pp. 21–26). IEEE (2016, October)

23. Khan, J.A., Shah, M.A.: Car Number Plate Recognition (CNPR) system using multiple template matching. In: Automation and Computing (ICAC), 2016 22nd International Conference on (pp. 290–295). IEEE (2016, September)

24. Babu, K.M., Raghunadh, M.V.: Vehicle number plate detection and recognition using bounding box method. In: Advanced Communication Control and Computing Technologies (ICACCCT), 2016 International Conference on (pp. 106–110). IEEE (2016, May)

25. Agarwal, A., Goswami, S.: An efficient algorithm for automatic car plate detection & recognition. In: Computational Intelligence & Communication Technology (CICT), 2016 Second International Conference on (pp. 644–648), IEEE (2016, February)

Image Pixel Prediction from Neighborhood Pixels Using Multilayer Perceptron

Ayonya Prabhakaran and Rajarshi Pal

Abstract Predicting a pixel value in an image with the help of neighborhood pixels is important for several tasks, like image compression, reversible data hiding. The paper proposes a novel approach to predict the pixel intensities in an image using a multilayer perceptron. Experimental results show that the proposed method is more efficient than several popular pixel prediction methods.

1 Introduction

Image pixel prediction methods predict a pixel's intensity on the basis of the neighborhood pixel intensities. Pixel prediction has a wide range of applications like image compression and reversible data hiding. These applications require that the error between the predicted pixel and original pixel must be as low as possible. A low prediction error boosts the entire set of applications wherever pixel prediction is used.

A good number of pixel prediction schemes are available in the literature. The median edge detector (MED) predictor which is used in [1, 2] uses only three causal pixels from the neighborhood for prediction. MED is also used in JPEG-LS standard [3]. Even in recent papers like, Gui et al. [4] the predictor used is the gradient adjusted predictor (GAP) which was proposed in CALIC algorithm [5]. GAP outperforms the MED detector as the edges' existence and strength are taken into account. GAP uses seven causal pixels from the neighborhood prediction. In [6], the gradient edge detector (GED) is proposed that combines the advantages of both the GAP and the MED predictor. Hence, the GED predictor outperforms the GAP and MED predictor and uses only a single threshold value that is preset. Jaiswal et al. [7] proposed an

A. Prabhakaran
Indian Institute of Information Technology, Allahabad, Allahabad, India
e-mail: ihm2014502@iiita.ac.in

R. Pal (✉)
Institute of Development & Research in Banking Technology, Hyderabad, India
e-mail: iamrajarshi@yahoo.co.in

© Springer Nature Singapore Pte Ltd. 2019
J. C. Bansal et al. (eds.), *Soft Computing for Problem Solving*, Advances in Intelligent Systems and Computing 817, https://doi.org/10.1007/978-981-13-1595-4_17

efficient prediction technique that is used for watermarking wherein for the prediction of certain type of pixels, watermarked pixels are used, making the method inefficient for pixel prediction alone as it increases the complexity of the pixel prediction. In [8] again, two sets of pixels are used where the pixels of one set are watermarked by using the other set for prediction. The prediction of the first set is done with original pixels and for the second set using the modified pixels using a rhombus predictor. In 2012, Dragoi and Coltuc [9] have proposed the modified rhombus interpolation scheme wherein the interpolated pixel is computed as the average of the horizontal pixels or of the vertical pixels or of the entire set of four pixels. Olu et al. [10] and Li et al. [11] are variations of the rhombus predictor wherein the prediction on the rhombus context is computed by partial differential equations (PDE), and in [11], the rhombus context is extended to the full 3×3 window and the central pixel is estimated as 75% of the average of the horizontal/vertical pixels and 25% of the average of the diagonal pixels. In [12], a least squares (LS) prediction is proposed in which the solution minimizes the sum of squares of the prediction error using the LS predictor. But, the predictors are required to be embedded into the watermarked image in order to be available at detection limiting the number of predictors in block-based prediction schemes. In [13], the pixel value is predicted by calculating the mean of the four neighboring pixels (in the North, South, East, and West directions). Since rough prediction methods reduce the hiding capacity, Lu et al. [14] proposed a method to perform reversible data hiding based on multiple predictors, i.e., GAP, MED, and INP using asymmetric histograms. INP is an image interpolation method to enlarge the image proposed in [15] in 2012.

Existing techniques, as discussed above, attempted to propose some mathematical relation between center pixel's intensity and those of its neighborhood pixels. But establishing an ideal mathematical model which can be generalized over every pixel and its neighborhood is a challenging task. Hence, there has been always a scope for taking this research forward. Considering the difficulty in generalizing a mathematical relation which can suit on a majority of image pixels and their neighborhood, this paper proposes a multilayer perceptron (one kind of artificial neural network)-based implementation of the pixel prediction. Existing techniques did not exploit the property that ANNs can actually learn from observing data sets, and hence, it can be used for pixel prediction using the neighborhood property of images. According to the method proposed in this paper, if the neighborhood of the pixel to be predicted contains edge pixels, then the network is neither trained or tested on these pixels. The proposed method gives lower values of mean square errors in comparison with other notable methods in the literature.

The rest of the paper is organized as follows: In Sect. 2, the proposed method of pixel prediction using the neighborhood pixels is described. This is followed by the Sect. 3 detailing the design of the experiment. Section 4 contains the experimental results and the observations on the results. Concluding remarks have been drawn in Sect. 5.

2   Proposed Method

This paper attempts to predict intensity value of a pixel using the intensities of other pixels in the neighborhood. It can be understood (and experimentally verified in later part of this paper) that prediction error will be high if an edge passes through the local neighborhood of size $a \times a$. A better accuracy can be achieved in relatively smoother regions in the image. Hence, the proposed method follows the steps as mentioned here.

1. At first, edge pixels are identified.
2. A pixel is considered for either training or testing of the prediction scheme if there is no edge pixel in the $a \times a$ window surrounding the pixel. Thus, a training data set is created considering all such pixels and their local neighborhood.
3. A multilayer perceptron (MLP) is modeled to contain an input layer, one or two hidden layers depending on the architecture that will be chosen, and an output layer. Any one of the following three architectures can be used to obtain the best results wherein a 5×5 window is used. Hence, the input layer contains 24 nodes. The output layer contains one node and the configuration of the hidden layer(s) is any one among the following:

   a. one hidden layer with 48 nodes, or
   b. one hidden layer with 16 nodes, or
   c. two hidden layers with 8 nodes in each layer.

4. Thus, the trained multilayer perceptron can then be used to predict pixels.

The proposed approach is presented in detail as follows:

2.1   Edge Pixel Detection

Since the training and testing data sets should not contain edge points, the Canny edge detection algorithm is used to detect the edges. The Canny edge detector requires two threshold values that are set manually, for hysteresis thresholding. At first, the input image is smoothed using a Gaussian filter of kernel size 7×7 in order to remove noise. Then, Otsu's method is applied to obtain a threshold value. The obtained Otsu's threshold value is set as the higher threshold for the Canny detector and the lower threshold is set as 0.9 times the higher threshold. The output of the Canny edge detection is a binary edge image as presented in Fig. 1.

2.2   Data set Creation

A sliding window of size $a \times a$ is moved to every pixel in the input image in raster scan order. Intensity values of the pixels in the window are examined. If any of the

Fig. 1 A sample input image and the corresponding edge image

pixels in the window is an edge pixel, then the window is not considered for training and testing of the prediction scheme. Otherwise, the $(a \times a - 1)$ neighboring pixels are used to predict the center pixel. The training or test data is created by considering the neighboring pixel intensities as input data and the center pixel as desired output.

2.3 Multilayer Perceptron

The architecture of the multilayer perceptron is described here. As per the proposed model, a local neighborhood in a 5×5 window is used to predict the center pixel value. Therefore, the input layer contains 24 nodes. The output layer contains only a single node to output the predicted value of the center pixel's intensity. The hidden layer is proposed to be either of the following:

1. one hidden layer with 48 nodes, or
2. one hidden layer with 16 nodes, or
3. two hidden layers with 8 nodes in each layer.

No definite guideline is available in the literature regarding number of hidden layers and number of nodes in each of these hidden layers in a multilayer perceptron. But normal practices assume 1–2 hidden layers and number of nodes in each hidden layer is found to vary from 2/3rd to twice of the number of input nodes. Hence, the proposed method try out few such variations of the architecture. As demonstrated through experimental results, each of the above three multilayer perceptrons demonstrates equal performance in predicting the center pixel.

The gradient descent algorithm is used for training with the Adam optimizer. The rectifier or ReLU activation function(i.e. $f(x) = max(0, x)$ or the ramp function) is used and mean square error (MSE) is the loss that is monitored. The MSE values are calculated by finding the sum of the squared differences between the predicted pixel intensity and original pixel intensity over the data set. The initial learning rate is set to a high value and it decays over the number of epochs so that network converges. Once these network parameters have been set, the network can be trained until convergence. The number of epochs is also set to a large value (so that the network is not under-trained) and the network is trained in batches of N samples. A batch approximates the distribution of the input data better than a sample (single input). The weights are updated after each batch by the process of backpropagation. The larger the batch, the better the approximation; but the longer it will take to process the data. The network stops training when convergence is attained on monitoring the MSE values on the training data. The architecture of the network, the weights, and the training configuration (loss, optimizer) are then stored on the disk in order to use it later for validation and testing.

3 Design of the Experiment

This section mentions the images which are used for the experiments and describes how the training and test data sets are created.

Experiments are carried out with 8 images of size 512×512 namely *Mandrill, Peppers, Tiffany, Airplane, Lake, House, San Diego (Miramar NAS)* and *Lena* from the USC-SIPI Image Database as shown in Fig. 2. Out of these 8 images, 2 images (the Mandrill and an aerial view of San Diego) contain a large number of edge pixels.

The experimental setup is detailed as follows. At first, pixels from 7 out of 8 images are considered to construct the training data. But every pixel in these images does not contribute to the training set. At first, the pixels having edge pixels in its 5×5 neighborhood are excluded. Then, remaining pixels are divided into 9:1 ratio between training and test set. These pixels are traversed in raster scan order and are grouped into a set of 10 pixels. One pixel from every 10 pixels is part of the test set. This test set is referred to in subsequent discussions as *test1*. This experiment is repeated 10 times wherein each of these 10 iterations, a different pixel becomes part of the test set.

Pixels from the left out image (1 out of 8 images) constitute the second test set (*test2*). Pixels having edges in their 5×5 neighborhood are not considered in this case too. The difference between two test sets are: (i) *test1* is comprised of pixels from the images which are used to generate training data set too. (ii) *test2* is generated

Fig. 2 Database of eight images

from pixels of a different image. The entire process is repeated 8 times. Then, a total of $10 \times 8 = 80$ experiments are conducted involving slightly different training and test data sets in each case.

The performance of the proposed method is compared by evaluating the mean square errors (MSEs) of prediction.

4   Experimental Results and Discussion

In general, most practical models of multilayer perceptron contain one or two hidden layers, while the number of nodes in each hidden layer varies between half- of or two-thirds the number of nodes in the input layer to not more than twice the number of nodes in the input layer. Besides the three architectures mentioned in Sect. 2.3, the experiments have also been performed on the following configurations:

1. 3×3 neighborhood window and one hidden layer with 16 nodes.
2. 3×3 neighborhood window and two hidden layers with 8 nodes in each layer.
3. 5×5 neighborhood window and two hidden layers with 24 nodes in each layer.

If a 3×3 window is used, then the number of nodes in the input layer is 8. In the case of the 5×5 window, the number of nodes in the input layer is 24. The learning rate has been initialized to a value of 0.05, and the number of epochs is set to 500 with a batch size of 10,000. An early stopping is enforced, where the training stops on the condition if three successive epochs have their MSE differences less than 0.001 implying the model of the network has converged.

A second implementation following the same experimental setup is also performed wherein edge pixel removal detection and removal have not been performed so as to check the efficiency of the proposed method.

For each of the six MLP architectures (as described earlier), MSEs are computed separately for *test1* and *test2*, resulting in 160 MSE values (80 for *test1* and 80 for *test2*) for one architecture. To test the similarity of the six architectures, *Z-Test for Two Means* is performed with the null and alternative hypothesis as follows:

$$H_0 : \mu_1 = \mu_2 \tag{1}$$

$$H_a : \mu_1 \neq \mu_2 \tag{2}$$

where μ_i is the mean of the MSE values of the ith architecture. H_0 in (1) implies that the pair of architectures taken for comparison are similar and H_a in (2) implies that they are not similar. The Z-Test is performed for all pairs of architectures, thus giving 15 pairs ($^6C_2$) to compare in total. If the null hypothesis (H_0) is accepted, then that pair of network architectures are considered equivalent and have similar performance. If H_0 is rejected, then H_a is true and the network with the lower value of mean of MSE is the better network as our aim is to get the lowest MSE possible. Table 1 summarizes the results obtained where the first four rows and columns represent the

Table 1 6 × 6 matrix of the results, where *Accept* means H_0 is accepted and *Reject* means H_a is accepted

No. of hidden layers	16	8 + 8	48	24 + 24	16 (3 × 3)	8 + 8 (3 × 3)
16						
8 + 8	Accept					
48	Accept	Accept				
24 + 24	Reject	Accept	Reject			
16 (3 × 3)	Reject	Reject	Reject	Reject		
8 + 8 (3 × 3)	Reject	Reject	Reject	Reject	Accept	

results for the 5 × 5 window and the last two rows and columns represent the results for the 3 × 3 window.

From Table 1, it can be observed that following architectures provide equivalent performance for this task:

1. The first set of architectures that are equivalent are:

   a. a 5 × 5 window and one hidden layer with 48 nodes,
   b. a 5 × 5 window and one hidden layer with 16 nodes and
   c. a 5 × 5 window and two hidden layers with 8 nodes per layer.

2. The second set of architectures that are equivalent are:

   a. 3 × 3 window with one hidden layer having 16 nodes and
   b. 3 × 3 window with two hidden layers having 8 nodes per layer.

From the mean values of the MSEs (i.e., the mean of the 80 iterations of the experiment) listed in Table 2, it can be concluded that the architectures in the first set have the least mean values and hence produce the best results. While using a 5 × 5 window, the number of neighborhood pixels to predict the center pixel is higher than the case of a 3 × 3 window.

Table 3 summarizes the results obtained when the neighborhoods having edge pixels were also considered to generate training and testing data sets, following the same experimental setup. From the mean values of the MSEs (the mean of the 80 iterations) listed in Tables 2 and 3, it can be concluded that the mean of the MSEs (where edge pixel is not excluded) is higher than those of the cases where edge pixels are excluded. This is because the neighborhood contains edge pixels which is not homogeneous and has drastic variation in the pixel intensities; thus, the MLP is unable to learn and predict the pixels with greater accuracy.

Performance of the proposed method of pixel prediction using the MLP is compared with existing methods in the literature. Table 4 summarizes these results, where the MSE values are calculated image-wise. The values for the proposed method given in the table as *MLP(1)*, *MLP(2)* and *MLP(3)* where (1), (2), and (3) are from the following set of architectures when the edge pixels are excluded:

Table 2 Mean of MSE values of the *test1* and *test2* as obtained on six architectures

Architecture/Mean of MSEs	*test1*	*test2*
5 × 5 window and one hidden layer with 48 nodes	12.64	17.58
5 × 5 window and one hidden layer with 16 nodes	12.74	17.40
5 × 5 window and two hidden layers with 8 nodes per layer	12.81	17.67
5 × 5 window and two hidden layers with 24 nodes per layer	13.16	17.95
3 × 3 window and one hidden layer with 16 nodes	17.81	24.47
3 × 3 window and two hidden layers with 8 nodes in each layer	17.82	24.52

Table 3 Mean of the MSEs of *test1* and *test2* as obtained on six architectures when edge pixel removal has not been performed

Architecture/Mean of MSEs	*test1*	*test2*
5 × 5 window and one hidden layer with 48 nodes	64.00	76.55
5 × 5 window and one hidden layer with 16 nodes	67.33	76.64
5 × 5 window and two hidden layers with 8 nodes per layer	70.96	80.29
5 × 5 window and two hidden layers with 24 nodes per layer	69.54	79.53
3 × 3 window and one hidden layer with 16 nodes	70.92	75.75
3 × 3 window and two hidden layers with 8 nodes in each layer	76.29	81.32

1. a 5 × 5 window and one hidden layer with 48 nodes,
2. a 5 × 5 window and one hidden layer with 16 nodes and
3. a 5 × 5 window and two hidden layers with 8 nodes per layer.

It is evident from the values of Table 4 that the proposed multilayer perceptron-based pixel prediction scheme outperforms several popular pixel prediction methods.

Table 4 Image-wise MSE of the proposed method versus existing methods

Predictor/Images	Mandrill	Peppers	Tiffany	Airplane	Lake	House	San Diego	Lena
MED [1]	431.25	66.86	47.67	45.49	123.15	67.71	337.14	51.01
GAP [4]	409.09	53.02	38.54	44.21	109.79	67.78	300.14	40.74
GED 8 [6]	480.31	63.63	49.46	59.42	142.40	88.75	373.82	59.31
Rhombus Avg. [8]	273.85	38.82	32.11	27.52	66.91	47.69	177.59	24.64
Imp. Rhombus [9]	309.59	33.40	28.21	23.55	69.04	44.63	190.41	23.77
MLP(1)	28.36	16.83	15.24	4.89	21.75	7.51	37.03	9.02
MLP(2)	27.75	15.94	15.84	5.11	21.05	7.59	36.80	9.11
MLP(3)	27.38	18.72	16.38	5.16	20.69	8.30	35.80	8.91

5   Conclusion

In this research, we have proposed a method to predict pixels based on the neighborhood pixel intensities using a multilayer perceptron. The method is tested extensively by using multiple architectures, varying the window size and also with inclusion and exclusion of the edge pixels. The 5 × 5 neighboring window using MLP architecture with one hidden layer having 16 nodes, one hidden layer having 48 nodes, and two hidden layers having 24 nodes per layer demonstrates the best performance among several other variations as tested. Even the proposed method demonstrates better performance as compared to several popular pixel prediction methods. This initial experiment is first of its kind in using artificial neural network models. But the success of the initial experiment is encouraging enough to carry out further investigation to try out various other models of artificial neural network for image pixel prediction.

References

1. Thodi, D.M., Rodriguez, J.J.: Expansion embedding techniques for reversible watermarking. IEEE Trans. Image Process. **16**(3), 721–729 (2007)
2. Hu, Y., Lee, H.K., Li, J.: DE-based reversible data hiding with improved overflow location map. IEEE Trans. Circuits Syst. Video Technol. **19**(2), 250–260 (2009)
3. Weinberger, M., Seroussi, G., Sapiro, G.: The LOCO-I lossless image compression algorithm: Principles and standardization into JPEG-LS. IEEE Trans. Image Process. **9**(8), 1309–1324 (2000)
4. Xinlu Gui, Xiaolong Li, Bin Yang: A high capacity reversible data hiding scheme based on generalized prediction-error expansion & adaptive embedding. Signal Process. **98**, 370–380 (2014)
5. Wu, X., Memon, N.: Context-based, adaptive, lossless image coding. IEEE Trans. Commun. **45**(4), 437–444 (1997)

6. Avramovi, A., Reljin, B.: Gradient edge detection predictor for image lossless compression. In: Proceedings ELMAR-2010, pp. 131–134. IEEE (2010)
7. Jaiswal, S.P., Au, O.C., Jakhetiya, V., Guo, Y., Tiwari, A.K., Yue, K.: Efficient adaptive prediction based reversible image watermarking. In: 2013 IEEE International Conference on Image Processing, VIC, pp. 4540–4544 (2013)
8. Sachnev, V., Kim, H.J., Nam, J., Suresh, S., Shi, Y.Q.: Reversible watermarking algorithm using sorting and prediction. IEEE Trans. Circuits Syst. Video Technol. **19**(7), 989–999 (2009)
9. Dragoi, C., Coltuc, D.: Improved rhombus interpolation for reversible watermarking by difference expansion. In: 2012 Proceedings of the 20th European Signal Processing Conference (EUSIPCO) (2012)
10. Ou, B., Li, X., Zhao, Y., Ni, R.: Reversible data hiding based on PDE predictor. J. Syst. Softw. **86**(10) 2700–2709 (2012)
11. Li, X., Li, B., Yang, B., Zeng, T.: General framework to histogram-shifting-based reversible data hiding. IEEE Trans. Image Process. **22**(6), 2181–2191 (2013)
12. Dragoi, I.C., Coltuc, D.: Local-prediction-based difference expansion reversible watermarking. IEEE Trans. Image Process. **23**(4), 1779–1790 (2014)
13. Sachnev, V., Kim, H.J., Nam, J., Suresh, S., Shi, Y.Q.: Reversible watermarking algorithm using sorting and prediction. IEEE Trans. Circuits Syst. Video Technol. **19**(7), 989–999 (2009)
14. Lu, T.-C., Chen, C.-M., Lin, M.-C., Huang, Y.-H.: Multiple predictors hiding scheme using asymmetric histograms. Multimed. Tools Appl. **76**(3), 3361–3382 (2017)
15. Lee, C.F., Huang, Y.L.: An efficient image interpolation increasing payload in reversible data hiding. Expert Syst. Appl. **39**(8) 6712–6719 (2012)

Classification of Histopathological Images Through Bag-of-Visual-Words and Gravitational Search Algorithm

Himanshu Mittal and Mukesh Saraswat

Abstract The automated quantification of different cell structures available in histopathological images is a challenging task due to the presence of complex background structures. Moreover, the tissues of different categories, namely epithelium tissue, connective tissue, muscular tissue, and nervous tissue have heterogeneous structure which limits the applicability of an algorithm to only a single class of tissue for the quantification analysis of histopathological images. Therefore, this paper introduces a novel method for categorization of histopathological images into the respective tissue category before quantification analysis. The proposed method uses SIFT method for feature extraction which are further processed by gravitational search algorithm to obtain optimal bag-of-visual-words. Moreover, support vector machine is trained on these bag-of-visual-words to classify the images into respective categories. The experimental results show that the proposed method outperforms the traditional K-means-based method for histopathological image classification.

Keywords Histopathological image classification · Bag-of-visual-words
Gravitational search algorithm · SIFT method

1 Introduction

Histopathological image analysis plays an important role in the drug development and disease identification. Generally, this analysis is performed manually by pathologists and has inherent some flaws such as highly time-consuming, biased in nature as the analysis depends on the knowledge and experience of a pathologist, and low qualitative report [1]. Therefore, automated histopathological image analysis has become an important area of research in medical imaging [2]. Histopathological images may

H. Mittal (✉) · M. Saraswat
Jaypee Institute of Information Technology, Noida, India
e-mail: himanshu.mittal224@gmail.com

M. Saraswat
e-mail: saraswatmukesh@gmail.com

© Springer Nature Singapore Pte Ltd. 2019   231
J. C. Bansal et al. (eds.), *Soft Computing for Problem Solving*, Advances in Intelligent
Systems and Computing 817, https://doi.org/10.1007/978-981-13-1595-4_18

be categorized into four different types of tissue images [3], namely 1. epithelium tissue—lines and covers surfaces, 2. connective tissue—protect, support, and bind together 3. muscular tissue—produces movement, and 4. nervous tissue—receives stimuli and conducts impulses. Recently, a good number of automated methods have been proposed in the field of histopathological quantification [1, 2]. These methods include the process of quantification of different types of cells and differentiate diseased image from normal image. However, no method has been proposed for the classification of histopathological images into their respective tissue class (epithelium, connective, muscular, nervous). This will help to make a generalize method for all the tissue images for their quantification. Therefore, this paper introduces a novel method for histopathological image classification into different tissue categories.

The accuracy of a classification problem is generally associated with the extraction of relevant features [4]. A number of methods have been proposed for extraction of features from histopathological images which can generally be classified into two main classes [5], namely statistics-based and learning-based methods. The statistics-based feature methods use shape, size, and distribution of the nuclei to represent the histopathological images. However, it has been observed that key nuclei are sometimes not considered while feature measurement due to their limited quantity, which may degrade the performance of a classifier. Some researchers have also used graph-based features or mixed features to represent the histopathological images. Furthermore, learning-based methods use different machine learning methods to extract the features such as auto-encoders [6, 7], restricted Boltzmann machines [8, 9], convolutional neural network (CNN) [10, 11], and many more [12–16]. In general, the learning-based features extraction methods are computational expensive methods.

Moreover, some researchers also used classical feature descriptors for representing the histopathological images such as scale-invariant feature transform (SIFT) [17], histogram of oriented gradient (HOG) [18] and local binary pattern (LBP) [19]. Kandemir et al. [20] used color histogram, SIFT, LBP, and a set of well-designed nuclear features to represent the histopathological images. Classical features have given considerable accuracy in histopathological image analysis. Recently, bag-of-visual-words (BoVW), which is generally used for document classification, has widely been used in many computer vision applications [21, 22]. This approach considers an image as a histogram of code words. The BoVW approach aims at identifying the visual patterns that are relevant to the whole image collection. This approach has shown robustness in terms of occlusion and affine transformations. Further, it is computational efficient too. However, the generation of BoVW is generally done using K-means clustering method which is sensitive toward the initial clusters [23]. In such cases, meta-heuristic methods have exhibited better performance [24–34]. Therefore, this paper uses gravitational search algorithm (GSA) to find the optimal BoVW.

GSA is introduced by Rashedi et al. [35] and has been widely used in the solving computationally intensive real-world problems [36–38]. This paper focuses on using GSA for generating codebook by performing clustering on the collection of extracted features with intra-class distance as the objective function. Selection of GSA has been made on the basis of its performance as compared to other meta-heuristic methods such as particle swarm optimization (PSO) [39] and differential evolution (DE) [40], especially in the clustering task [25, 41–43]. The accuracy of proposed method has been tested on the histopathological image dataset.

Rest of the paper is organized as follows: Section 2 describes the gravitational search algorithm (GSA). The proposed method is explained in Sect. 3. In Sect. 4, experimental results are compared and discussed. Finally, Sect. 5 concludes the paper.

2   Gravitational Search Algorithm

Gravitational search algorithm (GSA) [35] is a meta-heuristic method based on physical phenomenon of mass interactions. The GSA is based on Newton's law of gravity and law of motion. The algorithm considers each agent as an object and evaluates the performance in terms of masses. Each object applies force in the system according to law of gravity and changes its position according to law of motion. The number of objects exerting force at an iteration is represented by *Kbest* which controls the trade-off between exploration and exploitation [44]. The slower movement of heavier object corresponds to the exploitation and represents better solution. The fittest object is the heaviest mass and corresponding position represents the optimal solution of the problem at the end of the stopping criteria.

Consider a system of U objects in u-dimensional search space of GSA where the position of each object is depicted by Eq. 1.

$$X_i = (x_i^1, \ldots, x_i^d, \ldots, x_i^u), i = 1, 2, \ldots, U \tag{1}$$

Here, $x_i{}^d$ denotes the dth dimension of the ith object.

The total force $F_i^d(t)$ in dth dimension of ith object at tth iteration is defined as randomly weighted sum of dth components of the forces from other *Kbest* objects and is shown in Eq. (2).

$$F_i^d(t) = \sum_{j=1, j \neq i}^{Kbest} rand_j F_{ij}{}^d(t), \tag{2}$$

Here, $rand_j$ is a random number in the interval $[0, 1]$ and *Kbest* for the tth iteration is defined by Eq. (3).

$$Kbest(t) = final\_per + \left(\frac{1-t}{max\_it}\right) \times (100 - final\_per), \tag{3}$$

where max_it is the maximum number of iterations and $final\_per$ is the percent of objects which apply force to others. Equation (3) shows that the value of *Kbest* decreases linearly over iterations.

In Eq. (2), F_{ij} is the force of jth object on ith object and is computed by Eq. (4).

$$F_{ij}{}^d(t) = G(t)\frac{M_i(t) \times M_j(t)}{R_{ij}(t) + \varepsilon}(x_j^d(t) - x_i^d(t)) \qquad (4)$$

Here, $G(t)$ is a gravitational constant computed as Eq. (5) at iteration t, ε is a small constant, and $R_{ij}(t)$ is an Euclidean distance between two objects i and j.

$$G(t) = G(t_0) * exp\left(-\beta * \frac{t}{max\_it}\right) \qquad (5)$$

where $G(t_0)$ is the initial gravitational constant value and β corresponds to a constant.

In Eq. (4), M_i and M_j correspond to masses of objects i and j, respectively. For ith object with fitness value $(fit_i(t))$, M_i is calculated in every iteration t by Eq. (8).

$$M_i = M_{ii} = M_{gi}, i = 1, 2, \ldots, U \qquad (6)$$

$$m_i(t) = \frac{fit_i(t) - worst(t)}{best(t) - worst(t)}, \qquad (7)$$

$$M_i(t) = \frac{m_i(t)}{\sum_{j=1}^{N} m_j(t)}, \qquad (8)$$

where M_{gi} is gravitational mass and M_{ii} is inertia mass of an object i. $best(t)$ and $worst(t)$ are measured by Eq. (9) and Eq. (10), respectively, for minimization problem.

$$best(t) = \min_{j \in \{1, \ldots, U\}} fit_j(t) \qquad (9)$$

$$worst(t) = \max_{j \in \{1, \ldots, U\}} fit_j(t) \qquad (10)$$

The calculated force $F_i^d(t)$ and mass $M_i(t)$ are used to find the acceleration of ith object in the dth dimension according to law of motion as shown in Eq. (11).

$$a_i^d(t) = \frac{F_i^d(t)}{M_i(t)}, \qquad (11)$$

Now, the updated velocity and position of an object are calculated as Eq. (12) and Eq. (13), respectively.

$$v_i^d(t+1) = rand_i \times v_i^d(t) + a_i^d(t), \qquad (12)$$

$$x_i^d(t+1) = x_i^d(t) + v_i^d(t+1). \qquad (13)$$

Algorithm 1 Gravitational Search Algorithm (GSA)

Input: U objects having u dimensions. Assume the value of G_0, *final_per*, and β.
Output: Best solution having heaviest mass.
Randomly initialize the initial population of U objects;
Evaluate the fitness (*fit*) of each object;
Compute the mass M of each object by Eq. (8) and G by Eq. (5) ;
Set *Kbest* $= U$;
while stopping criteria is not satisfied **do**
    Compute the acceleration a of each object by Eq. (11);
    Compute the velocity v of each object by Eq. (12);
    Update the position of each object by Eq. (13);
    Evaluate the fitness *fit* for each object;
    Compute the mass M of each object by Eq. (8);
    Update G;
    Update *Kbest* as: $Kbest = final\_per + \left(\frac{1-t}{max\_it}\right) \times (100 - final\_per)$,
end while

As object's position corresponds to solution, heaviest object in the system will be the fittest agent and its position will represent the optimal solution of the problem at the end of stopping criteria. The pseudocode of the GSA is presented in Algorithm 1 [35].

3 Proposed Method

The proposed method uses bag-of-visual-words (BoVW) and gravitational search algorithm to classify the histopathological images into the respective classes, i.e., epithelium tissue, connective tissue, muscular tissue, and nervous tissue. The flow graph of the proposed method is depicted in Fig. 1. In the first step of the proposed method, scale-invariant feature transform (SIFT) is used to convert each image as a collection of feature vectors of 128 dimensions. The different vectors can be used in any order. These feature vectors are used to generate BOVW using GSA.

Like document classification, the bag-of-visual-words treat the image features as words termed as codewords. In other words, a codeword is a representative of several similar patches of images. The proposed method represents the image as a vector of occurrence counts of codewords. Generally, these code words are found by performing K-means clustering over all the feature vectors [45]. However, K-means clustering sometimes shows biased behavior due to initial clusters [23]. Therefore, this proposed method replaces the K-means clustering method by GSA which is a robust meta-heuristic method. GSA uses intra-class variance as a fitness function to

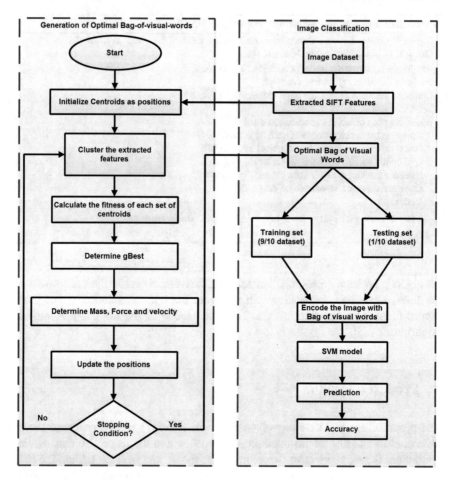

Fig. 1 The flow graph of the proposed method

find the optimal set of code words which serve as centers of the learned clusters. The number of clusters is known as codebook size. The proposed method then maps the learned code words with images and represents each image as a histogram of code words.

The final step of the proposed method includes the training of the support vector machine (SVM) with the generated BoVW. The trained SVM is further used to classify the test image into its respective class. The description of the proposed method is also depicted in Algorithm 2.

4 Experimental Results

The image dataset consists of four different histopathological categories, namely epithelium tissue, connective tissue, muscular tissue, and nervous tissue. Each image category contains 101 images. The images are taken from various publicly available sources [46, 47]. Each category contains images taken after various staining methods. The detailed information about the dataset is mentioned in Table 1. Moreover, Fig. 2 shows representative images from each category of histopathological image dataset. Furthermore, the dataset is divided into training and testing sets using stratified random sampling. For classification, multi-class SVM has been employed.

Algorithm 2 Proposed Approach

Extract features of each image.
Combine the features of all the images as a single matrix.
Identify N cluster centroids by performing GSA with intra-cluster variance as objective function.
The N cluster centroids define the visual dictionary.
Generate bag-of-visual-words for each image.
Train the classifier.
Extract the features of test image.
Generate the normalized term frequency of the test image.
Identify the label using trained classifier.

The performance of the proposed method has been compared with K-means-based method in terms of accuracy, precision, recall, and F-measure. Table 2 shows the performance values of accuracy, precision, recall, and F-measure. The best value is represented in bold. From the table, it is observed that the proposed method shows better performance for almost all performance parameters whereas K-means gives better results in only recall and precision values for connective tissues and muscle tissues, respectively. However, the proposed method outperforms the K-means-based method in terms of F-measure for all the histopathological categories. Moreover, the accuracy of proposed method is 51.6% which is better than K-means-based method.

Table 1 The different categories of image dataset and their staining methods

S. No.	Category	Staining	No. of images
1	Connective tissue	H&E, TRI, EL, TB, RET, CCY	101
2	Epithelial tissue	H&E, VG, MB, PAS/H&E	101
3	Muscle tissue	H&E, WHP, IH, HAFTEG	101
4	Nervous tissue	H&E, BC, ICC, VG, LFC, H&E/MB	101

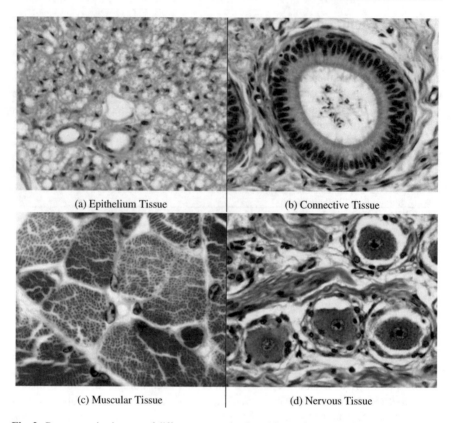

(a) Epithelium Tissue   (b) Connective Tissue

(c) Muscular Tissue   (d) Nervous Tissue

Fig. 2 Representative images of different categories from histopathological image dataset

Table 2 Comparative analysis of K-means-based method and proposed method

S. No.	Algorithm	Accuracy	Classes	Recall	Precision	F-measure
1	K-means	43.5	Muscle tissue	48.3	**68.1**	56.6
			Connective tissue	**61.2**	39.5	48.1
			Epithelial tissue	51.6	43.2	47.05
			Nervous tissue	12.9	23.5	16.6
2	GSA	**51.6**	Muscle tissue	**74.1**	57.5	**64.7**
			Connective tissue	45.1	**66.6**	**80.6**
			Epithelial tissue	**70.9**	**47.8**	**73.3**
			Nervous tissue	**16.1**	**29.4**	**20.8**

5 Conclusion

The proposed method introduces a novel method to classify the histopathological images into epithelium, connective tissue, muscular tissue, and nervous tissue classes. The method uses gravitational search algorithm for generating the optimal bag-of-visual-words (BoVW). For feature extraction, SIFT method has been used and these feature vectors are used by GSA for generating BoVW. Further, SVM is used to classify the images into respective categories. The performance has been compared with K-means-based method in terms of accuracy, recall, precision, and F-measure. The results depicted that the proposed method outperforms the considered method.

Acknowledgements Authors are thankful to Science and Engineering Research Board, Department of Science & Technology, Government of India, New Delhi for funding this work as part of the project (ECR/2016/000844).

References

1. Saraswat, M., Arya, K.: Automated microscopic image analysis for leukocytes identification: a survey. Micron **65**, 20–33 (2014)
2. Gurcan, M.N., Boucheron, L.E., Can, A., Madabhushi, A., Rajpoot, N.M., Yener, B.: Histopathological image analysis: a review. IEEE Rev. Biomed. Eng. **2**, 147–171 (2009)
3. B.L.F. JO ANN EURELL, Dellmanns textbook of veterinary histology (2006)
4. Saraswat, M., Arya, K.: Feature selection and classification of leukocytes using random forest. Med. Biol. Eng. Comput. **52**(12), 1041–1052 (2014)
5. Zheng, Y., Jiang, Z., Xie, F., Zhang, H., Ma, Y., Shi, H., Zhao, Y.: Feature extraction from histopathological images based on nucleus-guided convolutional neural network for breast lesion classification. Pattern Recognit. **71**, 14–25 (2017)
6. Cruz-Roa, A.A., Ovalle, J.E.A., Madabhushi, A., Osorio, F.A.G.: A deep learning architecture for image representation, visual interpretability and automated basal-cell carcinoma cancer detection. In: International Conference on Medical Image Computing and Computer-Assisted Intervention. pp. 403–410. Springer (2013)
7. Xu, J., Xiang, L., Liu, Q., Gilmore, H., Wu, J., Tang, J., Madabhushi, A.: Stacked sparse autoencoder (SSAE) for nuclei detection on breast cancer histopathology images. IEEE Trans. Med. Imaging **35**, 119–130 (2016)
8. Nayak, N., Chang, H., Borowsky, A., Spellman, P., Parvin, B.: Classification of tumor histopathology via sparse feature learning. In: 2013 IEEE 10th International Symposium on Biomedical Imaging (ISBI), pp. 410–413. IEEE (2013)
9. Chang, H., Nayak, N., Spellman, P.T., Parvin, B.: Characterization of tissue histopathology via predictive sparse decomposition and spatial pyramid matching. In: International Conference on Medical Image Computing and Computer-Assisted Intervention, pp. 91–98. Springer (2013)
10. Malon, C., Miller, M., Burger, H.C., Cosatto, E., Graf, H.P.: Identifying histological elements with convolutional neural networks. In: Proceedings of the 5th international conference on Soft Computing as Transdisciplinary Science and Technology, pp. 450–456. ACM (2008)
11. Hou, L., Samaras, D., Kurc, T.M., Gao, Y., Davis, J.E., Saltz, J.H: Patch-based convolutional neural network for whole slide tissue image classification. In: Proceedings of the IEEE Conference on Computer Vision and Pattern Recognition, pp. 2424–2433. IEEE (2016)
12. Zhou, Y., Chang, H., Barner, K., Spellman, P., Parvin, B.: Classification of histology sections via multispectral convolutional sparse coding. In: Proceedings of the IEEE Conference on Computer Vision and Pattern Recognition, pp. 3081–3088. IEEE (2014)

13. Srinivas, U., Mousavi, H.S., Monga, V., Hattel, A., Jayarao, B.: Simultaneous sparsity model for histopathological image representation and classification. IEEE Trans. Med. Imaging **33**(5), 1163–1179 (2014)
14. Vu, T.H., Mousavi, H.S., Monga, V., Rao, G., Rao, U.A.: Histopathological image classification using discriminative feature-oriented dictionary learning. IEEE Trans. Med. Imaging **35**, 738–751 (2016)
15. Arevalo, J., Cruz-Roa, A., Arias, V., Romero, E., González, F.A.: An unsupervised feature learning framework for basal cell carcinoma image analysis. Artif. Intell. Med. **64**(2), 131–145 (2015)
16. Xu, J., Luo, X., Wang, G., Gilmore, H., Madabhushi, A.: A deep convolutional neural network for segmenting and classifying epithelial and stromal regions in histopathological images. Neurocomputing **191**, 214–223 (2016)
17. Lowe, D.G.: Distinctive image features from scale-invariant keypoints. Int. J. Comput. Vis. **60**, 91–110 (2004)
18. Dalal, N., Triggs, B.: Histograms of oriented gradients for human detection. In: 2005 IEEE Computer Society Conference on Computer Vision and Pattern Recognition, CVPR, vol. 1, pp. 886–893. IEEE (2005)
19. Ojala, T., Pietikäinen, M., Harwood, D.: A comparative study of texture measures with classification based on featured distributions. Pattern Recognit. **29**, 51–59 (1996)
20. Kandemir, M., Hamprecht, F.A.: Computer-aided diagnosis from weak supervision: a benchmarking study. Comput. Med. Imaging Graph. **42**, 44–50 (2015)
21. Csurka, G. Dance, C., Fan, L. Willamowski, J., Bray, C.: Visual categorization with bags of keypoints. In: Workshop on statistical learning in computer vision, ECCV, vol. 1, pp. 1–2. Prague (2004)
22. Leung, T., Malik, J.: Representing and recognizing the visual appearance of materials using three-dimensional textons. Int. J. Comput. Vis. **43**, 29–44 (2001)
23. Jain, A.K.: Data clustering: 50 years beyond k-means. Pattern Recognit. Lett. **31**, 651–666 (2010)
24. Ahmed, H., Shedeed, H.A., Hamad, S., Tolba, M.F.: On combining nature-inspired algorithms for data clustering. In: Handbook of Research on Machine Learning Innovations and Trends, IGI Global, pp. 826–855 (2017)
25. Han, X., Quan, L., Xiong, X., Almeter, M., Xiang, J., Lan, Y.: A novel data clustering algorithm based on modified gravitational search algorithm. Eng. Appl. Artif. Intell. **61**, 1–7 (2017)
26. Tripathi, A.K., Sharma, K., Bala, M.: Dynamic frequency based parallel k-bat algorithm for massive data clustering (DFBPKBA). Int. J. Syst. Assur. Eng. Manag. II **I**, 1–9 (2017)
27. Chakraborty, A., Kar, A.K.: Swarm intelligence: a review of algorithms. In: Nature-Inspired Computing and Optimization, pp. 475–494. Springer (2017)
28. Bansal, N., Kumar, S., Tripathi, A.: Application of artificial BEE colony algorithm using Hadoop. In: 2016 3rd International Conference on Computing for Sustainable Global Development (INDIACom), pp. 3615–3619. IEEE (2016)
29. Anari, B., Torkestani, J.A., Rahmani, A.: Automatic data clustering using continuous action-set learning automata and its application in segmentation of images. Appl. Soft Comput. **51**, 253–265 (2017)
30. Pandey, A.C., Pal, R., Kulhari, A.: Unsupervised data classification using improved biogeography based optimization. Int. J. Syst. Assur. Eng. Manag. **III**, 1–9 (2017)
31. Pandey, A.C., Rajpoot, D.S., Saraswat, M.: Twitter sentiment analysis using hybrid cuckoo search method. Inf. Process. Manag. **53**, 764–779 (2017)
32. Pal, R., Pandey, H.M.A., Saraswat, M.: BEECP: Biogeography optimization-based energy efficient clustering protocol for HWSNs. In: 2016 Ninth International Conference on Contemporary Computing (IC3), pp. 1–6. IEEE (2016)
33. Kulhari, A., Pandey, A., Pal, R., Mittal, H.: Unsupervised data classification using modified cuckoo search method. In: 2016 Ninth International Conference on Contemporary Computing (IC3), pp. 1–5. IEEE (2016)

34. Pandey, A.C., Rajpoot, D.S., Saraswat, M.: Data clustering using hybrid improved cuckoo search method. In: 2016 Ninth International Conference on Contemporary Computing (IC3), pp. 1–6. IEEE (2016)
35. Rashedi, E., Nezamabadi-Pour, H., Saryazdi, S.: GSA: a gravitational search algorithm. Inform. Sci. **179**(13), 2232–2248 (2009)
36. Shaw, B., Mukherjee, V., Ghoshal, S.: A novel opposition-based gravitational search algorithm for combined economic and emission dispatch problems of power systems. Int. J. Electr. Power Energy Syst. **35**, 21–33 (2012)
37. Niknam, T., Golestaneh, F., Malekpour, A.: Probabilistic energy and operation management of a microgrid containing wind/photovoltaic/fuel cell generation and energy storage devices based on point estimate method and self-adaptive gravitational search algorithm. Energy **43**, 427–437 (2012)
38. Yin, M., Hu, Y., Yang, F., Li, X., Gu, W.: A novel hybrid k-harmonic means and gravitational search algorithm approach for clustering. Expert Syst. Appl. **38**, 9319–9324 (2011)
39. Kennedy, J.: Particle swarm optimization. In: Encyclopedia of machine learning, pp. 760–766. Springer (2011)
40. Storn, R., Price, K.: Differential evolution-a simple and efficient heuristic for global optimization over continuous spaces. J. Glob. Optim. **11**, 341–359 (1997)
41. Hatamlou, A., Abdullah, S., Othman, Z.: Gravitational search algorithm with heuristic search for clustering problems. In: 2011 3rd conference on Data mining and optimization (DMO), pp. 190–193. IEEE (2011)
42. Sun, L., Tao, T., Chen, F., Luo, Y.: An optimized clustering method with improved cluster center for social network based on gravitational search algorithm. In: International Conference on Industrial IoT Technologies and Applications, pp. 61–71. Springer (2017)
43. Hatamlou, A., Abdullah, S., Nezamabadi-Pour, H.: A combined approach for clustering based on k-means and gravitational search algorithms. Swarm Evol. Comput. **6**, 47–52 (2012)
44. Mittal, H., Pal, R., Kulhari, A., Saraswat, M.: Chaotic kbest gravitational search algorithm (CKGSA). In: Proceedings of International Conference on Contemporary Computing (IC3) (2016)
45. Leung, T., Malik, J.: Representing and recognizing the visual appearance of materials using three-dimensional textons. Int. J. Comput. Vis. **43**(1), 29–44 (2001)
46. Blue histology (Accessed on 10/04/2017). http://www.lab.anhb.uwa.edu.au/mb140/
47. Sirinukunwattana, K., Raza, S.E.A., Tsang, Y.-W., Snead, D.R., Cree, I.A., Rajpoot, N.M.: Locality sensitive deep learning for detection and classification of nuclei in routine colon cancer histology images. IEEE Trans. Med. Imaging **35**, 1196–1206 (2016)

Sentiment Score Analysis and Topic Modelling for GST Implementation in India

Nidhi Singh, Nonita Sharma, Ajay K. Sharma and Akanksha Juneja

Abstract Sentiment analysis has been widely used as a powerful tool in the era of predictive mining. However, combining sentiment analysis with social network analytics enhances the predictability power of the same. This research work attempts to provide the mining of the sentiments extracted from Twitter social application for analysis of the current trending topic in India, i.e. Goods and Services Tax (GST) and its impact on different sectors of Indian economy. This work is carried out to gain a bigger perspective of the current sentiment based on the live reactions and opinions of the people instead of smaller, restricted polls typically done by media corporations. A variety of classifiers are implemented to get the best possible accuracy on the dataset. A novel method is proposed to analyse the sentiment of the tweets and its impact on various sectors. Further, the sector trend is also analysed through the stock market analyses and the mapping between the two is made. Furthermore, the accuracy of stated approach is compared with state-of-the-art classifiers like SVM, naïve Bayes and random forest and the results demonstrate accuracy of stated approach outperformed all the other three techniques. Along with this, topic modelling was also done to get a picture of trending topics that are linked to GST. LDA and text ranking algorithms were applied to get connected topics.

Keywords Sentiment analysis · Goods and services tax · Classification
Text mining · Topic modelling · Text rank · LDA · Support vector machine
Naïve Bayes classifier · Random forest

N. Singh (✉) · A. K. Sharma
National Institute of Technology Delhi, New Delhi 110040, Delhi, India
e-mail: 162211001@nitdelhi.ac.in

A. K. Sharma
e-mail: director@nitdelhi.ac.in

N. Sharma
Dr. B. R. Ambedkar National Institute of Technology Jalandhar, Jalandhar, India
e-mail: nonitasharma@nitdelhi.ac.in

A. Juneja
Jawaharlal Nehru University, New Delhi 110040, Delhi, India
e-mail: akankshajuneja.jnu@gmail.com

© Springer Nature Singapore Pte Ltd. 2019
J. C. Bansal et al. (eds.), *Soft Computing for Problem Solving*, Advances in Intelligent
Systems and Computing 817, https://doi.org/10.1007/978-981-13-1595-4_19

1   Introduction

Goods and Services Tax (GST) has been a trending topic on Twitter since the announcement made by the Indian government for its imposition in India. It is an indirect tax or value-added tax for the whole nation, which will make India one unified common market [1]. Earlier there used to be different types of taxes for state and central government (local body tax (LBT), value-added tax (VAT), service tax, etc.). GST, on the other hand, is a single tax on the supply of goods and services, right from the manufacturer to the consumer. The final consumer will thus bear only the GST charged by the last dealer in the supply chain, with set-off benefits at all the previous stages. Thus, the implementation of GST seems to be very promising for Indian markets since no cross-utilization of credit would be permitted. Hence, there is an imperative need to analyse the impact of GST on Indian economy using sentiment of common people.

Sentiment is a simple, view or opinion held or expressed by an individual or group of individuals. The process of identification or classification of sentiments in any sentence or text computationally is called sentiment analysis [2]. One of the techniques is to calculate the sentiment score of the sentence and classify it according to this. Sentiment score represents the numerical value of polarity of sentence. The sentiment polarity is a verbal representation of the sentiment. It can be "negative", "neutral" or "positive". The value less than zero represents the negative sentiment, a value greater than zero represents positive sentiment and value equal to zero represents neutral sentiment. The sentiment score can be calculated for any sentence, document, named entities, themes and queries.

To attain a large and varied dataset of recent public opinions on the GST, we have used Twitter to gain real-time access of the opinions across the country. Different classification techniques are used to better understand the dataset that is generated and then to analyse the sentiment accurately. Support vector machines that find a hyperplane as decision boundary are used in this manuscript for classification of tweets. Another common classifier for text categorization, i.e. naïve Bayes which works on probabilities, is also implemented. Then classification is boosted using ensemble method. The ensemble was implemented using random forests that construct a number of decision trees for classification. The methodology involves collection of tweets, pre-processing them and then using for analysis of sentiment. For analysis purpose, we have employed different classifiers and their accuracies are tabulated in the results section.

Topic modelling is another aspect of text mining. As there is data explosion, we need to analyse documents or sentences to find similarity among them. In topic modelling, we extract similarity out of documents and convert them into graphs. There are various models to perform this modelling. It uses concepts of machine learning, natural language processing and statistics to derive the model. The most commonly used models are latent Dirichlet analysis (LDA) and text ranking algorithm. In LDA model, the document is viewed as mixture of different topics in the given corpus. The model proposes that each word in the document is attributable to one of the document's topics. In text ranking, we use graph-based algorithms that find linkages between different topics across a given set of documents. The most prevalent topic is highly connected in the graph. We have used these techniques to find the topics that are reacted to us GST in our collected tweets.

This manuscript proceeds as follows. Related work is reviewed and discussed to provide the theoretical background and foundation for our study in Sect. 2. We then describe the data and discussion on sentiment analysis procedures in Sect. 3. In Sect. 4, we present our results about the topic. Finally, we conclude the paper by discussing study implications and suggesting future research directions.

2 Literature Review

Wilson et al. [3] have described the method to find the contextual polarity. In contextual polarity, the words comprising a phrase are to be considered and then the sentiment is found for the complete phrase. This method has to be used for a large dataset where each sentence is complex. Complex sentence here refers to a sentence consisting of more than two phrases.

Alessia et al. [4] have described firstly about the types of approaches to do sentiment classification along with their advantages and disadvantages. Then it dealt with the tools with respect to the techniques discussed. This paper also discusses the emerging domains where sentiment analysis can be done.

Barbara and Schindler [5] discussed that social media has been exploding as a category of online discourse where people create content, share and discuss in communication network. From a business and marketing perspective, they noticed that the media landscape has dramatically changed in the recent years, with traditional now supplemented or replaced by social media. In contrast to the content provided by traditional media sources, social media content tends to be more "human being" oriented.

Yu et al. [6] discussed the use of social media and trends that are available on social media instead of conventional media. Since the major proportion of our population uses social media, so a huge amount of information is generated on any issue. This information can be used to carry out analysis and research. This paper discussed the sentiment analysis of social networking data for stock market analysis.

Prabowo and Thelwall [7] discussed the use of multiple classifiers in sentiment analysis. This paper combined rule-based classification, supervised learning and machine learning into a new combined method. This method was then tested on datasets and accuracy, and recall was shown to be increased compared to existing methods.

Hiroshi et al. [8] have discussed automated sentiment analysis. Various approaches have been applied to predict the sentiments of words, expressions or documents. These are natural language processing (NLP) and pattern-based machine learning algorithms, such as naive Bayes (NB), maximum entropy (ME), support vector machine (SVM) and unsupervised learning.

3 Methodology

3.1 Introduction

The framework for this research work is presented here in this section. The process starts with Twitter search API. From this API, we get keys and tokens to retrieve the tweets in real-time scenario (Fig. 1).

The data used in this work is real-time data in the form of tweets. It is collected using the Twitter search API that allows developers to extract tweets programmatically. The tweets were collected here for the period of 30 days, i.e. from 11 July 2017 to 15 September 2017 using the "twitteroauth" version of public API[1] by Williams (2012). This was the period right after the launch of GST. On an average, everyday there were around 3000 tweets. The tweets were extracted on daily basis and stored in comma separated value (CSV) files. Hence, the data that was under consideration comprised of more than 1,00,000 tweets. With each tweet, there comes metadata like user id, name, tweet id, text, longitude, latitude. For the analysis in this work, we use text, user id and name.

Once an issue pops up on Twitter, there are numerous hashtags created from the same issue. Similarly for the GST, many hashtags were used. So we selected the best five hashtags out of all and used them as Twitter handles to collect tweets from Twitter. The most used hashtags are GST, GSTBill, gstrollout and GST for New India [9]. The tweets are generated in three given ways: original, re-tweets and replies. Original tweets are any messages posted on Twitter. Re-tweets are reposting

Fig. 1 Framework of the sentiment analysis

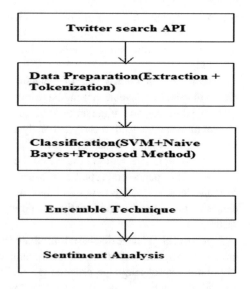

---

[1] https://dev.twitter.com/rest/public/search.

Fig. 2 Data extraction
procedure

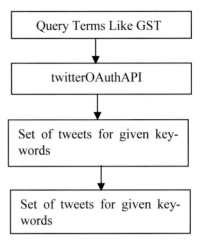

of another person's tweet. Replies are answers to some queries or a form chat. These tweets were retrieved from the Twitter as JSON objects and then converted into textual format for analysis. There are pre-defined functions to convert into text as shown in Fig. 2.

Once we get text from tweet, it needs to be filtered to remove unnecessary information as the nature of tweets is random and casual. In order to filter out these useless data, the Standford Natural Language Processing (SNLP group 2015) is used.[2] It is an open-source tool that gives the grammatical relations between words in a sentence as output. Then process of tokenization is carried out. Tokenization is breaking up of sentences into words, phrases, keywords or tokens. A token is an instance of sequence of characters that are grouped together as a useful semantic unit for processing. This breaking up is as per the application requirement. These tokens are then used for the text mining and parsing purposes. There are inbuilt methods for tokenization in natural language processing (NLP) toolkits like Natural Language Toolkit, Apache Lucene and Solr, Apache OpenNLP.

3.2 Classification

Classification is the process of categorization of new patterns with respect to the old patterns on which it is trained where class labels are already known. There are two kinds of classifiers: lazy and active. Lazy (e.g. naïve Bayes) are those which don't build any model until testing starts. Active (e.g. decision trees) are those which create a model as training data is presented to them.

Support vector machine (SVM) is an active classifier that measures the complexity of hypothesis based on margin that separates the plane and not number of features.

---

[2]https://nlp.stanford.edu/projects/DeepLearningInNaturalLanguageProcessing.shtm.

The input and output patterns for the SVM are 0/1(0–negative/1–positive). Sentences or tweets as retrieved are not suitable for classification. The transformation needs to be done so that the input is recognized by the classifier. For this, we have already discussed the pre-processing/tokenization of tweets. There are two kinds of SVM classifiers: linear SVM classifier and nonlinear SVM classifier. In the linear classifier model that we used, we assumed that training examples are plotted in 2D space. These data points are expected to be separated by an apparent gap. It predicts a straight hyperplane dividing two classes. The primary focus while drawing the hyperplane is on maximizing the distance from hyperplane to the nearest data point of either class. The drawn hyperplane is called maximum-margin hyperplane.

Naive Bayes classifiers have the ability to classify any type of data: text, networks features, phrases, parse trees, etc. In naïve Bayes classifier, a set of words as bigrams or trigrams can also be used. This classifier takes a small set of positive and negative words and finds the probability of each tweet with respect to these words and then finds the conditional probability. Below is the formula for calculating the conditional probability.

$$P(X|C) = \frac{(P(C|X) * P(C))}{P(X)} \tag{1}$$

where

$P(C)$ is the probability of class C. This is known as the prior probability.
$P(X)$ is the probability of the evidence (regardless of the class).
$P(X|C)$ is the probability of the evidence given it belongs to class C.
$P(C|X)$ is the probability of the class C true, and it is an evidence

The probability that a document belongs to a class C is given by the maximum of class probability $P(C)$ multiplied by the products of the conditional probabilities of each word for that class. It is also called as maximum a posteriori (MAP). It is given as:

$$MAP(C) = max(P(X|C)) \tag{2}$$

This is used to calculate the probability of tweet with respect to each class, i.e. positive and negative.

Ensemble is a technique of combining two or more algorithms of similar or dissimilar types called base learners. This is done to increase the robustness. For each tweet, the classification process is carried out by every classifier in the ensemble (it may also be a classifier of the same type, but trained on a different learning sample). Then the results of all classifiers are aggregated as the final ensemble classifier. This can be done in three ways: (a) majority-based voting, (b) weighted voting and (c) rank-based methods. In the case of tie, randomly anyone is chosen. Here or classification of tweets, we have used random forests as an ensemble technique. In random forest, a specified number of decision trees are constructed for classification on the partitioned datasets.

Pseudocode:

Input: Dataset D (set of Tweets extracted from a particular handle)

Training:
1. Select "m" words from every tweet randomly.
2. From these "m" words, calculate the splitting point "s".
3. Break it into child nodes using split points calculated.
4. Repeat steps 1 through 3 till threshold is reached.
5. Build "n" trees using steps 1–4.

Testing:
1. Select "m" words randomly and using created decision tree to predict the outcome.
2. Find the number of votes for each class C.
3. Find the maximum voted for class.

3.3 Proposed Method

The approach followed here is to count the positive and negative words in each domain-specific tweet and assign a sentiment score. It is given as:

$$Sentiment\ Score = Pc - Nc \tag{3}$$

where Pc represents positive word count and Nc represents negative word count.

This way, we can ascertain how positive or negative a tweet is

Pseudocode:

Input: Dataset D (set of Tweets extracted from a particular handle)

---

1. Create a list of domain/sector-specific words (e.g., auto, FCMG)
2. Create a list of domain-specific positive words.
3. Create a list of domain-specific negative words.
4. Apply functions to remove punctuation symbols, extra spaces, tabs, digits, RT, etc.
5. Convert obtained text to lowercase text.
6. Filter the tweet according to the domain.
7. Calculate the sentiment score for each domain-specific tweet.
    i. Count the number of the positive words in sentence as p.
    ii. Count the number of the negative words in sentence as n.
    iii. Calculate the sentiment score as: Score $= p - n$
8. Categorize the tweets into three categories:
    i. Positive Class where Score > 0
    ii. Negative Class where Score < 0
    iii. Neutral Class where Score $= 0$.

3.4 Topic Modelling

Topic modelling is a technique that is frequently used to mine data in text mining. It is tool to find semantic structures in a document or sentence. These models are called as probabilistic models as it uses probability and statistics to find the latent semantic structures. Practically, we try to fit the parameters of model to accommodate the data corpus completely using techniques like maximum likelihood fit. Another assumption that we follow here is that is distributed in known probabilistic distributions (Fig. 3).

Text rank is a graph-based clustering algorithm to the rank of all the vertices in a given graph. A node is ranked higher than its peers if it has more connections to other nodes. The graph with highest degree of connection is the most important topic in the given set of documents. To find the rank, we use the following formula:

$$R(Vi) = (1 - d) + d * \frac{R(Vj)}{out(Vj)} \tag{4}$$

where

$R(Vi)$ is the rank of ith vertex in given graph G(V, E)
D is the damping factor
$out(Vj)$ Id the out degree of the vertex jth that is connected to ith

The formula is used in the algorithm along with other steps to give the output as graph. The pseudocode of the algorithm is given as:

Latent Dirichlet analysis is approach that teases out the topics that are significant in the document. It is used to find the themes out from the given set of documents. It creates a probabilistic model with a generative process. The LDA model discovers

Fig. 3 Topic modelling procedure

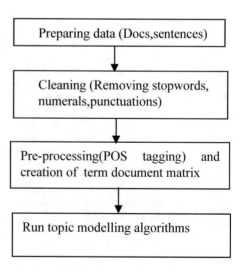

Input: Dataset D (set of Tweets extracted from a particular handle)

Code:
- Select the text units that are related to each other and put them in graph.
- Find the relation that connects various vertices to each other in the given graph.
- Iterate the graph-based ranking algorithm until the convergence.
- Sort the vertices based on their scores.

Output: Graph G (V, E)

the different topics that different documents represent and amount of information contained in them. In LDA, we view every document as mixture of topics that follow any probability distribution. Then we sample topics out of them and find their conditional probabilities. The pseudocode is given as:

Input: Dataset D (set of Tweets extracted from a particular handle)

Code:
- For each tweet t, for each word w;
- Calculate P(context c | tweet t) that is ratio of words is tweet t that are assigned to a given context c.
- Calculate P(word w| context c) that is ratio of assignments to context c, overall tweets t that comes from word w.
- Reassign word w a new context c' where we choose context c' with P(context c'| tweet t)* P(word w| context c')
- This generative model gives the probability that context c' generated word w.
- Repeat these steps a large number of times till steady state is reached.

Output: Documents and important terms in a list.

4 Results and Discussions

In this research, we have constructed our own dataset of about 1 lakh tweets for 30 days using the Twitter search API. On a daily basis, around three thousand tweets were generated. After the pre-processing that was discussed in Sect. 3, the classification algorithms and other analysis were done. The hashtags used are GST, GSTBill, gstrollout and GST for New India. The R language was used for the analysis and classification part (Table 1).

Table 1 Classification results

Classifier	Accuracy (%)
SVM	64.34
Naïve Bayes	58.44
Sentiment score (proposed method)	68.75
Random forests	65.21

4.1 Accuracy

In this manuscript, we have used four different classification techniques: SVM, naïve Bayes, sentiment score and random forests. The accuracy was found using

$$Accuracy = \frac{(TP + TN)}{(TP + TN + FP + FN)} \tag{5}$$

where TP, FP, TN, FN being number of true positives, false positives, true negatives and false negatives, respectively.

We have tabulated their results as follows:

As it can be inferred from the table, the accuracy of all classifiers lies in the range of 60–70%. The highest accuracy was found in proposed method, i.e. sentiment score and lowest in the naïve Bayes classifier. SVM and random forest have performed fairly well.

4.2 Text Models

The different models were created and results are shown here. Text ranking algorithm gave the output as graph of connected nodes in Fig. 4. This graph shows the topics that are linked to each other in the given set of tweets. Since this was economic term, the related terms that were found out are demonetization, sectors of economy, transport, entertainment, etc. Some graph can exist as disconnected unit from the bigger graph as they did not have associations with the terms of larger graph. It is more visually appealing techniques as it outputs the graph which can be interpreted easily.

Another technique that was applied was LDA. It gave the list of themes that can be found in the given set of documents and it gave a table showing the mapping. Table 2 was generated as result for the set of ten documents. It can be clearly inferred from the results that GST was the main theme of all the ten documents followed by demonetization, economy, education, sector-wise growth, etc. It was an economic term and hence the term associated with it and is prevalent in the result set (Table 2).

Fig. 4 Graph from text ranking algorithm

Table 2 Results of LDA

Document no.	Topics
Document 1	GST, Education, Demonetization
Document 2	Economy, GST, Costs
Document 3	Education, GST, Common man
Document 4	Sectors, Demonetization, GST
Document 5	Black money, GST, Demonetization,
Document 6	GST, Transport, Sector
Document 7	GST, Logistics, costs
Document 8	GST, Sector, common man
Document 9	GST, Demonetization, Sector
Document 10	GST, Entertainment, Sector

5   Conclusions

In summary, we have performed a classification of sentiment towards the launch of new taxing policy in India, i.e. GST. It was analysed that overall the attitude of Indian population was neutral towards it. The analysis was also done gender-wise, region-wise. Apart from this, time series analysis was done. It has shown an increase in positive attitude and neutral attitude. Further, sector-wise analysis was also done. The topic modelling techniques were also applied on the tweets, and results were shown.

As a future work, historical data analysis of GST can be done using tweets from before enforcement of the bill till the bill was enforced successfully. It can give a more a clear picture of the opinions transformations during this period. Other classifiers may also be tested on this dataset as it is large. If the accuracy is not improved, then boosting and bagging can be utilized to enhance the performance. This analysis can also be used in other domains like politics, product reviews.

References

1. Dani, S.: A research paper on an impact of goods and service tax (GST) on Indian economy. Bus Eco. J. **7**, 264 (2016). https://doi.org/10.4172/2151-6219.1000264
2. Liu, B.: Sentiment analysis and opinion mining. Synth. Lect. Hum. Lang. Technol. **5**(1), 1–167 (2012)
3. Wilson, T., Wiebe, J., Hoffmann, P.: Recognizing contextual polarity in phrase-level sentiment analysis. In: Proceedings of the Conference on Human Language Technology and Empirical Methods in Natural Language Processing. Association for Computational Linguistics (2005)
4. Alessia, D., et al.: Approaches, tools and applications for sentiment analysis implementation. Int. J. Comput. Appl. **125**(3) (2015)
5. Barbara, B., Schindler, R.M.: Internet forums as influential sources of consumer information. J. Interact. Mark. **15**(3), 31–40 (2001). ISSN 1094–9968. http://dx.doi.org/10.1002/dir.1014
6. Yu, Y., Duan, W., Cao, Q.: The impact of social and conventional media on firm equity value: a sentiment analysis approach. Decis. Support Syst. **55**(4), 919–926 (2013). ISSN 0167–9236. http://dx.doi.org/10.1016/j.dss.2012.12.028
7. Prabowo, R., Thelwall, M.: Sentiment analysis: A combined approach. J. Informetr. **3**(2), 143–157 (2009)
8. Hiroshi, K., Tetsuya, N., Hideo, W.: Deeper sentiment analysis using machine translation technology. In: Proceedings of the 20th International Conference on Computational Linguistics (COLING 2004), pp. 494–500. Geneva, Switzerland 23–27 Aug 2004
9. Trending Hashtags. https://www.hashtagify.com. Accessed 15 Sept 2017

Numerical Investigation of Flexural Properties of Curved Layer FDM Parts

Xiongbin Peng, Biranchi Panda, Akhil Garg, H. Guan and M. M. Savalani

Abstract The printing approach of curved layer fused deposition modelling (CLFDM) is gaining popularity for its superior part properties when compared to those obtained from conventional FDM approach. Past studies reveal that the study of effects of fill gap (FG) on these two approaches (CLFDM and conventional FDM) with different printing directions needs a thorough numerical investigation. Therefore, the present work introduces numerical modelling based on genetic programming for investigating the flexural strength of the fabricated parts from FDM and CLFDM. It was found that the GP-based flexural strength models are able to generalize both processes satisfactorily. It was also noticed that the GP-based flexural strength models for horizontal printed parts predict with higher accuracy than those for parts printed in vertical directions. Experiments were performed to validate the GP models.

Keywords Fill gap · Raster angle · Flexural strength · CLFDM
Genetic programming

1 Introduction

3D printing technology has gained significant attention in recent years because of its ability to manufacture complex shaped prototypes from computer-aided design (CAD) data automatically without much need for the human intervention [1–3]. This

X. Peng · A. Garg (✉)
Intelligent Manufacturing Key Laboratory of Ministry of Education,
Shantou University, Shantou, China
e-mail: akhil@stu.edu.cn

B. Panda
Department of Industrial Design, National Institute of Technology, Rourkela, Rourkela, India

H. Guan · M. M. Savalani
Department of Industrial and Systems Engineering, Hong Kong
Polytechnic University, Kowloon, Hong Kong

© Springer Nature Singapore Pte Ltd. 2019
J. C. Bansal et al. (eds.), *Soft Computing for Problem Solving*, Advances in Intelligent
Systems and Computing 817, https://doi.org/10.1007/978-981-13-1595-4_20

technology has been extensively used in architecture design and industrial engineering and medical applications such as for organ printing. Mironov et al. [4] have used 3D printing processes to study tissue engineering questions in organ transplantation crisis. However, the present technology enables the solid models to be printed layer by layer, and the properties of the parts are anisotropic in nature. This nature of 3D printing technology offers limit to its diversified applications. Researches have explored the procedures to counter this problem and enhance the mechanical properties. Literature [5–10] studies have reported that the printing parameters have a significant influence on part mechanical property. For instance, Bellini and Güçeri [11] proposed the experimental procedure and tested the effect of raster angles on mechanical properties such as the tensile strength and flexural strength of parts. Recently, the use of curved layer fused deposition modelling (CLFDM) that slices the thin shell 3D model and prints the model minimizes the staircase effect and shortens the printing time and saves printing materials. This technology is an emerging one, and studies have been reported which determine its superior mechanical properties over the planar layer-by-layer FDM approach [12–17]. The studies conducted measure the effect of raster angles (RAs) on the mechanical strength. As discussed, the anisotropic nature of the process results in variations in flexural strength values. Therefore, it would be interesting to study the interactive effect of FG and its printing directions (horizontal or vertical or criss-cross (CC) on the flexural strength of parts fabricated by CLFDM approach. The values of flexural strength achieved from the CLFDM process in these directions by varying FG will be compared to those obtained from the conventional layer-by-layer FDM approach. Further, the quantification of the flexural strength with respect to FG and the printing directions can be useful for the experts in prediction and monitoring of the CLFDM and FDM approaches. In this context, the numerical modelling based on genetic programming (GP) algorithm can be used to formulate the explicit and generalized models.

Therefore, the present work investigates the effects of varying the FG on the flexural strength of the parts fabricated by two technologies (conventional planar layer-by-layer FDM and CLFDM). CLFDM and FDM approaches are implemented by printing the parts in three different directions of horizontal, vertical and criss-cross (CC). In this work, the first set of samples is printed in horizontal direction with FG ranging from 0.5 to 1.1 mm and with interval of 0.2 mm. The second set of samples is printed in vertical direction with FG of 0.5, 0.9 and 1.1 mm. The third set of samples is printed using CLFDM with CC approach, and the corresponding FGs are 0.5, 0.7 mm. The data samples collected on the flexural strength values based on CLFDM and FDM approaches in these directions are then given as input to framework of GP algorithm to formulate the models representing the explicit relationship between the flexural strength and FG. The statistical analysis is then performed on the formulated models to evaluate its extrapolation ability. The following section discusses the experimental procedures for measuring the flexural strength from CLFDM set-up in three directions.

2   Experimental Design of CLFDM and Data Acquisition

In this section, the details of the experiment conducted on CLFDM machine for measuring the flexural strength are discussed. The sample dimension chosen can be referred in works done by Huang and Singamneni [14]. In this work, as a result of

Fig. 1   CLFDM sample dimension (mm)

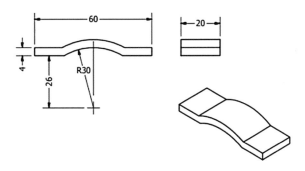

Fig. 2   Flexural strength with respect to fill gap (FG) for printed part in horizontal direction

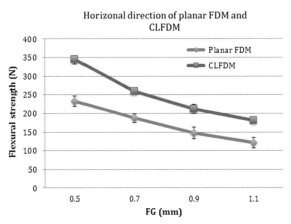

Fig. 3   Flexural strength with respect to fill gap (FG) for printed part in vertical direction

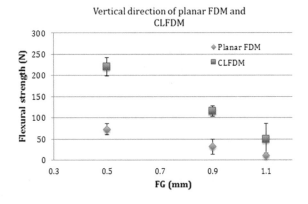

the jig and fixture requirement, an additional 10 mm for the length of the samples was extended. Therefore, the length of the part considered in this work increased from 50 to 60 mm. Figure 1 shows the whole dimension of the sample.

A total of 80 samples are fabricated. In order to clearly note the samples, a notation is introduced. An INSTRON 5566 material testing equipment was used to do a three-point bending test. Five samples were tested for each parameter, and the loading speed was set at 5 mm/sec based on the preliminary trials and the ISO 1209.

The flexural strength is studied as a function of fill gap for two approaches: Planar FDM and CLFDM. The comparison between the two FDM approaches is introduced. Figure 2 shows that with an increase in FG, the flexural strength of parts fabricated by both CLFDM and conventional planar FDM decreases. The parts fabricated by CLFDM could withstand almost 30.9% more flexural force than these printed by conventional layer-by-layer FDM. The results are found to be in good agreement with the actual findings by Singamneni et al. [13]. Figure 3 shows the effects of different FG on parts flexural strength printed in vertical direction by both conventional layer-by-layer FDM and CLFDM.

3   Flexural Strength Model Formulation Using GP Algorithm

3.1   Data Acquisition and GP Algorithm Description

In total, 36 data samples were collected from the experimental procedure of FDM and CLFDM in horizontal and vertical directions. The regression splines were used to extrapolate and interpolate the collected points so that the 36 samples are sufficient for training and testing of models formulated from GP algorithm. It should be noted that the learning capability of the algorithm depends on the selection of training and testing data, which is obtained by randomly selecting it from the given data. About 80% of the samples were used as a set of training data while the remaining were used as the testing data. The data is then fed into the paradigm of GP cluster for the training of models. Genetic programming (GP) (Fig. 4) is an evolutionary approach that mimics the process of biological evolution [18, 19]. The mathematical models in GP are laid on symbolic regression—a type of analysis that searches the space of mathematical expressions to find the best-fit model of a given dataset. Usually, these models or programs are represented by tree structures [20].

The GP simulations are performed in MATLAB R2010 with the parameters such as population size, number of generations, depth of the tree, tournament size, number of iterations set at 400, 100, 7, 5 and 8, respectively. The settings are shown by screenshot for GP code written in M-file (Fig. 5). The crossover, mutation and direct reproduction probabilities are taken as 0.85, 0.1 and 0.05, respectively. The best GP models (Eqs. A1–A4 in Appendix) are selected based on the minimum mean absolute percentage error (MAPE), and their performance is discussed in Sect. 4.

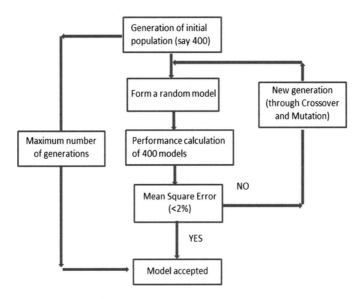

Fig. 4 Schematic of GP algorithm

Fig. 5 Screenshot of M-file code of GP

Table 1 Statistical metrics of the GP-based flexural strength models

Models	R^2		RMSE (%)		MAPE (%)	
	Training phase	Testing phase	Training phase	Testing phase	Training phase	Testing phase
CLFDM_GP	0.98	0.95	1.39	16.98	0.49	8.01
FDM_GP	0.99	0.98	0.42	3.39	0.17	2.90
CLFDM_GP	0.96	0.88	12.17	16.45	6.31	11.29
FDM_GP	0.94	0.86	5.22	18.10	8.10	17.82

$$MAPE(\%) = \frac{1}{n} \sum_i \left| \frac{A_i - M_i}{A_i} \right| \times 100 \tag{1}$$

where M_i and A_i are the predicted and actual values, respectively; $\overline{M_i}$ and $\overline{A_i}$ are the average values of the predicted and actual values, respectively; and n is number of training samples.

3.2 Statistical Analysis of the GP-Based Flexural Strength Models

The GP models (Eqs. A1–A4 in Appendix) for flexural strength are evaluated against the experimental data as discussed in Sect. 2. The three performance measures (correlation coefficient (R2), root mean square error (RMSE), and multi-objective error (R2 + RMSE/R2)) evaluate the performance of the GP models. Table 1 clearly shows that the GP-based flexural strength models for FDM and CLFDM parts in horizontal directions have very good training accuracy with higher values for the coefficient of determination and lower values of MAPE and RMSE. Similarly, on the testing data the GP-based flexural point models have shown higher generalization performance. The curves of the predicted GP models and actual experimental values are shown in Fig. 6. This clearly shows that the flexural strength values for the horizontal printed CLFDM (Fig. 6a) and FDM (Fig. 6b) parts obtained from the GP models are in well agreement with the actual values and found to be better than those of vertical printed parts [9]. Similarly, the performance of the GP-based flexural strength models for vertical printed CLFDM and FDM parts is also acceptable.

4 Conclusion

The present work addresses the motivation behind conducting the numerical investigation on study of fill gap on the flexural strength of the FDM and CLFDM printed parts in horizontal, vertical and CC directions. The numerical procedure based on

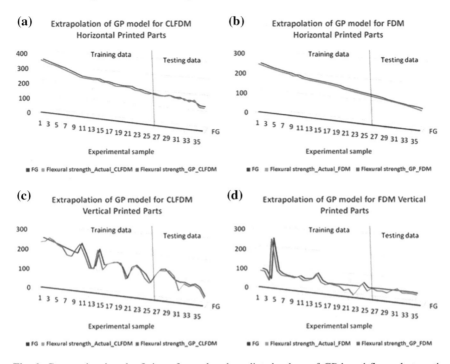

Fig. 6 Curves showing the fitting of actual and predicted values of GP-based flexural strength models, **a** and **b** horizontal, **c** and **d** vertical printed parts from CLFDM and FDM

evolutionary approach of GP is proposed to formulate the functional relationships between the flexural strength and the FG for the parts printed in horizontal and vertical directions. These explicit and generalized relationships of flexural strength can be used by expert's offline to predict and monitor the flexural strength, thus saving the experimental resources. It was also found that the models based on GP could predict the flexural strength of horizontal printed parts with accuracy higher than those of vertical printed parts. Future work for authors is to consider additional input layer thickness and propose a generic GP approach which could predict the flexural strength of vertical printed parts accurately.

Acknowledgements We would like to thank the Hong Kong Polytechnic University under the project code: G-YM77 titled "An Innovative Approach to Additive Manufacturing" for their support. This study is supported by Shantou University Scientific Research Foundation (NTF 16002, NTF 16011). Authors would also like to acknowledge Guangdong Sailing Plan Talent fund of the year 2016. Authors also like to acknowledge Guangdong University Youth Innovation Talent Project (2016KQNCX053) Supported by Department of Education of Guangdong Province.

Appendix

Flexural strength Horizontal_CLFDM_GP

$$= -1611.6967 + (-14.945) * (\cos(\cos((\tan((x1) + ((16.361923))))$$
$$+ ((-18.517307))))) + (193.3434) * (\tan((\tanh((x1) + ((-18.517307)))) * (plog(x1))))$$
$$+ (-5.6196) * (\sin(\exp(\exp(x1)))) + (64.4466) * (plog(\tan((\sin(\sin(\tanh(x1)))) * (plog(\tanh(x1)))))$$
$$+ (1.4019) * (((plog(\tan(x1))) - (\tan(((15.521232)) - (x1)))) * (plog(\tan(((-18.517307))$$
$$* (x1))))) + (-57.8947) * (plog(\tan((\exp(\cos(\cos(x1)))) * (plog(\tanh((x1) + ((16.361923))))))))); \quad (A1)$$

Flexural strength Horizontal_FDM_GP

$$= 23.6759 + (3.9216). * (\sin(\cos(\tan((plog(\tan(x1))) + (\exp(x1))))))$$
$$+ (-55.6616). * (plog((\tan(x1)). * ((\cos(\tan(\cos(x1)))) + (x1)))) + (63.1144.$$
$$* (plog(((-0.816101)). * ((\cos(\tan(\cos(x1)))) + ((-17.537172)))))$$
$$+ (-0.15459). * (\tan(\tanh(\cos(\exp(((x1) - ((-17.537172))) + (\cos(x1)))))))$$
$$+ (18.2054). * ((\sin(\exp(\tan(\exp((-8.889970)))))) - (x1))$$
$$+ (2.4745). * (\sin(\tanh(\tan(((((x1) - ((-13.964471)))$$
$$- (\cos((15.506033)))). * (\sin(\sin(x1)))))))); \quad (A2)$$

Flexuralstrength Vertical_CLFDM_GP

$$= -223.2723 + (362.7024) * (\exp(\sin(\exp(\exp(\tan(((0.157757)) * (x1)))))))$$
$$+ (0.018109) * ((((\tan(\tan((x1) * (x1)))) * (\exp(\sin(x1)))) * (\exp(\exp(\sin((x1) * (x1))))$$
$$* (\tan(\exp(plog(\tan(\exp(x1)))))) + (0.14521) * (\tan(\tan(((\tanh(x1)) + (\exp(x1)))$$
$$- (\exp(\sin((x1) * (x1)))))) + (31.4346) * (\cos((\exp(\tan(\tan((x1) * (x1)))) * (\exp(\cos((\cos(x1))$$
$$- (plog(x1)))))) + (24.5903) * (\cos((\tan(\tan((x1) * (x1)))) * (\exp(\sin(x1)))))$$
$$+ (0.85622) * (\tan(((x1) + (\tan(x1))) + (\tan(\exp(\tan(((-10.254681)) - (x1))))))); \quad (A3)$$

Flexuralstrength Vertical_FDM_GP

$$= 31.4187 + (2.6007). * (\tan((\exp((\tan(\tanh(x1))) + (\tanh(x1)))) - (plog(\tanh(\tanh(\cos(x1)))))))$$
$$+ (-23.0296). * (\sin(plog(\tan(x1)))) + (-28.9959). * (\tanh((\cos(x1)) - (\exp(\tanh(\exp(x1))))))$$
$$+ (-9.2956). * (((\tanh(\tan(\sin(\tanh(x1))))) - (\tan((\tanh((x1)+(x1))). * (\cos(\tan(x1)))))).$$
$$* ((\sin(\tan(plog(\cos(x1))))). * ((((\exp(x1)) - (\tan(x1))). * (\tan((x1) + ((10.438387)))))$$
$$- (\sin(plog(((-5.811520)). * (x1)))))))) + (7.487). * (\sin((\sin(((15.712123)). * (x1))). * (x1)))$$
$$+ (15.894). * (\sin((((\cos((x1) + (x1))) - ((\exp((11.195960)))$$
$$- (((19.548163)) + (x1))). * (\sin(\tan(x1)))) + ((\sin(\sin(\exp(x1)))). * (plog(\tan(x1)))))); \quad (A4)$$

References

1. Panda, B., Tan, M.J., Gibson, I., Chua, C.K.: The disruptive evolution of 3D printing. In: Chua, C.K., Yaong, W.Y., Tan, M.J., Liu, E., Tor, S.B. (eds) Proceedings of 2nd International Conference Progress Additive Manufacturing (Pro-AM 2016). National Research Foundation, Singapore, pp. 152–157 (2016)
2. Gibson, I., Rosen, D.W., Stucker, B.: Additive manufacturing technologies, vol. 238. Springer, New York (2010)
3. Mahapatra, S.S., Panda, B.N.: Benchmarking of rapid prototyping systems using grey relational analysis. Int. J. Serv. Oper. Manag. **16**(4), 460–477 (2013)

4. Mironov, V., et al.: Organ printing: computer-aided jet-based 3D tissue engineering. Trends Biotechnol. **21**(4), 157–161 (2003)
5. Panda, B., Paul, S.C., Tan, M.J.: Anisotropic mechanical performance of 3D printed fiber reinforced sustainable construction material. Mater. Lett. **209**, 146–149 (2017)
6. Panda, B.N., Bahubalendruni, R.M., Biswal, B.B., Leite, M.: A CAD-based approach for measuring volumetric error in layered manufacturing. Proc. Inst. Mech. Eng. Part C J. Mech. Eng. Sci. **231**(13), 2398–2406 (2017)
7. Sood, A.K., Ohdar, R.K., Mahapatra, S.S.: Parametric appraisal of mechanical property of fused deposition modelling processed parts. Mater. Des. **31**(1), 287–295 (2010)
8. Thrimurthulu, K., Pandey, P.M., Reddy, N.V.: Optimum part deposition orientation in fused deposition modeling. Int. J. Mach. Tools Manuf. **44**(6), 585–594 (2004)
9. Panda, B., Paul, S.C., Mohamed, N.A.N., Tay, Y.W.D., Tan, M.J.: Measurement of tensile bond strength of 3D printed geopolymer mortar. Measurement **113**, 108–116 (2018)
10. Hutmacher, D.W., et al.: Mechanical properties and cell cultural response of polycaprolactone scaffolds designed and fabricated via fused deposition modeling. J. Biomed. Mater. Res. **55**(2), 203–216 (2001)
11. Bellini, A., Güçeri, S.: Mechanical characterization of parts fabricated using fused deposition modeling. Rapid Prototyping J. **9**(4), 252–264 (2003)
12. Guan, H.W., et al.: Influence of fill gap on flexural strength of parts fabricated by curved layer fused deposition modeling. Proc. Technol. **20**, 243–248 (2015)
13. Singamneni, S., et al.: Modeling and evaluation of curved layer fused deposition. J. Mater. Process. Technol. **212**(1), 27–35 (2012)
14. Huang, B., Singamneni, S.B.: Curved layer adaptive slicing (CLAS) for fused deposition modelling. Rapid Prototyping J. **21**(4), 354–367 (2015)
15. Huang, B., Singamneni, S.: Curved layer fused deposition modeling with varying raster orientations. Appl. Mech. Mater. **446**, 263–269 (2014)
16. Klosterman, D.A., et al.: Development of a curved layer LOM process for monolithic ceramics and ceramic matrix composites. Rapid Prototyping J. **5**(2), 61–71 (1999)
17. Chakraborty, D., Reddy, B.A., Choudhury, A.R.: Extruder path generation for curved layer fused deposition modeling. Comput. Aided Des. **40**(2), 235–243 (2008)
18. Koza, J.R.: Genetic Programming II: Automatic Discovery of Reusable Programs. MIT USA (1994)
19. Panda, B.N., Shankhwar, K., Garg, A., Jian, Z.: Performance evaluation of warping characteristic of fused deposition modelling process. Int. J. Adv. Manuf. Technol. **88**(5–8), 1799–1811 (2017)
20. Garg, A., Vijayaraghavan, V., Lam, J.S.L., Singru, M.P., Liang, G.: A Molecular Simulation Based Computational Intelligence Study of a Nano-machining Process with Implications on its Environmental Performance, Swarm and Evolutionary Computation, vol. 21, pp. 54–63 (2015)

Small Signal Stability Enhancement of Power System by Modified GWO-Optimized UPFC-Based PI-Lead-Lag Controller

Narayan Nahak, Soumya Ranjan Sahoo and Ranjan Kumar Mallick

Abstract A major decisive task in power system stability enhancement is optimal setting of damping controller parameters. This work proposes small signal stability enhancement of power system using UPFC-based optimal PI-lead-lag controller, whose parameters are optimized by modified Grey Wolf Optimizer technique. Lead-lag structure has been very much popular in UPFC damping controller design but in this work the efficacy of lead-lag controller has been improved by proportional–integral (PI) structure. The modified GWO technique proposed here has been compared with GWO-optimized lead-lag controller and PSO, DE-optimized PI-lead-lag controller to justify its supremacy. ITAE criterion is selected for minimization problem considering an increase in input mechanical power to generator. The system eigenvalues, speed and line power deviations subjected to disturbance in power system show that the proposed PI-lead-lag controller performs better than conventional lead-lag controller and is much better in comparison to other optimization techniques to tune the controller parameters for enhancing stability of power system.

Keywords FACTS · UPFC · Modified GWO · Damping controller
Small signal stability

1 Introduction

Small signal stability is a challenging task for a recent power system network interconnected by weak tie lines and continuously subjected to different disturbances. This stability issue is addressed here in term of low-frequency oscillations in the

N. Nahak · S. R. Sahoo
Department of Electrical Engineering, Siksha 'O' Anusandhan University,
Bhubaneswar 751030, India

R. K. Mallick (✉)
Department of Electrical and Electronics Engineering, Siksha 'O' Anusandhan University,
Bhubaneswar 751030, India
e-mail: rkm.iter@gmail.com

© Springer Nature Singapore Pte Ltd. 2019
J. C. Bansal et al. (eds.), *Soft Computing for Problem Solving*, Advances in Intelligent
Systems and Computing 817, https://doi.org/10.1007/978-981-13-1595-4_21

range of 0.2–3 Hz observed in an extended PS (power system) network [1]. Power system stabilizer (PSS) has been very popular to damp these oscillations but, with variation in operating conditions, the performance of PSS may vary [2]. With the inclusion of power semiconductor technology, FACTS devices are becoming more popular for enhancing stability and improving controllability of PS [3–6]. As per researches, the PSS based on FACTS devices is more superior in comparison to conventional PSS [7–9]. UPFC is a versatile member of FACTS family with three controllable variables, which are magnitude, phase angle of series injected voltage and reactive current drawn by shunt voltage source converter (VSC). Wang [10] had addressed linear Heffron-Phillips transfer function model with UPFC for small signal stability assessment. But a systematic process to design UPFC-based damping controller has been reported in [11]. In this work, the optimal controller parameters were designed by conventional approach. However, the most important decisive task is online tuning of controller parameter and which can be better obtained by a suitable optimization technique [12]. So the next important task is selection of a suitable optimization tool to handle the problem. The evolution in nature has given impulse to different promising optimization techniques employing population-based search. PSO is a simple, efficient and robust optimization technique and already being used to optimize damping controller parameter [13, 14]. But, it is prone to premature convergence and while handling heavy constrained optimization problem. It may get trapped in local optima. The differential evolution (DE) technique has been implemented for optimal location parameter setting and damping controller parameter tuning based on SSSC and reactive power management [15, 16].

Recently, metaheuristics algorithms are gaining much popularity because they are simple and can be easily implemented and inspired by natural phenomenon like behaviour of animals. Also, they need less parameters to tune with straightforward property [17]. The GWO has been reported in [17], which has been influenced by the idea and strategy of grey wolves in their hunting process. GWO technique has been used in [18] to design damping controller with lead-lag controller. Different hybrid techniques are also proposed in current literatures to design damping controller like hGA-GSA in [19] and DE-GWO in [20] where lead-lag controllers and dual controllers have been implemented. The GWO can provide good balance in exploration and exploitation, but it may get trapped in local optima due to dependency on certain mechanism [21]. Hence, the performance of GWO has been modified in [21], where the searching agent's movement in searched space has been modified. This modified version of GWO has been utilized here to design damping controller, and again, the earlier lead-lag controller has been modified with PI controller for optimum efficacy of controller. In this work, modified GWO has been proposed to tune UPFC-based PI-lead-lag controller.

2 Power System with UPFC

Here, a single-machine infinite bus (SMIB) system is considered for study as given in Fig. 1 [10]. The UPFC is installed in one of the parallel lines. IEEE ST1A excitation system is taken for the generator. The UPFC has two voltage source converters, VSC-E and VSC-B, which are connected in parallel and in series with the line, respectively. The VSC-E has modulation index and phase angle m_E and δ_E, respectively. VSC-B has modulation index and phase angle m_B, δ_B, respectively. m_E, δ_E, m_B and δ_B are the four controller variables of UPFC.

2.1 Nonlinear Modelling of Power System

Neglecting resistance associated with power system, nonlinear model can be presented by the following equations [10, 20]. All the notations are taken from [20].

$$\dot{\omega} = \left(\frac{P_i - P_e - D\Delta\omega}{M}\right) \tag{1}$$

$$\dot{\delta} = \omega_0(\omega - 1) \tag{2}$$

$$\dot{E}'_q = (-E_q + E_{fd})/T'_{d0} \tag{3}$$

$$\dot{E}_{fd} = [-E_{fd} + K_a(V_{ref} - V_t)]/T_a \tag{4}$$

$$\dot{V}_{dc} = \frac{3m_E}{4C_{dc}}\left(I_{Eq} \sin \delta_E + I_{Ed} \cos \delta_E\right) + \frac{3m_B}{4C_{dc}}\left(I_{Bq} \sin \delta_B + I_{Bd} \cos \delta_B\right) \tag{5}$$

Equation which represents balance of real power among series converter and shunt converter of UPFC is mentioned in Eq. (6) as

$$\text{Re}(V_{\mathbf{B}}I_{\mathbf{B}}^* - V_E I_E^*) = 0 \tag{6}$$

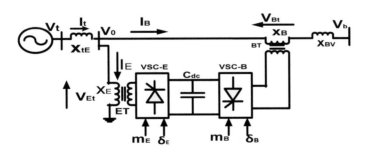

Fig. 1 SMIB system under study

2.2   Linear Modelling for Dynamic Study

For small signal stability assessment, the nonlinear model of power system can be linearized about the initial operating point as given by following equations [10].

$$\Delta\dot{\delta} = \omega_0\Delta\omega \tag{7}$$

$$\Delta\dot{\omega} = \left(\frac{-\Delta P_e - D\Delta\omega}{M}\right) \tag{8}$$

$$\Delta\dot{E}_q = (-\Delta E_q + \Delta E_{fd})/T_{d0} \tag{9}$$

$$\Delta\dot{E}_{fd} = \left[-\Delta E_{fd} + K_a(\Delta V_{ref} - \Delta V_t)\right]/T_a \tag{10}$$

$$\Delta V_{dc} = K_7\Delta\delta + K_8\Delta E_q' - K_q\Delta V_{dc} + K_{ce}\Delta m_E + K_{c\delta e}\Delta\delta_E + K_{cb}\Delta m_B + K_{c\delta b}\Delta\delta_B \tag{11}$$

where

$$\Delta P_e = K_1\Delta\delta + K_3\Delta E_q' + K_{pd}\Delta V_{dc} + K_{pe}\Delta m_E + K_{p\delta E}\Delta\delta_E + K_{pb}\Delta m_B + K_{p\delta b}\Delta\delta_B \tag{12}$$

$$\Delta E_d = K_4\Delta\delta + K_3\Delta E_q' + K_{qd}\Delta V_{dc} + K_{qe}\Delta m_E + K_{q\delta_E}\Delta\delta_E + K_{qb}\Delta m_B + K_{q\delta_B}\Delta\delta_B \tag{13}$$

$$\Delta V_t = K_5\Delta\delta + K_6\Delta E_q' + K_{vd}\Delta V_{dc} + K_{ve}\Delta m_E + K_{v\delta E}\Delta\delta_E + K_{vb}\Delta m_B + K_{v\delta b}\Delta\delta_B \tag{14}$$

3   Heffron-Phillips Power System Model Including UPFC

The Heffron-Phillips transfer function model is given in Fig. 2, which is a linear model including UPFC. The constants of model being calculated with respect to initial operating condition [11], provided in the appendix. In this model, $[\Delta U]$ is the input column vector, and $[K_{pu}]$, $[K_{vu}]$, $[K_{qu}]$, $[K_{cu}]$ are row vectors. The $[\Delta U]$ vector is given by $[\Delta U] = [\Delta m_E \ \Delta\delta_E \ \Delta m_B \ \Delta\delta_B]^T$.

Input vectors are given by, $[K_{pu}] = [K_{pe} \ K_{p\delta e} \ K_{pb} \ K_{p\delta b}]$, $[K_{vu}] = [\ K_{ve} \ K_{v\delta e} \ K_{vb} \ K_{v\delta b}]$ $[K_{qu}] = [\ K_{qe} \ K_{q\delta e} \ K_{qb} \ K_{q\delta b}]$, $[K_{cu}] = [\ K_{ce} \ K_{c\delta e} \ K_{cb} \ K_{c\delta b}]$.

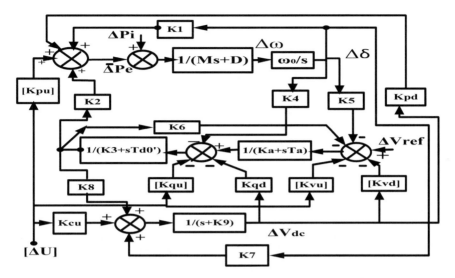

Fig. 2 Modification of Heffron-Phillips model including UPFC

4 Proposed PI-Lead-Lag Damping Controller Structure

The structure of the proposed PI-lead-lag damping controller is shown in Fig. 3 including time delay. The PI controller can provide fast error response and drive the system towards zero steady-state error, so the lead-lag controller can enhance stability or speed of response and provide smaller oscillations [16]. So here a PI controller is connected in series with lead-lag controller for enhancing stability of power system. The block D shows time delay, which may vary depending on whether local or remote signal available. In this work, delay time of 20 ms is considered. K1 and K2 are the gains of proportional and integral controller. Kp is the gain, and T1, T2 are time constants of lead-lag controller. All the parameters, K1, K2, Kp, T1 and T2, can be optimized by modified GWO and other optimization methods to compare the results.

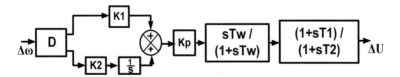

Fig. 3 PI-lead-lag controller with sensor and time delay

5 Objective Function

To design the damping controller, the main objective is reducing speed deviation when subjected to any disturbance. For this purpose, ITAE criterion [12] has been considered here, which is integral time of absolute error. For objective function, a step increase of 10 per cent in input mechanical power has been considered. The objective function is represented by J taking into account deviations in speed, dc bus voltage and real power deviation as:

$$J = \int_0^{tsim} t|\Delta\omega|dt + \int_0^{tsim} t|\Delta V_{dc}|dt + \int_0^{tsim} t|\Delta P_e|dt \qquad (15)$$

Now, the problem of optimization is minimizing J so that the controller parameters K1, K2, Kp, T1 and T2 should be within their minimum and maximum limiting values. The t_{sim} is simulation time. Typically, low-frequency oscillations lie in the range of 0.1 to 2 Hz. Here, the damping is provided by PI controller gains and phase compensation is provided by lead-lag controller. Hence, the proportional gains range is taken between 1 and 100 and integral gain as well as range of time constants is taken between 0.01 and 1 in this work.

6 PSO Algorithm Overview

It is a popular swarm intelligence-type technique [13, 14], which is population-based. It is a simple and efficient technique. The procedure to get optimal solution for a problem in PSO is on the basis of employing particles and allowing them to move in search space with multiple dimensions. The position of a particle is being updated by its personal experience and experience of its neighbour. PSO can provide challenging performance even in comparison to genetic algorithm for optimizing a problem. In [14], PSO was used to design UPFC controller for damping oscillations.

7 DE Algorithm Overview

It is an evolutionary technique of optimization [15, 16], where the searching process is being guided by distance as well as direction from current population. Here, a trial vector is obtained by target and difference vector. The trial vector is produced by crossover process which is performed by mainly parent vector and mutated vector. In [15], parameter setting of UPFC has been addressed for security of power system.

8 Grey Wolf Optimizer (GWO)

GWO is a recently revealed, population-based metaheuristic technique [17]. This technique is being imitated by the attitudes of grey wolves to make a plan and hunt a prey. To implement the hunting process, they are put in a rank, which are α, β, δ and ω. And the process is lead by α group followed by β, δ and finally by ω group. The process of hunting is mathematically modelled by considering the position of α as the most fittest solution followed by β, δ and rest solution by ω group. Three major steps in GWO are encircling the prey, hunting and attacking the prey. The flow chart of GWO is given in Fig. 4.

At starting, grey wolves encircle the prey which is given as [17].

$$\vec{D} = \left| \vec{C} \cdot \vec{X}_P(t) - \vec{X}(t) \right| \tag{16}$$

$$\vec{X}(t+1) = \vec{X}_p(t) - \vec{A} \cdot \vec{D} \tag{17}$$

Here, t represents current iteration. The coefficient vectors being $\vec{A}$, $\vec{C}$. The $\vec{X_P}$. and $\vec{X}$ represent the position vector of victim and grey wolf, respectively.

$\vec{A}$ and $\vec{C}$ can be calculated by Eqs. 18 and 19 as

$$\vec{A} = 2 \cdot \vec{a} \cdot \vec{r}_1 - \vec{a} \tag{18}$$

$$\vec{C} = 2 \cdot \vec{r}_2 \tag{19}$$

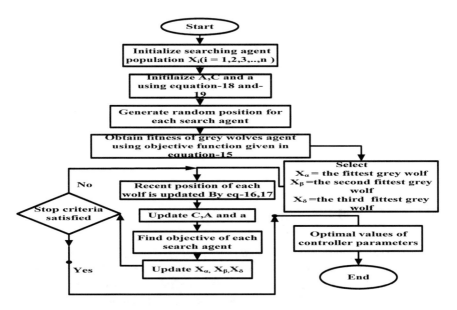

Fig. 4 Flow chart of GWO technique

where $\vec{a}$ component decreases from 2 to 0 in linear manner during an iteration, whereas $r1$ and r_2 are the random vector between [0, 1].

The position of grey wolf can be updated by Eqs. (16) and (17).

The hunting process can be expressed by following equations:

$$\vec{D}_\alpha = \left| \vec{C}_1 \cdot \vec{X}_\alpha - \vec{X} \right|, \vec{D}_\beta = \left| \vec{C}_2 \cdot \vec{X}_\beta - \vec{X} \right|, \vec{D}_\delta = \left| \vec{C}_3 \cdot \vec{X}_\delta - \vec{X} \right| \qquad (20)$$

$$\vec{X}_1 = \vec{X}_\alpha - \vec{A}_1 \cdot (\vec{D}_\alpha), \vec{X}_2 = \vec{X}_\beta - \vec{A}_2 \cdot (\vec{D}_\beta), \vec{X}_3 = \vec{X}_\delta - \vec{A}_3 \cdot (\vec{D}_\delta) \qquad (21)$$

The best position of the victim or prey can be obtained by taking the average values of positions of α, β and δ wolves as:

$$\vec{X}(t+1) = \frac{\vec{X}_1 + \vec{X}_2 + \vec{X}_3}{3} \qquad (22)$$

9 Modified GWO

Although the GWO can have good balancing in exploration and exploitation, still it is only dependent on vector C. So there is probability of trap in local optima for GWO but in modified GWO as represented in [21] the parameter A vector can also be used to find D'_α, D'_β and D'_δ, in which case magnitude of A may be less or more than one. The movement of searching agents depends on α, β, δ or on the agent chosen, which depends on random vector A.

Hence in the modified GWO, a new strategy is implemented to obtain D'_α, D'_β and D'_δ to avoid trapping in local optima. Therefore, this modified GWO is proposed here for tuning the controller parameters. The updated positions D'_α, D'_β and D'_δ are represented by given equations as:

$$\vec{D}'_\alpha = \left| \vec{C}_1 \cdot \vec{X}_{p1} - \vec{X}_{p3} \right|, \vec{D}'_\beta = \left| \vec{C}_2 \cdot \vec{X}_{p2} - \vec{X}_{p1} \right|, \vec{D}'_\delta = \left| \vec{C}_3 \cdot \vec{X}_{p3} - \vec{X}_{p1} \right| \qquad (23)$$

$$\vec{X}'_1 = \vec{X}_\alpha - \vec{A}_1 \cdot (\vec{D}'_\alpha), \vec{X}'_2 = \vec{X}_\beta - \vec{A}_2 \cdot (\vec{D}'_\beta), \vec{X}'_3 = \vec{X}_\delta - \vec{A}_3 \cdot (\vec{D}'_\delta) \qquad (24)$$

$$\vec{X}'(t+1) = \frac{\vec{X}'_1 + \vec{X}'_2 + \vec{X}'_3}{3} \qquad (25)$$

10 Results and Discussion

The objective of this work is optimal design of UPFC-based PI-lead-lag controller to damp oscillations in power system, for which the parameters of controllers have been optimized by PSO, DE and modified GWO techniques. The objective function considered here is ITAE criterion as given in Eq. 15 for minimization. The input to controller is speed deviation, obtained by raising the input mechanical power to

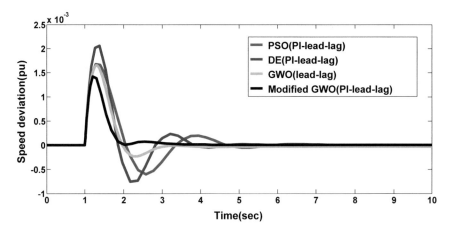

Fig. 5 Dynamic response of $\Delta\omega$ with mB-based controller

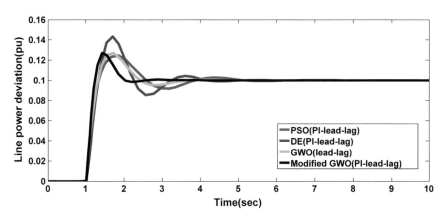

Fig. 6 Line power deviation with mB controller

generator by 10 per cent. The output of controller is the action to be performed to meet the objective. In this study, m_B and δ_E control actions of UPFC are considered as these are the best actions to design damping controller. Here, Pe = 0.8, Qe = 0.17 and line reactance Xe = 0.5. The speed deviations and line power deviations are shown in Fig. 5 and Fig. 6, respectively, for m_B-based controller and Fig. 7 and Fig. 8, respectively, for δ_E-based controller. In the figures, the results are also compared with GWO-optimized lead-lag controller [18]. The optimized controller parameters given in Tables 1 and 2 show the system eigenvalues with all controllers.

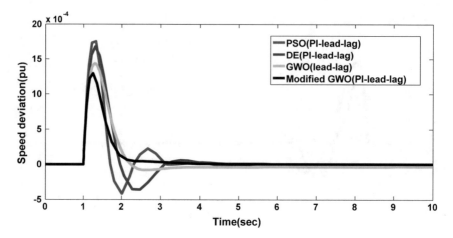

Fig. 7 Dynamic response of $\Delta\omega$ with δE-based controller

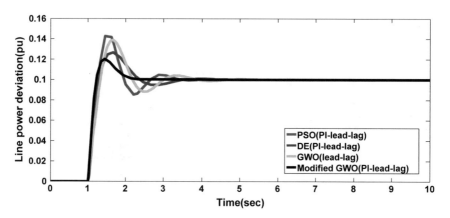

Fig. 8 Line power deviation with δE-based controller

11 Conclusion

- In this work, modified GWO-optimized PI-lead-lag controller based on UPFC is proposed to damp small signal oscillations in power system to enhance stability.
- The ITAE criterion has been considered for minimization problem. The disturbance taken is step increase in prime mover input power to generator.
- From the system eigenvalues, response of speed and line power deviation, subjected to disturbance, it was observed that the proposed controller performance is better than conventional lead-lag controller and the efficacy of modified GWO technique is also much better than standard PSO, DE and GWO techniques to optimize the controller to damp oscillations and enhancing stability of power system.

Table 1 Optimized parameters of controller

Control action	Optimization technique	K1	K2	Kp	T1	T2
m_B control action	PSO	68.7448	0.4401	71.5787	0.6139	0.3946
	DE	62.9513	0.7829	80.0249	0.6359	0.8420
	Modified GWO	59.998	0.6006	86.6941	0.3034	0.6442
δ_E control action	PSO	62.333	0.8445	85.6784	0.2302	0.9187
	DE	30.9834	0.4037	54.9337	0.3727	0.2705
	Modified GWO	60	0.3644	92.9071	0.4341	0.7791

Table 2 System eigenvalues with all controllers

PSO	DE	Modified GWO
m_B control action		
$-98.4012, -42.7122$	$-98.3574, -46.7947$	$-98.3412, -44.1182$
-8.2739	$-1.2412 \pm 2.4325i$	$-3.3199 \pm 0.43195i$
$-0.8217 \pm 2.4716i$	$-2.6207, -1.2588$	$-1.3877, -0.1046$
$-1.8297, -0.1004$	$-0.1004, -0.0053$	$-0.0053, -0.0000$
$-0.0053, -0.0000$	-0.0000	0
δ_E control action		
$-98.3283, -47.0825$	$-98.3407, -43.9614$	$-98.3486, -42.0500$
$-1.7216 \pm 4.9279i$	$-6.5425,$	$-4.1477 \pm 0.9227i$
$-1.7557, -0.8006$	$-1.4928 \pm 2.7983i$	$-1.8603, -1.0508$
$-0.1050, -0.0053$	$-2.1925, -0.1014$	$-0.1055, -0.0053$
$-0.0000, 0$	$-0.0053, -0.0000, 0$	$-0.0000, 0$

Appendix

(All the data are in per unit unless mentioned except constants)
Single-machine infinite bus test system data

$C_{dc} = 1$, $H = 4 MJ/MVA$, $Ka = 100$, $Ta = 0.01$, $T_{d0} = 5.044\,s$, $D = 0$, $\delta_0 = 47.13^0$, $V_b = 1$,

$V_{dc} = 2$, $V_t = 1$, $X_B = X_E = 0.1$, $X_{BV} = 0.3$, $X_d = 1$, $X_E = 0.1$, $X'_d = 0.3$, $X_q = 0.6$, $Xe = 0.5$

References

1. Kundur, P.: Power System Stability and Control. McGraw-Hill, New York (1994)
2. Keri, A.J.F., Lombard, X., Edris, A.A.: Unified power flow controller: modeling and analysis. IEEE Trans. Power Deliv. **14**(2), 648–654 (1999)
3. Li, G., Lie, T., Shrestha, G., Lo, K.: Implementation of coordinated multiple facts controllers for damping oscillations. Int. J. Electr. Power Energy Syst. **22**, 79–92 (2000)
4. Eslami, M., Shareef, H., Khajehzadeh, M.: Optimal design of damping controllers using a new hybrid artificial bee colony algorithm. Int. J. Electr. Power Energy Syst. **52**, 42–54 (2013)
5. Eslami, M., Shareef, H., Mohamed, A., Khajehzadeh, M.: Particle swarm optimization for simultaneous tuning of static var compensator and power system stabilizer. Przeglad Elektrotechniczny **87**, 343–347 (2011)
6. Abido, M.: Parameter optimization of multimachine power system stabilizers using genetic local search. Int. J. Electr. Power Energy Syst. **23**, 785–794 (2001)
7. Zhang, P., Messina, A.R., Coonick, A., Cory, B.J.: Selection of locations and input signals for multiple SVC damping controllers in large scale power systems. In: Proceedings of IEEE power engineering society winter meeting, pp. 667–70. (1998). paper IEEE-0-7803-4403-0
8. Farsangi, M.M., Song, Y.H., Lee, K.Y.: Choice of FACTS devices control inputs for damping inter area oscillations. IEEE Trans. Power Syst. **19**(2), 1135–1143 (2004)
9. Zhao, Q., Jiang, J.: Robust SVC controller design for improving power system damping. IEEE Trans Energy Convers. **10**, 201–209 (1995)
10. Wang, H.F.: A unified model for the analysis of FACTS devices in damping power system oscillations—part III: unified power flow controller. IEEE Trans. Power Deliv. **15**(3), 978–983 (2000)
11. Tambey, N., Kothari, M.L.: Damping of power system oscillations with unified power flow controller (UPFC). IEE Proc. Gener. Trans. Distrib. **150**, 129–140 (2003)
12. Taher, S.A., Hemmati, R., Abdolalipou,r A., Akbari, S.: Comparison of different robust control methods in the design of decentralized UPFC controllers. Int. J. Electr. Power Energy Syst. **43**, 173–84 (2012)
13. Ali, T., Al-Awami,Y.L., Abdel-Magid, M.A., Abido.: A particle-swarm-based approach of power system stability enhancement with unified power flow controller. Int. J. Electr. Power Energy Syst. **29**, 251–259 (2007)
14. Shayeghi, H., Shayanfar, H.A., Jalilzadeh, S., Safari, A.: Design of output feedback UPFC controller for damping of electromechanical oscillations using PSO. Energy Conver. Manag. **50** 2554–2561 (2009)
15. Shaheen, H.I., Rashed, G.I., Cheng, S.J.: Optimal location and parameter setting of UPFC for enhancing power system security based on differential evolution algorithm. Electr. Power Energy Syst. 33 94–105 (2011)
16. Sakra, W.S., EL-Sehiemya, R.A., Azmyb, A.M.: Adaptive differential evolution algorithm for efficient reactive power management. Appl. Soft Comput. **53** 336–351 (2017)
17. Mirjalili, S., Mirjalili, S.M., Lewis, A.: Grey Wolf Optimizer. Adv. Eng. Softw. **69**, 46–61 (2014)
18. Mallick, R.K., Nahak, N.: Grey wolves-based optimization technique for tuning damping controller parameters of unified power flow controller. In: IEEE International Conference on Electrical, Electronics, and Optimization Techniques, ICEEOT 2016, pp. 1458–1463
19. Khadanga, R.K., Satapathy, J.K.: A new hybrid GA–GSA algorithm for tuning damping controller parameters for a unified power flow controller. Electr. Power Energy Syst. **73**, 1060–1069 (2015)
20. Nahak, N., Mallick, R.K.: Damping of power system oscillations by a novel DE-GWO optimized dual UPFC controller. Eng. Sci. Technol. Int. J. **20**(4), 1275–1284 (2017)
21. Muangkote, N., Sunat, K., Chiewchanwattana, S.: An improved grey wolf optimizer for training q-Gaussian radial basis functional-link nets. In: International Computer Science and Engineering Conference (ICSEC) (2014)

A Robust Accelerated PSO MPPT for Photovoltaic System

Sarat Kumar Sahoo⊙**, M. Balamurugan**⊙**, Piyush Kumar Mishra, Kamakhya Krupa Mishra and Manas Ranjan Meher**

Abstract This paper focuses on the study and implementation of optimization techniques based on Maximum Power Point Tracking (MPPT) algorithm for photovoltaic (PV) fed DC–DC boost converter for constant voltage operation at the load side. It consists of PV array which is connected to the load through a DC–DC boost converter. The voltage and current are sensed from the converter at the input side by using voltage and current sensor those are given to MPPT block, which calculates the most suitable duty cycle, and it is fed to the boost converter for operation at the maximum peak point. Initially, the output is observed for conventional Fixed Duty Cycle, this method fails to converge for varying environmental conditions, and then advanced optimization techniques like particle swarm optimization (PSO) and accelerated particle swarm optimization (APSO) are used, and results are compared with the previous methods. The software simulation is carried out in MATLAB and then implemented in hardware using dSPACE DS1103 real-time controller interface.

Keywords Photovoltaic · MPPT · Matlab · dSPACE · APSO

S. K. Sahoo · M. Balamurugan (✉) · P. K. Mishra · K. K. Mishra · M. R. Meher
School of Electrical Engineering, VIT University, Vellore, India
e-mail: balamurugan.mano@vit.ac.in

S. K. Sahoo
e-mail: sksahoo@vit.ac.in

P. K. Mishra
e-mail: piyushkumar.mishra@yahoo.com

K. K. Mishra
e-mail: mishra.kamakhya@yahoo.com

M. R. Meher
e-mail: manasranjan1204@yahoo.com

© Springer Nature Singapore Pte Ltd. 2019
J. C. Bansal et al. (eds.), *Soft Computing for Problem Solving*, Advances in Intelligent Systems and Computing 817, https://doi.org/10.1007/978-981-13-1595-4_22

1 Introduction

The main concern in the power segment is the routine increase in the power supply demand. Due to lack of adequate resources, the power demand using the non-renewable energy sources is not fulfilled. Therefore, renewable energy resources which come from natural sources such as sunlight, wind, tidal waves, and geothermal heat are considered to fulfill the demand [1–4].

The conventional energy sources like coal, petroleum, hydro, nuclear sources was considered more even though it causes serious environmental problems when there was less existence of non-conventional energy sources like wind, solar, geothermal, wind and tidal sources [5]. The waste from these sources affects the surrounding, and moreover, these energy sources are not going to last longer as these are non-renewable. The use of nuclear energy source can be dangerous if not handled properly. In India, power transmission to the geographical remote area like the Himalaya ranges, the ranges in northeastern part, the desserts in the west, the Western Ghats is not suitable using conventional energy sources. So something other than the conventional energy sources is required [6–8]. As a result of this, the non-conventional energy sources are taken into consideration. The major advantage of these resources is that they are renewable and eco-friendly. These resources can be used to fulfill the growing demand for electricity since they are reliable. Even though the installation of these resources is more costly, researchers are round the clock to make this form better and more cost-effective. There is a bright future of these renewable energy resources, and it will play a vital role in world energy structure [9].

Solar energy is commonly accessible that has made it potential to operate it appropriately. Therefore, it can be utilized to provide power to the rural areas where the accessibility of grid is low. An additional benefit of utilizing solar energy is the convenient process wherever essential [10–12]. In order to solve the energy crisis, a proficient method has to be established in which power has to be taken out from the inward solar radiation. The power conversion efficiency is not up to the need in present times because of the changing environmental condition. Therefore, progressive technique called the Maximum Power Point Tracking (MPPT) algorithm which led to the growth in the effectiveness of the process of the solar modules and thus is operative in the area of application of renewable energy source [13–15].

Various conventional techniques have been applied to find the best possible way for tracking the maximum power. Different techniques have shown improved results in the output characteristics leading to finding its use in day-to-day applications. Initially, it started with simple conventional techniques such as Perturb and Observe (P&O) and Incremental Conductance method, but they have shown less efficiency due to its fluctuating output characteristics. Moreover, they were unable to operate at the global maximum point for partial shading of PV. So to overcome such drawbacks, advanced optimization techniques such as PSO and APSO have been used in this process to achieve the desired outcomes.

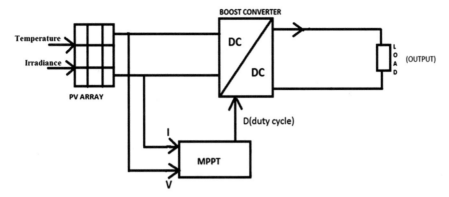

Fig. 1 Block diagram of PV fed DC–DC boost converter

2   System Description

The voltage generated by the PV panel is less; so to step up the voltage, DC–DC boost converter is utilized. Because of different temperature and irradiation obtained in different environmental conditions, the solar panel does not give the same output always.

Since electrical equipment cannot operate properly in varying voltage conditions, MPPT technique is used to obtain the constant maximum output from the array. The block diagram of PV fed boost converter is revealed in Fig. 1. Advanced optimization algorithms like PSO and APSO are used in this paper. By sensing the input voltage and current from the PV array, the best duty cycle is calculated using the MPPT algorithm and the same duty is given to the boost converter to generate the maximum output.

3   MPPT Algorithms

A solar panel converts about 30–40% of the solar insolation into electrical energy. Maximum Power Point Tracking technique is utilized to progress the productivity of the solar panel. As per the Maximum Power Transfer Theorem, the power production of a circuit is extreme when the source impedance of the circuit is same as the load impedance. Therefore, the difficulty of tracing the determined power fact decreases to an impedance matching problem. From the output side, boost converter is coupled with the solar panel to enhance the voltage at the output. By varying the pulse width of the boost converter properly, one can match the source impedance with the load impedance.

Different conventional MPPT methods like Fixed Duty Cycle, P&O, and Incremental Conductance methods have already been implemented, but because of certain drawbacks in the conventional techniques advanced optimization techniques like

PSO and APSO have been implemented in this project. To show the superiority of this method, the results of APSO method have been compared with the Fixed Duty Cycle method and Pulse Width Modulation (PWM) method.

3.1 PSO

Particle swarm optimization (PSO) is a universal optimization technique put forward originally by Kennedy and Eberhart in 1995 [14, 15]. This method has been established based on the swarm intelligence. While the birds are roaming to find the food, some of the bird can find the food very easily because of the smell of the food. Then the bird shares the information with the other bird about the location of food [16].

As per the PSO algorithm, bird is compared to solution swarm, and when the bird moves from one place to other place is called the developmental stage, and the information is called as the optimal solution and the source of food is called as the most optimal solution.

In swarm optimization, the particle position is mostly affected by the optimist position and its surrounding. When the particle is surrounded by the whole particle surrounding, then the position is called as the whole optimist solution [17].

$$V_{i(k+1)} = w * V_{i(k)} + c_1 * r_1 * (P_{best\_i} - X_{i(k)}) + c_2 * r_2 * (G_{best} - X_{i(k)}) \quad (1)$$

$$X_{i(k+1)} = X_{i(k)} + V_{i(k+1)} \quad (2)$$

where

$i = 1, 2, \ldots, N$ (particle number)
$K = 1, 2, 3, \ldots$ (iteration number)
$X =$ Current position of particle
$V =$ Velocity of particle
$w =$ Inertia constant
$c_1, c_2 =$ Acceleration constant
$P_{best,i} =$ Personal best of ith particle
$G_{best} =$ Global best of all the particle

The narrow surrounding is used in the algorithm which is called as the partial PSO (PPSO) algorithm. Each particle can be exposed to its present position and speed, the most optimist position of each individual and the surrounding. In the PPSO algorithm, the position and speed of every particle will vary according to (1) and (2). The velocity of the particle is calculated using (1), and the position of the particle has been updated by adding the current position with the velocity of the particle.

3.2 APSO

The conventional PSO has used both the global best and individual best for convergence; this may create some randomness in the solution. Therefore, accelerated particle swarm optimization (APSO) algorithm is a modest form of the PSO algorithm which utilizes only the global best to accelerate the convergence of this algorithm. Therefore, in APSO the velocity vector is produced by a simple equation which is written in (3) and (4). It is important to point out that the velocity vector is not mentioned in (5) because there is no deal with the initialization of voltage vectors. APSO is very simple because of the absence of velocity vector.

$$V_{j(t+1)} = V_{j(t)} + \alpha * \varepsilon + \beta * (G_{best} - X_{j(t)}) \tag{3}$$

$$X_{j(t+1)} = X_{j(t)} + V_{j(t+1)} \tag{4}$$

$$X_{j(t+1)} = (1 - \beta) * X_{j(t)} + \beta * G_{best} + \alpha * \varepsilon \tag{5}$$

where

$j = 1, 2, \ldots, N$ (particle number),
$t = 1, 2, 3 \ldots$ (iteration number)
$\alpha = \varepsilon^{\wedge} t$
$\varepsilon = $ Constant $[0, 1]$
$\beta = $ Speed of convergence $(0-1)$
$G_{best} = $ Global best of all the particle.

Pseudocode for APSO Algorithm.

Step 1: All the parameters (α, β, ε) are set, and the objective function is defined.
Step 2: The particle positions are initialized, and fitness evaluation is done.
Step 3: Gbest is calculated from the fitness evaluation.
Step 4: The position of the particle is modified using iteration equation.
Step 5: The boundary conditions are checked, and values of positions are verified.
Step 6: All the convergence criteria are verified. If the convergence criteria are not met, then proceed to step 3.
Step 7: The final Gbest values will give the best duty cycle.

4 Simulation Results

The simulation of the entire system has been simulated in MATLAB/Simulink environment. The constraints of the PV array and boost converter are listed in Table 1.

The I-V and P-V characteristics for different insolation levels such as 0.1 kW/m^2, 0.5 kW/m^2, and 1 kW/m^2 are shown in Fig. 2 and Fig. 3, respectively. The output characteristics of power and voltage waveforms of Fixed Duty Cycle method have

Table 1 Simulation parameters

S. no.	Components	Quantity/Capacity
1	PV panel	250 W
2	No. of panels connected in series	2
3	No. of panels connected in parallel	3
4	Cells per module	60
5	Open circuit voltage	37.62 V
6	Short circuit current	8.76 A
7	Voltage at maximum power	29.76 V
8	Current at maximum power	8.40 A
9	Inductor	30 μH
10	Capacitor	1000 μF
11	Resistor	67 Ω
12	Switching frequency	20 kHz

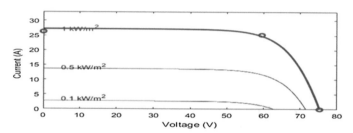

Fig. 2 I-V characteristics of PV string

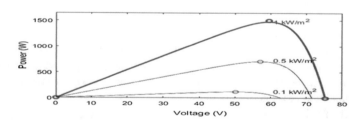

Fig. 3 P-V characteristics of PV string

been presented in Fig. 4 where the values of voltage and power are 240 V and 864.10 W, respectively.

The output characteristics of power and voltage waveforms of Fixed Duty Cycle method have been shown in Fig. 5, where the values of voltage and power are 199.9 V and 705.2 W, respectively.

The output characteristics of power and voltage waveforms of Fixed Duty Cycle method have been presented in Fig. 6, where the values of voltage and power are 271.5 V and 1100 W, respectively.

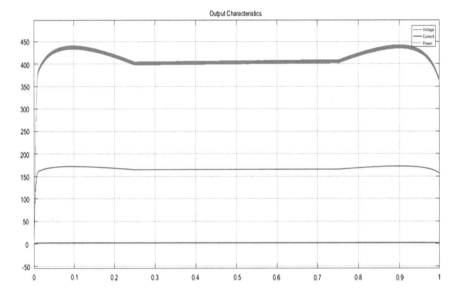

Fig. 4 Output characteristics of PV using Fixed Duty Cycle method

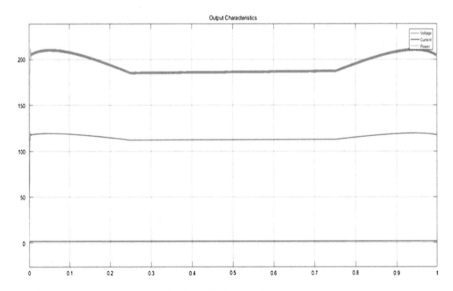

Fig. 5 Output characteristics of PV using PWM method

The output characteristics of power and voltage waveforms of Fixed Duty Cycle method have been shown in Fig. 7, where the values of voltage and power are 282 V and 1187 W, respectively.

The superiority of different algorithms has been measured in terms of duty cycle and compilation time. Therefore, the comparison has been made in terms of duty

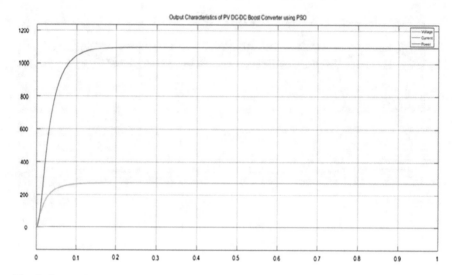

Fig. 6 Output characteristics of PV using PSO method

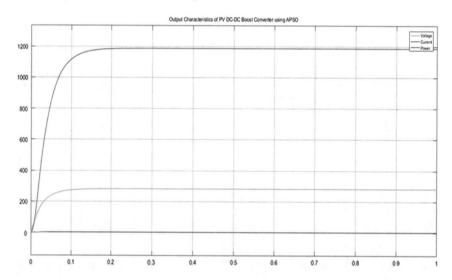

Fig. 7 Output characteristics of PV using APSO method

cycle and compilation time parameters for both PSO and APSO algorithms which are listed in Table 2. From the table, one can say that the APSO algorithm has shown improved performance in terms of compilation time when compared to PSO.

The efficiency of different types of MPPT algorithms has been compared which is listed in Table 3. While looking at the data, APSO has shown better efficiency and superior performance when compared to other methods.

Table 2 Comparison of PSO and APSO in terms of duty cycle and compilation time

Voltage from PV (V)	PSO		APSO	
	Duty cycle	Compilation time (s)	Duty cycle	Compilation time (s)
115	0.4250	4.211331	0.4254	3.390627
125	0.3750	4.351143	0.4048	4.143533
135	0.3250	4.390656	0.3571	3.696438
145	0.2750	4.490427	0.3095	3.790236
155	0.2250	4.560881	0.2619	3.876345
165	0.1750	4.703826	0.2143	3.942198
175	0.1250	4.790627	0.1667	4.054328
185	0.0750	4.879011	0.1697	4.194732
195	0.0250	4.982989	0.0714	4.273654

Table 3 Comparison of pulse generation methods

Pulse generation method	Output voltage (V)	Output power (W)	System efficiency (%)	Inference
Fixed duty cycle	240	864.10	57.60	Not suitable for varying environmental conditions
PWM	199.9	705.2	52.5	Due to the use of modulating signals, the efficiency is low
PSO	271.5	1100	73.33	Oscillations are very less, and efficiency is better
APSO	282	1187	79.13	Efficiency is very high, and the waveforms are free from oscillations

5 Experimental Results

The implementation of the boost converter in real time which is done by using the dSPACE DS1103 real-time interface control is shown in Fig. 8.

In the hardware arrangement, the PV module is coupled to the DC–DC boost converter. dSPACE DS1103 is used for the control system and data acquisition implementation with the help of digital signal processor card on the PC. The voltage is measured with the differential probe, while the PV current is measured with current

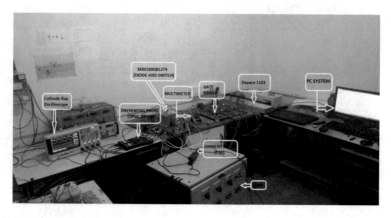

Fig. 8 Experimental setup

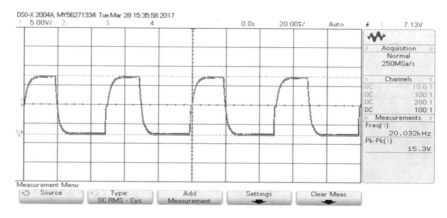

Fig. 9 Duty cycle of boost converter using APSO method

probe. The analog measured magnitudes of the PV voltage and current which are fed to A/D converter of the DS1103 to be utilized in the Simulink MPPT control block. The dSPACE A/D channel signal must be in the range from −10 V to +10 V.

APSO algorithm has shown better performance in simulation. Therefore, APSO method has been chosen for implementation in real time. The circuit connections are set up, and dSPACE is interfaced with MATLAB. Then the 3A load is provided by the load bank, and the load across it is found to be 67.8 Ω. Then the pulse generated for the switch obtained in DSO is shown in Fig. 9.

The input and output characteristics of boost converter are shown in Fig. 10. PV voltage, PV current, and the boost converter output voltage and current are displayed in the four channels of DSO. From the waveform, one can clearly understand that the voltage generated by PV string is 66 V with the help of APSO algorithm the voltage has been boosted to 124 V by varying the duty cycle of the boost converter. The Channel 3 represents the output voltage (blue line) which has shown fewer

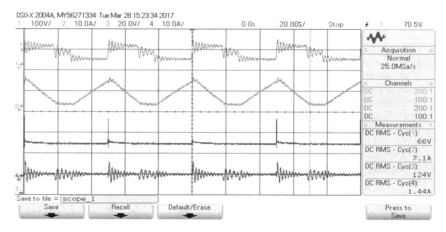

Fig. 10 Waveforms of boost converter using APSO method

distortions even when the PV voltage has high distortions as shown in Channel 1. Therefore, by using APSO method one can obtain the maximum power from PV string.

6   Conclusion

In this paper, a robust-based APSO MPPT algorithm has been demonstrated. The simulation has been done using MATLAB/Simulink for varying environmental conditions. Based on the simulation results, one can say that the APSO method has shown better performance when compared with Fixed Duty Cycle, PWM, and PSO methods. The experimental implementation of PV interfaced boost converter has been developed in the laboratory. The pulse has been generated with the help of APSO algorithm which has been given to the switch while interfaced with dSPACE DS1103. The experimental results established that the projected algorithm has shown decent performance and it can be implemented for high-power applications.

Acknowledgements The authors would like to acknowledge the financing support from Department of Science and Technology (DST), Government of India, Project No. DST/TSG/NTS/2013/59. This work has been carried out in School of Electrical Engineering, VIT University, Vellore, India.

References

1. M. Kasper, D. Bortis, T. Friedli, J.W. Kolar: Classification and comparative evaluation of PV panel integrated DC-DC converter concepts. In: Proceedings on 15th International IEEE Power

Electronics Motion Control Conference. IEEE, Serbia (2012). https://doi.org/10.1109/epepem c.2012.6397403

2. Liu, Y.H., Huang, J.W.: A fast and low cost analog maximum power point tracking method for low power photovoltaic systems. Sol. Energy **85**, 2771–2780 (2011)
3. P. Kumar, R. K. Pachauri, Y.K. Chuahan: Duty ratio control schemes of DC-DC boost converter integrated with solar PV system. In: Proceedings of International Conference on Energy Economics and Environment, pp. 1–6. IEEE, Noida (2015)
4. Seyedmahmoudian, M., Mekhilef, S., Rahmani, R., Yusof, R., Renani, E.: Analytical modeling of partially shaded photovoltaic systems. Energies **6**, 128–144 (2013)
5. Kalpana, Ch., Babu, Ch., Surya Kumari, J.: Design and Implementation of different MPPT algorithms for PV system. Int. J. Sci. Eng. Technol. Res. (IJSETR), **2**(10), 1926–1923 (2013)
6. Logeswaran, T., SenthilKumar, A.: A review of maximum power point racking algorithms for photovoltaic systems under uniform and non-uniform irradiances. Energy Proc. **54**, 228–235 (2014)
7. Ali, A.N.A., Saied, M.H., Mostafa, M.Z., Abdel- Moneim, T.M.: A survey of MPPT techniques of PV systems. In: Proceedings of Energytech, pp. 1–17. IEEE, USA (2012)
8. Sera, D., Member, S., Teodorescu, R., Member, S., Hantschel, J., Knoll, M.: Optimized Maximum Power Point Tracker for Fast-Changing Environmental Conditions. IEEE Trans. Industr. Electron. **55**, 2629–2637 (2008)
9. Jiang, Y., Abu Qahouq, J.A., Haskew, T.A.: Adaptive step size with adaptive-perturbation-frequency digital MPPT controller for a single-sensor photovoltaic solar system. IEEE Trans. Power Electron. **28**, 3195–3205 (2013)
10. Rahim, N.A., Che Soh, A., Radzi, M.A.M., Zainuri, M.A.A.M.: Development of adaptive perturb and observe-fuzzy control maximum power point tracking for photovoltaic boost DC-DC converter. IET Renew. Power Gener. **8**, 183–194 (2014)
11. Elgendy, M.A., Zahawi, B., Atkinson, D.J.: Assessment of perturb and observe MPPT algorithm implementation techniques for PV pumping applications. IEEE Transactions on Sustainable Energy **3**, 21–33 (2012)
12. Patel, H., Agarwal, V.: Maximum Power Point Tracking Scheme for PV Systems Operating Under Partially Shaded Conditions. IEEE Trans. Industr. Electron. **55**(4), 1689–1698 (2008)
13. Subha, R., Himavathi, S.: Accelerated particle swarm optimization algorithm for maximum power point tracking in partially shaded PV systems. In: Proceedings of 3rd International Conference on Electrical Energy Systems (ICEES), pp. 232–236. IEEE, Chennai (2016)
14. Liu, Y.H., Huang, S.C.: A particle swarm optimization-based maximum power point tracking algorithm for PV systems operating under partially shaded conditions. IEEE Trans. Energy Convers. **27**(4), 1027–1035 (2012)
15. Suryavanshi, R., Joshi, D.R., Jangamshetti, S.H.: Jangamshetti: PSO and P&O based MPPT technique for SPV panel under varying atmospheric conditions. In: Proceedings of International Conference on Power, Signals, Controls and Computation (EPSCICON). https://doi.org/10.1 109/epscicon.2012.6175270. Kerala (2012)
16. Selvapriyanka, P., Vijayakumar, G.: Particle swarm optimization based MPPT for PV system under partial shading conditions. Int. J. Innovat. Res. in Sci. Eng. Technol. **3**(1), 856–861 (2014)
17. Liu, Y.H., Chen, J.H., Huang, J.W.: Global maximum power point tracking algorithm for PV systems operating under partially shaded conditions using the segmentation search method. Sol. Energy **103**, 350–363 (2014)

Learned Invariant Feature Transform and Extreme Learning Machines for Face Recognition

A. Vinay, Nishanth S. Hegde, S. K. Tejas, Naveen V. Patil, S. Natarajan
and K. N. Balasubramanya Murthy

Abstract In this paper, we propose a novel face recognition pipeline based on key-point detection and classification. The proposed approach makes use of the recent advances in the field of deep learning through a recently proposed keypoint detector and descriptor called learned invariant feature transform (LIFT). We also incorporate extreme learning machines (ELMs) for the purpose of classification. The descriptors are aggregated using vector of locally aggregated descriptors (VLADs). This approach is tested extensively in comparison with other well-known descriptors on databases like ORL, Faces94, and Grimace and has been proved to outperform other descriptors in most cases. This provides a fast and accurate algorithm for face recognition.

A. Vinay · N. S. Hegde (✉) · S. K. Tejas · N. V. Patil · S. Natarajan
K. N. Balasubramanya Murthy
PES University, Bangalore, India
e-mail: hegde.nishanth@gmail.com

A. Vinay
e-mail: a.vinay@pes.edu

S. K. Tejas
e-mail: tejaskasetty@gmail.com

N. V. Patil
e-mail: naveenpatil26@gmail.com

S. Natarajan
e-mail: natarajan@pes.edu

K. N. Balasubramanya Murthy
e-mail: vice.chancellor@pes.edu

© Springer Nature Singapore Pte Ltd. 2019
J. C. Bansal et al. (eds.), *Soft Computing for Problem Solving*, Advances in Intelligent
Systems and Computing 817, https://doi.org/10.1007/978-981-13-1595-4_23

1 Introduction

Person identification using face recognition is one the most challenging task in field of computer vision. It is studied extensively due its immense practical utilization scenarios [1]. Generally, face recognition is a more suited method of person identification because of it's non-intrusive nature [2]. Face recognition in computer vision is the process of analyzing an image of a person's face and recognizing the person in the image. The approach for this problem traditionally has been to detect the keypoints in the face and extracting a feature vector that numerically describes them. Utilizing a classifier, these features are recognized to characterize a face existing in the database.

Some of the challenges in the process of face recognition are pose, scale, orientation, illumination, facial expression. Over the time, there have been several algorithms developed to handle these complexities [3–5]. Face recognition has seen a significant development in the recent years with advent of artificial neural networks [1, 6]. The existing technologies for feature description like scale-invariant feature transform (SIFT), speeded-up robust features (SURF), oriented FAST and rotated brief (ORB) do not extensively take advantages of recent developments in deep learning.

Learned invariant feature transform (LIFT) is a recently developed deep network architecture for keypoint detection, orientation estimation, and feature description [7]. LIFT has proven to outperform the state-of-the-art algorithms in feature description like SIFT, SURF, ORB, and the like with respect to the number of keypoints matched irrespective of changes in viewpoint and illumination. We utilize this feature descriptor to extract features from the images. The features thus obtained are invariant to scale and orientation, thereby increasing the accuracy in the identification process.

Due to the fact that the traditional neural networks are too slow to train because of backpropagation, we use extreme learning machines which are much faster than traditional backpropagation feedforward neural networks for classification [8].

2 Related Work

Previously, the work related to feature descriptors for face recognition have not considered LIFT, and we would like to explore the uses of LIFT, along with our training pipeline for the purpose of face recognition (Fig. 1).

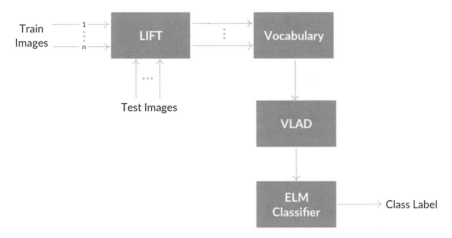

Fig. 1 Overview of pipeline

2.1 Descriptors

2.1.1 Scale-Invariant Feature Transform

SIFT is an algorithm proposed by D. Lowe which computes descriptors for the keypoints detected in an image, and these descriptors are invariant to scaling, rotation, and partially invariant to change in illumination and 3D camera viewpoint [9]. The algorithm is effective in detecting features even in the presence of occlusion, clutter, or noise. SIFT has been proved to be effective and has shown a good performance in recognizing objects, scenes, and faces.

By using SIFT features for constructing discriminative local features, and support vector machines as classifier, there are face recognition methods which have shown satisfying results on databases which are considered to be a benchmark for face recognition [10].

2.1.2 Speeded-Up Robust Features

SURF is a speeded-up version of SIFT [5]. Along with approximation of Laplacian of Gaussian (LoG) with difference of Gaussian for the scale-space. SIFT approximates LoG with box filter. With the help of integral images, the convolution of box filter can be easily calculated which is a key advantage of this approximation, and this can be done in parallel for different scales.

2.1.3   Oriented FAST and Rotated BRIEF

ORB is one of the another feature descriptors which have a lower computation cost and good matching performance. Its an amalgamation of FAST detector and brief descriptor with a lot modifications for enhancing the performance [4].

Each of the descriptors is been used in various applications over the years. ORB being the most recent one also stands out compared to the other two in various scenarios as per the results of a comparative study done on these descriptors [11].

2.1.4   LIFT

LIFT is an abbreviation for learned invariant feature transform. It uses a four-branched Siamese network, each comprising three CNNs. One each to detect, estimate orientation, and to compute feature descriptors. The branches of the Siamese networks take as input image patches from different viewpoints. The network is trained to effectively detect keypoints in an image and compute the feature descriptor. In the testing pipeline, the authors of LIFT have developed an end-to-end network comprising of the CNNs mentioned above to compute the descriptors. LIFT was tested to accurately detect pictures of places. We have drawn inspiration from this to test the feature descriptor's accuracy for the task of face recognition and to test the invariance of the descriptor to the pose of the subject and illumination on the image.

LIFT is similar to SIFT with respect to the dimensionality of the descriptor. It provides us with a 128-dimensional vector for each keypoint detected within an image. SURF provides a 64-dimensional descriptor and ORB, a 32-dimensional descriptor.

Hence, we will be comparing LIFT against these descriptors in the above-mentioned pipeline.

2.2   Vector of Locally Aggregated Descriptors

VLAD was originally developed by Jegou et al. as an extension over bag-of-words model to preserve more discriminative properties in the aggregated feature vector [12]. In bag-of-words (BoW) model, we first train a codebook,

$$C = \{c_1, c_2, \ldots c_k\} \tag{1}$$

where k represents the number of clusters formed using k-means clustering and c_i represent the centroid of the cluster. Aggregated VLAD vector of an image is constructed by summing up the differences between local descriptors and the cluster centers, followed by L2 normalization [13]. Thus, the VLAD vector for a image can be represented as follows:

$$v_{ij} = \sum_{x|x=NN(c_i)} (x_j - c_{ij}) \tag{2}$$

2.3 Extreme Learning Machines

Neural networks are very widely used for regression or classification problems. These networks however require to be trained extensively to produce state-of-the-art results. Perhaps, the most widely used method employed to train these networks are the gradient-based methods where the weights are randomly initialized at first and then learned through an iterative algorithm called backpropagation algorithm [14].

Extreme learning machines on the other hand do not use the aforementioned gradient-based techniques to learn the weights. The technique implemented is as follows. The input layer is initialized with random weights, and the weights for the output layer are calculated by finding the pseudo-inverse of a matrix which can be expressed as a function of the inputs and the weights of the hidden layer. This eliminates the need for backpropagation as the weights are computed keeping the outputs in mind.

The most simplest model of an ELM with W_{ho} hidden to output weight matrix, W_{ih} input to hidden weight matrix, and some activation function σ can be represented as follows:

$$Y = W_{ho}\sigma(W_{ih}x) \tag{3}$$

3 Proposed Work

Our work uses LIFT feature descriptor which incorporates the immense advances in the field of deep learning to detect keypoints and compute the respective feature descriptors which are invariant to scale, orientation, and illumination [7]. The descriptors of an image are aggregated into a single descriptor using VLAD. We also explore the use of ELM as a classifier to extract the advantages that it poses in terms of speed of training [8].

3.1 VLAD

We use VLAD to aggregate the various feature descriptors extracted from the images in the previous step to describe them systematically in a single vector. The bag-of-words model is trained by using k-means clustering algorithm, and such a model will have k code words in its codebook. After aggregating the feature vectors into one global descriptor per image, we are presented with a dataset of dimensions

$n \times (k \times 128)$ where n is the number of training images and k is the number of clusters.

3.2 ELM

For the purpose of classifying globally aggregated row vector of 128 elements per image into the corresponding class labels, we have chosen ELM as a classifier. We use a single hidden-layer feedforward network (SLFN ELM) with multi-quadric transfer function as the transfer function. The number of hidden nodes were empirically tuned to achieve maximum classification accuracy for each dataset.

We also explore the usage of random vector functional link networks instead of ELM for classification. In this type of neural networks, input nodes are first connected to enhancement nodes, and the feature vector, which is concatenation of direct features, and enhanced features are fed to the output nodes. Such random initializations has been proved to improve generalization [15].

The main advantage of this pipeline is the training time. Though large convolutional neural networks (CNNs) can be trained for the task of face recognition with high accuracy, the main problem with that would be the long training period. In case, a new class has to be incorporated for classification, it would require a fresh training of the network. This can be avoided in our algorithm where we propose a method that takes substantially lesser time for training without significantly trading-off on accuracy. Although the LIFT feature descriptor involves CNNs, these networks will not need repeated retraining.

The whole process can be summed up in Algorithm 1.

Algorithm 1 Face recognition using ELM

---

procedure FEATURES
    $Y \leftarrow$ Array for transformed features
    for image $i = 1 \ldots n$ **do**
        $F_i \leftarrow \text{FE}(i)$ ▷ FE is a feature extractor like LIFT/SIFT.
        Cluster F into k clusters
        $Y_i \leftarrow$ VLAD transform the clusters
    return Y
procedure LIFT(*Image*)
    Detect keypoints(*image*)
    $keypoints_o \leftarrow \forall keypoints$ Estimate orientation
    $descriptors \leftarrow \forall keypoints_o$ Compute descriptors
    return *descriptors*
procedure CLASSIFY(Y)
    $W_{ih} \leftarrow$ Random initialization
    Learn the weights of W_{ho}
    Predict class labels $\forall i$ in Y

Table 1 Results on ORL dataset

Metrics	LIFT + ELM	LIFT + RVFL	SIFT	SURF	ORB
Accuracy	**98.75**	**98.75**	97.5	95.00	93.75
Precision	**98.61**	**98.61**	97.22	94.44	95.13

Table 2 Results on Faces94 dataset

Metrics	LIFT + ELM	LIFT + RVFL	SIFT	SURF	ORB
Accuracy	93.09	93.58	**94.24**	93.25	90.13
Precision	94.48	94.8	**95.73**	95.05	92.81

4 Results

The framework is benchmarked on ORL faces, Faces94, Grimace datasets. ORL faces are a standard benchmarking database for semi-supervised algorithms, consisting of 10 images for each of 40 different subjects. This dataset has images of varying illumination and facial expressions. The size of each image is 92 by 112 pixels, with 256 gray levels per pixel [16]. Faces94 images consist of 153 individuals with 20 images per individual. The image size is 180 by 200 pixels. These images have large variations in facial expressions. Background is of green color [17]. Grimace dataset consists of 360 images of 18 individuals with 20 images per individual. The image size is 180 by 200 pixels. These images have variations in facial expressions, and the images are all captured in a low-light setting. Background is generally plain with very little variations in lighting. There are large variations in head orientation [18].

Features of each image are extracted and aggregated. The images are segregated into two training and testing sets with the test set being 20% of the dataset. The ELM learns to recognize faces from the training set and is presented with the test set for measuring metrics like accuracy and precision of the model. The results for the database with respect to these metrics are calculated and tabulated as shown in Tables 1, 2 and 3. In the tables below, LIFT, SIFT, SURF, and ORB refer to the pipeline with the respective descriptor used in the first step.

$$Accuracy = \frac{TP + TN}{TP + FP + TN + FN} \tag{4}$$

$$Precision = \frac{TP}{TP + FP} \tag{5}$$

From Fig. 2, we can observe the performance of the descriptors under consideration for a arbitrary image from the Faces94 dataset. We can easily observe that the keypoints detected by LIFT are more concentrated on features of the face. The number of keypoints detected out of the face is considerably low compared to SURF. The SURF keypoints show no clear distinctions between the sharp features on the

Table 3 Results on Grimace dataset

Metrics	LIFT + ELM	LIFT + RVFL	SIFT	SURF	ORB
Accuracy	**95.83**	**95.83**	94.44	93.05	89.47
Precision	**97.88**	**97.88**	96.11	96.06	91.48

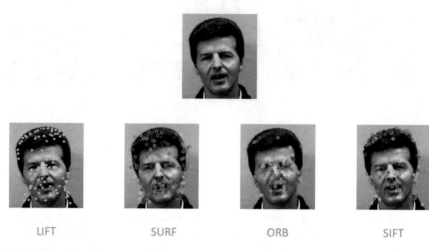

Fig. 2 Keypoints detected on a face taken from Faces94 dataset

face; that is, the distribution of keypoints is uniform over the face irrespective of features and also shows a good number of keypoints on the background.

The results of ORB seem to be focused on the eye region and very few keypoints in other regions. The results of SIFT, however, seem to be very well distributed. It seem to strike a good balance between the number of points per facial feature and seem to detect all facial features. The results for SIFT from the Face94 dataset reinforce the observations from the keypoint distribution.

5 Discussions and Conclusion

From the results shown in Tables 1, 2 and 3, it can be observed that our pipeline LIFT + VLAD + ELM generally outperforms the same with SIFT, SURF, and ORB as descriptors in most datasets except for Faces94 dataset, where SIFT is slightly better. This can be accounted due to very less change in orientation of the faces in that dataset. The performance of RVFL is usually on par with ELM, but in case of Faces94, it is seen that RVFL performs slightly better than ELM, though SIFT with ELM achieves better accuracy. Also, since the LIFT pipeline is based on a neural network, it takes more time than other algorithms for feature extraction. For cases where accuracy and precision are of high priority, LIFT can be used although it is slow. LIFT might not be suitable for cases where speed is of at most importance. Other feature detectors like ORB which is specifically designed for speed can be used while having a slight trade-off in accuracy.

References

1. Wang, Y., Bao, T., Ding, C., Zhu, M.: Face recognition in real-world surveillance videos with deep learning method. In: 2017 2nd International Conference on Image, Vision and Computing (ICIVC), pp. 239–243 (2017). https://doi.org/10.1109/ICIVC.2017.7984553
2. Zhao, W., Chellappa, R., Phillips, P.J., Rosenfeld, A.: Face recognition: a literature survey. ACM Comput. Surv. 35(4), 399–458 (2003). https://doi.org/10.1145/954339.954342
3. Lowe, D.G.: Object recognition from local scale-invariant features. In: Proceedings of the Seventh IEEE International Conference on Computer Vision, vol. 2, pp. 1150–1157 (1999). https://doi.org/10.1109/ICCV.1999.790410
4. Rublee, E., Rabaud, V., Konolige, K., Bradski, G.: ORB: an efficient alternative to SIFT or SURF. In: 2011 International Conference on Computer Vision, pp. 2564–2571 (2011). https://doi.org/10.1109/ICCV.2011.6126544
5. Bay, H., Tuytelaars, T., Van Gool, L.: SURF: Speeded Up Robust Features. Springer, Berlin, Heidelberg, pp. 404–417 (2006). https://doi.org/10.1007/11744023_32
6. Wan, L., Liu, N., Huo, H., Fang, T.: Face recognition with convolutional neural networks and subspace learning. In: 2017 2nd International Conference on Image, Vision and Computing (ICIVC), pp. 228–233 (2017). https://doi.org/10.1109/ICIVC.2017.7984551
7. Yi, K.M., Trulls, E., Lepetit, V., Fua, P.: LIFT: Learned Invariant Feature Transform. Springer International Publishing, Cham, pp. 467–483 (2016). https://doi.org/10.1007/978-3-319-46466-4_28
8. Huang, G.-B., Zhu, Q.-Y., Siew, C.-K.: Extreme learning machine: theory and applications. Neurocomputing 70(13), 489–501 (2006). ISSN 0925-2312. https://doi.org/10.1016/j.neucom.2005.12.126, http://www.sciencedirect.com/science/article/pii/S0925231206000385
9. Lowe, D.G.: Distinctive image features from scale-invariant keypoints. Int. J. Comput. Vis. 60(2), 91–110 (2004). https://doi.org/10.1023/B:VISI.0000029664.99615.94
10. Zhang, L., Chen, J., Lu, Y., Wang, P.: Face recognition using scale invariant feature transform and support vector machine. In: 2008 The 9th International Conference for Young Computer Scientists, pp. 1766–1770 (2008). https://doi.org/10.1109/ICYCS.2008.481
11. Isik, S.: A comparative evaluation of well-known feature detectors and descriptors. Int. J. Appl. Math. Electron. Comput. 3(1), 16 (2014)
12. Jegou, H., Douze, M., Schmid, C., Perez, P.: Aggregating local descriptors into a compact image representation. In CVPR 2010—23rd IEEE Conference on Computer Vision & Pattern Recognition. IEEE Computer Society, San Francisco, United States, pp. 3304–3311 (2010). https://doi.org/10.1109/CVPR.2010.5540039
13. Arandjelovic, R., Zisserman, A.: All About VLAD. In: 2013 IEEE Conference on Computer Vision and Pattern Recognition, pp. 1578–1585 (2013). https://doi.org/10.1109/CVPR.2013.207
14. Rumelhart, D.E., Hinton, G.E., Williams, R.J.: Parallel Distributed Processing: Explorations in the Microstructure of Cognition, vol. 1. MIT Press, Cambridge, MA, USA, Chapter Learning Internal Representations by Error Propagation, pp. 318–362 (1986). http://dl.acm.org/citation.cfm?id=104279.104293
15. Zhang, L., Suganthan, P.N.: A comprehensive evaluation of random vector functional link networks. Inf. Sci. (2016). https://doi.org/10.1016/j.ins.2015.09.025
16. Simple Faces Dataset (2002). http://www.cl.cam.ac.uk/research/dtg/attarchive/facedatabase.html

17. Spacek, L.: Collection of Facial Images: Faces94 (2007). http://cswww.essex.ac.uk/mv/allfaces/faces94.html
18. Spacek, L.: Collection of Facial Images: Grimace (2007). http://cswww.essex.ac.uk/mv/allfaces/grimace.html

Prediction of NO to NO$_2$ Conversion Efficiency with NTP-Based Diesel Exhaust Treatment Using Radial Basis Functions

Srikanth Allamsetty⊕ and **Sankarsan Mohapatro**⊕

Abstract Non-thermal plasma (NTP) technique is one of the most outstanding techniques which gives superior results in reducing the concentration of NO$_X$ from diesel exhaust. However, NO to NO$_2$ conversion reactions are predominant while NTP alone is used for exhaust treatment, with no adsorbents/catalysts. The NO to NO$_2$ conversion efficiency depends on various electrical and physical parameters, and very few studies have been carried out to predict it with the variation in those parameters using soft computing techniques. In the present study, experiments were conducted with variations in voltage, flow rate, discharge gap width, and initial NO$_X$ concentration and observed the changes in the concentrations of NO and NO$_2$. An approach has been made using MATLAB and radial basis functions (RBFs) to predict the NO to NO$_2$ conversion efficiency of this particular experimental setup. Predicted values are well matched with the experimental values and as such provide strong support to use this method for such prediction problems.

Keywords Non-thermal plasma · Dielectric barrier discharge · Diesel engine exhaust · NO removal · NO conversion · Prediction · Radial basis functions

1 Introduction

Non-thermal plasma (NTP) technique is being widely used for reducing the concentration of NO$_X$ from diesel exhaust. Researchers throughout the world have been coming up with experimental studies to make it convenient for practical applications of pollution control. However, there is no much research carried out to predict the efficiency of NTP technique using soft computing techniques. One of the major constraints in applying this technology for vehicle pollution control is estimating the operating parameter values to treat the real diesel exhaust. To conquer this, prediction

S. Allamsetty (✉) · S. Mohapatro
DEI Lab, School of Electrical Sciences, Indian Institute of Technology Bhubaneswar,
Bhubaneswar, India
e-mail: sa22@iitbbs.ac.in

© Springer Nature Singapore Pte Ltd. 2019
J. C. Bansal et al. (eds.), *Soft Computing for Problem Solving*, Advances in Intelligent
Systems and Computing 817, https://doi.org/10.1007/978-981-13-1595-4_24

of DeNO$_X$ efficiency as well as NO to NO$_2$ conversion efficiency based on various electrical and physical parameters is necessary.

There are so many experimental studies [1–11] conducted by various researchers to learn the effects of various operating parameters such as voltage, frequency, flow rate, engine load, temperature, additives. But there are very few studies carried out to predict the removal efficiency with changes in operating parameters. A study was performed in [12] to model the NO$_X$ emission of a coal-fired boiler using artificial neural network (ANN), least squares support vector machine (LSSVM), and partial least squares (PLS) methods, and the results were compared to know the best method toward prediction accuracy and time consumption. Twenty-eight various variables were selected as the inputs with the NO$_X$ emission as the only output. This study concluded that ANN gives the best training accuracy and PLS is least time-consuming. The work in [13] investigated the influence of compression ratio on the emissions of a diesel engine using biodiesel-blended diesel fuel. This study concluded that an increase in compression ratio led to decrease in the CO and hydrocarbon emissions while NO$_X$ emissions increase. The desirability approach of the response surface methodology was used to predict the response parameters such as brake-specific fuel consumption, brake thermal efficiency, CO, hydrocarbon, and NO$_X$.

A kinetic model was developed in [14] to characterize the chemical reactions taking place in a pulsed streamer corona discharge reactor. The electron density for the chemical reactions taken place in the reactor was obtained by fitting a model to the experimental data. Here, NO concentration was considered as function of the applied electric field and the reactor residence times. Later, the model was used to predict the concentrations of other species, i.e., O$_3$, NO$_2$, N$_2$O, and byproducts of ethylene decomposition.

Various data-fitting problems were considered in [15] to understand the relation of NO concentration with power input and voltage applied. A control value is obtained from applied voltage, and NO concentration was fitted against it using a polynomial equation. Then, NO concentration was fitted against deposited energy density (ratio of power input to the flow rate) and also to a form based on the hyperbolic *tanh* function superimposed on a sloping baseline.

Here, in this present study, an approach is made to predict NO to NO$_2$ conversion efficiency (η_{NO-NO_2}) of a particular experimental setup, using MATLAB and RBF, with variations in electrical and physical parameters such as voltage (v), flow rate (fr), discharge gap width (dg), and initial NO$_X$ concentration (ic). Test data has been taken by conducting experiments using the setup explained in the Sect. 2, and that data has been analyzed in the Sect. 3. Problem formulation using RBF has been explained in Sect. 4 with the support of an algorithm. The prediction results have been discussed in Sect. 5, and a conclusion based on the present study has been provided in Sect. 6.

2 Experimental Setup

Figure 1 represents the experimental setup used for this present study. Various components of this setup and procedure followed to get the required data have been explained in this section.

A 5 kVA diesel generator set is loaded with 3 kW bulb load to get the exhaust required for the experiments. This raw exhaust is made to pass through the tubes consisting of steel wool, drierite, and particulate filter (Make: Ultrafilter; Model: EG0020) to remove soot, particulate matter, moisture, and oil mist before passing through the plasma reactor. Flow controller is placed just before the reactor chamber to monitor and vary the flow rate of the exhaust.

A high voltage AC test set (Make: Rectifiers and Electronics) is used to give high voltage required for the process. The voltage of this test set is variable from 0 to 30 kV. A voltage divider (Make: IWATSU, Model: HV-P60A, DC to 50 MHz, within −3 dB) of attenuation 2000:1 ±5% and a digital storage oscilloscope (Make: RIGOL; Model: DS 1074: 70 MHz) are used for electrical parameters measuring purpose.

The plasma reactor, made up of borosil glass, has been used for exhaust treatment with an inner diameter of 20 mm and outer diameter of 24 mm. It is facilitated with inlet and outlet for getting the gas into it and leaving it back after treatment. Aluminum foil is wrapped between the inlet and outlet of the tube for a length of 280 mm to form the ground electrode. Three numbers of rod-type electrodes, made

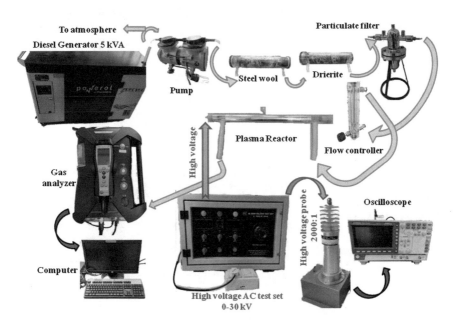

Fig. 1 Block diagram of experimental setup used during the experiments

up of stainless steel and with three different diameters, i.e., 3, 4 and 5 mm, are used as high-voltage electrodes.

Gas analyzer (Make: Testo; Model: Testo-350) is used to measure the concentrations of various pollutants. Real-time measurement mode of this analyzer enables the measurement process to be carried out continuously throughout the duration of experiments.

3 Experimental Results

As mentioned earlier, experiments are conducted with variation in voltage, flow rate, discharge gap width, and initial NO_X concentration. The data obtained from the experiments has been analyzed in this section with respect to each parameter considered as follows.

3.1 Variation in Voltage

Irrespective of the energization, whether it is AC or DC or pulse power supply, NO_X removal efficiency would increase with the increase in voltage while maintaining all other parameters constant. Here in this present study also, NO to NO_2 conversion efficiency has also been found to be increasing with the increase in voltage to some extent. But in some cases, for example, in cases a and c, this efficiency started decreasing with further increment in voltage as shown in Fig. 2. This happened as the total NO_X concentration is reduced due to the increased discharge power to the reactor, and thus, increased number of micro-discharges took place.

Fig. 2 Change in NO to NO_2 conversion efficiency with variation in voltage. Case: a. fr: 4 lpm, dg: 8 mm and ic: 155 ppm; Case: b. fr: 4 lpm, dg: 8.5 mm and ic: 297 ppm; Case: c. fr: 6 lpm, dg: 7.5 mm and ic: 179 ppm; Case: d. fr: 8 lpm, dg: 8 mm and ic: 171 ppm

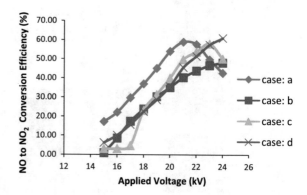

3.2 Variation in Flow Rate

When the flow rate is increased, gas residence time in the reactor chamber would decrease. So increase in flow rate causes a reduction in number of chemical reactions taking place in the reactor. Thus, NTP effect would be less on the exhaust gas and NO to NO$_2$ conversion efficiency slightly decreases as shown in Fig. 3. In this figure, maximum NO to NO$_2$ conversion efficiency that can be achieved with a particular setup has been plotted with variation in flow rate. However, decrement in NO$_X$ removal efficiency would be more with increase in flow rate. In this case, voltage, discharge gap width, and initial concentration are maintained constant.

3.3 Variation in Discharge Gap Width

Discharge gap width has been varied by changing the electrode diameter while maintaining the reactor dimensions same. Here in this study, three numbers of electrodes, of diameters 3, 4 and 5 mm, are used. As the inner diameter of the reactor is 20 mm, these electrodes are allowing discharge gaps of 8.5 mm, 8 mm, and 7.5 mm, respectively. When the discharge gap width decreases, electric field in the reactor would get more intensified at a lesser voltage. Intense electric field supports more chemical reactions to take place, and thus, the NO to NO$_2$ conversion efficiency increases as shown in Fig. 4.

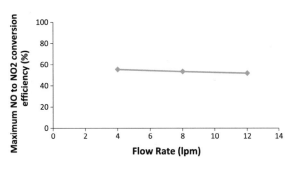

Fig. 3 Change in NO to NO$_2$ conversion efficiency with variation in flow rate

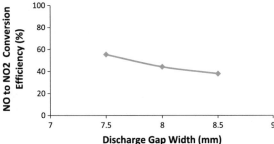

Fig. 4 Change in NO to NO$_2$ conversion efficiency with variation in discharge gap width

Fig. 5 Change in NO to
NO₂ conversion efficiency
with variations in initial
NO$_X$ concentration and
applied voltage

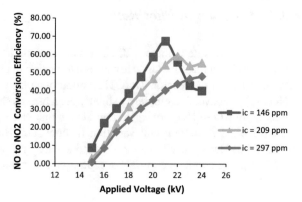

3.4 Variation in Initial NO$_X$ Concentration

Initial NO$_X$ concentration varies with the load on the diesel engine. In the present
study, 5 kVA diesel engine is loaded up to 3 kW. When the initial concentration is
less, reactions corresponding to NO$_X$ removal also take place leading to reduced NO
to NO₂ conversion efficiency at higher applied voltage. Figure 5 shows the change
in NO to NO₂ conversion efficiency with variations in initial NO$_X$ concentration and
applied voltage.

4 Problem Formulation Using RBF

An RBF network, to represent the predictive model shown in Fig. 6, is constructed in
MATLAB. Voltage, flow rate, discharge gap width, and initial NO$_X$ concentrations
are taken as independent variables and the NO to NO₂ conversion efficiency as the
dependent variable. In this study, RBF network is trained using experimental data
obtained from conducting experiments using the setup described in the previous
section. This data has been tabulated in MS Excel, and that table is called using
MATLAB programming to take the set of measured input and output variables. An
algorithm representing the step-by-step procedure for predicting the NO to NO₂
conversion efficiency has been given in Fig. 7.

Experimental data considered for training the RBF network in this study consisted
of 130 data sets, i.e., N. The training of the RBF network is composed of two stages,
i.e., finding the centers c_i, and then finding the weights w_i.

The selection of centers is done using K-means clustering algorithm. The number
of centers is decided in advance, i.e., 7, and each center c_i is supposed to be representative of a group of data points. The initial centers are randomly chosen from
the training data set. Then the following steps are repeated till a fixed number of
iterations.

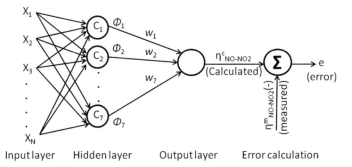

Fig. 6 RBF network

1. Find the nearest c_i to x_j; $(i = 1, 2, \ldots, 7)$. Suppose that this is found to be c_k.
2. $c_k^{new} = c_k^{old} + \eta(x_j - c_k^{old})$ where $\eta > 0$ is a small number called learning rate.
3. Set c_k^{new} as c_k^{old}.

This process is repeated until there are no more changes in all the centers when compared to the previous iteration. The predicted NO to NO$_2$ conversion efficiency can be calculated using the following formula.

$$\eta_{NO-NO_{2j}} = \sum_{j=1}^{h} \Phi_j w_j;$$

$$\Phi_j = \Phi(\|x - c_j\|)$$

Here Φ is Gaussian radial function, i.e., $\Phi(z) = e^{-z^2/2\sigma^2}$.

Centers and weights are updated using following equations based on gradient descent learning method.

$$c_{ij}(t+1) = c_{ij}(t) + \eta_1\left(\eta_{NO-NO_{2j}} - \eta_{NO-NO_2}\right)w_i\frac{\Phi_i}{\sigma^2}(x_j - c_{ij})$$

$$w_i(t+1) = w_i(t) + \eta_2\left(\eta_{NO-NO_{2j}} - \eta_{NO-NO_2}\right)\Phi_i$$

5 Results and Discussion

The experimental data provided to the training of the RBF network is of 130 sets. Each set of data contains the values of four independent variables, i.e., voltage, flow rate, discharge gap width, and initial NO$_X$ concentrations; and one dependent variable, i.e., NO to NO$_2$ conversion efficiency. Predicted NO to NO$_2$ conversion efficiency has been obtained after updating the centers and weights until there is no

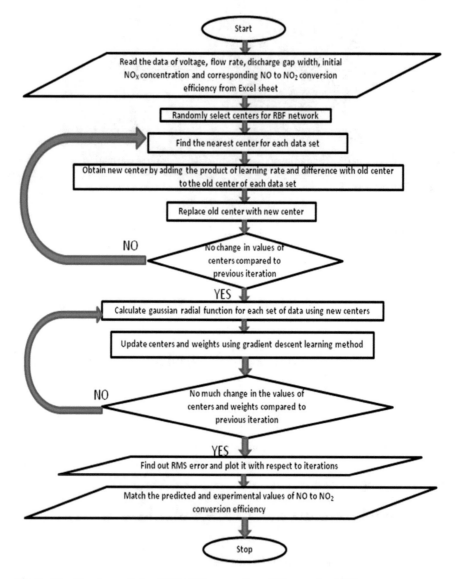

Fig. 7 Algorithm for predicting NO to NO$_2$ conversion efficiency using RBF

much change in them compared to the previous iteration. The RMS error is found to be reducing to 0.06 in 500 iterations as shown in Fig. 8.

It can be said that the predicted values are well following the experimental values of NO to NO$_2$ conversion efficiency from the residual plots shown in Fig. 9 for the training data. Residuals have been standardized first using their standard deviation in order to compare them at different data points. As the standardized residuals are not following any pattern, it can be said that the model is working well for the prediction.

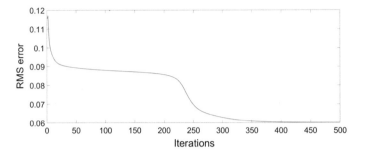

Fig. 8 Change in RMS error with iterations

Fig. 9 Residual plots for the training data. **a** Experimental values versus predicted values of conversion efficiency. **b** Standardized residuals versus predicted values of conversion efficiency

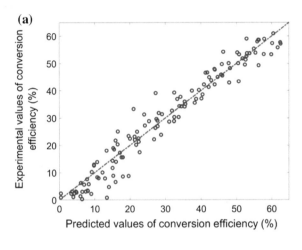

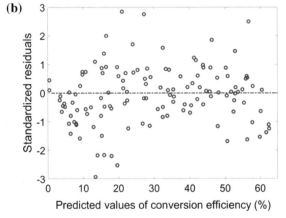

The final values of centers and weights can be considered as optimal values, and they can be further used for predicting NO to NO$_2$ conversion efficiency. Test data of ten numbers of sets has been given and obtained predicted efficiency values. The

Table 1 Test data and predicted NO to NO_2 conversion efficiencies

S. No.	Voltage	Flow rate	Discharge gap width	Initial NO_X concentration	Measured NO to NO_2 conversion efficiency	Predicted NO to NO_2 conversion efficiency
1	23	16	8.5	197	32.63	34.26
2	20	6	7.5	212	34.22	38.89
3	19	8	8	171	28.65	35.17
4	19	4	7.5	209	39.31	36.41
5	22	8	8.5	281	42.28	39.24
6	19	4	7.5	186	35.96	35.93
7	19	4	8	155	45.07	40.56
8	21	8	7.5	213	37.13	42.13
9	22	6	7.5	212	50.05	47.41
10	21	8	8	171	45.52	46.21

difference between experimental and predicted values of conversion efficiency for this test data can be seen in Fig. 10. Here, the RMS error is 0.07, and in terms of percentage, it is 3.7%. The test data along with predicted NO to NO_2 conversion efficiencies has been given in Table 1.

6   Conclusion

In this present study, NO to NO_2 conversion efficiency of a particular experimental setup has been predicted using RBF. Experimental data has been obtained with variations in voltage, flow rate, discharge gap width, and initial NO_X concentration. Predicted NO to NO_2 conversion efficiency has been obtained with updated centers and weights of RBF network and found it well matching with the experimental values. The RMS error is 0.06 during the training and 0.07 during testing. When comes to percentage by which experimental and predicted NO to NO_2 conversion efficiencies are varying, it is 3.7% for the random test data. Accuracy of this model would be increased with increase in the number of experimental data sets available. This method can be further modified to predict NO_X removal efficiency or to determine optimal values of various parameters of an experimental setup. Hence, developing a more accurate model with an objective of obtaining optimal operating parameter values for achieving maximum NO_X removal efficiency will be our future work.

Fig. 10 Residual plots and prediction results for the test data. **a** Experimental values versus predicted values of conversion efficiency. **b** Standardized residuals versus predicted values of conversion efficiency. **c** Bar graph for experimental and predicted conversion efficiencies

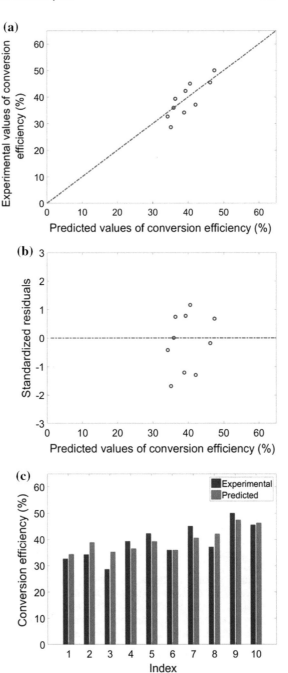

References

1. Bhattacharyya, A., Rajanikanth, B.S.: Performance of helical and straight-wire corona electrodes for NOx abatement under AC/Pulse energizations. Int. J. Plasma Environ. Sci. Technol. **7**(2), 148–156 (2013)
2. Du, B.X., Liu, H.J., Wang, X.H., Wang, K.F.: Application of dielectric barrier discharge in the removal of NO_X from diesel exhaust. In: CEIDP 2009 Annual Report, Conference on Electrical Insulation and Dielectric Phenomena, pp. 642–645. IEEE, USA (2009)
3. Babaie, M., Davari, P., Zare, F., Rahman, M., Rahimzadeh, H., Ristovski, Z., Brown, R.: Effect of pulsed power on particle matter in diesel engine exhaust using a DBD plasma reactor. Plasma Sci. **41**(8), 2349–2358 (2013)
4. Song, C.-L., Bin, F., Tao, Z.-M., Li, F.-C., Huang, Q.-F.: Simultaneous removals of NOx, HC and PM from diesel exhaust emissions by dielectric barrier discharges. J. Hazard. Mater. **166**(1), 523–530 (2009)
5. Namihira, T., Tsukamoto, S., Wang, D., Hori, H., Katsuki, S., Hackam, R., Akiyama, H., Shimizu, M., Yokoyama, K.: Influence of gas flow rate and reactor length on NO removal using pulsed power. IEEE Trans. Plasma Sci. **29**(4), 592–598 (2001)
6. Namihira, T., Tsukamoto, S., Wang, D., Katsuki, S., Hackam, R., Akiyama, H., Uchida, Y., Koike, M.: Improvement of NO_X removal efficiency using short-width pulsed power. IEEE Trans. Plasma Sci. **28**(2), 434–442 (2000)
7. Chirumamilla, V.R., Hoeben, W.F.L.M., Beckers, F.J.C.M., Huiskamp, T., Pemen, A.J.M.: Experimental investigation on NO_X removal using pulsed dielectric barrier discharges in combination with catalysts. In: 22nd International Symposium on Plasma Chemistry, Belgium, pp. 1–2 (2015)
8. Wang, P., Cai, Y.X., Li, X.H., Wang, J., Zhang, L.: Study on harmful emissions of diesel engine with non-thermal plasma assisted catalyst technology. In: 4th Internationl Conference on Bioinformatics Biomedical Engineering ICBBE, pp. 1–4. IEEE, USA (2010)
9. Mohapatro, S., Srikanth, A., Apeksha, M., Nikhil, Kumar S.: Study of nano second pulse discharge based nitrogen oxides treatment using different electrode configurations. High Volt. **2**(2), 60–68 (2017)
10. Wang, T., Sun, B.M., Xiao, H.P., Zeng, J.Y., Duan, E.P., Xin, J., Li, C.: Effect of reactor structure in DBD for nonthermal plasma processing of NO in N_2 at ambient temperature. Plasma Chem. Plasma Process. **32**(6), 1189–1201 (2012)
11. Wang, Z.: Reaction mechanism of NO_X destruction by non-thermal plasma discharge. Doctor of Philosophy Thesis, Clark Atlanta University (1999)
12. You L., Jizhen L., Tingting Y., Yuguang N.: Comparison of artificial neuro network, least squares support vector machine and partial least squares modelling on NOx emission. In: Power and Energy Engineering Conference (APPEEC). IEEE, USA (2012)
13. Sivaramakrishnan, K., Ravikumar, P.: Optimization of operational parameters on performance and emissions of a diesel engine using biodiesel. Int. J. Environ. Sci. Technol. **11**(4), 949–958 (2014)
14. Sathiamoorthy, G., Kalyana, S., Finney, W.C., Clark, R.J., Locke, B.R.: Chemical reaction kinetics and reactor modeling of NOx removal in a pulsed streamer corona discharge reactor. Ind. Eng. Chem. Res. **38**, 1844–1855 (1999)
15. Peter, A., Gorry, J., Christopher, W., Jinhui, W.: Adaptive control for NOx removal in nonthermal plasma processing. Plasma Process. Polym. **4**(5), 556–562 (2007)

Comparative Analysis of Optimum Capacity Allocation and Pricing in Power Market by Different Optimization Algorithms

Ashok Parmar and Pranav Darji

Abstract The paper presents the optimum allocation of capacity and capacity pricing between two individual electricity markets, which are interconnected and have different designs. Among them, one market has energy-only market, whereas other has capacity-plus-energy market. Generation companies (GenCos) optimally allocate capacity in different markets in such a way that it maximizes their overall revenue. Similarly, independent system operator (ISO) purchases capacity and energy in such a way that it minimizes their purchase cost. Both these problems are stochastic and nonlinear optimization problems. It could not be solved by classical method. Therefore, three meta-heuristic optimization algorithms such as Dragonfly, Moth-flame, and Whale optimization algorithms are used to solve both problems. Furthermore, both problem solutions are illustrated by a numerical example and capacity allocation and capacity pricing results are compared with the literature. Additionally, the comparative analysis is done for the capacity price, GenCos' revenue, and the ISO's purchase cost by using above-mentioned algorithms under different market conditions.

Keywords Generation company (GenCoC) · Independent system operator (ISO)
Dragonfly algorithm (DA) · Whale optimization algorithm (WOA) · Moth-flame
optimization algorithm (MFOA)

A. Parmar (✉) · P. Darji
Department of Electrical Engineering, Sardar Vallabhbhai National Institute
of Technology, Surat 395007, Gujarat, India
e-mail: d16el004@eed.svnit.ac.in

P. Darji
e-mail: pranav@eed.svnit.ac.in

© Springer Nature Singapore Pte Ltd. 2019   311
J. C. Bansal et al. (eds.), *Soft Computing for Problem Solving*, Advances in Intelligent
Systems and Computing 817, https://doi.org/10.1007/978-981-13-1595-4_25

1 Introduction

In the power system domain, every market has a different design; such few markets are working as energy-only market, whereas others are working as a capacity-plus-energy market. Energy-only market is only procured right amount of energy to fulfill load demand and rated in MWH or KWH, whereas capacity market is procured installed and available capacity for long-term resources adequacy and operational reliability. It is rated on MW-months, MW-year, or MW-years. All power markets are interconnected with surrounding regional state markets or country markets. In the USA, all regional markets are interconnected with nearby markets. The UK market is interconnected with Scotland and Ireland markets [1–5].

Cross-border and inter-state trades facilitate efficient utilization of available resources and increased social welfare. The excess capacity in any market is transferred to another market during peak time to fulfill peak demand of both the markets. In cross-border trades, consumers are able to get low-cost electricity if interconnections are unconstrained. Foreign entries in any market lead to competition and encourage existing player to perform well [2, 6, 7].

Generally, in any interconnected markets, capacity recourses of both markets are strategically participated in different markets to increase their revenue. Therefore, it is an important issue, how generation companies can optimally allocate their capacity to different markets to maximize their total revenue by selling different products like capacity and energy. Similarly, ISOs of both markets are trying to purchase least-cost capacity as well energy by placing offers to both markets. Therefore, GenCos' capacity allocation strategies depend on the ISO's purchase strategies and vice versa. Both these issues are discussed in the context of existing, differently designed, interconnected markets such as the PJM market and Midwest ISO [2, 6, 8, 9].

In the literature, optimum capacity placement and sizing problems are discussed to achieve different objectives such as loss reduction, reliability, and voltage profile enhancement of distribution system in conventional environment, whereas we are presented GenCos' revenue maximization and ISO's purchase cost minimization under deregulated environment by optimum capacity allocation between two neighboring power markets [10–12].

Main objectives are how GenCos can optimally allocate their capacity between two markets to maximize its overall revenue and how the ISO can minimize their overall purchase cost by adopting optimum purchase strategies [12]. GenCos' revenue and ISO's purchase cost are derived in Sect. 2. The optimization model and different algorithms are presented in Sect. 3. Result analysis is carried out in Sect. 4.

2 Problem Formulation

It is assumed that two nearby power markets are interconnected. One of them is recognized as an internal market, and it has energy-plus-capacity market, whereas the other is recognized as an external market, and it has energy-only market [1, 12].

Similar to the PJM market, the internal market is adopted two CRMs such as capacity market and capacity obligation. All LSEs are purchased their obliged capacity from the capacity market according to the predicted demand of their customers. For simplicity, few assumptions are made as under [1, 12];

- The forecasted day-ahead market prices of the external market are higher than the internal market's forecasted day-ahead market prices.
- The external market is always in capacity-deficient condition, and any generation companies removed from the internal market can give up its total energy to the external market.
- The ISO purchases capacity and energy on behalf of LSEs and is behaving as a single buyer. Furthermore, it is assumed that fixed and variable cost of all the generators are same. Therefore, any extra profit earned by any generators as a difference of the market clearing price and the marginal cost of generation is same. Thus, all generator companies are treated as generation company C (GencoC), and its capacity is the total available capacity of all generation companies of the internal market.

2.1 GenCoC's Revenue

To maximize the revenue, GenCoC has decided to allocate available capacity in following ways [12].

1. Total capacity of GencoC is $G_t = G_1 + G_2 \cdot G_1$ capacity is allocated to the internal market. Furthermore, G_1 consists of G_{1e} and $G_{1ee} \cdot G_{1e}$ is completely sold to internal market as capacity and energy resources, whereas G_{1ee} is sold to internal market as capacity resources and allocated as energy resources to external market. When the capacity scarcity takes place in the internal market, ISO of the internal market can recall G_{1ee} or part of G_{1ee} capacity.
2. G_2 is removed from of the internal market and dispatched to the external market. Here, in any situation, the internal market's ISO does not have right to recall G_2.

GencoC can earn revenue in following ways [12].

1. Selling capacity to internal capacity market;

$$x_{1g} = M_c * \min(G_1, G_r) \tag{1}$$

where $\min(G_1, G_r)$ is the auctioned capacity in internal market (MW), M_c is the capacity price ($/MW-day), and G_r is the ISO's capacity obligation (MW).
2. Selling energy to the internal energy-only market;

$$x_{1e} = \sum_{t=1}^{t=24} M_{1e}(t) * \min(D_1(t), G_{1e}) \tag{2}$$

where $D_1(t)$ is forecasted internal market load (MW), and $M_{1e}(t)$ is day-ahead forecasted energy prices of internal energy-only market (\$/MWh).

3. Act as a capacity recourses of the internal market but selling energy to external energy-only market or selling energy to internal energy-only market by the recall;

$$x_{1ee} = \sum_{t=t1}^{t=t2} G_{1ee} * (M_{2e}(t) - T_p) + \sum_{t=t1}^{t=t2} \{\min(G_{1ee}, \max(D_1(t) - G_{1e}, 0))$$
$$* (M\,1ee(t) + T_p - M_{2e}(t) - Pen_2(t))\} \tag{3}$$

$$Pen_2(t) = M_{2e}(t) * 0.1 \tag{4}$$

where $M_{2e}(t)$ is day-ahead energy price of external market (\$/MWh), T_p is transmission price (\$/MWh), $M_{1ee}(t)$ is real-time price of internal market during specific trading period, $Pen_2(t)$ is the penalty of the external market when capacity is recalled by the internal market (\$/MWh), and $t_1 - t_2$ is the peak time of trading period. For simplicity, T_p is kept constant. $Pen_2(t)$ is 10% of day-ahead market prices of the external market. The $\min(G_{1ee}, \max(D_1(t) - G_{1e}, 0))$ is the recalled capacity (MW).

4. Removed from internal capacity market and selling energy to external market;

$$x_{2e} = \sum_{t=t1}^{t=t2} G_2 * (M_{2e}(t) - T_p) \tag{5}$$

$$G_2 = G_t - G_{1e} - G_{1ee} \tag{6}$$

where $G_t = G_1 + G_2$ is the total capacity and $G_1 = G_{1e} + G_{1ee}$.

From all above equations, total revenue earned by GenCoC is mentioned in (7). Where G_{1e}, G_{1ee}, and M_c are the variables and constraints given by (8) and (9). T_c is the transmission capacity between internal and external markets (MW).

$$x(G_{1e}, G_{1ee}, M_c) = x_{1g} + x_{1e} + x_{1ee} + x_{2e} \tag{7}$$

$$G_t - T_c \le G_{1e} \le G_t \tag{8}$$

$$0 \le G_{1ee} \le G_t - G_{1e} \tag{9}$$

2.2 ISO's Purchase Cost

ISO of internal market can obtain necessary capacity and energy by purchasing it in the following way [12].

1. Purchasing capacity from internal capacity market;

$$y_{1g} = M_{cap} * \max(0, G_r - G_1) + M_c * \min(G_1, G_r) \tag{10}$$

where M_{cap} is called capacity market price-cap (\$/MW-day).

2. Purchasing energy from the internal energy-only market;

$$y_{1e} = \sum_{t=1}^{t=24} M_{1e}(t) * \min(D_1(t), G_{1e}) \tag{11}$$

3. Purchasing capacity by recalling capacity from the external market;

$$y_{1ee} = \sum_{t=t1}^{t=t2} \min(G_{1ee}, \max(D_1(t) - G_{1e}, 0)) * (M_{1ee}(t) + T_p) \tag{12}$$

4. Purchasing energy from the external market;

$$y_{2e} = \sum_{t=t2}^{t=t1} \min(T_c, \max(0, D_1(t) - G_1)) * (M_{2e}(t) + T_p) \tag{13}$$

So that, the ISO's total purchase cost is expressed as follows.

$$y(G_{1e}, G_{1ee}, M_c) = y_{1g} + y_{1e} + y_{1ee} + y_{2e} \tag{14}$$

where constraint in addition to Eqs. (8) and (9) is as under

$$G_r \leq G_1 \tag{15}$$

3 Optimization Model

3.1 GenCoC's Revenue Maximization and ISO's Purchase Cost Minimization

GenCoC's revenue maximization and ISO's revenue minimization problems are tested with DA, MFOA, WOA, SCA, MVO, and PSO algorithms. It is found that DA, MFOA, WOA, and PSO are presented best result under different scenarios. To analyze the effect of only advance algorithms, comparative analysis of three algorithms (DA, MFOA, and WOA) is presented here. Additionally, the heuristic algorithm preserves the all search space information of the previous iteration, whereas in evolutionary algorithm, previous round search space information is discarded when new generation form. Both these problems are stochastic and nonlinear optimization problems, and it could not be solved by analytical method. The parameters selected to solve this problems are as follows [13–15]. With reference to all these parameters, Dragonfly, Moth-flame, and Whale optimization algorithms are used to optimize below-mentioned objective functions. It is subjected to constraints mentioned in Eqs. (8), (9) and (15).

- Numbers of iteration = 30, Nos of search agent = 70.
- Numbers of variables = 3, $x(1) = G_{1e}$, $x(2) = G_{1ee}$, $x(3) = M_c$.

$$f(x) = Maximize\{x(G_{1e}, G_{1ee}, M_c)\} \qquad (16)$$

$$f(y) = Minimize\{y(G_{1e}, G_{1ee}, M_c)\} \qquad (17)$$

3.2 Different Optimization Algorithms (DA, MFOA, WOA)

The Dragonfly algorithm is meta-heuristic optimization algorithm, and it was introduced by Mirjalili [13]. It is inspired by the natural swarming behavior of Dragonflies. There is two swarming behaviors of Dragonflies—static and dynamic. In the static behavior, all Dragonflies formed a small group and fly in different directions to cover a large area to search the food. This is called exploration. In the dynamic behavior, all Dragonflies form a big group and migrate in a specific direction and it is called exploitation. Dragonflies do the random movement in search space is the main characteristic of static swarming behavior. Dragonflies are considered as search agents, and variables are the positions and velocity of Dragonflies in search space [13].

The Moth-flame optimization algorithm is also a meta-heuristic optimization algorithm, and it is inspired by navigation method of Moths. It was introduced by Mirjalili [14]. Moths follow a specific navigation method to fly in the night. Moths have always maintained a fixed angle with respect to the moon during flying, and to fly a long run in a straight way in the night. This is called transverse orientation. The moon is so far from the earth, and therefore, maintaining fixed angle with the moon is ensuring straight run. Moths are misguided by the artificial source of light, and they try to maintain fixed angle with respect to the artificial source of light for straight fly, but this light is very close to moths, and this behavior misleads them for a spiral fly around the light, and finally they reach to light. Therefore, moth finally flies toward light through changing their position in the spiral path. Moths are considered as search agents, and variables are the positions of moth search space [14].

The Whale optimization algorithm is another meta-heuristic optimization algorithm, and it was introduced by Mirjalili and Lewis [15]. It is inspired by humpback whale's bubble-net hunting strategy and mimics the hunting behavior of humpback whales. Humpback whale is one of the biggest mammals, and they adopt two actions of hunting herds of fish or krill. One of them is called upward spiral, whereas other is known as double loops. In an upward spiral, whales go deep into the sea up to 15 m and are producing bubbles in spiral-shape path or 9-shape path to encircle prey and swim toward the surface. They prefer to foraging herds at surface. This interesting hunting pattern is called bubble-net feeding method. The whale is considered as search agents, and variables are the positions of whale in search space [15].

4 Result Analysis

4.1 Numerical Example

To implement problems, the following data are considered [12].

- Total installed capacity of GencoC (G_t): 59637 MW.
- Capacity obligation of ISO (G_r): 57557 MW.
- Capacity market price-cap (Mcap): 350 \$/MW-day.
- Available transmission capacity (T_c): 3500 MW.
- Transmission price (T_p): 18 \$/MWh.
- The peak load energy contract duration: $t_1 = 5{:}00$ to $t_2 = 20{:}00$.

Forecasted load demand of the internal market (D_1), day-ahead energy prices of the internal energy-only market (M_1e), day-ahead energy prices of the external market (M_2e), and real-time energy prices in the internal market (M_1ee) are presented in Table 1 [12].

Table 1 Load and market prices data

t (h)	$M_{1e}(t)$ (\$/MWh)	$M_{1ee}(t)$ (\$/MWh)	$M_{2e}(t)$ (\$/MWh)	$D_1(t)$ (MW)	t (h)	$M_{1e}(t)$ (\$/MWh)	$M_{1ee}(t)$ (\$/MWh)	$M_{2e}(t)$ (\$/MWh)	$D_1(t)$ (MW)
0	25	24	25	38,000	13	90	340	370	53,000
1	24	23	24	37,200	14	125	270	330	54,000
2	23	22	23	36,000	15	145	450	480	54,900
3	23	22	23	34,000	16	180	140	200	55,700
4	23	22	23	33,000	17	155	125	175	55,700
5	24	23	24	32,500	18	140	100	150	54,500
6	24	23	24	34,000	19	100	75	125	53,000
7	24	23	24	36,700	20	75	51	75	52,000
8	38	23	38	39,300	21	75	50	50	51,000
9	48	45	48	41,800	22	75	48	49	50,000
10	50	50	50	45,000	23	70	32	29	45,000
11	65	80	83	48,500	24	25	25	22	41,000
12	75	110	115	51,000					

A. Parmar and P. Darji

4.2 Effect of the Recall Probability, Cost of Recall, and Load Forecasting (LF) Error on M_c and GenCoC's Revenues

GenCoC normally prefers to sells energy to external market. But due to T_c constraints, GenCoC has to sell minimum $(G_t - T_c)$ to the internal market. Therefore, internal market capacity G_1 is greater than or equal to $G_t - T_c$. GenCoC has two options to optimally allocate capacity between two markets. (1) GenCoC removed some of its capacity from the internal market and sold energy to the external energy market. (2) GenCoC has allocated their all capacities to the internal market as capacity resources to get capacity payment and sell energy to the external energy market and face recall possibility.

Table 2 Effect of T_c on M_c, GencoC's revenue, and ISO's purchase cost

Sr. no.	T_c (MW)	G_{1e} (MW)	G_{1ee} (MW)	M_c (\$/MW day)	GeCoC's revenue (\$)	ISO's purchase cost (\$)
Reference paper (Genetic algorithm)						
1	2700	56,936	1127.4	109.59	–	–
2	3100	56,537	1524.8	133.7	–	–
3	3500	56,137	1891	149.69	–	–
4	3900	55,738	2209.9	168.2	–	–
5	4300	55,339	2561.5	189.32	–	–
Dragonfly optimization algorithm						
1	2700	56,936	1135.8	108.9	96815671.0	91459437.1
2	3100	56,535	1518.7781	134.0	98987410.3	92973186.2
3	3500	56,134	1956	151.9	100756250.3	93778984.2
4	3900	55,736	2194.44	167.7	102450108.9	94,838,033
5	4300	55,336	2424.81644	191.1	104422816.7	96,207,025
Moth-flame optimization algorithm						
1	2700	56,936	1133.3	110.0	96815671.0	91419722.77
2	3100	56,535	1541.9528	134.5	98987410.8	92743533.00
3	3500	56,134	1881.185	151.9	100756250.0	93635091.7
4	3900	55,736	2176.19151	167.4	102450108.9	94751697.5
5	4300	55,336	2358.72785	192.6	104422816.7	95890461.5
Whale optimization algorithm						
1	2700	56,936	1129	110.9	96815671.3	91327056
2	3100	56,535	1550	134.2	98987410.0	92714179.7
3	3500	56,134	1956	151.9	100756250.0	93738694.3
4	3900	55,736	2225	167.7	102450108.9	94849544.4
5	4300	55,336	2599	191.1	104422816.7	95988308.4

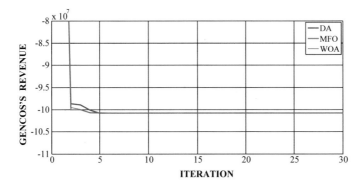

Fig. 1 GenCoC's revenue for T_c: 3500 MW

The recall probability is changed by changing T_c. If T_c is increased, GenCoC can dispatch more capacity to the external energy market. Therefore, the probability of D_1 exceeds, the $G_t - T_c$ is increased, and the ISO requires to recall more capacity to fulfill load requirement. If capacity G_{1ee} is recalled from the external market, GenCoC faces recall penalties. Therefore, more capacity payment is required to pay to compensate this loss and capacity cost will be increased. GenCoC's total revenue is increased if the total capacity allocation is increased, and it equals to a difference between two market energy prices minus recall penalties. All results are presented in Table 2. Additionally, GenCoC's revenue is also found for each value of T_c, which is not presented in Ref. [12]. It is found that GenCoC's revenue is remarkably increased with increasing T_c from 2700 to 4300 MW. All algorithms are conversed in different iterations but provided the same amount of GenCoC's revenue. GenCoC's revenue for 3500 MW value of Tc is presented in Fig. 1.

If capacity payment is not made to GenCoC, the price difference between two markets is regarded as an opportunity cost at which GenCoC sell energy to internal market, and it is called the cost of the recall. Hence, the cost of the recall is equal to the price difference between markets. Here, M_{1e} and M_{1ee} are kept constant and M_{2e} are increased or decreased to observe the effect of the cost of the recall on M_c and GenCoC's revenue. If M2e is increased, the cost of the recall is increased, and as a result, M_c and GenCoC's revenue are also gradually increased. Furthermore, results are found using DA, WOA, and MFOA and compared with Ref. [12]. It is found that different algorithms have given best result with a different scenario. MFOA has provided the best result for 90% values of M2e, whereas DA has provided the best result for 95% value of M_{2e}. For remaining three values of M_{2e}, WOA has provided best results. All results are presented in Table 3, and GenCoC's revenue graphs for 100% value of M_{2e} are presented in Fig. 2.

In power system, it is most difficult to predict load with exact precision and there is always the possibility of error. If real-time load (D_1) is less than predicted value, available capacity $G_t - T_c$ is greater than D_1 and hence there were less probability of recall and capacity prices. If real-time load ($D1$) is greater than predicted load

Table 3 Effect of the cost of recall on M_c and GenCoC's revenue

	Reference paper (Genetic algorithm)		Dragonfly optimization algorithm		Moth-flame optimization algorithm		Whale optimization algorithm	
Sr. no.	Change of M_{2e} (%)	M_c ($/MWh)	M_c ($/MWh)	GenCoC's revenue	M_c ($/MWh)	GenCoC's revenue	M_c ($/MWh)	GenCoC's revenue
1	−10%	94.69	99.1	95526568.7	110.9	96205741.3	95.6	95,325,119
2	−5%	121.63	123.1	97,212,901	120.5	96510706.0	122.3	97166856.1
3	0	149.69	151.3	99,140,974.1	148.3	96815671.3	151.3	99140974.1
4	5%	172.72	170.0	100,522,255	168.8	97120636.3	173.2	100706437.4
5	10%	201.16	187.6	101,840,223	202.4	97425601.7	199	102,496,373

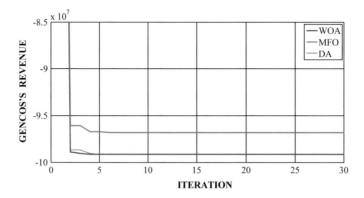

Fig. 2 GenCoC's revenue for M_{2e}:100 %

value, D_1 is greater than available capacity $G_t - T_c$. Thus, probabilities of recall and capacity prices are also increased. Capacity prices and GenCoC's revenues are found using DA, WOA, and MFOA and compared with Ref. [12] for each value of LF error. It is found that different algorithms have given the best result with a different scenario. WOA provided the best result for 0% LF error, whereas DA provided the best result for other values of the LF error. It is found that GenCoC's revenue is increased with the LF error increases. All results are presented in Table 4, and GenCoC's revenue for 0% value of the LF error is presented in Fig. 3.

4.3 Effect of the Recall Probability, Cost of Recall, and Load Forecasting (LF) Error on M_c and ISO's Purchase Cost

ISO is always preferred to buy energy and capacity from the internal market to minimize total purchase cost. However, due to T_c constraints, GenCoC has to sell minimum $G_t - T_c$ to the internal market and remaining capacity is assigned to the external market to earn more profit.

How probability of recall is changed was discussed in Sect. 4.2. If the probability of recall is increased, capacity costs and the ISO's purchase cost are increased. ISO's purchase cost for each value of optimum capacity is presented in Table 2. Furthermore, ISO's purchase cost is also found for each value of T_c, which is not presented in Ref. [12]. It is found that the ISO's purchase cost is remarkably increased with increasing T_c, and different algorithms are given the best result with a different scenario. WOA has provided the best result for 2700 and 3100 MW values of T_c, whereas MFO algorithm has provided the best result for remaining values of Tc.

What is the cost of the recall and how it is changed was discussed in previous Sect. 4.2. If M_{2e} is increased, the cost of the recall is increased, and as a result, Mc and ISO's purchase cost are also gradually increased. ISO's purchase cost is found

Table 4 Effect of load forecasting error on M_c and GenCoC's revenue

Sr. no.	Reference paper (Genetic algorithm)		Dragonfly optimization algorithm		Moth-flame optimization algorithm		Whale optimization algorithm	
	Load forecasting error	M_c ($/MWh)	M_c ($/MWh)	GenCoC's revenue	M_c ($/MWh)	GenCoC's revenue	M_c ($/MWh)	GenCoC's revenue
1	−20%	115.28	116.2	80098543.4	110.9	79793491.3	114	79,971,918
2	−10%	138.28	137.7	89847108.9	136.2	88304581.3	137.6	89841353.2
3	0	149.09	149.9	99060394.0	148.3	96815671.3	150.3	99083417.1
4	10%	153.21	154.2	104762150.5	152.8	102,296,252	153.8	104710338.8
5	20%	165.41	168.3	107,626,908	167.58	104,358,782	166	107495956.0

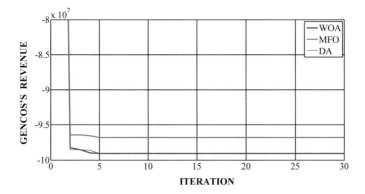

Fig. 3 GenCoC's revenue for forecasting error: 0%

using DA, WOA, and MFO algorithm and compared with Ref. [12]. It is found that different algorithms have given the best result with a different scenario. DA and MFO algorithm have provided a same best result for 90, 100, and 105% values of M_{2e}. For remaining values of M_{2e}, MFOA and WOA have provided best results. All results are presented in Table 5.

The effect of the LF error on capacity prices and the ISO's purchase cost are found using DA, MFOA, and WAO, and it compared with Ref. [12]. It is found that different algorithms have given best result with a different scenario. WOA has provided the best result for +10% values of LF error, whereas DA is provided the best result for remaining values of the LF error. It is found that the ISO's purchase cost is increased with the LF error increases. All results are presented in Table 6.

5 Conclusion

The way generation companies can increase their revenue with efficient utilization of capacity and the way the ISO can reduce their purchase cost by adopting optimum purchase strategies are equally important. Here, mathematical modeling of the GenCoC's revenue maximization problem and the ISO's purchase cost minimization problem is presented and solved by three optimization algorithms such as DA, MFOA, and WOA. Optimized maximum values of GenCoC's revenue and optimized minimum values ISO's purchase cost with optimum capacity allocation and pricing are found. Additionally, it presented a comparative analysis of the effect of the cost of the recall, the probability of recall and the LF error on the capacity prices (M_c), GenCoC's revenue and the ISO's purchase cost using three optimization algorithms. Furthermore, it is found that capacity price, the maximum value of Gen-CoC's revenue, and the minimum value of ISO's purchase cost are increased with the probability of recall, the cost of the recall, and the LF error. Different algorithms are

A. Parmar and P. Darji

Table 5 Effect of the cost of recall on M_c and ISO's purchase cost

Reference paper (Genetic algorithm)			Dragonfly optimization algorithm		Moth-flame optimization algorithm		Whale optimization algorithm	
Sr. no.	Change of M_{2e} (%)	M_c ($/MWh)	M_c ($/MW day)	ISO purchase cost ($)	M_c ($/MWh)	ISO purchase cost ($)	M_c ($/MWh)	ISO purchase cost ($)
1	−10%	94.69	93	90,463,701	93	90,463,701	93.2	90475212.4
2	−5%	121.63	121.1	92081052.0	120.5	92023495.1	121.2	92086808.4
3	0	149.69	148.4	93635091.7	148.8	93635091.7	148.2	93640847.4
4	5%	172.72	170.4	94901345.1	170	94901345.1	172.1	94958902.7
5	10%	201.16	187.6	96616544.3	201.1	96685612.7	198.2	96599277.2

Table 6 Effect of load forecasting error on M_c and ISO's purchase cost

Sr. no.	LF error (%)	Reference paper (Genetic algorithm) M_c ($/MW day)	Dragonfly optimization algorithm M_c ($/MW day)	ISO purchase cost ($)	Moth-flame optimization algorithm M_c ($/MW day)	ISO purchase cost ($)	Whale optimization algorithm M_c ($/MW day)	ISO purchase cost ($)
1	−20%	115.28	114	74,650,218	110.9	74684752.2	114.3	74667485.1
2	−10%	138.28	137.3	84502386.1	137	84525408.9	137.6	84519653.2
3	0	149.09	148.6	93663870.2	148.7	93669625.9	151.2	93813518.4
4	10%	153.21	154.0	103372145.8	152.8	103,353,151	152.1	103258182.9
5	20%	165.41	164.0	104005272.8	166.7	113646782.2	164.2	113474111.28

converged in different numbers of iteration and provided best result with different conditions, except the effect of the recall probability on GenCoC's revenue, where all algorithms have provided the same result. Additionally, results are compared with the available literature (Genetic algorithm). Finally, it is concluded that among these three algorithms, no one is superior to others with respect to the discussed problems.

References

1. Gore, O., Vanadzina, E., Viljainen, S.: Linking the energy-only market and the energy-plus-capacity market. Util. Policy **38**, 52–61 (2016)
2. Vasileva, E., Viljainen, S., Sulamaa, P., Kuleshov, D.: RES support in Russia: impact on capacity and electricity market prices. Renew. Ener. **76**, 82–90 (2015)
3. Bowring, J.: Capacity markets in PJM. Econ. Ener. Environ. Policy **2**, 47–64 (2013)
4. Ott, A.L.: Experience with PJM market operation, system design, and implementation. IEEE Trans. Power Syst. **18**, 528–534 (2003)
5. Thomas, S.: The Wholesale Electricity Market in Britain–1990-2001, Public Service International Research Unit. PSIRU), University of Greenwich, London (2001)
6. Bhagwat, P.C., de Vries, L.J., Hobbs, B.F.: Expert survey on capacity markets in the US: lessons for the EU. Util. Policy **38**, 11–17 (2016)
7. Logan, D.M., Dean, P., Edward, F.: RPM Implementation Review (2013)
8. Cramton, P., Stoft, S.: A capacity market that makes sense. Electr. J. **18**, 43–54 (2005)
9. Rau, N.S.: The need for capacity markets in the deregulated electrical industry- a review. In: Power Engineering Society 1999 Winter Meeting, vol. 1, pp. 411–415. IEEE (1999)
10. Anwar, A., Pota, H.: Optimum capacity allocation of DG units based on unbalanced three-phase optimal power flow. In: Power and Energy Society General Meeting, pp. 1–8. IEEE (2012)
11. Moradi, M.H., Abedini, M.: Optimal multi-distributed generation location and capacity by genetic algorithms. In: Power Engineering and Optimization Conference (PEOCO), 2010 4th International, pp. 440–444 (2010)
12. Wang, Y., Wong, K.P., Chung, C.Y., Wen, F.S.: Determination of appropriate price level in installed capacity market. IET Gener. Transm. Distrib. **1**, 127–132 (2007)
13. Mirjalili, S.: Dragonfly algorithm: a new meta-heuristic optimization technique for solving single-objective, discrete, and multi-objective problems. Neural Comput. Appl. **27**, 1053–1073 (2016)
14. Mirjalili, S.: Moth-flame optimization algorithm: a novel nature-inspired heuristic paradigm. Knowl. Based Syst. **89**, 228–249 (2015)
15. Mirjalili, S., Lewis, A.: The whale optimization algorithm. Adv. Eng. Softw. **95**, 51–67 (2016)

Improved Grey Wolf Optimizer Based on Opposition-Based Learning

Shubham Gupta and Kusum Deep

Abstract Swarm intelligence (SI)-based algorithms are very popular optimization techniques to deal with complex and nonlinear optimization problems. Grey wolf optimizer (GWO) is one of the newest and efficient algorithms based on hunting activity and leadership hierarchy of grey wolves. To avoid the slow convergence and stagnation problem in local optima, in this paper, opposition-based learning (OBL) is incorporated in GWO for the population initialization as well as for the iteration jumping. In this strategy, opposite numbers have been used to deal with the problem of slow convergence. The proposed algorithm is named as opposition-based explored grey wolf optimizer (OBE-GWO). To evaluate the performance of OBE-GWO, it is tested on some well-known benchmark problems. The experimental analysis concludes the better performance of OBE-GWO.

Keywords Meta-heuristics · Swarm intelligence · Opposition-based learning Grey wolf optimizer

1 Introduction

Meta-heuristic optimization techniques such as genetic algorithm (GA) [1], particle swarm optimization (PSO) [2], differential evolution (DE) [3], biogeography-based optimization (BBO) [4] and tabu search (TS) [5] are some popular techniques to solve complex real-life optimization problems that are widely used in engineering. These meta-heuristics are widely applicable to solve real-world optimization problems because of their simplicity, flexibility, derivative-free structure, etc. Swarm Intelligence is a very popular subarea in the field of meta-heuristics. PSO, BBO, Spi-

S. Gupta (✉) · K. Deep
Department of Mathematics, Indian Institute of Technology Roorkee, Roorkee 247667,
Uttarakhand, India
e-mail: g.shubh93@gmail.com

K. Deep
e-mail: kusumfma@iitr.ac.in; kusumdeep@gmail.com

© Springer Nature Singapore Pte Ltd. 2019
J. C. Bansal et al. (eds.), *Soft Computing for Problem Solving*, Advances in Intelligent
Systems and Computing 817, https://doi.org/10.1007/978-981-13-1595-4_26

der monkey optimization (SMO) [6] are some well-known techniques in this field of swarm intelligence. Grey wolf optimizer (GWO) [7], based on leadership hierarchy of grey wolves, is one of the newest optimizer algorithms in the field of swarm intelligence.

GWO is developed by Mirjalili et al. [7] in 2014 by simulating the hunting activity and leadership behaviour of wolves. As in GWO, three leaders lead to the process of searching the prey; therefore, it explores the whole search space which helps in finding the solution of the problem. From that time, some improvement has been done to improve its performance and used to solve many applications problems of engineering and computer science [8–12].

The concept of opposition-based learning (OBL) was given by Tizhoosh [13]. OBL is used in meta-heuristics to enhance the diversity of search process. In the present paper, to deal the problem of premature convergence which is the result of stagnation of the solution in local optima, GWO based on OBL strategy has been proposed. In this work, opposite numbers have been introduced to improve the exploration of the search space, so that the stagnation problem of the solution can be ignored.

It has been found that opposite numbers are better approximate to the optimal solution as compared to the purely random numbers. In this paper, the use of opposite numbers is analysed by replacing them with random ones. The paper presents a comprehensive analysis of convergence and benefits of opposite numbers.

The rest of the paper is as follows: Section 2 provides an overview of GWO. In Sect. 3 proposed improved version of GWO has been presented. In Sect. 4 experimental analysis has been done, and finally, Sect. 5 concludes the paper.

2   Overview of GWO

In 2014, Mirjalili et al. [7] designed GWO by mimicking the hunting behaviour and leadership discipline of the grey wolf pack. In grey wolf pack, there are four classes of wolves based on the mental intelligence and ability of managing the pack. Alpha is the dominant wolf of the pack and belongs to first class of the pack. They are responsible for all the major decisions within the pack. Beta wolf (second-class wolf) is the wolf that passes the important messages of alpha to other wolves. Beta wolf can take the position of alpha in case of the absence of alpha wolf. Delta wolves are the lowest class wolf that has the permission to eat food in the last. Elders and protector of the pack belongs to the third class of the pack, and they take care of the wolves from natural and external damages. This leadership hierarchy is the main characteristic of the wolf pack that helps in finding the prey. Grey wolves find their food (prey) by following the steps

   I.   chasing the prey
  II.   encircling the prey, and
 III.   attacking towards the prey.

This behaviour of wolves can be modelled mathematically [7] to solve any optimization problem as follows:

A. In any optimization problem, to simulate the social behaviour, the top three wolves are considered as alpha, beta and delta wolves (leaders of the pack).

B. To simulate the encircling behaviour, in [7], the following equations are proposed

$$y(t+1) = y_P(t) - A \cdot d \tag{1}$$

$$d = |C \cdot y_P(t) - y(t)| \tag{2}$$

$$A = 2 \cdot a \cdot rand_1 - a \tag{3}$$

$$a = 2 - 2 \cdot (t/\text{max\_iterations}) \tag{4}$$

$$C = 2 \cdot rand_2 \tag{5}$$

where $rand_1$ and $rand_2$ are uniformly distributed random vectors within the interval $(0, 1)$, y_P and y are prey positions and wolf positions, respectively, and t is the iteration count.

C. To simulate the last step of hunting that is to attack on a prey, it is considered that alpha, beta and delta wolf have enough knowledge regarding the prey; therefore, the preys' position can be approximated with the help of these leading wolves to update the state of the wolves. The following equations are used to update the position of wolf:

$$y_1 = y_\alpha(t) - A_\alpha \cdot d_\alpha \tag{6}$$

$$y_2 = y_\beta(t) - A_\beta \cdot d_\beta \tag{7}$$

$$y_3 = y_\delta(t) - A_\delta \cdot d_\delta \tag{8}$$

$$y(t+1) = (y_1 + y_2 + y_3)/3 \tag{9}$$

where

$$d_\alpha = |C_\alpha \cdot y_\alpha(t) - y(t)| \tag{10}$$

$$d_\beta = |C_\beta \cdot y_\beta(t) - y(t)| \tag{11}$$

$$d_\delta = |C_\delta \cdot y_\delta(t) - y(t)| \tag{12}$$

To find the optima for a particular optimization problem, (A), (B) and (C) forms a one iteration of search process. By continuing these iterations optima of the optimization problem can be found.

In the above equations, the vectors A and C are responsible for exploration as well as for exploitation in each iteration. $|A| < 1 \, or \, |C| < 1$ leads to the exploitation, and $|A| > 1 \, or \, |C| > 1$ leads to the exploration of the search space. The framework of the algorithm is provided in Algorithm 1.

3   Improved Grey Wolf Optimizer Based on OBL

In original GWO, each wolf updates their position with the help of leaders, namely alpha, beta and delta, but in some cases, GWO is trapped in local minima due to lack of diversity. Therefore in this paper, opposite numbers have been introduced in GWO to enhance the diversity of searching wolf in GWO and to avoid the premature convergence. Opposition-based learning shows their impact in many meta-heuristics [14–18] to solve real-life problems. Therefore, the opposition-based learning is effective when the population tends to stagnate or drift. In this case, the opposition-based allows an extra sampling in a possibly unexplored area of the domain [19]. This consideration was also made in [20].

Before employing opposite numbers in GWO, opposite numbers must be defined. Algorithm 1: Grey Wolf Optimizer (GWO)

1.	Initialization uniformly distributed wolf population W_p
2.	Calculate the fitness of each wolf
3.	Initialize the parameters $\{a, A, C\}$
4.	Select the leaders **alpha, beta** and **delta** for the search process
5.	while $l <$ max_iterations (termination criteria)
6.	**for each wolf**
7.	update the position by equation (1) to (12)
8.	**end for**
9.	update the leaders alpha, beta and delta
10.	update the algorithm parameters $\{a, A, C\}$
11.	$l = l + 1$
12.	**end while**

Opposite Number

Let $x \in [a, b]$ be a real number, then the opposite number y of x is defined as

$$y = a + b - x.$$

Similarly, this definition can be extended for the higher dimension [13].

A perturbation rate or oppositional probability (OP) is predefined in the algorithm, which decides that when to introduce OBL in the algorithm. The incorporated OBL phase in the proposed improved GWO can be summarized as follows:

In this phase, a uniformly distributed random number $rand()$ is generated; if $rand() \leq OP$, then the opposite number-based population O_p is generated as

$\boldsymbol{for}\ i = 1: wolf\ population\ (W_p)$
$\boldsymbol{for}\ j = 1: d$
$\qquad O_p(i,j) = a + b - y(i,j)$
$\boldsymbol{end}$
$\boldsymbol{end}$
where $y \in W_p$

In the present paper, the modified algorithm GWO with OBL is named as OBE-GWO. The framework of the proposed OBE-GWO is presented in Algorithm 2. Algorithm 2: Opposition-based explored grey wolf optimizer (OBE-GWO)

1. Initialization uniformly distributed wolf population W_p
2. Calculate the fitness of each wolf
3. Initialize the parameters $\{a, A, C\}$
4. Generate another population O_p by OBL
5. Select W_p fittest solutions from a set $\{W_p, O_p\}$ to form a wolf population
6. Select the leaders **alpha**, **beta** and **delta** for the search process
7. while $l < $ max_iterations (termination criteria)
8. **for each wolf**
9.   update the position by equation (1)
10. **end for loop**
11. update the leaders alpha, beta and delta
12. update the algorithm parameters $\{a, A, C\}$
    %opposition phase
13. **if** $rand() < OP$
14. Generate population O_p by OBL
15. Select W_p fittest wolf as a wolf population from a set $\{W_p, O_p\}$
16. Update the leaders
17. **end if**
18. $l = l + 1$
19. **end while loop**

4 Experimental Setup

Some well-known benchmark test beds are used for the evaluation of the proposed OBE-GWO algorithm. The characterization and search bounds of these test problems are represented in Table 1. All the reported functions are of minimization type. From P1 to P7, the functions are unimodal and scalable, from P8 to P13, functions are scalable and multimodal and functions from P14 to P23 are non-scalable and fixed-dimensional multimodal.

S. Gupta and K. Deep

Table 1 Results on set of problems from P1 to P23

Problem	OBE-GWO		GWO	
	Average	STD	Average	STD
P1	**2.44E−32**	**1.11E−31**	**7.17E−28**	**1.33E−27**
Rank (±)	+		−	
P2	**1.95E−20**	**3.15E−20**	**7.57E−17**	**5.92E−17**
Rank (±)	+		−	
P3	**4.01E−20**	**2.19E−19**	**5.77E−05**	**2.41E−04**
Rank (±)	+		−	
P4	**9.23E−15**	**3.27E−14**	**9.50E−07**	**1.71E−06**
Rank (±)	+		−	
P5	29.06	11.56	**27.12**	0.84
Rank (±)	−		+	
P6	**0.55206**	0.32139	0.77110	0.43763
Rank (±)	+		−	
P7	**6.61E−04**	**6.46E−04**	**1.84E−03**	**9.66E−04**
Rank (±)	+		−	
P8	−4876.38	1580.64	**−6122.07**	764.68
Rank (±)	−		+	
P9	**2.86399**	11.8266	2.98046	3.570393
Rank (±)	=		=	
P10	**1.15E−14**	**3.96E−15**	**1.05E−13**	**1.84E−14**
Rank (±)	+		−	
P11	**0**	0	0.005396	0.010005
Rank (±)	+		−	
P12	0.038471	0.030937	0.064855	0.094445
Rank (±)	+		−	
P13	0.626366	0.248764	0.72232	0.185166
Rank (±)	+		−	
P14	4.334733	3.697561	**3.61588**	3.500087
Rank (±)	−		+	
P15	0.005715	0.008984	0.005052	0.008592
Rank (±)	−		+	
P16	**−1.0316**	**3.79E−05**	**−1.0316**	**6.78E−16**
Rank (±)	=		=	
P17	**0.39789**	**1.69E−16**	**0.39789**	**1.69E−16**
Rank (±)	=		=	
P18	3.000027	**4.5E−05**	3.00005	**6.3E−05**
Rank (±)	+		−	
P19	−3.86125	0.002829	−3.86178	0.002207
Rank (±)	=		=	

(continued)

Table 1 (continued)

Problem	OBE-GWO		GWO	
	Average	STD	Average	STD
P20	−3.23812	0.08847	−3.26028	0.082178
Rank (±)	=		=	
P21	−9.54868	1.886677	**−9.64456**	1.545962
Rank (±)	=		=	
P22	**−10.4025**	0.000281	−10.116	1.563574
Rank (±)	+		−	
P23	**−10.5358**	0.000373	−10.2909	1.334039
Rank (±)	+		−	

The experiment is conducted on these problems 30 times, and MATLAB 2010a with 4 GB system has been used for experimentation. The number of iterations is kept same as in original GWO [7]. For the fair comparison, the same population of wolves have been taken. The obtained results are reported in Table 1. In this table, average and standard deviation of fitness (objective function value) has been presented. For OBE-GWO, OP = 0.1 (opposition probability) has been taken.

4.1 Results and Discussion

In Table 1, the obtained results are presented after implementing OBE-GWO on reported test problems. From the table, it can be easily observed that OBE-GWO outperforms GWO in most of the problems. From the functions P8 to P23, results are very effective compared to original GWO. This shows that incorporation of opposite number improves the exploration ability of GWO as the multimodal functions are useful to check the exploration ability of any search algorithm. Therefore, opposition=based learning (OBL) is very effective in exploring the unexplored search regions and in preventing the solution from trapping in local optima. In this way, by enhancing the exploration ability, an attempt for a proper balance between exploration and exploitation is established in the present paper.

In Table 1, a rank is mentioned based on Wilcoxon signed rank test based on '+' and '−' sign. Thus, the statistical analysis concludes the superior performance of OBE-GWO compared to original GWO. The distribution of objective function values for test problems is presented in Fig. 1.

5 Conclusion

In this paper, the opposition-based learning strategy has been employed for the initialization of the population and for the iteration jumping also. The proposed algorithm OBE-GWO has been tested on some famous benchmark beds to analyse the perfor-

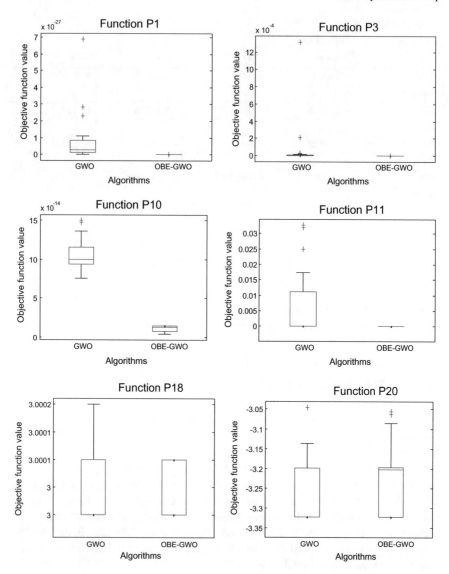

Fig. 1 Box plot for the distribution of objective function values in different test problems

mance. The obtained results on these test beds indicate that incorporation of opposite numbers helps in GWO to prevent from the problem of premature convergence due to the stagnation in local solutions and enhances the diversity of the algorithm.

For the future work, proposed OBE-GWO can be used for solving real-life optimization problems. Also, the discrete version of OBE-GWO can be proposed so that the discrete optimization problem can be solved.

Appendix

Test Problems

---

1. Unimodal problems

$P1(x) = \sum_{j=1}^{D} x_j^2$
$dim = 30$
$x \in [-100, 100]^D$
$P1_{min} = 0$

$P2(x) = \sum_{j=1}^{D} |x_j| + \prod_{j=1}^{n} x_j$
$dim = 30$
$x \in [-10, 10]^D$
$P2_{min} = 0$

$P3(x) = \sum_{j=1}^{D} (\sum_{k-1}^{j} x_k)^2$
$dim = 30$
$x \in [-100, 100]^D$
$P3_{min} = 0$

$P4(x) = max_j\{|x_j| : j \in [1, D]\}$
$dim = 30$
$x \in [-100, 100]^D$
$P4_{min} = 0$

$P5(x) = \sum_{i=1}^{D-1}[(x_j - 1)^2 + 100(x_{j+1} - x_j^2)^2]$
$dim = 30$
$x \in [-30, 30]^D$
$P5_{min} = 0$

$P6(x) = \sum_{i=1}^{D}([0.5 + x_j])^2]$
$dim = 30$
$x \in [-100, 100]^D$
$P6_{min} = 0$

$P7(x) = \sum_{i=1}^{D} rand[0, 1) + jx_j^4$
$dim = 30$
$x \in [-1.28, 1.28]^D$
$P7_{min} = 0$

Multimodal problems

$P8(x) = \sum_{i=1}^{D} -x_j \sin\left(\sqrt{|x_j|}\right)$
$dim = 30$
$x \in [-500, 500]^D$
$P8_{min} = 12569.487$

$P9(x) = \sum_{i=1}^{D} -x_j \sin\left(\sqrt{|x_j|}\right)$
$dim = 30$
$x \in [-5.12, 5.12]^D$
$P9_{min} = 0$

$P10(x) = -e^{\frac{1}{D}\sum_{j=1}^{D}\cos(2\pi x_i)} - 20e^{-0.2\sqrt{\frac{1}{D}\sum_{j=1}^{D} x_j^2}} + 20 + e$
$dim = 30$
$x \in [-32, 32]^D$
$P10_{min} = 0$

$P11(x) = 1 + \frac{1}{4000}\sum_{j=1}^{D} x_j^2 - \prod_{j=1}^{D} \cos(x_j/\sqrt{j})$
$dim = 30$
$x \in [-600, 600]^D$
$P11_{min} = 0$

$P12(x) =$
$\sum_{j=1}^{D} w(x_j, 10, 100, 4) +$
$\pi/D\,[10\sin(\pi z_1) + \sum_{j=1}^{D-1}(z_j - 1)^2\{1 + 10\sin^2(\pi z_{j+1})\} + (z_D - 1)^2]$
$z_j = (x_j + 5)/4$
$w(x_j, b, K, n) = \begin{cases} K(x_j - b)^n & \text{if } x_j > b \\ 0 & \text{otherwise} \\ K(-x_j - b)^n & \text{if } x_j < -b \end{cases}$
$dim = 30$
$x \in [-50, 50]^D$
$P12_{min} = 0$

$P13(x) = \sum_{j=1}^{D} w(x_j, 5, 100, 4) + 0.1\,[sin^2(3\pi x_1) + \sum_{j=1}^{D-1}(x_j - 1)^2\{1 + sin^2(3\pi x_j + 1)\} + (x_j - 1)^2\{1 + sin^2(2\pi x_D + 1)\}]$
$dim = 30$
$x \in [-50, 50]^D$
$P13_{min} = 0$

Fixed dimension multimodal problems

$P14(x) = 1/\left(\frac{1}{500} + \sum_{i=1}^{25} \frac{1}{i + \sum_{k=1}^{2}(x_k - a_{ki})^6}\right)$

$dim = 2$

$x \in [-5.12, 5.12]^D$

$P14_{min} = 1$

$P15(x) = \sum_{j=1}^{11} \left[a_i - \frac{x_1(b_j^2 + b_j x_2)}{b_j^2 + b_j x_3 + x_4}\right]^2$

$dim = 4$

$x \in [-5, 5]^D$

$P15_{min} = 0.00030$

$P16(x) = 4x_1^2 - 2.1x_1^4 + (x_1^6/3) + x_1 x_2 - 4x_2^2 + 4x_2^4$

$dim = 2$

$x \in [-5, 5]^D$

$P16_{min} = -1.0316$

$P17(x) = 10\left[1 + \left(1 - \frac{1}{8\pi}\right)\cos x_1\right] + (x_2 - \frac{5.1}{4\pi^2}x_1^2 + \frac{5}{\pi}x_1 - 6)^2$

$dim = 2$

$x \in [-5, 5]^D$

$P17_{min} = 0.398$

$P18(x) = \left[1 + (x_1 + x_2 + 1)^2(19 - 14x_1 + 3x_1^2 - 14x_2 + 6x_1 x_2 + 3x_2^2)\right] \cdot$
$\left[30 + (2x_1 - 3x_2)^2 \cdot (18 - 32x_1 + 12x_1^2 + 48x_2 - 36x_1 x_2 + 27x_2^2)\right]$

$dim = 2$

$x \in [-2, 2]^D$

$P18_{min} = 3$

$P19(x) = -\sum_{j=1}^{4} c_j \exp\left(-\sum_{k=1}^{3} b_{jk}(x_k - q_{jk})^2\right)$

$dim = 3$

$x \in [1, 3]^D$

$P19_{min} = -3.86$

$P20(x) = -\sum_{j=1}^{4} c_j \exp\left(-\sum_{k=1}^{6} b_{jk}(x_k - q_{jk})^2\right)$

$dim = 6$

$x \in [0, 1]^D$

$P20_{min} = -3.32$

$P21(x) = -\sum_{j=1}^{5}[(X - b_j)(X - b_j)^T + c_j]^{-1}$

$dim = 4$

$x \in [0, 10]^D$

$P21_{min} = -10.1532$

$P22(x) = -\sum_{j=1}^{7}[(X - b_j)(X - b_j)^T + c_j]^{-1}$

$dim = 4$

$x \in [0, 10]^D$

$P22_{min} = -10.4028$

$P23(x) = -\sum_{j=1}^{7}[(X - b_j)(X - b_j)^T + c_j]^{-1}$

$dim = 4$

$x \in [0, 10]^D$

$P23_{min} = -10.5363$

Box Plot
See Fig. 1.

References

1. Holland, J.H.: Adaptation in Natural and Artificial Systems: An Introductory Analysis with Applications to Biology, Control, and Artificial Intelligence. MIT press (1992)
2. Eberhart, R., Kennedy, J.: A new optimizer using particle swarm theory. In: Proceedings of the Sixth International Symposium on Micro Machine and Human Science, 1995. MHS'95, pp. 39–43. IEEE (1995)
3. Storn, R., Price, K.: Differential evolution–a simple and efficient heuristic for global optimization over continuous spaces. J. Global Optim. **11**(4), 341–359 (1997)
4. Simon, D.: Biogeography-based optimization. IEEE Trans. Evol. Comput. **12**(6), 702–713 (2008)
5. Glover, F.: Tabu search—part I. ORSA J. Comput. **1**(3), 190–206 (1989)
6. Bansal, J.C., Sharma, H., Jadon, S.S., Clerc, M.: Spider monkey optimization algorithm for numerical optimization. Memetic Comput. **6**(1), 31–47 (2014)
7. Mirjalili, S., Mirjalili, S.M., Lewis, A.: Grey wolf optimizer. Adv. Eng. Softw. **69**, 46–61 (2014)
8. Mirjalili, S.: How effective is the Grey Wolf optimizer in training multi-layer perceptrons. Appl. Intell. **43**(1), 150–161 (2015)
9. Song, H.M., Sulaiman, M.H., Mohamed, M.R.: An application of grey wolf optimizer for solving combined economic emission dispatch problems. Int. Rev. Modell. Simul. (IREMOS) **7**(5), 838–844 (2014)
10. Gupta, S., Deep, K.: A novel random walk grey wolf optimizer. Swarm Evol. Comput. (2018). https://doi.org/10.1016/j.swevo.2018.01.001
11. Gupta, S., Deep K.: Random walk grey wolf optimizer for constrained engineering optimization problems. Comput. Intell. https://doi.org/10.1111/coin.12160.
12. Lal, D.K., Barisal, A.K., Tripathy, M.: Grey wolf optimizer algorithm based fuzzy PID controller for AGC of multi-area power system with TCPS. Proc. Comput. Sci. **92**, 99–105 (2016)
13. Tizhoosh, H.R.: Opposition-based learning: a new scheme for machine intelligence. In: International Conference on Computational Intelligence for Modelling, Control and Automation, 2005 and International Conference on Intelligent Agents, Web Technologies and Internet Commerce, vol. 1, pp. 695–701. IEEE (2005)
14. Wang, H., Zhijian, W., Rahnamayan, S., Liu, Y., Ventresca, M.: Enhancing particle swarm optimization using generalized opposition-based learning. Inf. Sci. **181**(20), 4699–4714 (2011)
15. Rahnamayan, S., Tizhoosh, H.R., Salama, M.M.A.: Opposition-based differential evolution. IEEE Trans. Evolut. Comput. **12**(1), 64–79 (2008)
16. Gao, W., Liu, S.: Improved artificial bee colony algorithm for global optimization. Inf. Process. Lett. **111**(17), 871–882 (2011)
17. Ergezer, M., Simon, D., Du, D.: Oppositional biogeography-based optimization. In: IEEE International Conference on Systems, Man and Cybernetics, 2009. SMC 2009, pp. 1009–1014. IEEE (2009)
18. Dinkar, S.K., Deep, K.: Opposition based Laplacian ant lion optimizer. J. Comput. Sci. (2017). https://doi.org/10.1016/j.jocs.2017.10.007
19. Iacca, G., Neri, F., Mininno, E.: Opposition-based learning in compact differential evolution. Appl. Evolut. Comput. 264–273 (2011)
20. Neri, F., Tirronen, V.: Recent advances in differential evolution: a survey and experimental analysis. Artif. Intell. Rev. **33**(1–2), 61–106 (2010)

Spectrum Sensing for Fading Wireless Channel Using Matched Filter

Suresh Dannana, Babji Prasad Chapa and Gottapu Sasibhushana Rao

Abstract Ever-increasing use of wireless applications is exerting pressure on the limited, insufficient, and expensive licensed spectrum. Actually, because of allocation of fixed spectrum, more portion of spectrum is underutilized. Spectrum sensing can be used for the efficient and effective use of the radio spectrum. It detects the unused spectrum channels in cognitive radio network. In cognitive radio, spectrum sensing techniques such as energy detection, matched filter detection have been used. In this paper, receiver operating characteristics (ROC) of matched filter and energy detector are compared at -10 and -15 dB signal-to-noise Ratio (SNR) levels in fading wireless channels. From the results obtained it is found that matched filter system performed well over the ROC of energy detector at lower values of SNR.

1 Introduction

At present, accessing of spectrum is the key issue in wireless communication. Fixed spectrum is allotted to primary users (PU) for exclusive use, while other users (secondary users) are not allowed to access this spectrum even when it is idle. Cognitive radio (CR) [1] is proposed to enhance the spectrum efficiency. In CR, secondary users (SU) are allowed to access the spectrum when primary user is idle. Before accessing the spectrum, SU sense the spectrum whether it is idle or not. Sensing techniques detect the idle spectrum. Maximizing the probability of detection (PD) [2–4] without sacrificing much probability of false alarm (PFA) while minimizing the hardware and the time to sense the spectrum is the main purpose of the spectrum sensing. Energy

S. Dannana (✉) · B. P. Chapa · G. S. Rao
Department of Electronics and Communication Engineering, A.U.C.E (A) Andhra University,
Visakhapatnam 530003, Andhra Pradesh, India
e-mail: suress445@gmail.com

B. P. Chapa
e-mail: babjiprasad.ch@gmail.com

G. S. Rao
e-mail: sasigps@gmail.com

© Springer Nature Singapore Pte Ltd. 2019
J. C. Bansal et al. (eds.), *Soft Computing for Problem Solving*, Advances in Intelligent
Systems and Computing 817, https://doi.org/10.1007/978-981-13-1595-4_27

Fig. 1 A fading wireless channel

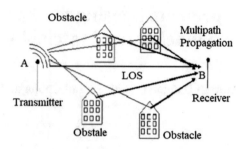

detector [5–7] is a simple sensing technique, which detects the PU. It performs better at higher values of SNR. However, in wireless channel, signal usually suffers from fading [8]. The presence of obstacles in the surroundings of transmitter and receiver acts as reflectors, which is shown in Fig. 1. Receiver collects the same signal from all the reflectors, as a result superposition of signal is occurred. The signal from all the paths undergoes different attenuation, delay and phase. This can cause interference of a signal which lowers the SNR. For such fading channels matched filter detector [9, 10] gives better sensing performance to detect the primary users.

2 System Model

Consider a fading wireless channel described by the model below:

$$\bar{y} = p\bar{x} + n$$

$\bar{x}$ is a transmitted pilot vector with an average power E, i.e.,

$$\|\bar{x}\|^2 = E$$

where $\|\bar{x}\|$ is norm of a vector $\bar{x}$.

p   is channel coefficient, which controls the power at the receiver.
p   is complex channel coefficient $p = a + jb$.
n   is noise, which follows the Gaussian distribution with mean zero, and variance σ_n^2.
$\bar{y}$ is the signal at the receiver by various paths. This multipath propagation introduces fading (interference of signals).

2.1 Matched Filter

Consider binary hypothesis testing for spectrum sensing.

$$H_0 : \bar{y} = \bar{n}$$

$$H_1 : \bar{y} = p\bar{x} + \bar{n} \tag{1}$$

Let the signal, $\bar{m} = p\bar{x}$
Therefore, under hypothesis

$$H_1 : \bar{y} = \bar{m} + \bar{n}$$

If the channel characteristics are known then the matched filter-based detection is suitable to sense the primary user. It requires less time to sense.
The resulting outputs are

$$\text{under hypothesis } H_0 : \tilde{y} = \frac{m}{\|m\|} y = \frac{m^H}{\|m\|} n = \tilde{n} \tag{2}$$

$$\text{under hypothesis } H_1 : \tilde{y} = \frac{m^H}{\|m\|} y = \frac{m^H}{\|m\|}(m + n) = \|m\| + \tilde{n} \tag{3}$$

Here, $\tilde{n}$ is a Gaussian random variable with mean zero and variance σ_n^2. It results that under hypothesis H_0, $\tilde{y}$ is a Gaussian random variable with mean zero and variance σ_n^2.

Under hypothesis H_1, $\tilde{y}$ is a Gaussian random variable with mean $\|m\|$ and variance σ_n^2. The same can be shown in Fig. 2.

From Fig. 2, it is observed that the optimum threshold can be $\|m\|/2$.

i.e., if $\tilde{y} > \|m\|/2$. choose H_1

$$\tilde{y} < \|m\|/2. \text{ choose } H_0$$

Probability of detection (PD) is evaluated under H_1

$$PD = Q\left(\frac{k - |p|\|\bar{x}\|}{\sigma_n/\sqrt{2}}\right) \tag{4}$$

Probability of false alarm (PFA) is evaluated under H_0

$$PFA = Q\left(\frac{k}{\sigma_n/\sqrt{2}}\right) \tag{5}$$

Fig. 2 Decision threshold

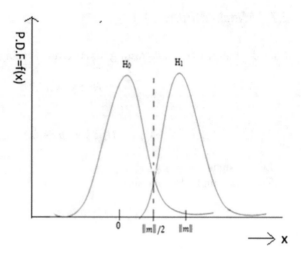

$Q(u)$ is an error function, which gives error probability. It is a monotonically decreasing function.

$$Q(u) = \int\limits_{u}^{\infty} \frac{1}{\sqrt{2\pi}} e^{\frac{-x^2}{2}} dx$$

2.2 Energy Detector

Energy detection-based spectrum sensing is the method to detect the primary user. If the characteristics of a channel are unknown, this method is well suitable to detect the primary user. However, it does not distinguish the signal and noise. The resulting outputs are

$$\tilde{y} = \frac{\bar{x}^H}{\|\bar{x}\|} \bar{y} \tag{6}$$

Under hypothesis H_1: $\|y\|^2 \geq k$
Under hypothesis H_0: $\|y\|^2 < k$
Probability of detection is

$$PD = Q_{\aleph_{2r}^2} \left(\frac{k}{0.5\left(\|\bar{x}\|^2 + \sigma_n^2\right)} \right) \tag{7}$$

Probability of false alarm is

$$PFA = Q_{\aleph_{2r}^2}\left(\frac{k}{0.5(\sigma_n^2)}\right) \tag{8}$$

where $Q_{\aleph_{2r}^2}(x)$ denotes the complementary cumulative distribution function (CCDF) of a $\aleph_{2r}^2$ random variable with 2r degree of freedom.

3 Results

The receiver operating characteristics (ROC) illustrates the probability of detection and probability of false alarm for the given threshold. The ROC for wireless fading channel are plotted in Figs. 3 and 4, which show the performance of the spectrum sensing. Here, ROC of a matched filter and ROC of the energy detector are compared. Average power of a transmitted signal is 1.414, and SNR of signals is taken as -10 and -15 dB. In this simulation, 50,000 iterations are performed where 80 threshold values that ranged from -10 to 30 are considered. Number of pilot symbols are 50. From the results, it can be observed that the matched filter detector shows better performance than the energy detector for low SNR values. Since the signal characteristics are known, matched filter gives better probability of detection. In these plots, the simulated results in both the cases are also compared with theoretical values.

4 Conclusion

In this paper, a matched filter-based spectrum sensing technique is proposed for primary user detection in a fading wireless channel. The receiver operating characteristics are observed for different SNR values of -10 and -15 dB. The receiver operating characteristics of the proposed matched filter are compared with energy detection method. The matched filter-based detection shows better performance for lower SNR values. It is also observed that the matched filter shows better probability of detection for lower values of PFA. However, matched filter sensing technique requires prior information of primary user for better probability of detection.

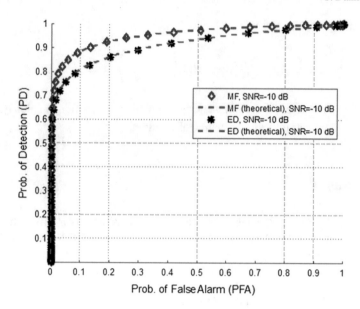

Fig. 3 Receiver operating characteristics for SNR= −10 dB

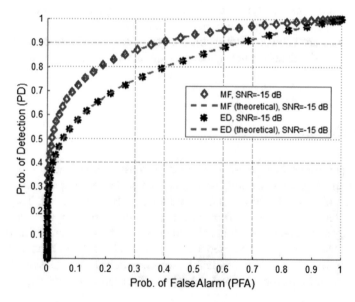

Fig. 4 Receiver operating characteristics for SNR= −15 dB

References

1. Haykin, S.: Cognitive radio: brain-empowered wireless communications. IEEE J. Sel. Areas Commun. **23**(2), 201–220 (2005)
2. Kay, S.M.: Fundamentals of Statistical Signal Processing, Estimation Theory, vol. I. Prentice hall (1993)
3. Kay, S.M.: Fundamentals of Statistical Signal Processing, Detection Theory, vol. I. Prentice hall (1998)
4. Van Trees, H.L.: Detection, Estimation, and Modulation theory. Wiley (2004)
5. Ali, A., Hamouda, W.: Spectrum monitoring using energy ratio algorithm for OFDM-based cognitive radio networks. IEEE Trans. Wirel. Commun. **14**(4), 2257–2268 (2015)
6. Xuping, Z., Jianguo, P.: Energy-detection based spectrum sensing for cognitive radio, pp. 944–947 (2007)
7. Plata, D.M.M., Reátiga, Á.G.A.: Evaluation of energy detection for spectrum sensing based on the dynamic selection of detection-threshold. Procedia Eng **35**(2012): 135–143
8. Ling, F.: Matched filter-bound for time-discrete multipath Rayleigh fading channels. IEEE Trans. Commun. **43**(234), 710–713 (1995)
9. Xuping, Z., Jianguo, P.: Energy-detection based spectrum sensing for cognitive radio, pp. 944–947 (2007)
10. Ling, F.: Matched filter-bound for time-discrete multipath Rayleigh fading channels. IEEE Trans. Commun. **43**(234), 710–713 (1995)

Energy Detection-Based Spectrum Sensing for MIMO Wireless Channel with Different Antenna Configurations

Babji Prasad Chapa, Suresh Dannana and Sasibhushana Rao Gottapu

Abstract The spectrum sensing acquired importance due to ever-growing demands for the high data rates. It is one of the prominent functions in the cognitive radio, 4G, and beyond 4G cellular systems. Different facets of spectrum sensing methods are examined from a cognitive radio and multi-input and multi-output (MIMO) systems perspective. The major task of cognitive radio is to identify the unused spectrum and thereby allocating that spectrum to the required users without causing interference to the other users, and the important task of 4G system is to provide user demanding bandwidth by detecting the existence of the user in order to allocate the unused subcarrier channels to the required users. Hence, spectrum sensing is the key feature of cognitive radio. In this paper, a new energy detection-based spectrum sensing for MIMO wireless channel is proposed. The proposed technique performs better in case of higher order antenna configurations compared to lower order.

1 Introduction

Spectrum is a very precious commodity, and the authorized spectrum is accessed by those who own the spectrum [1]. Cognitive radio is a tempting technology which addresses the spectrum underutilization problem. Orthogonal frequency division multiplexing (OFDM) is one of the most frequently used technologies in current wireless communication systems, which has the potential of fulfilling the requirements of cognitive radios inherently or with small changes. The cognitive radio has key features like spectrum awareness and spectrum intelligence. Spectrum intelligence is an ability to learn the spectrum environment and adapt transmission parameters. The operation of cognitive radio starts by sensing the surrounding electromagnetic

B. P. Chapa (✉) · S. Dannana · S. R. Gottapu
Department of Electronics and Communication Engineering, A.U.C.E (A),
Andhra University, Visakhapatnam 530003, Andhra Pradesh, India
e-mail: babjichapa@gmail.com

S. Dannana
e-mail: suress445@gmail.com

© Springer Nature Singapore Pte Ltd. 2019
J. C. Bansal et al. (eds.), *Soft Computing for Problem Solving*, Advances in Intelligent
Systems and Computing 817, https://doi.org/10.1007/978-981-13-1595-4_28

environment in order to detect the existing vacant spectrum or spectrum holes. The cognitive radio then adapts radio operating transmission parameters in order to exploit the detected spectrum opportunities in the best possible way. Hence, the spectrum sensing is the most important component for the establishment of cognitive radio.

2   Energy Detection Model

The model has been set up for energy detection based on binary hypothesis testing problem, in which hypothesis 0 or null hypothesis (H_0) and hypothesis 1 (H_1) are the user absence and user presence, respectively. Spectrum holes can be identified when H_0 is true. The energy detector measures the energy of received signal over a specified time duration and bandwidth. The measured value is then compared with appropriate selected threshold value to determine the presence or the absence of the signal [2, 3] (Fig. 1).

2.1   Spectrum Sensing for MIMO Wireless Channel

The binary hypothesis testing problem for spectrum sensing in the MIMO wireless channel is given as follows:

$$H_0 : y = q \tag{1}$$
$$H_1 : y = Kx + q \tag{2}$$

where y is the received signal, x is transmitted signal, q is i.i.d Gaussian noise with zero mean and variance σ^2 and K is the channel matrix. Furthermore, transmitted signal and noise samples are uncorrelated and independent. If the channel matrix K is unknown, one can employ energy detection [4, 5]. Now compare the output $\|y\|^2$ with a suitable threshold Λ to yield the detector. The hypothesis H_1 is chosen if $\|y\|^2 \geq \Lambda$; otherwise, null hypothesis is chosen.

The probability of false alarm and probability of detection can be calculated as follows:

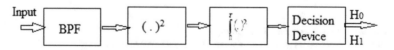

Fig. 1 Block diagram of energy detection model

Considering H_0, the probability of false alarm is

$$P_{FA} = \Pr(\|y\|^2 > \Lambda) \tag{3}$$

In order to formulate the above expression in terms of complementary cumulative distribution function

$$P_{FA} = \Pr\left(\frac{\|y\|^2}{\sigma^2/2} > \frac{\Lambda}{\sigma^2/2}\right) \tag{4}$$

Above expression seems to be chi-squared distributed random variable with two degrees of freedom. The probability of false alarm will be

$$P_{FA} = Q_{\chi^2_{2r}}\left(\frac{\Lambda}{\sigma^2/2}\right) \tag{5}$$

Considering H_1, the probability of detection is

$$P_D = \Pr\left(\|y\|^2 > \Lambda\right) \tag{6}$$

In order to formulate above expression in terms of complementary cumulative distribution function,

$$P_D = \Pr\left(\frac{\|y\|^2}{\frac{1}{2}(P + \sigma^2)} > \frac{\Lambda}{\frac{1}{2}(P + \sigma^2)}\right) \tag{7}$$

Above expression seems to be chi-squared distributed random variable with two degrees of freedom. The probability of detection is

$$P_D = Q_{\chi^2_{2r}}\left(\frac{\Lambda}{\frac{1}{2}(P + \sigma^2)}\right) \tag{8}$$

3 Results

The efficiency of an energy detector for spectrum sensing can be evaluated by simulating using MATLAB version R2016b. Monte Carlo (MC) simulations are used to assess the energy detector performance for MIMO wireless channel. The receiver performance can be determined by illustrating the receiver operating characteristics. Plotting of receiver operating characteristics is done keeping one parameter constant and varying the other parameter. The detection probability versus false alarm probability curves is plotted for various signals to noise ratios.

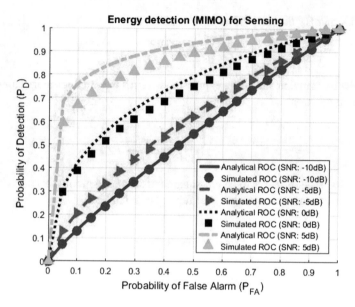

Fig. 2 Receiver operating characteristics for energy detection over the MIMO channel for 2×2 antenna configuration

Table 1 Detection probability for a given false alarm probability over MIMO channel for 2×2 antenna configuration

Signal to noise ratio (SNR in dB)	False alarm probability	Detection probability
5	0.1	0.7598 (Analytical)
5	0.1	0.6790 (Simulated)
0	0.1	0.4211 (Analytical)
0	0.1	0.3920 (Simulated)
−5	0.1	0.2062 (Analytical)
−5	0.1	0.2062 (Simulated)
−10	0.1	0.1330 (Analytical)
−10	0.1	0.1330 (Simulated)

Signal to noise ratio affects the detection performance and is measured by using an energy detector. Figure 2 shows the detection performance of the detector simulating over an MIMO wireless channel for 2×2 antenna configuration. Here, the signal to noise ratio is set at different values like -10, -5, 0, and 5 dB, the number of MC sample points is set to 10,000, and results are plotted between false alarm and detection probability. Analytical and simulated results show that the detection probability is more when the signal to noise ratio is high at a given false alarm probability (Table 1).

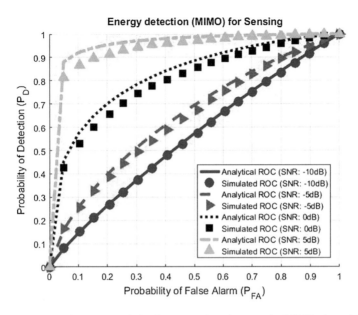

Fig. 3 Receiver operating characteristics for energy detection over the MIMO channel for 4 × 4 antenna configuration

Table 2 Detection probability for a given false alarm probability over the MIMO channel for 4 × 4 antenna configuration

Signal to noise ratio (SNR in dB)	False alarm probability	Detection probability
5	0.1	0.9205 (Analytical)
5	0.1	0.8657 (Simulated)
0	0.1	0.5714 (Analytical)
0	0.1	0.5408 (Simulated)
−5	0.1	0.2486 (Analytical)
−5	0.1	0.2486 (Simulated)
−10	0.1	0.1455 (Analytical)
−10	0.1	0.1455 (Simulated)

In Fig. 3, for different values of the signal to noise ratios like −10, −5, 0, and 5 dB, the receiver operating characteristics are plotted for 4 × 4 antenna configuration. Analytical and simulated results show that the detection probability is high for high signal to noise ratio at a given false alarm probability (Table 2).

In Fig. 4, for different values of the signal to noise ratios like −10, −5, 0, and 5 dB, the receiver operating characteristics curves are plotted for 6 × 6 MIMO antenna configuration. It shows the plot of probability of false alarm versus probability of

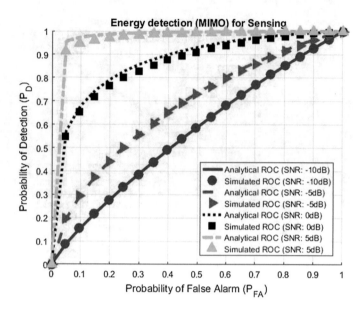

Fig. 4 Receiver operating characteristics for energy detection over the MIMO wireless channel for 6×6 antenna configuration

Table 3 Detection probability for a given false alarm probability over the MIMO wireless Channel for 6×6 antenna configuration

Signal to noise ratio (SNR in dB)	False alarm probability	Detection probability
5	0.1	0.9737 (Analytical)
5	0.1	0.9428 (Simulated)
0	0.1	0.6793 (Analytical)
0	0.1	0.6508 (Simulated)
−5	0.1	0.2921 (Analytical)
−5	0.1	0.2921 (Simulated)
−10	0.1	0.1585 (Analytical)
−10	0.1	0.1585 (Simulated)

detection. Analytical and simulated results show that the detection probability is high for the signal to noise ratio of 5 dB at a given probability of false alarm (Table 3).

In Fig. 5, for different values of the signal to noise ratios like $-10, -5, 0$, and 5 dB, the receiver operating characteristics curves are plotted for 8×8 MIMO channel. It shows the plot of probability of false alarm versus probability of detection. Analytical and simulated results show that the detection probability is high for the signal to noise ratio of 5 dB at a given probability of false alarm (Table 4).

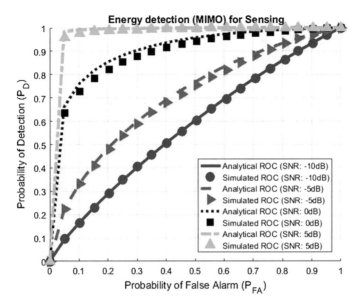

Fig. 5 Receiver operating characteristics for energy detection over the MIMO wireless channel for 8×8 antenna configuration

Table 4 Detection probability for a given false alarm probability over the MIMO wireless channel for 8×8 antenna configuration

Signal to noise ratio (SNR in dB)	False alarm probability	Detection probability
5	0.1	0.9766 (Analytical)
5	0.1	0.9766 (Simulated)
0	0.1	0.7596 (Analytical)
0	0.1	0.7262 (Simulated)
−5	0.1	0.3302 (Analytical)
−5	0.1	0.3302 (Simulated)
−10	0.1	0.1634 (Analytical)
−10	0.1	0.1634 (Simulated)

4 Conclusions

This study imparts the behavior of the energy detection technique for the MIMO wireless channel with different antenna configurations. The energy detector performance in finding an unoccupied spectrum was measured. Simulations are carried out to examine the performance of the signal detector over MIMO wireless channel with different antenna configurations. The detection performance of the energy detector over MIMO channel with different antenna configurations can be improved by proper selection of threshold value, and detection probability can be increased by increasing the SNR. In case of MIMO with 2×2 antenna configuration, the detection probability at a given false alarm is better for SNR of 5 dB among selected values. On simulating the detector over the MIMO wireless channel with 4×4 antenna configuration, it is observed that the detection probability is high for the signal to noise ratio of 5 dB and also the detection performance is improved in 4×4 case (0.9205) compared to 2×2 case (0.7598) antenna configuration. When the antenna configuration is changed to 6×6, the detection probability is more at high SNR and detection probability is even better compared to the previous two cases. By simulating the energy detector over 8×8 MIMO wireless channel, it is observed that detection probability is high at SNR of 5 dB. Finally, it is observed that the detection probability is better in case of the 8×8 MIMO wireless channel (0.9766) when compared to other 2×2 (0.7598), 4×4 (0.9205), and 6×6 (0.9737) antenna configurations.

References

1. Federal Communication Commission, Spectrum Policy Task Force Report, ET Docket No. 02-155, Nov 2002
2. Digham, F.F., Alouini, M.S., Simon, M.K.: On the energy detection of unknown signals over fading channels. IEEE Trans. Commun. **55**(1), 21–24 (2007)
3. Tandra, R., Sahai, A.: SNR walls for signal detection. IEEE J. Sel. Topics Signal Process **2**(1), 4–17 (2008)
4. Yucek, T., Arslan, H.: A survey of spectrum sensing algorithms for cognitive radio applications. IEEE Commun. Surv. Tutor. **11**, 116–130 (2009)
5. Amich, A., Imran, M.A., Tafazolli, R., Cheraghi, P.: Accurate and efficient algorithms for cognitive radio modeling applications under the i.n.i.d. paradigm. IEEE Trans. Veh. Technol. **64**(5), 1750–1765 (2015)

Performance Evaluation of Fuzzy C Means Segmentation and Support Vector Machine Classification for MRI Brain Tumor

B. Srinivas and G. Sasibhushana Rao

Abstract The medical imaging field has its significance with an increase in the demand for automatic and efficient diagnosis in a brief time period. In this paper, the Fuzzy C-means (FCM) algorithm is used to identify the tumor and extract it and parameters like segmented area, Mean Squared Error (MSE) and Peak Signal to Noise Ratio (PSNR) are found. The daubechies three level decomposition of Discrete 2-D Wavelet Transform (DWT) is used to extract the coefficients of wavelets for the Magnetic Resonance (MR) image and then its dimensionality is reduced by Principle Component Analysis (PCA) algorithm. Gray level co-occurrence matrix of these coefficients are found and then thirteen statistical features are extracted from the given MR image. The extracted features describe the input image's texture and the structural information of the intensity. These extracted features of the training set of images constitute the training feature database. This database is used for training and a test input image is given to Support Vector Machine (SVM) classifier to classify 105 MR brain tumor images into either benign or malignant class. From the obtained experimental results, the proposed SVM classifier had an accuracy of 98.82%, 100% of sensitivity, 97.83% of specificity, and 1.17% of error rate.

Keywords MR brain tumor segmentation and classification · Fuzzy C-means (FCM) algorithm · Discrete wavelet transform (DWT) · Principle component analysis (PCA) algorithm · Support vector machine (SVM) classifier

B. Srinivas (✉)
Deptartment of ECE, MVGR College of Engineering(A),
Visakhapatnam 535005, India
e-mail: srinivas.b@mvgrce.edu.in

G. Sasibhushana Rao
Department of ECE, Andhra University College of Engineering(A),
Visakhapatnam 530003, Andhra Pradesh, India
e-mail: sasigps@gmail.com

© Springer Nature Singapore Pte Ltd. 2019
J. C. Bansal et al. (eds.), *Soft Computing for Problem Solving*, Advances in Intelligent Systems and Computing 817, https://doi.org/10.1007/978-981-13-1595-4_29

1   Introduction

One of the leading cause of death in people with various age groups is brain diseases. These diseases can be of different types such as cerebrovascular disease (stroke), neoplastic diseases (brain tumor), infectious diseases, and degenerative diseases. Almost all of the brain diseases cause serious issues in the human brain and sometimes prompt to death. It is very important to detect such diseases early, which gives assistance to the radiologist for better treatment and also reduces the risk rate of the disease. As a consequence, the usage of computer and IT is now spread across all medical areas, like cancer related research, heart diseases, brain tumors etc. Magnetic Resonance Imaging (MRI) [1, 2] is a powerful imaging technique that is used in the area of the surgical and clinical environment. Its characteristics include superior soft tissue differentiation, high spatial resolution, and contrast. They give great information for biomedical research and disease diagnosis. It uses non harmful ionizing radiation to patients. Through visual interpretation, Radiologist examines MR images to determine the presence of abnormal tissue [3]. Due to a shortage of radiologists and a huge quantity of MR Images have made the work labor intensive, costly and often incorrect. Hence for analysis and classification of medical images, there is a requirement to develop an automated systems. In this regarding, a large research work for MR medical images exits.

1.1   Segmentation

A study of cluster or segmentation technique [3, 4] based on clustering assembles a set of entities in a manner that entities in the identical cluster have a superior degree of alikeness to each compared to the other clusters. Clusters are defined as contiguous regions of more than one-dimensional space comprising comparative points of high density, alienated from other exemplary regions comprising moderate points of low density. In image breakdown, clustering is the order of arrangement of pixels conferring to more or less features like intensity. Under hard clustering, data elements fit into one cluster simply and the membership value of belongingness to a cluster is precisely one. Under soft clustering, elements of data fit into greater than the single cluster and the membership value of belongingness to the cluster varies from 0 to 1.

1.2   Features Extraction

Extraction of a feature [5] is a practice to discover the significant features from MR images that are utilized to apprehend the images simpler. These input images data set are altered into packed down kind is understood as feature extraction. It will shrink the task for an additional process such as the classification of the image. Features, if carefully chosen are symbolic of the maximum appropriate info that the image has to propose for a broad categorization of a lesion. Extraction of feature techniques analyzes images and objects to extract the utmost noticeable features that are symbolic of the countless object classes. These features are used as contributions to classifiers [6] that commit them to the class that they symbolize. The aim of the feature extraction is to diminish the novel data by ascertaining assured features that distinguish between different input plans. The feature extracted must offer the features of the input class to the classifier by bearing in mind the report of the related image properties into a set of feature vectors. In the offered methodology, Shape Intensity Index features: Variance, Mean, Standard Variance, Kurtosis, Skewness, Correlation, Contrast, Homogeneity, and Entropy are extracted, which name the structure proof of shape, intensity, and texture.

1.3 Support Vector Machine (SVM)

It is a supervised learning methodology and a decent tool for the analysis and classification of data. It encompasses a quick learning speed even in huge data. It is employed for 2 or a lot of classes of classification problems [7, 8]. It primarily relies on the theory of division planes. A division plane is one which separates between a set of data having different class memberships. It is used to detect and classify the MR brain tumor. The tumor class presented in the image is classified by the SVM algorithm [5, 9]. Basically, it uses two parts namely training part and testing part.

2 Algorithms

In this section, all the algorithms Fuzzy C means (FCM) clustering for tumor segmentation, DWT for wavelet coefficients, PCA for the dimensionally reduced feature set and SVM for classification are explained in detail.

2.1 Fuzzy C Means (FCM) Clustering Algorithm

FCM algorithm is one kind of clustering method, which was introduced by Dunn, was enhanced by Bezdek and was titivated further by M. Matteucci. During segmentation process, only local information is considered in the FCM algorithm [10]. The membership function is permitted for each information focuses (data points) directly related to each group focus (cluster center), based on the distance between the group focus and information focuses. The membership function and group focuses are upgraded after each cycle.

n   Number of information focuses
v_q Group focuses
m   Fuzziness index m $\in$ [1, ∞]
K   Number of group focuses
μ_{pq} Membership function of information focuses to group focuses
d_{pq} The Euclidean distance between pth information focuses and qth group focuses

The main FCM objective function is to minimize

$$G(u, v) = \sum_{p=1}^{n} \sum_{q=1}^{k} \left(\mu_{p,q}\right)^m \left\|X_p - v_q\right\|^2 \tag{1}$$

where $\left\|X_p - v_q\right\|^2$ is the Euclidean distance between pth information focuses and qth group focuses.

Steps for FCM [11] algorithm:
$S = \{s_1, s_2, s_3 \ldots, s_x\}$ is the information focuses set and $D_c = \{d_1, d_2, d_3 \ldots, v_v\}$ is the set of group focuses.

Step (i) Arbitrarily select k group focuses.

Step (ii) Function of Fuzzy membership μ_{pq} is calculated as $\mu_{pq} = \dfrac{1}{\sum_{r=1}^{k} \left(\frac{d_{pq}}{d_{pr}}\right)^{\frac{2}{m-1}}}$

Step (iii) Calculate Fuzzy centers $v_q = \dfrac{\sum_{p=1}^{n} \left(u_{pq}\right)^m x_p}{\sum_{p=1}^{n} \left(u_{pq}\right)^m}$

Step (iv) Until the G is reached to minimum value or $\|U_{r+1} - U_r\| < E$, Repeat 2 and 3 Steps where,

r   Iteration step
E   Termination criterion is in the range of [0, 1]
U (μ_{pq})+c is the matrix of fuzzy membership.

2.2 Discrete Wavelet Transform (DWT)

DWT is one kind of linear transformation on the input data vector to transform into another vector. The length of that transformed vector is an integer number to the power of 2. DWT differentiates data into different components [5] of frequency named as Low Low (LL), Low High (LH), High Low (HL) and High High (HH). All the frequency components are considered with a proper resolution that is coordinated to their scales. In DWT, the digital signal is represented in time-scale utilizing digital filtering algorithms. The signal which is to be evaluated is allowed to go through filters that are having diverse cutoff frequencies at various scales. This signal and the localized basis functions, which are defined as the shifted and scaled versions of some already fixed mother wavelets are compared with each other. The main characteristic feature of the wavelets is that localized frequency information about a function of a signal is provided, which is further more advantageous for classification to classify the tumor. The continuous domain wavelet transform of $x(t)$ is well-defined as

$$W_\psi(r, s) = \int x(t) \times \psi_{r,s}(t)dx \qquad (2)$$

where $\psi_{r,s}(t) = \frac{1}{\sqrt{|r|}}\psi\left(\frac{t-r}{s}\right)$, and the wavelet $\psi_{r,s}(t)$ is obtained from the main mother wavelet $\psi(t)$ by considering a dilation and b translation parameters which are positive and numbers. DWT can be obtained by restraining a and b parameters to the above Eq. (2). This DWT is expressed as shown below:

$$DWT_{X[n]} = d_{j,k} = x_{[n]}r_j * 2_{j,k}, a_{j,k} = x_{[n]}s_j * 2_{j,k} \qquad (3)$$

The coefficients $d_{j,k}$ represents the details of the components while the coefficients $a_{j,k}$ represents the components that are approximated within $x[n]$ corresponding to the wavelet function. As per the above Eq. (3), the functions s(n) and r(n) points to low pass and high pass filter coefficients respectively, whereas j refers to wavelet scale and k refers to translation factor. Example for Daubechies multiresolution analysis, there exists a connection between the wavelet of Daubechies and a 2 band Daubechies filter bank. BY apply this methodology to MR brain images, the DWT is applied individually on every dimension. Resulting in four subbands like LL, LH, HH, HL images. The DWT decomposition technique [12] next stage is applied to the first LL subband. This LL subband can be considered as an image approximation cofficients. The remaining subbands such as LH, HL, HH can be considered as an image detail cofficients. The higher is the level of the decomposition, the more is the compactness of the approximation components. Hence, wavelets give a straight forward path of a hierarchical framework for information interpretation of the image. In this paper, decomposition can be performed by three levels using Daubechies wavelet for extracting the features.

2.3 Principal Component Analysis (PCA)

PCA [13] is one kind of statistical process to get a compacted number of dimensions in the information lattice that is useful for acquiring the most extreme conceivable quantity of variance. The whole methodology depends on the topic that gives a huge dataset, here the set is examined for connections between the independent points contained in the set. It basis vectors is nothing but the given input data covariance matrix Eigen vectors. This can be advantageous in exploring the multivariations data analysis as the new dimensions which are termed as the principal components PCs of the dataset. A reduced dimension is obtained by means of selecting the PCs related to the most elevated Eigen values.

3 Methodology

The software is developed to identify a tumor in the Brain MR Image. Firstly, a random brain MRI image is taken and converted to a Grayscale image. It is then segmented on the basis of FCM clustering method. In this method, 3 clusters centroid values are taken on basis of which the clustering takes place. In FCM, the centroid values lie between 0 and 255 since the Grayscale values range from 0 to 255. In Fuzzy clustering, each point has a likelihood of belonging to each group, instead of totally belonging to only one group as it is the situation in the K-means algorithm. Find the parameters like Segmented area, Mean Square Error (MSE), Peak Signal to Noise Ratio (PSNR). A wavelet transform is a great apparatus to include extraction of features from the MR brain images since it permits images analysis at different levels of resolution in view of its multi-resolution diagnostic property. Therefore the Brain MRI image is now converted to the single level discrete 2-D wavelet transform and perform the principal component analysis operation. This returns a new set of features matrix which has reduced dimensionally. This dataset is transformed to Gray level co-occurrence matrix and the statistical features are extracted from the given brain MR image. About 105 MR brain images are used using SVM classifier for classifying the images into either benign or malignant.

4 Performance Measures

This section segmentation performance measures are used in evaluating the performance of the proposed unsupervised clustering algorithm for segmentation of brain tumor and classifier performance metrics are used to evaluating the accuracy of the classifier.

4.1 Segmentation Performance Measures

The following five performance measures are used in evaluating the performance of the proposed unsupervised clustering algorithms for segmentation of brain tumor.

MSE (mean squared error): The process of squaring the differentiated values are indicated by MSE [14]. The average of the sum of the squares of the errors is called as MSE, is obtained by subtraction of the input and the segmented images (Fig. 1).

MSE can be expressed as

$$\text{MSE} = \frac{1}{mn} \sum_{a=0}^{m-1} \sum_{b=0}^{n-1} [R(a,b) - s(a,b)]^2 \qquad (4)$$

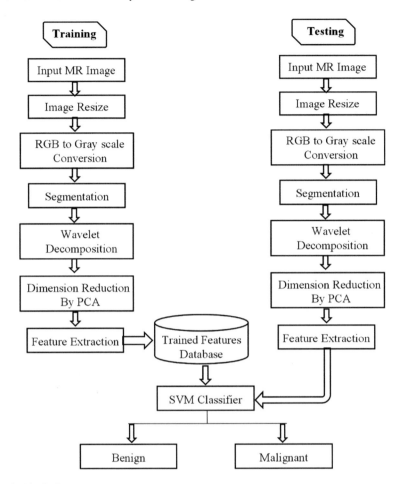

Fig. 1 Block diagram

where R(a, b) is the input image, S(a, b) is the segmented image, 'm' and 'n' denotes the number of rows and columns in the input MR brain image. To get better PSNR, the obtained value of MSE should be low for the segmented image.

PSNR (peak signal to noise ratio): Noise immunity of an image is indicated by peak signal to noise ratio [14]. More PSNR means that the MR brain image has very less interference due to noise. In the input MR brain image, the highest pixel value is denoted as a max_i. The MSE values represent the PSNR. The algorithm produced PSNR values, which is in the range of 40 and 100 dB, and so it is less sensitive to noise.

Usually, PSNR can be expressed as

$$PSNR = 10\,log_{10}\left(\frac{MAX_i^2}{MSE}\right) = 20\,log_{10}\left(\frac{MAX_i}{\sqrt{MSE}}\right) \tag{5}$$

Execution Time or Processing (s) [14] can be expressed as the required time for the system to complete the execution of the program.

Segmented Area (pixel) of an image is defined as the number of ones in the segmented image.

4.2 Performance Evaluation of Classifier

The following performance metrics are used to evaluate the classifier performance [15] like sensitivity, specificity, accuracy, and error rate.

Sensitivity: It can be defined as the true positive rate.

$$Sensitivity = \frac{TP}{TP + FN} \times 100$$

Specificity: It can be defined as the true negative rate.

$$Specificity = \frac{TN}{TN + FP} \times 100$$

Accuracy: It cab be defined as the number of images which are classified correctly to the total number of input images.

$$Accuracy = \frac{TP + TN}{TP + TN + FP + FN} \times 100$$

Error rate: It can be defined as the number of images which are classified incorrectly to the total number of images.

$$Errorrate = \frac{FP + FN}{TP + TN + FP + FN} \times 100$$

where *True Positive (TP)*: Malignant rightly classified as Malignant, *True Negative (TN)*: Benign rightly classified as Benign. *False Positive (FP)*: Benign wrongly classified as Malignant. *False Negative (FN)*: Malignant wrongly classified as Benign.

5 Experimental Results

The implementation of the proposed segmentation and Classification algorithms for a brain tumor is done by using the software MatlabR2013a. The experiments are performed in Intel Core i5 CPU 2.5 GHz processor has a RAM of 8 GB. The dataset of MR images is taken from the open data source http://www.cancerimagearchive.net/display/public/collections. For the purpose of segmentation, 14 MR brain images are taken. The MR images considered for the experimentation have a default size

Table 1 Performance measures of segmentation

Image no.	MSE	PSNR	P.Time (s)	Segmented area (pixels)
1	0.2437	54.2631	16.214	3563
2	0.0975	58.2408	7.012	3849
3	0.1984	55.1554	13.954	3134
4	0.1012	58.0779	18.866	1196
5	0.3923	52.1941	9.370	2572
6	0.0545	60.7648	13.564	872
7	0.0545	60.7648	12.844	1021
8	0.0259	63.9936	10.820	830
9	0.0548	60.7470	9.889	1038
10	0.0464	61.4609	10.425	441
11	0.0445	61.6496	9.395	496
12	0.0152	66.3195	12.985	671
13	0.0684	59.7802	8.594	602
14	0.0522	60.9499	10.304	674

of 200×200. The segmentation results of the FCM for 14 MRI brain images are shown in Fig. 2, which consists of an original MR image, Clustered image1, Clustered image2, Clustered image3, and segmented image of Brain.

An original MR image is converted to gray scale image from RGB image, Where Clustered Image is the image obtained when centroids are chosen near to background, brain image, and tumor image pixel value of original image value.

The initial fuzzy partition matrix is first generated and calculated the initial fuzzy cluster center values $cc1 = 10.41$, $cc2 = 84.43$, $cc3 = 173.32$ are taken as an initial cluster centroids. In each iteration, the cluster centers and the membership grade point are updated and the best location for the clusters is obtained by minimizing the objective function. This process is stopped when the highest number of iterations are attained or when the improvement in the objective function between two consecutive iterations is less than specified minimum improvement. The final cluster center values are $ccc1 = 3.8650$; $ccc2 = 84.5062$; $ccc3 = 172.5919$.

The fifth image (v) of Fig. 2 shows the required segmented image that is obtained by applying area opening operation on the clustered image3. The results of MR image segmentation are shown in Fig. 2 in a stepwise fashion.

The performance parameters like MSE, PSNR value (dB), Processing Time (s) and Segmented Area of the MRI Brain tumor image no. 4 in Fig. 2 using the proposed methodology are calculated. The obtained performance parameters are shown in Table 1.

The wavelet coefficients from the MR image is determined using Daubechies 3 level decomposition of 2-D DWT and its dimensionality is then reduced by using PCA algorithm. These coefficients are then converted to a Gray level co-occurrence matrix.

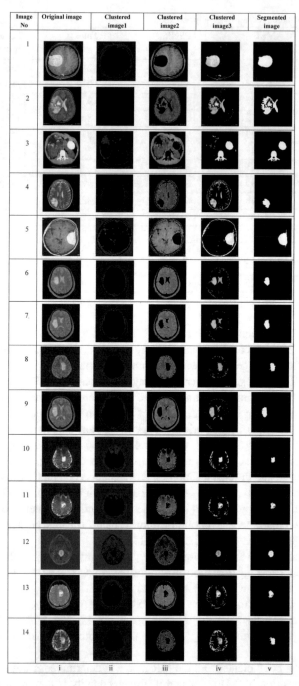

Fig. 2 (i) Original image (ii) Clustered image1, (iii) Clustered image2, (iv) Clustered image3, and (v) Segmented image

Table 2 Extraction of features for a brain MR image

S. no.	Features	Values
1	Mean	0.0057
2	Standard deviation	0.0896
3	Entropy	2.6622
4	RMS	0.0898
5	Variance	0.0080
6	Smoothness	0.9551
7	Kurtosis	13.0402
8	Skewness	1.3124
9	Inverse difference movement (IDM)	1.2778
10	Contrast	0.2925
11	Correlation	0.1584
12	Energy	0.7588
13	Homogeneity	0.9330

Table 3 Performance metrics

S. no.	Testing classes	Target classes	
		Malignant (Positive) 40	Benign (Negative) 45
1	Malignant	TP (39)	FP (1)
2	Benign	FN (0)	TN (45)

The extraction of the following statistical features from the given MR image is done. They are Mean, Variance, Standard Variance, Median Intensity, Skewness, Smoothness, Inverse Difference Movement (IDM), Kurtosis, Contrast, Correlation, Entropy, Energy, and Homogeneity. The above extracted features describe the image's structure information of intensity and the texture. These extracted features are shown in Table 2.

In Feature extraction, 13 features are extracted from each image. There are 2 phases of the classification process, i.e., the training and the testing. In the training, the known data i.e. 13 features of 10 benign and 10 malignant images are given to the classifier for training. In the testing phase, 85 images are given as input to the classifier and the proposed SVM algorithm is used for classification purpose. From the Table 3, the proposed SVM classifier, 45 images are classified as benign tumor correctly one malignant tumor is misclassified as benign tumor out of that 40 images.

Sensitivity, specificity, accuracy, and error rate of the SVM classifier are shown in Table 4, The Proposed SVM classifier had 98.82% of classification accuracy.

Table 4 SVM classifier performance

S. no.	Performance parameter	Percentage (%)
1	Sensitivity	100
2	Specificity	97.83
3	Accuracy	98.82
4	Error rate	1.17

6 Conclusions

In this paper, an FCM algorithm is applied to segment the input Brain MR images and SVM for classification of the given test images into benign and malignant classes respectively. The segmentation results of the proposed algorithm for 14 MR brain images are obtained and performance parameters like MSE, Processing Time (s), PSNR value (dB), are obtained and the segmented area is calculated using the proposed methodology and shown in the Table 1. The statistical features are extracted by wavelet decomposition followed by PCA algorithm for dimensionality reduction. These extracted features of the training set of images constitute the training feature database. This database is used when a test input image is given to classifying the brain tumor into either benign or malignant i.e., Brain tumor classification. 105 Brain MR Images are used to diagnose such as Benign, and Malignant based on the proposed supervised learning SVM classification algorithm. SVM classifier gave an accuracy of 98.82%, 100% of sensitivity, 97.83% of specificity, and 1.17% of error rate in experimental results.

References

1. Sheethal, M.S., Kannan, B., Varghese, A., Sobha, T.: Intelligent classification technique of human brain MRI with efficient wavelet based feature extraction using local binary pattern. In: 2013 International Conference on Control Communication and Computing (ICCC), pp. 368–372. IEEE (2013)
2. Thara, K.S., Jasmine, K.: Brain tumour detection in MRI images using PNN and GRNN. In: International Conference on Wireless Communication, Signal Processing and Networking (WiSPNET), pp. 1504–1510. IEEE (2016)
3. Selvathi, D., Ram Prakash, R.S., Thamarai Selvi, S.: Performance evaluation of kernel based techniques for brain MRI data classification. In: International Conference on Conference on Computational Intelligence and Multimedia Applications, 2007, vol. 2, pp. 456–460. IEEE (2007)
4. Nabizadeh, N., Kubat, M.: Brain tumors detection and segmentation in MR images: Gabor wavelet vs. statistical features. Comput. Elect. Engg. **45**, 286–301 (2015)
5. Singh, A.: Detection of brain tumor in MRI images, using combination of fuzzy c-means and SVM. In: 2015 2nd International Conference on Signal Processing and Integrated Networks (SPIN), pp. 98–102. IEEE (2015)

6. Selvaraj, H., Thamarai Selvi, S., Selvathi, D., Ramkumar, R.: Support vector machine based automatic classification of the human brain using MR image features. Intl. J. Comput. Intell. Appl. **6**(3), 357–370 (2006)
7. Vanitha, L., Venmathi, A.R.: Classification of medical images using support vector machine. In: International Conference on Information and Network Technology, Singapore, vol. 4, pp. 63–67 (2011)
8. Prasad, P.S., Rao, B.P.: Condition monitoring of 11 kV overhead power distribution line insulators using combined wavelet and LBP-HF features. IET Gener. Transm. Distrib. **11**(5), 1144–1153 (2016)
9. Nandpuru, H.B., Salankar, S.S., Bora, V.R.: MRI brain cancer classification using support vector machine. In: 2014 IEEE Students' Conference on Electrical, Electronics and Computer Science (SCEECS), pp. 1–6. IEEE (2014)
10. Raghtate, G.S., Shankar, S.S.: Modified fuzzy C means with optimized ant colony algorithm for image segmentation. In: 2015 International Conference on Computational Intelligence and Communication Networks (CICN), pp. 1283–1288. IEEE (2015)
11. Telrandhe, S.R., Pimpalkar, A., Kendhe, A.: Detection of brain tumor from MRI images by using segmentation and SVM. In: World Conference on Futuristic Trends in Research and Innovation for Social Welfare (Startup Conclave), pp. 1–6. IEEE (2016)
12. Lahmiri, S., Boukadoum, M.: Brain MRI classification using an ensemble system and LH and HL wavelet sub-bands features. In: 2011 IEEE Third International Workshop on Computational Intelligence in Medical Imaging (CIMI), pp. 1–7. IEEE (2011)
13. Sengur, A.: An expert system based on principal component analysis, artificial immune system and fuzzy k-NN for diagnosis of valvular heart diseases. Comput. Biol. Med. **38**(3), 329–338 (2008)
14. Vishnuvarthanan, G., Rajasekaran, M.P., Subbaraj, P., Anitha V.: An unsupervised learning method with a clustering approach for tumor identification and tissue segmentation in magnetic resonance brain images. Appl. Soft Comput. **38**, 190–212 (2016)
15. Machhale, K., Nandpuru, H.B., Kapur, V., Kosta, L.: MRI brain cancer classification using hybrid classifier (SVM-KNN). In: 2015 International Conference on Industrial Instrumentation and Control (ICIC), pp. 60–65. IEEE (2015)

Study of Real-Coded Hybrid Genetic Algorithm (RGA) to Find Least-Cost Ration for Non-pregnant Dairy Buffaloes

Ravinder Singh Kuntal, Radha Gupta, Duraisamy Rajendran and Vishal Patil

Abstract In Mandya District of Karnataka, the cost of milk per liter was more in case of buffaloes compared to local cows due to high fat content and high nutritive value of buffaloes' milk than its counterpart. Based on earlier research, it was clear that the productivity of the buffaloes maintained by different dairy farms was lower. Therefore, there is a need to focus on two important aspects of dairy farming: One to increase the milk productivity of buffaloes and the other one to minimize the diet cost by upgrading the scientific dairy farming practices. Though several techniques are in use for animal diet formulation but a successful application of soft computing technique to improve the quality of the solution is always preferred as the rigidity of the functions in LPP can be easily handled. Therefore, to meet the nutrient requirements at lowest cost, we have developed a hybrid real-coded genetic algorithm (RGA) for formulating the least-cost ration for dairy buffaloes. This technique is better than old conventional techniques, in the sense that it does not break if the inputs are modified and provides better results over complex problem even if it is linear programming model. The linear programming model is developed from primary data collected from NIANP and as per the standards of ICAR. Next, the developed algorithms RGA and Hybrid RGA are executed and compared with other least-cost feed formulation techniques in non-pregnant dairy buffalo weighing 450 kg and yielding 10 L milk with 6% of fat content as a model and considering standard nutrient requirement on dry matter basis. Further, goal programming model (GP model) has been developed as there are two high priority objectives (out of eight goals), i.e., least-cost and dry matter intake, to be achieved simultaneously, if possible. This GP model is also solved by hybrid RGA showing that four goals out of eight are fully achieved. It could be concluded that real-coded genetic algorithm (RGA) with hybrid function

R. S. Kuntal (✉) · V. Patil
Department of Mathematics, Jain University, Bengaluru 562112, India
e-mail: ravindercertain@gmail.com

R. Gupta
Department of Mathematics, Dayananda Sagar University, Bengaluru, India

D. Rajendran
ICAR-National Institute of Animal Nutrition and Physiology,
Adugodi, Bengaluru 560030, India

© Springer Nature Singapore Pte Ltd. 2019
J. C. Bansal et al. (eds.), *Soft Computing for Problem Solving*, Advances in Intelligent
Systems and Computing 817, https://doi.org/10.1007/978-981-13-1595-4_30

369

can effectively be used to economize the total mixed ration cost such that the feed requirements of the animals are met without any nutrients deficiency.

Keywords Dairy feed · Least cost · Real-coded genetic algorithm
Goal programming

1 Introduction

In dairy farming, feeding cost accounts for about 70% of the total operation cost. Therefore, different diet plan is required for different categories of buffaloes in which the most critical part is to formulate the accurate low-cost balanced diet. While formulating the ration, it is necessary to understand the nutrient requirements of buffaloes at various stages. As per the 19th Livestock Census Report of 2003, out of total livestock in country, buffaloes are 21.2%, whereas since 1997–2003 buffaloes population has increased by 8.9%. Female buffalo population has increased by 7.99% over the previous census, and the total number of female buffaloes is 92.5 million in 2012. The buffalo population has increased from 105.3 million to 108.7 million showing a growth of 3.19%. Also, the number of milking buffaloes has increased from 111.09 million to 118.59 million, an increase of 6.75% [1].This resulted in most of the farmers not feeding their buffaloes properly due to high feed cost and unavailability of proper feedstuffs in their respective regions, which effects in productivity of milk yield [2]. Most of the fibrous feed available to buffaloes is deficient in essential nutrient like protein. The quantity as well as quality of feedstuffs decreases in dry season where farmers face major difficulty to provide proper low-cost balanced ration to their animals in milk [3–7]. Therefore, it is necessary that such lactating buffaloes should be supplemented with balanced ration, which is nutritionally rich and should be of low cost [8–11]. Since 1991, many researchers studied the feeding practices of dairy cattle, in which small farmers have limited resources for feeding and they are not having proper knowledge as well as resources to provide low-cost balanced ration [12]. Study also reveals that in most of the country animal diet is imbalanced by overfeeding energy, protein, and minerals. Hence, there is a large scope to improvement in the nutritional status by adopting improved feeding practices [13–15]. In addition, the dairy sector in India is majorly depending upon buffalo milk because of their contribution to total milk production (49%), as it is known that buffalo milk is rich in fat and solid content [16]. Livestock industry plays one of the major roles in development of Indian Economy as the share of livestock sector in agriculture GDP is increased from 13.88% in 1980–81 to 29.20% in 2012–13 [16]. As per the Annual Report of Department of Animal Husbandry, Dairying, and Fisheries Ministry of Agriculture, Government of India (2015–16) livestock is one of the important contributors in Indian economy, where India stands first in the world in milk production (155.5 million tonnes) [17]. Livestock also contributes to 4% of the national gross domestic product (GDP) [18]. A field study on effects of feeding a balanced ration on milk production, nitrogen use efficiency (NUE), is conducted on 7090 lactating cows and 4534 lactating buffaloes, whereby using local feed resources a balanced least-cost ration can be obtained [19]. Therefore, by considering the economic importance and difficulties of Indian farmers and the importance of buffalo rearing, an improvement in feeding practice is required, which results in effective milk production by optimizing the utilization of available feedstuffs and also attempt to minimize the feed cost of ration. Straight writing computer programs are a standout among the most normally utilized strategies followed in numerous business and non-business bolster plan programs; however, Rehman and Romero [20, 21] call attention to that LP have numerous restrictions while detailing apportions practically speaking. The presumptions in the LP strategy confine target capacity to be single and requirements to be settled—right-hand side (RHS). This implies the diminishment of an objective programming model comprises of limitations and set of objectives, which are organized a few times. The goal of objective writing computer programs is to discover the arrangement, which fulfills the limitations and approach the expressed objectives of separate issue. Hypothetically, objectives could be fulfilled totally, incompletely, or in some outrageous cases, some of them may likewise not be met [22, 23]. This viciousness is estimated utilizing positive and negative deviation factors that are characterized for every objective independently, normally known as finished or under-accomplishment of the objective. Since the target capacity of the WGP plan limits the whole of aggregate deviation from set objectives, the acquired outcome may yield trade-off arrangement between opposing objectives [24]. Evolutionary algorithms consist of genetic algorithm, genetic programming and their hybrid functions [25, 26] and EA highly depends upon its operators [27]. Furuya et al. in 1997 used genetic algorithms in which the ratio of ingredients has evolved. In this paper, he used new form of crossover and mutation in which mutation was generated by a combination of uniform random number and normal distribution random number and the mutation rate was very high (12) he also solves the nonlinear constraint which showed that EA is a good technique for diet problem [28]. Sahman et al. used GA to find least-cost diet for a livestock, which results in good solution with few constraints [29]. Since many researcher in past addressing the issue of diet problems using EA [30–36]. In this study, we made an effort to formulate low-cost ration for non-pregnant buffaloes weighing 450 kg at third lactating period with ten liter-milk yield containing 6% fat and solving using a real-coded hybrid genetic algorithm and extended our approach to nonlinear weighted goal programming by taking the square root of sum of the squares of the deviations, to formulate the ration cost of non-pregnant buffaloes weighing 450 kg at third lactating period with ten liter milk yield containing 6% fat.

2   Problem Description

The present study computes the balanced least-cost ration for buffalo of body weight 450 kg in third lactation period, where buffalo need ration for body maintenance with ten liter milk production (6% fat content). Nutrient requirements of buffalo is calcu-

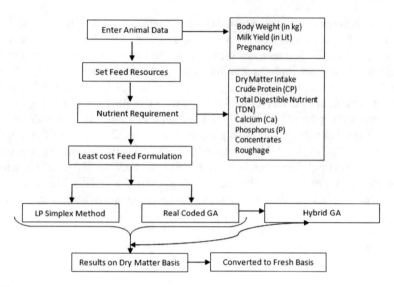

Fig. 1 Methodology to calculate least-cost balanced ration

lated by Excel-based computer programme developed by NIANP, Bangalore, as per the Indian Council of Agricultural Research-ICAR 2013 and NRC 2001 standard. A linear programming model is introduced for specified condition of buffalo. Though there is a well-defined method to solve LPP, evolutionary algorithm is preferable over them as these algorithms efficiently deal with mathematical rigidity of the constraints and can handle large dimensional problems and also gives multiple solutions. Since the number of variables represented as feedstuffs is more in number and the constraints are very rigid, we opted to develop a real-coded hybrid genetic algorithm to solve the ration formulation problem for buffaloes. Various versions of this algorithm are created, and a comparative study is planned to obtain suitable technique, which is farmer friendly and can be used at farm level (Figs. 1, 2 and 3).

3 Data Collection and Data Handling

Ration has to be formulated and needs to be updated in regular basis to avoid overfeeding. The common guidelines for diet formulation is the Indian Council of Agricultural Research [37], National Research Council's (NRC 2001), this provides the nutrient requirements for cow and buffaloes at different conditions. Hence, Excel-based computer programme developed by NIANP, Bangalore, provides actual dry matter intake, nutrient requirement values for optimizing the cost, it is difficult to optimize the cost while considering the buffaloes at different condition, so we obtained

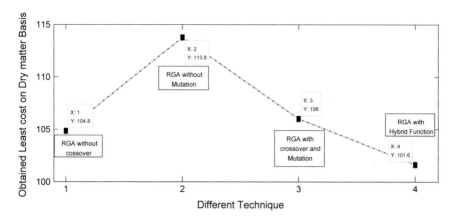

Fig. 2 Least-cost obtained by different techniques on DM basis

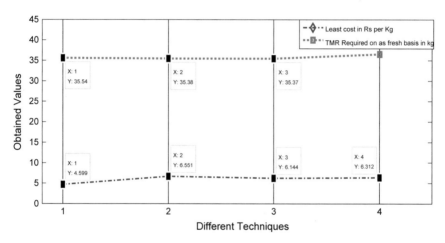

Fig. 3 Least cost in Rs. per kg versus TMR required on as fresh basis

a help from qualified nutritionist while formulating the diet. Balanced diet requires frequent analyses of feedstuffs and their cost. Table 1 gives the selected feedstuffs for diet formulation with their nutrient values. Some amount of variations in milk production can happen due to feeding practices, availability of feed resources, or animal grouping factor.

Roughages and concentrates which are rich in digestibility should be used with correct proportion depending upon lactation period, quantity of milk with high percentage of fat, so depending upon condition roughage-concentrate ration can vary. To obtain the least-cost feed, we have some constraints for each nutrient and it is unique for each animal. The minimum and maximum level of crude protein (CP 10.4–11%), total digestible nutrient (TDN 48.6–51.1%), calcium (0.4–0.5%), phosphorus (0.16–0.18%), these nutrient requirements can be met through different pro-

Table 1 Nutrient requirements for non-pregnant Buffalo weighing 450 kg and yielding ten liter milk with 6% fat

Animal	Total dry matter intake required (DMI in kg)	Crude protein (CP in kg)	Total digestible nutrient (TDN in kg)	Calcium (Ca in kg)	Phosphorus (P in kg)
	Min-Max	Min-Max	Min-Max	Min-Max	Min-Max
Buffalo	16.42–17.24	1.7158–1.8016	7.9857–9.1835	0.0680–0.0748	0.0270–0.0405

portion of roughage (40–80%), and concentrates (20–70%) on total dry matter intake for milking buffalo (Table 2).

4 Linear Programming Model

Objective Function:

$$\text{Min(z)} : \sum_{i=1}^{21} C_i x_i$$

Constraints:

$$16.42 \, \text{kg} \leq \sum_{i=1}^{21} x_i \leq 17.42 \, \text{kg}$$

$$1.7158 \, \text{kg} \leq \sum_{i=1}^{21} CP_i \leq 1.8016 \, \text{kg}$$

$$7.9857 \, \text{kg} \leq \sum_{i=1}^{21} TDN_i \leq 9.1835 \, \text{kg}$$

$$0.0680 \, \text{kg} \leq \sum_{i=1}^{21} Ca_i \leq 0.0748 \, \text{kg}$$

$$0.0270 \, \text{kg} \leq \sum_{i=1}^{21} Ph_i \leq 0.0405 \, \text{kg}$$

$$6.568 \, \text{kg} \leq \sum_{i=1}^{8} Rough_i \leq 13.136 \, \text{kg}$$

$$3.224 \, \text{kg} \leq \sum_{i=9}^{21} Conc_i \leq 11.494 \, \text{kg}$$

Table 2 Nutrient content of feedstuffs on dry matter basis

Variables	Feedstuffs	DM (in %)	CP (in kg)	TDN (in kg)	Ca (in kg)	P (in kg)	Cost C_i (in Rs./-)
Roughages							
x_1	Paddy straw	90	0.0513	0.45	0.0018	0.0008	5
x_2	CO-4 grass	20	0.08	0.52	0.0038	0.0036	3
x_3	Maize fodder	20	0.1086	0.58	0.0053	0.0014	3
x_4	Co Fs 29 sorghum fodder	90	0.0823	0.52	0.003	0.0025	3
x_5	Ragi straw	90	0.06	0.42	0.0058	0.0025	3.5
x_6	Berseem	20	0.158	0.66	0.0144	0.0014	2
x_7	Wheat straw	90	0.033	0.55	0.003	0.0006	2
x_8	Maize stover	90	0.048	0.58	0.0053	0.0014	1.5
Concentrates							
x_9	Maize	90	0.09	0.85	0.0053	0.0041	17
x_{10}	Soya DOC	90	0.46	0.7	0.0036	0.01	38
x_{11}	Copra DOC	90	0.27	0.7	0.002	0.009	23
x_{12}	Cotton DOC	90	0.35	0.7	0.0031	0.0072	23
x_{13}	Wheat bran	75	0.16	0.75	0.01067	0.00093	17
x_{14}	Gram chunies	90	0.1645	0.7	0.0028	0.0054	14
x_{15}	Cottonseed	90	0.17	1.1	0.003	0.0062	21
x_{16}	Chickpea husk	90	0.16	0.65	0.004	0.0141	10
x_{17}	Concentrate mix type I	90	0.22	0.7	0.005	0.0045	17
Minerals							
x_{18}	Calcite	97	0	0	0.36	0	4
x_{19}	MM	90	0	0	0.32	0.06	50
x_{20}	DCP	90	0	0	0.24	0.16	28
x_{21}	Salt	90	0	0	0	0	5

Search Space:

$$0.806 \leq x_1 \leq 4.03, 0.806 \leq x_2 \leq 4.03, 0.806 \leq x_3 \leq 4.03,$$
$$0.806 \leq x_4 \leq 4.03, 0.806 \leq x_5 \leq 4.03, 0.806 \leq x_6 \leq 3.224,$$
$$0.806 \leq x_7 \leq 3.224, 0.806 \leq x_8 \leq 4.03, 0.806 \leq x_9 \leq 3.224,$$
$$0 \leq x_{10} \leq 4.03, 0 \leq x_{11} \leq 4.03, 0 \leq x_{12} \leq 3.2240,$$
$$0 \leq x_{13} \leq 3.224, 0.1612 \leq x_{14} \leq 0.806, 0.1612 \leq x_{15} \leq 0.806,$$
$$0.0806 \leq x_{16} \leq 0.806, 0.806 \leq x_{17} \leq 3.224, 0 \leq x_{18} \leq 0.16120,$$
$$0 \leq x_{19} \leq 0.0806, 0 \leq x_{20} \leq 0.03224, 0.12896 \leq x_{21} \leq 0.1612$$

Goal programming model 1

$$\text{Min } z = \sqrt{\begin{array}{l} p_1\left(d_{\text{cost}}{}^+\right)^2 + p_2\left(d_{\text{DM}}{}^+\right)^2 + p_3\left(d_{\text{CP}}{}^+\right)^2 + p_4\left(d_{\text{TDN}}{}^+\right)^2 + p_5\left(d_{\text{Ca}}{}^+\right)^2 + \\ p_6\left(d_{\text{Ph}}{}^+\right)^2 + p_7\left(d_{\text{Rough}}{}^+\right)^2 + p_8\left(d_{\text{Conc}}{}^+\right)^2 \end{array}}$$

Subjected to :

$$\text{Goal 1(Least Cost)} : \sum_{i=1}^{21} C_i x_i + d_{\text{cost}}{}^- - d_{\text{cost}}{}^+ = 101.6073$$

$$\text{Goal 2(Dry Matter)} : \sum_{i=1}^{21} x_i + d_{\text{DM}}{}^- - d_{\text{DM}}{}^+ = 16.42 \text{ kg}$$

$$\text{Goal 3(Crude Protein)} : \sum_{i=1}^{21} CP_i + d_{\text{CP}}{}^- - d_{\text{CP}}{}^+ = 1.7158 \text{ kg}$$

$$\text{Goal 4(Total Digestible Nutrient)} : \sum_{i=1}^{21} TDN_i + d_{\text{TDN}}{}^- - d_{\text{TDN}}{}^+ = 9.1835 \text{ kg}$$

$$\text{Goal 5(Calcium)} : \sum_{i=1}^{21} Ca_i + d_{\text{Ca}}{}^- - d_{\text{Ca}}{}^+ = 0.0748 \text{ kg}$$

$$\text{Goal 6(Phosphorus)} : \sum_{i=1}^{21} Ph_i + d_{\text{Ph}}{}^- - d_{\text{Ph}}{}^+ = 0.0405 \text{ kg}$$

$$\text{Goal 7(Roughages)} : \sum_{i=1}^{8} Rough_i + d_{\text{Rough}}{}^- - d_{\text{Rough}}{}^+ = 13.136 \text{ kg}$$

$$\text{Goal 8(Concentrates)} : \sum_{i=9}^{21} Conc_i + d_{\text{Conc}}{}^- - d_{\text{Conc}}{}^+ = 3.2840 \text{ kg}$$

Priorities	Goals	Value
p_1	Min Least Cost	0.9
p_2	Min DM	0.8
p_3	Min CP	0.7
p_4	Min TDN	0.6
p_5	Min Ca	0.5
p_6	Min Ph	0.4
p_7	Min Rough	0.3
p_8	Min Conc	0.2

Goal programming model 2

$$\text{Min } z = \sqrt{p_1\left(d_{cost}^{+}\right)^2 + p_2\left(d_{DM}^{+}\right)^{2^2}}$$

Subjected to:

$$\text{Goal 1 (Least Cost): } \sum_{i=1}^{21} C_i x_i + d_{cost}^{-} - d_{cost}^{+} = 101.6073$$

$$\text{Goal 2 (Dry Matter): } \sum_{i=1}^{21} x_i + d_{DM}^{-} - d_{DM}^{+} = 16.42\,\text{kg}$$

Priorities	Goals	Value
p_1	Min Least Cost	0.9
p_2	Min DM	0.8

5 Development of Heuristic Approach (Hybrid GA)

Genetic algorithm is heuristic-based technique, which is based on evolution theory. GA works on solution space instead of state space, where it builds new solutions based on existing ones. GA combines good portions of solution just as nature does by combining DNA of living beings. To use GA, first we create initial population then we decide the "gene" representation, we choose default population type "double vector" to represent gene due to its flexibility. After selecting the representation of genes, it undergoes three main operators of GA such as selection, crossover, and mutation operators to create next generation from current generations. In brief, a selection operator select parents that combine to populate at next generation, crossover operator also combines two parents to create new offspring for the next generation and mutation operator randomly changes the individuals to create new offspring. MATLAB provides gaoptimset to create or modify the GA option structure. MATLAB does not provide every method available in the literature but provides lot of options to find the optimal solution.

5.1 Objective Function, Decision Variables, Representation and Plots

Firstly, we have provided the objective function, i.e., the function that calculates minimum cost of ration of each member of population. In our linear model, we have 21 decision variables and 14 constraints from which we have to find the minimum cost; hence, we set the no of variables to 21. We used gaoptimset for representation of population. As population, type specifies the type of input to the fitness function so we have restricted the input to population as "double vector" as we since we do not have complex decision variables. We used some plot options in gaoptimset such as @gaplotbestf, @gaplotdistance, @gaplotrange to see the performance of population in each generation, where gaplotbestf plots the best-fitness value in each generation, gaplotdistance plots the average distance between individuals by taking 20 samples at every generation. It calculates the distance of each sample and stores in choice and then calculates the average distance by using mean square error. The performance of GA will be affected by the diversity of initial population, if the distance between populations is large, the diversity is large, if the distance is low, and the diversity is less. More number of trials are required to experiment the right amount of diversity because if the diversity is too high or low, GA will not perform well and solution may stuck to local minimum. We have created the population in the range of lower and upper bounds and tried to perform GA to search in that space. Population size determines the size of population at each generation; hence, we have selected four times the number of variables to perform in every generation. Obviously, by increasing the population size GA will perform better to search more points and more likely to provide better solution. In this case, we have rigid constraints and bounds; therefore, by increasing the population size it will take long time for GA to generate population.

5.2 Crossover, Selection, Mutation, and Elitism

The selection procedure decides how an individual is selected to become parents. We used tournament selection procedure where an individual can be selected more than once as a parent. Tournament selection process chooses each parent by tournament size player at random and then choosing the best-fit individual from that set to be a parent. We have set tournament size as two. In crossover option, algorithm combines two parents to form a new offspring for the next generation. Crossover heuristic returns an offspring because it moves from worst parent to past best parent. The default value of ratio is 1.2. If P1 and P2 are parents and P1 has better fitness then offspring $= P2 + 1.2 * (P1 - P2)$. Mutation decides how algorithm makes small changes in the individuals randomly to create the mutation offsprings. It is an important parameter of algorithm as it provides diversity that allows GA to search in broader space. We have linear constraints and bounds hence; adaptive feasible mutation option generates a direction that is adaptive with respect to the last successful or unsuccessful generation. The feasible region is bounded by the constraints and inequality constraints. A step length is chosen along each direction so that linear constraints and bounds are satisfied.

5.3 Global Versus Local Minima

After specifying above genetic algorithm options for linear models, genetic algorithm sometimes returns a local minimum instead of global minimum, i.e., a point where the objective function value is less than the nearby points but possibly greater than the distant point in solution space. Therefore, to overcome this deficiency of genetic algorithm we have introduced hybrid command "fmincon" inside genetic algorithm, in which we allow GA to find the valley that contains global minimum and after last generation, it takes the last value of GA as the initial value of fmincon to converge quickly. Another way to make GA explore the wider range of points is to increase the diversity of the population, and it can be done by setting initial range of population. However, we have rigid constraints and bounds so we want to search the point in the specified lower and upper bounds only.

Fmincon is gradient-based search technique, which works on problem, which has continuous constraints and objective function and must have first derivative. In MATLAB, fmincon uses sequential quadratic programming (SQP) subproblem where SQP methods are representation of nonlinear programming methods. Schittkowski [38] has implemented and tested a version that outperforms every other tested method in terms of efficiency, accuracy of solution for large number of test problems. Based on the work of Biggs [39], Han [40], and Powell [41, 42] it has seen that the method mimic Newton's method for constrained problem. At every iteration, an approximation is made of the Hessian of the Lagrangian function using quasi-Newton method, which is used to generate a quadratic programming (QP) problem whose solution is used to form a search direction for a line search problem. The major overview of SQP is to be found in Fletcher [43], Gill et al. [44], Powell [41], and Hock and Schittkowski [45]. In this view, we have used SQP algorithm for fmincon solver, where SQP algorithm (identical to SQP-legacy algorithm) is same as active-set algorithm but has a different implementation. Moreover, interior point algorithm can also be used but SQP has faster execution time and less memory usage then SQP-legacy and interior point algorithm. Nocedal and Wright explain the basic SQP algorithm [26].

6 Result in Dry Matter and Fresh Basis

See Tables 3, 4, 5 and 6.

7 Results and Discussions

The results obtained by various techniques, viz., RGA without crossover, RGA without mutation, RGA with crossover and mutation, and RGA with hybrid function for least-cost ration (which is run 10 times up to 5000 generation) are presented in Table 3 for dairy buffalo. LP model for dairy buffalo has twenty-one variables, which are too complex for finding an optimal solution. RGA is also a heuristic technique, which works on a principle of survival of best fit and tournament selection. Adaptive feasibility mutation and heuristic crossover are used along with elitism to solve LP model for dairy buffalo. Though RGA without crossover, RGA without mutation, RGA with crossover, and mutation gives near optimal answer, we provide solution using RGA with hybrid function to avoid solution being stuck in local minima. As farmers need total mixed ration to feed dairy buffalo on "as fresh basis," we have converted the least-cost and total mixed ration obtained by all technique to "as fresh basis" and the results are given in Table 4. As per Table 3, this category of animals requires about 16.42 kg of dry matter from various kinds of feeds, which should contain 1715.8 g of protein, 9183.5 g of TDN, 74.8 g of calcium, and 40.5 g of phosphorus. These nutrient requirements can be met from 13.13 kg of roughage and 3.28 kg of concentrates on dry matter. Its corresponding total mixed ration (TMR) cost on dry matter basis is Rs. 101.6703, i.e., Rs. 6.192 per kg.

"As fresh basis" is a feed nutrient content with moisture included. After converting to "as fresh basis" feedstuff required for dairy buffalo is approximately 36.5 kg per day amounting Rs. 4.5 per kg using RGA with hybrid function. When the expected nutrient requirement was tried to achieve using RGA without crossover, without mutation, and with crossover and mutation the least-cost ration obtained was Rs. 4.59 per kg, Rs. 4.93 per kg, and Rs. 4.67 per kg, respectively. The detailed analysis is of ration is showed in Tables 3 and 4 which exactly meets the requirement of dry matter, CP, TDN, calcium, and phosphorus. Requirement of roughage: concentrates are also met with in permissible range.

In addition, results obtained by GP model 1 and 2 are shown in Tables 4 and 5 which reveal that from GP 1 goal 5 and 8 is fully achieved and Goal 1, 2, 3, 4, 6, 7 are underachieved. The least cost obtained by GP 1 is Rs. 100.79 on DM basis but DMI, and CP was not satisfied. Whereas by GP model 2, our highly priorities goal 1 and 2 are completely satisfied and rest all goals are either underachieved or overachieved,

Table 3 Least-cost ration formulated by various techniques for non-pregnant dairy buffalo on dry matter basis

	Minimum value obtained in ten runs for 5000 generations			RGA with hybrid function
Feedstuffs	RGA without crossover	RGA without mutation	RGA with crossover and mutation	
Roughages (In kg)				
Paddy straw	3.7907	3.8114	3.187	2.363
CO-4 grass	0.8758	0.806	1.3551	0.806
Maize fodder	2.3054	2.7951	2.0736	3.086
Co Fs 29 sorghum fodder	1.3639	0.806	1.112	2.266
Ragi straw	1.7162	1.5121	2.0725	2.197
Berseem	1.2667	0.806	0.9759	0.806
Wheat straw	0.9611	1.601	0.8703	0.806
Maize stover	0.8572	0.806	1.3561	0.806
Concentrates (In kg)				
Maize	0.8060	0.806	0.8061	0.806
Soya DOC	0.0001	0.5647	0.0057	0.000
Copra DOC	0.0015	0	0.0101	0.000
Cotton DOC	1.1233	0.4301	1.2101	0.951
Wheat bran	0.0003	0	0.0114	0.000
Gram chunies	0.1624	0.4549	0.1652	0.161
Cottonseed	0.1613	0.1612	0.1612	0.161
Chickpea husk	0.0855	0.0806	0.0911	0.081
Concentrate mix type I	0.8076	0.806	0.8258	0.963
Minerals (In kg)				
Calcite	0.0017	0.0117	0.0007	0.000
MM	0.0000	0	0	0.000
DCP	0.0000	0	0	0.000
Salt	0.1322	0.1612	0.129	0.161
Constraints (In kg)				
Dry matter intake (DMI)	16.419	16.42	16.419	16.42
Crude protein (CP)	1.7148	1.7148	1.7151	1.7158
Total digestible nutrient (TDN)	9.175	9.1835	9.1816	9.1835
Calcium (Ca)	0.0758	0.0746	0.075	0.0748
Phosphorus (P)	0.0388	0.0391	0.0412	0.0405
Roughage	13.137	12.9436	13.0025	13.1360
Concentrates	3.282	3.4765	3.4165	3.2840
Least cost (z)	104.8323	113.7639	106.0145	101.6073

Table 4 Least-cost ration formulated by various techniques for non-pregnant dairy buffalo on "as fresh basis"

Feedstuffs	RGA without crossover	RGA without mutation	RGA with crossover and mutation	RGA with hybrid function
Paddy straw	4.211889	4.234889	3.541111	2.625556
CO-4 grass	4.379	4.03	6.7755	4.03
Maize fodder	11.527	13.9755	10.368	15.43
Co Fs 29 sorghum fodder	1.515444	0.895556	1.235556	2.517778
Ragi straw	1.906889	1.680111	2.302778	2.441111
Berseem	6.3335	4.03	4.8795	4.03
Wheat straw	1.067889	1.778889	0.967	0.895556
Maize stover	0.952444	0.895556	1.506778	0.895556
Maize	0.895556	0.895556	0.895667	0.895556
Soya DOC	0.000111	0.627444	0.006333	0
Copra DOC	0.001667	0	0.011222	0
Cotton DOC	1.248111	0.477889	1.344556	1.056667
Wheat bran	0.0004	0	0.0152	0
Gram chunies	0.180444	0.505444	0.183556	0.178889
Cottonseed	0.179222	0.179111	0.179111	0.178889
Chickpea husk	0.095	0.089556	0.101222	0.09
Concentrate mix type I	0.897333	0.895556	0.917556	1.07
Calcite	0.001753	0.012062	0.000722	0
MM	0	0	0	0
DCP	0	0	0	0
Salt	0.146889	0.179111	0.143333	0.178889
Total mixed ration on "as fresh basis" in kg	35.5405415	35.3822285	35.3746994	36.5144444
Least cost in Rs. on "as fresh basis"	163.445699	174.679247	165.425231	164.572778
Least cost (in Rs. per kg)	4.59885225	4.93692045	4.67637135	4.50705961

Table 5 Least cost and deviation value solved by hybrid RGA for GP model 1

Feedstuff	Values
On dry matter basis analysis	
Paddy straw	3.5806
CO-4 grass	0.1175
Maize fodder	0.0194
Co Fs 29 sorghum fodder	0.1334
Ragi straw	1.0355
Berseem	2.7540
Wheat straw	2.6429
Maize stover	0.5415
Maize	1.3333
Soya DOC	1.0162
Copra DOC	0.0146
Cotton DOC	0.0240
Wheat bran	0.0000
Gram chunies	0.0107
Cottonseed	0.0000
Chickpea husk	0.0214
Concentrate mix type I	0.0000
Calcite	0.0000
MM	0.0000
DCP	0.0000
Salt	0.8638
Deviations	
d_{cost}^-	0.8115
d_{cost}^+	0.0000
d_{DM}^-	2.3113
d_{DM}^+	0
d_{CP}^-	0.2942
d_{CP}^+	0.0000
d_{TDN}^-	1.5173
d_{TDN}^+	0
d_{Ca}^-	0.0000
d_{Ca}^+	0
d_{Ph}^-	0.0118
d_{Ph}^+	0.0000

(continued)

Table 5 (continued)

Feedstuff	Values
d_{Rough}^{-}	2.3113
d_{Rough}^{+}	0.0000
d_{Conc}^{-}	0.0000
d_{Conc}^{+}	0.0000
Constraints	
Dry matter intake (DMI)	14.1088
Crude protein (CP)	1.4216
Total digestible nutrient (TDN)	7.6663
Calcium (Ca)	0.0748
Phosphorus (P)	0.0287
Roughage	10.8248
Concentrates	3.2840
Least cost on DM basis	100.7965
As fresh basis analysis	
Total mixed ration on "as fresh basis" in kg	26.91883333
Least cost in Rs. on "as fresh basis"	135.0132778
Least cost (in Rs. per kg)	5.015569438

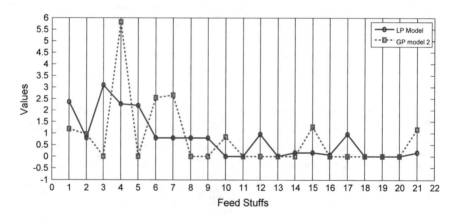

Fig. 4 Comparison of feedstuffs obtained by LP model and GP model 2

but the least cost obtained was Rs. 100.6090 on DM basis which is same as LP model. The amount of feedstuff is very less compared to LP model, which is truly an advantage to farmers because of the limitation of feedstuff (see Fig. 4).

As a farmer's point of view, the TMR required by the farmer by RGA hybrid function is 36.51 kg per day amounting Rs. 4.50, whereas by GP model 2 TMR required is 31.90 kg per day amounting 4.50 approx, which is less compared to

Table 6 Least cost and deviation value solved by hybrid RGA for GP model 2

Feedstuffs	Values
On dry matter basis analysis	
Paddy straw	1.1775
CO-4 grass	0.9571
Maize fodder	0.0001
Co Fs 29 sorghum fodder	5.7964
Ragi straw	0.0000
Berseem	2.5561
Wheat straw	2.6487
Maize stover	0.0000
Maize	0.0000
Soya DOC	0.8576
Copra DOC	0.0000
Cotton DOC	0.0001
Wheat bran	0.0000
Gram chunies	0
Cottonseed	1.2705
Chickpea husk	0.0000
Concentrate mix type I	0
Calcite	0.0000
MM	0
DCP	0
Salt	1.1559
Deviations	
d_{cost}^{-}	0.0001
d_{cost}^{+}	0
d_{DM}^{-}	0.0000
d_{DM}^{+}	0
d_{CP}^{-}	0.0032
d_{CP}^{+}	0.0997
d_{TDN}^{-}	0.2440
d_{TDN}^{+}	1.5679
d_{Ca}^{-}	0.0055
d_{Ca}^{+}	0.2868
d_{Ph}^{-}	0.0663
d_{Ph}^{+}	0.7203

(continued)

Table 6 (continued)

Feedstuffs	Values
$d_{Rough}{}^{-}$	0.0216
$d_{Rough}{}^{+}$	0.0182
$d_{Conc}{}^{-}$	0.2544
$d_{Conc}{}^{+}$	0.0095
Constraints	
Dry matter intake (DMI)	16.4200
Crude protein (CP)	1.7158
Total digestible nutrient (TDN)	9.1835
Calcium (Ca)	0.0748
Phosphorus (P)	0.0405
Roughage	13.1359
Concentrates	3.2841
Least cost on DM basis	101.6090
As fresh basis analysis	
Total mixed ration on "as fresh basis" in kg	31.90727778
Least cost in Rs. on "as fresh basis"	143.947
Least cost (in Rs. per kg)	4.511415891

the RGA hybrid function. However, for better output for farmers we need a further discussion with qualified cattle nutritionist.

8   Conclusion

The present study addresses the use of real-coded genetic algorithm with hybrid function as a tool to provide good quality feed mix to the dairy buffalo for better health and milk production. All the techniques, viz., RGA without crossover, RGA without mutation, RGA with crossover, and mutation and RGA with hybrid function for least-cost ration are performing equally. However, RGA without crossover and RGA without mutation operator provide near to optimal answer but solutions seem to be get stuck in local minima; hence, it is proved that real-coded genetic algorithm (RGA) with hybrid function provide optimal solution and this method can also be used for ration formulation to find least-cost feedstuff in dairy buffalo. However, fixing of constraints and the use of code for making software is considered while choosing the techniques for making least-cost feed formulation. Further, detailed research with various species of animals and with different physiological needs may require to fine tune the techniques for farmer use. We are able to economize the TMR cost such that the feed requirements of the animals are met without any nutrient deficiency.

References

1. Govt of India, ministry of agriculture department of animal husbandry, dairying and fisheries, Krishi Bhavan, 19 livestock census-2012 all India report, New Delhi
2. Garg, M.R.: FAO 2012. Balanced feeding for improving livestock productivity—increase in milk production and nutrient use efficiency and decrease in methane emission. In: FAO Animal Production and Health Paper No. 173. Rome, Italy
3. Afolayan, M.O., Afolayan, M.: Nigeria oriented poultry feed formulation software requirements. J. Appl. Sci. Res. **4**, 1596–1602 (2008)
4. Hertzler, G.: Dynamically optimal and approximately optimal beef cattle diets formulated by nonlinear programming. West. J. Agr. Econ. **13**, 7–17 (1987)
5. Goswami, S.N., Chaturvedi, A.: Least cost diet plan of cows for small dairy farmers of Central India. Afr. J. Agric. Res. **8**, 5989–5995 (2013)
6. Almasad, M., Altahat, E., AL-Sharafat, A.: Applying linear programming technique to formulate least cost balanced ration for white eggs layers in Jordan. Int. J. Empir. Res. **1**, 112–120 (2011)
7. Waugh, F.: The minimum-cost dairy feed. J Farm Econ. **33**, 299–310 (1951)
8. Olorunfemi Teitope, O.S.: Linear programming approach to least cost ration formulation for poults. Inf. Technol. J. **6**, 294–299 (2007)
9. Oladokun, V.O., Johnson, A.: Feed formulation in Nigerian poultry farms: a mathematical programming approach. Am. J. Sci. Ind. Res. **3**, 14–20 (2012)
10. Kuntal, R.S., Gupta, R., Rajendran, D., Patil, V.: Application of real coded genetic algorithm (RGA) to find least cost feedstuffs for dairy cattle during pregnancy. Asian J. Anim. Vet. Adv. **11**, 594–607 (2016)
11. Sreenivas, B., Ramappa, P.: Performance of livestock sector in India (with reference to Bovine population). Curr. Agric. Res. J. **4**(1), 108–113 (2016)
12. Leng, R.A.: Feeding strategies for improving milk production of dairy animals managed by small-farmers in the tropics. In: Speedy, A., Sansoucy, R. (eds.) Feeding Dairy Cows in the Tropics. Proceedings of the FAO Expert Consultation held in Bangkok, Thailand, p. 82 (1991)
13. Mudgal, V., Mehta, M.K., Rane, A.S., Nanavati, S.: A survey on feeding practices and nutritional status of dairy animals in Madhya Pradesh. Indian J. Anim. Nutr. **20**(2), 217–220 (2003)
14. Afrouziyeh, M., Shivazad, M., Chamani, M., Dashti, G.: Use of non-linear programming to determine the economically optimal energy density in laying hens diet during phase 1. Afr. J. Agric. Res. **5**, 2270–2777 (2010)
15. Al-Deseit, B.: Least-cost broiler ration formulation using linear programming technique. J. Anim. Vet. Adv. **8**, 1274–1278 (2009)
16. Islam, M.M., Anjum, S., Modi, R.J., Wadhwani, K.N.: Scenario of livestock and poultry in india and their Contribution to national economy. Int J Sci Environ Technol. **5**(3), 956–965 (2016)
17. Annual Report 2016–17, Department of Animal Husbandry, Dairying and Fisheries, Ministry of Agriculture and Farmers Welfare, Govt. of India
18. Angadi, U.B., Anandan, S., Gowda, N.K.S., Rajendran, D., Devi, L., Elangovan, A.V., Jash, S.: "Feed assist"—an expert system on balanced feeding for dairy animals. In: AGRIS on-line Papers in Economics and Informatics, vol. 8, no. 3, pp. 3–12. ISSN 1804-1930. https://doi.org/10.7160/aol.2016.080301
19. Garg, M.R., Sherasia, P.L., Bhanderi, B.M., Makkar, H.P.S.: Nitrogen use efficiency for milk production on feeding a balanced ration and predicting manure nitrogen excretion in lactating cows and buffaloes under tropical conditions. Anim. Nutr. Feed Technol. **16**, 1–12 (2016)

20. Rehman, T., Romero, C.: Multiple-criteria decision-making techniques and their role in livestock ration formulation. Agri. Sys. **15**, 23–49 (1984)
21. Rehman, T., Romero, C.: Goal programming with penalty functions and livestock ration formulation. Agri. Sys. **23**, 117–132 (1987)
22. Zhang, F., Roush, W.B.: Multiple-objective (Goal) programming model for feed formulation: an example for reducing nutrient variation. Poult. Sci. **81**, 182–192 (2002)
23. Zgajnar, J., Juvancic, L., Kavcic, S.: Combination of linear and weighted goal programming with penalty function in optimization of daily dairy cow ration. Agric. Econ. Czech (10), 492–500 (2009)
24. Gupta, R., Kuntal, R.S., Ramesh, K.: Heuristic approach to goal programming problem for animal ration formulation. Int. J. Eng. Innov. Technol. (IJEIT) **3**(4), 414–422 (2013)
25. Rehman, A.R.: Evolutionary algorithms with average crossover and power heuristics for aquaculture diet formulation. Ph.D. thesis, University Utara Malaysia, Sintok, Malaysia (2014)
26. Nocedal, J., Wright, S.J.: Numerical Optimization, 2nd edn. Springer Series in Operations Research, Springer (2006)
27. Koda, M.: Chaos search in Fourier amplitude sensitivity test version. J. Inf. Commun. Technol. **11**, 1–16 (2012)
28. Tozer, P.R., Stokes, J.R.: A multiobjective programming approach to feed ration balancing and nutrient management. Agri. Syst. **67**, 201–215 (2001)
29. Şahman, M.A., Çunkaş, M., İnal, Ş., İnal, F., Coşkun, B., Taşkiran, U.: Cost optimization of feed mixes by genetic algorithms. Adv. Eng. Softw. **40**(10), 965–974 (2009)
30. Ghosh, S., Ghosh, J., Pal, D.T., Gupta, R.: Current concepts of feed formulation for livestock using mathematical modeling. Anim. Nutr. Feed Technol. **14**, 205–223 (2014)
31. Powell, M.J.D.: Variable metric methods for constrained optimization. In: Bachem, A., Grotschel, M., Korte, B. (eds.) Mathematical Programming: The State of the Art, pp 288–311. Springer (1983)
32. Sexena, P., Chandra, M.: Animal diet formulation models: a review (1950–2010). Anim. Sci. Rev. 189–197 (2011)
33. Sexena, P.: Comparison of linear and non-linear programming techniques for animal diet. Appl. Math. **1**, 106–108 (2011)
34. Radhika, V., Rao, S.B.N.: Formulation of low cost balanced ration for livestock using Microsoft excel, Wayamba J. Anim. Sci. 38–41 (2010)
35. Schittkowski, K.: NLQPL: A FORTRAN-subroutine solving constrained nonlinear programming problems. Ann. Oper. Res. **5**, 485–500 (1985)
36. Furuya, T., Satake, T., Minami, Y.: Evolutionary programming for mix design. Comput. Electron. Agric. **18**(2–3), 129–135 (1997)
37. Nutrient Requirements of Animals-Cattle and buffalo (ICAR-NIANP). ISBN: 978-81-7164-139-9 (2013)
38. Singh, K.P., Singh, I.: Buffalo diversity in India: breeds and defined populations. Dairy Year Book (2014–15), pp. 33–36 (2015)
39. Biggs, M.C., Constrained Minimization Using Recursive Quadratic Programming, Towards Global Optimization. In: Dixon, L.C.W., Szergo, G.P. (eds.) North-Holland, pp. 341–349 (1975)
40. Han, S.P.: A globally convergent method for nonlinear programming. J. Optim. Theory Appl. **22**, 297 (1977)
41. Powell, M.J.D.: The convergence of variable metric methods for nonlinearly constrained optimization calculations. In: Mangasarian, O.L., Meyer, R.R. Robinson, S.M. (eds.) Nonlinear Programming 3. Academic Press (1978)
42. Powell, M.J.D.: A fast algorithm for nonlinearly constrained optimization calculations. In: Watson, G.A. (ed.) Numerical Analysis. Lecture Notes in Mathematics, vol. 630. Springer (1978)
43. Fletcher, R.: Practical Methods of Optimization. Wiley (1987)

44. Gill, P.E., Murray, W., Wright, M.H.: Practical Optimization. Academic Press, London (1981)
45. Hock, W., Schittkowski, K.: A comparative performance evaluation of 27 nonlinear programming codes. Computing **30**, 335 (1983)
46. Sebastian, C., Akinnifesi, F.K., Oluyede, C.A., Gudeta, S., Simon, M., France, M.T.: A simple method of formulating least-cost diets for smallholder dairy production in sub-Saharan Africa. Afr. J. Biotech. **7**, 2925–2933 (2008)

Multimetrics-Based Objective Function for Low-power and Lossy Networks Under Mobility

Shridhar Sanshi and C. D. Jaidhar

Abstract Due to the popularity of Low-power and Lossy Networks (LLN), numerous low-power device applications are emerging and driving the need for an efficient routing protocol. Recently, the Internet Engineering Task Force (IETF) standardized the IPv6 Routing Protocol for Low-power and Lossy Networks (RPL). To route the packets, the RPL constructed a Directed Acyclic Graph (DAG) rooted towards the DAG root using the Objective Function (OF). However, the OF supported by the standard RPL did not yield better performance, since it used a single metric. Therefore, an OF based on Multimetrics (MMOF), which combines multiple routing metrics for Static Router Nodes (SRNs) and Mobile Nodes (MNs), has been proposed in this work. From the simulation results, it was observed that the proposed MMOF showed better performance compared with other existing OFs of the RPL.

Keywords Preferred parent node · Mobility · Routing metrics
Received signal strength indicator · Expected transmission count
Residual energy

1 Introduction

LLN consist of embedded devices characterized by its small size, constrained power, limited memory, and low processing capabilities. These resource-restrained devices can only handle a limited amount of data, and its considerable number of nodes are

S. Sanshi (✉) · C. D. Jaidhar
Department of Information Technology, National Institute
of Technology Karnataka, Surathkal, Karnataka, India
e-mail: it15f03.sanshi@nitk.edu.in

C. D. Jaidhar
e-mail: jaidharcd@nitk.edu.in

© Springer Nature Singapore Pte Ltd. 2019
J. C. Bansal et al. (eds.), *Soft Computing for Problem Solving*, Advances in Intelligent
Systems and Computing 817, https://doi.org/10.1007/978-981-13-1595-4_31

battery powered [18]. Due to its embedded nature, these low-cost devices are subject to a high variance of environmental factors, interference and noise.

The Wireless Sensor Network (WSN) is a subtype of the LLN [14], which consists of several sensor nodes. These nodes sense the surrounding environment and gather application-specific information, which they then send to the collector node over the Internet for the purpose of processing. The routing protocol for such Lossy Networks, where the transmitted messages are often lost, must be designed carefully so that the routing is performed effectively and efficiently.

In order to route packets over the LLN, the IETF has created the RPL [19], which is capable of fulfilling the precise requirements of the LLN. The RPL is simply a IPv6 distance vector routing protocol designed for the LLN. It constructs DAG rooted towards the DAG root using control messages and the OF. The routes constructed using OF are optimized routes to the destination, while the RPL maintains multiple routes to the destination. The OF uses a routing metric to construct the DAG, and the default routing metric used in the RPL is the Expected Transmission Count (ETX) [4]. However, for some applications where energy is crucial, the OF that considers only the ETX does not yield better performance. Therefore, routing metrics should be used based on the application.

Due to the popularity of the LLN, many applications like industrial automation, warehouse, healthcare monitoring require mobility support and most of the applications demand transmission of data in real time. Hence, the RPL protocol should offer support for mobility, and also at the same time, provide guaranteed reliability to transmit the data during the mobility of the node. The OF, which plays a central role in the RPL, must consider the mobility of the node while selecting the Preferred Parent Node (PPN). However, the default OF used in the RPL does not consider mobility and uses a single routing metric. Therefore, in this paper, MMOF is proposed for SRNs and MNs. For the SRNs, link quality, Residual Energy (RE) and Distance (Dist) metrics were considered for selecting a PPN that has high energy, minimum distance, and high-quality link. For the MNs, link quality, Dist and direction of the MN were considered, as the position of the MN plays an important role in selecting the PPN.

The remaining sections of this paper are structured as follows: Section 2 briefly discusses the RPL and the OF considered for the RPL. In Sect. 3, the proposed MMOF is presented, followed by an evaluation of the MMOF using the Cooja simulator in Sect. 4, and finally, in Sect. 5, the conclusion is presented.

2 Background

2.1 Overview of RPL

The IETF designed RPL to route the packets from source to destination for a network consists of a large number of low-power devices. The RPL is a simple, distance vector routing protocol specially intended for the LLN. The RPL organizes the large network into a tree structure called the DAG for each sink node. The DAG is rooted towards one node, usually the DAG root (sink node), by forming a Destination-Oriented DAG (DODAG). The DODAGs are identified by a unique DODAG Identifier in the network. It determines the routes between the sink node and any node in the network based on the node rank with respect to the DODAG root. The rank of each node is determined based on the OF, which indicates the metrics (link quality, hop count or other parameters) and constraints among the nodes (minimum energy of the node) in use.

The RPL classifies the nodes into three categories based on their role in the network. The sink node is the DODAG root, which is connected to the outside world through the Internet or any other technology. Router nodes are the intermediate nodes between the sink node and the leaf nodes and have the ability to forward data traffic from the other nodes. Another category of nodes are the leaf nodes placed at the edge of the DODAG topology; they do not have the capability to forward data from the other nodes.

In order to construct and maintain the DODAG, the RPL makes use of the following control messages:

- DODAG Information Object (DIO) message contains all the necessary information to calculate the rank of a node.
- DODAG Information Solicitation (DIS) message is used to receive DIO messages from neighbouring nodes.
- Destination Advertisement Object (DAO) message is used to create downward routes between the nodes.
- DAO Acknowledgement (DAO-ACK) is used to acknowledge reception of any DAO message.

Initially, the DODAG root advertises its presence by broadcasting DIO messages to its neighbours. The node calculates the rank and decides whether or not to join the DODAG. Then, each node broadcasts its information to its neighbours, and this process is repeated until the DIO message reaches the leaf node. The node creates a candidate parent list, which has the potential to become PPN from the DIO messages. With the information in the DIO message, the node calculates the OF for each parent node in the list and selects one node as the PPN to forward the data traffic towards the sink node.

Each node sends a DIO message based on the trickle algorithm [16]. This algorithm controls the frequency of DIO messages in the network. The algorithm increases its interval value if it receives a consistent message, otherwise, the interval becomes shorter, and DIO messages are transmitted frequently in order to stabilize the network. The interval timer is bound by lower-value I_{min} and upper-value I_{max}, where I_{min} is the minimum interval defined in milliseconds by a base-2 value (e.g., $2^{10} = 1024$ ms), and $I_{max} = I_{min} \times 2^{I_{doubling}}$. $I_{doubling}$ is used to limit the number of times the I_{min} can double. Suppose, $I_{doubling}$ is 8 then, the maximum interval given by $I_{max} = 1024 \times 2^8 = 26{,}2144$ ms. For a more detailed explanation of the trickle algorithm, Ref. [16].

2.2 RPL Objective Functions

An OF defines the routing metrics and constraints used to determine the path towards any node and to select the parents in the DODAG. To select one among the available parents, the path cost for each parent should be computed. The path cost is calculated using an OF that translates the routing metrics and constraints into a scalar value. Based on the application requirements, different routing metrics or constraints can be defined to optimize the application performance.

The standard RPL defines two OFs, namely Minimum Rank with Hysteresis Objective Function (MRHOF) [8] and Objective Function Zero (OF0) [17]. The MRHOF is based on the additive metric aimed to select a good quality link towards the sink node. It uses the ETX as a routing metric, which is defined as the expected transmissions required for the node to deliver a packet successfully. The OF0 is based on the hop count, which is defined as the distance from the node to the DODAG root in terms of intermediate nodes (hops). The hop count of the DODAG root is zero, which then increases down the link towards the leaf nodes. Once the PPN is selected, the node calculates its rank and path cost and disseminates this data through a DIO message.

2.3 Objective Functions Proposed for RPL in the Literature

Several OFs are defined in the literature to improve energy consumption, minimize end-to-end delay and maximize reliability or to prolong network lifetime. The present work concentrates on the OF defined for RPL to improve its performance.

In [3], the authors proposed a new metric based on energy consumption by nodes along the path towards the sink node. Each node sends energy consumed along the path in a DIO message, which is used to determine the path cost. The node with least energy consumption is selected as the PPN. However, a node with very low energy may be present along the path. In [13], the authors proposed a routing metric by considering the nodes remaining energy. To calculate the remaining energy of a node, a well-known battery theoretical model was used. However, this did not consider link quality while selecting the PPN.

Taking into consideration only the ETX causes uneven distribution of energy and reduces network lifetime. Similarly, taking only energy into consideration causes packet loss in the network. In order to tackle these problems, the authors [2] combined the ETX and energy routing metrics and analysed the performance of the RPL. Similarly, in [11], the authors proposed a new OF based on expected lifetime routing metric to prolong network lifetime. The RPL constructed the DODAG by considering amount of traffic, link quality and remaining energy. However, they did not consider the distance metric, which is important in the mobile scenario.

In [9], the authors proposed a new routing metric to minimize the overall delay in reaching the DODAG root. They also modified the ContikiMAC to support the different sleeping periods of the nodes. However, the protocol suffered from packet loss and required retransmission of the packets. In [7], the authors proposed an OF based on fuzzy logic for static and mobile LLN by considering ETX, end-to-end delay, hop count and link quality. The simulation results showed better performance as compared with the standard RPL. However, they did not consider the distance routing metric.

In the present work, ETX, RE and Dist routing metrics have been considered for the SRNs, and ETX, Dist and Distance variation (Dist_var) routing metrics for the MNs. The proposed OF is discussed in detail in Sect. 3.

3   Objective Function Based on Multimetrics for RPL

In this section, the MMOF for the SRNs and MNs and routing metrics used to calculate the new OF are discussed in detail.

3.1   Objective Function for Static Router Node

Since the OF plays a central role in creating the DAG structure, designing an OF for the RPL becomes absolutely necessary. Therefore, a new OF (MMOF) was devised that combined multiple routing metrics to select the PPN. The routing metrics used in the MMOF are Dist, ETX and RE. The MMOF is calculated as per Eq. (1).

$$MMOF = \alpha \frac{ETX_{curr}}{ETX_{max}} 100 + \beta (100 - \frac{RE_{curr}}{E_{total}} 100) + \gamma \frac{Dist_{curr}}{Dist_{max}} 100 \quad (1)$$

where, α, β and γ are weights assigned to the routing metrics ETX, RE and Dist, respectively, such that $\alpha + \beta + \gamma = 1$. The weights can be altered according to the priority of the metrics. ETX_{curr} is the current link quality, ETX_{max} is the maximum value that an ETX can take, RE_{curr} is the remaining energy of the node, E_{total} is the energy of the node during deployment, $Dist_{curr}$ is the current distance from the node, and $Dist_{max}$ is the maximum distance in which a node can communicate with another node. The calculation of each routing metric is explained in the subsections below.

Expected Transmission Count-Based Routing Metric ETX is the number of expected transmissions required by a node to successfully deliver the packet to its destination. The value of ETX is calculated by measuring the probability that a packet reaches the neighbour (D_f) successfully and the probability that an acknowledgement packet is successfully received (D_r) as given in Eq. (2).

$$ETX_{curr} = \frac{1}{D_f \times D_r} \quad (2)$$

Residual Energy-Based Routing Metric RE is the amount of energy remaining before the node runs out of its energy. It is calculated by subtracting the amount of energy depleted from the total energy of the node as given in Eq. (3).

$$RE_{curr} = E_{total} - E_{depletion} \qquad (3)$$

where E_{total} is the initial energy of the node during deployment and $E_{depletion}$ is the energy consumed by the node. The energy depletion $E_{depletion}$ of the node is calculated as the sum of the energy consumed during processing, communicating and sensing. As per [6], it is calculated as in Eq. (4).

$$E_{depletion} = V \times (I_{ap}T_{ap} + I_{lp}T_{lp} + I_{tx}T_{tx} + I_{rx}T_{rx} + \sum_i I_{ni}T_{ni}) \qquad (4)$$

where V is a supply voltage, I_{ap}, I_{lp}, I_{tx}, I_{rx} and I_{ni} is the current required by the microcontroller in active mode, microcontroller in low-power mode, transmission mode of the communicating device, receiving mode of the communication device, and sensors, respectively and T_{ap}, T_{lp}, T_{tx}, T_{rx}, and T_{ni} are the activity duration of the microcontroller in active mode, microcontroller in low-power mode, transmission mode of the communicating device, receiving mode of the communication device, and sensors, respectively.

Distance-Based Routing Metric The distance routing metric reduces the transmission range of the node so that the node with minimum distance is selected as the PPN to avoid long-distance communication. The received signal strength diminishes with the distance raised to the nth power, where n is the path loss exponent, which depends on the environment. Therefore, the route with the minimum distance is generally more energy efficient than one with a longer distance.

The distance is determined using the RSSI value based on the empirical model. The advantage of using the RSSI value is that it uses less costly hardware and can be implemented in low-power devices.

In this work, the Tmote Sky provided by the Cooja Simulator has been used [12]. Tmote Sky is a built-in IEEE 802.15.4 compliant CC2420 radio, which implements physical layer functions and also a few medium access control layer functions. Taking the IEEE 802.15.4 characteristics and low-power operations into account, the CC2420 provides a reliable wireless communication. The CC2420 has a built-in RSSI register, which provides an 8-bit digital value (RSSI_VAL) averaged over 8 symbol periods [10]. The status bit (RSSI VALID) signifies when the value is adequate. The RSSI register value is determined and continuously updated for each symbol, when the RSSI VALID bit is set.

To adjust the error in RSSI_VAL due to variation in antenna radiation pattern, a known RSSI offset is added to the RSSI_VAL, which is not a function of the environment or a position of the node, and its value is discovered empirically during system development from the front end gain. The $RSSI$ value is computed as in Eq. (5) as per [15].

$$RSSI = RSSI\_VAL + RSSI\_OFFSET \quad [dBm] \tag{5}$$

where $RSSI\_VAL$ is the 8-bit register value and $RSSI\_OFFSET$ is set to approximately -45 dBm [10]. The $RSSI\_OFFSET$ is added to compensate an error in $RSSI\_VAL$. For example, if the value in the RSSI register is -40 dBm, then the $RSSI$ is approximately -85 dBm. The distance is calculated using the relation between $RSSI$ value and the power of the received signal as per [15]:

$$RSSI \propto 10 \log(\frac{1}{D^n})$$

$$RSSI = -10n \log(D) + C \tag{6}$$

where D is the distance between MN and the parent node, 'n' is the path loss exponent factor, and C is fixed constant. Equation (6) can be rewritten as:

$$RSSI = -m \log(D) + C \tag{7}$$

where 'm' is the slope of the linear equation between $\log(D)$ and $RSSI$. From Eqs. (6) and (7), the path loss exponent factor is calculated as:

$$n = \frac{m}{10} \tag{8}$$

Algorithm 1: Preferred Parent Node Selection()

$PreferredParent = NULL$
$MAX\_OF = Maximum\_Value$
if $Candidate\_Parent\_Table == TRUE$ **then**
   | **foreach** $CP \in Candidate\_Parent\_Table$ **do**
   | | **if** CP.OF $<MAX\_OF$ **then**
   | | | MAX_OF = CP.MMOF
   | | | PreferredParent = CP
   | | **end**
   | **end**
end
if *PreferredParent* **then**
   | Send(DAO message to PreferredParent)
else
   | Broadcast DIS control message
end

3.2 Objective Function for Mobile Node

The MN has the ability to change its location in the network. Due to the mobility, if the MN is effectively distant from its parent node then the route towards the parent node becomes inconsistent. Therefore, mobility has to be given more priority over other routing metrics. A new routing metric $Dist\_Var$ is defined to identify the change in position of the MN over a period of time. The MMOF for the MN is calculated using multiple routing metrics, i.e. ETX, $Dist$ and $Dist\_Var$ as given in Eq. (9).

$$MMOF = \alpha \frac{ETX_{curr}}{ETX_{max}} 100 + \beta \frac{Dist_{curr}}{Dist_{max}} 100 + \gamma \frac{Dist\_Var}{Dist_{max}} 100 \qquad (9)$$

where, α, β and γ are weights assigned to ETX, Dist and Dist_Var, respectively, such that $\alpha + \beta + \gamma = 1$. The weights can be altered according to the priority of the metrics. ETX_{curr} is link quality, ETX_{max} is maximum value that an ETX can take, $Dist_{curr}$ is the distance from the node, $Dist_{max}$ is the maximum distance that a node can communicate and this routing metric is given highest priority among other routing metrics, and $Dist\_Var$ is the distance variation of the node over a period of time. ETX_{curr} and $Dist_{curr}$ are calculated according to Eqs. (2) and (7), respectively, while $Dist\_Var$ is calculated using the formula given in Eq. (10).

$$Dist\_Var = Dist_{t2} - Dist_{t1} \qquad (10)$$

where $Dist_{t1}$, $Dist_{t2}$ are the distances from the node calculated at the time of receiving successive DIO messages. $Dist_{t1}$ and $Dist_{t2}$ are determined using the RSSI as per Eq. (7). Figure 1 shows the MMOF for RPL in the Contiki operating system in which the program logic uses the proposed MMOF defined for SRNs and MNs in

Fig. 1 MMOF for RPL in Contiki operating system

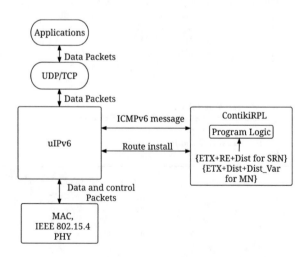

order to select the PPN. Algorithm 1 shows the procedure involved in selecting the PPN. For selecting the PPN, the node determines the path cost for each candidate parent using the MMOF and selects the candidate parent, which has the least path cost as the PPN.

4   Simulation Results and Discussion

In this section, the performance of the proposed MMOF was evaluated using the Cooja simulator provided by Instant Contiki 2.6 [5, 12]. The proposed MMOF was compared with the MRHOF [8], OF0 [17], OFE [13], and OFDE [9]. In this simulation, 1 sink node, 30 SRNs and 9 MNs were used in random topology. Routing metric *Dist* was given higher priority compared with the other metrics, and therefore, it was set to 0.4 and the others were set to 0.3. In random topology, the SRNs were placed randomly such that at least one SRN was in the communication range of another SRN, similar to an actual environment. The detailed simulation parameters used in the work are shown in Table 1. To generate mobility traces, the BonnMotion software was used [1], and the parameters are shown in the Table 2. The simulation results were obtained by varying the number of packet transmissions and trickle timer in random topology.

Reason for varying Traffic: Packets get lost when the MN moves away from the communication range of the PPN and requires retransmission of such packets to be delivered successfully to the destination. To analyse the performance of the MMOF, the packet transmission rate was set to 1 packet per 2 min, 1 packet per minute, 2 packets per minute and 3 packets per minute.

Table 1 Simulation parameters

Simulation parameter	Values
Radio model	Unit disk graph model
Transmission range	30 m
Mote type	Tmote Sky
Network size	100 m × 100 m
Node density	SRN: 30, MN: 9
Mobility model	Random waypoint
Simulation time	600 s

Table 2 BonnMotion software parameters

Settings	Area	Maximum speed	Nodes	Interval time
Value	$10{,}000\,\mathrm{m}^2$	1.5 m/s	9	10 s

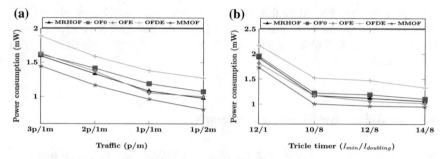

Fig. 2 Power consumption with respect to **a** traffic and **b** trickle timer

Reason for varying Trickle timer: Since the PPN is updated based on the OF obtained from the DIO message, which in turn broadcasts based on trickle timer, the trickle timer was varied to analyse the performance of the MMOF. At first, $[I_{min}, I_{doubling}]$ was fixed to $[12, 1]$ then $[I_{min}, I_{doubling}]$ was set to $[10, 8]$, $[12, 8]$, and $[14, 8]$. To measure the performance of the MMOF, power consumption, PDR and latency were considered as the evaluation metrics.

4.1 Analysis of Power Consumption

Figure 2 consumption in random topology with varying numbers of packet trans- mission (Fig. 2a) and trickle timer (Fig. 2b). In general, the power consumption is directly dependent on the number of packets transmitted over the network. As seen in Fig. 2, the power consumption decreased as the number of data packets and control packets were reduced.

The OFDE and OFE consumed more power as they required retransmission of the packets. In the case of OF0 and MRHOF, the PPN was selected irrespective of its distance to the PPN, consuming more power. In contrast, the MMOF consumed less power compared with other OFs, as it selected a node at minimum distance from the PPN and also as high ETX value that reduced the number of retransmissions.

4.2 Analysis of Packet Delivery Ratio

Figure 3 shows a slight decrease in PDR with varying number of packet transmissions (Fig. 3a) and trickle timer (Fig. 3b). In general, the PPN is updated based on the OF determined from the DIO message. Therefore, the decrease in DIO messages reduces the PDR as the packets are transmitted to an inconsistent PPN.

The OFDE, OFE and OF0 ignore the link quality and show less PDR. The MRHOF shows better PDR. However, if the node is distant and moving away from the PPN,

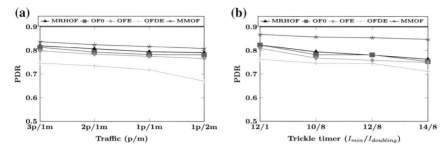

Fig. 3 PDR with respect to **a** traffic and **b** trickle timer

then the packets get lost. In contrast, the PPN is selected based on the position and link quality in MMOF, thus outperforming the other OFs.

4.3 Analysis of Latency

Figure 4 shows latency with varying number of packet transmissions (Fig. 4a) and trickle timer (Fig. 4b). In general, the latency increases if more number of packets are injected over the network, since each node requires some amount of time to process the received packet. As seen in Fig. 4, the latency is high initially and gradually decreases as a number of packets are injected into the network.

The OFE and MRHOF have higher latency as compared with the proposed MMOF. However, the MMOF has higher latency compared with OF0 and OFDE. While selecting the PPN, the MMOF does not consider the end-to-end delay to the root node.

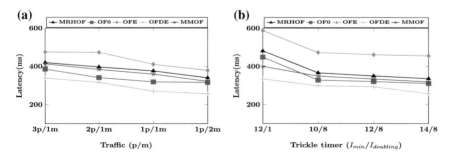

Fig. 4 Latency with respect to **a** traffic and **b** trickle timer

5 Conclusion

In this work, an MMOF was proposed by combining different routing metrics for SRNs and MNs. This MMOF minimizes the energy consumption of the nodes, while selecting a PPN based on the combination of minimum distance, maximum energy, high-quality link and direction of the node. The performance of the MMOF was evaluated and compared with the MRHOF, OF0, OFE and OFDE. The simulation results showed that the MMOF reduced power consumption, increased PDR compared with the other OFs, and smaller latency compared with the MRHOF and OFE.

References

1. Aschenbruck, N., Ernst, R., Gerhards-Padilla, E., Schwamborn, M.: Bonnmotion: a mobility scenario generation and analysis tool. In: Proceedings of the 3rd International ICST Conference on Simulation Tools and Techniques, p. 51. ICST (Institute for Computer Sciences, Social-Informatics and Telecommunications Engineering) (2010)
2. Chang, L.H., Lee, T.H., Chen, S.J., Liao, C.Y.: Energy-efficient oriented routing algorithm in wireless sensor networks. In: 2013 IEEE International Conference on Systems, Man, and Cybernetics, pp. 3813–3818 (2013)
3. Demicheli, F., Ferrari, G., Gonizzi, P.: Design, Implementation and Evaluation of an Energy RPL Routing Metric: Study of an Energy Efficient Routing Metric. LAP Lambert Academic Publishing (2014)
4. Draves, R., Padhye, J., Zill, B.: Routing in multi-radio, multi-hop wireless mesh networks. In: Proceedings of the 10th Annual International Conference on Mobile Computing and Networking MobiCom'04, pp. 114–128. ACM, New York, NY, USA (2004)
5. Dunkels, A.: Contiki Operating System (2012). http://www.contiki-os.org
6. Dunkels, A., Osterlind, F., Tsiftes, N., He, Z.: Software-based on-line energy estimation for sensor nodes. In: Proceedings of the 4th Workshop on Embedded Networked Sensors EmNets'07, pp. 28–32. ACM, New York, NY, USA (2007)
7. Gaddour, O., Koubâa, A., Abid, M.: Quality-of-service aware routing for static and mobile IPV6-based low-power and lossy sensor networks using RPL. Ad Hoc Netw. **33**, 233–256 (2015)
8. Gnawali, O.: The Minimum Rank with Hysteresis Objective Function (2012)
9. Gonizzi, P., Monica, R., Ferrari, G.: Design and evaluation of a delay efficient RPL routing metric. In: 2013 9th International Wireless Communications and Mobile Computing Conference (IWCMC), pp. 1573–1577 (2013)
10. Instruments, T.: Cc2420 datasheet. Reference SWRS041B (2007)
11. Iova, O., Theoleyre, F., Noel, T.: Improving the network lifetime with energy-balancing routing: application to RPL. In: 2014 7th IFIP Wireless and Mobile Networking Conference (WMNC), pp. 1–8 (2014)
12. Jevtić, Miloš and Zogović, Nikola and Dimić, Goran.: Evaluation of wireless sensor network simulators. In: Proceedings of the 17th Telecommunications Forum (TELFOR 2009), Belgrade, Serbia. pp. 1303–1306 (2009)
13. Kamgueu, P.O., Nataf, E., Djotio, T.N., Festor, O.: Energy-based metric for the routing protocol in low-power and lossy network. In: SENSORNETS, pp. 145–148 (2013)
14. Ko, J., Terzis, A., Dawson-Haggerty, S., Culler, D.E., Hui, J.W., Levis, P.: Connecting low-power and lossy networks to the internet. IEEE Commun. Mag. **49**(4), 96–101 (2011)
15. Kumar, P., Reddy, L., Varma, S.: Distance measurement and error estimation scheme for rssi based localization in wireless sensor networks. In: 2009 Fifth International Conference on Wireless Communication and Sensor Networks (WCSN), pp. 1–4 (2009)

16. Levis, P., Patel, N., Culler, D., Shenker, S.: Trickle: a self-regulating algorithm for code propagation and maintenance in wireless sensor networks. In: Proceedings of the 1st Conference on Symposium on Networked Systems Design and Implementation NSDI'04, USA, vol. 1, pp. 2–2 (2004)
17. Thubert, P.: Objective function zero for the routing protocol for low-power and lossy networks (RPL) (2012)
18. Tripathi, J., de Oliveira, J.C., Vasseur, J.P.: A performance evaluation study of rpl: Routing protocol for low power and lossy networks. In: 2010 44th Annual Conference on Information Sciences and Systems (CISS), pp. 1–6 (2010)
19. Winter, T.: RPL: IPv6 routing protocol for low-power and lossy networks (2012)

Analysis of User's Behavior Using Markov Model

X. Arputha Rathina, M. Ponnavaikko, K. M. Mehata and M. S. Kavitha

Abstract This paper proposes to model the dynamic interaction of the complex emotions excited by audiovisual stimuli. The behavior of the user is modeled through Markov chain procedure, and the transition probabilities are calculated. A parameter variation study of the dynamical behavior is undertaken to examine the scope of chaos and stability, thus leading to the derivation of the extent of stability. The arch motive of the paper lies in the prediction of chaos and stability of emotional dynamics from the instantiation of emotion in facial expression stimulated by external audiovisual impetus. Such prediction may help in the early discovery of mental disorders.

Keywords Image processing · Emotional intelligence · Markov model
Lyapunov exponent

1 Introduction

Emotional intelligence is a wide arena which has evolved over time growing out of a large number of theories. The theories have gained their scope from understanding of various emotions. An emotion is basically a state which is got through the arousal of variety of feelings, behaviors, and thoughts.

X. Arputha Rathina (✉) · K. M. Mehata
B. S. Abdur Rahman Crescent Institute of Science and Technology, Chennai, India
e-mail: xarathna@gmail.com

K. M. Mehata
e-mail: mehatakm@gmail.com

M. Ponnavaikko
Bharath Institute of Science and Technology, Chennai, India
e-mail: ponnav@gmail.com

M. S. Kavitha
Kyungpook National University, Daegu, South Korea
e-mail: mkkavi14@gmail.com

© Springer Nature Singapore Pte Ltd. 2019 405
J. C. Bansal et al. (eds.), *Soft Computing for Problem Solving*, Advances in Intelligent
Systems and Computing 817, https://doi.org/10.1007/978-981-13-1595-4_32

Fig. 1 Multi-emotional interactions

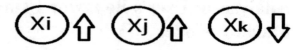

Many physiologists attribute the hormonal and physiological activity of the human body to be the main cause of the creation of emotion. Emotions are evolved from four phases as believed by the psychologists such as cognition, evaluation, motivation, and feeling.

Cognitive component matches a given situation to an already-registered situation and sends a message to the component of evaluation to evaluate the emotion from a suggested set. This has a great significance in determining the impact of emotion as a positive or a negative one. The motivational component allows a subject to maintain or bring change in his emotional state. The feeling component includes the sensation of the body and the various moods.

Emotion detection was once regarded as an unworthy entity incapable of being a topic of research. Later theories experimented with emotions and revealed that emotions were valuable component of the brain and its expression provides an opportunity to detect the physiological nature of mankind [1, 2].

Humans compare their current situations with previous situations, and when the level of current stimulation far exceeds the level they have experienced long enough to get accustomed to it, arousal of emotion takes place. Experimental observations reveal that arousal of common/simple emotions, such as sadness, happiness, disgust, fear, love, hatred, and even sexual desire, supports the above phenomenon.

1.1 Communication Between Emotions

Emotions can be triggered by means of suitable audiovisual stimulus. There exist two types of complex emotions—competitive and cooperative emotions. When two simple emotions arise at the similar time, they produce an emotion which is either one of the complex types. A person's change in the physiological state either accelerates or retards the arousal of emotions.

Let X_i, X_j, and X_k be three states that describe the concentration of individual emotions [1]. Suppose X_j for some j co-operates with X_i, while X_k, for some k, competes with X_i. In other words, the growth rate of X_i will be accelerated with an increase in X_j, but will be decelerated with an increase in X_k (Fig. 1).

1.2 Applications of Emotional Stability Identification

Stability analysis of a subject may serve as a valuable tool for dealing with patients who are suffering from mental disorders such as trauma or epilepsy. Further, such stability identification is used to assess stress tolerance of a person at the time of interviews or in normal walk of life.

1.3 Markov Model Prediction

Prediction of the stability assessment is done through a Markov model which assumes a Markov property. The model is defined for stochastic nature of the process and satisfies memorylessness and stationary properties.

2 Methodologies

2.1 Existing System

Identification of chaos and stability was done initially using conventional methods such as electroencephalogram (EEG), functional magnetic resonance imaging (fMRI), and positron emission tomography (PET scan).There is also an ongoing research which employs fuzzy logic, neural networks, and many statistical methods [3, 2]. These methods are found to be tedious and time-consuming, making it obvious for a need of a better method.

2.2 Proposed System

Emotion detection is carried out using two methods—one by mathematical analysis method using Lyapunov exponent [3] and the other by employing a Markov model. Lyapunov exponent models a graph which is used to give the result of positive or negative stability, whereas the Markov model predicts the percentage of stability using the idea of transition probabilities [4].

The overall architecture diagram is shown in Fig. 2. Stimuli video is created by combining various emotion triggering video clips. The users are allowed to watch the stimuli video. The user's response to the stimuli video is recorded using a webcam. The recorded user's behavior is used as the input to the system. The system analyzes the facial features and the areas occupied by the eye and mouth regions. Using the eye opening and mouth opening, the emotions are analyzed using Lyapunov analysis.

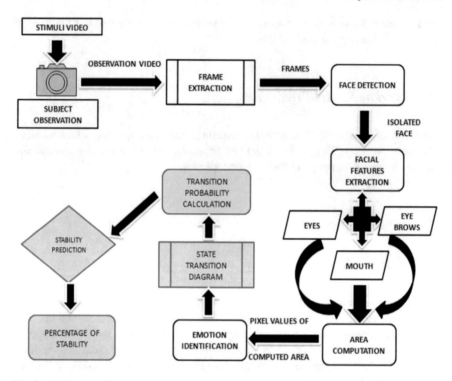

Fig. 2 Architecture diagram

3 Methods

3.1 Stimuli Video Creation

Different emotion triggering video clips, each of which evokes a particular emotion, are combined into a single long video, thereby creating the required stimuli video (Fig. 3).

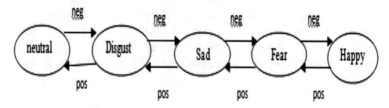

Fig. 3 Emotion triggering sequence

3.2 Observation Video Collection

The user is made to watch the stimuli video, and the emotions exhibited by the user are recorded into a single video. This recorded video is given to the system as input.

3.3 Frame Extraction

The observation video is segmented into multiple frames, and the required frames for face detection are extracted. Initially, one frame per emotion was extracted. Later, the observation was that subjects did not universally respond at the same instance to the stimuli video. Hence, five frames per emotion were taken, thereby giving a wider window of opportunity for capturing the instance of emotion. It also simplified the analysis.

3.4 Face Detection and Isolation

The face is identified from a set of selected frames and isolated to a separate image file by using image cropping function.

YCbCr color model is used which plots a red line-box around a face detected based on the skin color. Two or more faces might be detected when a similar pattern is observed in the frame, but this problem is resolved by processing the face closest to the camera.

3.5 Facial Feature Extraction

From the cropped image file containing the face and the pixel locations of facial features (eyebrows, eyes, and mouth) are tracked. Smiling detection algorithm [5] is used for this extraction which postulates that each of the facial features is symmetrically located at a proportionate distance from the forehead. This simplified the task of tracking the facial features.

3.6 Area Computation

Using the pixel locations, bounding boxes are drawn around the facial features and their areas are computed based on edge detection algorithm which converts a colored image into a black-and-white image, where edges of each facial feature are clearly identified. Edges are displayed in white.

For computing the area of each eye, it locates the first pixel from the top, which is upper eyelid and the first pixel from the bottom, which is the lower eyelid. Then, it computes the distance between these two pixels, which gives the width of the eye. This is the eye opening (EO).

For computing the area of the mouth, it finds the length/width and height between the upper and lower lips and finds its product which gives the area of the mouth. This is the mouth opening (MO).

For computing the position of the eyebrow, identification is based on the region where the eyebrow pixels are located. Discreetly we give the values 1 and 0 for upper and lower regions, respectively, where the eyebrows are found.

3.7 Emotion Identification

The areas of facial features obtained from the image are compared with the reference table, and the corresponding emotions are identified.

3.8 Stability Analysis

The area values of the facial features are substituted in the Lyapunov exponent equation, and chaos is predicted using the result theoretically. A graph is plotted for each value obtained after substitution, and the final results are obtained.

The probabilities are defined for each transition for different emotional states, and the state transition diagram is drawn for each observation video using the concept of the Markov model. By calculating the total probabilities, the percentage of the emotion exhibition of a subject can be predicted with precision (Fig. 4).

4 Implementation

4.1 Development Process

The application is first started by the creation of the stimuli video. This video is a collection of emotion triggering clips appended together to make the subject react in different manners. The observation video, which is the video that records these reactions, is done next. The video is then analyzed, and frames are extracted. The faces of the subject are extracted from each frame, and facial features are isolated. Based on the areas of the facial features, emotions are identified, and finally, stability is analyzed.

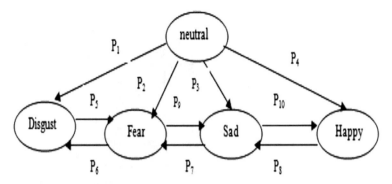

Fig. 4 State transition diagram

4.2 Stability Analysis

The stability of a subject is inferred from the result of the Lyapunov equation and by Markov model. Theoretically, when using the Lyapunov equation, the area values of each facial feature are substituted into it and the results are tabulated. The time difference values are then calculated and tabulated. A graph is then charted using these time difference values for each emotion. If the graph remains mostly on the negative side of the x-axis for all values, the person is considered to be unstable [2, 3]. Another type of analysis is done by using the transition probabilities of the Markov model. The probabilities are defined for each transition depending on the priority of the emotions traversal in the stimuli video [4]. The calculation of probabilities is done on the basis of the formula

$$P_1 = \frac{P_1}{P_1 + P_2 + \cdots + P_n}$$

where

P_1 Required probability
$P_1 + P_2 + \cdots + P_n$ Total probability

Using this, all the necessary probabilities are calculated. These probabilities are used in the corresponding state transition diagram to give an estimate of the percentage of the subject's stability.

4.3 Face Detection

For detecting the emotion of the subject, the face and the facial features have to be identified. The face detection technique identifies the face. This process involves the skin color detection and isolation. It combines motion detection by spatiotemporal

Fig. 5 Face detection

filtering with an appearance-based face model in the form of a neural net. Color is used as a cue for detection, and tracking is described. Color provides a computationally efficient yet effective method which is robust under rotations in depth and partial occlusions. It can be combined with motion and appearance-based face detection. Human skin forms a relatively tight cluster in color space even when different races are considered (Fig. 5).

4.4 Emotion Triggering

In order to arouse mixed emotions, audiovisual stimulus is used to cause the arousal of different emotions. In other words, persistence of more than one emotion can be maintained by exciting subjects with one base emotion, superimposed with other relatively short-duration emotions. For instance, a base emotion of happiness, when perturbed by a burst of disgust and fear in time-succession, represents a coexistence of multiple emotions for a certain time frame. After the mixed emotions are synthesized, three important facial attributes, such as mouth opening (MO), eye opening (EO), and eyebrow constriction (EBC), are determined from the facial expression of the subject. This needs segmentation of eye, mouth, and eyebrow regions and their localization. The exact range of these three adjectives depends on individual's personality, culture and upbringing, and country/state where he/she belongs to. In order to keep the measurements error-free, we normalized the measurements by dividing the actual measurement of MO, EO, and EBC by its largest experimental value for the given subject [2, 3].

Table 1 Emotions based on facial feature areas

Emotion transition	Eye opening and eyebrow constriction	Mouth opening
Neutral -> disgust	Increases	Increases
Neutral -> fear	Increases	Decreases
Neutral -> happy	Decreases	Increases
Neutral -> sad	Equal	Equal

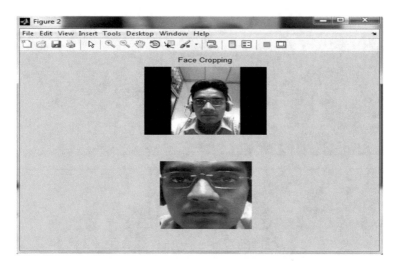

Fig. 6 Face cropping

4.5 Facial Feature Extraction

Human skin maintains a standard range in hue, saturation, and value, thereby providing a simple means to segment skin region from the faces. After the skin region is segmented, next we go for segmentation of eye, eyebrows, and lip region. Lip region in a colored image can be easily segmented because of its distinctive color profile that segregates it from the skin color. Segmentation of the eye region is performed in two phases. Firstly, we obtain the third dark-most subregion in the monochrome version of a given facial image, and then, we grow the subregion, so as to identify the region containing dark eyeballs. Similarly, segmentation of the eyebrow region is performed by determining the second dark-most region from top of the face, starting from the hair region. A localization algorithm is also undertaken to localize the segmented regions in the image. After segmentation and localization, we determine the eye opening (EO), mouth opening (MO), and eyebrow constriction (EBC) from the counting of pixels of our interest. Based on the size range of the features, the emotion can be identified as given in Table 1 [2, 3] (Figs. 6 and 7).

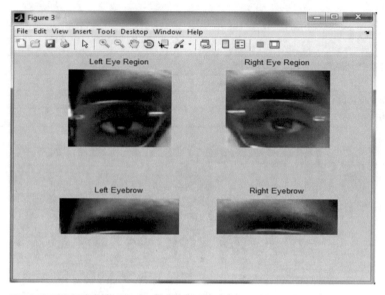

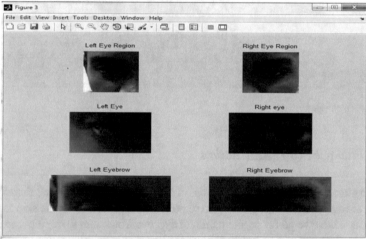

Fig. 7 Eye and eyebrow detection

5   Experimentation and Testing

5.1   Theoretical Testing

Lyapunov exponent is normally calculated for difficult weight setting of the discrete dynamics. When Lyapunov exponent is found to be positive, the dynamics is said to have strong chaotic behavior. Harter et al. considered different parameter sets of

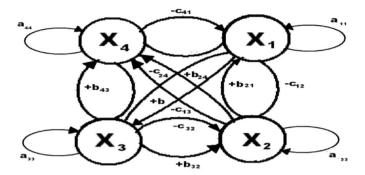

Fig. 8 State diagram for Lyapunov equation

the discrete dynamics and plotted the Lyapunov exponents against various weights scaling in the range 0–1. Their graphical demonstration envisages that for most weight scaling, the Lyapunov exponent is positive indicating the positive behavior of the dynamics [2].

The state diagram for the Lyapunov equation is given as (Fig. 8).

The Lyapunov equation is given as follows:

$$\frac{dX_i}{dt} = a_{ii}X_i\left(1 - \frac{X_i}{K}\right) + \sum_{\exists j} b_{ji}X_i\left(1 - exp\left(-\beta_{ji}X_j\right)\right) - \sum_{\exists k} c_{ki}X_i\left(1 - exp(-\lambda_{ik}X_k)\right)$$

(1)

where $i = 1, 2, ..., 4, j = 1, 2, ..., 4$.

X_i—self-growth emotional state, a_{ii}—inertial coefficient that regulates the self-growth of X_i. $(1 - X_i/K)$—controlling term that selects the sign of growth rate of a_{ii}. When $X_i < k$, the first term is positive; when $X_i = k$, the first term is zero; otherwise, it is negative. The second term represents the cooperation between X_i and X_j. The third term represents the competition of X_i with growing X_k. Parameters β_{ji} and λ_{ik} control the growth of X_i and X_j.

The difference equation values for various time frames are obtained and plotted against time. If the line remains above zero, then the subject is said to be stable, satisfying the condition for stability,

$$dx/dt > 0;$$

(2)

The duration of stimuli video is 3 min, and details of the frames at 50, 261, 580, 696, 725, 783, 2726, 2755, 2958, 3190, 3364, and 3480 were taken. The details involve mouth opening, eye opening, and eyebrow information. These mouth opening $x_2(0)$ and eye opening $x_1(0)$ are substituted.

Table 2 shows the sample parameters of facial extract of a subject (in pixels).

Substituting the values in the Eq. (1), we get positive or negative values as results. For instance, we know that:

Table 2 Parameters of facial extract of a subject

Feature	T0	T1	T2	T3	T4
EO	32	54	69	102	51
MO	1222	897	780	1110	876

$$dx/dt = a_{11}X_1(1 - X_2/k); \tag{3}$$

$$k = 1000, X_{21}(1) = 30, X_1(0) = 32, X_2(0) = 1222; X_{21}(1) = 1, \quad \lambda_{21} = 0.0045 \tag{4}$$

$$=> X_{21}(1) = X_1(0) + a_{11}X_1(0) + a_{11}X_1(0)(1 - X_2(0)/k); \tag{5}$$

$$=> 30 = 32(1 + a_{11}(1 - 32/1000); \tag{6}$$

$$=> a_{11} = -0.064; \tag{7}$$

$$=> dx_1/dt = a_{11}X_1(1 - X_1/k) - c_{12}X_1(1 - \exp(-\lambda_{21}X_2(2)) \tag{8}$$

$$=> X_1(2)/X_1(1) = 1 + a_{11}(1 - X_1(1)/1000) - c_{12}(1 - \exp(-\lambda_{21}X_2(2)); \tag{9}$$

Substituting the values for the variables,

$$C_{12} = -0.062; \tag{10}$$

Using the values of a_{11} and c_{12} in the primary equation, we get,

$$X_1(2)/X_1(1) = 1.0001. \tag{11}$$

Similarly, when we apply the equation for other scenarios of dx_i/dt, we get

$$X_1(3)/X_1(2) = 0.9939; \tag{12}$$
$$X_1(4)/X_1(3) = 0.93792; \tag{13}$$
$$X_1(5)/X_1(4) = 0.9377; \tag{14}$$
$$X_1(6)/X_1(5) = 1.000; \tag{15}$$
$$X_1(7)/X_1(6) = 0.9939; \tag{16}$$
$$X_1(8)/X_1(7) = 0.9381; \tag{17}$$
$$X_1(9)/X_1(8) = 0.9381; \tag{18}$$
$$X_1(10)/X_1(9) = 0.90; \tag{19}$$
$$X_1(11)/X_1(10) = 1.0001; \tag{20}$$
$$X_1(12)/X_1(11) = 1; \tag{21}$$

When a graph is plotted with these values with dx/dt on the x-axis and time on the y-axis, we get a graph like this (Fig. 9).

Since dx/dt remains above 0 completely, the subject is assumed to be stable by the condition of stability.

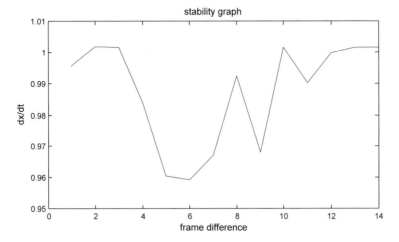

Fig. 9 Graph showing stability state (dx/dt $\geq$ 0)

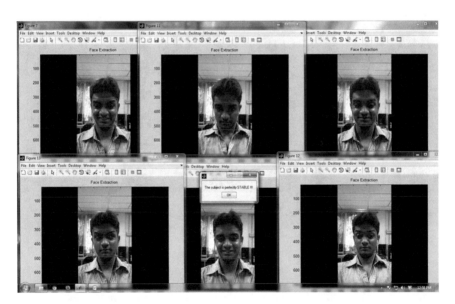

Similarly, the graph is obtained for a chaotic person as below [2], where the values are either 0 or –ve (Fig. 10).

The duration of stimuli video is 3 min, and details of the frames at 50, 261, 580, 696, 725, 783, 2726, 2755, 2958, 3190, 3364, and 3480 were taken. The details involve mouth opening, eye opening, and eyebrow information. These mouth opening and eye opening are substituted in the above equations, and finally with a help of a graph, we have found whether the person is in NORMAL state or in CHAOTIC state (Fig. 11).

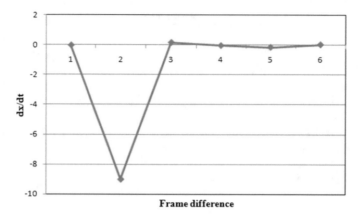

Fig. 10 Graph showing chaotic state (dx/dt $\leq$ 0)

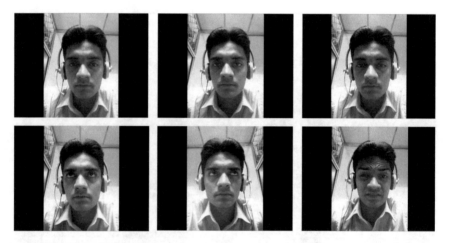

Fig. 11 Images at specified frames

5.2 Theoretical Testing Using Markov Model

Similar to the Lyapunov exponent, Markov model is said to model the dynamic behavior of the emotions using transition probabilities. The transition probabilities are defined for each transition from one emotional state to another in a defined period of time. The transitions are numbered based on the likeliness of transition of the states. For example, when a transition from a state S1 to state S2 has a higher likeliness of transition, the probability is labeled above 0.5 and the rest of the lower probability transitions are divided equally, making the total cost of probability to be 1 (Fig. 12).

By calculating the total probability from a set of transitions, the score of the person can be calculated. Thus, the percentage of the stability can be calculated

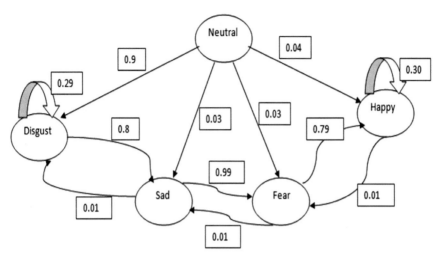

Fig. 12 Transition diagram with probabilities

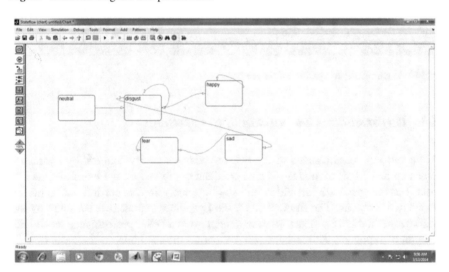

Fig. 13 Transition diagram of emotional states

easily using the score. The transition diagram is drawn using the emotion exhibited in the frame. The emotion shown in each frame is calculated by using the reference value of the mouth opening, eye opening, and the eyebrow constriction obtained in the initial implementation. Stability is assessed depending on the percentage. Figure 13 represents the transition diagram of emotional states of a stable person.

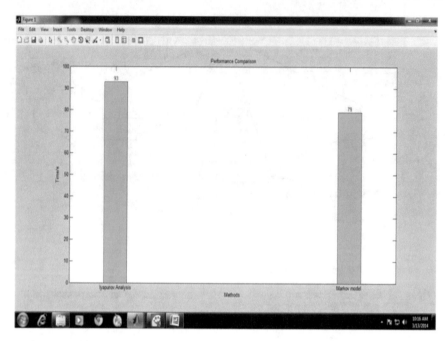

Fig. 14 Performance comparison of the methods

5.3 Performance Comparison of the Methods

At the end, the performance of the two methods—stability analysis by Lyapunov exponent and Markov model—is analyzed and compared on the basis of time criteria. For the given stimuli video, the stability analysis was carried out using the above two methods. The time taken by the Lyapunov exponent is 92 s and by the Markov method is 78 s. Thus, Markov model gives a better performance result than Lyapunov exponent. Furthermore, on evaluating the methods on the basis of efficiency, Markov model shows the clear interpretation of stability on the grounds of percentage, whereas Lyapunov exponent gives only two results—stable or unstable. In conclusion, Markov model serves as a more efficient method of the analysis of stability (Fig. 14).

5.4 Test Cases

If the face in the observation video is not clearly centered and visible, it leads to the out-of-bound error while extracting the frames from the video. This impedes the further analysis of the frames. The other test cases are subjects with head cover and spectacles (Figs. 15 and 16).

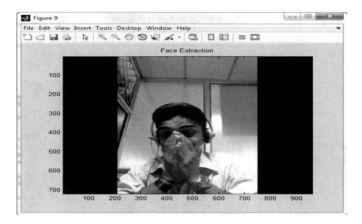

Fig. 15 Out-of-bound error

Fig. 16 Face detection with head cover

The system detects the face when the object has head covers, but when the subject has objects obstructing the view of their eyes such as spectacles, it leads to an altered computation of the eye area. When the spectacle is frameless, the system works similar to that of the subject without an obstacle (Fig. 17).

The system also gives the out-of-frame error message when the subject is not in the capturing region or not focused on the stimuli video. The success rate of the system was calculated using the ratio of number of successful cases to the number of cases taken for testing.

Fig. 17 Face out of frame

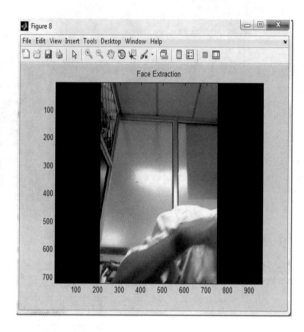

6 Conclusion

This paper aims to accurately identify the user's response to the stimuli video. The system is highly useful in order to identify psychological deficiencies of people who have recently suffered a mental disorder. This system is however only used as a *simple diagnostic tool* and cannot be used as a substitute for medical tools such as EEG and fMRI [1]. fMRI is the functional magnetic resonance imaging/functional MRI, which can be used to measure the brain activity by detecting associated changes in the blood flow.

6.1 Future Enhancements

In future, this application can be extended onto a mobile platform to develop an application using android technology, where the front camera of the mobile phone can be used to record the reactions of the subject, while it displays the stimuli video on its screen. The processing can be done remotely using a server-based system. This way the results can be obtained remotely without the need of heavy equipment to be carried around. It can also be made openly available as a commercial application to mobile application stores to benefit a greater population.

Acknowledgements
• Statement of Ethical Approval
B. S. Abdur Rahman Crescent Institute of Science and Technology; Bharath Institute of Science and Technology, Chennai, India; Kyungpook National University, South Korea, approved our study.
• Statement of Consent
All authors consent to publish this article, and there is no any conflict of interest.

References

1. Chakraborty, A., Konar, A.: Emotional Intelligence: A Cybernetic Approach. Springer, Heidelberg (2005)
2. Sanjeetha Sanofer, R., Arputha Rathina, X.: Chaos & stability in dynamic behavior of multi emotion interactions. In: ACEEE Prof of International Conference on Recent trends in Signal processing, Image processing & VLSI, ICtrSiv (2014)
3. Ghosh, M., Chakraborty, A., Konar, A., Nagar, A., Janarthanan, R.: Chaos and Stability in the Dynamic Behavior of the Multi-emotion Interactions, 978-1-4244-2963-9/08. IEEE (2008)
4. Shukla, D., Singhai, R.: Analysis of Users Web Browsing Behavior Using Markov chain Model (2011)
5. Rai, P., Dixit, M.: Smile detection via Bezie curve of mouth interest points. Int. J. Adv. Res. Comput. Sci. Softw. Eng. (2013)

Analysing Image Compression Using Generative Adversarial Networks

Amit Adate, Rishabh Saxena and B. Gladys Gnana Kiruba

Abstract Most image compression algorithms rely on custom-built encoder–decoder pairs, and they lack flexibility and are differing to the data being compressed. In this paper, we have elaborated on the notion of generative compression by implementing various compression techniques using a generative adversarial network. We have also analysed the compression approaches that are implemented using deep learning. Our experiments are performed on the handwritten digits database and are yielding progressive results with both conditional and quantifiable benchmarks.

1 Introduction

Traditional image compression techniques rely on a variety of different ways to compress and decompress images which can be compared using image compression metrics to figure out if the compressed image is economically feasible to export using a certain technique. These compression techniques involve varying degrees of transformations on the base image to either produce two types of traditional image compression techniques, namely lossy and lossless.

A. Adate · R. Saxena (✉) · B. Gladys Gnana Kiruba
Vellore Institue of Technology, Vellore, India
e-mail: rishabh.saxena2014@vit.ac.in; cjrishabhsaxena@gmail.com

A. Adate
e-mail: adateamit.sanjay2014@vit.ac.in

B. Gladys Gnana Kiruba
e-mail: gladys@vit.ac.in

© Springer Nature Singapore Pte Ltd. 2019
J. C. Bansal et al. (eds.), *Soft Computing for Problem Solving*, Advances in Intelligent Systems and Computing 817, https://doi.org/10.1007/978-981-13-1595-4_33

1.1 Lossy Compression

Lossy image compression techniques refer to the methods which can used to compress images by reducing the total amount of information present in any image. This causes the images to be compressed with a form of data loss that is mostly invisible to the human vision, for which the compression technique has been designed for, but sometimes does produce some degree of unwanted noise in the structure. Loss image compression uses techniques like Transform Encoding [1] which uses information about the image prior to compression and then discards any information that does not suit the intended purpose; Sub-Band Encoding [2] which breaks the image signal into multiple separate signals and encodes each one differently, where the encodings can be same for all signals or different for each sub-signal depending on preference of encoding; Block Truncate Encoding [3] which divides the image into certain "blocks" and uses a quantizer to reduce grey levels in each block while preserving the mean and standard deviation; Vector Quantization [4] that uses the distribution in the form of vectors to prototype the compression in an image; and Fractal Encoding [5] which uses the Hutchinson operator to compress images by keep the same domain while encoding the image where the image is scanned for blocks and then hashed to a function in the same domain.

These methods follow one of the three operations to compress images. These methods usually involve predicting neighbouring pixel values and to subtract that image for the residual image to form an image that is compressed in size as compared to the original image, transforming the image into a less redundant form that can later be encoded/decoded into more readable format in a more relevant information environment; and quantization compresses the data bits that belong to a range of values into a single quantum of data and reduces the overall pixel range data that is needed to encode data.

1.2 Lossless Compression

Another paradigm used for data compression is a lossless form of data compression that, as the name suggests, is opposite to the lossy data compression previously mentioned. In this compression methodology, the data is compressed by preserving the pixel data and using other more sophisticated data methods to reduce size. These techniques include Statistical Encoding [6] which is a compression technique that uses entropy coding which recognizes statistical patterns in the data to be compressed and takes advantage of these patterns to carry out the compression; Bit Plane Encoding [7] which takes the most significant bit from the binary values of the pixel intensities and compresses them in a plane of bit positions; Predictive Encoding [8] which encodes characters by their frequency of occurrence and whether they will appear ahead or not; and Run Length encoding [9] which runs along the data and stores similar data in the form of count and data instead of the entire repetitive data.

1.3   Recent Trends

In a more recent scenario, compression techniques have improved to accommodate newer technologies and provide better performance. One metric for performance measurement is the compression ratio which is a widely accepted scoring format that uses a ratio between the size of the original image and the compressed image.

A recent breakthrough in image compression techniques comes from the wavelet transformation techniques which use multiwavelets to transform the image from one spatial domain to the other where the latter is less spatially intensive to store. Martin et al. [10] have shown that multiwavelet transformations that selectively choose specific desirable properties in an image such as orthogonality and symmetry are taken and processed by wavelet function. Their limitations include not being able to choose all the desirable properties in an image which can cause a loss of data, hence they come under lossy image compression, and can cause perturbations in an image due to loss of the aforementioned properties.

Another recent image compression technique introduced by Weinberger et al. [11] called the LOCO-I or the Low Complexity Lossless Compression for Images is an image compression technique that is a variation of the JPEG-LS [12] compression technique for continuous images. It is based on a simple fixed context model which can attain complex universal techniques for capturing high-order dependencies. It can effectively code low-entropy images for near-lossless compression. The obvious drawback for the algorithm is its inability to reach compression levels that are attainable by lossy compression techniques.

Though they enjoy the best of both worlds, this paper dives into a more unique compression techniques that can work on near-lossless quality domain and still present a compression ratio that is comparable to any lossy compression model. As such, it is a difficult task to perform both these operations in a single model, hence we used a more generative approach to the compression mechanism.

1.4   Deep Learning

Before diving into the generative field, it should be noted that neural network-based compression techniques have a pre-existing history which can be related to the concept of various deep learning models such as sparse autoencoders and sparse coding [13], denoising autoencoders [14], deconvolutional networks [15], restricted Boltzmann machines [16], deep Boltzmann machines [17], generative adversarial networks [18] and variational autoencoders [19].

A more detailed look at variational autoencoders used for conceptual compression [20] shows that they can be used to store a data distribution in a latent variable that can be extracted later by a distribution that can be used for classifying objects. This approach can be used to store data in a latent variable that can later be decoded as a compressed image. Since a single pass cannot be used in such architectures, Gregor et al. proposed a recurrent feedback mechanism called DRAW that takes the image information, stores it in a latent variable and corrects itself during reconstruction using that variable. This algorithm is then combined with a convolutional architecture for the first network to form a Convolutional DRAW [20] that uses the following at one level of convolution:

Input x, reconstruction r, reconstruction error ε, the state of the encoder recurrent net h^e, the state of the decoder recurrent net h^d and latent variable z. The variables h^e, h^d and r are recurrent (passed between different time steps) and are initialized with learned biases. Then at each time step $t \in \{1, T\}$, convolutional DRAW performs the following updates:

$$\epsilon = x - r$$

$$h^e = Rnn(h, \epsilon, h^e, h^d)$$

$$z\,q = q(z|h^e)$$

$$p = p(z|h^d)$$

$$h^d = Rnn(z, h^d, r)$$

$$r = r + Wh^d$$

$$L_t^z = KL(q|p)$$

After training their model on ImageNet, they reached at a compression ratio that was comparable to that of JPEG2000 [21] which shows promise in using neural engines for image compression.

1.5 Generative Adversarial Networks

In this paper, we analyse a more robust and efficient generative algorithm designed on the concept of generative adversarial networks, which uses adversarial learning to train a pair of networks known as the generator and discriminator that learn by using optimizing their loss functions in conjunction with one another to improve both networks until they reach minima. This method is similar to the aforementioned one but uses a more recent and powerful training method known as the adversarial learning method. This allows for a more efficient network that can generate more accurate and less lossy images of the image set that it was trained on. Another major advantage of such networks is their resilience to noisy channels that cause bit errors in the images being compressed (Fig. 1).

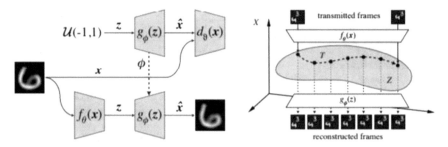

Fig. 1 Architecture for generative compression [22]

2 Experiment

This paper works to implement a model that uses a combination of an autoencoder and a generative adversarial network. We work on the previous work of Santurkur et al. [22] which uses a similar architecture to generate compressed images by using the generator function from the adversarial network as the input function of the autoencoder.

A DCGAN is first pre-trained, and its generator is then used as the decoder for the autoencoder. The input to the encoder of the autoencoder are raw images which are then turned into latent-variable vectors and used as the input for the generator function which then generates the compressed images.

Our implementation of the model differs slightly where we used shared weights for the generator and the Discriminator while training the initial DCGAN. We used various optimization techniques to improve the training and convergence of the network. We used Batch Normalization [23] while training the network to provide a faster and more efficient way to train the network while reaching a high number of epochs. We also implemented dropout [24] to reduce overfitting in the model.

We trained on the MNIST database due to its widely available results and taking into consideration computational limits of hardware while testing a new architecture. We used 28×28 inputs in the first layer of the network along with 100 hidden layer units in the vanilla encoder of the autoencoder.

We then trained a DCGAN using shared layers containing a generator function $g_\phi : Z \to X$ which is trained along with the discriminator function $d_\vartheta : X \to [1, 0]$.

We then trained the autoencoder using the function $f_\theta : X \to Z$ where X is the state space used by the generator function of the shared-DCGAN while Z is the state space of the uniform prior $\mu[-1, 1]$ that is used to decode the image with the generator. This allows the generator and the encoder to work in conjunction to provide with the compressed image. We then train the encoder to reduce the noise using $L(x, g_\phi(f_\theta(x)))$ that produces better compressed outputs. The encoder produces a N-latent-variable vector, where N is 100, that is used by the generator to create the final output (Fig. 2).

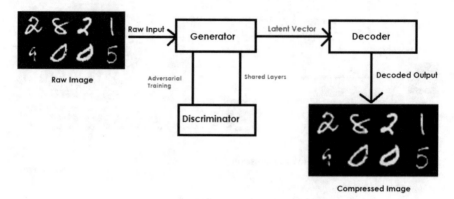

Fig. 2 Diagram for the architecture using generator and decoder of the network

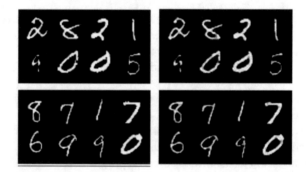

Fig. 3 (Left to Right) Original MNIST database images, compressed images using shared layer generative compression technique

To view our experiments and images generated: https://github.com/amitadate/gan-noise.

We trained the network with a combined loss function of the pixel loss and the perceptual loss [25] which encompasses both the structure and the texture of the image being compressed. The pixel loss function allowed the autoencoder to create a latent space vector which was similar to the original image while the perceptual loss allows the network to learn the similarities of the images being compressed [26] (Fig. 3).

3 Results

We concluded that the network learns to compress images as a compression ratio of 1.0309 with a RSNR of 50 where the input images were MNIST images taken from the test set. We implemented the graph for the following bits per pixels as shown in the Fig. 4.

Fig. 4 Realistic SNR versus compression ratio for the network

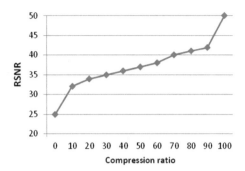

Comparing this data to those of previously used compression techniques like JPEG2000, TIFF and PNG, we see a significant difference in the data reduction as well as better quality of the compress images was observed. This allows for the architecture to be of significant value in compression since the shared layers all for a smoother training experience.

A more viable way for implementing such a technique would be to use cached models on the sender and the receiver side, which would store low-dimensional object embeddings in it's manifold, containing common images as it's dataset. This could replace traditional compression techniques for a more human-like image transferring system.

References

1. Raid, A.M., Khedr, W.M., El-dosuky, M.A., Ahmed, W.: JPEG image compression using discrete cosine transform—a survey. CoRR, arXiv:1405.6147 (2014)
2. Roy, S., Gupta,D.B., Chaudhuri, S.S., Banerjee, P.K.: Studies and implementation of subband coder and decoder of speech signal using Rayleigh distribution. In: Emerging Trends in Computing and Communication, pp. 11–25. Springer (2014)
3. Delp, E., Mitchell, O.: Image compression using block truncation coding. IEEE Trans. Commun. **27**(9), 1335–1342 (1979)
4. Nasrabadi, N.M., King, R.A.: Image coding using vector quantization: a review. IEEE Trans. Commun. **36**(8), 957–971 (1988)
5. Miar Naimi, H., Salarian, M.: A fast fractal image compression algorithm using predefined values for contrast scaling. CoRR, arXiv:1501.04140 (2015)
6. Liu, S., Zhang, Z., Qi, L., Ma, M.: A fractal image encoding method based on statistical loss used in agricultural image compression. Multimed. Tools Appl. **75**(23), 15525–15536 (2016)
7. Chakraborty, S., Jalal, A.S., Bhatnagar, C.: An efficient bit plane X-OR algorithm for irreversible image steganography. CoRR, arXiv:1410.3117 (2014)
8. Choi, M., Tani, J.: Predictive coding for dynamic visual processing: Development of functional hierarchy in a multiple spatio-temporal scales RNN model. CoRR, arXiv:1708.00812 (2017)
9. Zaman, N., Pippenger, N.: Asymptotic analysis of run-length encoding. CoRR, arXiv:1504.04070 (2015)
10. Martin, M.B., Bell, A.E.: New image compression techniques using multiwavelets and multiwavelet packets. IEEE Trans. Image Process. **10**(4), 500–510 (2001)

11. Weinberger, M.J., Seroussi, G., Sapiro, G.: The loco-i lossless image compression algorithm: principles and standardization into JPEG-ls. IEEE Trans. Image Process. **9**(8), 1309–1324 (2000)
12. Rane, S.D., Sapiro, G.: Evaluation of JPEG-ls, the new lossless and controlled-lossy still image compression standard, for compression of high-resolution elevation data. IEEE Trans. Geosci. Remote Sens. **39**(10), 2298–2306 (2001)
13. Andrew N.G.: Sparse autoencoder. CS294A Lect. Notes, **72**(2011), 1–19 (2011)
14. Vincent, P., Larochelle, H., Bengio, Y., Manzagol, P.-A.: Extracting and composing robust features with denoising autoencoders. In: Proceedings of the 25th International Conference on Machine Learning, pp. 1096–1103. ACM (2008)
15. Zeiler, M.D., Krishnan, D., Taylor, G.W., Fergus, R.: Deconvolutional networks. In: 2010 IEEE Conference on Computer Vision and Pattern Recognition (CVPR), pp. 2528–2535. IEEE (2010)
16. Sutskever, I., Hinton, G.E., Taylor, G.W.: The recurrent temporal restricted Boltzmann machine. In: Advances in Neural Information Processing Systems, pp. 1601–1608 (2009)
17. Salakhutdinov, R., Hinton, G.: Deep Boltzmann machines. In: Artificial Intelligence and Statistics, pp. 448–455 (2009)
18. Radford, A., Metz, L., Chintala, S.: Unsupervised representation learning with deep convolutional generative adversarial networks, arXiv:1511.06434 (2015)
19. Pu, Y., Gan, Z., Henao, R., Yuan, X., Li, C., Stevens, A., Carin, L.: Variational autoencoder for deep learning of images, labels and captions. In: Advances in Neural Information Processing Systems, pp. 2352–2360 (2016)
20. Gregor, K., Besse, F., Rezende, D.J., Danihelka, I., Wierstra, D.: Towards conceptual compression. In: Advances in Neural Information Processing Systems, pp. 3549–3557 (2016)
21. Taubman, D., Marcellin, M.: JPEG2000 Image Compression Fundamentals, Standards and Practice: Image Compression Fundamentals, Standards and Practice, volume 642. Springer Science & Business Media (2012)
22. Santurkar, S., Budden, D.M., Shavit, N.: Generative compression. CoRR, arXiv:1703.01467 (2017)
23. Ioffe, S., Szegedy, C.: Batch normalization: accelerating deep network training by reducing internal covariate shift. In: International Conference on Machine Learning, pp. 448–456 (2015)
24. Srivastava, N., Hinton, G.E., Krizhevsky, A., Sutskever, I., Salakhutdinov, R.: Dropout: a simple way to prevent neural networks from overfitting. J. Mach. Learn. Res. **15**(1), 1929–1958 (2014)
25. Johnson, J., Alahi, A., Fei-Fei, L.: Perceptual losses for real-time style transfer and super-resolution. In: European Conference on Computer Vision, pp. 694–711. Springer (2016)
26. Long, J., Shelhamer, E., Darrell, T.: Fully convolutional networks for semantic segmentation. In: Proceedings of the IEEE Conference on Computer Vision and Pattern Recognition, pp. 3431–3440 (2015)

ANFIS Modeling of Boiling Heat Transfer over Tube Bundles

Abhilas Swain and Mihir Kumar Das

Abstract The article describes the application of artificial intelligence technique artificial neuro-fuzzy inference system (ANFIS) to predict the flow boiling heat transfer coefficient for distilled water on individual row in plain tube bundles. The variation of row-wise heat transfer coefficients is discussed with respect to the operating conditions such as mass flux, heat flux, and pitch to distance. A semi-empirical correlation is also formulated to predict the flow boiling Nusselt number taking the Peclet number, Froude number, and pitch-to-diameter ratio as inputs. The experimental data are predicted with $\pm 15\%$ accuracy by the semi-empirical correlation, whereas the ANFIS model is capable to predict within a maximum error of $\pm 10\%$.

Keywords First keyword · Second keyword · Third keyword

1 Introduction

Boiling heat transfer is applied widely in different types of industries such as thermal power plants, chemical, refrigeration, nuclear food processing, and other allied industries. The huge application of boiling heat transfer is due to high rate of heat transfer involved than other single-phase heat transfer processes. The flow boiling over is a highly nonlinear complex phenomenon. The reason behind this is that the process is greatly affected by various influential parameters such as properties of heater surface, material properties of heater, heater configuration, heater dimensions, operating parameters, and thermo-fluid properties of the liquid [1]. Furthermore, the process becomes further complex when flow boiling happens over tube bundles. This is due to the involvement of additional affecting variables such as mass flux, pitch-to-

A. Swain (✉)
School of Mechanical Engineering, KIIT Deemed to be University, Bhubaneswar 751024, India
e-mail: abhilas.swain@gmail.com

M. K. Das
School of Mechanical Sciences, IIT Bhubaneswar, Jatni 752050, India
e-mail: mihirdas@iitbbs.ac.in

© Springer Nature Singapore Pte Ltd. 2019
J. C. Bansal et al. (eds.), *Soft Computing for Problem Solving*, Advances in Intelligent Systems and Computing 817, https://doi.org/10.1007/978-981-13-1595-4_34

diameter ratio, and bundle configuration. Thus to formulating, mathematical model to predict the heat transfer coefficients taking all the influencing factors into account is a difficult task.

In the last two decades, significant advancements have been achieved in two differ- ent artificial intelligence subjects, which are fuzzy logic and neural networks. These techniques are useful in modeling nonlinear systems and processes. The adaptive neuro-fuzzy inference system (ANFIS) proposed by Jang [1] is a combined method- ology of fuzzy inference systems (FIS) and artificial neural network (ANN). Thus, it has advantages of fuzzy behavior and high learning competency.

The ANN has been already utilized to model the single-phase and two-phase heat transfer by Wang et al. [2] and Scalabrin et al. [3, 4]. The ANFIS is also adopted by Na [5] and Zaferanlouei et al. [6] for modeling the critical heat flux (CHF). Moreover, it is also used for modeling heat transfer coefficient in case of pool boiling by Das and Kishore [7, 8]. Recently, Tahseen et al. [9] implemented the ANFIS to model the heat transfer coefficient over a bundle for convection with inline arrangement. The data used for this modeling are obtained from numerical analysis. Razei et al. [10] have adopted ANFIS to model the natural convection occurring over V-shaped flat plate. From this literature, it can be inferred that the ANFIS method can model highly nonlinear-multifaceted process such as boiling heat transfer on account of its competences which embrace those demonstrated by the ANN and fuzzy inference system. The difference in the quantity and trend of boiling phase change heat transfer in different regimes, the variation of dependency on the various liquid properties, heater properties and operating conditions can be modeled by the use of this artificial intelligence methodology. ANFIS which in turn is essential for the effective design process of two-phase heat transfer equipment like shell and tube heat exchanger. The ANFIS is already tested and compared with other techniques. It is observed to predict the flow boiling row-wise HTC and bundle average HTC with sufficient accuracy [11–14]. Therefore, ANFIS is applied to predict the row-wise average convective boiling heat transfer coefficient over plain tube bundle in the present analysis.

2 Experimentation

The present experimental investigation is carried out through an experimental setup already well described in the articles Swain and Das [12, 13]. The setup has the facility to test the boiling heat transfer on tube bundles with different tube diameter and pitch to distances. The test vessel has a window onto which a PTFE sheet with the tubes can be attached to test the flow boiling behavior. Plain tube bundles with different pitch-to-diameter ratios are tested under different heat flux and mass fluxes. The detail process of data reduction and procedure is explained in Swain and Das [12, 13] (Fig. 1).

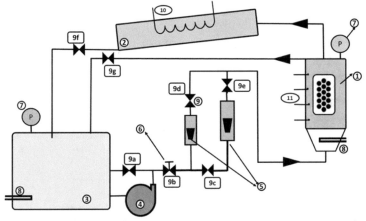

1-Test Vessel, 2-Condenser, 3-Reservoir, 4-Pump, 5-Rotameters, 6- Flow Control valves, 7-Pressure Gauge, 8-Heaters, 9-Shutoff valves, 10-Cooling water line,11- Thermocouples to measure the liquid temperature.

Fig. 1 Line diagram of the flow boiling experimental setup

3 ANFIS Methodology

The structure of ANFIS model resembles that of the multilayer feed-forward neural network which normally consists of neurons and directional links connecting the nodes. The major difference between ANN and ANFIS is seen in the quantity of nodes in the hidden layers. The number of nodes in ANFIS model is dependent on the number of inputs, whereas in case of ANN it is decided on the basis of nonlinearity of the physical problem to be modeled.

Generally, there are four layers besides the first and last layers of the ANFIS models. A single ANFIS model is generally applied for predicting single output using the inputs. For a physical problem with multiple outputs, different numbers of similar structures are to be considered to simulate the system. Further, it is to be noted that the kind of membership function and number of membership function applied to the inputs in the first layer are changed according to the nonlinearity in the problem.

Layer 1: This is the first layer where input data enter to the model. As membership functions are applied here, it is named as rule layer or membership function (MF) layer. The first layer is normally adaptive; that is, the parameters in these membership functions are changed in the learning process. The bell-type membership function used here is defined as follows.

$$\text{mfij}(x) = \frac{1}{1 + \left\{ ((x - c_i)/a_i)^2 \right\}^{b_i}} i, j = 1, 2, 3 \qquad (1)$$

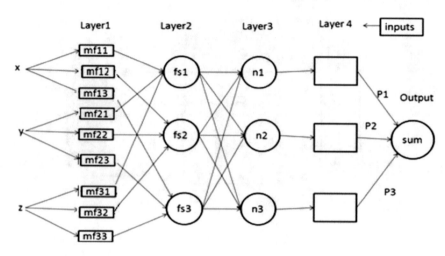

Fig. 2 Architecture of ANFIS model

The value of these factors ai, bi, and ci is changed iteratively to get suitable output through the training algorithm. Different membership functions may be used for different inputs. This layer brings the advantages of fuzzy logic into the model.

Layer 2: The output of the first layer is transferred to the next layer. There are no adaptive parameters in this second layer. The function of the first neuron of this layer is to multiply the output values of the first membership functions obtained from first layer for all the inputs. These multiplied values are called as the firing strengths of the rules. Therefore, the number of neurons or multiplied values is equal to the number of membership functions.

$$Wi = (mf\,1j)(mf\,2j)(mf\,3j) \quad \text{for } j = 1, 2, 3 \tag{2}$$

Layer 3: The function of the third layer is to normalize the multiplied firing strengths values obtained from the second layer. This is presented as follows.

$$W_i^n = W_i/(W_1 + W_2 + W_3) \tag{3}$$

Layer 4: The fourth layer neurons are adaptive in nature which means the parameters pi, qi, ri, and si are tuned by the training algorithm. The values obtained from the third layer are multiplied with a linear polynomial having the input data and the above-said parameters are defined as follows.

$$Pi = W_i^n (p_i x + q_i y + r_i z + s_i) \tag{4}$$

Layer 5: The function of last layer or so-called output node is to sum up all the values obtained from fourth layers to get the final output (Fig. 2).

4 Results and Discussions

The flow boiling heat transfer experiments are conducted under the operating parameters of mass flux, heat flux, and pitch-to-diameter ratio (P/D ratio). The trend of row-wise average heat transfer coefficients (HTC) is presented here with respect to the above-mentioned operating parameters. Figure 3 shows the variation of row average HTC with heat flux for different rows. This could be perceived from Fig. 1 that the HTC increases with increase in heat flux applied. Another observation from the graph is that the HTC is also increasing along the height for a fixed value of mass flux, heat flux, and P/D. Similar variations are observed for other values of mass flux and P/D.

Figure 4 displays the plot of row-wise heat transfer coefficient with mass flux for a given value of heat flux and P/D ratio. It can be observed that for the row average HTC does not vary significantly for initial range of mass flux from 20.25–77.16 kg/m^{2}s. However, after that for higher mass flux values the HTC values decreases gradually. The details of the reason for this variation are mentioned in Swain and Das [11]. Similar variations were also observed for the other heat flux and P/D ratios.

The preceding discussion demonstrated that the flow boiling HTC on individual rows depends on factors such as mass flux, heat flux, pitch-to-diameter ratio, and the row number of the tube bundle. An enhancement category correlation is framed taking Peclet number, Froude Number, and row number (n) for predicting the row-wise local flow boiling HTC. The flow boiling HTC is articulated as Nusselt number shown in below Eq. (5).

$$Nu = HTC * d_o/K_l \tag{5}$$

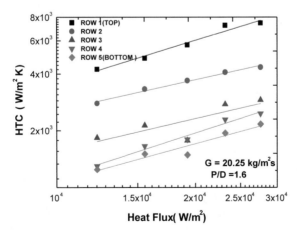

Fig. 3 Variation of row-wise heat transfer coefficient with heat flux for G = 20.25 kg/m^{2}s and P/D = 1.6

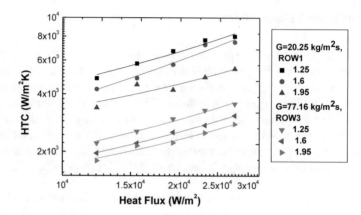

Fig. 4 Variation of row average HTC with heat flux for P/D = 1.25, 1.6, and 1.95 for rows = 3 and 5 and G = 20.25 and 77.16 kg/m²s

The Peclet number used for modeling the convection process is the product of Reynolds number and Prandtl number. Here for boiling heat transfer process, the Peclet number is the modified one given by following expression given in Eq. (6).

$$Pe = \frac{q}{\alpha \rho \lambda} \sqrt{\frac{\sigma}{g(\rho_l - \rho_v)}} \tag{6}$$

In current experimentations, five rows are instrumented to calculate the flow boiling HTC. The row number is designated as 'Nr' and its value varies from 1 (bottom row) to 5 (top row). The pitch-to-diameter ratio is also an important factor driving the two-phase flow in the tube bundles and therefore is considered as input in model. The heat input to the tube is assimilated into the correlations in terms of $(1 + Pe)$ and exponential form of row serial number to address the sliding bubble effect of the two-phase flow invading from lower tubes. The final formula of the correlation is structured as given in Eq. (7).

$$Nu_{rfb} = C_1 * \left(\frac{P}{D}\right)^{-0.37} * (1 + Pe)^{0.57} * Fr^{-0.042} * \exp(C_2 * Nr) \tag{7}$$

where C1 = 0.094 and C2 = 0.31, respectively.

Figure 5 shows a graph between the values of row-wise average Nusselt number determined from experiment and those determined from Eq. (7) for the saturated flow boiling of distilled water on a plain tube bundle. From this figure, this is observed that calculation match excellently with experimental values with most of the data is below an error of ±15%. The R-square value for the model is 0.9034.

The ANFIS model is created in the ANFIS toolbox of MATLAB platform considering the row average HTC as output and mass flux, heat flux, P/D ratio, and row number as input to the model. The model is able to predict within maximum error

Fig. 5 Comparison of predicted and experimental for row-wise average Nusselt number

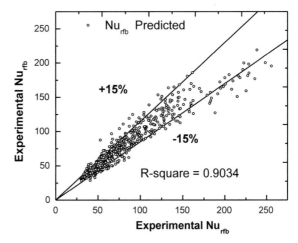

Fig. 6 Comparison of predicted and experimental for row-wise average Nusselt number

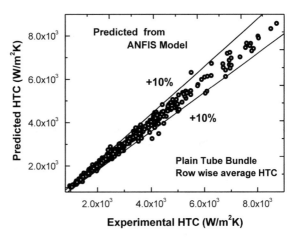

of $\pm 10\%$ the experimental data as shown in Fig. 6. Three membership functions of bell type are used to build the model. The default least square error method was used to train the model.

5 Conclusion

The variation of row-wise average HTC during flow boiling heat transfer over plain tube bundles is studied experimentally. The dependency of row average HTC on the operating parameters such as heat flux, mass flux, P/D ratio, and row number is modeled through conventional semi-empirical correlation and ANFIS technique. The ANFIS model is able to predict the data more accurately, i.e., within a maximum error of $\pm 10\%$.

References

1. Jang, J.: ANFIS: adaptive network-based fuzzy inference systems. IEEE Trans. Syst. Man Cybern. **23**, 665–685 (1993)
2. Wang, W.J., Zhao, L.X., Zhang, C.L.: Generalized neural network correlations for flow boiling heat transfer of R22 and its alternative refrigerants inside horizontal smooth tubes. Int. J. Heat Mass Transf. **49**(15–16), 2458–2465 (2006)
3. Scalabrin, G., Condosta, M., Marchi, P.: Modeling flow boiling heat transfer of pure fluids through artificial neural networks. Int. J. Therm. Sci. **45**(7), 643–663 (2006)
4. Scalabrin, G., Condosta, M., Marchi, P.: Flow boiling of pure fluids: Local heat transfer and flow pattern modeling through artificial neural networks. Int. J. Therm. Sci. **45**(8), 739–751 (2006)
5. Na, M.G.: DNB limit estimation using an adaptive fuzzy inference system. In: Nuclear Science Symposium, vol. 3, pp. 1708–1713. IEEE (2000)
6. Zaferanlouei, S., Rostamifard, D., Setayeshi, S.: Prediction of critical heat flux using ANFIS. Ann. Nucl. Energy **37**, 813–821 (2006)
7. Das, M., Kishore, N.: Adaptive fuzzy model identification to predict the heat transfer coefficient in pool boiling of distilled water. Expert Syst. Appl. **36**, 1142–1154 (2009)
8. Das, M., Kishore, N.: Determination of heat transfer coefficient in pool boiling of organic liquids using fuzzy modelling approach. Heat Transf. Eng. **31**(1), 45–58 (2010)
9. Tahseen, T.A., Ishak, M., Rahman, M.M.: Performance predictions of laminar heat transfer and pressure drop in an in-line flat tube bundle using an adaptive neuro-fuzzy inference system (ANFIS) model. Int. Commun. Heat Mass Transf. **50**, 85–97 (2014)
10. Rezaei, E., Karami, A., Yousefi, T., Mahmoudinezhad, S.: Modeling the free convection heat transfer in a partitioned cavity using ANFIS. Int. Commun. Heat Mass Transf. **39**(3), 470–475 (2012)
11. Swain, A., Das, M.K.: Flow boiling of distilled water over plain tube bundle with uniform and varying heat flux along the height of the tube bundle. Exp. Therm. Fluid Sci. **82**, 222–230 (2017)
12. Swain, A., Das, M.K.: Performance of porous coated 5×3 staggered horizontal tube bundle under flow boiling. Appl. Therm. Eng. **128**(5), 444–452 (2018)
13. Swain, A., Das, M.K.: Prediction of heat transfer coefficient in flow boiling over tube bundles using ANFIS. Heat Transf. Eng. **37**(5), 443–455 (2016)
14. Swain, A., Das, M.K.: Artificial intelligence approach for the prediction of heat transfer coefficient in boiling over tube bundles. Proceed. Inst. Mech. Eng. Part C: J. Mech. Eng. Sci. **228**(10), 1680–1688 (2013)

Prediction of Damage Level of Inner Conventional Rubble Mound Breakwater of Tandem Breakwater Using Swarm Intelligence-Based Neural Network (PSO-ANN) Approach

Geetha Kuntoji, Subba Rao, Manu and Eluru Nava Bharath Reddy

Abstract The conventional rubble mound breakwater is a coastal protective structure commonly used decades before which alone failed to withstand the deepwater wave and its energy, and suffered a catastrophic failure. Keeping in mind both the safe functioning of harbor and stability of the breakwater for the fast-growing economy of the country, different types of breakwaters are being developed to serve this purpose. Tandem breakwater is an innovative type of breakwater, which is a combination of main conventional rubble mound breakwater and submerged reef in front of it. One of the advantages of this breakwater is that most of the wave energy is dissipated and wave intensity is reduced by submerged reef and the smaller waves interact with main breakwater and ensure its stability. Experimental studies are laborious and time-consuming to conduct. Therefore, it is necessary to carry out the detailed study of tandem breakwater stability by making use of simple and alternate techniques using the experimental data. In the present study, an attempt is made to understand the suitability and applicability of PSO-ANN, a hybrid soft computing technique for predicting damage level of conventional rubble mound breakwater of tandem breakwater. Based on the experimental data available in Marine Structure Laboratory, NITK, Surathkal, India, soft computing models are developed. The performances of the models are evaluated using model performance indicators. Results obtained demonstrate that the proposed new approach can be used to predict the damage level of conventional rubble mound breakwater of tandem breakwater efficiently and accurately.

Keywords Breakwaters · Tandem breakwater · Damage level · Particle swarm optimization (PSO) · Artificial neural network (ANN)

G. Kuntoji (✉) · S. Rao · Manu · E. N. B. Reddy
Department of Applied Mechanics & Hydraulics, National Institute
of Technology Karnataka, Surathkal 575025, Karnataka, India
e-mail: geeta.kuntoji@gmail.com

© Springer Nature Singapore Pte Ltd. 2019

441

J. C. Bansal et al. (eds.), *Soft Computing for Problem Solving*, Advances in Intelligent
Systems and Computing 817, https://doi.org/10.1007/978-981-13-1595-4_35

1   Introduction

The construction of breakwater is closely related to developments of ports around the world over centuries. Experimental works have been carried out on different types of innovative breakwaters by Rao and Shirlal [1–3]. Designing coastal structures economically such as breakwaters is of significant task in coastal engineering. Rubble mound breakwater is the most accepted design for large harbors because of its flexibility due to rearrangement of armor stones with the impacting waves. The failure of conventional rubble mound breakwater basically happens close to the waterline, the material drooping down the incline, shaping an augmented toe. The waves break on the toe and spread over the bench before at last passing on the structure. It was perceived that this bench was truly abundance filler material and could be expelled and moved somewhat front with little effect on the structure execution, shaping a wave stilling bowl behind the toe. The outcome is a tandem breakwater, with an external submerged reef and an inward primary breakwater or non-overtopping rubble mound breakwater. The detached submerged reef acts as a wave attenuator and reduces the wave impacts on the conventional breakwater, thus minimizing the size of armor stones required for the given conditions. The conventional rubble mound breakwater with the presence of submerged reef in front of it possesses merits of reef by dissipating maximum wave intensity before reaching to main inner breakwater and makes it structurally stable, which eventually makes tandem breakwater as one of the economical-, stable-, and innovative-type protective structures [4, 5].

The tandem breakwater may not always provide the least cost alternative, as section costs are strongly related to material costs and water depths. However, the tandem breakwater has offered significantly reduced design risk since high return period events such as large storm waves are reduced to a more manageable level, reducing the chance of catastrophic failure of the main breakwater in storm events outside the design envelope [4].

Keeping in mind the end goal to reduce the cost and time in conducting the trial work and to overcome the uncertainties involved in mathematical and numerical modeling, soft computing techniques, viz. artificial neural network (ANN), are being applied to solve a wide variety of coastal engineering problems [6].

ANN resembles the biological neurons in human brain that pass information through them as stimuli. Artificial neural networks are trained using a set of data, and once the model learns from examples, predictions are done. The obtained results are then optimized using different optimization techniques such as PSO and FIS. Particle swarm optimization (PSO) is a population-based stochastic enhancement and optimization tool developed by [7, 8]. The present study focuses on application of artificial neural network in combination with particle swarm optimization (PSO) technique based on the experimental data of Rao and Shirlal [2], for predicting damage level of the conventional rubble mound breakwater of tandem breakwater.

2 Development of PSO-Based ANN Models

Experimental data (number of data set = 288) is collected and organized systematically such that consistency of the data is checked using time series plots after normalizing the data set between 0 and 1 (all ranges of data should be included in the training and testing sets). Data is then separated into two sets, 70% (201) for training and 30% (87) for testing models.

Non-dimensional analysis is performed to arrive at some of the dimensionless parameters. The dimensionless parameters that influence damage level (S) are relative wave steepness $\left(\frac{H_i}{gT^2}\right)$, the relative spacing $\left(\frac{X}{d}\right)$, stability number $\left(\frac{H}{\Delta D_{n50}}\right)$, relative crest widths $\left(\frac{B}{d}\right)$, $\left(\frac{B}{L_o}\right)$, relative crest height $\left(\frac{h}{d}\right)$, relative submergence $\left(\frac{F}{H_i}\right)$, relative water depth $\left(\frac{d}{gT^2}\right)$, of tandem breakwater which are considered in the present study as shown in Table 1.

2.1 Generating Hybrid PSO-ANN Model

The particle swarm optimization research toolbox given by George Evers (2010) in MATLAB is used to replace standard back-propagation algorithm using Levenberg—Marquardt algorithm (LMA) in ANN to develop PSO-ANN algorithm. While using the PSO-ANN in the training stage for predicting the damage level, the "trainlm" parameter is replaced by "trainpso" so that the number of hidden neurons and layers are optimized by PSO and error is measured using the RMSE as a fitness criterion. The statistical parameters are calculated for PSO-ANN models by varying number of hidden neurons to evaluate the model performance, and the network with best prediction is considered as best.

Table 1 Non-dimensional input parameters and their range used in the experiment

Sl. no.	Non-dimensional parameters	Notation	Range
1	Deepwater wave steepness parameter	H_i/gT^2	1.46×10^{-3} to 7.85×10^{-3}
2	Depth parameter	d/gT^2	0.004–0.018
3	Relative reef crest width	B/d	0.25–1.33
4	Relative reef crest width	B/L_o	0.01–0.114
5	Relative reef submergence	F/H_i	-0.312 to -1.5
6	Relative reef crest height	h/d	0.625–0.833
7	Relative reef spacing	X/d	2.5–13.33
8	Stability number	$H_i/\Delta D_{n50}$	2–3

The particle swarm optimization idea comprises of, at each time step, changing the speed of (accelerating) every particle toward its pbest (best position) and lbest (best locations) (general form of PSO). Increasing speed is weighted by an irregular term, with isolated arbitrary numbers being produced for accelerating toward pbest and lbest areas.

One more reason for adopting PSO as optimizer is the advantage and attractiveness of the PSO to modify the couple of parameters. One form, with slight variations, functions effectively in a wide assortment of uses. Optimization by PSO has been utilized for methodologies that can be utilized over an extensive variety of uses, and also for particular applications concentrated on a given prerequisite. Swarm intelligence concept began as a simulation of improved social and simplified system. The first goal was to graphically recreate the choreography of a bird block or fish school. In any case, it is discovered that the particle swarm model can be utilized as an optimizer.

3 Methodology

For the present study, experimental data is normalized (Eq. 1), which range from 0 to 1 and then classified into two sets: One set with 70% data is used to train the networks, and 30% data for testing the models. Both the training and testing data sets contain the full range of normalized data (0 to 1). The input parameters that influence output parameters' damage level (S) of conventional rubble mound breakwater of tandem breakwater are H_i/gT^2, X/d, F/H_i, B/L_o, $H_i/\Delta D_{n50}$, d/gT^2, h/d, and B/d considered to train PSO-ANN network (Fig. 1).

The normalization of data is achieved by the following equation:

$$X_{norm} = \frac{X - X_{min}}{X_{max} - X_{min}} \tag{1}$$

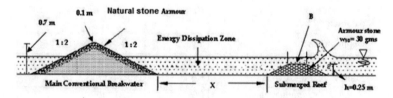

Fig. 1 Tandem breakwater experimental setup

where

X	Actual value
X_{norm}	Normalized value
X_{max}	Maximum value
X_{min}	Minimum value

3.1 PSO-ANN Algorithm

As explained before, PSO reenacts the practices of bird flocking. PSO gains the knowledge from the situation and utilizes it to tackle the problem of optimization. In PSO, each single arrangement is a "bird" in the search space which is known as "particle (swarm)." Each one of the particles has fitness values, which are assessed by the function of fitness to be optimized, and has velocities which coordinate the movement of the particles. The movement of the particles "y" in the search space takes after the present ideal particles. The methodology of the present study is illustrated in the flowchart given in Fig. 2.

The performances of the model are assessed by the following statistical measures: Root-mean-square error is calculated using the formula:

$$\text{RMSE} = \sqrt[2]{\frac{1}{N} \sum_{i=0}^{N} \left(P_{pi} - O_{pi}\right)^2} \tag{2}$$

Nash–Sutcliffe efficiency is calculated using the formula:

$$\text{NSE} = 1 - \frac{\sum_{i=1}^{N} \left(P_{pi} - O_{pi}\right)^2}{\sum_{i=1}^{N} \left(O_{pi} - \overline{O_{pi}}\right)^2} \tag{3}$$

Scatter index is calculated using the formula:

$$\text{SI} = \frac{\sqrt[2]{\frac{1}{N} \sum_{i=0}^{N} \left(P_{pi} - O_{pi}\right)^2}}{\overline{O_{pi}}} \tag{4}$$

Correlation coefficient is calculated using the formula:

$$\text{CC} = \frac{\sum_{i=1}^{N} (P_{pi} - \overline{P_{pi}})\left(O_{pi} - \overline{O_{pi}}\right)}{\sqrt{\sum_{i=1}^{N} \left(P_{pi} - \overline{P_{pi}}\right)^2} \sqrt{\sum_{i=1}^{N} \left(O_{pi} - \overline{O_{pi}}\right)^2}} \tag{5}$$

where O_{pi}—observed wave transmission, P_{pi}—predicted wave transmission, "n"—No. of data sets $\overline{O_{pi}}$, and $\overline{P_{pi}}$—average observed and predicted wave transmission, respectively.

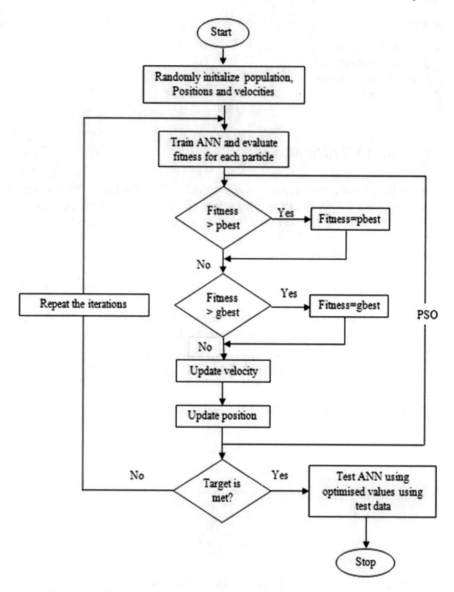

Fig. 2 Flowchart of the PSO-ANN used to predict damage level

4 Results and Discussion

In the present paper, PSO-tuned ANN models are generated and damage levels of conventional rubble mound breakwater of tandem breakwater are predicted. The performances of the model are evaluated using statistical measures, namely root-mean-square error (RMSE), correlation coefficient (CC), scatter index (SI), and Nash–Sutcliffe efficiency (NSE).

4.1 Performance of PSO-ANN Model

As mentioned, PSO function is called in place of LMA function. The statistical parameters RMSE, CC, NSE, and SI are calculated for various PSO-ANN models varying hidden neuron size, and the network with best prediction is considered as reliable and efficient model. Table 2 shows the results of different statistical parameters for various network architectures for predicting damage level (S).

From Table 2, it can be seen that hybrid PSO-ANN with (8-2-1) network in prediction of damage level is found to be better with CC = 0.769 for testing. Figure 4 shows scatter plots for damage level (S) predictions by PSO-ANN model with different hidden neurons. The efficiency of model is presented in terms of NSE $\cong$ 0.60 in both training and testing cases. The hidden neurons are varied from 1 to 5, and the results seem to remain same after increasing hidden neurons beyond 2. The decisions are made based on the statistical parameters, root-mean-square error, model efficiency, correlation coefficient, and scatter index of the predicted damage level with reference to the measured damage level.

Further, the results are found to be marginally close to the measured one and give only 75% correlation and 60% efficiency. This is due to the presence of zero damage levels (damage less than 5% is considered as zero damage) at some points. Damage to the main breakwater is prevented by placing the submerged reef in front of it. Based on the above discussion, it is clear that, because of the presence of zero damage, predicted results are considered to be better in terms of all the statistical measures used to evaluate the model performance.

From Fig. 3b, for PSO-ANN model with 8-2-1 network combination, the scatter index is less compared to PSO-ANN model with 1, 3, and 4 hidden neurons' combinations of PSO-ANN networks as shown in Fig. 3a, c, d, respectively. It is also observed that after the increase in hidden neurons beyond 4, the PSO-ANN model performance remains constant.

The distribution of predicted values of each PSO-ANN network with reference to the observed damage level of tandem breakwater is presented in the form of box–whiskers plot. The box plot representation shows the spread of the predicted damage level and the observed damage level of tandem breakwater. It is found that higher damage levels are predicted well compared to lower damage levels due to the presence of zero damage at certain locations. It is also observed that there are more outliers found in the lower quartile of PSO-ANN with (8-3-1) and (8-4-1) compared to PSO-ANN with (8-2-1) network. Therefore, the PSO-ANN model with (8-2-1) network predicts the higher damage level well in the upper quartile and close to the observed damage level compared to the remaining networks of PSO-ANN model as shown in Fig. 4.

Table 2 Statistical results of prediction of PSO-ANN model

Output	Network	RMSE		CC		SI		NSE	
		Train	Test	Train	Test	Train	Test	Train	Test
S	8-1-1	0.157	0.144	0.772	0.755	1.083	1.119	0.594	0.570
	8-2-1	**0.158**	**0.141**	**0.767**	**0.769**	**1.096**	**1.093**	**0.584**	**0.589**
	8-3-1	0.159	0.145	0.763	0.756	1.102	1.125	0.579	0.565
	8-4-1	0.160	0.147	0.761	0.744	1.103	1.142	0.579	0.552
	8-5-1	0.165	0.141	0.740	0.769	1.143	1.094	0.548	0.589

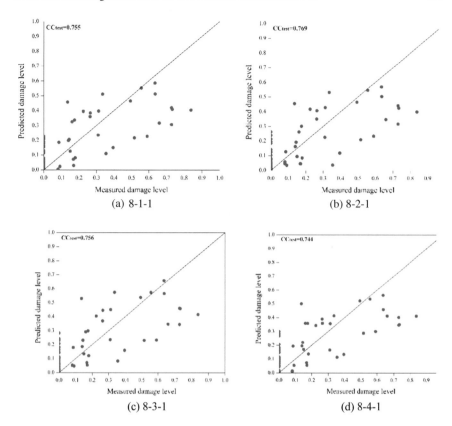

Fig. 3 Scatter plots of PSO-ANN model for damage level

The accuracy and reliability of the PSO-ANN model with (8-2-1) network in predicting damage level with reference to observed damage level during testing phase are illustrated in Fig. 5. From Figs. 5 and 3b, it is also noticed that PSO-ANN in combination with two hidden neurons and eight input parameters predicts the damage efficiently. The model is considered as reliable and alternative approach and can be used to predict damage level of tandem breakwater if the site conditions are favorable and to meet immediate needs.

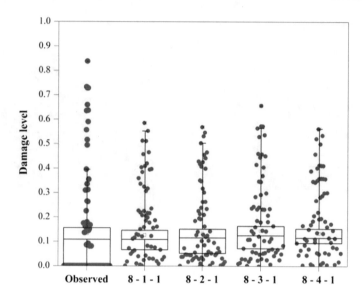

Fig. 4 Box–whiskers plot in damage level prediction by PSO-ANN

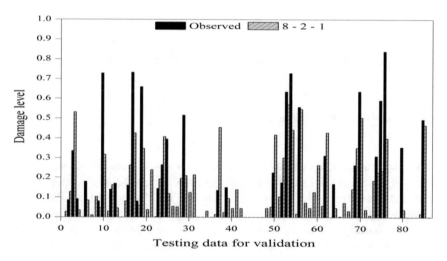

Fig. 5 Performance validation of PSO-ANN model in predicting damage level

4.2 Comparison and Validation

The authors also developed ANN model with LMA algorithm for predicting damage level of the conventional rubble mound breakwater. The predicted results are given in Table 3, in terms of statistical measures. The results of the ANN with LMA are found to be better compared to the PSO-ANN model.

Table 3 Statistical results of prediction of LMA-ANN model

Models	Network	RMSE		CC		SI		NSE	
		Train	Test	Train	Test	Train	Test	Train	Test
ANN1	8-1-1	0.144	0.163	0.808	0.695	1.00	1.264	0.653	0.451
ANN2	8-2-1	0.083	0.075	0.941	0.941	0.576	0.581	0.885	0.884
ANN3	8-3-1	0.069	0.072	0.959	0.947	0.481	0.555	0.920	0.894
ANN4	8-4-1	0.066	0.082	0.963	0.932	0.455	0.640	0.928	0.859
ANN5	**8-5-1**	**0.055**	**0.064**	**0.974**	**0.956**	**0.382**	**0.515**	**0.949**	**0.909**

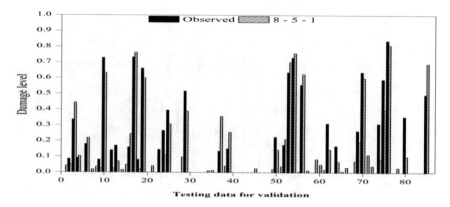

Fig. 6 Performance validation of LMA-ANN model for predicting damage level

From Table 3, it can be seen that ANN1 model gives a very poor result which can be seen from all the five statistical parameters. The network efficiency improves by increasing hidden neuron size to 2 and 3, and thereafter, it decreases for fourth neuron, but there is an increase again for hidden neuron size 5. The network architecture is observed to give high performance for all the four statistical parameters and hence is considered as optimum. The predictions of LMA-ANN value of CC = 0.956 for testing with five hidden neurons (ANN5) are found to be better compared to PSO-ANN with two hidden neurons as CC = 0.769. The value of NSE = 0.59 which is very less compared to the LMA-ANN. Although the RMSE value for the network ANN5 is 0.064 for testing, the error is more generalized in case of LMA-ANN compared to the PSO-ANN. Therefore, the LMA-ANN model with (8-5-1) network outperformed PSO-ANN with (8-2-1) network in predicting the damage level of the main breakwater as shown in Fig. 6.

5   Summary and Conclusion

In the present discussion, a hybrid PSO-ANN model with various hidden neurons is attempted. For selecting the accurate and reliable ANN hybrid model, different statistical measures are used. The results obtained by PSO-ANN and LMA-ANN models show that soft computing techniques like ANN can be used efficiently in combination with different optimization techniques. The particle swarm intelligence and LMA approach are used in the present study where these techniques will optimize weights and bias of the ANN model by varying the hidden neurons resulting in more generalized error.

Based on the statistical results, box plot, and comparison graphs, it is observed that LMA-ANN predicts the damage level accurately with better efficiency of NSE = 0.90 with lower RMSE and less scatter index than PSO-ANN. Hence, it can be applied successfully for the prediction of damage level. Therefore, PSO-based ANN approach is not suitable to predict the damage level of the inner conventional rubble mound breakwater of the tandem breakwater. LMA-based ANN can be utilized as an alternate and reliable tool for predicting the damage level influencing the stability of the tandem breakwater.

Acknowledgements The authors are thankful to the Director of National Institute of Technology Karnataka, Surathkal, India, and Head of the Department of Applied Mechanics and Hydraulics, NITK, Surathkal, for providing the necessary facilities to carry out the project and for granting the permission to publish this work.

References

1. Shirlal, K.G., Rao, S.: Ocean wave transmission by submerged reef—A physical model study. Ocean Eng. **34**, 2093–2099 (2007). https://doi.org/10.1016/j.oceaneng.2007.02.008
2. Rao, S., Shirlal, K.G.: Studies on the stability of conventional rubble mound breakwater defended by a seaward submerged reef (2004)
3. Binumol, S., Hegde, A.V., Rao, S.: Effect of water depth on wave reflection and loss characteristics of an emerged perforated quarter circle breakwater. Int. J. Ecol. Dev. **31**, 13–22 (2016)
4. Cox, J.C., Clark, G.R.: Design development of a tandem breakwater system for Hammond Indiana. In: Coastal Structure Breaking Conference, Thomas Telford, London, p. 10 (1992)
5. Gadre, M.R., Poonawala, I.Z., Kale, A.G., Kudale, M.D.: Rehabilitation of rubble mound breakwater. In: 3rd National Conference on Dock & Harbour Engineering, Karnataka Regional Engineering College (K.R.E.C), Surathkal, Srinivasnagar, India, p. 6 (1989)
6. Block, H.D., Knight, B.W., Rosenblatt, F.: Analysis of a four-layer series-coupled perceptron. II. Rev. Mod. Phys. **34**, 135–142 (1962). https://doi.org/10.1103/revmodphys.34.135
7. Kennedy, J., Eberhart, R.: Particle swarm optimization. In: Proceedings of the IEEE International Conference on Neural Networks, 1942–1948, vol. 4 (1995). https://doi.org/10.1109/icnn.1995.488968
8. Nikelshpur, D., Tappert, C.: Using particle swarm optimization to pre-train artificial neural networks: selecting initial training weights for feed-forward back-propagation neural networks. In: Proceedings of the Student-Faculty Research Day, CSIS, Pace University, New York, NY, USA, pp. C5.1–C5.7 (2013)

Swarm Intelligence-Based Support Vector Machine (PSO-SVM) Approach in the Prediction of Scour Depth Around the Bridge Pier

B. M. Sreedhara ⓘ **, Manu and S. Mandal**

Abstract The mechanism of scour around the bridge pier is a complex phenomenon, and it is very difficult to make a common method to predict or estimate the depth of scour hole. In this paper, a hybrid model is developed, combining support vector machine and particle swarm optimization (PSO-SVM) to predict scour depth around a bridge pier. The input parameters such as sediment size (d_{50}), the velocity of flow (U), and time (t) are used in the study to predict the scour depth. The models are developed with RBF, polynomial, and linear kernel functions, and the performances are evaluated using different statistical parameters such as CC, RMSE, NSE, and NMB. The predicted results are compared with measured scour depth. The predicted scour depth reveals that PSO-SVM with RBF kernel function model is found to be reliable and efficient in predicting the scour depth around bridge piers.

Keywords Bridge pier scour · Particle swarm optimization (PSO) · Support vector machine (SVM) · PSO-SVM

1 Introduction

Bridges play an important role in the society since they enable quick access to a river or any water body. Bridges facilitate transportation of goods and people and hence play a key role in the development of a region. The failure of bridges is a serious problem because of the high investment costs, safety problems in the event of a failure, and adverse effects on the economy of the region. Scour related to bridge hydraulics received much attention in the past decade, including its relation to flood hydrology and hydraulic processes. Scour means the lowering of the riverbed level

B. M. Sreedhara (✉) · Manu
Department of Applied Mechanics & Hydraulics, National Institute
of Technology Karnataka, Surathkal 575025, Karnataka, India
e-mail: shreedhar.am13f07@nitk.edu.in

S. Mandal
Department of Civil Engineering, PES University, Bangalore, India

© Springer Nature Singapore Pte Ltd. 2019
J. C. Bansal et al. (eds.), *Soft Computing for Problem Solving*, Advances in Intelligent
Systems and Computing 817, https://doi.org/10.1007/978-981-13-1595-4_36

Fig. 1 Mechanism of local scour (adapted from [1])

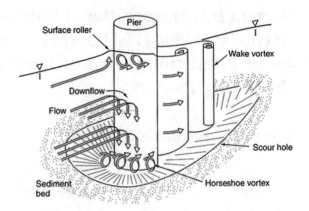

by water erosion, such that there is a tendency to expose the foundations of a bridge. The typical local scour mechanism is shown in Fig. 1.

Laboratory research has been the primary tool in defining the relations among variables affecting the depth of pier scour in recent years. Various studies have been performed in the laboratory tests to study the scour phenomenon around a bridge pier. Those experimental data were further used for numerical modeling and soft computing applications. The different soft computing techniques such as artificial neural network (ANN), fuzzy logic, support vector machine (SVM), linear regression, genetic programming (GP), group method of data handling (GMDH), and model tree approach have been used to predict the scour depth around the bridge pier using the experimental data, and the performances were compared with several empirical relations. In the present days, the researchers are focusing on the hybridization by combining individual's models with other optimization techniques to improve the prediction accuracy. There are numbers of optimization techniques such as particle swarm optimization (PSO), ant colony optimization (ACO), honey bee search from the literature, it was found that swarm intelligence-based algorithms which are today's one of the powerful tools of optimization techniques. It is observed from the literature that the hybrid approaches like ANFIS-ACO [2], ANFIS-LR [3], SVM-GA [4], GMDH-BP [5], PSO-ANN [6], ANFIS-PSO [7, 8], and other models are developed to solve scour-related and other problems. However, the application of PSO-SVM approach is not yet carried out in the prediction of scour depth and other scour-related problems. Therefore, in the current discussion, an effort is made to study and estimate the scour depth around the bridge pier using PSO-SVM, a hybrid approach.

2 Data Analysis

The data are taken from the doctoral thesis with the title "Evaluation of scour depth around bridge piers" done by Goswami Pankaj, submitted to Gauhati University in 2013 [9]. The laboratory data sets are taken with a 1000 mm wide, 1300 mm depth, and 19.25 m length of flume dimensions. Three input parameters, namely sediment size (d50), velocity (U), and time (t), are used to predict the scour depth around the bridge pier. The whole data set is divided randomly into training data (50%) and testing data (50%). The statistical parameters such as maximum, minimum, mean, and standard deviations are listed in Table 1.

3 Methodology

3.1 Particle Swarm Optimization-Based Support Vector Machine (PSO-SVM)

Support vector machine (SVM), one of the supervised learning methods, was introduced in 1992 by Vapnik [10]. Basically, SVM is a training algorithm for linear and nonlinear classification as well as regression. The basic idea of the SVM learning algorithm can be summarized in two steps. Firstly, the input space is transformed into a higher-dimensional linear feature space by a nonlinear transform function φ. Then, the optimal linear separating plane can be constructed in this higher-dimensional feature space [11]. The SVM develops a different hyperplane margin between the

Table 1 Statistical parameters

Data set	Variable	Max	Min	Mean	St. dev.
Training	Scour depth (mm)	118	55	84.262	14.166
	Sediment size, d_{50} (mm)	4.2	0.42	2.31	1.89
	Velocity, U (m/s)	0.351	0.184	0.261	0.0515
	Time, t (h)	6	0	3	2
Testing	Scour depth (mm)	115	54	84.512	14.206
	Sediment size, d_{50} (mm)	4.2	0.42	2.31	1.89
	Velocity, U (m/s)	0.351	0.184	0.216	0.0515
	Time, t (h)	6	0	3	2

Fig. 2 The separating
hyperplane (adapted from
[14])

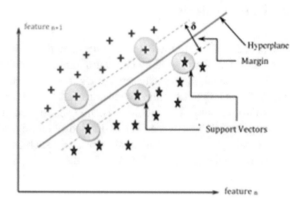

points in the feature space and amplifies edge between two informational indexes of two input points. It makes an effort of constructing a fit curve with a kernel function and is used on entire data points such that data points should lie between two largest marginal hyperplanes to minimize the error of regression as shown in Fig. 2 [12, 13].

Particle swarm optimization (PSO) is a population-based stochastic optimization technique motivated by social behavior, such as bird flocking and fish schooling. PSO was first proposed by Kennedy and Eberhart [15]. In PSO, each particle makes use of its individual memory and knowledge gained by the swarm as a whole to find the best solution. All the particles have fitness values, which are evaluated by fitness function to be optimized and have velocities which direct the movement of the particles. The best position of each particle is chosen from its own and neighboring particle experience in the process of movement of the particles. For every iteration, each particle is updated by the following two "best" values called pbest and gbest [16].

The PSO algorithm is defined by the direction and movement of each particle through the search space, by updating its velocity and position:

$$Vel_{j+1}^{i} = W_j Vel_j^{i} + C_1 rand_1 \left(pbest_j - pos_j^{i} \right) + C_2 rand_2 \left(gbest_j - pos_j^{i} \right) \quad (1)$$

$$pos_{j+1}^{i} = pos_{j+1}^{i} + Vel_{j+1}^{i} \quad (2)$$

where pos_j^{i} is the current position of the particle i with subscript j representing iteration count, Vel_j^{1} is the search velocity of the ith particle, C_1 and C_2 are the cognitive and social scaling parameters, $rand_1$ and $rand_2$ are the random numbers with interval [0, 1] applied to the ith particle, W_j is the particle inertia, $pbest_j$ is the best position found by the ith particle (personal best), and $gbest_j$ is the global best position found among all the particles in the swarm. The particle inertia controls the balance of global and local search abilities, where a larger W_j facilitates a global search. Particle i flutters toward a new position using Eqs. (1) and (2), which allow all particles in the swarm to update their $pbest_j$ and $gbest_j$.

The algorithm steps are as follows:

Step 1: Initialize PSO parameters particle positions and velocities;
Step 2: Evaluate the objective function values using initial particle positions and velocities;
Step 3: Update the optimum particle positions and global optimum particle position using fitness function;
Step 4: Update the position of each particle using its previous position and update the velocity vector using Eqs. (1) and (2);
Step 5: Repeat steps 2–4 until the stopping criteria are met.

3.2 Performance Analysis

Once the predictions of PSO-SVM are obtained, then the model predictions are assessed and validated using model performance indicators, as defined below.

1. Root-mean-square error (RMSE)

$$RMSE = \sqrt{\left(\frac{\sum_{i=1}^{N}(X_i - Y_i)^2}{N}\right)} \tag{3}$$

2. Normalized mean bias (NMB)

$$NMB = \sum_{i=1}^{N}\left(\frac{Y_i - X_i}{X_i}\right) = \left(\frac{\overline{Y_i}}{\overline{X_i}} - 1\right) \tag{4}$$

3. Nash–Sutcliffe coefficient (NSE)

$$NSE = 1 - \left(\frac{\sum_{i=1}^{N}(X_i - Y_i)^2}{\sum_{i-1}^{N}(X_i - \overline{X})^2}\right) \tag{5}$$

4. Correlation coefficient (CC)

$$CC = \frac{\sum_{i=1}^{N}(X_i - \overline{X}) \cdot (Y_i - \overline{Y})}{\sqrt{\sum_{i=1}^{N}(X_i - \overline{X})^2 \cdot \sum_{i=1}^{N}(Y_i - \overline{Y})^2}} \tag{6}$$

where

X Observed/measured values;
Y Predicted values;
$\overline{X}$ Average of actual data;
N Total number of data points.

4 Results and Discussion

The PSO-SVM models are developed to predict the scour depth around the bridge pier. The RBF, polynomial, and linear kernel functions are utilized as a part of this investigation. The experimental data used for this study are given in Table 1. For effective forecast, it is important to prepare the model for running the testing stage. The prediction accuracy of the models is reassessed in terms of performance analysis, and the obtained statistical parameters are given in Table 2 for both training and testing stages. The model's effectiveness and accuracy are assessed, and the results are validated by plotting the estimated results of PSO-SVM with measured scour depths as shown in Fig. 3 for RBF, Fig. 4 for polynomial, and Fig. 5 for linear kernel functions. From the plots, it is clear that the models are giving a good correlation in the training phase in comparing with the testing phase for all types of kernel functions. The obtained statistical parameters for all three kernel functions are listed in Table 2, and it is clear that the PSO-SVM with RBF kernel function model is performing better with higher CC=0.952, NSE=0.90, and lower RMSE=0.460 compared to other two kernel functions. The NMB value from Table 2 shows that all three kernel functions are performing under prediction for the testing phase.

The comparison of the measured scour depth to predicted scour depth from all three kernel functions is plotted in Fig. 6. It is shown that the predicted scour depth from PSO-SVM with RBF kernel functions well correlates with measured results.

Table 2 Statistical results of PSO-SVM models

PSO-SVM	CC		RMSE		NSE		NMB	
	Training	Testing	Training	Testing	Training	Testing	Training	Testing
RBF	0.959	0.952	4.04	4.56	0.92	0.90	−0.0022	−0.0131
Polynomial	0.953	0.949	4.31	4.55	0.91	0.90	0.0003	−0.0094
Linear	0.882	0.881	6.70	6.77	0.78	0.77	0.0061	−0.0026

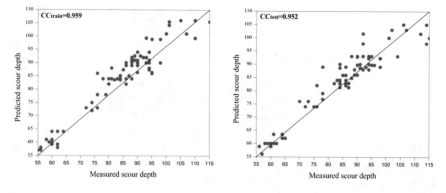

Fig. 3 Scatter plot of measured versus predicted scour depth from PSO-SVM models with RBF kernel function

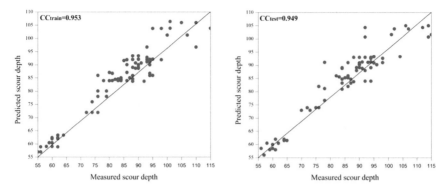

Fig. 4 Scatter plot of measured versus predicted scour depth from PSO-SVM models with polynomial kernel function

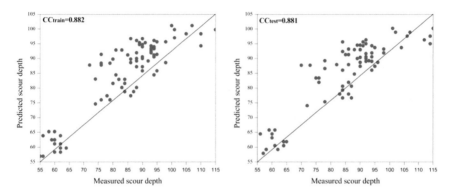

Fig. 5 Scatter plot of measured versus predicted scour depth from PSO-SVM models with linear kernel function

Box plots are also used to learn the spread of measured and predicted scour depth values. It is observed that predicted values from RBF and polynomial kernel functions showing the similar spread in comparison with measured results as shown in Fig. 7.

5 Conclusions

The application of PSO-SVM models in the prediction of scour depth around the bridge pier is discussed in this study. The RBF, polynomial, and linear kernel functions are used in the prediction. The predicted scour depth is compared and validated with experimentally measured scour depth. The study shows PSO-SVM with RBF kernel function performs better than polynomial and linear kernel function models and showed good correlation with experimental results. From the study, it can

Fig. 6 Comparison plot of measured with predicted (PSO-SVM) scour depth

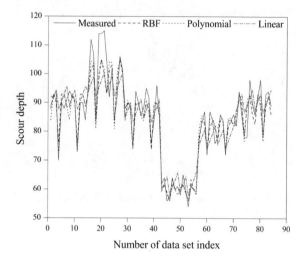

Fig. 7 Box plot of measured with predicted (PSO-SVM) scour depth

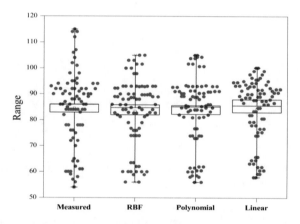

be concluded that PSO-SVM with RBF kernel function model could serve a better alternate for scour depth prediction.

Acknowledgements The authors would like to express their sincere gratitude to Dr. Goswami Pankaj, Gauhati University, for providing experimental data. Also, grateful to Director and Head of the department, Applied Mechanics and Hydraulics department, NITK, Surathkal for necessary support.

References

1. Abdalla, M.G.: A study on scour for irrigation canals in egypt, "case study: the first reach of El-ibrahimeya canal." Am. J. Eng. Technol. Manag. **1**, 65–77 (2016). https://doi.org/10.1164 8/j.ajetm.20160104.13
2. Cus, F., Balic, J., Zuperl, U.: Hybrid ANFIS-ants system based optimisation of turning parameters. J. Achiev. Mater. Manuf. Eng. **36**(1), 79–86 (2009)
3. Akib, S., Mohammadhassani, M., Jahangirzadeh, A.: Application of ANFIS and LR in prediction of scour depth in bridges. Comput. Fluids (2014). https://doi.org/10.1016/j.compfluid.20 13.12.004
4. Chou, J.S., Pham, A.D.: Hybrid computational model for predicting bridge scour depth near piers and abutments. Autom. Constr. (2014). https://doi.org/10.1016/j.autcon.2014.08.006
5. Najafzadeh, M., Barani, G.A.: Comparison of group method of data handling based genetic programming and back propagation systems to predict scour depth around bridge piers. Sci. Iran. (2011). https://doi.org/10.1016/j.scient.2011.11.017
6. Hasanipanah, M., Noorian-Bidgoli, M., Armaghani, J.D., Khamesi, H.: Feasibility of PSO-ANN model for predicting surface settlement caused by tunneling (2016). https://doi.org/10.1 007/s00366-016-0447-0
7. Annaty, M., Eghbalzadeh, A., Hosseini, S.: Hybrid ANFIS model for predicting scour depth using particle swarm optimization. Indian J. Sci. Technol. **8**, 326–332 (2015). https://doi.org/ 10.17485/ijst/2015/v8i
8. Basser, H., Karami, H., Shamshirband, S., Akib, S., Amirmojahedi, M., Ahmad, R., et al.: Hybrid ANFIS-PSO approach for predicting optimum parameters of a protective spur dike. Appl. Soft Comput. J. (2015). https://doi.org/10.1016/j.asoc.2015.02.011
9. Pankaj, G.: Evaluation of scour depth around bridge piers. Gauhati University (2013)
10. Cortes, C., Vapnik, V.: Support-vector networks. Mach. Learn. **20**, 273–297 (1995). https://do i.org/10.1023/A:1022627411411
11. Zhang, X., Guo, Y.: Optimization of SVM parameters based on PSO algorithm. In: 2009 Fifth International Conference on Natural Computation, vol. 1, pp. 536–539 (2009). https://doi.org/ 10.1109/icnc.2009.257
12. Mahjoobi, J., Mosabbeb, E.A.: Prediction of significant wave height using regressive support vector machines. Ocean Eng. **36**, 339–347 (2009). https://doi.org/10.1016/j.oceaneng.2009.0 1.001
13. Mandal, S., Rao, S., Harish, N., Lokesha: Damage level prediction of non-reshaped berm breakwater using ANN, SVM and ANFIS models. Int. J. Nav. Archit. Ocean Eng. **4**, 112–122 (2012). https://doi.org/10.3744/jnaoe.2012.4.2.112
14. Ivanciuc, O.: Applications of support vector machines in chemistry. Rev. Comput. Chem. **23**, 291–400 (2007). https://doi.org/10.1002/9780470116449.ch6
15. Kennedy, J., Eberhart, R.: Particle swarm optimization. In: 1995 IEEE International Conference on Neural Networks (ICNN'95), 1942–1948, vol. 4 (1995). https://doi.org/10.1109/icnn.1995. 488968
16. Harish, N., Mandal, S., Rao, S., Patil, S.G.: Particle swarm optimization based support vector machine for damage level prediction of non-reshaped berm breakwater. Appl. Soft Comput. J. **27**, 313–321 (2015). https://doi.org/10.1016/j.asoc.2014.10.041

A Multi-scale Retinex with Color Restoration (MSR-CR) Technique for Skin Cancer Detection

Prapti Pandey, Praneet Saurabh, Bhupendra Verma and Basant Tiwari

Abstract Image enhancement is one of the key concerns pertaining to better quality image photography captured through modern digital cameras. Probability of digital images getting compromised through lightning and weather conditions remains high. Due to these environmental limitations, many a time loss of information from images is reported. Major role of image amplification is to bring out hidden details of an image from the sample. It provides multiple options for enhancing the visual quality of images. This paper addresses problem of early skin cancer detection using image enhancement techniques and presents a multi-scale retinex with color restoration (MSR-CR) technique for skin cancer detection. The actual skin portion suffering from cancer is identified by comparing enhanced image with available ground truth image. Experimental result shows significant improvement over previously available techniques.

Keywords Skin cancer · Diseases · Melanoma · Color enhancement · Image enhancement · Retinex algorithm

1 Introduction

Advent of technology facilitated detection of various diseases. But all these techniques require precise detection. Skin cancer is such a disease where accuracy among images matters a lot and plays very significant role. Accuracy in diagnosis of skin

P. Pandey (✉) · P. Saurabh · B. Verma · B. Tiwari
Technocrats Institute of Technology, Bhopal 462021, Madhya Pradesh, India
e-mail: kanha.prapti24@gmail.com

P. Saurabh
e-mail: praneetsaurabh@gmail.com

B. Verma
e-mail: bkverma3@gmail.com

B. Tiwari
e-mail: basanttiw@gmail.com

© Springer Nature Singapore Pte Ltd. 2019
J. C. Bansal et al. (eds.), *Soft Computing for Problem Solving*, Advances in Intelligent Systems and Computing 817, https://doi.org/10.1007/978-981-13-1595-4_37

cancer helps in early and its precise detection and thereafter reduces the mortality rate. The skin disease does not affect the skin. It can also affect the quality of life. So, identifying the hidden pattern is very important for better diagnosis and enhances the quality of image. The abnormal growth of cells leads to cancer. Basically, damaged cell of mass and tissues or lumps are the one which become harmful. The lump or masses of tissue are known as tumors. The uncontrolled growth of tissue affects human body functions. The digestive, circulatory, and nervous systems get affected badly due to tumors [1]. Skin cancer is caused due to uncontrolled abnormal growth of skin cells. There is rapid multiplication of skin cells due to unrepaired DNA. The unrepaired DNA damages the skin cell in the form of mutation or genetic defects and forms malignant tumors. Ultraviolet rays (UVR) are the major cause for skin cancer, due to sunlight and artificial source like sun beds. In this paper, retinex theory is used for precise detection of cancer region, exact region. This paper is organized in the following way: Section 2 discusses the basic theory and related work, Sect. 3 presents proposed work, Sect. 4 introduces experimentation and results analysis, and in the end Sect. 5 contains the conclusion.

2 Basic Theory and Related Work

2.1 Basic Theory

Skin cancer is a state that is not desired and termed as dangerous as it effects skin directly and has severe consequences. Primarily, it is classified as malignant melanoma and non-melanoma skin cancers with another variation illustrated in Fig. 1.

Malignant melanoma is the prime or most prominent reason for the sores that leads to bleeding [1, 2]. Malignant melanoma matures from the cancerous development in pigmented skin injury. In this scenario, it is still curable if detected at the right time [3]. Many a time, laboratory sampling of skin results in inflammation that even widens in some specific cases. Detection of cancerous cells is based on images of the affected areas which in some cases suffer due to environmental variations and other external factors [4, 5]. These limitations in image capturing and processing are

Fig. 1 Block diagram of types of skin cancer

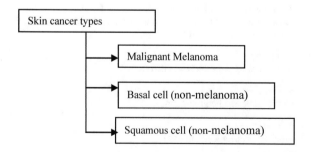

addressed in the proposed multi-scale retinex with color restoration (MSR-CR) for skin cancer detection. The actual skin portion suffering from cancer is identified by comparing enhanced image with available ground truth image.

2.2 Related Work

This section explores the various advancements in current state of the art in the domain of skin cancer detection system, concerning with a different method and technology. Choudhari and Biday [6] presented a mechanism for early detection of skin cancer. This system reduced mortality and morbidity. Some subsequent works have used image segmentation, feature extraction techniques and fused it with neural networks for better results. In another significant work, Mendonca et al. [7] illustrated a system that captured skin cancer images with mobile camera and subsequently demonstrated superior results in doubtful cases. Thereafter, Amarathunga and Ellawala [8] put forward a new technique that integrated image processing and data mining to diagnose and identify human skin for potential skin cancer very efficiently. In another critical work, Nasr-Esfahani et al. [9] fused graphical processing with deep learning for early skin cancer detection. Afterward, Premaladha and Ravichandran [10], Jaworek-Korjakowska [11], and Basturk et al. developed different effective systems that showed significant improvements. Later on, Basturk et al. [12] identified and stated melanoma as a serious cancer that results in grave consequences. In their work, it is diagnosed with interpretation of the dermoscopy images by the ABCD rule. In this explore, a deep neural network (DNN) is used as a new method for diagnosis of melanoma skin cancer.

All the existing methods focused on the extraction and classification of significant feature to produce result with black and white regions that actually suffered from the environmental complexities. This paper presents, Multi-scale retinex with color restoration that enhances images not only for the detection of skin cancer but also provide the better result of image limitations with color contrast. The proposed system identifies the low contrast, over-saturation, and non-uniform illumination. This work addresses that issue, proposes a method to identify the exact area of skin cancer, and separates it from unaffected or less affected area. Some of recent works using bio-inspired computing [13] have acquired attention as it proves to be helpful in realizing different goals [14, 15] of this domain [16].

3 Proposed Work

This section presents a multi-scale retinex with color restoration (MSR-CR) technique for skin cancer detection to prevail over the limitations.

3.1 *Multi-scale Retinex with Color Restoration*

The proposed multi-scale retinex with color restoration (MSR-CR) technique for skin cancer detection selects actual skin portion suffering from cancer, makes enhancements, and then compares the image with available ground truth image. In this method, image enhancement employs image improvement and image restoration. Restored color images are good in contrast as compared to the original images.

3.2 *Algorithm-1*

The presented algorithm strives to enhance the quality of melanoma image, in the pre-processing step as it takes out the additional information. This step helps in detecting skin cancer more accurately and precisely.

Step 1: take an input image. (Input: Image I).
Step 2: perform preprocessing of image:
    PI: Preprocess (I)   // PI: preprocess Image
Step 3: After pre-processing perform segmentation of image obtained in previous Process.
    SI: Segmentation (PI)   // SI: Segmented image
Step 4: now, take a loop for no. of Rows and Columns to find out the filtered image:
    Loop 1: M   // M: No. of Rows
    Loop 1: N   // N: No. of columns
    FI (M, N)   Median (SI, M, N) // FI: Filtered Image
    End loop.
    End loop.
Step 5: To identify the segmented color image, again take a looping process on Filtered image obtained in the above filtration process. And Apply retinex Algorithm (Multi-Scale retinex with color restoration) technique on image.
    Loop 1: M;   // M: No. of Rows
    Loop 1: N;   // N: No. of columns
    SCI (M, N)   MSR-CR (FI [M, N])
    End loop.
    End loop.
Step 6: Finally the skin cancer diagnosis system traced the actual treatment area Of skin cancer in originality (with enhance color of image).

4   Simulation and Result Analysis

4.1   *Experimentation*

This section covers the simulations and subsequent results analysis and then comparisons of proposed MSR-CR with the existing state of the art to establish the improvements.

4.2   *Simulation*

Numerical computation and visualization are performed using MATLAB. It provides an interactive atmosphere. MATLAB stands for matrix laboratory and possesses many toolboxes with inbuilt functions. Matrix is the basic constituent of MATLAB. Arrays are the fundamental data type of MATLAB.

4.3   *Assessment Methodology*

		(+Ve) detection	(-Ve) detection
Actual	Positive	A: True positive	B: False negative
	Negative	C: False positive	D: True negative

(i)   **True positive rate (TPR) or recall or sensitivity**: It is the fraction that tells the number of relevant instances over a combination of total number of instances.

$$TPR = \frac{True\ possitive}{True\ possitive + False\ negative} \tag{1}$$

(ii)   **Positive predictive value (PPV) or precision**: It is the fraction that states number of relevant instances among total number of retrieved instances.

$$PPV = \frac{True\ possitive}{True\ possitive + False\ possitive} \tag{2}$$

(iii)   **F-Measure**: It is defined as the measure of test accuracy as weighted harmonic mean of the PPV and TPR of the test. Therefore, harmonic mean of PPV and TPR is given as:

$$F\text{-measure} = \frac{2 * (positive\ predictive\ value * true\ positive\ rate)}{Positive\ predictive\ value + True\ positive\ rate} \tag{3}$$

4.4 Results Analysis

This section explains the experiments on the skin cancer images using MATLAB as it contained rich libraries which have many inbuilt functions that can be directly used for attaining different purposes. Performance results are compared with previous work using support vector machine (SVM) because SVM focused on the extraction and classification of significant feature to produce result in black and white regions. These results showed more accurate values under different performance parameters as compared to the method based on SVM. Experimental results also specified that MSR-CR successfully enhanced to a much better extent than SVM.

Table 1 shows the comparative results obtained under proposed MSR-CR and state of art based on SVM. MSR-CR reported better result of image limitations along with color contrast even in the cases of poor quality of images. This work also addressed the issue of exact identification of the region of skin cancer and separates it from unaffected or less affected area.

Table 2 shows the experimental results of the proposed MSR-CR with the state of art under different parameters of precision, recall, and f-measure. Experimental results show that the proposed work has achieved high precision value, high recall value, and less execution time in the MSR-CR against state of art. It has shown in table that MSR-CR algorithm is more accurate as compared to SVM.

Figure 2 shows the comparison of precision value of MSR-CR and support vector machine. Figure 2 very clearly shows high precision value as compared to SVM. Under different scenarios, image enhancement through MSR-CR yields more accurate result than support vector machine. Figure 3 shows the comparison of recall value of MSR-CR and SVM. From the results, it can be very easily concluded that MSR-CR performs better and yields superior results under different scenarios.

Table 1 Comparison with MSR-CR and state of the art [10]

S. No	Original image	Ground truth Image	Performance state of art [3]	Performance with MSR-CR
1				
2				
3				

Table 2 Comparison with MSR-CR and state of the art [10]

Images	Methods	Precision	Recall (TPR)	F-measure
1	MSR-CR algorithm	0.733,333	0.93617	0.82243
	State of art [10]	0.235174	0.881226	0.371267
2	MSR-CR algorithm	0.745455	0.82	0.780952
	State of art [10]	0.147702	0.5625	0.233969
3	MSR-CR algorithm	0.672131	0.931818	0.780952
	State of art [10]	0.239955	0.669782	0.353328
4	MSR-CR algorithm	0.816327	0.727273	0.769231
	State of art [10]	0.101373	0.585366	0.172817

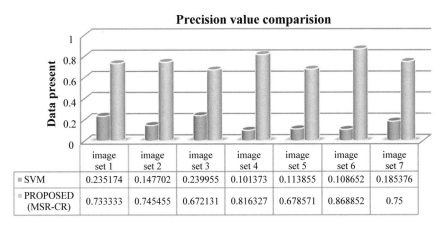

Precision value comparision

	image set 1	image set 2	image set 3	image set 4	image set 5	image set 6	image set 7
SVM	0.235174	0.147702	0.239955	0.101373	0.113855	0.108652	0.185376
PROPOSED (MSR-CR)	0.733333	0.745455	0.672131	0.816327	0.678571	0.868852	0.75

Fig. 2 Comparison of precision value from the proposed MSR-CR algorithm and SVM

5 Conclusion

This research work proposed and developed multi-scale retinex with color restoration (MSR-CR) technique for skin cancer detection that selects actual skin portion suffering from cancer, makes enhancements, and then compares the image with available ground truth image. MSR-CR processed the images quite well and reported better results as compared to the existing state of art. Standard MSR performed a mixture of local (via ratios) and global (via pigmented logarithms) contrast adjustment. In this research work, firstly detection of cancer with restoration (retinex) algorithm is done and then retinex theory is applied for enhancement of system performance. In

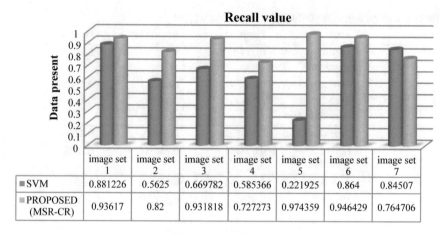

Fig. 3 Comparison of recall (TPR) value from the proposed MSR-CR algorithm and SVM

the part of retinex theory, MSR-CR algorithm is applied for enhancing the image quality so that better results can be obtained even in the cases of for poor quality of images.

References

1. Thangaraju, P., Mehala, R.: Novel classification based approaches over cancer diseases. Int. J. Adv. Res. Comput. Commun. Eng. **4**(3), 294–297 (2015)
2. Sheha, M.A., Mabrouk, M.S.: Automatic detection of melanoma skin cancer. Int. J. Comput. Appl. (0975–8887) 22–26 (2012)
3. National Cancer Institute: Melanoma and other skin cancer, US department of health and human service, NIH publication no. 10-7625 (2010)
4. Karargyris, A., Karargyis, O., Pantelopoulos, A.: An advanced image-processing mobile application for monitoring skin cancer. In: IEEE 24th International Conference on Tools with Artificial Intelligence, vol. 2, pp. 1–7 (2012)
5. Chauhan, S., Prasad, R., Saurabh, P., Mewada, P.: Dominant and LBP-based content image retrieval using combination of color, shape and texture features. In: Progress in Computing, Analytics and Networking. Advances in Intelligent Systems and Computing, vol. 710, pp. 235–243. Springer (2018)
6. Choudhari, S., Biday, S.: Artificial neural network for skin cancer detection. Int. J. Emerg. Trends Technol. Comput. Sci. **3**(5), 147–153 (2014)
7. Mendonca, T., Ferreira, P.M., Marques, J.S., Marcal, A.R., Rozeira, J.: PH2—A dermoscopic image database for research and benchmarking. In: 35th Annual International Conference of the IEEE Engineering in Medicine and Biology Society, pp. 5437–5440 (2013)
8. Amarathunga, A.A.L.C., Ellawala, E.P.W.C.: Expert system for diagnosis of skin diseases. Int. J. Sci. Technol. Res. **4**(1), 174–178 (2015)
9. Nasr-Esfahani, E., Samavi, S., Karimi, N., Soroushmehr, S.M.R., Jafari, M.H., Ward, K., Najarian, K.: Melanoma detection by analysis of clinical images using convolutional neural network. In: 38th Annual International Conference of the IEEE Engineering in Medicine and Biology Society, pp. 1373–1376 (2016)

10. Premaladha, J., Ravichandran, K.S.: Novel approaches for diagnosing melanoma skin lesions through supervised and deep learning algorithms. J. Med. Syst. 96 (2016)
11. Jaworek-Korjakowska, J.: Computer-Aided Diagnosis of Micro-Malignant Melanoma Lesions Applying Support Vector Machines, vol. 2016, 8 pp. Hindawi Publishing Corporation, Article ID 4381972 (2016)
12. Basturk, A., Yukesi, M.E., Badem, H., Caliskan, A.: Deep neural network based diagnosis system for melanoma skin cancer. In: 25th IEEE Signal Processing and Communications Applications Conference (SIU), pp. 1–4 (2017)
13. Saurabh, P., Verma, B.: An efficient proactive artificial immune system based anomaly detection and prevention system. Expert Syst. Appl. Elsevier **60**, 311–320 (2016)
14. Saurabh, P., Verma, B.: Cooperative negative selection algorithm. Int. J. Comput. Appl. (0975–8887) **95**(17), 27–32 (2014)
15. Saurabh, P., Verma, B, Sharma, S.: An immunity inspired anomaly detection system: a general framework a general framework. In: Proceedings of Seventh International Conference on Bio-Inspired Computing: Theories and Applications (BIC-TA 2012). Advances in Intelligent Systems and Computing, vol. 202, pp. 417–428. Springer (2012)
16. Saurabh, P., Verma, B, Sharma, S.: Biologically inspired computer security system: the way ahead. In: Recent Trends in Computer Networks and Distributed Systems Security. Communications in Computer and Information Science, vol. 335, pp. 474–484. Springer (2011)

Generative Power of Hexagonal Prusa Grammar Model Revisited

T. Kamaraj

Abstract Hexagonal Prusa grammar (hex-PG) is an appealing version of Prusa grammar model proposed recently for generating hexagonal picture patterns. This model features the hexagonal context-free derivation rules in parallel application mode. It has been already proved that the class of languages generated by hex-PG includes that of classical context-free hexagonal array grammars and is included in a recent class of languages generated by regional hexagonal tile rewriting grammars. In this paper, we study the expressive power of hex-PG by comparing it with other existing hexagonal array generating grammars involving parallel rewriting rules.

Keywords Hexagonal arrays · Hexagonal Prusa grammars · Pure grammars
Parallel contextual grammars

1 Introduction

Hexagonal picture patterns occur in several application areas, especially in picture processing and image analysis [5, 7]. To study the problem of hexagonal picture generation where pictures are considered as connected, finite arrays of symbols in a hexagonal grid, there has been continued interest in adapting techniques of formal string language theory for developing new grammar models [2, 3, 7, 9]. Hexagonal Kolam array grammars for generating hexagonal arrays on triangular grid, which can be treated as two-dimensional representation of 3D blocks, were constructed by Siromoney and Siromomey [7] in 1976 and were later generalized to hexagonal array grammars (HAG) [8]. Recently, Kamaraj and Thomas [3] introduced an isometric version of hexagonal array grammars called regional hexagonal tile rewriting grammars (RHTRG) based on the notion of regional partition of pictures, which is more general than HAG. On the other hand, Subramanian et al. proposed a non-isometric grammar model, pure 2D hexagonal context-free array grammars

T. Kamaraj (✉)
Department of Mathematics, Sathyabama University, Chennai 600119, India
e-mail: kamaraj_mx@yahoo.co.in

© Springer Nature Singapore Pte Ltd. 2019
J. C. Bansal et al. (eds.), *Soft Computing for Problem Solving*, Advances in Intelligent
Systems and Computing 817, https://doi.org/10.1007/978-981-13-1595-4_38

(pure-HCFG) [9, 10], which exhibits parallel rewriting rules using terminal symbols only. Differently, D. G. Thomas et al. introduced parallel contextual hexagonal array grammars (PCHAG) [11], in which array contexts are used to generate the hexagonal pictures.

Prusa grammar, recently introduced by Prusa [6] for generating rectangular pictures, exploits parallel application of rules and maintains a rectangular grid pattern in all stages of derivation. The hexagonal variant of this model, hexagonal Prusa grammar (hex-PG), is later attempted by Kamaraj and Thomas [4] with the motivation of framing hexagonal grammar device involving parallel rewriting rules to the full extent, which are very rare in the literature. In [4], hex-PG is compared with sequential and parallel hexagonal grammar models, HAG and RHTRG, and proved to be more general than context-free HAG but weaker than RHTRG. In this paper, we deepen the study on expressiveness of hex-PG by comparing it with hexagonal array generating models, extended pure-HCFG and z-direction external PCHAG, which involve rich application of parallel derivation rules.

In Sect. 2, we brief the notions of hexagonal pictures, languages and generating devices. In Sect. 3, we review the notions of hexagonal Prusa grammar and illustrate through examples. In Sect. 4, we produce the comparison results of hex-PG with other hexagonal array generating models.

2 Preliminaries

Let D be a finite alphabet of symbols. A hexagonal array (picture) H is a function from a set of hexagonal grid points of two-dimensional plane to D. The coordinates of hexagonal array elements are determined by the triad of triangular axes [2].

Let D^H denote the set of all hexagonal pictures over the alphabet D, and D^{+H} denote the set of all non-empty hexagonal pictures over D. Let D^* denotes the set of all strings in the three directions parallel to the triangular axes x, y and z over D, and D^+ denotes set of all non-empty strings in the x, y, z-directions. A hexagonal picture language L over D is a subset of D^{+H}.

We study hexagons of the shape

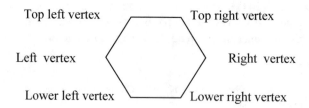

Given a hexagonal array b $\in$ D^{+H}, let (l, m, n) denote the size of b where ℓ, m, n denote the number of elements in the boundary of b respectively from top left vertex to left vertex parallel to x (axes)-direction, from top right vertex to right vertex along y (axes)-direction and from top left vertex to top right vertex in the z (axes)-direction. The coordinates are determined with origin as $(1, 1, 1)$ representing the top left vertex position.

A hexagonal array b of size (ℓ, m, n) can be symbolized as $[b_{ijk}]^{+(\ell,m,n)H}$, where b_{ijk} denote the symbol in b, having the coordinates (i, j, k) where $1 \leq i \leq \ell, 1 \leq j \leq m$, $1 \leq k \leq n$. D$^{+(\ell,m,n)H}$ denote the set of hexagonal arrays of size (ℓ, m, n) over D.

In the sequel, we adapt the notions of arrowhead, a non-convex hexagonal array and its catenations pertaining to hexagonal array growth as in [2, 8]. We can also use a compact notation, $x\langle y\rangle z$ to denote an arrowhead (in clock-wise) [8] in any of the six directions, where x, y, z are rectangular arrays of symbols with same number of rows.

We now recall the formal definition of pure 2D hexagonal context-free grammars [9, 10] and z-direction external parallel contextual hexagonal array grammars [11].

Definition 1 A pure 2D hexagonal context-free grammar (pure-HCFG) is an 8-tuple G = (D, P$_{tr}$, P$_{tl}$, P$_{lr}$, P$_{ll}$, P$_l$, P$_r$, A$_0$), where D is a finite alphabet, P$_{ur}$ = $\{c_{tr_i}/1 \leq i \leq m\}$; Each c_{tr_i}, $(1 \leq i \leq m)$, called a TR table, is a set of context-free rules of the form d $\rightarrow \alpha$, d $\in$ D, $\alpha \in$ D* such that any two rules of the form d $\rightarrow \alpha$, e $\rightarrow \beta$ in c_{tr_i}, we have $|\alpha| = |\beta|$ where $|\alpha|$ denotes the length of $\propto$, each of the other five components P$_{tl}$, P$_{lr}$, P$_{ll}$, P$_l$ and P$_r$ is similarly defined, A$_0 \subseteq$ D^{+H} is called set of axioms, a finite set of hexagonal pictures (arrow heads).

Derivation of pictures and languages are stated as follows:

For any two hexagonal arrays b_1, b_2, we write $b_1 \Rightarrow b_2$ if b_2 is obtained from b_1 by rewriting all the symbols simultaneously in an arrowhead of thickness one of b_1 by rules of a relevant table in P$_{ur} \cup$ P$_{ul} \cup$ P$_{lr} \cup$ P$_{ll} \cup$ P$_l \cup$ P$_r$. $\overset{*}{\Rightarrow}$ is the reflexive transitive closure of $\Rightarrow$. The hexagonal array language L(G) generated by G is the set $\{b/m_0 \overset{*}{\Rightarrow} b \in$ D^H, for some $m_0 \in$ A$_0\}$.

To increase the generative power of pure-HCFG, we associate a regular control language defined over an alphabet Γ and also consider D = D$_f \cup$ D$_c$ where D$_f$ is the alphabet of terminal symbols and D$_c$ is the set of intermediate symbols as in [1] and called as extended pure 2D hexagonal context-free grammars (epure-HCFG).

Definition 2 A z-direction external parallel hexagonal contextual array grammar (ze-PHCAG) is a tuple F = $\langle$D, E, C$_{xy}$, C$_{yx}$, C$_{xx}$, C$_{yy}$, ϕ_{xy}, ϕ_{xy}, ϕ_{xx}, $\phi_{yy}\rangle$ where D is a finite alphabet, E $\subseteq$ D^H is a finite hexagonal array language over D, called the base of F; C$_{xy}$ is a finite set of elements of the form $\begin{bmatrix} u_1 \\ u_2 \end{bmatrix} \#_{xy} \begin{bmatrix} v_1 \\ v_2 \end{bmatrix}$, where u_1, u_2 are of the size $1 \times m$ and v_1, v_2 are of the size $1 \times n$ for some $m, n \geq 0, m + n > 0$ and $\#_{xy}$ is a new symbol not belonging to D. The elements of C$_{xy}$ are called xy contexts of F. When $u_i = \lambda$ (empty array) for $i = 1, 2$. Then mentioning of u_1, u_2 in $\begin{bmatrix} u_1 \\ u_2 \end{bmatrix} \#_{xy} \begin{bmatrix} v_1 \\ v_2 \end{bmatrix}$ can be avoided. Likewise for the case when $v_i = \lambda$ for $i = 1, 2$.

In the similar notion, C_{yx} denotes a finite set of yx array contexts of the form $\begin{bmatrix} u_1 \\ u_2 \end{bmatrix}$

$\#_{yx} \begin{bmatrix} v_1 \\ v_2 \end{bmatrix}$, C_{xx} denotes a finite set of xx array contexts of the form $\begin{bmatrix} u_1 \\ u_2 \end{bmatrix} \#_{xy}$

$\begin{bmatrix} v_1 \\ v_2 \end{bmatrix}$, C_{yy} denotes a finite set of yy array contexts of the form $\begin{bmatrix} u_1 \\ u_2 \end{bmatrix} \#_{yy} \begin{bmatrix} v_1 \\ v_2 \end{bmatrix}$.

$\phi_{xy} : D^H \rightarrow 2^{C_{xy}}$, $\phi_{yx} : D^H \rightarrow 2^{C_{yx}}$, $\phi_{xx} : D^H \rightarrow 2^{C_{xx}}$, $\phi_{yy} : D^H \rightarrow 2^{C_{yy}}$ are partial mappings and are, respectively, called the xy, yx, xx, yy context choice functions. There exists a relation on D^H, denoted by $\Rightarrow_{ze}$ defined as follows, associated with the grammar.

For any b, c $\in D^H$, b $\Rightarrow_{ze}$ c if and only if c $= l$ b r, where l, r are left and right arrow heads, obtained by the xx, yx, xx, yy contextual operations corresponding to choice functions. $\overset{*}{\Rightarrow}_{ze}$ is a reflexive and transitive closure of the relation $\Rightarrow_{ze}$.

The hexagonal array language generated by ze-PHCAG grammar F is defined to be L[F] = {b $\in D^H$/$\exists$c $\in$ E such that c $\overset{*}{\Rightarrow}_{ze}$ b}.

3 Hexagonal Prusa Grammar

In this section, we recall formal definition of hexagonal Prusa grammar [4] and illustrate some simple examples of languages generated by these grammars.

Definition 3 A hexagonal Prusa grammar (hex-PG) is a tuple G $= \langle M, D, R, S \rangle$ where M is a finite set of non-terminals, D is a finite set of terminals, R : M $\rightarrow$ [(M $\cup$ D)$^{+H} \cup$ (M $\cup$ D)$^+$] and S $\in$ M is the start symbol.

The hexagonal picture language L(G$_A$) over D for every $A \in M$, defined by the following recursive productions.

(1) Terminal production: If A $\rightarrow$ Y is in R, and Y $\in$ (D$^{+H} \cup$ D$^+$), then Y $\in$ L(G$_A$).
(2) Mixed production: Let A $\rightarrow$ Y be in R with Y $\in \cup$ (M $\cup$ D)$^{+(\ell', m', n')H}$, $1 \leq \ell' \leq \ell$, $1 \leq m' \leq m$ and $1 \leq n' \leq n$ and $P_{ijk}(1 \leq i \leq \ell, 1 \leq j \leq m, 1 \leq k \leq n)$ be such that

    (i) if Y_{ijk} is terminal then $P_{ijk} = Y_{ijk}$
    (ii) if Y_{ijk} is non-terminal then $P_{ijk} \in$ L[G$_Y$]

Then, if P $= [P_{ijk}]^{+(l', m', n')H}$ is formed using either string catenation or arrow head catenation then P $\in$ L[G$_A$].

The set L[G$_A$] contains only those pictures that can be obtained by applying the productions (1) and (2). The hexagonal language L[G] generated by the grammar G is defined to be the language L[G$_S$].

Informally, the productions result in either hexagonal array of terminal symbols or hexagonal array of terminals and non-terminals.

Example 1 The language

$$L_1 = \left\{ \begin{array}{cccc} \begin{matrix} c & c \\ c & c & c, \\ c & c \end{matrix} & \begin{matrix} c & c & d \\ c & c & c & d, \\ c & c & d \end{matrix} & \begin{matrix} c & c & d & d \\ c & c & c & d & d, \\ c & c & d & d \end{matrix} & \cdots \end{array} \right\}$$

is generated by hex-PG $G_1 = \langle M, D, R, S \rangle$ where $M = \{Y, S\}$, $D = \{c, d\}$ and

$$R = \left\{ S \rightarrow SY, \quad S \rightarrow \begin{matrix} c & c \\ c & c & c, \\ c & c \end{matrix} \quad Y \rightarrow \begin{matrix} d \\ d \\ d \end{matrix} \right\}$$

By applying terminal productions

$$Y \rightarrow \begin{matrix} d \\ d, \\ d \end{matrix} \quad S \rightarrow \begin{matrix} c & c \\ c & c & c \\ c & c \end{matrix} \text{ we get } \begin{matrix} d \\ d \\ d \end{matrix} \in L[G_Y] \text{ and } \begin{matrix} c & c \\ c & c & c \\ c & c \end{matrix} \in L[G_S]$$

By applying mixed production $S \rightarrow SY$, we have $\begin{matrix} a & a & d \\ a & a & a & d \\ a & a & d \end{matrix} \in L[G_Y]$

Similarly if $\begin{matrix} c & c & d^k \\ c & c & c & d^k \\ c & c & d^k \end{matrix} \in L[G_S]$

on applying mixed production $S \rightarrow SY$, we have $\begin{matrix} c & c & d^{k+1} \\ c & c & c & a^{k+1} \\ c & c & d^{k+1} \end{matrix} \in L[G_S]$.

Example 2 The language

$$L_2 = \left\{ \begin{array}{cccc} \begin{matrix} 0 & 0 \\ 0 & 1 & 0, \\ 0 & 0 \end{matrix} & \begin{matrix} 0 & 0 & 0 \\ 0 & 1 & 1 & 0, \\ 0 & 0 & 0 \end{matrix} & \begin{matrix} 0 & 0 & 0 & 0 \\ 0 & 1 & 1 & 1 & 0, \\ 0 & 0 & 0 & 0 \end{matrix} & \cdots \end{array} \right\}$$

can be generated by the hex-PG

$G_2 = \langle M, D, R, S \rangle$ where $M = \{S, U, V, T\}$, $D = \{0, 1\}$ and

$$R = \left\{ S \rightarrow UT / \begin{matrix} 0 & 0 \\ 0 & 1 & 0, \\ 0 & 0 \end{matrix} \quad U \rightarrow UV / \begin{matrix} 0 & 0 \\ 0 & 1 & 0, \\ 0 & 0 \end{matrix} \quad V \rightarrow \begin{matrix} 0 \\ 1, \\ 0 \end{matrix} \quad T \rightarrow \begin{matrix} 0 \\ 0 \\ 0 \end{matrix} \right\}$$

From the terminal productions, $0 \begin{smallmatrix} 0 & 0 \\ 1 & 0 \\ 0 & 0 \end{smallmatrix} \in L[G_S]$, $\begin{smallmatrix} 0 \\ 1 \\ 0 \end{smallmatrix} \in L[G_V]$. Mixed productions: $U \rightarrow UV$, $S \rightarrow UT$ repeatedly yields the hexagonal pictures of L_2

Example 3 Let $G_3 = \langle M, D, R, S \rangle$, where $M = \{U, V, W, S\}$, $D = \{0, 1\}$ and

$$R = \left\{ S \rightarrow \begin{smallmatrix} 1 & U \\ W & S & 1 \\ 1 & V \end{smallmatrix}, \quad U \rightarrow \begin{smallmatrix} 0 & U \\ & 0 \end{smallmatrix} \middle/ 1, \quad V \rightarrow \begin{smallmatrix} 0 \\ 0 & V \end{smallmatrix} \middle/ 1, \right.$$

$$\left. W \rightarrow \begin{smallmatrix} 0 \\ W \\ 0 \end{smallmatrix} \middle/ 1, \quad S \rightarrow \begin{smallmatrix} 1 & 1 \\ 1 & 1 & 1 \\ 1 & 1 \end{smallmatrix} \right\}$$

generates the language L_3 of symmetrical hexagonal pictures over $\{0, 1\}$ with 1's in the diagonal positions and 0's in the other positions.

From the terminal production $S \rightarrow \begin{smallmatrix} 1 & 1 \\ 1 & 1 & 1 \\ 1 & 1 \end{smallmatrix}$, we have $\begin{smallmatrix} 1 & 1 \\ 1 & 1 & 1 \\ 1 & 1 \end{smallmatrix} \in L[G_S]$.

After applying the terminal and mixed productions with left-hand sides as U, V, W, we have

$0 1 \begin{smallmatrix} 0 \\ & \\ 0 \end{smallmatrix} \in L[G_U]$, $0 B \begin{smallmatrix} 0 \\ \\ 0 \end{smallmatrix} \in L[G_V]$ and $1 \begin{smallmatrix} 0 \\ \\ 0 \end{smallmatrix} \in L[G_W]$.

Again by applying mixed production $S \rightarrow \begin{smallmatrix} 1 & U \\ W & S & 1 \\ 1 & V \end{smallmatrix}$ we get

$$\begin{array}{ccccc} & 1 & 0 & 1 & \\ 0 & 1 & 1 & 0 & \\ 1 & 1 & 1 & 1 & 1 \\ 0 & 1 & 1 & 0 & \\ & 1 & 0 & 1 & \end{array} \in L[G_S].$$

If the whole application sequence of rules repeat for another time, it will yield

$$
\begin{array}{ccccccc}
1 & 0 & 0 & 1 & & & \\
0 & 1 & 0 & 1 & 0 & & \\
0 & 0 & 1 & 1 & 0 & 0 & \\
1 & 1 & 1 & 1 & 1 & 1 & 1 \in L[G_s]. \\
0 & 0 & 1 & 1 & 0 & 0 & \\
0 & 1 & 0 & 1 & 0 & & \\
1 & 0 & 0 & 1 & & &
\end{array}
$$

Example 4 The language L_4 of left arrowheads of width one over $D = \{x, y\}$ with

symbols forming palindromic words of the form

$$
\begin{array}{c}
x \\
\quad x \\
\quad\quad y \\
x \\
\quad y \\
\quad\quad x \\
\quad\quad\quad x
\end{array}
$$

is generated by the hex-PG, $G = \langle M, D, R, S \rangle$ with $M = \{S\}$, $D = \{x, y\}$,

$$
R = \left\{ S \to \begin{array}{c} x \\ S \\ x \end{array}, \quad S \to \begin{array}{c} y \\ S \\ y \end{array}, \quad S \to \begin{array}{c} x \\ x \\ x \end{array} \middle/ \begin{array}{c} x \\ y \\ x \end{array} \middle/ \begin{array}{c} y \\ x \\ y \end{array} \middle/ \begin{array}{c} y \\ y \\ y \end{array} \right\}
$$

4 Comparison Results

We use £[X] to denote the class of languages generated by the grammar X.

Theorem 1 *£[hex-PG] and £[epure-HCFG] are incomparable but not disjoint.*

Proof The hexagonal pictures of language L_2 given in Example 1 can be generated by a epure-HCFG, $G = (D, P_{ur}, P_{ul}, P_{lr}, P_{ll}, P_l, P_r, A_0)$, where $D = D_f \cup D_c$; $D_f = \{0, 1\}$, $D_c = \{a, b\}$,

$$
P_r = \{r_1, r_2\}; r_1 = \{a \to 0a, b \to 1a\}, r_2 = \{a \to 0, b \to 1\},
$$

$$
P_{ur} = P_{ul} = P_{lr} = P_{ll} = P_l = \varnothing \text{ and } A_0 = \left\{ \begin{array}{ccc} & a & 0 \\ 0 & b & 0 \\ & a & 0 \end{array} \right\}
$$

The regular control language associated with G is $r_1 * r_2$. Hence $£[hex - PG] \cap £[epure - HCFG] \neq \varnothing$

For incomparability: The language L_4 given in Example 4, of left arrowheads comprised of palindromic words, cannot be produced by any epure-HCFG, since by applying controlled table rules of epure-HCFG grammar, the thickness of arrowhead can be increased but not the length of the arrowhead.

On the other hand, if we consider the language $L_A \subseteq D^{+(l,m,n)H}$, where $l = m = 2$ and $n = 2k+3, k \geq 0$, consists of hexagonal arrays of alternate left(right) arrowheads with identical symbols over $\{1, 2\}$ of the form

$$
\begin{array}{ccccccccc}
2 & 1 & 2 & 1 & 1 & 2 & 1 & 2 & \\
2 & 1 & 2 & 1 & 2 & 1 & 2 & 1 & 2 \\
2 & 1 & 2 & 1 & 1 & 2 & 1 & 2 &
\end{array}
$$

This language can be generated by a epure-HCFG, $G = (D, P_{ur}, P_{ul}, P_{lr}, P_{ll}, P_l, P_r, A_0)$, where $D = D_f \cup D_c$; $D_f = \{1, 2\}$, $D_c = \{u, v\}$,

$$P_r = \{r_1, r_2\}; r_1 = \{v \rightarrow 1u, u \rightarrow 2v\}, r_2 = \{v \rightarrow 1, u \rightarrow 2\},$$
$$P_l = \{l_1, l_2\}; l_1 = \{v1 \leftarrow u, u2 \leftarrow v\}, l_2 = \{1 \leftarrow u, 2 \leftarrow v\}$$

$$P_{ur} = P_{ul} = P_{lr} = P_{ll} = \varnothing \text{ and } A_0 = \left\{ \begin{array}{ccc} & u & v \\ u & 2 & v \\ & u & v \end{array} \right\}$$

The regular control language associated with G is $(l_1 r_1) * l_2 r_2$. But the language L_A cannot be generated by any hex-PG. In order to build the pictures of L_A, a rule of hex-PG should be in the form of $S \rightarrow ASB$. But parallel application of rules may yield left and right arrowheads, but will not provide the required alternate pattern as the controlled of rules for A and B is not possible.

Theorem 2 *£[hex-PG] and £[ze-PHCAG] are incomparable but not disjoint.*

Proof The language L_1 in Example 1 can be generated by a ze-PHCAG, $G = \langle D, E, C_{xy}, C_{yx}, C_{xx}, C_{yy}, \phi_{xy}, \phi_{xy}, \phi_{xx}, \phi_{yy} \rangle$, where $D = \{c, d\}$, $E = \left\{ \begin{array}{ccc} & c & c \\ c & c & c \\ & c & c \end{array} \right\}$,

$$C_{xy} = \left\{ \begin{bmatrix} & c \\ c & \end{bmatrix} \#_{xy} \right\}, C_{yx} = \left\{ \begin{bmatrix} c & \\ & c \end{bmatrix} \#_{yx} \right\} \text{ and } C_{xx} = C_{yy} = \varnothing,$$

$$\phi_{xy} \left(\begin{bmatrix} & c \\ c^{k+1} & \end{bmatrix} \right) = \begin{bmatrix} d & \\ & d \end{bmatrix} \#_{xy}, \phi_{yx} \left(\begin{bmatrix} c^{k+1} & \\ & c \end{bmatrix} \right) = \begin{bmatrix} d & \\ & d \end{bmatrix} \#_{yx}$$

Hence $£[hex\text{-}PG] \cap £[ze\text{-}PHCAG] \neq \varnothing$. Now consider the language L_A in the proof of the Theorem 1. This language can be generated by a ze-PHCAG $G = \langle D, E, C_{xy}, C_{yx}, C_{xx}, C_{yy}, \phi_{xy}, \phi_{xy}, \phi_{xx}, \phi_{yy} \rangle$, where $D = \{1, 2\}, E = \left\{ \begin{array}{ccc} & 1 & 1 \\ 1 & 2 & 1 \\ & 1 & 1 \end{array} \right\}$,

$$C_{xy} = \left\{ \begin{bmatrix} & 2 \\ 2 & \end{bmatrix} \#_{xy} \begin{bmatrix} 2 & \\ & 2 \end{bmatrix}, \begin{bmatrix} & 1 \\ 1 & \end{bmatrix} \#_{xy} \begin{bmatrix} 1 & \\ & 1 \end{bmatrix} \right\},$$

$$C_{yx} = \left\{ \begin{bmatrix} 1 \\ & 1 \end{bmatrix} \#_{yx} \begin{bmatrix} 1 \\ & 1 \end{bmatrix}, \begin{bmatrix} 2 \\ & 2 \end{bmatrix} \#_{yx} \begin{bmatrix} 2 \\ & 2 \end{bmatrix} \right\} \text{ and } C_{xx} = C_{yy} = \varnothing,$$

$$\phi_{xy} \left(\begin{bmatrix} (12)^k 11(21)^k \\ (12)^{2k+1} 1 \end{bmatrix} \right) = \begin{bmatrix} 2 \\ & 2 \end{bmatrix} \#_{xy} \begin{bmatrix} 2 \\ & 2 \end{bmatrix},$$

$$\phi_{xy} \left(\begin{bmatrix} (21)^k (12)^k \\ (21)^{2k} 2 \end{bmatrix} \right) = \begin{bmatrix} 1 \\ & 1 \end{bmatrix} \#_{xy} \begin{bmatrix} 1 \\ & 1 \end{bmatrix},$$

$$\phi_{yx} \left(\begin{bmatrix} (12)^{2k+1} 1 \\ (12)^k 11(21)^k \end{bmatrix} \right) = \begin{bmatrix} 2 \\ & 2 \end{bmatrix} \#_{yx} \begin{bmatrix} 2 \\ & 2 \end{bmatrix},$$

$$\phi_{yx} \left(\begin{bmatrix} (21)^{2k} 2 \\ (21)^k (12)^k \end{bmatrix} \right) = \begin{bmatrix} 1 \\ & 1 \end{bmatrix} \#_{yx} \begin{bmatrix} 1 \\ & 1 \end{bmatrix},$$

The hexagonal picture language L_3 generated by hex-PG cannot be generated by any ze-PHCAG since from the definition it is clear that when we add an array context A to a hexagonal picture B, growth of B can happen only in the z-direction.

5 Conclusions

Hexagonal Prusa grammars exploit the parallel application of non-isometric rules to the full extent. Because of this feature, the growing array maintains a hexagonal grid-like structure, which gives more generative capacity than HAG, yet incomparable with other expressive models epure-HCFG and ze-PHCAG. But in future generative capacity of hex-PG model can be enhanced by associating regular or context-free control to the production rules.

References

1. Bersani, M.M., Frigeri, A., Cherubini, A.: On some classes of 2D languages and their relations. In: Aggarwal, J.K., et al. (eds.) IWCIA 2011. LNCS, vol. 6636, pp. 222–234, Springer, Heidelberg (2011)
2. Dersanambika, K.S., Krithivasan, K., Martin-Vide, C., Subramanian, K.G.: Local and recognizable hexagonal picture languages. Int. J. Pattern Recognit. Artif. Intell. 19(7), 853–871 (2005)
3. Kamaraj, T., Thomas, D.G.: Regional hexagonal tile rewriting grammars. In: Barneva, R.P., et al. (eds.) IWCIA 2012. LNCS, vol. 7655, pp. 181–195, Springer, Heidelberg (2012)
4. Kamaraj, T., Thomas, D.G.: Hexagonal Prusa grammar model for context-free hexagonal picture languages. In: Krishnan, G.S.S., et al. (eds.) ICC3 2013. Advances in Intelligent Systems and Computing, vol. 246, pp. 305–311. Springer, India (2014)
5. Middleton, L., Sivaswamy, J.: Hexagonal image processing: a practical approach. In: Advances in Computer Vision and Pattern Recognition Series. Springer (2005)

6. Prusa, D.: Two-dimensional Languages, Ph.D. Thesis, Charles University, Faculty of Mathematics and Physics, Czech Republic (2004)
7. Siromoney, G., Siromomey, R.: Hexagonal arrays and rectangular blocks. Comput. Graph. Image Process. **5**(3), 353–381 (1976)
8. Subramanian, K.G.: Hexagonal array grammars. Comput. Graph. Image Process. **10**(4), 388–394 (1979)
9. Subramanian, K.G., Ali, R.M., Geethalakshmi, M., Nagar, A.K.: Pure 2D picture grammars and languages. Discrete Appl. Math. **157**(16), 3401–3411 (2009)
10. Subramanian, K.G., Geethalakshmi, M., Nagar, A.K., Lee, S.K.: Two-dimensional picture grammar models. In: Proceedings of the 2nd European Modelling Symposium, Liverpool Hope University, England, pp. 263–267 (2008)
11. Thomas, D.G., Humrosia Begam, M., Gnanamalar David, N.: Parallel contextual hexagonal array grammars and languages. In: Proceedings of 13th International Workshop on Combinatorial Image Analysis. LNCS, vol. 5852, pp. 344–357. Springer, Heidelberg (2009)

Substructuring Waveform Relaxation Methods for Parabolic Optimal Control Problems

Bankim C. Mandal

Abstract We study in this paper Dirichlet–Neumann and Neumann–Neumann waveform relaxation methods for the parallel solution of linear-quadratic parabolic optimal control problems, originating from the examples of transient optimal heating with distributed control. Unlike in the case of single linear or nonlinear parabolic problem, we need to solve here two coupled parabolic problems that arise as a part of optimality system for the optimal control problem. We present the detail algorithms for the case of two non-overlapping subdomains and show conditional convergence properties in few special cases. We illustrate our findings with numerical results.

Keywords Dirichlet–Neumann · Neumann–Neumann · Waveform relaxation
Domain decomposition methods · Optimal control problems

1 Introduction

This work is primarily based on one of the active topics of recent years, namely space–time parallel methods for the numerical solution of partial differential equations (PDEs). We consider optimal control problems (OCP) where the constraint is given by a parabolic PDE: We formulate the DNWR and NNWR algorithms for linear-quadratic parabolic optimal control problems that originate from the examples of transient optimal heating with distributed control, or identification of a source of pollution in a waterbody.

To illustrate our ideas, we consider a simple model problem on a bounded spatial domain $\Omega \subset \mathbb{R}^d$ with an objective functional to be minimized as the following:

$$J(y, u) = \frac{1}{2} \int_0^T \int_\Omega \left(y(x, t) - \hat{y}(x, t) \right)^2 dx dt + \frac{\lambda}{2} \int_0^T \int_\Omega u^2(x, t) \, dx dt. \quad (1)$$

B. C. Mandal (✉)
School of Basic Sciences, Indian Institute of Technology Bhubaneswar,
Bhubaneswar 752050, Odisha, India
e-mail: bmandal@iitbbs.ac.in

© Springer Nature Singapore Pte Ltd. 2019
J. C. Bansal et al. (eds.), *Soft Computing for Problem Solving*, Advances in Intelligent
Systems and Computing 817, https://doi.org/10.1007/978-981-13-1595-4_39

The constraint is given by the parabolic PDE with distributed control,

$$\begin{aligned}
\partial_t y(x,t) - \nu \Delta y(x,t) &= u(x,t), \quad x \in \Omega, t \in [0,T], \\
y(x,0) &= y_0(x), \quad x \in \Omega, \\
y(x,t) &= 0, \qquad\quad x \in \partial\Omega \times [0,T].
\end{aligned} \tag{2}$$

For more such space–time problems, see [1, 21]. In our setting, $y(x,t)$ denotes the temperature at a particular point x and a particular time t, ν is the thermal conductivity of Ω, and $\lambda > 0$ is a regularization parameter. We assume $u, \hat{y} \in L^2(0,T; L^2(\Omega))$ to ensure a solution of the problem [13–15]. For simplicity, we also consider $U_{\mathrm{ad}} = L^2(0,T; L^2(\Omega))$ as the set of all feasible controls.

From the first-order optimality conditions (cf. [29]), we obtain a system of equations containing the (forward) state equation and the (backward) adjoint equation which are coupled by an optimality condition. So, one needs to focus on the development of fast and effective tools for their numerical solution. DD methods are often used as powerful parallel computing tools for large-scale optimization problems governed by PDEs. For DD methods applied to linear-quadratic elliptic optimal control problems, see [10, 15, 16]. On the other hand, for time-dependent problems, one can consider either time DD methods [13, 19] or spatial DD methods [2, 4, 14]. Heinkenschloss and Herty [14] presented the numerical behavior of a non-overlapping spatial DD method for the solution of linear-quadratic parabolic optimal control problems. But there is to our knowledge no systematic statement and analysis of the DN [3, 6, 25–27] and NN [5, 7, 20, 28] methods for optimal control problems in the literature, not even in the simple case of the heat equation in 1D or in regular 2D domains. In this article, we systematically introduce the DNWR and the NNWR [8, 9, 11, 17, 18, 22–24] algorithms for the model OCP (1)–(2), and present relevant numerical results. The problem formulation is done in Sect. 2. The algorithms are then presented in Sect. 3, with solution strategy in Sect. 4. Numerical illustration depicting convergence in few cases is presented in Sect. 5.

2   Optimality Condition and Formulation

To obtain the adjoint equation and the optimality condition for the OCP, we consider the Lagrangian [12],

$$L(y,u,p) = J(y,u) + \int_0^T \int_0^1 p(x,t)\left(u(x,t) + \nu\Delta y(x,t) - \partial_t y(x,t)\right) dx dt,$$

with p being the Lagrange multiplier or the adjoint variable. Then from the first-order optimality conditions (cf. [29]), the adjoint equation corresponding to the problem (1)–(2) becomes

$$-\partial_t p(x,t) - \nu \Delta p(x,t) = y(x,t) - \hat{y}(x,t), \quad x \in \Omega, t \in [0,T],$$
$$p(x,T) = 0, \qquad\qquad x \in \Omega, \qquad (3)$$
$$p(x,t) = 0, \qquad\qquad x \in \partial\Omega \times [0,T],$$

with the optimality condition

$$p(x,t) + \lambda u(x,t) = 0. \qquad (4)$$

It is well known that the backward heat equation is ill-posed, but the adjoint Eq. (3) is a terminal value problem, and thus not a backward heat equation. The problem is thus well-posed, which can be seen as follows: Set $U(x,t) = p(x, T - t)$. Then, U satisfies the following initial boundary value equation:

$$\partial_t U - \nu \Delta U = y(x, T - t) - \hat{y}(x, T - t) \text{ in } \Omega, \quad U(x,0) = 0 \text{ in } \Omega, \quad U(x,t) = 0 \text{ on } \partial\Omega,$$

for $t \in [0, T]$, and it is a well-posed problem. The above form is also more suitable for numerical implementation.

Now, we apply the DNWR and the NNWR methods to solve the underlying space–time PDEs (2) and (3). We only consider the error equations in the rest of our discussion; so we take $\hat{y}(x,t) = 0$ and $y_0(x) = 0$ in (1)–(4) and analyze convergence to zero, since the corresponding error equations coincide with these homogeneous equations.

3   DNWR and NNWR Algorithms

To explain the new algorithms, we assume for simplicity that the spatial domain $\Omega = (0,1)$ is partitioned into two non-overlapping subdomains $\Omega_1 = (0, \alpha)$ and $\Omega_2 = (\alpha, 1)$. We denote by y_i, p_i, u_i the restriction of y, p, u to $\Omega_i, i = 1, 2$.

The DNWR for the coupled PDEs consists of the following steps: Given two initial guesses $w_y^{[0]}(t)$ and $w_p^{[0]}(t)$ along the interface $\{x = \alpha\} \times [0, T]$, compute for iteration $k = 1, 2, \ldots$ with $t \in [0, T]$,

$$
\begin{aligned}
\partial_t y_1^{[k]} - \nu \Delta y_1^{[k]} &= u_1^{[k]}, & \text{in } \Omega_1, & \quad -\partial_t p_1^{[k]} - \nu \Delta p_1^{[k]} = y_1^{[k]}, & \text{in } \Omega_1, \\
y_1^{[k]}(x,0) &= 0, & \text{in } \Omega_1, & \quad p_1^{[k]}(x,T) = 0, & \text{in } \Omega_1, \\
y_1^{[k]}(0,t) &= 0, & & \quad p_1^{[k]}(0,t) = 0, & \\
y_1^{[k]}(\alpha,t) &= w_y^{[k-1]}(t), & & \quad p_1^{[k]}(\alpha,t) = w_p^{[k-1]}(t), &
\end{aligned}
\qquad (5)
$$

and

$$\partial_t y_2^{[k]} - \nu \Delta y_2^{[k]} = u_2^{[k]}, \qquad \text{in } \Omega_2, \quad -\partial_t p_2^{[k]} - \nu \Delta p_2^{[k]} = y_2^{[k]}, \qquad \text{in } \Omega_1,$$
$$y_2^{[k]}(x, 0) = 0, \qquad \text{in } \Omega_2, \qquad p_2^{[k]}(x, T) = 0, \qquad \text{in } \Omega_1,$$
$$\partial_x y_2^{[k]}(\alpha, t) = \partial_x y_1^{[k]}(\alpha, t), \qquad \partial_x p_2^{[k]}(\alpha, t) = \partial_x p_1^{[k]}(\alpha, t),$$
$$y_1^{[k]}(1, t) = 0, \qquad p_2^{[k]}(1, t) = 0,$$

$$(6)$$

together with the update conditions

$$w_y^{[k]}(t) = \theta_y y_2^{[k]}(\alpha, t) + (1 - \theta_y) w_y^{[k-1]}(t),$$
$$w_p^{[k]}(t) = \theta_p p_2^{[k]}(\alpha, t) + (1 - \theta_p) w_p^{[k-1]}(t),$$

$$(7)$$

where $\theta_y, \theta_p \in (0, 1)$ are two relaxation parameters.

The NNWR starts with two initial guesses $\vartheta_y^{[0]}(t)$ and $\vartheta_p^{[0]}(t)$ along the interface $\{x = \alpha\} \times [0, T]$ and then computes simultaneously for $t \in [0, T]$ for $i = 1, 2$ with $k = 1, 2, \ldots$

$$\partial_t y_1^{[k]} - \nu \Delta y_1^{[k]} = u_1^{[k]}, \qquad \text{in } \Omega_1, \quad \partial_t y_2^{[k]} - \nu \Delta y_2^{[k]} = u_2^{[k]}, \qquad \text{in } \Omega_2,$$
$$y_1^{[k]}(x, 0) = 0, \qquad \text{in } \Omega_1, \qquad y_2^{[k]}(x, 0) = 0, \qquad \text{in } \Omega_2,$$
$$y_1^{[k]}(0, t) = 0, \qquad y_2^{[k]}(\alpha, t) = \vartheta_y^{[k-1]}(t),$$
$$y_1^{[k]}(\alpha, t) = \vartheta_y^{[k-1]}(t), \qquad y_2^{[k]}(1, t) = 0,$$

and then after computing $\xi(t) = \left(\partial_x y_1^{[k]} - \partial_x y_2^{[k]} \right)\Big|_{x=\alpha}$, calculate

$$\partial_t \psi_1^{[k]} - \nu \Delta \psi_1^{[k]} = 0, \qquad \text{in } \Omega_1, \quad \partial_t \psi_2^{[k]} - \nu \Delta \psi_2^{[k]} = 0, \qquad \text{in } \Omega_2,$$
$$\psi_1^{[k]}(x, 0) = 0, \qquad \text{in } \Omega_1, \qquad \psi_2^{[k]}(x, 0) = 0, \qquad \text{in } \Omega_2,$$
$$\psi_1^{[k]}(0, t) = 0, \qquad \partial_x \psi_2^{[k]}(\alpha, t) = \xi(t),$$
$$\partial_x \psi_1^{[k]}(\alpha, t) = \xi(t), \qquad \psi_2^{[k]}(1, t) = 0,$$

and for the adjoint equation, we have

$$-\partial_t p_1^{[k]} - \nu \Delta p_1^{[k]} = y_1^{[k]}, \qquad \text{in } \Omega_1, \quad -\partial_t p_2^{[k]} - \nu \Delta p_2^{[k]} = y_2^{[k]}, \qquad \text{in } \Omega_2,$$
$$p_1^{[k]}(x, T) = 0, \qquad \text{in } \Omega_1, \qquad p_2^{[k]}(x, T) = 0, \qquad \text{in } \Omega_2,$$
$$p_1^{[k]}(0, t) = 0, \qquad p_2^{[k]}(\alpha, t) = \vartheta_p^{[k-1]}(t),$$
$$p_1^{[k]}(\alpha, t) = \vartheta_p^{[k-1]}(t), \qquad p_2^{[k]}(1, t) = 0,$$

and after computing $\eta(t) = \left(\partial_x p_1^{[k]} - \partial_x p_2^{[k]} \right)\Big|_{x=\alpha}$, we calculate

$$-\partial_t \varphi_1^{[k]} - \nu \Delta \varphi_1^{[k]} = 0, \qquad \text{in } \Omega_1, \quad -\partial_t \varphi_2^{[k]} - \nu \Delta \varphi_2^{[k]} = 0, \qquad \text{in } \Omega_2,$$
$$\varphi_1^{[k]}(x, T) = 0, \qquad \text{in } \Omega_1, \qquad \varphi_2^{[k]}(x, T) = 0, \qquad \text{in } \Omega_2,$$
$$\varphi_1^{[k]}(0, t) = 0, \qquad \partial_x \varphi_2^{[k]}(\alpha, t) = \eta(t),$$
$$\partial_x \varphi_1^{[k]}(\alpha, t) = \eta(t), \qquad \varphi_2^{[k]}(1, t) = 0,$$

with the update conditions

$$\vartheta_y^{[k]}(t) = \vartheta_y^{[k-1]}(t) - \theta_y \left(\psi_1^{[k]}(\alpha, t) - \psi_2^{[k]}(\alpha, t) \right),$$

$$\vartheta_p^{[k]}(t) = \vartheta_p^{[k-1]}(t) - \theta_p \left(\varphi_1^{[k]}(\alpha, t) - \varphi_2^{[k]}(\alpha, t) \right),$$

where $\theta_y, \theta_p \in (0, 1)$ are relaxation parameters.

4 Solution Procedure and Convergence Result

To analyze the convergence behavior of the DNWR method (5)–(7) (or for the NNWR method), we need to solve the coupled problems (5) and (6). But from (4), we have $u_i^{[k]} = -p_i^{[k]}/\lambda$ for $i = 1, 2$. So eliminating $p_1^{[k]}$ from (5), we get

$$\nu^2 \frac{\partial^4 y_1^{[k]}}{\partial x^4} - \frac{\partial^2 y_1^{[k]}}{\partial t^2} + \frac{1}{\lambda} y_1^{[k]} = 0, \tag{8}$$

with the initial conditions

$$(i)\ y_1^{[k]}(x, 0) = 0, \quad (ii)\ \left(\partial_t y_1^{[k]} - \nu \partial_{xx} y_1^{[k]} \right) \Big|_{(x,T)} = 0,$$

and the boundary conditions

$$(a)\ y_1^{[k]}(0, t) = 0, \qquad (b)\ \left(\partial_t y_1^{[k]} - \nu \partial_{xx} y_1^{[k]} \right) \Big|_{(0,t)} = 0,$$

$$(c)\ y_1^{[k]}(\alpha, t) = w_y^{[k-1]}(t),\ (d)\ \left(\partial_t y_1^{[k]} - \nu \partial_{xx} y_1^{[k]} \right) \Big|_{(\alpha,t)} = -w_p^{[k-1]}/\lambda.$$

We now use separation of variables to solve (8). For simplicity, we drop the iteration number in calculation. Let us suppose $y_1(x, t) = H(x)V(t)$. Therefore substituting y_1 back into Eq. (8), we obtain

$$\frac{H^{(4)}}{H} + \frac{1}{\lambda \nu^2} = \frac{1}{\nu^2} \frac{V^{(2)}}{V}, \tag{9}$$

with the initial and boundary conditions (considering only non-trivial solutions)

$$\text{IC}: \ (i)\ V(0) = 0, \quad (ii)\ H(x)V^{(1)}(T) = \nu H^{(2)}(x)V(T),\ \forall x \in \Omega_1$$

$$\text{BC}: (iii)\ H(0) = 0, \qquad (iv)\ H^{(2)}(0) = 0,$$

$$(v)\ H(\alpha)V(t) = w_y(t), \ (vi)\ H(\alpha)V^{(1)}(t) - \nu H^{(2)}(\alpha)V(t) = -w_p(t)/\lambda.$$

Note that the symbol $f^{(m)}$ denotes the mth derivative of the function f with respect to the corresponding variable. One can have the following possibilities:

Case 1 If both sides of (9) are equal to a constant $-\kappa^2$, then we get a solution set as:

$$H(x) = A_1 \sinh\left(\omega_1 x/\sqrt{2}\right) \cos\left(\omega_1 x/\sqrt{2}\right) + B_1 \cosh\left(\omega_1 x/\sqrt{2}\right) \sin\left(\omega_1 x/\sqrt{2}\right),$$
$$V(t) = C_1 \sin(\nu\kappa t),$$

together with $\omega_1^4 = \kappa^2 + \frac{1}{\lambda\nu^2}$ and the conditions (ii), (v), (vi). Now, the initial condition (ii) yields

$$\frac{H^{(2)}(x)}{H(x)} = -\frac{1}{\nu}\frac{V^{(1)}(T)}{V(T)} =: \beta_1,$$

and so one has $\beta_1 = \kappa \cot(\kappa\nu T)$ and $H^{(2)}(x) - \beta_1 H(x) = 0$. Therefore, the solution of (8) would be of the form

$$y_1(x, t) = \sum_{n\geq 1} \sin(\nu\kappa_n t) \left(A_n \sinh(\omega_{1,n} x/\sqrt{2}) \cos(\omega_{1,n} x/\sqrt{2}) \right.$$
$$\left. + B_n \cosh(\omega_{1,n} x/\sqrt{2}) \sin(\omega_{1,n} x/\sqrt{2}) \right).$$

Case 2 If the ratio in (9) is equal to zero, then we obtain a solution set as:

$$H(x) = A_2 \sinh\left(\omega_2 x/\sqrt{2}\right) \cos\left(\omega_2 x/\sqrt{2}\right) + B_2 \cosh\left(\omega_2 x/\sqrt{2}\right) \sin\left(\omega_2 x/\sqrt{2}\right),$$
$$V(t) = C_2 t,$$

together with $\omega_2^4 = \frac{1}{\lambda\nu^2}$ and the conditions (ii), (v), (vi). Also, the condition (ii) gives

$$\frac{H^{(2)}(x)}{H(x)} = -\frac{1}{\nu}\frac{V^{(1)}(T)}{V(T)} =: \beta_2,$$

and so we have $\beta_2 = \frac{1}{\nu T}$ and $H^{(2)}(x) - \beta_2 H(x) = 0$. Therefore, the solution of (8) would be of the form

$$y_1(x, t) = \sum_{n\geq 1} t \left(A_n \sinh(\omega_{2,n} x/\sqrt{2}) \cos(\omega_{2,n} x/\sqrt{2}) \right.$$
$$\left. + B_n \cosh(\omega_{2,n} x/\sqrt{2}) \sin(\omega_{2,n} x/\sqrt{2}) \right).$$

Case 3 If the ratio in (9) is equal to ϖ^2, then we get a solution set as:

$$H(x) = A_3 \sinh\left(\omega_3 x/\sqrt{2}\right) \cos\left(\omega_3 x/\sqrt{2}\right) + B_3 \cosh\left(\omega_3 x/\sqrt{2}\right) \sin\left(\omega_3 x/\sqrt{2}\right),$$
$$V(t) = C_3 \sinh(\nu\varpi t),$$

together with $\omega_3^4 = \frac{1}{\lambda \nu^2} - \varpi^2$ and the conditions (ii), (v), (vi). In addition, we obtain from (ii)

$$\frac{H^{(2)}(x)}{H(x)} = \frac{1}{\nu} \frac{V^{(1)}(T)}{V(T)} =: \beta_3,$$

and so we get $\beta_3 = \varpi \coth(\varpi \nu T)$ and $H^{(2)}(x) - \beta_3 H(x) = 0$. Therefore, the solution of (8) would be of the form

$$y_1(x,t) = \sum_{n \geq 1} \sinh(\nu \varpi_n t) \left(A_n \sinh(\omega_{3,n} x/\sqrt{2}) \cos(\omega_{3,n} x/\sqrt{2}) \right.$$

$$\left. + B_n \cosh(\omega_{3,n} x/\sqrt{2}) \sin(\omega_{3,n} x/\sqrt{2}) \right).$$

But it is difficult to get the closed-form solutions to proceed further analysis. We keep this for future work.

5 Numerical Illustration

We implement these two algorithms numerically for the model problem (1)–(4) with $\lambda = 1$, $\nu = 1$, $\hat{y}(x,t) = 0$, $y_0(x) = 0$, with a discretization using standard centered finite differences in space and backward Euler in time with $\Delta x = \Delta t = 4 \times 10^{-3}$. For the DNWR and the NNWR algorithms, we consider two cases: first the symmetric decomposition, where $\alpha = 1/2$, and then a domain decomposition with $\Omega_1 = (0, 0.6)$, $\Omega_2 = (0.6, 1)$. In all our numerical experiments, we choose $w_y^{[0]}(t) = w_p^{[0]}(t) = t^2$, $t \in (0, T]$ as the initial guesses. Figure 1 gives the numerically measured convergence curves of the DNWR method for $T = 2$ and for different values of the parameters θ_y, θ_p for $\alpha = 1/2$ on the left, and $\alpha = 0.6$ on the right. We see that for the symmetric case, we get linear convergence for all relaxation parameters θ_y, θ_p, except for $(\theta_y, \theta_p) = (0.5, 0.5)$, when we observe two-step convergence. But for the asymmetric decomposition, we cannot predict numerically the best value of

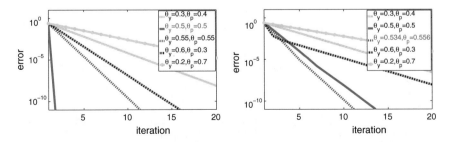

Fig. 1 Convergence of DNWR with various values of θ_y, θ_p for $T = 2$ for the symmetric subdomains on the left, and $\alpha = 0.6$ on the right

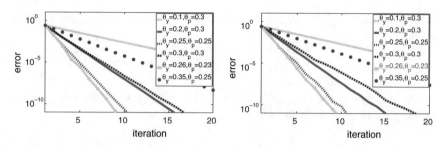

Fig. 2 Convergence of NNWR with various values of θ_y, θ_p for $T = 2$ for the symmetric subdomains on the left, and $\alpha = 0.6$ on the right

the parameters. In Fig. 2, we plot the convergence curves for $T = 2$ for the NNWR algorithm for $\alpha = 1/2$ on the left, and $\alpha = 0.6$ on the right. Unlike for the DNWR, we do not see any difference between the symmetric and asymmetric cases.

6 Concluding Remarks

We formulate the DNWR and the NNWR algorithms to solve the underlying coupled forward and backward space–time problems in parabolic optimal control problems. With numerical experiments, we showed two-step convergence, choosing a particular set of relaxation parameters for the ideal symmetric case in DNWR. For the NNWR method, we also see finite-step convergence, although we do not get two-step convergence even in the symmetric case.

Acknowledgements I would like to express my gratitude to Prof. Martin J. Gander and Prof. Felix Kwok for their constant support and stimulating suggestions.

References

1. Akçelik, V., Biros, G., Ghattas, O., Long, K.R., van Bloemen Waanders, B.: A variational finite element method for source inversion for convective-diffusive transport. Finite Elem. Anal. Des. **39**(8), 683–705 (2003)
2. Benamou, J.D.: Décomposition de domaine pour le contrôle optimal de systémes gouvernés par des équations aux dérivées partielles elliptiques. C.R. Acad. Sci. Paris Sér. I Math. **317**(2) (1993) 205–209
3. Bjørstad, P.E., Widlund, O.B.: Iterative methods for the solution of elliptic problems on regions partitioned into substructures. SIAM J. Numer. Anal. **23**(6), 1097–1120 (1986)
4. Bounaim, A.: On the optimal control problem of the heat equation: new formulation of the problem using a non-overlapping domain decomposition technique. Technical report, University of Oslo, Scientific Computing Group, Department of Informatics (2002)

5. Bourgat, J.F., Glowinski, R., Le Tallec, P., Vidrascu, M.: Variational formulation and algorithm for trace operator in domain decomposition calculations. In: Domain Decomposition Methods, Los Angeles, CA, 1988, pp. 3–16. SIAM, Philadelphia, PA (1989)
6. Bramble, J.H., Pasciak, J.E., Schatz, A.H.: An iterative method for elliptic problems on regions partitioned into substructures. Math. Comput. **46**(174), 361–369 (1986)
7. De Roeck, Y.H., Le Tallec, P.: Analysis and test of a local domain-decomposition preconditioner. In: Fourth International Symposium on Domain Decomposition Methods for Partial Differential Equations, Moscow, 1990, pp. 112–128. SIAM, Philadelphia, PA (1991)
8. Gander, M.J., Kwok, F., Mandal, B.C.: Dirichlet–Neumann and Neumann–Neumann waveform relaxation algorithms for parabolic problems. Electron. Trans. Numer. Anal. **45**, 424–456 (2016)
9. Gander, M.J., Kwok, F., Mandal, B.C.: Dirichlet–Neumann and Neumann–Neumann waveform relaxation for the wave equation. In: Dickopf, T., Gander, M.J., Halpern, L., Krause, R., Pavarino, L. (eds.) Domain Decomposition Methods in Science and Engineering XXII. Lecture Notes in Computational Science and Engineering, pp. 501–509. Springer International Publishing (2016)
10. Gander, M.J., Kwok, F., Mandal, B.C.: In: Domain Decomposition Methods in Science and Engineering XXIV. Lecture Notes in Computational Science and Engineering. Springer International Publishing (2017)
11. Gander, M.J., Kwok, F., Mandal, B.C.: Dirichlet–Neumann Waveform Relaxation Method for the 1D and 2D Heat and Wave Equations in Multiple Subdomains. arXiv:1507.04011
12. Gander, M.J., Kwok, F., Wanner, G.: Constrained optimization: from Lagrangian mechanics to optimal control and PDE constraints. In: Optimization with PDE Constraints. LNCSE, vol. 101, pp. 151–202 (2014)
13. Heinkenschloss, M.: Time-domain decomposition iterative methods for the solution of distributed linear quadratic optimal control problems. J. Comput. Appl. Math. **173**, 169–198 (2005)
14. Heinkenschloss, M., Herty, M.: A spatial domain decomposition method for parabolic optimal control problems. J. Comput. Appl. Math. **201**(1), 88–111 (2007)
15. Heinkenschloss, M., Nguyen, H.: Balancing Neumann–Neumann methods for elliptic optimal control problems. In: Kornhuber, R., Hoppe, R., Périaux, J., Pironneau, O., Widlund, O., Xu, J. (eds.) Domain Decomposition Methods in Science and Engineering, vol. 40, pp. 589–596. Springer, Heidelberg (2004)
16. Heinkenschloss, M., Nguyen, H.: Neumann–Neumann domain decomposition preconditioners for linear-quadratic elliptic optimal control problems. SIAM J. Sci. Comput. **28**(3), 1001–1028 (2006)
17. Hoang, T.T.P.: Space-time domain decomposition methods for mixed formulations of flow and transport problems in porous media. Ph.D. Thesis, University Paris 6, France (2013)
18. Kwok, F.: Neumann–Neumann waveform relaxation for the time-dependent heat equation. In: Erhel, J., Gander, M.J., Halpern, L., Pichot, G., Sassi, T., Widlund, O.B. (eds.) Domain Decomposition in Science and Engineering XXI, vol. 98, pp. 189–198. Springer-Verlag (2014)
19. Lagnese, J.E., Leugering, G.: Domain Decomposition Methods in Optimal Control of Partial Differential Equations. International Series of Numerical Mathematics, vol. 148. Basel (2004)
20. Le Tallec, P., De Roeck, Y.H., Vidrascu, M.: Domain decomposition methods for large linearly elliptic three-dimensional problems. J. Comput. Appl. Math. **34**(1), 93–117 (1991)
21. Lions, J.L.: Optimal Control of Systems Governed by Partial Differential Equations. Springer, Berlin (1971)
22. Mandal, B.C.: Convergence analysis of substructuring waveform relaxation methods for space-time problems and their application to optimal control problems. PhD thesis, University of Geneva (2014)
23. Mandal, B.C.: A time-dependent Dirichlet–Neumann method for the heat equation. In: Erhel, J., Gander, M.J., Halpern, L., Pichot, G., Sassi, T., Widlund, O.B. (eds.) Domain Decomposition in Science and Engineering XXI, vol. 98, pp. 467–475. Springer-Verlag (2014)

24. Mandal, B.C.: Neumann-Neumann waveform relaxation algorithm in multiple subdomains for hyperbolic problems in 1D and 2D. Numer. Methods Partial Differ. Equ. (2016)
25. Marini, L.D., Quarteroni, A.: A relaxation procedure for domain decomposition methods using finite elements. Numer. Math. **55**(5), 575–598 (1989)
26. Martini, L., Quarteroni, A.: An iterative procedure for domain decomposition methods: a finite element approach. In: Domain Decomposition Methods for PDEs, I, pp. 129–143. SIAM (1988)
27. Quarteroni, A., Valli, A.: Domain Decomposition Methods for Partial Differential Equations. Clarendon Press (1999)
28. Toselli, A., Widlund, O.: Domain Decomposition Methods—Algorithms and Theory. Springer Series in Computational Mathematics, vol.34. Springer, Berlin (2005)
29. Tröltzsch, F.: Optimal Control of Partial Differential Equations: Theory, Methods and Applications. Graduate Studies in Mathematics, vol. 112. American Mathematical Society (2010)

Solution of Optimization Problems in Fuzzy Background Using HVPSO Algorithm

Ashok Pal, Kusum Deep and S. B. Singh

Abstract The effectiveness of a HVPSO algorithm proposed by Pal (Decision making in crisp and fuzzy environments using particle swarm optimization, 2015 [1]) has focused to solve the problems of optimization in fuzzy background on eight test problems taken from the literature The problems of optimization in the present paper are solved by using the objectives like fuzzy goals and the fuzzy constraints, the same as the crisp constraints to satisfy a level of α of fuzzy parameters involved. A hybrid version particle swarm optimization (HVPSO) algorithm proposed (Pal, Decision making in crisp and fuzzy environments using particle swarm optimization, 2015 [1]) has been used to solve these optimization problems, and its performance is compared with MI-LXPM algorithm (Deep et al, Appl Math Comput 212(2):505–518, 2009 [2]) which is based on Genetic Algorithm. It has been observed that the performance of the HVPSO algorithm in solving the optimization problems in fuzzy background is as good as that of MI-LXPM.

Keywords Particle swarm optimization (PSO) · Standard particle swarm optimization (SPSO) FPP · HVPSO algorithm · MI-LXPM algorithm and fuzzy background

A. Pal (✉)
Department of Mathematics, Chandigarh University, Chandigarh, India
e-mail: ashokpmaths@gmail.com

K. Deep
Department of Mathematics, Indian Institute of Technology Roorkee, Roorkee 247667, Uttarakhand, India
e-mail: kusumdeep@gmail.com; kusumfma@iitr.ac.in

S. B. Singh
Department of Mathematics, Punjabi University, Patiala, India
e-mail: sbsingh69@yahoo.com

© Springer Nature Singapore Pte Ltd. 2019
J. C. Bansal et al. (eds.), *Soft Computing for Problem Solving*, Advances in Intelligent Systems and Computing 817, https://doi.org/10.1007/978-981-13-1595-4_40

495

1 Introduction

In the initial phase of the growth of fuzzy programming problem, the flexibility in constraints and the fuzziness in objectives are introduced into the usual mathematical programming problems to deal with a type of uncertainties, namely ambiguity/vagueness/imprecision.

Any FPP can be defined as

$$Fuzzy \ \min/\max f(x, \tilde{a}_i),$$

such that

$$x \in X(\tilde{b}) = \{x \in R^n \mid g_j(x, \tilde{b}_j) \le 0, \ j = 1, 2, \ldots, m\} \tag{1}$$

where $\tilde{a}_i = (\tilde{a}_{i1}, \tilde{a}_{i2}, \ldots, \tilde{a}_{ip})$ and $\tilde{b}_j = (\tilde{b}_{j1}, \tilde{b}_{j2}, \ldots, \tilde{b}_{jq})$ represent the vectors of fuzzy parameters involved in the objective function $f(x, \tilde{a}_i)$ and the constraint functions $g_j(x, \tilde{b}_j)$ ($\tilde{a}_{ir}$ and $\tilde{b}_{js}$ are assumed to be fuzzy numbers whose membership functions are $\mu_{\tilde{a}_{ir}}(a_{ir})$ and $\mu_{\tilde{b}_{js}}(b_{js})$, respectively). It is not necessary that all the parameters appearing in the problem be fuzzy. Some of the parameters can also be crisp. For the sake of convenience, in notation, constraints of the type $c_i \le x_i \le d_i$ which specify bounds on the values of the decision variables as well as integer restrictions, if any, imposed on the decision variables x_i for some or all indices i are not mentioned separately.

As per Tanaka et al. [3] and Zimmermenn [4], a fuzzy optimization problem considered as a goal optimization problem may be written in the subsequent form:

$$Max/Min(f_i(x)) \underset{\sim}{\ge} \ or \ \underset{\sim}{\le} c_i, \ i = 1, 2, \ldots, m$$

$$g_j(x) \underset{\sim}{\le} \ or \ \underset{\sim}{\ge} d_j, j = 1, 2, 3, \ldots, p \tag{2}$$

where $\underset{\sim}{\ge}$ is fuzzified form of $\ge$ and interpret as in essence greater than or equal to and $\underset{\sim}{\le}$ is fuzzified form of $\le$ and interpret as in essence smaller than or equal to. Here $f_i(x)$ and $g_j(x)$ are some fuzzy functions of variables $x = (x_1, x_2, \ldots, x_n)$ and c_i and d_j are fuzzy numbers.

If $m > 1$, then Eq. (2) is known as a multi-objective fuzzy optimization problem (MOFPP), otherwise it is called a single-objective fuzzy optimization problem (SOFPP). Fuzzy optimization problem (2) can be explained as finding $x \in R^n$ such that $f_i(x)$ is in essence greater than or equal to c_i and $g_j(x)$ is in essence smaller than or equal to $d_j \forall$ j. It is assumed that the fuzzy goal c_i is subjectively defined by the decision maker and it should reflect the decision maker's vagueness in understanding and/or describing the objective and the constraints. The fuzzy goals can be quantified based on the concept of fuzzy set.

Let X denotes a universal set. Then a fuzzy subset $\tilde{A}$ of X is defined by its membership function $\mu_{\tilde{A}} : X \to [0, 1]$, which assign to each element $x \in R$, a real number $\mu_{\tilde{A}}(x)$ in the interval [0, 1], where the value of $\mu_{\tilde{A}}(x)$ at x represents the grade of membership of x in $\tilde{A}$. A fuzzy subset $\tilde{A}$ can be characterized as a set of ordered pairs of element x and grade $\mu_{\tilde{A}}(x)$ and is often written $\tilde{A} = \{(x, \mu_{\tilde{A}}(x)) : x \in X\}$. When the membership function $\mu_{\tilde{A}}(x)$ contains only the two points 0 and 1, then $\mu_{\tilde{A}}(x)$ is the characteristic function $C_A : X \rightarrow \{0, 1\}$, and hence, $\tilde{A}$ is no longer a fuzzy subset, but an ordinary crisp set A.

The level of α set of a fuzzy set $\tilde{A}$ is taken as a usual set A_α for which the membership function degree exceeds the level of α, i.e.,

$$A_\alpha = \{x : \mu_A(x) \geq \alpha, \ \alpha \in [0, 1]\} \tag{3}$$

Each fuzzy goal in problem (2) can be taken as a fuzzy set in R^n as $\tilde{G}_i = (x, \mu_{\tilde{G}_i}(x))$ where for each $x \in R^n$, the related membership grade $\mu_{\tilde{G}_i}(x)$ shows the decision maker's level of fulfillment with respect to the degree of attainment of $f(x)$. Membership function $\mu_{\tilde{G}_i}(.)$ is given by

$$\mu_{\tilde{G}_i}(x) = \begin{cases} 0, & \text{if } f_i(x) \leq c_i', \\ u_i(f_i(x)), & \text{if } c_i' \leq f_i(x) \leq c_i, \\ 1, & \text{if } f_i(x) \geq c_i \end{cases} \tag{4}$$

where, is an increasing function with $u_i(c_i') = 0$, $u_i(c_i) = 1$. Similarly, fuzzy goal $\underset{\sim}{<}$ can be defined as a fuzzy set in R^n as $\tilde{C}_j = \{(x, \mu_{\tilde{C}_j}(x))\}$ where for each $x \in R^n$, the related membership grade $\mu_{\tilde{C}_j}(x)$ denotes the decision maker's level of fulfillment with respect to the degree of fulfillment for jth constraint. Membership function $\mu_{\tilde{C}_j}(.)$ is given by the term

$$\mu_{\tilde{C}_j}(x) = \begin{cases} 1, & \text{if } g_j(x) \leq d_j, \\ v_j(g_j(x)), & \text{if } d_j \leq g_j(x) \leq d_j', \\ 0, & \text{if } g_j(x) \geq d_j' \end{cases} \tag{5}$$

where v_j is an increasing function with $v_j(d_j') = 0$ and $v_j(d_j) = 1$.

Fuzzy optimization problem (2) now stated as fulfilling all the specific goals $\tilde{G}$ and $\tilde{C}_j \forall j$. The fuzzy conclusion to (2) now defines the fuzzy set where

$$\mu_{\tilde{D}}(x) = (*\mu_{\tilde{G}_i}(x)) * (*_j \mu_{\tilde{C}_j}(x)) \tag{6}$$

implies

$$\mu_{\tilde{D}}(x) = \min\{\mu_{\tilde{G}_i}(x), \mu_{\tilde{C}_j}(x) \forall j\} \tag{7}$$

where $*$ can be assumed to be some aggregation operator [20].

A variety of functional forms of u_i and v_j are currently being used in the literature, for instance, linear, piecewise linear, exponential, and hyperbolic [5–10]. A particular form of the membership function of a fuzzy goal is chosen by the decision maker to represent his level of satisfaction. For example, if the membership function $\mu_{\tilde{G}_i}(x)$ is linear, then the decision maker's level of fulfillment is linearly growing as $f(x)$ increases from c_i' to c_i. In actual fact, fuzzy optimization problems of type (2) are one of the applications of fuzzy set theory. A variety of successful applications of this methodology have been given in the literature [5, 11–18].

An appropriate expression of fuzzy number $\tilde{a}$ is a quadruple of parameters $(a^L, a^R, \underline{a}, \bar{a})_{LR}$ of its membership function:

$$\tilde{a}(x) = \begin{cases} 0, & \text{if } x < a^L - \underline{a} \\ L((a^L - x)/\underline{a}), & \text{if } a^L - \underline{a} \leq x < a^L \\ 1, & \text{if } a^L \leq x \leq a^R \\ R((x - a^R)/\bar{a}), & \text{if } a^R < x \leq a^R + \bar{a} \\ 0, & \text{if } x > a^R + \bar{a} \end{cases} \tag{8}$$

where $[a^L; a^R]$ is an interval of the most probable values, a^L and a^R are right and left positional points, $\underline{a}$ and $\bar{a}$ are (non-negative) right and left spreads of $\tilde{a}$. L and R are left and right reference functions that are decreasing on $(-\infty, +\infty)$ and $L(0) = R(0) = 1$, $L(1) = R(1) = 0$. $\tilde{a}$ is said to be L-R flat fuzzy number. If $a^L = a^R = a$, a is called the 'middle' value of $\tilde{a}$, and $\tilde{a}$ becomes a triangular fuzzy number, denoted as a triplet $(a, \underline{a}, \bar{a})_{LR}$. If these spreads of a fuzzy number are zero, then it becomes usual (crisp) number.

2 HVPSO Algorithm

N. H. Thanh and C. Mohan found a new test point p in the method of a 'Control Random Search Technique for Global Optimization problems using quadratic approximation [8].' In this random search technique, they choose $b_1 = L$, i.e., the point with the lowest function value, i.e., with the best optimal value from the uniformly generated random numbers in the range of 0 and 1 and two other distinct random points b_2 and b_3 which calculate the next test point say P using the quadratic approximation formula cited below

$$P = \frac{1}{2} * \frac{[f(b_1)(b_2^2 - b_3^2) + f(b_2)(b_3^2 - b_1^2) + f(b_3)(b_1^2 - b_2^2)]}{[f(b_1)(b_2 - b_3) + f(b_2)(b_3 - b_1) + f(b_3)(b_1 - b_2)]} \tag{9}$$

where P gives the extremal point of the quadratic curve passing through the points b_1, b_2, and b_3.

Deep and Bansal [19] proposed an algorithm named Quadratic approximation PSO (qPSO) in which the hybridization of PSO is performed with Quadratic Approximation operator (QA), by dividing the whole swarm into two associate swarms in such a way that the PSO operators are functional on one associate swarm, whereas the QA operator is applied on the other associate swarm, ensuring that both associate swarms are simplified using the global best particle of the entire swarm.

The HVPSO algorithm [1] in the present paper is a hybrid of the SPSO algorithm [21] and a random search technique with quadratic approximation formula [8] named hybrid version particle swarm optimization (HVPSO) algorithm.

The flow of the HVPSO algorithm is as under:

BEGIN:
Create and Initialize an n-dimensional swarm S
$\{x_i(t) : =1 \text{to } S\}$ *uniformly between 0 and 1.*
Assign w some value between 0.4 to 0.9 and
set c_1 *and* $c_2 = 2.0$.
For i=1 to S,
For d=1 to n,
Assign some value to P between 0 and 1.
If r (0, 1) < α , then
generate velocity vector using equation
standard PSO algorithm [21],
else generate it using equation of controlled
random search with quadratic approximation formula [8,19].
Calculate particle position as
$$x_i(t+1) = x_i(t) + v_i(t+1)$$
End- for-d;
calculate objective value of updated position; if needed,
change the historical information for P_i *and* P_g;
end-of-i;
Conclude if P_g *meets problem needs;*
End.

3   Discussion of Results

The hybrid edition of PSO method proposed in paper coded and compiled in C++ compiler on a Windows XP operating system with 2 GB RAM and a Intel Dual Core Processor for solving the optimization problems in fuzzy background. Method has been taken in solving eight test problems from the literature which are given in appendix. The fuzzy parameters used in solving the test problems have been taken on the basis of triangular membership functions, where L, M and R indicate left spread, model value and right spread of membership function values, respectively. The used fuzzy parameters are presented in Tables 1, 2, 3, and 4 and obtained results in Table 5.

Table 1 Values of fuzzy parameters used in Problems 1–5

Pb. no.	$\tilde{p}$	Fuzzy parameters taken [22]			Fuzzy parameters for which optimal solution achieved by HVPSO	
		Left spread (L)	Modal value (M)	Right spread (R)	$(\tilde{p})$	(α)
Pb1	$\tilde{a}_{11}$	3.8	4.0	4.3	4.28	0.963
	$\tilde{a}_{12}$	48.5	50	52	48.5	0.0
	$\tilde{b}_{11}$	0.9	1.0	1.1	1.093	0.963
	$\tilde{b}_{12}$	0.8	1.0	1.2	0.804	0.009
	$\tilde{b}_{13}$	0.85	1.0	1.15	0.89	0.144
Pb2	$\tilde{a}_{21}$	1.85	2.0	2.2	1.85	0.0
	$\tilde{a}_{22}$	18.2	20	22.5	18.2	0.0
	$\tilde{b}_{21}$	0.9	1.0	1.1	0.911	0.053
	$\tilde{b}_{22}$	0.8	1.0	1.2	1.10	0.751
	$\tilde{b}_{23}$	0.85	1.0	1.15	1.06	0.714
Pb3	$\tilde{a}_{31}$	2.9	3.0	3.5	2.9	0.0
	$\tilde{a}_{32}$	4.7	5.0	5.35	4.7	0.0
	$\tilde{b}_{31}$	0.9	1.0	1.1	1.10	1.0
	$\tilde{b}_{32}$	0.8	1.0	1.2	0.835	0.088
	$\tilde{b}_{33}$	0.85	1.0	1.15	0.85	0.0
Pb4	$\tilde{a}_{41}$	3.44	4	5.06	5.06	1.0
	$\tilde{a}_{42}$	2.78	3	3.33	3.33	1.0
	$\tilde{b}_{41}$	5.67	6	6.83	5.83	0.135
	$\tilde{b}_{42}$	6.94	7.5	8.67	6.94	0.0
	$\tilde{b}_{43}$	13.9	15	17.1	14.12	0.068
	$\tilde{b}_{44}$	9.44	10	11.56	9.44	0.0
	$\tilde{b}_{45}$	13.9	15	17.1	13.9	0.0
	$\tilde{b}_{46}$	8.74	9.3	10.33	1.59	0.0
Pb5	$\tilde{b}_{51}$	5.67	6	6.83	5.67	0.0
	$\tilde{b}_{52}$	6.94	7.5	8.67	6.94	0.0
	$\tilde{b}_{53}$	13.9	15	17.1	13.9	0.0
	$\tilde{b}_{54}$	9.44	10	11.56	9.904	0.219
	$\tilde{b}_{55}$	13.9	15	17.1	16.409	0.784
	$\tilde{b}_{56}$	8.74	9.3	10.33	10.33	1.0

Table 2 Values of fuzzy parameters used in Problem 6

$\tilde{p}$	Fuzzy parameters taken [22]			Fuzzy parameters for which optimal solution achieved by HVPSO	
	Left spread (L)	Modal value (M)	Right spread (R)	$(\tilde{p})$	(α)
$\tilde{a}_{61}$	-14.5	-14	-13.5	-14.5	0.0
$\tilde{a}_{62}$	-16.5	-16	-15.5	-16.5	0.0
$\tilde{a}_{63}$	3.7	4	4.4	3.724	0.034
$\tilde{a}_{64}$	4.5	5	5.5	4.501	0.001
$\tilde{a}_{65}$	6.4	7	7.5	6.4	0.0
$\tilde{b}_{61}$	-3.3	-3	-2.6	-2.601	0.999
$\tilde{b}_{62}$	-2.2	-2	-1.8	-2.2	0.0
$\tilde{b}_{63}$	-8.6	-8	-7.4	-7.434	0.972
$\tilde{b}_{64}$	1.8	2	2.2	2.2	0.999
$\tilde{b}_{65}$	1.7	2	2.4	1.732	0.08
$\tilde{b}_{66}$	-14.8	-14	-13.2	-14.592	0.130
$\tilde{b}_{67}$	-0.9	-0.5	-0.2	-0.202	0.997
$\tilde{b}_{68}$	0.8	1.0	1.2	1.106	0.766
$\tilde{b}_{69}$	-12.8	-12	-11.2	-11.205	0.997
$\tilde{b}_{610}$	6.5	7	7.5	7.42	0.92
$\tilde{b}_{611}$	3.6	4	4.4	3.602	0.002
$\tilde{b}_{612}$	-3.4	-3	-2.6	-3.4	0.0
$\tilde{b}_{613}$	-8.4	-8	-7.6	-8.39	0.015
$\tilde{b}_{614}$	-18.3	-17	-16	-17.902	0.173
$\tilde{b}_{615}$	-2.2	-2	-1.8	-2.2	0.001

3.1   Values of Fuzzy Parameters Used

See Tables 1, 2, 3, 4 and 5.

4   Conclusions

The effectiveness of a hybrid version particle swarm optimization (HVPSO) algorithm proposed by Pal [1] has been considered in the present paper in solving the problems of optimization in fuzzy background which are given in appendix. The

Table 3 Values of fuzzy parameters used in Problem 7

$\tilde{p}$	Fuzzy parameters taken [22]			Fuzzy parameters for which optimal solution achieved by HVPSO	
	Left spread (L)	Modal value (M)	Right spread (R)	$(\bar{p})$	(α)
$\tilde{a}_{71}$	4.5	5	5.5	4.5	0.0
$\tilde{a}_{72}$	2.7	3	3.3	2.708	0.014
$\tilde{a}_{73}$	6.5	7	7.5	6.566	0.066
$\tilde{a}_{74}$	−4.5	−4	−3.6	−4.5	0.0
$\tilde{a}_{75}$	−10.7	−10	−9.3	−10.68	0.014
$\tilde{b}_{71}$	−3.3	−3	−2.6	−2.672	0.897
$\tilde{b}_{72}$	−2.2	−2	−1.8	−2.182	0.046
$\tilde{b}_{73}$	−8.6	−8	−7.4	−7.401	0.999
$\tilde{b}_{74}$	1.8	2	2.2	2.197	0.992
$\tilde{b}_{75}$	1.7	2	2.4	1.710	0.025
$\tilde{b}_{76}$	−14.8	−14	−13.2	−13.30	0.938
$\tilde{b}_{77}$	−0.9	−0.5	−0.2	−0.2	1.0
$\tilde{b}_{78}$	0.8	1.0	1.2	1.2	1.0
$\tilde{b}_{79}$	−12.8	−12	−11.2	−11.235	0.978
$\tilde{b}_{710}$	6.5	7	7.5	7.5	1.0
$\tilde{b}_{711}$	3.6	4	4.4	3.6	0.0
$\tilde{b}_{712}$	−3.4	−3	−2.6	−3.4	0.0
$\tilde{b}_{713}$	−8.4	−8	−7.6	−7.885	0.644
$\tilde{b}_{614}$	−18.3	−17	−16	−16.306	0.876
$\tilde{b}_{715}$	−2.2	−2	−1.8	−2.16	0.100

optimization problems in this paper are solved by treating the objectives like fuzzy goals and the fuzzy constraints the same as the crisp constraints to satisfy a level of α of fuzzy parameters. The obtained results have been compared with the solutions obtained by Deep et al. [2] using a MI-LXPM algorithm. Our computations using hybrid version particle swarm optimization approach yield comparable results. Thus, hybrid version of particle swarm optimization technique can be used with equal confidence to solve the optimization problems in fuzzy environment also.

Table 4 Values of fuzzy parameters used in Problem 8

$\tilde{p}$	Fuzzy parameters taken [22]			Fuzzy parameters for which optimal solution achieved by HVPSO	
	Left spread (L)	Modal value (M)	Right spread (R)	$(\tilde{p})$	(α)
$\tilde{a}_{81}$	2.8	3	3.3	2.806	0.013
$\tilde{a}_{82}$	−12.8	−12	−11.2	−11.333	0.917
$\tilde{a}_{83}$	3.5	4	4.5	3.504	0.004
$\tilde{a}_{84}$	5.6	6	6.4	6.132	0.665
$\tilde{a}_{85}$	2.8	3	3.2	3.169	0.922
$\tilde{b}_{81}$	−3.3	−3	−2.6	−2.789	0.73
$\tilde{b}_{82}$	−2.2	−2	−1.8	−1.944	0.641
$\tilde{b}_{83}$	−8.6	−8	−7.4	−7.514	0.905
$\tilde{b}_{84}$	1.8	2	2.2	2.15	0.874
$\tilde{b}_{85}$	1.7	2	2.4	2.049	0.873
$\tilde{b}_{86}$	−14.8	−14	−13.2	−13.963	0.523
$\tilde{b}_{87}$	−0.9	−0.5	−0.2	−0.305	0.850
$\tilde{b}_{88}$	0.8	1.0	1.2	0.8	0.0
$\tilde{b}_{89}$	−12.8	−12	−11.2	−11.365	0.897
$\tilde{b}_{810}$	6.5	7	7.5	7.401	0.901
$\tilde{b}_{811}$	3.6	4	4.4	4.11	0.638
$\tilde{b}_{812}$	−3.4	−3	−2.6	−3.171	0.286
$\tilde{b}_{813}$	−8.4	−8	−7.6	−8.017	0.479
$\tilde{b}_{814}$	−18.3	−17	−16	−16.65	0.715
$\tilde{b}_{815}$	−2.2	−2	−1.8	1.907	0.733

Appendix

Problem 1 This problem is taken from [22] and has also been analyzed in [8].

$$\text{Min} f(x, \tilde{a}_1) = (x_1 + 5)^2 + \tilde{a}_{11} x_2^2 + 2(x_3 - \tilde{a}_{12})^2,$$

subject to

$$g_1(x, \tilde{b}_1) = \tilde{b}_{11} x_1^2 + \tilde{b}_{12} x_2^2 + \tilde{b}_{13} x_3^2 \leq 100,$$
$$0 \leq x_i \leq 10.0; \quad i = 1, 2, 3.$$

Table 5 Comparison of results HVPSO versus MI-LXPM based on optimal value of objective functions

Pb no.	HVPSO						MI-LXPM [2]
	Optimal value of objective function		Statistical data recorded				Optimal value of objective function
			SS	NR	ANE	SR (%)	
1	Min(f)	2989.50	50	30	238	100	2989.50
2	Min(f)	3484.97	50	30	317	100	3484.97
3	Min(f)	7142.50	50	30	325	100	7142.50
4	Max(f)	23.89	50	30	95,692	90	19.3580
5	Max(f)	18	50	30	820	100	16.6863
6	Min(f)	74.50	70	30	1,605,110	90	80.000
7	Min(f)	169	70	30	53,190	70	200.000
8	Min(f)	50	50	30	409,950	70	70.00

Where, NR denotes no. of runs, SS the swarm size, AFE the used average no. of functions evaluations, and SR the rate of success.

The membership functions for the fuzzy parameters used are triangular and listed in Table 1.

The optimal solution obtained by [22] is: $(f = 2989.5)$.

Problem 2 This problem is taken from [22] and has also been analyzed in [8].

$$\text{Min} f(x, \tilde{a}_2) = \tilde{a}_{21}(x_1 + 45)^2 + (x_2 + 15)^2 + 3(x_3 + \tilde{a}_{22})^2,$$
$$\text{subject to}$$
$$g_1(x, \tilde{b}_1) = \tilde{b}_{21}x_1^2 + \tilde{b}_{22}x_2^2 + \tilde{b}_{23}x_3^2 \le 100,$$
$$0 \le x_i \le 10.0; \quad i = 1, 2, 3.$$

The membership functions for the fuzzy parameters used are triangular and listed in Table 1.

The optimal solution obtained by [22] is $(f = 3484.97)$.

Problem 3 This problem is taken from [22] and has also been analyzed in [8].

$$\text{Min} f(x, \tilde{a}_3) = \tilde{a}_{31}(x_1 + 20)^2 + \tilde{a}_{32}(x_2 - 45)^2 + (x_3 + 15)^2,$$
$$\text{subject to}$$
$$g_1(x, \tilde{b}_1) = \tilde{b}_{31}x_1^2 + \tilde{b}_{32}x_2^2 + \tilde{b}_{33}x_3^2 \le 100,$$
$$0 \le x_i \le 10.0; \quad i = 1, 2, 3.$$

The membership functions for the fuzzy parameters used are triangular and listed in Table 1.

The optimal solution obtained by [22] is ($f = 7142.5$).

Problem 4 This problem is taken from [22] and has also been analyzed in [8].

$$\text{Max} \tilde{f} = \tilde{a}_{41} x_1 + \tilde{a}_{42} x_2 + 2x_4$$

subject to the constraints

$$3x_1 + 3x_3 + 4x_3 + 3x_4 \leq 18$$
$$9x_1 + 4x_3 + 4x_3 + 6x_4 \leq (35, \ 40/9)_{RR}$$
$$\tilde{b}_{41} x_1 + \tilde{b}_{42} x_3 + \tilde{b}_{43} x_3 + \tilde{b}_{44} x_4 \leq (60, \ 50/9)_{RR}$$
$$8x_1 + \tilde{b}_{45} x_3 + \tilde{b}_{46} x_3 + 5x_4 \leq (50, \ 60/9)_{RR}$$
$$0 \leq x_1, \ x_2, \ x_3 \leq 4; \ 0 \leq x_4 \leq 6.$$

The optimal solution reported in [22] is (f = 19.358).
The fuzzy parameters used in this problem are triangular and given in Table 1.

Problem 5 This problem is also taken from [8].

$$\text{Max} \tilde{f} = x_1 + x_3 + 3x_4$$

subject to the constraints

$$3x_1 + 3x_3 + 4x_3 + 3x_4 \leq 18$$
$$9x_1 + 4x_3 + 4x_3 + 6x_4 \leq (35, \ 40/9)_{RR}$$
$$\tilde{b}_{51} x_1 + \tilde{b}_{52} x_3 + \tilde{b}_{53} x_3 + \tilde{b}_{54} x_4 \leq (60, \ 50/9)_{RR}$$
$$8x_1 + \tilde{b}_{55} x_3 + \tilde{b}_{56} x_3 + 5x_4 \leq (50, \ 60/9)_{RR}$$
$$0 \leq x_1, \ x_2, \ x_3 \leq 4; \ 0 \leq x_4 \leq 6.$$

The optimal solution reported in K. P. Singh [22] is (f = 16.683 with $\alpha = 0$).
The fuzzy parameters used in this problem are triangular and given in Table 1.

Problem 6 This problem is taken from K. P. Singh [22] and studied by Nguyen [8].

$$\text{Min} f(x, \tilde{a}_6) = 7x_1^2 - x_2^2 + x_1 x_2 + \tilde{a}_{61} x_1 + \tilde{a}_{62} x_2 + 8(x_3 - 10)^2 + \tilde{a}_{63}(x_4 - 5)^2$$
$$+ (x_5 - 3)^2 + 2(x_6 - 1)^2 + \tilde{a}_{64} x_7^2 + \tilde{a}_{65}(x_8 - 11)^2 + 2(x_9 - 10)^2$$
$$+ x_{10}^2 + 45$$

subject to:

$$\tilde{b}_{61} - 4(x_2 - 3)^2 - 2x_3^2 + 7x_4 + \tilde{b}_{62}x_5x_6x_8 + 120 \geq 0,$$

$$-5x_1^2 + \tilde{b}_{63}x_2 - (x_3 - 6)^2 + \tilde{b}_{64}x_4 + 40 \geq 0,$$

$$-x_1^2 - 2(x_2 - 2)^2 + \tilde{b}_{65}x_1x_2 + \tilde{b}_{66}x_5 - 6x_5x_6 \geq 0,$$

$$\tilde{b}_{67}(x_1 - 8)^2 - 2(x_2 - 4)^2 - 3x_5^2 + 7x_4 + \tilde{b}_{68}x_5x_8 + 30 \geq 0,$$

$$3x_1 - 6x_2 + \tilde{b}_{69}(x_9 - 8) + \tilde{b}_{610}x_{10} \geq 0,$$

$$\tilde{b}_{611}x_1 + 5x_2 + \tilde{b}_{612}x_7 + 9x_8 \leq 105,$$

$$10x_1 + \tilde{b}_{613}x_2 + \tilde{b}_{614}x_7 + 2x_8 \leq 0,$$

$$-8x_1 + 2x_2 + 5x_9 + \tilde{b}_{615}x_{10} \leq 12,$$

$$-5.0 \leq x_i \leq 10.0; \quad i = 1, 2, \ldots, 10.$$

The membership functions for the fuzzy parameters are triangular and are listed in Table 2. The optimal solution of the problem as reported in the source [22] is $f = 80$.

Problem 7 This problem is taken from [22] and has also been analyzed in [8].

$$\mathrm{Min} f(x, \tilde{a}_7) = (x_1 - 5)^2 + a_{71}(x_2 - 12)^2 + 0.5x_3^4 + \tilde{a}_{72}(x_4 - 11)^2 + 0.2x_5^5$$
$$+ \tilde{a}_{73}x_6^2 + 0.1x_7^4 + \tilde{a}_{74}x_6x_7 + \tilde{a}_{75}x_6 - 8x_7 + x_8^2 + 3(x_9 - 5)^2$$
$$+ (x_{10} - 5)^2.$$

subject to:

$$\tilde{b}_{71} - 4(x_2 - 3)^2 - 2x_3^2 + 7x_4 + \tilde{b}_{72}x_5x_6x_8 + 120 \geq 0,$$

$$-5x_1^2 + \tilde{b}_{73}x_2 - (x_3 - 6)^2 + \tilde{b}_{74}x_4 + 40 \geq 0,$$

$$-x_1^2 - 2(x_2 - 2)^2 + \tilde{b}_{75}x_1x_2 + \tilde{b}_{76}x_5 - 6x_5x_6 \geq 0,$$

$$\tilde{b}_{77}(x_1 - 8)^2 - 2(x_2 - 4)^2 - 3x_5^2 + 7x_4 + \tilde{b}_{78}x_5x_8 + 30 \geq 0,$$

$$3x_1 - 6x_2 + \tilde{b}_{79}(x_9 - 8) + \tilde{b}_{710}x_{10} \geq 0,$$

$$\tilde{b}_{711}x_1 + 5x_2 + \tilde{b}_{712}x_7 + 9x_8 \leq 105,$$

$$10x_1 + \tilde{b}_{713}x_2 + \tilde{b}_{714}x_7 + 2x_8 \leq 0,$$

$$-8x_1 + 2x_2 + 5x_9 + \tilde{b}_{715}x_{10} \leq 12,$$

$$-5.0 \leq x_i \leq 10.0; \quad i = 1, 2, \ldots, 10.$$

The membership functions for the fuzzy parameters are triangular and are listed in Table 3. The optimal solution of the problem as reported in the source [22] is $f = 200$.

Problem 8 This problem is taken from [22] and has also been analyzed in [8].

$$\text{Min} f(x, \tilde{a}_8) = x_1^3 + (x_2 - 5)^2 + a_{81}(x_3 - 9)^2 + \tilde{a}_{82}x_3 + 2x_4^3 + \tilde{a}_{83}x_5^2$$
$$+ (x_6 - 5)^2 + \tilde{a}_{84}x_7^2 + \tilde{a}_{85}(x_7 - 2)x_8^2 - x_9x_{10} + 4x_9^3 + 5x_1$$
$$- 8x_1x_7$$

subject to:

$$\tilde{b}_{81} - 4(x_2 - 3)^2 - 2x_3^2 + 7x_4 + \tilde{b}_{82}x_5x_6x_8 + 120 \geq 0,$$
$$-5x_1^2 + \tilde{b}_{83}x_2 - (x_3 - 6)^2 + \tilde{b}_{84}x_4 + 40 \geq 0,$$
$$-x_1^2 - 2(x_2 - 2)^2 + \tilde{b}_{85}x_1x_2 + \tilde{b}_{86}x_5 - 6x_5x_6 \geq 0,$$
$$\tilde{b}_{87}(x_1 - 8)^2 - 2(x_2 - 4)^2 - 3x_5^2 + 7x_4 + \tilde{b}_{88}x_5x_8 + 30 \geq 0,$$
$$3x_1 - 6x_2 + \tilde{b}_{89}(x_9 - 8) + \tilde{b}_{810}x_{10} \geq 0,$$
$$\tilde{b}_{811}x_1 + 5x_2 + \tilde{b}_{812}x_7 + 9x_8 \leq 105,$$
$$10x_1 + \tilde{b}_{813}x_2 + \tilde{b}_{814}x_7 + 2x_8 \leq 0,$$
$$-8x_1 + 2x_2 + 5x_9 + \tilde{b}_{815}x_{10} \leq 12,$$
$$-5.0 \leq x_i \leq 10.0; \quad i = 1, 2, \ldots, 10.$$

The membership functions for the fuzzy parameters are triangular and are listed in Table 4.

References

1. Pal, A.: Decision making in crisp and fuzzy environments using particle swarm optimization. Ph.D. Thesis, Department of Mathematics, Punjabi University, Patiala-India (2015)
2. Deep, K., Singh, K.P., Kansal, M.L., Mohan, C.: A real coded genetic algorithm for solving integer and mixed integer optimization problems. Appl. Math. Comput. **212**(2), 505–518 (2009)
3. Tanaka, H., Okuda, T.O., Asai, K.: On fuzzy mathematical programming. J. Cybern. **34**, 37–46 (1974)
4. Zimmermann, H.J.: Fuzzy programming and linear programming with several objective functions. Fuzzy Sets Syst. **1**, 45–55 (1978)
5. Dhingra, A.K., Moskowitz, H.: Application of fuzzy theory to multiple objective decision making in system design. Eur. J. Oper. Res. **53**, 348–361 (1991)
6. Sakawa, M.: Fuzzy Sets and Interactive Multi-objective Optimization. Plenum Press, New York (1993)
7. Chen, J.E., Otto, K.N.: Construction membership functions using interpolation and measurement theory. Fuzzy Sets Syst. **73**, 313–327 (1995)
8. Nguyen, H.T., Some global optimization techniques and their use in solving optimization problems in crisp and fuzzy environments, Ph.D. Thesis, Department of Mathematics, University of Roorkee, Roorkee, India,1996
9. Mohan, C., Nguyen, H.T.: A controlled random search technique incorporating the simulated Annealing concept for solving integer and mixed integer global optimization problems. Comput. Optim. Appl. **14**, 103–132 (1999)
10. Sakawa, M., Yauchi, K.: An interactive fuzzy satisficing method for multi-objective non convex programming problems with fuzzy numbers through co evolutionary genetic algorithms. IEEE Trans. Syst. Man Cybern. Part B: Cybern. **31**, 459–467 (2001)

11. Zimmermenn, H.J.: Fuzzy Set, Decision Making and Expert System. Kluwer Acadamic Publisher, Dordrecht (1987)
12. Zimmermenn, H.J.: Fuzzy Set Theory and its Applications, 4th edn. Kluwer Acadamic Publisher, Dordrecht (2001)
13. Kacprzyk, J., Orlovski, S.: Optimization Models using Fuzzy Sets and Possibility Theory. D. Reidel, Dordrecht (1987)
14. Fedrizzi, M., Kacprzyk, J., Roubens, M.: Interactive Fuzzy Optimization. Lecture Notes in Economics and Mathematical Systems, vol. 368 (1991)
15. Delgado, D., Kacprzyk, J., Verdegy, J.L., Villa, M.A.: Fuzzy Optimization: Recent Advances. Physica Verlag, Germany (1994)
16. Bector, C.R., Chandra, S., Dutta, J.: Principle of Optimization Theory. Narosa Publishing house, New Delhi (2005)
17. Kahraman, C.: Fuzzy Applications in Industrial Engineering. Springer (2006)
18. Kahraman, C.: Fuzzy Multi-Criteria Decision Making: Theory and Applications with Recent Developments. Springer, Londan (2008)
19. Deep, K., Bansal, J.C.: Hybridization of particle swarm optimization with quadratic approximation. OPSEARCH 46(1), 1–22 (2009)
20. Zadeh, L.A.: Fuzzy algorithms. Inf. Control 12, 92–102 (1968)
21. Clerc, M.: Confinements and biases in particle swarm optimisation. Technical report, Open archive HAL (2006). http://hal.archives-ouvertes.fr/hal-00122799
22. Singh, K.P.: Multi-criteria decision making techniques for engineering and management problems. Ph.D. Thesis, IIT Roorkee-India (2009)

An Analytical Comparative Approach of Cloud Forensic Tools During Cyber Attacks in Cloud

Shaik Khaja Mohiddin, Suresh Babu Yalavarthi and Venkatesh Kondragunta

Abstract Cloud forensics is a product which has evolved with the combination of digital and computer forensics. Digital evidences involved in digital crime play an important role, and when this digital crime is carried out in cloud, then it is called as cloud crime. In a crime scene where it is important to isolate the entire scenario from the surroundings in order to collect the exact data at the time of crime, but the same is a little bit difficult and a challenging task to recover the information from a cloud crime, because here the investigator is connected to various cloud resources which may or may not be under his control and this may sometime lead against the privacy issues of cloud users. In this technological era where technology is changing the lifestyles of people, trends in various sectors either related to business, education, engineering, medical, scientific, and many other such kind of areas, with this vast growth of cloud in various areas, have opened a door for the intruders to carry out their mischief tasks, which may lead to various challenges regarding data breaches various other ways which lead to the disturbance of data or resources in the cloud. In digital forensics, it is evident to collect the information from a scene where Cloud forensics is a strategic solution to all such kind of things, where it is important to prove that how the disturbance or mischief task was carried out with a proof. Cloud computing is a boon for small-scale industries, with exponential growth of technology and the resources being required for the establishment of small-scale industries, and being very cost-effective which they are not affordable for such cases, cloud computing provides the resources to these industries so that they can even withstand the tough competition among themselves as well as from the big established companies. These

S. Khaja Mohiddin (✉)
Department of CSE, Acharya Nagarjuna University, Guntur, Andhra Pradesh, India
e-mail: mail2mohiddin@gmail.com

S. B. Yalavarthi
JKC College, Guntur, Andhra Pradesh, India
e-mail: Yalavarthi_s@yahoo.com

V. Kondragunta
Sr. Technology Consultant/Technology Lead Deloitte Consulting LLP., Atlanta, USA
e-mail: Kondragunta.v@hotmail.com

© Springer Nature Singapore Pte Ltd. 2019   509
J. C. Bansal et al. (eds.), *Soft Computing for Problem Solving*, Advances in Intelligent
Systems and Computing 817, https://doi.org/10.1007/978-981-13-1595-4_41

things made the popularity of Cloud along with efficient and effective computing paradigm made cloud computing a popular thing.

Keywords Cloud forensics (CF) · Cloud crime (CC) · Forensic tool kit (FTK) Cloud service provider (CSP)

1 Introduction

A remarkable increase in variety, volume, velocity of data in cloud computing has been increasing drastically for the past years, and this will go to further increase in coming future. However, the concept of cloud computing is not unfamiliar, due to the recent observed cyber attacks on the cloud repositories have revealed a huge loss with respect to public and social security as well. Beside high security measures taken by the cloud service providers, they are being exploited by the criminals. Beside the huge development and deployments of cloud computing services utilized by various organizations on daily basis, there is an increasing demand for the forensic investigation which has increased the concept of research in this field. Digital forensics deals with preserving, analyzing documenting the digital evidences which are being collected from the digital devices which may fall from any one of these categories, such as smart phone, desktop, laptop, hard disk, digital camera, and many other such storage devices [1]. In cloud forensic investigations, it is very important to analyze the flow of data, and this can be categorized into three stages, with respect to the client devices, i.e., the data at rest, data in transit, and with respect to the servers as data at rest, thereby it becomes necessary to carry out with static and dynamic analysis in order to trace out the proper reason for the loss of modified data, if the data is being modified then at what credential was the data modified and also if the data is being lost, we have to bring back the lost data. When a crime happens in a cloud, at that time the data may not be in a required format in order to collect the digital evidence, in such cases it become necessary to decode it either by the application or by the operating system, so one has to carry out both empirical and comprehensive investigations to get expected results from both client side and on cloud servers. With the current existing tools, the results are inbuilt and incomplete [2].

In a cloud environment where the cloud services are provided to the customers by the cloud providers based on that, a general classification of cloud services can be given as of the following three types which are shown in Fig. 1.

The services offered by the cloud are + being made classes of as Software as a Service (SaaS), Infrastructure as a Service (IaaS), Platform as a Service (PaaS).

IaaS: It deals with on-demand infrastructural resource provisioning with respect to virtual machines. Examples are Amazon EC2 [3], GoGrid [4].

PaaS: It refers to providing resource layers for platform it includes software development frameworks and operating systems, e.g., Google App Engine [5], Force.com.

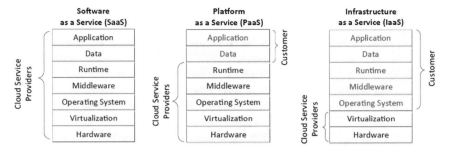

Fig. 1 General service models offered by the cloud

SaaS: It refers to providing applications on demand on the Internet, e.g., salesfor ce.com [6], Rackspace [7].

The main reason for the development of digital forensic is that criminals are also using technology to carryout digital crime, so it becomes necessary to collect the digital evidence from the crime location as soon as possible, because sometimes these evidences may be lapsed [8]. Various methods which are involved in digital forensics as discussed in [9] which involves identification, collection, acquisition, preservation, examination, and analysis.

2 Need to Have Forensic-Enabled Cloud

As there is no accurate direct tool in cloud which can be applied during the investigations, all the existing tools are being derived from digital or computer forensics, as there are certain constraints in the cloud itself which enables the need for a cloud forensic systems; its requirement can be categorized into functional and non-functional requirements, which was also stressed by [10–13] (Table 1).

In functional requirements, F1 specifies that one should follow a common unique standard digital forensic process which should be carrying out the same number of steps and in the same manner whomever might be dealing with the investigation in the cloud. In F2, there should be certain specific predefined steps that are designed in a well worse order so that any investigator during the investigation in cloud should follow these standard steps in order to reach the required goal; F3 deals with the concept of collaboration such that as the data in the cloud or the resources related to the cloud are not placed at a same location so different investigators are investigating the concepts of cloud forensics at different places in the cloud which are located at different locations. At final stages, the reports collected by them should be combined in order to get common report which should give a required conclusion. F4 specifies that the task or the steps for the implementation of cloud forensics is very forbidding thing, so some of the steps during the cloud investigation should be predefined, so that the process goes in a smooth way.

Table 1 Specifying various functional and non-functional requirements for forensic cloud

Requirements

Functional (F)	Non-functional (NF)
F1. Implementing a standard digital forensic process in the cloud	**NF1**. Flexibility [14]
F2. A cloud investigator should be guided through a standard digital forensic process	**NF2**. Ease of use and efficiency
F3. Collaboration should be allowed between all the existing jurisdictional by agencies in any cloud investigation	**NF3**. Scalability
F4. Semi-automated	**NF4**. Security [15]
	NF5. Audit ability
	NF6. Investigating agency should be given maximum control of the service stack

In non-functional requirements, NF1 specifies that while investigation by the cloud investigator in a cloud, he comes across certain regional, national, and international standards so it would be better to follow a flexible way which gives importance to any kind of standard which is encountered during the investigation. NF2 specifies that during the investigation, an investigator comes across a system with which he has to deal, so the new system should not be complex and it should not allow the user to take a long time, i.e., it should give a friendly interface, otherwise time for the investigation will increase which is going to waste the valuable time. NF3 specifies that during the investigation, a voluminous information is going to be gathered due to which depending on the need the collected information can be extracted either less or more, NF4 specifies that the collected data should be protected further from the intruders so that intruders should not be successful in any way to change the integrity of the collected data. NF5 specifies that the collected data in order should be in an auditable form. NF6 specifies that sometimes investigator might compromise the collected evidence so the investigating system should be very strong in order to protect such kind of activities form the other side.

3 Steps to Be Carried Out for Cloud Forensics Process

Cloud forensic process is carried out in the following steps as:

Incident Confirmation: It is the first stage of investigation where the incident is confirmed either by intruder detection alarms, snap shots which are being collected during the investigation process by these things one can confirm that the incident has been carried out.

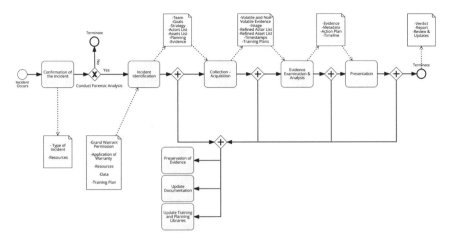

Fig. 2 Process for cloud forensic investigation

Incident Identification: After confirmation of the occurrence of the incident, it is necessary to confirm the identity of the incident, i.e., how it has carried out, what was the loss, how much the client is going to be affected with this loss.

Collection–Acquisition: Clues should be collected properly either directly or indirectly, because they lead the forensic investigation so immediately after the cloud crime has been carried out; it is advised to be fast in order to collect the data, there might be chances where an intruder may sometime try forcefully to delete the clues during the crime.

Examination–Analysis: Examination of the collected data should be carried out in a good way either by using the default defined process or the new initiated process, which ever might be the thing but the analysis should be carried out in a correct way in order to get the relevant results (Fig. 2).

Presentation: After collection of all the necessary information related to investigation, it should be presented in a well manner, as the clues or information which is collected is related to technical background and it has to be explained to the people who are not aware of the technical concepts, i.e., it has to be explain to a judge during the court process, principles which are being carried out, so that the people who are related with it should get a complete picture of the thing that is being carried out and they get an idea about the cloud crime, and how it has been done.

Concurrent Activities: Along with the investigation process simultaneously, the collected evidence is preserved, investigation-related documents are being updated, and along with that training and planning libraries are also updated.

4 An Overview of the Existing Cloud Forensic Tools

There exist certain limitations for digital forensics which can be defined by [15, 16], which exists in the areas of legal aspect, volume of data, capacity of tools, forensic analysis automation and visualization; these existing limitations make further need for the new cloud forensic tool which has to carry out the necessary things (Table 2).

Table 2 List of different forensic tools which are being used during an investigation [17]

S. no.	Tool name	Purpose
1	Digital forensics frameworks	It is an open source. It is very popular digital dedicated platform, and it can be easily handled by non-experts as well as experts also. It comes with GPL license. It is used for digital chain of custody. Even hidden and deleted files can also be recovered
2	Open computer forensics architecture	It is an open-source forensic framework, it is used for storing data and it uses PostgreSQL, and it works on Linux platform
3	CAINE	Computer-aided investigative environment is an open source; it is helpful in integrating software modules from existing software tools which are carried out in an user-friendly manner
4	X-Ways forensics	It runs on almost all available Windows versions, digital examiners consider it as an advanced stand, it is used in various dimensions such as cloning, disk imaging, bulk hash calculations, and it also supports maximum existing file formats
5	SANS investigative forensics toolkit (SIFT)	It is a operating system forensic used for multipurpose which has all required tools to be used in digital forensic process which has an inbuilt platform on Ubuntu
6	EnCase	This tool has a forensic platform which is utilized for multipurpose, it gathers information from different devices very fast, and also this tool also produces report-based evidence. It is a paid tool
7	Registry recon	This is a paid tool, it is well known for registry analysis, it gathers the registry info from evidence and again rebuilds the registry, and it can be used for registry rebuild for both previous and current Windows
8	Sleuth Kit (+Autopsy)	It is used in the forensic analysis of computers, it is a Windows- and Unix-based tool, and it has many tools which are being used for analyzing disk images, file systems depth analysis
9	Libforensics	It is available with various demo tools in order to extract information from different types of evidences, and this tool was developed by python. It consists of various libraries which are required during digital forensic applications
10	Volatility	It is used for analysis in malware, it is a memory-based forensic framework, this tool is helpful in extracting information from the processes which are running, and from Windows crash also one can extract the information

(continued)

Table 2 (continued)

S. no.	Tool name	Purpose
11	WindowsSCOPE	It is used for volatile memory analysis, and it also works as reverse engineering tool, utilizing this tool one can analyze Windows virtual and physical memory and kernel drivers
12	Corner's toolkit	This tool is used to recover the data from the devices which are working based on Unix operating systems
13	Oxygen forensic suite	Using this tool, one can gather the information from mobile, one can also recover calendar information, call logs, contacts, and messages and also deleted messages, and the reports which are being generated by this tool are very easy to analyze
14	Bulk extractor	This tool is very fast as it does not follow file system structure during the extraction of data from files; it is helpful for the scanning of disk images, for the extraction of useful information from files
15	Xplico	The data from the applications which uses network and Internet protocols can be extracted using this tool, It is an open-source tool for forensic analysis, it supports many protocols such as TCP, UDP, SIP, HTTP, IMAP, it also supports both ip6 and ip4, and SQLite database is used for the storage of output
16	Mandiant Redline	It is used for file and memory analysis, when a process is running on the host it collects the information, and it is also helpful to gather meta data, registry data, and Internet history for building up a report
17	Computer online forensic evidence extractor (COFEE)	Forensic experts in computers use this tool kit, and as it was developed by Microsoft, it collects evidences within the Windows systems. It can be installed in the system with the help of external hard disk or pen drive, as it is very fast so it finishes the analysis within 20 min
18	P2eXplorer	It is an image mounting tool, and on the hard disk these images are mounted and they are then analyzed by file explorer; here more images can be mounted within short span of time
19	PlainSight	With Linux distribution, it is a CD-based Knoppix, and it is useful in gathering information related to internet history, extracting password hashes, checking of Windows firewall, data carving, inspecting USB device usage
20	XRY	Developed by Micro Systemation, it is helpful for recovery of crucial and analyzed data from the mobile; this tool is a combination of both hardware and software
21	HELIX3	It is an incident responsive CD-based digital forensic suite, it also includes hex editors, tools for password cracking, data carving, and it is also available in free version
22	Cellebrite UFED	It is very helpful to collect information with high accuracy on mobile data, and using this tool one can also get up to date extraction of mobile data; the designing of UFED field series is done such that it unifies the workflow between lab and the field

(continued)

Table 2 (continued)

S. no.	Tool name	Purpose
24	FTK imager	This tool is used for the examination of folders and files which are being stored in network drives, DVDs/CDs, hard drives; it is also helpful in creating MD5 or SHA1 hash files, and one can also recover deleted files even from Recycle Bin
25	DEFT	It is a Linux-based CD which consists of number of open-source and freely available forensic tools; this is helpful in hashing, data recovery, network forensic, and mobile forensics
26	Bulk extractor	This tool is helpful for scanning of directory of files, disk images, e-mail address, credit card number, various ZIP file and URLs

5 Conclusion

Though there exist several digital forensic tools, which are helpful during the investigation but there limitation is up to a desktop, laptop, smart phone, or any kind of smart devices, utilizations of these tools give results but not to the satisfactory levels so it is necessary to have the an individual dedicated cloud-based computer tool and which can be utilized in a proper way to enhance and to get good results during the forensic investigations. So there is a need of a proposal model, which is expected to reach to that extent and it also need certain things to be enhanced and has to be implemented to get the desired results. The data on which investigation have to be carried out is fragile, i.e., it is in a changed format and also here investigator has to carry out the investigation on data that are not on a crime scene it is situated on remote servers, and accessing such kind of data on remote servers where other clients data is also resides may cause breach to their privacy and may sometimes lead to the change of existing data which there by looses data integrity concept. In our paper, we have traced out the comparisons of various available digital forensic tools and the necessity for the existence of separate tool, which is dedicated with respect to the cloud in order to carry out the forensic investigations. In the paper, we have also traced out the importance and the necessity for cloud forensic tool.

References

1. Ting, Y.-H., et al.: Design and implementation of a cloud digital forensic laboratory. In: Symposium on Cryptography and Information Security (2013)
2. Roussev, V., Ahmed, I., Barreto, A., McCulley, S., Shanmughan, V.: Cloud forensics–tool development studies and future outlook. Digital Investig. **18**, 79–95 (2016)
3. Amazon Elastic Computing Cloud. aws.amazon.com/ec2
4. Cloud Hosting: CLoud Computing and Hybrid Infrastructure from GoGrid. http://www.gogrid.com
5. Google App Engine. http://code.google.com/appengine
6. Salesforce CRM. http://www.salesforce.com/platform

7. Dedicated Server, Managed Hosting, Web Hosting by Rackspace Hosting. http://www.racksp ace.com
8. Mishra, A.K., et al.: Cloud forensics: state-of-the-art and reasearch challenges. In: International Symposium on Cloud and Services Computing (2012)
9. Mohiddin, S.K., Yalavarthi, S.B., Sharmila, S.: A complete ontological survey of cloud forensic in the area of cloud computing. In: Deep, K., et al. (eds.) Proceedings of Sixth International Conference on Soft Computing for Problem Solving. Advances in Intelligent Systems and Computing, vol. 547. Springer, Singapore (2017)
10. Microsoft: NW3C-Microsoft COFEE (2014). https://cofee.nw3c.org
11. Mitrokotsa, A., et al.: Intrusion detection in MANET using classification algorithms: the effects of cost and model selection. Ad Hoc Netw. **11**(1), 226–237 (2013). ISSN: 15708705. https://d oi.org/10.1013/j.adhoc.2012.05.006
12. Volatility: Volatility Introduction—volatility—Introduction to volatility—An advanced memory forensics framework (2014)
13. Slay, J., et al.: Advances in Digital Forensics V, Volume 306 of IFIP Advances in Information and Communication Technology, pp. 37–47. Springer, Berlin, Heidelberg (2009)
14. Craiger, P., et al.: Advances in Digital Forensics IV, Volume 285 of IFIP the International Federation for Information Processia. Springer, US (2008)
15. Al Fahdi, M., et al.: Challenges of digital forensic—a survey of researchers and practitioners attitudes and opinions. In: Information Security for South Africa, pp. 1–8. IEEE (Aug 2013)
16. Henry, P., et al.: The SANS survey of digital forensics and incident response. Technical report, SANS (July 2013)
17. https://resources.infosecinstitute.com/computer-forensics-tools

Prediction of Inverse Kinematics for a 6-DOF Industrial Robot Arm Using Soft Computing Techniques

Golak Bihari Mahanta, B. B. V. L. Deepak, M. Dileep, B. B. Biswal and S. K. Pattanayak

Abstract In this twenty-first century, due to the heavy demand for high quality and great accuracy product from the customer, a large number of industries nowadays shifted their focus toward the installation of the robotic arm in their assembly line for faster production of the product. One of the most challenging problems of the robotic system is the inverse kinematics which deals to find the joint angles for the given robotic configurations. When the DOF of the system increases, it is very difficult to calculate the precise result with the help of the analytical methods and also the computing time taken for solving the analytical methods of the robotic systems is more. So, to achieve a better result for the inverse kinematics problem various intelligent and nontraditional techniques are used in recent years. In this article, authors have presented the application of soft computing techniques to obtain the inverse kinematics of Kawasaki RS06L 6-DOF robotic manipulator for a pick and place operation. For validating and checking the efficiency of the proposed approaches, a comprehensive study has been conducted among the techniques such as artificial bee colony (ABC), firefly algorithm (FA), invasive weed optimization (IWO), and particle swarm optimization (PSO). In a simulation environment, a thorough study conducted to find the inverse kinematics of the proposed system before the actual experimentation, and obtained results are fed into the Kawasaki Controller for the pick and place task which shows efficient results.

G. B. Mahanta (✉) · B. B. V. L. Deepak · M. Dileep · B. B. Biswal
Industrial Design Department, NIT Rourkela, Rourkela 769008, Odisha, India
e-mail: golakmahanta@gmail.com

B. B. V. L. Deepak
e-mail: bbv@nitrkl.ac.in

M. Dileep
e-mail: mdileepkumar48@gmail.com

B. B. Biswal
e-mail: bbbiswal@nitrkl.ac.in

S. K. Pattanayak
Department of Mechanical Engineering, NIT Silchar, Silchar 788010, Assam, India
e-mail: sujupat@gmail.com

© Springer Nature Singapore Pte Ltd. 2019
J. C. Bansal et al. (eds.), *Soft Computing for Problem Solving*, Advances in Intelligent Systems and Computing 817, https://doi.org/10.1007/978-981-13-1595-4_42

Keywords Inverse kinematics · Metaheuristics · Artificial bee colony
Firefly algorithm · Invasive weed algorithm · Particle swarm optimization

1 Introduction

In this cutting-edge world, robotics is a modern field of advanced innovation that
pushes the traditional engineering boundaries toward the modern era of automation.
For a thoughtful understanding of the complexity of robots and their applications in
everyday life requires learning of electrical, mechanical engineering, basic science,
and software. Numerous industrial tasks have been carried out such as pick and
place operation, assembly operation at the manufacturing plant, welding and spray
painting in automobile industry, army, home, as well as in a medical application for
surgery by several advanced robotic systems [1]. If the robotic manipulators work
uncontrollably or inaccurately, they can cause harm to the working environment. If
a painting robotic manipulator in an automobile plant painting a car does not have a
precise and consistent controller, and the robot not able to paint at the desired place.
For solving the above-said problem, it is necessary to derive the proper working
parameters for the industrial manipulators. Formulations of the forward kinematics
equations are relatively easier than that of its counterpart inverse kinematics equation
[2]. The formulation of inverse kinematics equation of the robotic systems leads to
the time-varying nonlinear equations as well as transcendental functions which leads
to a computational difficulty for solving those developed equations using geometric,
iterative, or algebraic method.

Mathematical model, Jacobian, geometric, analytical, and pseudo-inverse are
some of the traditional approaches frequently used for solving the inverse kinematics problem of the robotic systems effectively. Although these methods have the
tendency to solve the inverse kinematics problem of the robotic systems, it inherits
some limitations which triggers researcher to look toward the savvy or soft computing
methods to discover the inverse kinematic arrangements. Some of the well-known
soft computing techniques which explored to solve for various engineering problem
named as genetic algorithm (GA), particle swarm optimization (PSO), ant colony
optimization (ACO), artificial neural network (ANN), hybrid ANN, fuzzy logic,
hybrid fuzzy, ANFIS, and much more.

In this article, four powerful commonly used soft computing techniques called artificial bee colony (ABC), firefly algorithm (FA), invasive weed optimization (IWO),
and particle swarm optimization (PSO) have been employed for solving IK problem
in a 6-DOF industrial robot (Kawasaki RS06L). The remaining of this article organized as literature review, soft computing technique, proposed methodology, results
and discussion, and conclusion.

2 Literature Review

To find the inverse kinematics problem of a robotic system is one of the most challenging jobs. It is time-consuming for the real controller of the manipulator to solve the complex equations. There are abundant number of methodologies used to solve the inverse kinematics issue among which few traditional techniques such as analytical solution [3, 4], iterative solution [5–7], geometric solution [2, 8, 9], quaternion algebra [10], theory of screws [11], optimization techniques [12, 13], cyclic coordinate descent method [14], and Monte Carlo method [15]. Although all the above traditional techniques can solve the inverse kinematics problem, these algorithms have their limitations when it comes to the higher degree of freedom. The above problem leads to exploring toward the nontraditional optimization techniques such as genetic algorithm [16–18], simulated annealing [19], and harmony search [20]. Some of the other well-known soft computing techniques which explored to solve the inverse kinematics are fuzzy logic, neural network, artificial neural network, ANFIS, hybrid neural network [21–24], etc. At the same time, much focus toward the use of the nature-inspired algorithm for resolving the inverse kinematics problem of the industrial manipulator as well as other problems related to the engineering field. The most commonly used algorithms for solving the real-world problems are artificial bee colony (ABC) [25], firefly algorithm (FA) [26], particle swarm optimization (PSO) [27–30].

3 Soft Computing Techniques

In this study, four well-known soft computing techniques are used and described below.

3.1 Artificial Bee Colony (ABC)

ABC algorithm is one of the well-known optimization metaheuristic algorithms which is based on the intelligent foraging behavior of a honey bee developed by Dervis Karabogain in 2005. The algorithm uses common parameters like a maximum number of cycles and population size to control the outcome of optimization. ABC algorithm comprises of three groups of bees named as: employed bees whose main purpose is to collect nectar from the food source, onlooker's bees whose sole purpose is to looks on the proper working of the proceedings and remain on the dance floor and gathers information about the position and amount of nectar present at a food source, and scout's bees randomly searched for the food sources. Parameters used for the simulation operation as follows:

Population size (Colony Size) = 30, number of onlooker bees = 30, acceleration coefficient upper bound = 1.

3.2 Invasive Weed Optimization (IWO)

Invasive weed optimization is developed from the inspiration based on the natural behaviors of weeds in colonize and its best way for getting the appropriate place for the growth and reproduction. The algorithm mainly has four stages. The algorithm starts with the first phase by initializing the initial population candidates of the algorithm over a specified search area, and a countable number of seeds have been spread over the predefined area. The second phase of the algorithm starts with *reproduction phase* in which each seeds is allowed to develop a flowering plants according its fitness value. The third stage of the algorithm deals with the *spatial dispersal* where randomness and adaptation will be provided into it. Generated seeds at the previous stages have been randomly distributed over a specified dimensional search space according to the normally distributed random numbers with mean value equal to zero. Normally distributed random numbers with mean value zero is used to disperse the seeds for ensuring the distribution of the seeds near the parent plants. The developed seeds from the parents have been randomly distributed around their parents with a decreasing variation and grow to new plants according to its fitness. The fourth and final phase of the algorithm is the competitive exclusion which allows to find maximum number of plants according to the fitness of each seed. This whole process continues until it achieved either the maximum set iteration or the treated the plant as the optimal solution having the best optimal value.

Parameters considered: initial population = 30, maximum population = 35, minimum number of seeds = 0, maximum number of seeds = 5, variance reduction exponent = 2, initial value of standard deviation = 0.5, final value of standard deviation = 0.001.

3.3 Particle Swarm Optimization (PSO)

This algorithm was developed by Kennedy and Eberhart based on the collective behavior of various insect colonies, birds, and animal societies search technique as well as the synchronous flocking behavior. It is one of the most famous and versatile technique. To find out an optimal solution for their work, the particles reorganize their flying according to their own experience and the flying experience of their companions. In gbest (Global Best) the particles are attracted toward the best solution found by a member of the swarm, whereas in lbest (Local Best) is attracted to the best option obtained among its neighborhood. As the gbest model suggests, the particles in the population travel to their best position and toward the best particle position in the swarm. For a given position, velocity and fitness value, the particles will update their own best value (i.e., personal best value) if there is an improvement in the fitness value. The above process is carried out for every particle, and the position and fitness value of the best particle in entire swarm population is updated as the global best value. Then the velocities of all particles are updated using their previous velocity, personal best value, and the global best value and provide a new position of the particles. To enhance the exploration of the search space, mutation or local search can be used for a particular group of particles in the population. A predetermined stopping criterion is set to terminate the iterations. Parameters considered as follows:

Population size (Swarm Size) = 30, inertia weight = 1, inertia weight damping ratio = 0.99, personal learning coefficient = 1.5, global learning coefficient = 2.0.

3.4 Firefly Algorithm (FA)

Xin-She Yang was taken the inspiration of the flashing pattern and behavior of fireflies to develop a nature-inspired algorithm which is named as firefly algorithm (FA). There are so many species of firefly around two thousand produces bioluminescence due to which a flashing light is shown at the abdomen of the firefly. All the firefly produces their pattern of flashes, and the main purpose of their flashing is to attract a mate (communication) and to attract potential prey and produce an amazing sight in the tropical areas during summer. The flashing can also be used to send information between fireflies. The firefly's characteristics are defined by Xin-She Yang [26] as:

All fireflies are unisex for attracted to remaining fireflies regardless of their sex.

- Attractiveness is proportional to their brightness thus for any two flashing fireflies the low brighter one will move toward the higher brighter one.
- The attractiveness is directly proportional to the brightness, and they both decrease as their distance increases. If no one is brighter than a specified firefly, it moves randomly.
- The brightness or light intensity of a firefly is determined by the landscape of the objective function to be optimized.

Parameters considered for this study as follow:

Number of fireflies (Swarm Size) = 30, light absorption coefficient = 0.8, attraction coefficient base value = 0.2, mutation coefficient = 0.2, mutation coefficient damping ratio = 0.997, uniform mutation range = 0.9.

4 Mathematical Modeling of the Robotic System

Mathematical modeling of the proposed robotic system is very much required to deal with the inverse kinematics problem of the robotic systems. By knowing the link lengths and joint positions of the robotic system, it is easy to compute the forward kinematics equation which is nothing but to find the end effector positions. For a

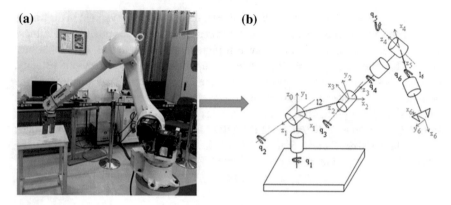

Fig. 1 a Kawasaki RS06L, **b** schematic diagram of the Kawasaki Robot

Table 1 D–H parameter of the robotic system

Link	a_i	Alpha (α_i)	d_i	θ_i
1	0	90°	0	θ_1
2	l_2	0°	0	θ_2
3	0	+90°	0	θ_3
4	0	−90°	l_4	θ_4
5	0	90°	0	θ_5
6	0	0°	l_6	θ_6

higher DOF robotic systems, the formulation of IK problem leads to complexity. The most successful way to represents the proposed robotic system is by using the Denavit–Hartenberg (D–H) model. Denavit–Hartenberg (D–H) model can represent the robot links and joints for any robot configuration, regardless of its sequence and complexity. D–H parameters of the considered system (Kawasaki RS06L) and its line diagram are illustrated as follows (Fig. 1 and Table 1).

The forward kinematics equations for the robotic system is

$$
{}^0T_1 = \begin{bmatrix} \cos\theta_1 & -\cos\alpha_1 \times \sin\theta_1 & \sin\alpha_1 \times \sin\theta_1 & a_1 \times \cos\theta_1 \\ \sin\theta_1 & \cos\alpha_1 \times \cos\theta_1 & -\sin\alpha_1 \times \cos\theta_1 & a_1 \times \sin\theta_1 \\ 0 & \sin\alpha_1 & \cos\alpha_1 & d_1 \\ 0 & 0 & 0 & 1 \end{bmatrix}
$$

$$
{}^1T_2 = \begin{bmatrix} \cos\theta_2 & -\cos\alpha_2 \times \sin\theta_2 & \sin\alpha_2 \times \sin\theta_2 & a_2 \times \cos\theta_2 \\ \sin\theta_2 & \cos\alpha_2 \times \cos\theta_2 & -\sin\alpha_2 \times \cos\theta_2 & a_2 \times \sin\theta_2 \\ 0 & \sin\alpha_2 & \cos\alpha_2 & d_2 \\ 0 & 0 & 0 & 1 \end{bmatrix}
$$

$$^{2}T_{3} = \begin{bmatrix} \cos\theta_3 & -\cos\alpha_3 \times \sin\theta_3 & \sin\alpha_3 \times \sin\theta_3 & a_3 \times \cos\theta_3 \\ \sin\theta_3 & \cos\alpha_3 \times \cos\theta_3 & -\sin\alpha_3 \times \cos\theta_3 & a_1 \times \sin\theta_3 \\ 0 & \sin\alpha_3 & \cos\alpha_3 & d_3 \\ 0 & 0 & 0 & 1 \end{bmatrix}$$

$$^{3}T_{4} = \begin{bmatrix} \cos\theta_4 & -\cos\alpha_4 \times \sin\theta_4 & \sin\alpha_4 \times \sin\theta_4 & a_4 \times \cos\theta_4 \\ \sin\theta_4 & \cos\alpha_4 \times \cos\theta_4 & -\sin\alpha_4 \times \cos\theta_4 & a_4 \times \sin\theta_4 \\ 0 & \sin\alpha_4 & \cos\alpha_4 & d_4 \\ 0 & 0 & 0 & 1 \end{bmatrix}$$

$$^{4}T_{5} = \begin{bmatrix} \cos\theta_5 & -\cos\alpha_5 \times \sin\theta_5 & \sin\alpha_5 \times \sin\theta_5 & a_5 \times \cos\theta_5 \\ \sin\theta_5 & \cos\alpha_5 \times \cos\theta_5 & -\sin\alpha_5 \times \cos\theta_5 & a_5 \times \sin\theta_5 \\ 0 & \sin\alpha_5 & \cos\alpha_5 & d_5 \\ 0 & 0 & 0 & 1 \end{bmatrix}$$

$$^{5}T_{6} = \begin{bmatrix} \cos\theta_6 & -\cos\alpha_6 \times \sin\theta_6 & \sin\alpha_6 \times \sin\theta_6 & a_6 \times \cos\theta_6 \\ \sin\theta_6 & \cos\alpha_6 \times \cos\theta_6 & -\sin\alpha_6 \times \cos\theta_6 & a_6 \times \sin\theta_6 \\ 0 & \sin\alpha_6 & \cos\alpha_6 & d_6 \\ 0 & 0 & 0 & 1 \end{bmatrix}$$

The final position of the end effector is determined by the following Eq. (1)

$$^{0}T_{6} = {}^{0}T_{1} \times {}^{1}T_{2} \times {}^{2}T_{3} \times {}^{4}T_{5} \times {}^{5}T_{6} \tag{1}$$

The end effector position is represented by (X, Y, Z) in Cartesian coordinate system and the target point value is calculated from the below-provided equation.

$$X = {}^{0}T_{6}(1, 4); Y = {}^{0}T_{6}(2, 4); Z = {}^{0}T_{6}(3, 4); \tag{2}$$

Knowing the target point of the system (A, B, C), we can formulate the fitness function of the proposed 6-DOF robotic system as follows:

$$F_{fitness} = \sqrt{(X - A)^2 + (Y - B)^2 + (Z - C)^2} \tag{3}$$

5 Proposed Methodology

The problem of inverse kinematics of a 6-DOF Kawasaki robotic manipulator is studied and simulated by four different metaheuristics techniques which include ABC, FA, IWO, PSO, and work is focused on minimizing the square of the Euclidian

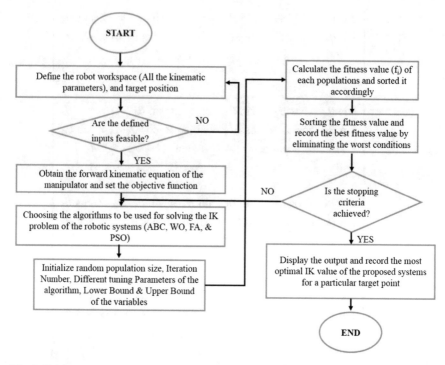

Fig. 2 Proposed methodology for finding the IK using soft computing techniques

distance between the target position and the nearest point achieved by the algorithm. The objective function (fitness function) for the 6-DOF robotic manipulator has defined an Eq. 1. The behavior and performance of each algorithm are analyzed based on the result obtained from the fitness function and the convergence time (Fig. 2).

Various tuning parameters are needed for each soft computing techniques, and the result obtained is heavily depends on the parameters of the systems. All the considered parameters are described in the Sect. 3.

6 Results and Discussion

This section of the research paper presents the outcome from the simulation and experimental setup for the considered robotic systems. The simulations of the proposed methodology to find out the IK problem of 6-DOF robotic system has been implemented using a personal computer with Windows 7 power by Intel Core i7, 3.54 GHz processor and 8 GB RAM. The programming has been done using MATLAB 2016a for the simulation study of the proposed method. As the soft computing methods are considered the random value for the initial population, the output result

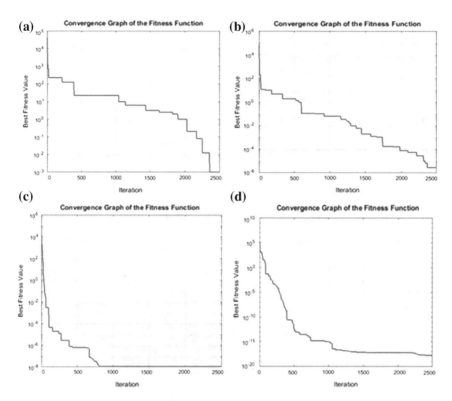

Fig. 3 Convergence graph of the fitness function, **a** ABC, **b** IWO, **c** PSO, **d** FA

may vary time to time. Hence, to overcome such difficulty and to avoid any ambiguous behavior simulations are carried for 100 times and the best solution for each method selected. In this study, four well-known soft computing such as ABC, IWO, PSO, and FA considered for the simulation study and the best method is obtained after comparing the performances of each technique based on the positional error between the target and achieved point as well as the time needed for the simulation.

All the simulation carried out for 2500 iterations (Fig. 3).

In Table 2, the obtained results of the different joints angle of the 6-DOF Kawasaki Robot for a fixed target point of the end effector (350, 400, 540) has been recorded. The same target point is fixed for all the metaheuristics algorithm as well as keeping the iteration number 2500. Table 2 also compares the positional error as well as the computational time for the different proposed algorithms.

The results for each method in this table are ordered as highest position error to the lowest positional error. From Table 2, it is clearly seen that firefly algorithm dominate other soft computing techniques and lowest positional error obtained. Table 2 shows the best output result for the firefly algorithm and the error is in 1.65×10^{-18} mm. Time taken for the experiment is shown in Fig. 4, and it shows that for 2500 iteration PSO taking less time. Firefly techniques followed PSO in time duration taking 29 s.

Table 2 Predicated IK value for the proposed system

Soft computing techniques	Predicted value of the joint angles θ_1 to θ_6 for the target $(350, 400, 540)_{xyz}$ (in degree)	Positional error (mm)	Time (s)
ABC	(67.5, 142.8, −55.7, 40.33, 38.8, −165.5)	7.89×10^{-3}	34
IWO	(42, 110.8, −89.9, 71.8, 30.4, −123.7)	5.98×10^{-5}	39
PSO	(67.6, 87.8, −77.3, 43.3, 39.2, −142)	2.5×10^{-8}	25
FA	(52.5, 122.8, −59.76, 50, 30.8, −185.5)	1.65×10^{-18}	29

Fig. 4 Time comparison to perform 2500 iteration for each soft computing technique

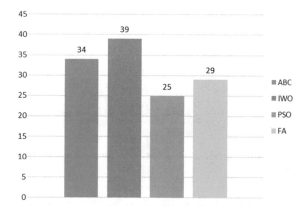

Only 4 s difference between PSO and FA for the whole simulation but the positional error much higher in case of PSO compared to FA. Hence, FA is recommending to find the IK value of the joint trajectory of the Kawasaki Industrial Robot.

After obtaining the joint parameters from the simulation study, it is fed into the Kawasaki E-Controller for performing the pick and place operation. Kawasaki E-Controller based on the AS programming language provided by the Kawasaki Company. The performed experiment for pick and place is shown in Fig. 5.

7 Conclusions

In this work, a thorough investigation of application of soft computing technique to predict the IK problem of industrial manipulator carried out. So far different methods are used to solve the IK problem of robotic manipulation but very less investigation in

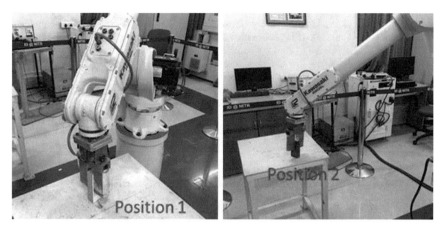

Fig. 5 Pick and place work from position 1 to the position 2

the real industrial manipulator. A well-organized comparative study conducted and based on the obtained performance firefly algorithm came as the winner. In future, several other recent developed soft computing techniques can be investigated as well as hybridization with other intelligence techniques will be done and investigated to find out IK for higher DOF industrial robotic arm.

References

1. Royakkers, L., Van Est, R.: A literature review on new robotics: automation from love to war. Int. J. Soc. Robot. **7**(5), 549–570 (2015)
2. Kucuk, S., Bingul, Z.: Robot Kinematics—Forward and Inverse Kinematics (2007)
3. D'Souza, A., Vijayakumar, S., Schaal, S.: Learning inverse kinematics. In: Proceedings of the 2001 IEEE/RSJ International Conference on Intelligent Robots and Systems. Expanding Society Role Robotics Next Millennium (Cat. No. 01CH37180), vol. 1, no. 1, pp. 298–303 (2001)
4. Lu, Z., Xu, C., Pan, Q., Zhao, X., Li, X.: Inverse Kinematic Analysis and Evaluation of a Robot for Nondestructive Testing Application, vol. 2015 (2015)
5. Baillieul, J.: Kinematic programming alternatives for redundant manipulators. In: Proceedings of the 1985 IEEE International Conference on Robotics and Automation, vol. 2, pp. 722–728 (1985)
6. Deo, A.S., Walker, I.D.: Adaptive non-linear least squares for inverse kinematics. In: Proceedings of the 1933 IEEE International Conference on Robotics and Automation, pp. 186–193 (1993)
7. Tevatia, G., Schaal, S.: Inverse kinematics for humanoid robots. In: Proceedings of the IEEE International Conference on Robotics and Automation, vol. 1, pp. 294–299 (2000)
8. Fu, Z., Yang, W., Yang, Z.: Solution of inverse kinematics for 6R robot manipulators with offset wrist based on geometric algebra. J. Mech. Robot. **5**(3), 310081–310087 (2013)
9. Han, L., Rudolph, L.: A unified geometric approach for inverse kinematics of a spatial chain with spherical joints. In: Proceedings of the IEEE International Conference on Robotics and Automation, pp. 4420–4427 (2007)

10. Aydm, Y., Kucuk, S.: Quaternion based inverse kinematics for industrial robot manipulators with euler wrist. In: 2006 IEEE International Conference on Mechatronics, ICM, pp. 581–586 (2006)
11. Chen, Q., Zhu, S., Zhang, X.: Improved inverse kinematics algorithm using screw theory for a Six-DOF robot manipulator. Int. J. Adv. Robot. Syst. 1 (2015)
12. Al-Faiz, M.Z., Ali, A.A., Miry, A.H.: Human arm inverse kinematic solution based geometric relations and optimization algorithm. Int. J. Robot. Autom. 2(4), 245–255 (2011)
13. Kumar, S., Sukavanam, N., Balasubramanian, R.: An optimization approach to solve the inverse kinematics of redundant manipulator. Int. J. Inf. Syst. Sci. 6(4), 414–423 (2010)
14. Kenwright, B.: Inverse kinematics—cyclic coordinate descent (CCD). J. Graph. Tools 16(4), 177–217 (2012)
15. Courty, N., Arnaud, E.: Sequential Monte Carlo Inverse Kinematics (2007)
16. Derman, L.J.: Solving the Inverse Kinematics Problem with Genetic Algorithms
17. Momani, S., Abo-Hammour, Z.S., Alsmadi, O.MK.: Solution of inverse kinematics problem using genetic algorithms. Appl. Math. Inf. Sci. 10(1), 225–233 (2016)
18. Tabandeh, S., Clark, C., Melek, W.: A genetic algorithm approach to solve for multiple solutions of inverse kinematics using adaptive niching and clustering, pp. 1815–1822 (2006)
19. Dutra, M.S., Salcedo, I.L., Margarita, L., Diaz, P.: New technique for inverse kinematics problem using simulated annealing. Eng. Optim. 1–5 (2008)
20. Ghafil, H.N.: Inverse acceleration solution for robot manipulators using harmony search algorithm. Int. J. Comput. Appl. 144(6), 1–7 (2016)
21. Alavandar, S., Nigam, M.J.: Inverse kinematics solution of 3 DOF planar robot using ANFIS. Int. J. Comput. Commun. Control. (Suppl. Issue Proc. ICCCC 2008) 3, 150–155 (2008)
22. Duka, A.-V.: Neural network based inverse kinematics solution for trajectory tracking of a robotic arm. Procedia Technol. 12, 20–27 (2014)
23. Duka, A.-V.: ANFIS based solution to the inverse kinematics of a 3 DOF planar manipulator. Procedia Technol. 19, 526–533 (2015)
24. Feng, Y., Wang, Y.-N., Yang, Y.-M.: Inverse kinematics solution for robot manipulator based on neural network under joint subspace. Int. J. Comput. Commun. Control 7(3), 459–472 (2012)
25. Abou-Shaara, H.F.: The foraging behavior of honey bees, Apis mellifera: a review. Vet. Med. (Praha) 59(1), 1–10 (2014)
26. Apostolopoulos, T., Vlachos, A.: Application of the firefly algorithm for solving the economic emissions load dispatch problem. Int. J. Comb. 2011, 1–23 (2011)
27. AL Rasheed, M.R.A.: A Modified Particle Swarm Optimization and its Application in Thermal Management of an Electronic Cooling System, p. 143 (2011)
28. Amouri, A., Mahfoudi, C., Zaatri, A., Merabti, H.: A new approach to solve inverse kinematics of a planar flexible continuum robot. AIP Conf. Proc. 1618, 643–646 (2014)
29. Bhushan, B., Pillai, S.: Particle swarm optimization and firefly algorithm: performance analysis. Adv. Comput. Conf. 746–751 (2013)
30. Coello, C.A., Reyes-Sierra, M.: Multi-objective particle swarm optimizers: a survey of the state-of-the-art. Int. J. Comput. Intell. Res. 2(3), 287–308 (2006)

Optimized Design Parameters for the Bidirectional Isolated Boost DC-DC Converter Using Particle Swarm Optimization

Bhatt Kunalkumar, Ram Avtar Gupta and Nitin Gupta

Abstract This paper describes dual active bidirectional buck–boost converter for battery charging/discharging application. The dual active bridge converter can charge and discharge the battery level with the wide voltage variation. When bidirectional DC-DC converter operates in the boost mode, an alternating path must be provided to reduce the circulating current to minimize the conduction losses. The RCD (a combination of resistance, capacitance, and diode) is the best way to provide the path for the circulating current which creates the voltage spike in the system. RCD snubber circuit provides an alternating path for circulating current; it also creates the I^2R losses in the system though it requires the accurate design parameters which can trade between voltage spike versus system losses. The particle swarm optimization (PSO) technique has been widely used for tuning purpose. The novel approach has been used in this paper to obtain the optimized design parameters using this PSO to reduce the copper losses in the system which is caused by RCD parameters. So PSO has been implemented to identify the best possible solution for RCD to the trade-off between losses and voltage spikes. The bidirectional converter system battery voltage has been considered in the wide range 60–320 V, and the output voltage has been considered 650 V with 20 kHz switching frequency. The loss analysis has been carried out with the optimal solution obtained from PSO and without the optimal solution of the system. The simulation result proves that optimal solution obtained from the PSO has lower losses compared to the optimal solution provided by genetic algorithm (GA).

Keywords Isolated boost · Bidirectional DC-DC converter · Dual active bridge converter · Particle swarm optimization (PSO)

B. Kunalkumar (✉) · R. A. Gupta · N. Gupta
Department of Electrical Engineering, Malaviya National Institute of Technology (MNIT), Jaipur, India
e-mail: kunalgcet04@gmail.com

R. A. Gupta
e-mail: ragupta.ee@mnit.ac.in

N. Gupta
e-mail: nitineed@gmail.com

© Springer Nature Singapore Pte Ltd. 2019   531
J. C. Bansal et al. (eds.), *Soft Computing for Problem Solving*, Advances in Intelligent Systems and Computing 817, https://doi.org/10.1007/978-981-13-1595-4_43

1 Introduction

Nowadays, the recent development of electric vehicle (EV) plays a significant role in the green global environment. EV DC-DC converter requires a high gain with the high energy efficiency [1]. The converter requires high power density as well as high conversion efficiency. The converter should also have the bidirectional power flow capability for energy saving purpose. The dual active bridge converters are popularly used for this application [2–4]. However, the converter needs to utilize one of the control schemes which are frequency modulation technique or phase-shift control methods. The frequency modulation techniques are widely adopted by the resonant converters [5–7]. The frequency modulation techniques have their limitations.

The phase-shift modulation technique is widely adopted for the dual active bridge converter. The phase-shift modulation technique cannot provide a constant voltage at one end with the wide voltage variation at the other end. Therefore, isolated boost converter is mandatory to meet the demand of the wide voltage variation with the high power density [8].

The isolated boost bidirectional converter suffers from high voltage spike and conduction loss due to the current difference between source inductance and leakage inductance of the transformer at the low-voltage side. The alternative path must be provided to this current to overcome the problem mentioned above. The most common way is to provide the RCD (a combination of resistance, capacitance, and diode) such a way that the current difference will flow through this path and reduce the conduction losses and voltage spike, but it also adds the copper losses in the system [9]. There should be a trade-off between the copper losses and voltage spike while designing the RCD snubber circuit. The design of RCD snubber requires utmost care and preciseness.

Optimizations techniques are being searched for optimization purposes. The different approaches for fuzzy control, optimal control, or sliding mode control can be used for optimization purpose. The neural networks and fuzzy logic-based controllers are accurate and robust. Optimal control has been used to achieve the finest response from the controller. Hybrid techniques, PSO, and genetic algorithms (GA) can be used to optimize the controller parameters.

The different algorithms have been used to tune the proportional integrator (PI) or proportional integrator–differentiator (PID) for the different controlling algorithms [10]. There are lots of studies based on tuning of the parameters of the PID controller which includes GA, differential evolution (DE) algorithm [11], and PSO algorithm [12–14]. In this paper, PSO algorithm has been used to design of RCD parameters of the circuit as it requires the utmost care and accuracy.

This paper is divided into the following section. Section 2 explains the particle swarm optimization. Section 3 depicts isolated boost bidirectional DC-DC converter. Section 4 explores the design of RCD snubber circuit. Section 6 analyzes the simulation results and loss analysis. Section 7 describes the conclusion of the paper.

2 Practical Swarm Optimization (PSO)

The PSO technique is inspired by the social behavior of swarms of birds and fish schools. The PSO method has many similarities with other computational technique like genetic algorithm (GA), but PSO does not have evolutionary operators such as crossover and mutation. The PSO has a particle which represents the solution. There is a velocity related to each particle which is constantly updated and adjusted according to the populations. The best possible solution can be obtained by the changing the position of each particle with some velocities. The fitness function evaluates the performance of the particle and population performance with each iteration. It increases or decreases the velocity of each particle in the direction best possible solution (P_{best}) and best global solution (g_{best}) as shown in Fig. 1.

The most important aspect of the topology is how the particles communicate with each other. There are many topologies where there are different ways to communicate with each other.

The velocity of each particle is defined by Eq. (1)

$$V_i = V_{i-1} + C_1 r_1 (P_{best_i} - x_{i-1}) + C_2 r_2 (g_{best_i} - x_{i-1}) \tag{1}$$

$$x_i = x_{i-1} + V_i \tag{2}$$

where V_i is the velocity, x_i is the position of the particle, and c_1 and c_2 are the learning factors. The particle is referred as a possible solution of particle swarm optimization problem which is an individual of a pool of solutions. The particle is obtained from the current position as well as previous best position obtained in the past with optimal solution. A particle has two parameters; each particle is optimized by continuous comparison with the best value obtained in the past. The value of *Pbest* and *gbest* is constantly updated based on the comparison.

Fig. 1 Graphical evolution of particle evolution

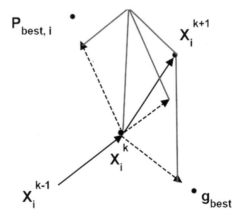

3   Isolated Boost Bidirectional DC-DC Converter

Isolated boost bidirectional DC-DC converter allows the power transfer in both the direction, and the isolated boost bidirectional DC-DC converter operates in the two modes of operation—(I) step-up mode and (II) step-down mode. The step-up mode has two sub-mode of operation—(a) energy storing mode and (b) energy transferring mode (Fig. 2).

The low-voltage side (V_{out}) has the wide voltage range variation which needs to be boosted up to provide the desired voltage level at high-voltage side (V_{in}). The necessary boost operation is achieved with the help of inductor same as the boost converter. The energy transferring mode will be the same as the normal full bridge converter with the phase-shifted topology. The problem occurs during the transition between boost mode energy transferring mode in step-up conversion. The circulation current caused because of the nonideality of the transformer will flow through the switches and contributes to the conduction losses as well as high voltage spike across the switches. Some auxiliary circuits must be provided to provide an alternate path to this current. In this paper, RCD clamp circuit has been used to provide this alternative path to the circuit.

$$V_{out} = n V_{in} D \tag{3}$$

$$V_{in} = \frac{n V_{out}}{(1 - D)} \tag{4}$$

The input/output equation in the step-down mode of the equation can be given by Eq. (3). The input/output equation in the step-down mode of the equation can be given by Eq. (4).

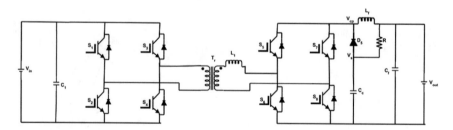

Fig. 2 Isolated boost bidirectional DC-DC converter

4 Design of RCD Snubber

With considering the initial condition $V_s = V_{cp}$ and $I_s = I_o$, the average current through capacitor should be zero in the steady-state condition. The time period obtained from ringing frequency can be given as follows:

$$\Delta t = \frac{L_1 I_r(t_1)}{V_{cp} - V_d} \tag{5}$$

The current flowing through D_5 is given by:

$$\int_{t1}^{t2} I_r(t) dt = \frac{I_r(t_1)}{2} \Delta t = \frac{1}{2} L_1 * I_r^2(t_1) \left(\frac{1}{V_{cp} - V_d} \right) \tag{6}$$

By solving above equation, we will get

$$I_r^2(t_1) = \frac{C}{L_1} V_{cp}(2V_d - V_{cp}) \tag{7}$$

The resistance value can be given by Eq. (8)

$$R = \frac{T(V_{cp} - V_p)(V_{cp} - V_d)}{C * V_{cp}(2V_d - V_{cp})} \tag{8}$$

The diode rating can be found out from Eq. (7). The clamping capacitor value should be large enough to give constant voltage tank with V_{cp}. The compromise must be made between V_{cp} and R for power loss. The optimization method has been used to minimize the power loss and to give the optimum value from particle swarm optimization [9] (Figs. 3 and 4).

5 Optimized Parameters for RCD Design

This research is conducted to develop optimized design parameters for RCD snubber circuit to minimize the power losses. The identification of minimizing fitness function can be made with the help of the optimized values of the parameters.

The accurate fitness function is required to improve the efficiency of the system. The chosen fitness function is the following:

$$F = \min \frac{[2 * (V_{cp} - V_o) * T * (V_{cp} - V_d)(V_{cp} - V_o)]}{R * C * V_{cp} * (2V_d - V_{cp})} \tag{9}$$

where V_{cp} and R are the variables with other parameters constant for the given rating—PSO: a swarm size of 50, an iteration number of 100. The PSO has been

(a)

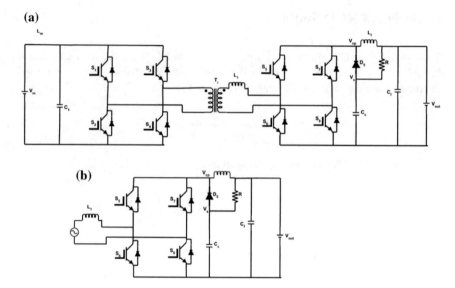

(b)

Fig. 3 a Isolated boost bidirectional DC-DC converter with RCD snubber, **b** simplified structure of isolated bidirectional DC-DC converter

Table 1 Constant parameters for PSO algorithm

Parameters	Value
V_o^*	310 V
C^*	500 pf
V_d	360 V
T	50 μS

Note V_o is chosen based on high voltage though variation in output voltage

chosen as it is a single-objective optimization technique for continuous optimization problem. Researchers have found out that DE provides the best possible solution, but time consumption is more in the DE. While according to the research by researches, PSO provides the second best solution with less time consumption. So PSO has been chosen in this paper [14] (Table 1).

6 Simulation Result

The simulation result proves that with the wide voltage variation 48–320 V, it is able to achieve the 650 V at the input side in the step-up mode of conversion. The dual active bridge converter operates during the energy transferring mode in step-up operation. The simulation has been done with the input voltage (60 V) with 25 A

Fig. 4 Flowchart for designing RCD snubber using PSO

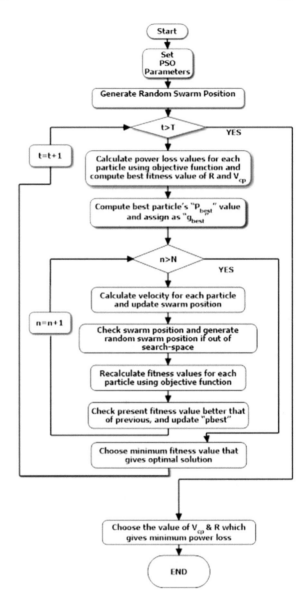

constant charging current to maintain the DC link voltage constant of 650 V on the secondary side. The voltage across primary side of transformer in dual active bridge converter can be represented by the Fig. 5a. The voltage across secondary side of transformer can be represented by Fig. 5b. The Fig. 5c represents the battery voltage and battery current. The Fig. 5 represents the charging mode for 310 V, 25 A battery rating. The Fig. 6 represents the charging mode for 48 V, 25 A battery rating. The Fig. 7 represents the discharge mode of operation during discharging mode with

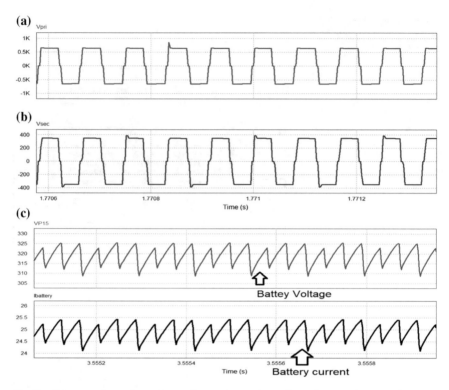

Fig. 5 a Primary voltage of transformer, **b** secondary voltage of transformer, **c** battery voltage (310 V) and battery current (25 A) during charging mode

Table 2 Simulation parameters

List of components	Value
Source inductance	1 mH
Transformer turns ratio (N2: N1)	0.55
Leakage inductance	100 μH
Input voltage	620–660 V
Output voltage	48–320 V
Output current	25 A
Switching frequency	20 kHz
Clamping capacitor	1000 μF

310 V battery voltage. The Fig. 8 represents the discharge mode of operation during discharging mode with 48 V battery voltage (Figs. 5, 6, 7, and 8 and Table 2).

Tables 3, 4, 5, and 6 depict that all losses described in the table increase with a decrease in resistance (R). The RCD snubber loss is very much high compared to all the losses for the lower value of the R. Tables 3, 4, 5, and 6 conclude that the

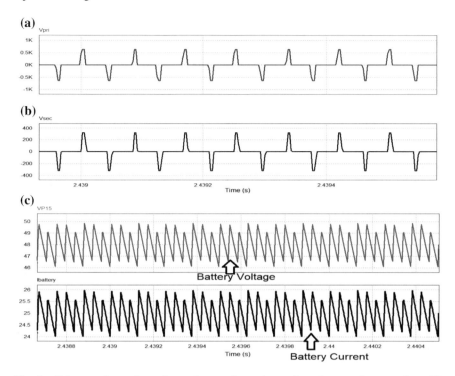

Fig. 6 **a** Primary voltage of transformer, **b** secondary voltage of transformer, **c** battery voltage (60 V) and battery current (25 A) during charging mode

lower values of R cannot be chosen as it results in very poor efficiency. The higher value of resistance increases voltage spikes. The value of R should be chosen from the optimized parameters obtained from the PSO. It can be concluded from the table that the RCD snubber loss reduces with the increase in the value of R. The higher value of R also increases the voltage spikes which may damage the switches. That is, the reason optimization technique has been utilized to draw an optimal solution for RCD parameters. The GA and PSO have been applied to identify the optimal solution of the RCD parameters. The GA method provides the optimal solution of $R = 375 \ \Omega$. The value of the resistance obtained from the PSO is $R = 400 \ \Omega$. Both the results are obtained with the condition that voltage spike should not be more than 125% of the rated value of the converter. The loss comparison between both the optimal solutions has been explained in Table 7. The value of the resistance $R = 400 \ \Omega$ has been selected for this RCD parameter design. The different advance swarm optimization can be applied to the same parameter design.

Table 3 Loss analysis in 310 V, 25 A charging mode

Charging mode 310 V, 25 A

Types of losses	R = 50	R = 100	R = 200	R = 500	R = 1000
Inverter switching loss (W)	125.81	122.767	106.38	122.97	129.12
Inverter diode loss (W)	2.6635	4.94	2.6532	4.088	5.10
RCD snubber loss (W)	105.72	65.14	26.33	13.13	10.78
Rectifier losses (W)	87.48	90.04	92.148	93.91	94.75

Table 4 Loss analysis in 310 V, 25 A discharging mode

Discharging mode 310 V, 25 A

Types of losses	R = 50	R = 100	R = 200	R = 500	R = 1000
Inverter switching loss (W)	156.86	151.11	150	146.44	129.12
Inverter diode loss (W)	3.86	3.89	3.92	1.83	2.09
RCD snubber loss (W)	233.48	157.20	105.64	76.13	74.69
Rectifier losses (W)	21.57	23.16	24.54	24.01	23.80

Table 5 Loss analysis in 60 V, 25 A charging mode

Charging mode 60 V, 25 A

Types of losses	R = 50	R = 100	R = 200	R = 500	R = 1000
Inverter switching loss (W)	68.17	58.49	52.46	45.90	44.86
Inverter diode loss (W)	3.52	4.13	4.18	4.632	4.70
RCD snubber loss (W)	764.23	608.13	287.64	158.47	83.46
Rectifier losses (W)	45.471	49.51	52.462	54.84	55.88

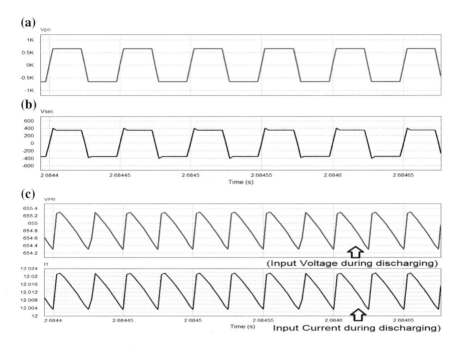

Fig. 7 **a** Primary voltage of transformer, **b** secondary voltage of transformer, **c** input voltage and input current during discharging mode (310 V, 25 A)

Table 6 Loss analysis in 60 V, 25 A discharging mode

Discharging mode 60 V, 25 A					
Types of losses	R = 50	R = 100	R = 200	R = 500	R = 1000
Inverter switching loss (W)	135.34	138.17	139.83	142.18	189.50
Inverter diode loss (W)	2.156	2.42	2.2465	5.37	16.45
RCD snubber loss (W)	752.4	529	502	25.92	156.91
Rectifier losses (W)	7.86	10.69	42	20.94	21.48

7 Conclusion

The optimization technique-based isolated boost bidirectional DC-DC converter with RCD snubber has been described in this paper. The current difference is created due to source inductance and leakage inductance of the transformer. The RCD snubber provides an alternative path for minimizing the voltage spike caused due to this

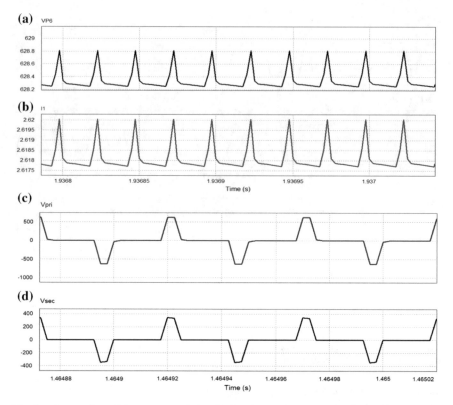

Fig. 8 **a** Input voltage, **b** input current during discharging mode, **c** primary voltage of transformer, **d** secondary voltage of transformer during discharging mode 60 V, 25 A battery rating

Table 7 RCD loss analysis in 60 V, 25 A discharging mode

Comparison between GA and PSO

Types of losses	R=375	R=400
310 V, 25 A charging mode	21.75	15.24
310 V, 25 A discharging mode	81.56	79.25
60 V, 25 A charging mode	150.35	147.91
60 V, 25 A discharging mode	184.09	160

mismatch current. The design of RCD snubber has been explained in the paper. The PSO algorithm has been used to identify the optimized value of the RCD snubber circuit. The optimal value obtained from the PSO is R = 400 Ω, and V_{cp} = 375 V. Optimized value obtained from PSO gives affordable losses (less than 5%) with less than 120% peak voltage including spikes in discharge mode of operation, while optimized value obtained from PSO gives 110% voltage spikes with 10% losses in discharge mode of conversion. The RCD snubber loss is less than 2% in 310 V, 25 A charging/discharging mode from the values obtained from PSO. The RCD snubber loss is less than 13% in 310 V, 25 A charging/discharging mode from the value obtained from PSO. The optimized value obtained from PSO gives a better result and lower losses compared to the optimized value obtained from GA approach.

References

1. Nymand, M., Andersen, M.A.E.: High-efficiency isolated boost DC-DC converter for high-power low-voltage fuel-cell applications. IEEE Trans. Ind. Electron. **57**(2), 505–514 (2010)
2. Huiqing, W., Jie, C.: Control and efficiency optimization of dual active–bridge DC/DC converter. In: Proceedings of IEEE International Conference on Power Electronics, Drives and Energy Systems (PEDES), Trivendram, pp. 1–6 (2016)
3. Du, Y., Lukic, S., Jacobson, B., Huang, A.: Review of high power isolated bi-directional DC-DC converters for PHEV/EV DC charging infrastructure. In: Proceedings of the IEEE Energy Conversion Congress and Exposition (ECCE), Phoenix, pp. 553–560 (2011)
4. Zhao, S.B., Song, Q., Liu, W., Sun, W.: Overview of dual-active-bridge isolated bidirectional DC-DC converter for high-frequency-link power-conversion system. IEEE Trans. Power Electron. **29**(8), 4091–4106 (2014)
5. Guo, Z., Sun, K., Wu, T.F., Li, C.: An improved modulation scheme of current-fed bidirectional DC-DC converters for loss reduction. IEEE Trans. Power Electron. **99** (2017)
6. Sun, X., Wu, X., Shen, Y., Li, X., Lu, Z.: A current-fed isolated bidirectional DC–DC converter. IEEE Trans. Power Electron. **32**(9), 6882–6895 (2017)
7. Jang, Y., Jovanovic, M.M.: A new PWM ZVS full-bridge converter. IEEE Trans. Power Electron. **22**(3), 987–994 (2007)
8. Jang, Y., Jovanovic, M.M.: A new PWM ZVS full-bridge converter. In: Proceedings of Annual IEEE Applied Power Electronics Conference and Exposition (APEC), Dallas, Texas, pp. 331–337 (2006)
9. Song-Yi, L., Chern-Lin, C.: Analysis and design for RCD clamped snubber used in output rectifier of phase-shift full-bridge ZVS converters. IEEE Trans. Ind. Electron. **45**(2), 358–359 (1998)
10. Zhu, L.: A novel soft-commutating isolated boost full-bridge ZVS-PWM DC-DC converter for bidirectional high power applications. In: Proceedings of Power Electronics Specialists Conference (PESC), Jeju, Korea, pp. 2141–2146 (2004)
11. Marsala, G., Ragusa, A.: A design method of zero ripple input current DC-DC converter assisted by constrained GA&PSO algorithms to minimize power losses. In: Proceedings of International Conference on Electrical Machines and Systems (ICEMS), Chiba, pp. 1–6 (2016)
12. Hoque, M.M., Hannan, M.A., Mohamed, A.: Optimal CC-CV charging of the lithium-ion battery for charge equalization controller. In: Proceedings of International Conference on Advances in Electrical, Electronic and Systems Engineering Putrajaya, pp. 610–615 (2016)
13. Ferreiro, J.B.L., et al.: Evaluation of a particle swarm optimization controller for DC-DC boost converters. In: Proceedings of International Conference on Compatibility and Power Electronics (CPE), Costa da Caparica, pp. 179–184 (2015)
14. Ab Wahab, M.N., Nefti-Meziani, S., Atyabi, A.: A comprehensive review of swarm optimization algorithms (Ed. Catalin Buiu). PLoS ONE **10**(5) (2015)

Fine-Tuned Constrained Nelder–Mead SOMA

Dipti Singh and Seema Agrawal

Abstract A number of population-based constrained optimization techniques are offered in the literature. Usually, a least amount of 50 population size is required to solve constrained optimization problems proficiently that results in large computation cost also. In this paper, a hybrid optimization technique, constrained Nelder–Mead self-organizing migrating algorithm (C-NMSOMA), has been projected that works with 20 population size only. It works with aim not only to handle constraints but also to condense the computation cost also. To confirm the claim, an experiment has been conducted for a set of ten constraint optimization problems by varying the population size from 20 to 100. C-NMSOMA works best with population size 20. To show its effectiveness over other algorithms, a convincing comparison has been made between the best results available by these algorithms and results obtained by C-NMSOMA using population size 20 only. Experimental results demonstrate that the presented algorithm C-NMSOMA is a robust optimization technique that yields feasible and viable solutions in lesser number of function evaluations using lesser population size. It combines the features of self-organizing migrating algorithm (SOMA) and Nelder–Mead (NM) simplex search. NM simplex search has been used as a crossover operator to produce new individuals in the solution space. A constraint handling technique based on preserving the feasibility of solutions with initialized feasible population has been adopted.

Keywords Self-organizing migrating algorithm · Nelder–Mead simplex search method · Constrained optimization problems

D. Singh
Department of Applied Sciences, Gautam Buddha University, Greater Noida, India
e-mail: diptipma@rediffmail.com

S. Agrawal (✉)
Department of Mathematics, S.S.V. College, Hapur, India
e-mail: seemagrwl7@gmail.com

© Springer Nature Singapore Pte Ltd. 2019   545
J. C. Bansal et al. (eds.), *Soft Computing for Problem Solving*, Advances in Intelligent
Systems and Computing 817, https://doi.org/10.1007/978-981-13-1595-4_44

1 Introduction

In general, a constrained optimization problem can be formulated mathematically as follows:

$$\text{Minimize} \quad f(x)$$
$$\text{Subject to:} \quad g_i(x) \geq 0, \quad i = 1, \dots, m$$
$$h_j(x) = 0, \quad j = 1, \dots, l$$
$$x_{min} \leq x \leq x_{max}$$

where $\{x = (x_1, x_2, \dots, x_n)^T\}$ is a vector solution, and m and l are the number of inequality and equality constraints, respectively.

Most of the constrained optimization problems are complex in nature. It is quiet complicated to solve these problems using traditional algorithms (TAs) such as Lagrange multiplier method [1], penalty function method [2, 3], sequential quadratic programming method [4], generalized reduced gradient method [5], etc. TAs require usually one or two initial guesses to work with that results faster convergence but chances of getting trapped to local optima are more. To beat this problem, evolutionary algorithms (EAs), such as evolutionary programming [6, 7], genetic algorithm [8], simulated annealing [9], particle swarm optimization [10], differential evolution algorithm [11], harmony search algorithm [12], gravitational search algorithm [13], water cycle algorithm [14], mine blast algorithm [15], have been used widely. These algorithms work with multiple guesses. These multiple guesses are usually known as population of EAs. A large number of population sizes may result in more exploration but also increases the computation cost also. To resolve a constrained optimization problem, there is a need of low computation cost technique that has good exploration quality also. This paper primarily focuses on two objectives: first handle the constraints resourcefully and second decrease the computation cost.

Quite a large number of constraint handling techniques have been developed in the literature to solve nonlinear constrained optimization problems. They can be classified as follows: a simple and most admired class of methods is penalty function-based methods which are easy to apply [16, 17]. In these methods, individuals out of feasible domain are penalized by using a penalty parameter. The constrained optimization problems are transformed into unconstrained optimization problems by appending penalty function to the objective function. In order to get good results while using penalty function methods, penalty parameter should be fine-tuned. Very small value of the penalty parameter provides infeasible solution [18], while large value of penalty parameter may generate the feasible solution deviated from optimality.

In another method, feasible solutions are favoured over infeasible one. A dominance-based selection scheme, free from fine tuning of penalty parameter to handle the constraints, was given by Coella and Mezura-Montes [19, 20]. Deb proposed three simple rules called feasibility-based rules [21] to select the feasible solutions over infeasible ones. The main disadvantage of this method is that it may effect in premature convergence due to enormous pressure on selection of feasible solutions.

Third method is based on initial feasibility of solution. In this approach, a population of feasible solutions has been initialized randomly; then these solutions are updated within the feasible section [22]. Particle swarm optimization (PSO) has proved its efficiency for highly constrained optimization problems using this feasibility preserving method [23]. Some other penalty-free constraint handling techniques can be found in [24–27].

Constrained optimization problems can also be solved capably using hybrid algorithms such as a hybrid method combining tabu search and simplex search [28], a combined Nelder–Mead simplex search and genetic algorithm [29], hybrid simulated annealing and direct search method [30], a hybrid method combining genetic algorithm and Hooke–Jeeves method [31], self-organizing migrating genetic algorithm (C-SOMGA) [32]. In this paper, a hybrid method (C-NMSOMA) with initial population of feasible solutions and preserving them using feasibility rules has been used for solving constrained optimization problems. Here C before NMSOMA has been used for constraint technique. Earlier NMSOMA-M has been used to solve large-scale unconstrained optimization problem [33]. It has been shown that to solve 1000-dimensional problem, it requires 10 population sizes only. C-NMSOMA works with less population size and hence requires lesser number of function evaluations. A fine tuning of population size has been done by varying the population size from 20 to 100. A comparison between the results obtained by C-NMSOMA and best results available by other algorithms has been made for these ten problems. The results for these problems are available by particle swarm optimization [34], an evolutionary algorithm-based pattern search approach (EA-PS) [35], constrained optimization with PSO (COPSO) [36], modified artificial bee colony algorithm (SB-ABC) [37], hybrid real-coded genetic algorithm with particle swarm optimization (RGA-PSO) algorithm [38], an improved electromagnetism-like mechanism algorithm (ICEM) [39], genetic algorithm with Hooke–Jeeves method (GAHJ) [31], improved backtracking search algorithm (IBSA) [40], constraint handling differential evolution (CDHE) algorithm [41], hybrid Nelder–Mead with PSO (NM-PSO) [42], water cycle algorithm (WCA), harmony search algorithm (HS).

The rest of the paper is structured as follows: In Sect. 2, the methodology of proposed algorithm, C-NMSOMA, is offered. In Sect. 3, the experimental results are discussed. Finally, the paper concludes with Sect. 4 portraying the conclusions of the proposed study.

2   Hybrid C-NMSOMA Algorithm

The algorithm C-NMSOMA is a hybrid constraint handling approach based on the functioning of SOMA combining NM simplex search crossover operator to generate new individuals in the solution space. It is based on preserving feasible solutions using feasibility rules with randomly initialized feasible solutions. The methodology of C-NMSOMA is set as follows:

The constraint violation function can be evaluated as follows:

$$\psi(x) = \sum_{i=1}^{m} G_i \big[g_i(x) \big]^2 + \sum_{j=1}^{l} \big[h_j(x) \big]^2 \tag{1}$$

where G_i is the heaviside operator such that $G_i = 0$ for $g_i(x) \geq 0$ and $G_i = 1$ for $g_i(x) < 0$.

The value of $\psi(x)$ indicates the amount of constraint violation. Value of $\psi(x)$ is 0 for the individuals inside the feasible region, and value of $\psi(x)$ is greater than 0 for the individuals outside the feasible region.

The proposed algorithm C-NMSOMA works as follows: a population of feasible solutions is generated uniformly in the search space randomly and is evaluated using fitness function, i.e. objective function. This population is sorted in increasing order of the objective function value. The individual with highest fitness value is chosen as leader, and the individual with worst objective function value is chosen as an active individual. In a generation also called migration loop, active individual travels towards leader in n steps $\left(n = \frac{path\,length}{step\,size} \right)$ of defined length. Each step gives the new position of active individual. In this way, for each active individual, a new population of size n is created. The movement of active individual towards leader is given as:

$$X_{i,j}^{MLNew} = X_{i,j,start}^{ML} + \left(X_{L,j}^{ML} - X_{i,j,start}^{ML} \right).t.PRTVector_j \tag{2}$$

where t is step size and ML is actual migration loop.

$X_{i,j}^{MLNew}$ is the new position of active individual.

$X_{i,j,start}^{ML}$ is the original position of active individual.

$X_{L,j}^{ML}$ is the position of the leader.

The movement of active individual towards leader is perturbed randomly by PRT (Perturbation) parameter. It is defined in the range $(0, 1)$.

This new population is sorted in increasing order of the fitness value. Now constraint violation function is evaluated for the best position of the new sorted population. If $\psi(x) = 0$, active individual is replaced with the current position, and if $\psi(x) > 0$ it moves to next best position of the sorted population. If there is no feasible solution accessible, then active individual remains the similar. The best and worst individual from the population are again chosen. Now a new solution is created using Nelder–Mead crossover operator using following steps:

Step1: choose parameters $\gamma > 1$, $\beta > 0$;

Step2: create an initial simplex with randomly selected three vertices;

find x_h (the worst point), x_l (the best point), x_g (next to the worst point);

evaluate their function values f_h, f_l, f_g; the worst point x_h is reflected with respect to the centroid (x_c) of other two points;

$$x_r = 2x_c - x_h. \text{(reflection)}$$
if $f_r < f_l$
$$x_{new} = (1 + \gamma)x_c - \gamma x_h. \text{(expansion)}.$$
else if $f_r >= f_h$
$$x_{new} = (1 - \beta)x_c + \beta x_h. \text{(contraction)}.$$
else if $f_g < f_r < f_h$
$$x_{new} = (1 + \beta)x_c - \beta x_h. \text{(contraction)}.$$
evaluate f_{new} and replace x_h by x_{new}. $\qquad(3)$

If this new solution satisfies constraint violation function, then this solution is acknowledged only if it is superior to active individual and is replaced with active individual. This process is continued till the termination criterion is contented. Flow chart of C-NMSOMA is given in Fig. 1.

3 Experimental Results

Computational cost is an important criterion to assess the effectiveness of an algorithm. It can be considered as number of function evaluations or computational time required to converge to the minima. Since population-based algorithm requires population to work with, its computational cost is highly affected by this factor. This has to be fine-tuned very carefully as exploration of an algorithm is highly dependent on it. Less population size may result in premature convergence or trapped in local optimal solution, and large population size results in high computational cost. There is a need of low computational cost algorithms that not only explore the search domain efficiently but also reduce the cost. This paper presents a hybrid constraint variant of SOMA (C-NMSOMA) that works with a population size of 20 only and has good exploration qualities also. First fine tuning of population size has been made by varying it from 20 to 100. The best optimal solution (best), worst optimal solution (worst), number of function evaluations (NFES) and computational time have been recorded for ten problems given in Table 1. C-NMSOMA works best with population size 20. In order to compare the performance of varying population size from 20 to 100, the value of performance index [34] is computed for these five population sizes. This performance index gives weighted importance to best, worst and NFES. Three cases arise based on the weights given to best, worst and NFES, and the graphs corresponding to each of the cases are shown in Fig. 2 (case i), Fig. 3 (case ii) and Fig. 4 (case iii). From the figures, it is evidently observed that the rate of performance index for population size 20 is high in contrast of other population size.

This paper also claims its effectiveness over other existing algorithms. To validate it, results of these problems taken by C-NMSOMA have been compared with the best results available by other algorithms and have been presented

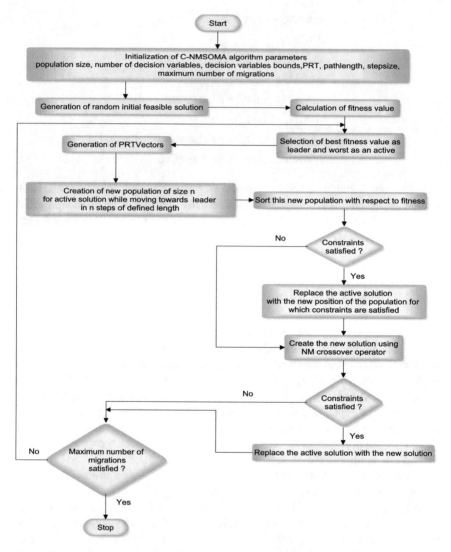

Fig. 1 Flow chart of C-NMSOMA

in Tables 2, 3, 4, 5, 6, 7, 8, 9, 10 and 11. C-NMSOMA is providing more accurate solutions at low computational cost. Throughout all experiments, the initial parameters for C-NMSOMA are set as follows: population size 20, PRT 0.3, path length 3.0, step size 0.11, β 0.8 and γ 1.8.

Table 1 Statistical results of C-NMSOMA for all the constrained problems (P01–P10) for different population sizes

Problems	Pop size	Best	Worst	NFES	Com. time (s)
P01	20	−15	−14.9955	69,042	0.16
	40	−15	−14.9966	110,834	0.27
	60	−15	−14.7529	114,906	0.285
	80	−14.9997	−14.9835	117,513	0.322
	100	−14.9995	−14.96	144,561	0.359
P02	20	−0.812033	−0.742144	69,327	1.981
	40	−0.810448	−0.718897	147,547	2.636
	60	−0.805297	−0.708436	126,321	4.446
	80	−0.807375	−0.666367	138,161	4.898
	100	−0.775971	−0.659132	150,000	4.987
P03	20	−1.0025	−0.101722	1299	0
	40	−1.00695	−0.0467949	2395	0
	60	−1.00923	−0.153777	3979	0
	80	−1.00851	−0.11574	6023	0
	100	−1.00697	−0.00371996	7564	0
P04	20	−30665.5	−30664.8	18,679	0.078
	40	−30665.5	−30664.9	46429	0.219
	60	−30665.5	−30664.5	49,880	0.249
	80	−30665.5	−30664.2	66,961	0.281
	100	−30665.5	−30662.3	80,564	0.702
P05	20	−0.0959963	−0.0083953	2701	0.015
	40	−0.0950052	−0.0169091	8069	0.016
	60	−0.0959401	−0.0165022	9393	0.016
	80	−0.0959026	−0.0003346	10,039	0
	100	−0.0945302	−0.00192626	15,248	0.016
P06	20	0.749303	0.762977	505	0
	40	0.75017	0.79962	973	0.015
	60	0.749919	0.955894	1656	0.062
	80	0.749309	1.01436	2537	1.02
	100	0.74	0.74001	30,139	1.19
P07	20	−6.04699	−6.04695	150,000	0.452
	40	−6.04583	−6.04539	150,000	0.689
	60	−6.04409	−6.04119	150,000	1.188
	80	−6.04116	−6.02364	150,000	1.326
	100	−6.03466	−5.966446	150,000	1.56

(continued)

Table 1 (continued)

Problems	Pop size	Best	Worst	NFES	Com. time (s)
P08	20	−5.508	−5.50767	9491	0.032
	40	−5.50798	−5.50654	14,937	0.063
	60	−5.50795	−5.49895	27,039	0.078
	80	−5.50795	−5.50226	30,343	0.110
	100	−5.50795	−5.50724	32,502	0.172
P09	20	10122.5	10122.5	150,000	0.234
	40	10122.5	10122.5	150,000	0.624
	60	10122.5	10122.5	150,000	0.718
	80	10122.5	10122.5	150,000	0.827
	100	10122.5	10122.5	150,000	0.905
P10	20	13.5908	13.5911	5077	0.016
	40	13.5908	13.5909	10,401	0.031
	60	13.5908	13.5911	12,566	0.062
	80	13.5908	13.5911	16,506	0.093
	100	13.5908	13.5911	22,463	0.203

Fig. 2 Case i

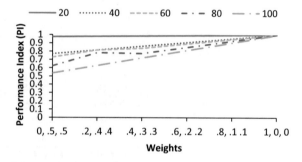

Fig. 3 Case ii

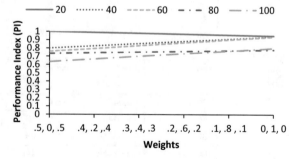

Fig. 4 Case iii

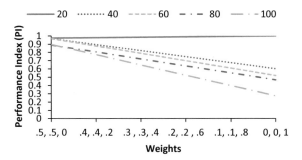

Table 2 Comparison of the results of C-NMSOMA and other published algorithms for constrained problem P01

Methods	Best optimal solution	NFES	Pop size
PSO [34]	−14.9987	–	20
EA−PS [35]	−14.989	65,000	208
COPSO [36]	−15	90,800	100
C-SOMGA [32]	−14.99225	30,992	20
SB-ABC [37]	−15	–	40
RGA-PSO [38]	−15	–	–
AIA-PSO [38]	−15	–	–
ICEM [39]	−15	–	100
GAHJ [31]	−15	–	260
IBSA [40]	−15	–	80
C-NMSOMA	**−15**	**20,670**	**20**

Table 3 Comparison of the results of C-NMSOMA and other published algorithms for constrained problem P02

Methods	Best optimal solution	NFES	Pop size
CHDE [41]	−0.803619	348,000	60
COPSO [36]	−0.803619	142,900	100
C-SOMGA [32]	−0.77542	1.35468	20
ICEM [39]	−0.803619	–	100
GAHJ [31]	−0.623867	–	400
IBSA [40]	−0.803614	–	80
C-NMSOMA	**−0.803832**	**62,385**	**20**

Table 4 Comparison of the results of C-NMSOMA and other published algorithms for constrained problem P03

Methods	Best optimal solution	NFES	Pop size
CHDE [42]	−1.00	348,000	60
COPSO [36]	−1.000005	315,100	100
C−SOMGA [32]	−0.88794	3603	20
WCA [14]	−0.999981	103,900	50
ICEM [39]	−1.0005	–	100
GAHJ [40]	−1.0031596	–	200
IBSA [41]	−1.012555	–	80
C-NMSOMA	**−1.00765**	**1392**	**20**

Table 5 Comparison of the results of C-NMSOMA and other published algorithms for constrained problem P04

Methods	Best optimal solution	NFES	Pop size
PSO [34]	−30665.500	–	20
CHDE [41]	−30665.539	348,000	60
HS [12]	−30665.500	–	–
EA-PS [35]	−30665.540	12,679	208
COPSO [36]	−30665.538672	59,600	100
NM-PSO [42]	−30665.5386	–	274
SB-ABC [37]	−30665.539	–	40
WCA [14]	−30665.5386	18,850	50
RGA-PSO [38]	−30665.539	–	100
AIA-PSO [38]	−30665.539	–	100
ICEM [39]	−30665.539	–	100
GAHJ [31]	−30665.547	–	260
IBSA [40]	−30665.539	–	80
C-NMSOMA	**−30665.500**	**22,768**	**20**

4 Conclusion

In this paper, a constrained handling optimization technique C-NMSOMA has been presented. It is a hybrid scheme combining the features of SOMA with NM search method as crossover operator. C-NMSOMA embedded with constraint handling method based on preserving the feasibility solution with initialized feasible population has been used to solve ten constrained benchmark optimization problems. The efficiency of the proposed method has been compared with the efficiency of various other optimization methods. The main feature that distinguishes it with other methods is that it provides solution at low cost and work with small population size.

Table 6 Comparison of the results of C-NMSOMA and other published algorithms for constrained problem P05

Methods	Best optimal solution	NFES	Pop size
CHDE [41]	0.095825	348,000	60
COPSO [36]	0.095825	3600	100
C-SOMGA [32]	0.08816	605	20
NM-PSO [42]	0.095825	–	43
SB-ABC [37]	0.095825	–	40
ICEM [39]	0.095825	–	100
GAHJ [31]	0.095825	–	40
IBSA [40]	0.095825	–	80
C-NMSOMA	**0.0958507**	**5461**	**20**

Table 7 Comparison of the results of C-NMSOMA and other published algorithms for constrained problem P06

Methods	Best optimal solution	NFES	Pop size
PSO [34]	0.75	–	20
CHDE [41]	0.749	348,000	60
COPSO [36]	0.749999	31,5000	100
C-SOMGA [32]	0.82519	559	20
ICEM [39]	0.7499	–	100
GAHJ [31]	0.7500150	–	40
IBSA [40]	0.7499	–	80
C-NMSOMA	**0.749477**	**737**	**20**

Table 8 Comparison of the results of C-NMSOMA and other published algorithms for constrained problem P07

Methods	Best optimal solution	NFES	Pop size
RGA-PSO [38]	−6.0417	–	100
AIA-PSO [38]	−6.0466	–	100
C-NMSOMA	**−6.04623**	**150,000**	**20**

Table 9 Comparison of the results of C-NMSOMA and other published algorithms for constrained problem P08

Methods	Best optimal solution	NFES	Pop size
COPSO [36]	−5.508013	14,900	100
C-NMSOMA	**−5.508**	**3038**	**20**

Table 10 Comparison of the results of C-NMSOMA and other published algorithms for constrained problem P09

Methods	Best optimal solution	NFES	Pop size
RGA-PSO [38]	−10122.4732	–	100
AIA-PSO[[38]	−10122.4852	–	100
C-NMSOMA	**−10122.5**	**150,000**	**20**

Table 11 Comparison of the results of C-NMSOMA and other published algorithms for constrained problem P10

Methods	Best optimal solution	NFES	Pop size
HS [12]	13.590845	–	–
C-SOMGA [32]	13.59610	–	20
C-NMSOMA	**13.5908**	**840**	**20**

The statistical results have proved its efficiency of solving variety of constraints problems.

Appendix

Constrained Problems

The first six functions are well-known benchmark taken from Runarsson and Yao [43].

Problem 01

$$\text{Minimize } f(x) = 5\sum_{i=1}^{4} x_i - 5\sum_{i=1}^{4} x_i^2 - \sum_{i=5}^{13} x_i$$

$$\text{subject to: } g_1(x) = 2x_1 + 2x_2 + x_{10} + x_{11} - 10 \leq 0$$

$$g_2(x) = 2x_1 + 2x_3 + x_{10} + x_{12} - 10 \leq 0$$

$$g_3(x) = 2x_2 + 2x_3 + x_{11} + x_{12} - 10 \leq 0$$

$$g_4(x) = -8x_1 + x_{10} \leq 0$$

$$g_5(x) = -8x_2 + x_{11} \leq 0$$

$$g_6(x) = -8x_3 + x_{12} \leq 0$$

$$g_7(x) = -2x_4 - x_5 + x_{10} \leq 0$$

$$g_8(x) = -2x_6 - x_7 + x_{11} \leq 0$$

$$g_9(x) = -2x_8 - x_9 + x_{12} \leq 0$$

$0 \le x_i \le 1, i = 1, .., 9; 0 \le x_i \le 100, i = 10, 11, 12; 0 \le x_{13} \le 1.$
The global minimum optimal solution is:

$$\bar{x} = (1, 1, 1, 1, 1, 1, 1, 1, 1, 3, 3, 3, 1), f(\bar{x}) = -15$$

Problem 02

$$\text{Maximize } f(x) = \left| \frac{\sum_{i=1}^{n} \cos^4(x_i) - 2 \prod_{i=1}^{n} \cos^2(x_i)}{\sqrt{\sum_{i=1}^{n} i x_i^2}} \right|$$

$$\text{Subject to: } g_1(x) = 0.75 - \prod_{i=1}^{n} x_i \le 0$$

$$g_2(x) = \sum_{i=1}^{n} x_i - 7.5\,n \le 0$$

where $n = 20$ and $0 \le x_i \le 10, i = 1, \ldots, n$.
The global maximum optimal solution is:

$$f(\bar{x}) = 0.803619.$$

Problem 03

$$\text{Maximize } f(x) = \left(\sqrt{n}\right)^n \prod_{i=1}^{n} x_i$$

$$\text{Subject to: } h(x) = \sum_{i=1}^{n} x_i^2 - 1 = 0$$

where $n = 10$ and $0 \le x_i \le 1, i = 1, \ldots, n$.
The global maximum optimal solution is:

$$\bar{x} = \left(\frac{1}{\sqrt{n}}, \ldots, \frac{1}{\sqrt{n}}\right), f(\bar{x}) = 1.$$

Problem 04

Minimize $f(x) = 5.3578547 x_3^2 + 0.8356891 x_1 x_5 + 37.293239 x_1 - 40792.141$

Subject to: $g_1(x) = 85.334407 + 0.0056858 x_2 x_5 + 0.0006262 x_1 x_4 - 0.0022053 x_3 x_5 - 92 \leq 0$

$g_2(x) = -85.334407 - 0.0056858 x_2 x_5 - 0.0006262 x_1 x_4 + 0.0022053 x_3 x_5 \leq 0$

$g_3(x) = 80.51249 + 0.0071317 x_2 x_5 + 0.0029955 x_1 x_2 + 0.0021813 x_3^2 - 110 \leq 0$

$g_4(x) = -80.51249 - 0.0071317 x_2 x_5 - 0.0029955 x_1 x_2 - 0.0021813 x_3^2 + 90 \leq 0$

$g_5(x) = 9.300961 + 0.0047026 x_3 x_5 + 0.0012547 x_1 x_3 + 0.0019085 x_3 x_4 - 25 \leq 0$

$g_6(x) = -9.300961 - 0.0047026 x_3 x_5 - 0.0012547 x_1 x_3 - 0.0019085 x_3 x_4 + 20 \leq 0$

$78 \leq x_1 \leq 102, 33 \leq x_2 \leq 45, 27 \leq x_i \leq 45, i = 3, 4, 5.$

The global minimum optimal solution is:

$\bar{x} = (78.0, 33.0, 29.995256025682, 45.0, 36.775812905788), f(\bar{x}) = -30665.539.$

Problem 05

$$\text{Maximize } f(x) = \frac{\sin^3(2\pi X_1)\sin(2\pi X_2)}{X_1^3(X_1+X_2)}$$

$$\text{Subject to: } g_1(x) = x_1^2 - x_2 + 1 \leq 0$$

$$g_2(x) = 1 - x_1 + (x_2 - 4)^2 \leq 0$$

$$0 \leq x_i \leq 10, \quad i = 1, 2.$$

The global maximum optimal solution is:

$$\bar{x} = (1.2279713, 4.2453733), f(\bar{x}) = 0.095825.$$

Problem 06

$$\text{Minimize } f(x) = x_1^2 + (x_2 - 1)^2$$

$$\text{Subject to: } h(x) = x_2 - x_1^2 = 0$$

$$-1 \leq x_i \leq 1, \quad i = 1, 2.$$

The global minimum optimal solution is:

$$\bar{x} = \left(\pm\frac{1}{\sqrt{2}}, \frac{1}{2}\right), f(\bar{x}) = 0.75.$$

Problem 07 This problem has been taken from handbook of test problems in local and global optimizations [44].

$$\text{Minimize } f(x) = \left(-x_1 - x_5 + 0.4\,x_1^{0.67}x_3^{-0.67} + 0.4\,x_5^{0.67}x_7^{-0.67}\right)$$

$$\text{Subject to: } g_1(x) = 0.05882\,x_3\,x_4 + 0.1\,x_1 \leq 1$$

$$g_2(x) = 0.05882\,x_7\,x_8 + 0.1\,x_1 + 0.1\,x_5 \leq 1$$

$$g_3(x) = 4\,x_2\,x_4^{-1} + 2\,x_2^{-0.71}x_4^{-1} + 0.05882\,x_2^{-1.3}x_3 \leq 1$$

$$g_4(x) = 4\,x_6\,x_8^{-1} + 2\,x_6^{-0.71}x_8^{-1} + 0.05882\,x_6^{-1.3}x_7 \leq 1$$

$$0.01 \leq x_i \leq 10, \ i = 1, 2, \ldots, 8.$$

The optimal solution of the problem is:

$$\bar{x} = (6.4225,\ 0.6686,\ 1.0239,\ 5.9399,\ 2.2673,\ 0.5960,\ 0.4029, 5.5288)$$
$$f(\bar{x}) = -6.0482.$$

Problem 08 This problem was proposed by Mezura and Coello [45].

$$\text{Minimize } f(x) = -x_1 - x_2$$

$$\text{Subject to: } g_1(x) = -2\,x_1^4 + 8\,x_1^3 - 8\,x_1^2 + x_2 - 2 \leq 0$$

$$g_2(x) = -4\,x_1^4 + 32\,x_1^3 - 88\,x_1^2 + 96\,x_1 + x_2 - 36 \leq 0$$

$$0 \leq x_1 \leq 3, 0 \leq x_2 \leq 4$$

The global minimum solution is:

$$\bar{x} = (2.3295, 3.17846), f(\bar{x}) = -5.50796.$$

Problem 09

$$\text{Minimize } f(x) = 5.3578\,x_3^2 + 0.8357\,x_1x_5 + 37.2392\,x_1$$

$$\text{Subject to: } g_1(x) = 0.00002584\,x_3x_5 - 0.0006663\,x_2x_5 - 0.0000734\,x_1x_4 - 1 \leq 0$$

$$g_2(x) = 0.000853007\,x_2x_5 + 0.00009395\,x_1x_4 - 0.00033085\,x_3 - 1 \leq 0$$

$$g_3(x) = 0.00024186\,x_2x_5 + 0.00010159\,x_1x_2 + 0.00007379\,x_3^2 - 1 \leq 0$$

$$g_4(x) = 1330.3294\,x_2^{-1}x_5^{-1} - 0.42\,x_1x_5^{-1} - 0.30586\,x_2^{-1}x_3^2x_5^{-1} - 1 \leq 0$$

$$g_5(x) = 2275.1327\,x_3^{-1}x_5^{-1} - 0.2668\,x_1x_5^{-1} - 0.40584\,x_4x_5^{-1} - 1 \leq 0$$

$$g_6(x) = 0.00029955\,x_3x_5 + 0.00007992\,x_1x_3 + 0.00012157\,x_3x_4 - 1 \leq 0$$

$$78 \leq x_1 \leq 102, 33 \leq x_2 \leq 45, 27 \leq x_i \leq 45, \ i = 3, 4, 5.$$

The global minimum optimal solution is:

$$\bar{x} = (78.0, 33.0, 29.998, 45.0, 36.7673), f(\bar{x}) = 10122.6964.$$

Problem 10

$$\text{Minimize}\ \ f(x) = \left(x_1^2 + x_2 - 11\right)^2 + \left(x_1 + x_2^2 - 7\right)^2$$
$$\text{Subject to:}\ g_1(x) = 4.84 - (x_1 - 0.05)^2 - (x_2 - 2.5)^2 \ge 0$$
$$g_2(x) = x_1^2 + (x_2 - 2.5)^2 - 4.84 \ge 0$$
$$0 \le x_i \le 6,\, i = 1, 2.$$

The global minimum optimal solution is:

$$\bar{x} = (2.246826, 2.381865), f(\bar{x}) = 13.59085.$$

References

1. Bertsekas, D.P.: Constrained optimization and lagrange multiplier methods. In: Linear Network Optimization: Algorithms and Codes. M.I.T. Press London LTD. (1991)
2. Homaifar, A., Qi, C.X., Lai, S.H.: Constrained optimization via genetic algorithms. Simulation **62**, 242–253 (1994)
3. Fletcher, R.: An ideal penalty function constrained optimization. IMA J. Appl. Math. **15**, 319–342 (1975)
4. Boggs, P.T., Tolle, Jon W.: Sequential quadratic programming. Acta Numer. **4** (1995)
5. Lasdon, L.S., Fox, R.L., Ratner, M.W.: Nonlinear optimization using the generalized reduced gradient method. RAIRO-Oper. Res. Rech. Oper. **8**, 73–103 (1974)
6. Dou, J., Wang, X.J.: An efficient evolutionary programming. In: International Symposium on Information Science and Engineering, ISISE'08, pp. 401– 404 (2008)
7. Michalewicz, Z.: Genetic algorithms, numerical optimization and constraints. In: Echelman, L.J. (ed.) Proceedings of the Sixth International Conference on Genetic Algorithms, pp. 151–158 (1995)
8. Deb, K.: An efficient constraint handling method for genetic algorithms. Comput. Methods Appl. Mech. Eng. **186**, 311–338 (2000)
9. Kirkpatrick, S., Gelatt Jr., C.D., Vecchi, M.P.: Optimization by simulated annealing. Science **220**, 671–680 (1983). https://doi.org/10.1126/science.220.4598.671
10. Eberhart, R., Kennedy, J.: A new optimizer using particle swarm theory. In: Proceedings of the Sixth International Symposium on Micro Machine and Human Science (1995)
11. Lampinen, J.: A constraint handling approach for the differential evolution algorithm. In: Proceedings of the 2002 Congress on Evolutionary Computation (CEC'2002), IEEE Service Center, Piscaaway, New Jersey, pp. 1468–1473 (2002)
12. Yang, X.S.: Harmony search as a metaheuristic algorithm. In: Geem, Z.W. (ed.) Music-Inspired Harmony Search Algorithm. Studies in Computational Intelligence, vol. 191, pp. 1–14. Springer Berlin, Heidelberg (2009)
13. Rashedi, E., Nezamabadi-pour, H., Saaryazdi, S.: GSA: a gravitational search algorithm. Inf. Sci. **179**, 2232–2248 (2009)
14. Eskander, H., Sadollah, A., Bahreininejad, A., Hamdi, M.: Water cycle algorithm-A novel metaheuristic optimization method for solving constrained engineering optimization problems. Comput. Struct. **110–111**, 151–166 (2012)
15. Eskander, H., Sadollah, A., Bahreininejad, A., Hamdi, M.: Mine blast algorithm-A new population based algorithm for solving constrained engineering optimization problems. Appl. Soft Comput. **13**, 2592–2612 (2013)

16. Michalewicz, Z., Attia, N.F.: Evolutionary optimization of constrained problems. In: Proceedings of Third Annual Conference on Evolutionary Programming, World Scientific, River Edge, NJ, pp. 998–1008 (1994)
17. Joines, J.A., Houck, C.R.: On the use of non-stationary penalty functions to solve nonlinear constrained optimization problems with GAs. In: Proceedings of the First IEEE Conference on Evolutionary Computation, IEEE World Congress on Computational Intelligence, Orlando, FL, USA, pp. 587–602 (1994)
18. Smith, A.E., Coit, D.W.: Constraint handling techniques-penalty functions. In: Handbook of Evolutionary Computation, Institute of Physics Publishing and Oxford University Press, Bristol, U.K., Chapter C 5.2 (1997)
19. Coello, C.A.C.: Theoretical and numerical constraint handling techniques used with evolutionary algorithms: a survey of the state of the art. Comput. Methods Appl. Mech. Eng. **191**, 1245–1287 (2002)
20. Mezura-Montes, E., Coello, C.A.C.: A simple evolution strategy to solve constrained optimization problems. In: Cantú-Paz, E. et al. (eds.) Genetic and Evolutionary Computation—GECCO 2003. Lecture Notes in Computer Science, vol. 2723. Springer, Berlin, Heidelberg (2003)
21. Mezura-Montes, E., Coello, C.A.C.: Adding a diversity mechanism to a simple evolutionary strategy to solve constrained optimization problems. In: Proceedings of IEEE International Congress on Evolutionary Computation (CEC'2003). IEEE Neural Network Society, vol. 1, pp. 6–13 (2003)
22. Deb, K.: An efficient constraint handling method for genetic algorithms. Comput. Methods Appl. Mech. Eng. **186**, 311–338 (2000)
23. Hu, X., Eberhart, R.: Solving constrained nonlinear optimization problems with particle swarm optimization. In: Proceedings of 6th World Multiconference on Systemics, Cybernetics and Informatics (SCI) (2002)
24. Sun, C., Zeng, J., Pan, J.: A new method for constrained optimization problems to produce initial values. In: Chinese Control and Decision Conference (CCDC'09), pp. 2690–2692 (2009)
25. Coello, C.A.C., Mezura-Montes, E.: Constraint-handling in genetic algorithms through the use of dominance based tournament selection. Adv. Eng. Inf. **16**, 193–203 (2002)
26. Deb, K., Agarwal, S.: A niched-penalty approach for constraint handling in genetic algorithms. In: Artificial Neural Nets and Genetic Algorithms. Springer, Vienna (1999)
27. Akhtar, S., Tai, K., Ray, T.: A socio-behavioural simulation model for engineering design optimization. Eng. Optim. **34**, 341–354 (2002)
28. Chelouah, R., Siarry, P.: A hybrid method combining continuous tabu search and Nelder-Mead simplex algorithms for the global optimization of multiminima functions. Eur. J. Oper. Res. **161**, 636–654 (2005)
29. Durand, N., Alliot, J.: A combined Nelder-Mead simplex and genetic algorithm. In: Proceedings of the Genetic and Evolutionary Computation Conference GECCO'99, vol. 99, pp. 1–7 (1999)
30. Hedar, A., Fukushima, M.: Derivative-free filter simulated annealing method for constrained continuous global optimization. J. Global Optim. **35**, 521–549 (2006)
31. Long, Q., Wu, C.: A hybrid method combining genetic algorithm and hook-jeeves method for constrained global optimization. J. Ind. Manag. Optim. **10**, 1279–1296 (2014)
32. Deep, K., Dipti: A self organizing migrating genetic algorithm for constrained optimization. Applied Mathematics and Computation, vol. 198, pp. 237–250 (2008)
33. Singh, D., Agrawal, S.: Self organizing migrating algorithm with Nelder-Mead crossover and log-logistic mutation for large scale optimization. In: Acharjya, D., Dehuri, S., Sanyal, S. (eds.) Computational Intelligence for Big Data Analysis. Adaptation, Learning, and Optimization, vol. 19, pp. 143–164. Springer (2015)
34. Pulido, G.T., Coello, C.A.C.: A constraint-handling mechanism for particle swarm optimization. In: IEEE Congress on Evolutionary Computation (CEC'2004), vol. 2, pp. 1396–1403 (2004)
35. Datta, R., Costa, M.F.P., Deb, K., Gaspar-Cunha, A.: An evolutionary algorithm based pattern search approach for constrained optimization. In: IEEE Congress on Evolutionary Computation (CEC), pp. 1355–1362 (2013)

36. Aguirre, A.H., Zavala, A.E.M., Diharce, E.V., Rionda, S.B.: COPSO: constrained optimization via PSO algorithm. In: Center for Research in Mathematics (CIMAT), Technical report No. I-07-04/22-02-2007
37. Stanarevic, N., Tuba, M., Bacanin, N.: Modified artificial bee colony algorithm for constrained problems optimization. Int. J. Math. Models Methods Appl. Sci. **5**, 644–651 (2011)
38. Wu, J.-Y.: Solving constrained global optimization problems by using hybrid evolutionary computing and artificial life approaches. Math. Probl. Eng. **2012**, Article ID 841410, 36 pp (2012). https://doi.org/10.1155/2012/841410
39. Zhang, C., Li, X., Gao, L., Wu, Q.: An improved electromagnetism-like mechanism algorithm for constrained optimization. Expert Syst. Appl. (Elsevier) **40**, 5621–5634 (2013)
40. Zhao, W., Wang, L., Yin, Y., Wang, B., Wei, Y., Yin, Y.: An improved backtracking search algorithm for constrained optimization problems. In: Buchmann, R., Kifor, C.V., Yu, J. (eds.) Knowledge Science, Engineering and Management. KSEM'2014. Lecture Notes in Computer Science, vol 8793, pp. 222–233. Springer (2014)
41. Mezura-Montes, E., Coello, C.A.C., Tun-Morales, E.I.: Simple feasibility rules and differential evolution for constrained optimization. In: Monroy, R., Arroyo-Figueroa, G., Sucar, L.E., Sossa, H. (eds.) In: MICAI 2004: Advances in Artificial Intelligence. MICAI 2004. Lecture Notes in Computer Science, vol. 2972. Springer, Berlin, Heidelberg (2004)
42. Zahara, E., Kao, Y.T.: Hybrid Nelder-Mead simplex search and particle swarm optimization for constrained engineering design problems. Expert Syst. Appl. (Elsevier) **36**, 3880–3886 (2009)
43. Runarson, T.P., Yao, X.: Stochastic ranking for constrained evolutionary optimization. IEEE Trans. Evol. Comput. **4**, 284–294 (2000)
44. Floudas, C.A., Pardalos, P.M., Adjiman, C.S., Esposito, W.R., Gümüs, Z.H., Harding, S.T., Klepeis, J.L., Meyer, C.A., Schweiger, C.A.: Handbook of Test Problems in Local and Global Optimization. Kluwer, Boston, Mass, USA (1999)
45. Mezura-Montes, E., Coello, C.A.C.: Useful infeasible solutions in engineering optimization with evolutionary algorithms. In: Gelbukh, A., de Albornoz, Á., Terashima-Marín, H. (eds.) MICAI 2005: Advances in Artificial Intelligence. MICAI'2005. Lecture Notes in Computer Science, vol. 3789. Springer, Berlin, Heidelberg

Hybrid System for MPAA Ratings of Movie Clips Using Support Vector Machine

Gagan Vishwakarma and Ghanshyam Singh Thakur

Abstract A hybrid system to assign Motion Picture Association of America (MPAA) rating to the movie clip by combining fundamentals of content-based video retrieval system (CBVR), with the assistance of discrete cosine transform (DCT) and finally applying support vector machine (SVM) for adding intelligence to the processing framework to rate movie clip have been proposed in this paper. In the conventional CBVR system, keyframe plays an important role, and instead of it, the proposed approach does not depend on the few keyframes and shows better performance due to the incorporation of DCT to extract related data which represent low recurrence in each block. At that point, the obtained DCT coefficients are further utilized as features for the classification procedure performed by exploiting multiclass SVM. Motion picture initially extracted into clips to assign a rating to individual clips, the most prominent rating among all individual clips concluded as a final rating. Intelligence and accuracy of the proposed framework depend on SVM training performed by extracted training data corresponding to particular rating specified by the user applying his human intelligence and common sense. It will make SVM as intelligent as users provided training data set to assign an appropriate rating to the potential movie clip.

Keywords Image/video retrieval · Classification · Machine learning · SVM

1 Introduction

Recently, foreign language movies commonly termed as 'Hollywood movies' are gaining massive popularity in India, so the scope of this situation is Indian scenario. When planning to watch such a movie with family members, may cause a condition of

G. Vishwakarma (✉) · G. S. Thakur
Computer Application Department, Maulana Azad National Institute of Technology,
Bhopal, Madhya Pradesh, India
e-mail: gagan.v.20@gmail.com

G. S. Thakur
e-mail: ghanshyamthakur@gmail.com

© Springer Nature Singapore Pte Ltd. 2019
J. C. Bansal et al. (eds.), *Soft Computing for Problem Solving*, Advances in Intelligent Systems and Computing 817, https://doi.org/10.1007/978-981-13-1595-4_45

dilemma that "Is content suitable to watch in the family?" As most of the so-called Hollywood movies or dubbed movies based on characters, scenario, lifestyle and culture belong to another country than India such as USA, China, Japan or UK and there is a vast cultural, moral and social difference among these countries and India. It creates a massive mismatch between content suitability to an Indian audience and content suitability rating allotted to them. Motion Picture Association of America (MPAA) ratings are globally accepted; they are responsible for rating the movies according to its content suitability for a particular audience. MPAA is an organization as similar to Censor Board Film Corporation (CBFC) of India. Independent division of MPAA known as Classification and rating administration (CARA) consists of an MPAA rating board issues ratings for movies exhibited and distributed in the United States, it does not decide the content nor does it assess the quality or social value of motion pictures. By issuing rating, it looks to educate parents of the level of particular material, for example, language, violence, drug, sex, adult activity and so on that parents may discover unseemly to view by their children. They review the elements of a movie such as script, scenes or any fragment of the movie and vote on the film's rating [11]. Studies show that there is no automation to rate movies [11, 12]. Rating system still depends upon the board members for evaluating the movies. This phenomenon produces a motivation towards the research work in this field. Ratings adopted by MPAA are as follows [12].

- **G-Rating (since 1968-present)**: Movies rated as G are suitable for "general audience" which means content depicting no nudity, sexuality, profanity, drug usage or violence, i.e. non-objectionable to any parents for their children and suitable to watch for any age group.
- **PG-Rating (since 1972-present)**: Movies rated as PG suggested parental guidance if being watched by children. Content depicting in the movie may not be suitable for children and require clarification and guidance of parents.
- **PG-13 Rating (since 1984-present)**: Movies rated as PG-13 suggested parents to be strictly cautious because content depicting in the movie may be unseemly for youngsters under age 13.
- **R-Rating (since 1968-present)**: Movies rated as R are strictly not suitable for children under the age 17; they are not permitted to watch in public without parents going with them.
- **NC-17 Rating (since 1990-present)**: Movies rated as NC-17 are strictly for adults only. Children cannot be allowed to watch them under any circumstances.

As previously motioned, unavailability of any automatic, computerized system to perform this task is a problem. Because rating of movies is a crucial process, it requires human intelligence and common sense to analyze content and rate it accordingly [11, 12].

2 Related Work

Conventional approaches were developed to enhance the retrieval accuracy of content-based image retrieval (CBIR) system. One way to achieve it is by utilizing extracted image features from the compressed data stream. Another direction is opposite of the classical scheme by obtaining an image descriptor from the first image; this retrieval scheme uncomplicatedly extracts image features from the compressed stream without depending on the decoding procedure. Modified retrieval scheme intended to lessen the time calculation for feature extraction or generation because most of the images are already transformed to the compressed domain before writing back to the memory device [4]. Accordingly, the past literature [15] proposed a new pattern-based indexing approach to enhance content-based video retrieval by approximate optimal solution. A video is made out of a succession of various shots made from keyframes. Due to co-relations of these consecutive shots, the video retrieval system, naturally, needs to consider the temporal continuity of shots. In other words, two video cuts are comparable if the subsequences of both are comparable. As a result of the complex video contents, the multifaceted nature of CBVR is significantly superior to the content-based image retrieval (CBIR).

2.1 Discrete Cosine Transform (DCT)

The DCT can be used to transform an image from spatial domain to frequency domain. The DCT demonstrates the strong mathematical ability to operate on any dimension as the image has an extra complexity of having two-dimension (2D) surfaces; that is why 2D DCT transform is required [14]. The 2D M × N DCT is defined as follows

$$f(x, y) = \sum_{u=0}^{M-1} \sum_{v=0}^{N-1} \propto (u) \propto (v) C(v, u) \cos\left[\frac{\pi(2x+1)u}{2M}\right] \cos\left[\frac{\pi(2y+1)v}{2N}\right]$$

where

$$\propto (u) = \begin{cases} \frac{1}{\sqrt{M}}, & u = 0 \\ \sqrt{\frac{2}{M}}, & u = 1, 2, \ldots, M-1 \end{cases}$$

$$\propto (v) = \begin{cases} \frac{1}{\sqrt{N}}, & v = 0 \\ \sqrt{\frac{2}{N}}, & v = 1, 2 \ldots, N-1 \end{cases}$$

2.2 Support Vector Machine (SVM)

SVM is a supervised machine learning concept which works on the basis of statistical learning theory; it also utilizes the guideline of structural risk minimization as an alternative of empirical risk minimization, initially proposed by Vapnik [2]. The SVM procedure finds a separating hyperplane also called decision boundary, with the maximum margin between two classes of data, and in this manner creating the enormous conceivable distance between the separating hyperplane and the elements around it. Two critical components of the implementation of SVM are the methods of mathematical programming and kernel functions. Exploiting the facility of kernel function gives us a nonlinear separating plane among original feature space, without increasing the computation cost if computation of kernel function maintained equivalent to the interior product [2].

Suppose we are given i training data $x1, \ldots, xi$ that are vectors in some space $X \in RN$ [3]. We are also given their labels $y1, \ldots, yi$ where $yi \in \{-1, +1\}$. For the optimal hyperplane, it can be characterized by a weight vector w and a bias b such that

$$w^t x + b = 0$$

Let x is an unknown point, to classify it, the decision function can be defined as

$$f(x) = \sin(w^t x + b) \text{ where } w = \sum_{i=1}^{N} a_i y_i x_i$$

The number of support vector is denoted as N; support vectors are x_i. The class label is y_i to which x_i belongs, and b is set by the particular conditions where a_i satisfies

$$\max L_d(a) = \sum_{i=1}^{N} a_i \sum_{i,j} a_i a_j y_i y_j x_i x_j$$

$$\text{Subject to} \sum_{i=1}^{N} a_i y_i = 0$$

2.3 Multiclass SVM

For classification issues with multiple classes, diverse methodologies are produced to choose whether the given data have a place with individual predefined classes or not. The most widely recognized methods are those that join a few binary classifiers and utilize a summarization technique to conclude the last classification decision [5]. Some popular schemes are termed as one-against-all [8], one-against-one [1, 10],

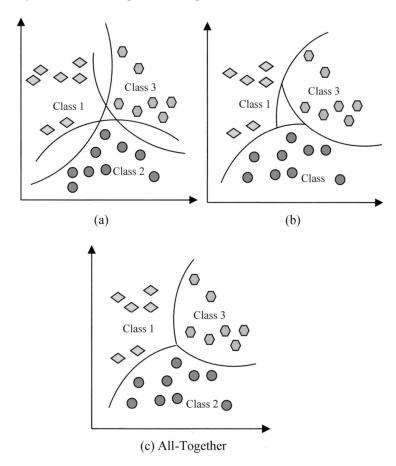

Fig. 1 Comparison of multiclass boundaries

directed acyclic graph (DAG) [13] and half-against-half method [9]. Another quite sophisticated approach is that endeavours to manufacture one support vector machine that isolates all classes in the meantime. In the following section, Fig. 1 depicts various multiclass SVM approaches and compares the decision boundaries for three classes.

3 Proposed Method

After reading and reviewing many research papers finally, we stay on the present topic of this article. Many research works going on the educational and commercial data set avail on the various websites for research point of view, but researchers are oversight and the very own and common issues came in front of us on a day-to-day

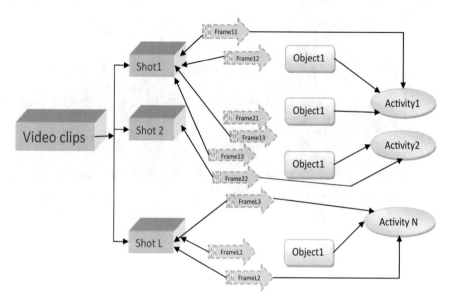

Fig. 2 Preprocessing of the movie to be rated

life. The reason behind it is we were unable to think outside the box; i.e., every second person in the world may not be very much fond of, but at some extent would like to watch movies in his leisure moments. In this paper, we proposed a technique as of hybrid system to rate the film with MPAA rating, according to the content presented in the film. Automation works on the concept of CBVR to mine the information, process with the assistance of discrete cosine transform (DCT) and finally applying support vector machine (SVM) for adding intelligence and rate the film.

3.1 Divide Movie into Scenes

In the present scheme, the primary object is the movie rating; practically, it is quite difficult to rate the whole film because on an average movie is at least 1 h long, and it is a collection of thousands of shots. To solve the present problem, we divide the film into small clips. Among all the clips, each clip may be a collection of at least 2–3 shots/scenes. Figure 2 shows an initial preprocessing process of the movie which gives us a group of frames extracted from the video clip. Now, we need to cluster frames belonging to each shot.

3.2 Clustering of Frames and Event Detection

Two kinds of transitions were experienced during analysis of continuous frames of a video. First one is abrupt transitions which were due to a large amount of change among neighbouring frames and another one is gradual transition which was due to smooth changes among neighbouring frames. Shot boundaries detection is done by finding out abrupt transitions between consecutive frames. For the detection of sudden transitions and events detection in a video sequence, the objects tracking must be performed first. An algorithm to consolidating the parameter of a Gaussian mixture model to separate the foreground and background features of the video then send extracted features to train error back propagation neural network so that it can efficiently identify the object and perceive the activity of the object [7] partially adapted to perform Preprocessing of the movie as shown in Fig. 2. By consolidating activity of the object and shot boundary, clusters of frames have been formed those belong to shot of the clip.

3.3 Combining SVM with DCT

Our proposed methodology begins by breaking the movie into clips, then shots, then sequence of frames which have been used to derive visual features. Overall working on the proposed system with the combination of SVM with DCT is summarized in following Fig. 3 illustrating the following steps

- Divide image into 8 × 8 blocks of pixels.
- Seeking from left to right and top to bottom, apply DCT to each block.
- The zigzag pattern will be applied to an individual block of DCT coefficients to produce a row of 64 elements.
- Top left 3 × 3 block is considered as DCT coefficient, and remaining elopements are discarded.
- Support vector machine learning is involved to specific values of each row in getting the final output.

Overall working of the proposed method is divided into two parts. The first-half is the training of SVM which is the most crucial part of the present work. The accuracy of the results depends on the training of SVM; proper training tends us toward a correct result. The second-half of working is rating which utilizes the training of SVM to rate movies as efficient as a human can rate.

4 Implementation

A large corpus of training and testing images was required as the basis for implementing and evaluating the image-based application. The data are organized into image

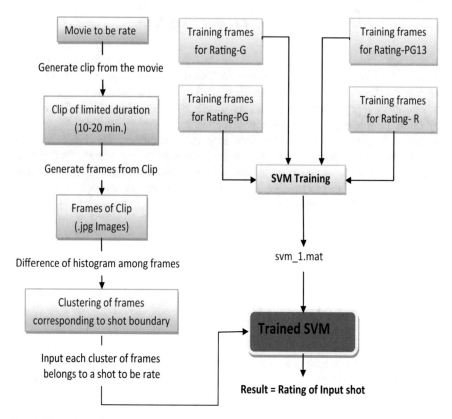

Fig. 3 Flow chart of rating procedure by a proposed system

sequences of subjects expressing events. The proceeding section contains details of training data set used for SVM training according to each rating with a description of an image sequence which is being utilized for rating and generate correct results.

Dataset

To successfully achieve the objective of the present framework, a large data set is required for training of SVM and for rating by using the trained SVM to deliver it. The dataset was created by the user for either training or rating. Table 1 presents a detailed description of training data set collected by using various movies for the training of SVM for specific rating and event. Similarly, we created a data set for movies to be the rate which also has a sequence of images extracted from clips of the movies.

Table 1 Ratings result in comparison by MPAA

Name of the movie	Training for the event	Number of images	Actual rating from IMDB
For rating G			
Cloudy with a Chance of Meatballs	Animated face	21	PG
Inception	Meeting at outdoor	155	PG-13
Tinker Bell And The Lost Treasure	Animation with bright colours	75	G
Stanly's Tiffin Box	Various activities of Children	332	G
For rating PG			
Nanny McPhee	Various activities of children	201	PG
Meet the Robinsons	Scary object and boy falling	280	G
Inkheart	group of man and woman talking	199	PG
For rating PG-13			
Clash of the Titans	Scared face, fog, smoke, and falling	187	PG-13
The Ugly Truth	Alcohol drinking and picturization at the pub	166	R
Firewall	Gun	61	PG-13
Daybreakers	Smoking	35	R
For rating PG-R			
Daybreakers	Burning human	103	R
Resident Evil	Partial female nudity	75	R
Love actually	Lovemaking	85	R
The Doomsday	A woman struggling with a man	190	R

4.1 Results

To demonstrate results in the form of graph and for simplifying the processing and identification of each rating during SVM training and classification, we assign them numerical values. Rating G is denoted as numerical value 1; rating PG is indicated as numerical value 2, Rating PG-13 meant as numerical value 3, and Rating R is denoted as numerical value 4. Before analyzing any movie, we divide it into small clips. Similarly, for results evaluation, we take seven movies and generate three clips from each video then convert each clip into a sequence of frames, this is our test data for films to be the rate.

Table 2 Ratings result in comparison by MPAA

Movie name	Clip id	Actual rating	Predicted rating
A Bug's Life	ABL 1	1	1
	ABL 2	1	2
	ABL 3	1	1
Rio	RIO1	1	1
	RIO2	1	1
	RIO3	1	2
Next Avengers: Heroes of Tomorrow	NA1	2	1
	NA2	2	2
	NA3	2	2
Cheaper by the Dozen	CBD1	2	2
	CBD2	2	2
	CBD3	2	1
The Proposal	TP1	3	4
	TP2	3	4
	TP3	3	3
New Moon	NM1	3	4
	NM2	3	3
	NM3	3	1
Resident Evil-Extinction [2007]	RE1	4	4
	RE2	4	4
	RE3	4	3

After applying the proposed rating procedure on test data, details of the movie, it's actual rating (from IMDB) and predicted rating showing in Table 2. Summary of experiments and result comparison demonstrated in following Fig. 4.

Performance Evaluation

Decision support accuracy metrics are popularly used as reversal rate, weighted errors, receiver operating characteristics. [6] **Precision** is the fraction of recovered cases that are relevant, while **recall** is the fraction of comparing cases that are extracted. Both precision and recall are therefore in the light of comprehension and measure of relevance. In much less ambiguous terms, high recall means that an algorithm returned the vast majority of the relevant results. High precision says that an algorithm returned more relevant results than irrelevant. A measure that joins precision and recall is the harmonic mean of precision and recall, the traditional **F-measure** or balanced **F-Score**.

$$F\text{-}Score = 2 * \frac{\text{Precision} * \text{Recall}}{\text{Precision} + \text{Recall}}$$

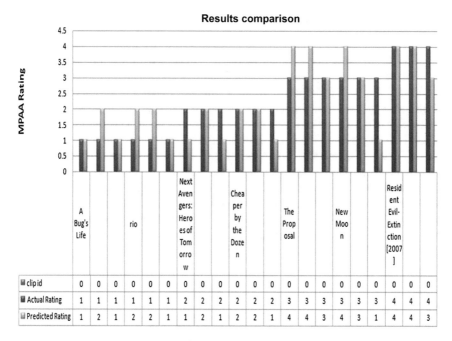

Fig. 4 Comparison of actual versus predicted rating results

Calculation:

$$Precision = \frac{Number\ of\ relevent\ documents\ retrieved}{Total\ number\ of\ documents\ retrieved}$$

$$Precision = \frac{12}{21} = 0.57$$

$$Recall = \frac{Number\ of\ relevent\ documents\ retrieved}{Total\ number\ of\ relevant\ documents}$$

$$Recall = \frac{12}{12} = 1$$

$$F\text{-}Score = 2 * \frac{0.57 * 1}{0.57 + 1} = 0.72$$

The F-Score can be considered as a measure of uniqueness of mean and their relation to the variability within the individual sample. A more significant value of F-Score reflects that the differences between the methods are not just coincidence but the results of the actual experiment. The significance of F-Score is 0.72 means; training by present training data set may produce 72% correct rating. This factor may vary with the change in the training dataset.

5 Conclusion

In this paper, we presented a framework for classification of movie clips using features from scene analysis. The events and the activities are detected using a framework based on SVM by the addition of the conventional video retrieval system. We added intelligence to the processing framework which helps us to computerize a critical task which needs severe and dedicated human involvement. The present framework did not go to replace human but improving his processing power and efficiency for achieving the function of movie rating. Presently, we implemented a hybrid system means a combination of various methods and approaches such as extracting frames from movie clip manually then dividing all extracted frames into the scene by detecting shot boundary using differences of histogram then applying these sequence of frames into SVM and finally getting a rating of the movie clip. In future, we may improve this work by combining all the methods and approaches into a single automated system as an enhancement of hybrid system.

References

1. Amine, A., Ghouzali, S., Rziza, M., Aboutajdine, D.: Investigation of Feature Dimension Reduction based DCT/SVM for Face Recognition, pp. 188–193 (2008)
2. Araki, K.: A SVM-based personal recommendation system for TV programs. In: 2006 12th International Multi-Media Model Conference, pp. 401–404. https://doi.org/10.1109/mmmc.2 006.1651358 (2006)
3. Chen, G., Wang, T.-J., Gong, L.-Y., Herrera, P.: Multi-class support vector machine active learning for music annotation. Int. J. Innov. Comput. **6**, 921–930 (2010)
4. Guo, J.M., Prasetyo, H.: Content-based image retrieval using error diffusion block truncation coding features. IEEE Trans. Image Process. **24**, 1010–1024 (2015). https://doi.org/10.1109/T IP.2014.2372619
5. Habib, M.S., Kalita, J.: Scalable Biomedical Named Entity Recognition: Investigation of a Database-Supported SVM Approach
6. Isinkaye, F.O., Folajimi, Y.O., Ojokoh, B.A.: Recommendation systems: principles, methods, and evaluation. Egypt Inform. J. **16**, 261–273 (2015). https://doi.org/10.1016/j.eij.2015.06.005
7. Jain, S., Vishwakarma, G., Jain, Y.K.: An artificial approach of video object action detection by using gaussian mixture model. Int. J. Eng. Trends Technol. **42**, 49–55 (2016). https://doi.org/10.14445/22315381/IJETT-V42P212
8. Jain, S., Vishwakarma, G., Kumar, Y.: Random forest classifier based on variable precision rough set theory. Int. J. Comput. Appl. **169**, 1–5 (2017). https://doi.org/10.5120/ijca2017914 866
9. Lei, H., Govindaraju, V.: Half-against-half multi-class support vector machines. Mult. Classif. Syst. 156–164 (2005). https://doi.org/10.1007/11494683_16
10. Milgram, J., Cheriet, M., Sabourin, R.: "One Against One" or "One Against All": Which One is Better for Handwriting Recognition with SVMs? (2006)
11. National Association of Theatre Owners (2010) Classification and Rating Rules
12. Perez, C.A.: Digital Commons@Georgia southern a content analysis of the MPAA rating system and its evolution a content analysis of the MPAA rating system and its evolution
13. Platt, J., Cristianini, N., Shawe-Taylor, J.: Large margin DAGs for multiclass classification. Int. Conf. Neural Inf. Process. Syst. 547–553 (2000)

14. Robinson, J., Kecman, V.: Combining support vector machine learning with the discrete cosine transform in image compression. IEEE Trans. Neural Netw. **14**, 950–958 (2003). https://doi.org/10.1109/TNN.2003.813842
15. Su, J., Huang, Y., Yeh, H., Tseng, V.S.: Expert systems with applications effective content-based video retrieval using pattern-indexing and matching techniques. Expert Syst. Appl. **37**, 5068–5085 (2010). https://doi.org/10.1016/j.eswa.2009.12.003

Empirical Study on Features Recommended by LSVC in Classifying Unknown Windows Malware

S. L. Shiva Darshan and C. D. Jaidhar

Abstract Modern malware has greatly evolved and become sophisticated with the capability to evade existing detection techniques. To defend against an advanced class of malware, behaviour-based malware detection technique has emerged as an essential complement. The major challenging task in this technique is to identify significant features from the original features' set. The main objective of this work was to explore the effectiveness of the linear support vector classification (LSVC) in choosing prominent features from an original feature set derived from the Cuckoo sandbox generated behaviour reports. In this work, the proposed malware detection system (MDS) utilizes the Cuckoo sandbox to obtain runtime behaviour report of the Windows executable file to be examined. From the report, features are extracted, and then LSVC is applied onto the extracted features to recognize crucial features, which boost the detection ability of the MDS. The efficiency of the proposed MDS was evaluated using real-world malware samples with tenfold cross-validation tests. The experimental results demonstrated that the proposed MDS is proficient in accurately detecting malware and benign executable files by attaining a detection accuracy of 98.429% with the sequential minimal optimization (SMO) classifier.

1 Introduction

Malicious software is widely known as malware and is developed with the intention to disrupt the normal operations of the target host without the consent of the host user. Today's modern sophisticated malware is highly powerful and has the capability

S. L. Shiva Darshan (✉) · C. D. Jaidhar
Department of Information Technology, National Institute of Technology Karnataka,
Surathkal, India
e-mail: it15f02.shivadarshan@nitk.edu.in

C. D. Jaidhar
e-mail: jaidharcd@nitk.edu.in

© Springer Nature Singapore Pte Ltd. 2019
J. C. Bansal et al. (eds.), *Soft Computing for Problem Solving*, Advances in Intelligent
Systems and Computing 817, https://doi.org/10.1007/978-981-13-1595-4_46

to tamper the file systems, registers and make it difficult for the users to login into their host system. Thus, the malware is no longer used to just damage or intrude on computer network systems, but also has become the primary tool for criminals to operate a profitable business. The extensive propagation of malware makes efficient and effective detection of unknown malware a challenging task. Therefore, a malware detection technique that detects unknown malware quickly and rapidly is essential to mitigate today's malware attacks.

Majority of the anti-malware defensive solutions extensively depend on the signature-based detection technique [23]. However, this technique exhibits its deficiency in spotting malware that uses obfuscation techniques [12]. Consequently, behavioural-based detection technique emerges as an essential complement to the signature-based detection technique. Generally, behaviour-based malware detection technique utilizes the sandbox to monitor and capture activities that occur during the execution of the target file, in particular, the sequence of system calls or application programming interface (API) calls. Then, it analyses the acquired report to conclude whether the target file is benign or malware. However, most of the modern malware first identify the sandbox when they are being analysed and then decide whether to instigate an attack on the targeted system or not. This indicates that even behaviour-based detection techniques are inadequate and fail to balance between false positive rate (FPR) and malware detection rate.

To build an efficient and robust machine learning-based malware detection system, identifying crucial features from the original feature set of a large size is essential. This process is referred to as the feature selection technique. In order to identify and eradicate noisy features from the original feature set, LSVC is employed. The prime objective is to measure the effectiveness of the features recommended by the LSVC in classifying unknown malware.

The major contributions of the proposed work are summarized as follows:

- In this work, the proposed MDS was designed, implemented, and evaluated using publicly available real-world malware samples. The proposed MDS is capable of identifying malware and benign Windows executable files precisely by using dynamic features suggested by the LSVC.
- The proposed MDS was evaluated with two different types of dynamic features, namely (1) API calls and (2) API calls with their category (category+API calls), to know which type of features provide better malware detection rate.
- To measure the malware detection rate of the MDS, tenfold validation tests were conducted. The obtained experimental results demonstrate that the MDS is able to attain detection accuracy of 98.429% with 0.016 FPR for API calls as the dynamic feature.

The rest of this paper is organized as follows: in Sect. 2, we review earlier research works related to malware identification. Section 3 elucidates an overview of the proposed malware detection system. The obtained empirical results are presented in Sect. 4. Finally, the conclusion is summarized in Sect. 5.

2   Related Work

A noteworthy amount of malware detection approaches have been proposed in the direction of malware detection [8, 18]. Malware identification approaches are mainly classified into two types: (1) static feature-based detection techniques and (2) dynamic feature-based detection techniques.

2.1   Static Feature-Based Malware Detection Techniques

A number of static feature-based MDSs have been proposed in the literature [1, 2, 19, 20, 22]. However, the limitations of these techniques are: First and foremost limitation being that sophisticated malware can make use of obfuscation techniques such as compression, encryption, code reordering, garbage instruction insertion to bypass the MDS [17]. Moser et al. [12] demonstrated in their work that static features are insufficient to spot unknown malware. Dynamic feature-based MDS has evolved as a promising solution to address the limitations of static feature-based MDS.

2.2   Dynamic Feature-Based Malware Detection Techniques

This type of detection approach examines the runtime behaviour of the input file. A number of dynamic feature-based MDSs have been proposed in the literature [5, 9]. Salehi et al. [16] proposed an MDS (MAAR) that treats API calls and their arguments as features to recognize the malware. The authors demonstrated in their work that MAAR achieves an accuracy of 99.4% with FPR lesser than one. Qiao et al. [13] proposed an automated malware analysis framework, which extracts API call sequences from the report generated by the dynamic analysis tool and then transforms them into byte-based sequential data. The main advantage of this framework is that it supports local deployment and reduces storage size and computational cost, thereby achieving high precision for malware clustering. An automatic malware examination framework based on behaviour traits using machine learning was proposed [15]. The authors attempted to form an effective malware detection based on the invoked system call sequence, which provides timely defence against the malware. In a similar direction, other works were also proposed [7, 10, 21].

To the best of our knowledge, LSVC has not been explored as a feature selector to select dynamic features given by the Cuckoo sandbox report. Thus, in this work, the prime aim is to investigate the effectiveness of the dynamic features suggested by the LSVC in identifying unknown malware.

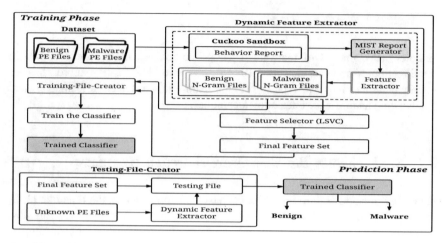

Fig. 1 The architecture of the proposed MDS

3 Architecture Overview

An overview of the proposed MDS is shown in Fig. 1. It utilizes dynamic malware analysis technique to spot malware. It mainly consists of two phases: (1) training phase and (2) prediction phase.

3.1 Training Phase

In the training phase, a set of benign and malware Windows portable executable (PE) files are executed one at a time on the Cuckoo sandbox to gather runtime behavioural report of the PE files. Each runtime behavioural report is pre-processed separately to obtain API calls and their category. Each gathered data is treated as an individual feature. In order to remove noisy features from a set of acquired features, the LSVC is employed as feature selector. The features thus recommended are used as final features to prepare a training file, which is essential to train the classifier. The main components of the training phase are: dynamic feature extractor, feature selector, final feature set, and training-file-creator.

3.1.1 Dynamic Feature Extractor

The prime task of the dynamic feature extractor is to observe and record the execution behaviour of the PE, while it is being executed on a controlled monitoring environment. Its main sub-components are: (1) Cuckoo sandbox, (2) malware instruction set (MIST) report generator, and (3) feature extractor.

```
{
        "category": "system",
        "status": 1,
        "stacktrace": [],
        "api": "LdrLoadDll",
        "return_value": 0,
        "arguments": {
                "basename": "comctl32",
                "module_address": "0x773d0000",
                "flags": 0,
                "module_name": "comctl32.dll"
        },
        "time": 1491435510.263125,
        "tid": 1124,
        "flags": {}
}
```

Fig. 2 Snippet of API call recorded by Cuckoo sandbox in JSON file format

3.1.2 Cuckoo Sandbox

The Cuckoo sandbox is a Windows executable file runtime behaviour acquiring tool [4]. It is used to obtain a runtime behavioural report of the executable file. It records the details onto the behavioural report in JavaScript Object Notation (JSON) file format. The behaviour report also called as the dynamic analysis report [3]. The Cuckoo sandbox captures the API calls and classifies them into several categories on the basis of the type of operation performed which includes `network, process, system, services, registry, misc, crypto, file, resources,` etc. [11]. Figure 2 shows a snippet of the recorded API call `"LdrLoadDll"` categorized under `system` category during the execution of `"01B43C0C8D620E8B88D846E4C9287CCD.bin"`. Moreover, the category of an API call in terms of the type of operation performed is a better indicator to understand the actions taken by the executable file. So, the focus is to extract details such as category API call with corresponding arguments from the JSON format and then represent the acquired particulars in the MIST format.

3.1.3 MIST Report Generator

The behavioural analysis report obtained in the JSON file format requires high storage and more processing time. In order to address this issue, the MIST report generator is used in this work [15]. It extracts system-level behaviour such as category, API calls and their arguments from JSON format and organizes them in different levels of blocks as shown in Fig. 3. MIST is the preferred format because it uses a smaller file size and reduces the processing time.

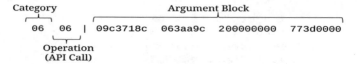

Fig. 3 MIST representation of API call reported in JSON file format shown in Fig. 2

3.1.4 Feature Extractor

The job of the feature extractor is to extract API calls (operation field) and their category from the MIST report (see Fig. 3). In this work, the feature extractor is implemented using the Python programming language. The extracted data is represented in overlapping substrings, which are obtained based on the sliding window approach known as N-grams [14] and are extensively used in information retrieval. Its major drawback is that the set of all the gathered N-grams (features) from a set of byte string of malware and benign PE files is enormous, and applying them directly for classification techniques degrades the classifier performance. Therefore, selection of the most relevant N-grams plays a crucial role. In order to choose prominent features, the feature selector is employed.

3.2 Feature Selector

LSCV [2, 6] is a wrapper-based feature selection technique. It is used to identify the most significant features from a set of gathered features. All the LSVC recommended features are treated as final features because there can be no further elimination of the features. The final features are used to prepare a training file as well as testing files to measure the efficiency of the classifiers. Suppose p, q, d, and n is the input vector, class label, dimension, and number of samples, respectively. Training data (p_i, q_i) is separated by hyperplane decision function D(p) based on the value of x and y. Where $p \in P^m$, $q_i \in \{+1, -1\}$ and $i = 1, 2, 3, ..., n$. The decision function D(p) is given by Eq. 1.

$$D(p) = \{x^T p\} + y = \sum_{i=1}^{n} x_i p_i + y \tag{1}$$

where $x = [x_1, x_2, \ldots, x_n]^T$ is the weight vector of the hyperplane. The training data with value $q_i = +1$ gets classified under D(p) > 0, and conversely, when the training data with value $q_i = -1$ gets classified under D(p) < 0.

For the LSVC, we mainly used the recommended default parameters provided in sklearn.svm.LinearSVC python library. Accordingly, the value of the penalty parameter 'C' was set to 1. The value of 'penalty' specifies the norm used in the penalization, i.e. 'l2' and the value of 'dual' is preferred to be false.

3.3 Training-File-Creator

Training-file-creator creates a training file essential to train the classifiers. It parses the benign and malware N-gram files corresponding to the training dataset with the final features in order to create a training file.

3.4 Prediction Phase

In the prediction phase, the main task of the testing-file-creator is to create a testing file necessary to appraise the predictive performance of the trained classifiers. It makes use of the final features and N-grams of the testing file to deliver a testing file. The generated testing file is sent to the trained classifier, which classifies whether the test input file is benign or malware.

4 Experimental Results

To evaluate the performance of the proposed MDS, all the experiments were conducted on a host system that had Ubuntu 14.04.5 LTS operating system, Intel i7-3770 CPU 3.40 GHz, and 8 GB RAM.

The experimental data consists of 200 benign and 200 malware PE files. The benign PE files include the Windows system files, which were collected from a freshly installed Windows virtual machine. The Windows malware PE files used in this experimental work were downloaded from the public source VirusShare.[1] Six different machine learning classifiers such as the SMO, simple logistic, logistic, J48, random forest, and random tree, which were available in WEKA[2] tool, were adopted. These classifiers were used throughout the experiments in tenfold cross-validation with default parameters settings.

4.1 Evaluation Metrics

The MDS with high detection rate and minimum FPR is proficient in identifying the unknown malware. The effectiveness of the MDS and the detection performance of the classifiers were estimated using five evaluation metrics such as the true positive rate (TPR), FPR, precision, recall, and accuracy (see Eq. 2).

---

[1] https://virusshare.com/.

[2] http://www.cs.waikato.ac.nz/ml/weka/.

$$Accuracy = \frac{(TP + TN)}{(TP + TN + FP + FN)} \quad TPR = \frac{TP}{TP + FN}$$

$$FPR = \frac{FP}{FP + TN} \quad Precision = \frac{TP}{TP + FP} \quad Recall = \frac{TP}{TP + FN} \quad (2)$$

where true positive (TP) indicates that the malware PE file is correctly classified as malware; true negative (TN) represents that the benign executable is precisely classified as benign; false positive (FP) indicates that the benign executable is misclassified as malware; and false negative (FN) indicates that the malware executable is misclassified as benign.

4.2 Results and Discussions

In this work, the prime aim of the proposed MDS was to explore the accurate detection and categorizations of malicious executables using the final features suggested by the LSVC. Two different types of dynamic features, namely API calls and API calls with their category for different N-grams of sizes 3, 4, and 5 bytes were used to measure the accuracy. Two stages of experiments were conducted. The first stage comprised of API calls sequence (N-grams) as features. The second stage consisted of category and API call together as a single element. The sequence of a combination of category and API as a feature (N-gram) were taken into consideration. The features suggested by the LSVC in both the experiments were used to appraise the efficiency of the classifier separately. Finally, a comparative analysis was made to know the impact of the LSVC recommended features.

During the training file creation phase, the Cuckoo sandbox stored the monitored behaviour of the executable file in JSON format, and the obtained data was preprocessed to convert into MIST format. From the MIST format file, the API call sequence was extracted, and then consecutive API calls were grouped to prepare N-grams of different sizes such as 3, 4, and 5 bytes. Subsequently, a combination of category and API call was extracted as a pair from the MIST format file and a separate N-gram file was prepared for each individual MIST file. Duplicate N-grams were removed and further LSVC was applied as feature selector to choose the distinct features and ignore the noisy features. The LSVC assigns a separate score to each individual feature and then selects a set of features based on their score.

First, the experiment was conducted with concern to the monitored API calls present in the MIST files. Initially, the length of N-gram as 3 bytes was selected to prepare N-grams considering all the benign and malware N-gram files. Further, we consolidated all N-grams of size 3 bytes belonging to benign and malware class individually. However, 4133 and 6555 distinct N-grams related to benign and malware class, respectively, were obtained. All these benign and malware N-grams were combined, and the duplicates were eliminated to obtain unique N-grams. Consequently, 7645 N-grams as were attained features. Moreover, it was impractical to consider all these 7645 N-grams as features to prepare a training file required to train the classifier. Therefore, the relevant features (90) were selected as final features by using the LSVC. Finally, a training file was constructed to train the classifier using the final features with the N-gram files corresponding to the training samples. Similarly, experiments were carried out for N-grams of 4 and 5 bytes, where the LSVC suggested 186 (4 byte N-grams) best features out of 26,143 N-grams and 124 (5 byte N-grams) significant features out of 35,226 N-grams to prepare the final features required to train the classifier.

In another set of experiments, both the category and API calls present in the MIST files were focused on to know their effectiveness. With respect to this, consecutive three pairs of category and API call were grouped to form N-grams of size 3 bytes. After removing duplicate N-grams, 16,188 N-grams were attained from all N-gram files and the LSVC was applied to choose the distinct features. The LSVC recommended 134 N-grams as prominent features and these were used to train the classifier. Correspondingly, experiments were conducted for N-grams of size 4 bytes (4 pairs of category and API call) and 5 bytes (5 pairs of category and API call). In this case, the LSVC advised 156 best features among 29,821 N-grams of size 4 bytes and 221 predominant features among 42,742 N-grams of size 5 bytes.

4.3 Analysis of MDS Based on Evaluation Metrics

From the experimental observations, the detection rate and FPR obtained for N-grams of size 3 bytes are depicted in Fig. 4a and Table 1. In particular, the SMO classifier achieved maximum accuracy of 96.923% with 0.031 FPR for combination of category+API calls features. Further, the simple logistic classifier also performed well by producing an accuracy of 93.846% with FPR 0.062. Relatively, the overall performance of other classifiers such as the logistic, J48, random forest, and random tree reported lowest accuracy and their corresponding values are 88.717% with FPR 0.113, 87.692% with FPR 0.123, 91.794% with FPR 0.082 and 88.717% with FPR 0.113.

The accuracy and FPR achieved by the different classifiers for N-grams of size 3 bytes with features API calls are shown in Fig.4a and Table 1. It can be seen from Fig. 4a that the highest accuracy of 98.429% with 0.016 FPR was yielded by the SMO classifier. However, the performance of the other classifiers was not remarkable. The second highest accuracy of 95.549% was accomplished by the Logistic classifier with 0.045 FPR. The simple logistic classifier recorded low accuracy of 94.764% with 0.052 FPR. Moreover, other classifiers such as the J48, random forest, and random tree underperformed by achieving least accuracy of 87.958% with 0.120 FPR, 93.193% with 0.068 FPR, and 87.696% with 0.123 FPR, respectively.

Comparative analysis was made between the values accomplished by the evaluation metrics on two different N-gram features such as the API calls and category+API calls of size 3 bytes. The highest TPR of 0.984, Precision of 0.985, and Recall of 0.984 were recorded by the SMO classifier when only API calls were considered as N-gram features. The other classifiers like the simple logistic, J48, random forest, and

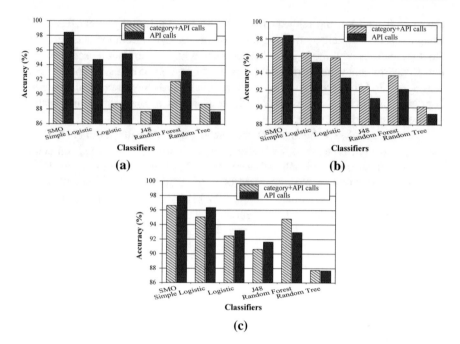

Fig. 4 Accuracy obtained considering API calls and API calls with their category for N-grams of different sizes **a** 3 bytes, **b** 4 bytes, and **c** 5 bytes

Table 1 Comparison analysis of evaluation metrics obtained for N-grams of size 3 bytes

Classifiers	TPR		FPR		Precision		Recall	
	C+API	API calls	C+API	API calls	C+API	API calls	C+API	API calls
SMO	0.969	0.984	0.031	0.016	0.969	0.985	0.969	0.984
Simple logistic	0.938	0.948	0.062	0.052	0.939	0.948	0.938	0.948
Logistic	0.887	0.955	0.113	0.045	0.888	0.957	0.887	0.955
J48	0.877	0.880	0.123	0.120	0.877	0.880	0.877	0.880
Random forest	0.918	0.932	0.082	0.068	0.922	0.932	0.918	0.932
Random tree	0.887	0.877	0.113	0.123	0.888	0.878	0.887	0.877

C+API: category+API calls

random tree performed poor and resulted in achieving least accuracy as demonstrated in Table 1.

Experiments were also performed by considering N-grams size of 4 bytes and 5 bytes. Accordingly, final features were constructed for both types of dynamic features API and category with API calls. As illustrated in Fig. 4b and Table 2, the

Table 2 Comparison analysis of evaluation metrics obtained for N-grams of size 4 bytes

Classifiers	TPR		FPR		Precision		Recall	
	C+API	API calls	C+API	API calls	C+API	API calls	C+API	API calls
SMO	0.982	**0.984**	0.018	**0.016**	0.982	**0.984**	0.982	**0.984**
Simple logisitic	0.964	0.953	0.036	0.047	0.964	0.953	0.964	0.953
Logistic	0.958	0.935	0.042	0.065	0.960	0.936	0.958	0.935
J48	0.924	0.911	0.076	0.089	0.925	0.914	0.924	0.911
Random forest	0.938	0.921	0.063	0.079	0.939	0.924	0.938	0.921
Random tree	0.901	0.893	0.099	0.107	0.902	0.894	0.901	0.893

C+API: category+API calls

maximum detection accuracy of 98.429% with 0.016 FPR was accomplished for the final features comprised of only API calls. However, a combination of final features consisting of category+API calls also achieved nearly equivalent detection rate of 98.177% with FPR 0.018 when compared with features of API calls. The second highest accuracy recorded was 96.335% with FPR 0.036 by the simple logistic classifier for the final features consisting of category+API calls as N-gram features, correspondingly, the same classifier was successful in achieving an almost identical accuracy of 95.288% with 0.047 FPR by final features of API calls. The logistic classifier attained least accuracy of 95.833% with 0.042 FPR for final features comprising of both category+API calls, but, it gained nearest accuracy of 93.455% with FPR 0.065 for final features built with only API calls. The performance of other classifiers such as the J48, random forest, and random tree was not remarkable.

The highest values of TPR, precision, and recall signifies better MDS. Therefore, the experiment results tabulated in Table 2 provides details of the comparative analysis made for the final features (N-grams size 4 bytes) consisting of only API calls and the combination of both category+API calls. Relatively, the nearest value to one was achieved for final features made with API calls and as proof-of-concept, the SMO classifier achieved the highest TPR with 0.984, precision with 0.984, and recall with 0.984.

The last experiment was performed on N-grams of size 5 bytes and the performance of the classifiers for both types of features such as API calls and category+API calls was observed. Figure 4c and Table 3 provides details of the detection accuracy and FPR achieved by the different classifiers. It was observed that the SMO classifier was successful in achieving highest accuracy of 97.905% with 0.021 FPR for final features comprised of API calls. Meanwhile, the same classifier performed well for the features of category+API calls and gained nearly equivalent accuracy of 96.614% with 0.034 FPR when compared with features of API call type. Further, the simple logistic classifier attained second highest accuracy of 96.355% with 0.037 FPR for final features consisting of API calls, and for the final features of category+API calls,

Table 3 Comparison analysis of evaluation metrics obtained for N-grams of size 5 bytes

Classifiers	TPR		FPR		Precision		Recall	
	C+API	API calls	C+API	API calls	C+API	API calls	C+API	API calls
SMO	0.966	0.979	0.034	0.021	0.967	0.980	0.966	0.979
Simple logistic	0.951	0.963	0.049	0.037	0.953	0.964	0.951	0.963
Logistic	0.924	0.932	0.076	0.068	0.925	0.933	0.924	0.932
J48	0.906	0.916	0.094	0.084	0.908	0.917	0.906	0.916
Random forest	0.948	0.929	0.052	0.071	0.951	0.932	0.948	0.929
Random tree	0.878	0.877	0.122	0.123	0.878	0.878	0.878	0.877

C+API: category+API calls

it yielded an accuracy of 95.052% with 0.049 FPR. The least accuracy obtained was 93.193% with 0.068 FPR by the logistic classifier for final features of API calls, whereas for the final features of category+API calls, it recorded an accuracy of 92.447% with 0.076 FPR. The other classifiers such as the J48, random forest, and random tree underperformed by achieving less accuracy as shown in Table 3.

The detection performance of the classifiers was also analysed for N-grams of size 5 bytes using other evaluation metrics as tabulated in Table 3. The highest TPR, precision, and recall were achieved by the SMO classifier with 0.979, 0.980, and 0.979, respectively, and were witnessed for the final features consisting of API calls as N-grams. However, the performance of the classifiers was not appreciable for the final features comprised of category+API calls.

The experiments were computed substantially and analysed to decide which N-gram size dynamic features encourage the classifier to achieve better accuracy. In this direction, the obtained and analysed results proved that the feature API calls alone exhibited promising results for N-grams of size 4 bytes. Further, it was crucial to have LSVC as feature selector to get the best compact set of features and better accuracy. Accordingly, the accuracy of 98.429% was achieved by the SMO classifier with few features recommended by the LSVC for the final features consisting of API calls. Meanwhile, experiments were also conducted by combining category+API calls, which resulted in less accuracy.

5 Conclusion

In this paper, the Cuckoo sandbox was used to analyse the execution time behaviour of the Windows executable files. In order to measure the effectiveness of the LSVC in selecting prominent features from the original feature set, it was applied onto two different types of features set such as (1) API calls and (2) API calls with their category. From the experiments, it was observed that the SMO classifier achieved detection accuracy of 98.429% with the final features (N-grams of size 4 bytes) comprising of API calls as recommended by the LSVC. Thus, the empirical results manifested that the LSVC is proficient in recommending the best N-gram features, which boosted the detection rate of the classifier.

References

1. Bai, J., Wang, J., Zou, G.: A malware detection scheme based on mining format information. Sci. World J. (2014)
2. Belaoued, M., Mazouzi, S.: A real-time pe-malware detection system based on chi-square test and pe-file features. In: IFIP International Conference on Computer Science and Its Applications_x000D_, pp. 416–425. Springer (2015)
3. Firdausi, I., Erwin, A., Nugroho, A.S., et al.: Analysis of machine learning techniques used in behavior-based malware detection. In: 2010 Second International Conference on Advances in Computing, Control and Telecommunication Technologies (ACT), pp. 201–203. IEEE (2010)
4. Guarnieri, C., Tanasi, A., Bremer, J., Schloesser, M.: Automated malware analysis-cuckoo sandbox (2012)
5. Juwono, J.T., Lim, C., Erwin, A.: A comparative study of behavior analysis sandboxes in malware detection. In: International Conference on New Media (CONMEDIA), p. 73 (2015)
6. Kawaguchi, N., Omote, K.: Malware function classification using apis in initial behavior. In: 2015 10th Asia Joint Conference on Information Security (AsiaJCIS), pp. 138–144. IEEE (2015)
7. Kim, J., Lee, S., Youn, J.M., Choi, H.: A study of simple classification of malware based on the dynamic api call counts. In: International Conference on Computer Science and its Applications, pp. 944–949. Springer (2016)
8. Kolter, J.Z., Maloof, M.A.: Learning to detect and classify malicious executables in the wild. J. Mach. Learn. Res. 7, 2721–2744 (2006)
9. Kruegel, C., Kirda, E., Bayer, U.: Ttanalyze: a tool for analyzing malware. In: Proceedings of the 15th European Institute for Computer Antivirus Research (EICAR 2006) Annual Conference, vol. 4 (2006)
10. Lengyel, T.K., Maresca, S., Payne, B.D., Webster, G.D., Vogl, S., Kiayias, A.: Scalability, fidelity and stealth in the drakvuf dynamic malware analysis system. In: Proceedings of the 30th Annual Computer Security Applications Conference, pp. 386–395. ACM (2014)
11. Miller, C., Glendowne, D., Cook, H., Thomas, D., Lanclos, C., Pape, P.: Insights gained from constructing a large scale dynamic analysis platform. Dig. Invest. 22, S48–S56 (2017)
12. Moser, A., Kruegel, C., Kirda, E.: Limits of static analysis for malware detection. In: Twenty-Third Annual Conference Computer Security Applications, 2007. ACSAC 2007, pp. 421–430. IEEE (2007)
13. Qiao, Y., Yang, Y., He, J., Tang, C., Liu, Z.: CBM: free, automatic malware analysis framework using api call sequences. In: Knowledge Engineering and Management, pp. 225–236. Springer (2014)
14. Raff, E., Zak, R., Cox, R., Sylvester, J., Yacci, P., Ward, R., Tracy, A., McLean, M., Nicholas, C.: An investigation of byte n-gram features for malware classification. J. Comput. Virol. Hack. Tech. 1–20 (2016)
15. Rieck, K., Trinius, P., Willems, C., Holz, T.: Automatic analysis of malware behavior using machine learning. J. Comput. Sec. 19(4), 639–668 (2011)
16. Salehi, Z., Sami, A., Ghiasi, M.: Maar: robust features to detect malicious activity based on api calls, their arguments and return values. Eng. Appl. Artif. Intel. 59, 93–102 (2017)

17. Santos, I., Brezo, F., Ugarte-Pedrero, X., Bringas, P.G.: Opcode sequences as representation of executables for data-mining-based unknown malware detection. Inf. Sci. **231**, 64–82 (2013)
18. Schultz, M.G., Eskin, E., Zadok, F., Stolfo, S.J.: Data mining methods for detection of new malicious executables. In: 2001 IEEE Symposium on Security and Privacy, 2001. S&P 2001. Proceedings, pp. 38–49. IEEE (2001)
19. Shabtai, A., Moskovitch, R., Elovici, Y., Glezer, C.: Detection of malicious code by applying machine learning classifiers on static features: a state-of-the-art survey. Inf. Sec. Tech. Report **14**(1), 16–29 (2009)
20. Siddiqui, M., Wang, M.C., Lee, J.: Data mining methods for malware detection using instruction sequences. In: Proceedings of the 26th IASTED International Conference on Artificial Intelligence and Applications, AIA '08, pp. 358–363. ACTA Press, Anaheim, CA, USA (2008). http://dl.acm.org/citation.cfm?id=1712759.1712825
21. Tsyganok, K., Tumoyan, E., Babenko, L., Anikeev, M.: Classification of polymorphic and metamorphic malware samples based on their behavior. In: Proceedings of the Fifth International Conference on Security of Information and Networks, pp. 111–116. ACM (2012)
22. Vinod, P., Laxmi, V., Gaur, M.S.: Scattered feature space for malware analysis. Adv. Comput. Commun. 562–571 (2011)
23. Ye, Y., Li, T., Adjeroh, D., Iyengar, S.S.: A survey on malware detection using data mining techniques. ACM Comput. Surv. (CSUR) **50**(3), 41 (2017)

An Improvised Competitive Swarm Optimizer for Large-Scale Optimization

Prabhujit Mohapatra⊙, Kedar Nath Das⊙ and Santanu Roy⊙

Abstract In this paper, an improvised competitive swarm optimizer (ICSO) is introduced for large-scale global optimization (LSGO) problems. The algorithm is fundamentally inspired by the competitive swarm optimizer (CSO) algorithm which neither remembers the personal best position nor global best position to update the particles. In CSO, a pair-wise competition mechanism was introduced, where the particle that loses the competition is updated by learning from the winner and the winner particles are simply passed to the next generation. The proposed algorithm introduces a new tri-competitive mechanism strategy to improve the solution quality. The algorithm has been performed on different dimensions of CEC2008 benchmark problems. The empirical results and analysis have shown better overall performance for the proposed ICSO than the CSO and many state-of-the-art meta-heuristic algorithms.

Keywords Competitive swarm optimizer · Evolutionary algorithms
Large-scale global optimization · Particle swarm optimization
Swarm intelligence

1 Introduction

Particle swarm optimization (PSO) is one of the popular swarm intelligence techniques [1] for solving global optimization problems. The algorithm is based on swarm behaviors of animals such as birds, fishes, and ants. Due to easy understanding and implementation, PSO has shown rapid growth in admiration over the last few decades. In PSO, each particle is able to memorize its previous best position, which is termed as 'personal best', whereas the best position that has been found by the whole swarm is known as 'global best.' To find the global optimal solution of

P. Mohapatra (✉) · K. N. Das · S. Roy
National Institute of Technology, Silchar, Silchar 788001, Assam, India
e-mail: prabhujit.mohapatra@gmail.com

© Springer Nature Singapore Pte Ltd. 2019
J. C. Bansal et al. (eds.), *Soft Computing for Problem Solving*, Advances in Intelligent
Systems and Computing 817, https://doi.org/10.1007/978-981-13-1595-4_47

the optimization problem, the particles learn from both the personal best and global best positions.

Although the algorithm has perceived rapid progresses over the past few decades, still its performance is unsatisfying when the problem is high dimensional, multimodal, and non-separable [2]. In order to address the issue and enhance the performance, several different variants of PSO have been proposed over the time. These variants are primarily classified into different categories such as adaptive control strategy of parameters in PSO [3–6], introduction of topological structure in neighborhood control to enhance swarm diversity of PSO [7, 8], multi-swarm PSO [9]. But for most of these PSO variants, the computational cost increases exponentially with the introduction of new mechanisms and new operators. Another major concern is about the strong effect of the global best position on the convergence speed, which is principally accountable for the premature convergence. To address this issue, Liang [8] introduced a new PSO variant deprived of the global best term and the update strategy banks on only on the particle best. A different way to address the issue of premature convergence is to get freed of both the factors, i.e., personal best and global best positions which was proposed with a multi-swarm framework [10]. Here, the updating strategy is carried out by a pair-wise competition mechanism between particles of two swarms. Afterward, the concept of the competitive mechanism has been introduced in several evolutionary algorithms [11, 12]. The weak particles of one swarm learn from stronger ones of other swarm, and the strong particles are self-motivated by the previous experiences to produce better solutions. Inspired by these works, a multi-swarm evolutionary framework based on a feedback mechanism [13] was introduced. There the feedback mechanism is supplemented with a convergence strategy and a mutation strategy. The convergence strategy helps to converge, whereas the mutation strategy carries the diversity to the population. Both the strategies synchronize to provide a good balance between exploration and exploitation. Subsequent to this idea, another method called competitive swarm optimizer (CSO) [14, 20] exposed the use of the competitive mechanism between particles within one single swarm. After each competition, the loser particle learns from the winner particle instead of from personal best or global best positions. The practice of pair-wise competitive mechanism and involvement of neither the particle best term nor the global best term makes the algorithm unique and conceptually different than other meta-heuristic algorithms. It has been popularly used to solve LSGO problems after its introduction.

Motivated by the working principle of CSO, a simple yet effective modification on the partition of a population is being proposed in this paper. Unlike CSO, the proposed improvised CSO (ICSO) suggests a new tri-competitive scenario based on the fitness differences. The tri-competition divides the whole population into three groups, namely winner, superior loser, and inferior loser groups. In each instance, the particles in the superior loser group are guided by the particles of the winner group whereas the particles in the inferior loser group are guided by the particles of the superior loser group. The basic idea is to allow the average and below average solutions to converge toward good solutions in a cascade manner. As a result, an improved rate of convergence is expected through a better rate of exploration in the search space. The paper is organized as follows. The motivation and proposition of the proposed algorithm are presented in Sect. 2. In Sects. 3 and 4, the experimental studies and analysis of results are carried out, respectively. Finally, the conclusion is drawn in Sect. 5.

2 Motivation and Proposition

2.1 Motivation

Competitive swarm optimizer (CSO) updates half of the population swarms during the simulation, which results in high diversity and slow rate of convergence. Motivated by CSO and aiming at the faster rate of convergences, a new tri-competitive scenario is introduced here. The tri-population concept helps not only in balancing exploration and exploitation, but also helps to maintain the diversity. The upgraded 2/3rd particles help to improve the convergence process, whereas the rest 1/3rd particle passed directly to the next generation to retain the necessity of swarm diversity.

2.2 Proposition

Let us consider the following problem to be minimized:

$$min f = f(x) \; s.t. \; x \in X.$$

where $X \in R^n$ is the feasible solution set and n is the search dimension. First, the problem is solved by randomly initializing a swarm $P(t)$ containing m particles, where m is known as the swarm size and t is known as the generation number. In each generation, the particles in $P(t)$ are sorted into $\frac{m}{3}$ triplets (assuming m is multiple of three). Then, a competition is made between the three particles in each triplet resulting in one winner and two losers. The superior loser is symbolized as l_1 and the inferior as l_2. Eventually, through this selection process, there will be a $K\left(= \frac{m}{3}\right)$ number of distinct competitions possible. Therefore, three distinct groups, namely winner group, superior loser group, and inferior loser group, will be formed, each of size K. Let $x_{w,k}(t), x_{l_1,k}(t), x_{l_2,k}(t)$ and $v_{w,k}(t), v_{l_1,k}(t), v_{l_2,k}(t)$ represent the position and velocity of the winner and two losers, respectively, in the kth round of competition ($k = 1, 2, \ldots, K$) at iteration t. The selection strategy of particles under tri-competition along with their inherited learning strategy is presented in Fig. 1.

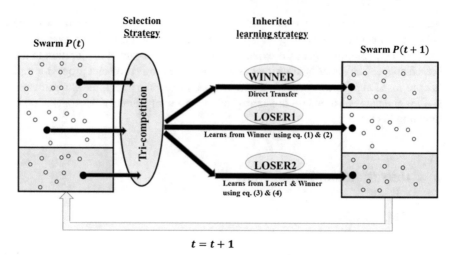

Fig. 1 Tri-competitive mechanism in ICSO

After each competition, both the loser particle's velocity and position are updated using the mechanism as follows:

$$v_{l_1,k}(t+1) = R_1(k,t)v_{l_1,k}(t) + R_2(k,t)\big(x_{w,k}(t) - x_{l_1,k}(t)\big)$$
$$+ \varphi_1 R_3(k,t)\big(\overline{x_k(t)} - x_{l_1,k}(t)\big) \tag{1}$$

$$v_{l_2,k}(t+1) = R_4(k,t)v_{l_{21},k}(t) + R_{52}(k,t)\big(x_{l_1,k}(t) - x_{l_2,k}(t)\big)$$
$$+ \varphi_2 R_6(k,t)\big(\overline{x_w(t)} - x_{l_2,k}(t)\big) \tag{2}$$

$$x_{l_1,k}(t+1) = x_{l_1,k}(t) + v_{l_1,k}(t+1) \tag{3}$$

$$x_{l_2,k}(t+1) = x_{l_2,k}(t) + v_{l_2,k}(t+1) \tag{4}$$

where $R_1(k,t)$ to $R_6(k,t) \in [0,1]^n$ are six randomly generated vectors after the competition and learning strategies in generation t. Here, $\overline{x_k(t)}$ and $\overline{x_w(t)}$ are the mean position values of the relevant particles and winners, respectively. The φ_1 and φ_2 are the parameters that control the influence of the mean positions. The first part of both Eqs. (1) and (2) $R_1(k,t)v_{l,k}(t)$, $R_4(k,t)v_{w,k}(t)$ are similar to the inertia term in the canonical PSO. But here in the proposed method, the inertia term ω is replaced by randomly created vectors. The second part $R_2(k,t)\big(x_{w,k}(t) - x_{l,k}(t)\big)$ is called the cognitive component after Kennedy and Eberhart. Here, the particle that loses the competition learns from its competitor unlike from its personal best in PSO. It does not store the personal best or global best position. The pseudocode of the ICSO is summarized in Algorithm 1.

Algorithm 1 The pseudo code of the Improved Competitive Swarm Optimizer (ICSO). U Represent a set of particles. The terminal condition is the total number of fitness evaluations.

1. $t = 0$;
2. Randomly initialize $P(0)$;
3. **While** *terminal condition* is not satisfied **do**
4.   calculate the fitness of all particles in $P(t)$;
5. $U = P(t), P(t+1) = \emptyset$
6.   **While** $U \neq \emptyset$ **do**
7.   Randomly choose three particles $X_1(t), X_2(t)$ and $X_3(t)$ from U such that $f(X_1(t)) \leq f(X_2(t)) \leq f(X_3(t))$
8.   Assign$X_w(t) = X_1(t), X_{l1}(t) = X_2(t)$ and $X_{l2}(t) = X_3(t)$;
9.   Add $X_w(t)$ into $P(t+1)$;
10.   Update $X_{l1}(t), X_{l2}(t)$ to $X_{l1}(t+1)$ and $X_{l2}(t+1)$ by eqn. (1-4) and add to P(t+1) ;
11.   Remove $X_1(t), X_2(t), X_3(t)$ from U ;
12.   **End while**
13.   t = t + 1;
14. **End while**

3 Experimental Setup

Here, a set of experiments is performed on the set of benchmark functions presented in CEC2008 special session on LSGO problems. There are seven benchmark functions available in the CEC2008 from which the functions f_1, f_4, and f_6 are separable functions and the rest are non-separable ones. But as the dimension increases, f_5 becomes separable function. In all the experiments, the population size m is varied from 200 to 1000 whereas the social components φ_1 and φ_2 are varied in the range of 0–0.3. The range of these parameters is taken same as in CSO. The experiments are simulated on a PC with an Intel Pentium-P6100 2.00 GHz CPU, Microsoft Windows 7 32-bit operating system, and MATLAB R2013a. The results are averaged over 25 different runs, and for each independent run, the maximum number of function evaluations (FEs) is set to be $5000 * n$ for the CEC2008 benchmark problem. It is needless to mention that n is the dimension of the search space.

The experiment is conducted to compare the proposed method ICSO with CSO and other recently proposed state-of-the-art algorithms for large-scale optimization on 100D and 500D CEC2008 benchmark functions. The compared algorithms for the benchmark include CCPSO2 [15], MLCC [16], sep-CMA-ES [17], EPUS-PSO [18], and DMS-PSO [19]. The criteria proposed in the respective benchmarks have been accepted in this experiment too.

4 Analysis of the Results

In this section, the proposed ICSO is employed to solve the set of CEC2008 LSGO problems under the parameter setting defined in the previous section. Considering three different cases, viz. 100D and 500D, the optimum solutions obtained by ICSO are compared with its original version CSO and some other state-of-the-art algorithms participated in the CEC2008 LSGO competition. To analyze these results, different types of statistical measures are adopted in this section as follows.

- *Mean and standard deviation*: In this section, the proposed algorithm is initially engaged to evaluate the best, median, worst, mean, and standard deviation of the optimization errors of different dimensions CEC2008 benchmark functions. Then, these results along with the respective results of the CSO and state-of-the-art algorithms participated in CEC2008 competition are reported in Table 1. From these statistical values reported in Table 1, it is observed that the dominant performance of ICSO is very much noticeable on 100D functions. In general regardless of search dimension, DMS-PSO always provides best results for the functions f_1 and f_5, whereas for the function f_2 ICSO algorithm is best suitable. However, when the dimension increases to 500 the proposed algorithm ICSO is best suitable for f_6 as well. Similarly, MLCC is perfectly matchable for the functions f_4 and f_7 in most cases and sep-CMA-ES is for the function f_3. One more observation is that although these algorithms are dedicated to produce optimum solution for the above particular functions, they are also capable of producing the optimum solution for other functions in different cases. But as the dimension increases to 500, these algorithms stick to their dedicated functions. It is also observed that the algorithms CCPSO2 and EPUS-PSO are incapable to yield optimum solutions for any of the functions present in the benchmark. The most remarkable observation is the performance of ICSO. Apart from f_2 and f_6, the algorithm produces very closed result with the other functions too. To find the significant difference among these results, the following statistical tests have been performed.
- *t test*: Further, in order to calculate the t values against each function, two algorithms at a time are needed to be considered. The t values are calculated and are reported in the above-said table along with mean and standard deviation. The overall best mean and least Std. for each function in the benchmark are highlighted with boldface letters. In the above-mentioned table, the t values with a significance level $\alpha = 0.05$ are listed in each row of respective algorithms. A negative t value represents relative small optimization error for the proposed algorithm, whereas positive t value represents the relative high optimization error.
  When the ICSO algorithm performs significantly better than other algorithm, the respective t value is highlighted and is considered as a win. In case of a tie, the corresponding t value is tilde in italic. The last column of the table counts the number of wins, ties, and losses symbolically as $w/t/l$ expression. The algorithms with high win values are again highlighted with bold letters. The higher number of wins than the number of ties and losses represent very significant difference between the proposed algorithm and the competitor algorithm. From these statis-

Table 1 Comparison of all the algorithms on the CEC2008 benchmark functions of 100 and 500 dimensions

Algorithms	Tests	f_1		f_2		f_3		f_4	
		100D	500D	100D	500D	100D	500D	100D	500D
ICSO	Mean	**0.00E+00**	3.07E−26	**2.04E−01**	**2.43E+01**	9.31E+01	4.91e+02	**0.00E+00**	1.71E+02
	Std.	**0.00E+00**	2.56E−27	**4.25E−03**	**5.41E+00**	1.56E−01	2.05E+01	**0.00E+00**	1.42E+01
	t-values	–	–	–	–	–	–	–	–
CSO	Mean	2.64E−27	4.71E−23	3.35E+01	2.65E+01	3.90E+02	5.39E+02	5.60E+01	3.25E+02
	Std.	4.09E−28	3.28E−24	5.38E+00	7.55E+00	5.53E+02	5.14E+01	7.48E+00	1.36E+01
	t-values	**−3.23E+01**	**−1.97E+01**	**−3.09E+01**	**−9.33E+00**	**−2.68E+00**	**−4.34E+00**	**−1.14E+01**	**−5.50E+01**
CCPSO2	Mean	7.73E−14	7.73E−14	6.08E+00	5.79E+01	4.23E+02	7.24E+02	3.98E−02	3.98E+02
	Std.	3.23E−14	3.23E−14	7.83E+00	4.21E+01	8.65E+02	1.54E+02	1.99E−01	1.99E−01
	t-values	**−1.20E+01**	**−1.20E−01**	**−3.75E+00**	**−3.96E+00**	**−1.91E+00**	**−7.50E+00**	***−1.00E+00***	1.39E+02
MLCC	Mean	6.82e−14	4.29E−13	2.53e−01	6.66E+01	1.50e+02	9.24E+02	**4.38E−13**	**1.79E−11**
	Std.	2.32E−14	3.31E−14	8.73E+00	5.69E+00	5.72E+01	1.72E+02	**9.21E−14**	**6.31E−11**
	t-values	**−1.47E+01**	**−6.48E+01**	**−1.44E+01**	**−2.69E+01**	**−4.97E+00**	**−1.25E+01**	**−2.38E−01**	1.39E+02
Sep-CMA-ES	Mean	9.02E−15	2.25E−14	2.31E+01	2.12E+02	**4.31E+00**	**2.93E+02**	2.78E+02	2.18E+03
	Std.	5.53E−15	6.10E−15	1.39E+01	1.74E+01	**1.26E+01**	**3.59E+01**	3.43E+01	1.51E+02
	t-values	**−8.16E+00**	**−1.84E+01**	**−8.24E+00**	**−5.15E+01**	3.52E+01	2.39E+01	**−3.49E+01**	**−5.88E+01**
EPUS-PSO	Mean	7.47E−01	8.45E+01	1.86E+01	4.35E+01	4.99E+03	5.77E+04	4.71E+02	3.49E+03
	Std.	1.70E−01	6.40E+00	2.26E+00	5.51E−01	5.35E+03	8.04E+03	5.94E+01	1.12E+02
	t-values	**−2.20E+01**	**−6.60E+01**	**−4.07E+01**	**−1.77E+01**	**−4.58E+00**	**−3.56E+01**	**−3.64E+01**	**−1.37E+02**
DMS-PSO	Mean	**0.00E+00**	**0.00E+00**	3.64E+00	6.89E+01	2.83E+02	4.67E+07	1.82E+02	1.60E+03
	Std.	**0.00E+00**	**0.00E+00**	7.30E−01	2.01E+00	9.40E+02	5.87E+06	2.16E+01	1.04E+02
	t-values	*0.00E+00*	6.00E+01	**−2.35E+01**	**−3.86E+01**	*−1.01E+00*	**−3.98E+01**	**−3.31E+01**	**−5.74E+01**

(continued)

Table 1 (continued)

Algorithms	Tests	f_5 100D	f_5 500D	f_6 100D	f_6 500D	f_7 100D	f_7 500D	w/t/l 100D	w/t/l 500D
ICSO	Mean	**0.00E+00**	1.11E−16	2.48E−14	**3.55E−14**	−1.48E+03	−7.06E+03	–	–
	Std.	**0.00E+00**	0.00E+00	1.67E−15	**2.62E−15**	5.47E+00	2.26E+01		
	t-values	–	–	–	–	–	–		
CSO	Mean	**0.00E+00**	2.22E−16	1.20E−14	2.08E−13	−1.47E+03	−7.03E+03	**6/1/0**	**6/0/1**
	Std.	**0.00E+00**	0.00E+00	1.52E−15	7.08E−14	1.27E+01	4.55E+01		
	t-values	*–*	6.25E+23	−4.61E+00	−1.22E+01	−3.62E+00	−2.95E+00		
CCPSO2	Mean	3.45E−03	1.18E−03	1.44E−13	5.34E−13	−1.50E+03	−7.23E+03	**5/0/2**	**4/1/2**
	Std.	4.88E−03	4.61E−03	3.06E−14	8.61E−14	1.04E+01	4.16E+01		
	t-values	*−3.53E+00*	*−1.28E+00*	−1.94E+01	−1.06E+01	0.00E+00	1.80E+01		
MLCC	Mean	3.40E−14	2.12E−13	1.11E−13	5.34E−13	**−1.54E+03**	**−7.43E+03**	**6/0/1**	**5/0/2**
	Std.	1.16E−14	2.47E−13	7.86E−15	7.01E−13	**2.52E+00**	**8.03E+00**		
	t-values	*−1.47E+01*	*−4.29E+00*	−5.36E+01	−3.56E+00	4.98E+01	7.71E+01		
Sep-CMA-ES	Mean	2.96E−04	7.88E−04	2.12E−01	2.15E−01	−1.39E+03	−6.37E+03	**5/1/1**	**5/1/1**
	Std.	1.48E−03	2.82E−03	4.02E−01	3.10E−01	2.64E+01	7.59E+01		
	t-values	*−1.00E+00*	*−1.40E+00*	−2.64E+02	−3.47E+02	−1.67E+01	−4.36E+01		
EPUS-PSO	Mean	3.72E−01	1.64E+00	2.06E+00	6.64E+00	−8.55E+02	−3.51E+03	**7/0/0**	**7/0/0**
	Std.	5.60E−02	4.69E−02	4.40E−01	4.49E−01	1.35E+01	2.10E+01		
	t-values	*−3.32E+01*	*−1.75E+02*	−2.34E+01	−7.39E+01	−2.15E+02	−5.75E+02		
DMS-PSO	Mean	**0.00E+00**	**0.00E+00**	**0.00E+00**	2.00E+00	−1.14E+03	−4.19E+03	3/4/0	**5/0/2**
	Std.	**0.00E+00**	**0.00E+00**	**0.00E+00**	9.66E−02	8.48E+00	1.29E+01		
	t-values	*0.00E+00*	7.52E+42	*0.00E+00*	−1.04E+02	−1.68E+02	−5.51E+02		

Table 2 Average ranking and best count for the algorithm in solving CEC2008

Algorithms	100D		500D	
	Average ranking	Best	Average ranking	Best
ICSO	**1.42**	**4**	**2.00**	**2**
CSO	3.57	1	3.00	0
CCPSO2	4.00	0	3.71	0
MLCC	3.14	1	3.71	2
Sep-CMA-ES	4.14	1	5.00	1
EPUS-PSO	5.71	0	6.14	0
DMS-PSO	2.85	3	4.42	2

tical results, it is clear that the proposed algorithm ICSO has significantly better overall performance in comparison with all other algorithms on 100D and 500D CEC2008 benchmark functions.

- *Average ranking according to Friedman test*: The relative ranking of each algorithm is computed according to the mean performance of each function, and then, the average ranking is computed through all the functions in the benchmark. For example, let us consider two algorithms X and Y in a benchmark of three functions. The function values for each of the algorithm may be given by (0.00E+00, 1.53E+02, 8.24E+03) and (1.23E+01, 4.18E+02, 1.12E+01), respectively. The relative ranking would be $Rank_X = (1, 1, 2)$ and $Rank_Y = (2, 2, 1)$,, whereas their average ranking would be: $R_X = 1.33$ and $R_Y = 1.67$. The average ranking computed through all the functions is reported in Table 2. It is observed from Table 2 that ICSO attains the best ranking and supersedes others including CSO for all dimensions under consideration.
- *Best count test*: The best count of an algorithm is the number of functions for which the algorithm provides the best results as compared to the rest algorithms. For each algorithm, the best count is reported just right to the average ranking under different dimensions in Table 2. The highest count of ICSO over others indicates that it outperforms others everywhere.
- *Convergence test*: The convergence graphs of all the functions for the 100D and 500D are plotted in Fig. 2. One can conclude that for the 100D benchmark ICSO converges very fast for two separable functions f_4 and f_7. But for the 500D scenario functions f_2 and f_7 provide fast convergence. Another observation is that although for some cases the convergence speed of the proposed algorithm is not as fast as CSO at the beginning, later it implements relatively more consistently to constantly recover the solution quality. In 100D, except the functions f_1 and f_6 for the remaining cases ICSO produces improved results than the original inspired algorithm CSO. From this information, one can conclude that the leading algorithm ICSO is worthy enough to be considered as the perfect revision to the novel algorithm CSO. From this figure, it can be concluded that sooner or later ICSO converges closer toward optimal solution as compared with CSO. In few cases, where ICSO initially could not beat CSO gradually could do it later.

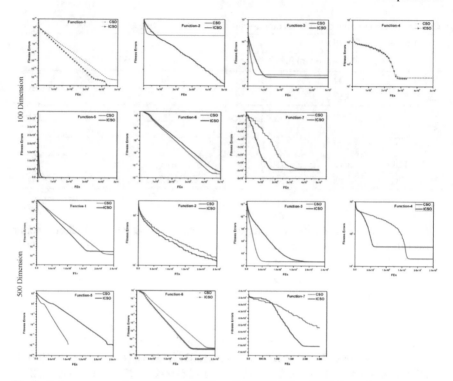

Fig. 2 Convergence graphs of CSO and ICSO on CEC2008 100 and 500 dimension

5 Conclusion

In this paper, a unique algorithm called as improved competitive swarm optimizer (ICSO) is presented. The algorithm neither practices particle best nor global best in the approach to modernize the positions of the particles. However, it adds an additional strategy to the CSO algorithm by updating the winner particles. The empirical experiments and investigations are accompanied by different dimensions of CEC2008 benchmark problems. The algorithm has revealed supremacy performance on CSO, and many meta-heuristic algorithms particularly aimed to resolve the large-scale optimization problems.

References

1. Kennedy, J., Eberhart, R.: Particle swarm optimization. In: Proceedings of the IEEE International Conference on Neural Networks, vol. 4, pp. 1942–1948. IEEE (1995)
2. Yang, Y., Pedersen, J.O.: A comparative study on feature selection in text categorization. In: Proceedings of International Conference on Machine Learning, pp. 412–420. Morgan Kaufmann Publishers (1997)

3. Kennedy, J., Eberhart, R.: Parameter selection in particle swarm optimization. In: Evolutionary Programming VII, pp. 591–600. Springer (1998)
4. Hu, M., Wu, T., Weir, J.D.: An adaptive particle swarm optimization with multiple adaptive methods. IEEE Trans. Evol. Comput. 17(5), 705–720 (2013)
5. Robinson, J., Sinton, S., Rahmat-Samii, Y.: Particle swarm, genetic algorithm, and their hybrids: optimization of a profiled corrugated horn antenna. In: Proceedings of IEEE Antennas and Propagation Society International Symposium, pp. 314–317. IEEE (2002)
6. Shelokar, P., Siarry, P., Jayaraman, V.K., Kulkarni, B.D.: Particle swarm and ant colony algorithms hybridized for improved continuous optimization. Appl. Math. Comput. 188(1), 129–142 (2007)
7. Kennedy, J., Mendes, R.: Population structure and particle swarm performance. In: Proceedings of IEEE Congress on Evolutionary Computation, pp. 1671–1676. IEEE (2002)
8. Liang, J.J., Qin, A., Suganthan, P.N., Baskar, S.: Comprehensive learning particle swarm optimizer for global optimization of multimodal functions. IEEE Trans. Evol. Comput. 10(3), 281–295 (2006)
9. Liang, J., Suganthan, P.: Dynamic multi-swarm particle swarm optimizer. In: Proceedings of IEEE Swarm Intelligence Symposium, pp. 124–129. IEEE (2005)
10. Kennedy, J.: Bare bones particle swarms. In: Proceedings of IEEE Swarm Intelligence Symposium, pp. 80–87. IEEE (2003)
11. Goh, C., Tan, K., Liu, D., Chiam, S.: A competitive and cooperative co-evolutionary approach to multi-objective particle swarm optimization algorithm design. Eur. J. Oper. Res. 202(1), 42–54 (2010)
12. Whitehead, B., Choate, T.: Cooperative-competitive genetic evolution of radial basis function centers and widths for time series prediction. IEEE Trans. Neural Netw. 7(4), 869–880 (1996)
13. Cheng, R., Sun, C., Jin, Y.: A multi-swarm evolutionary framework based on a feedback mechanism. In: Proceedings of IEEE Congress on Evolutionary Computation, pp. 718–724. IEEE (2013)
14. Ran, C., Yaochu, J.: A competitive swarm optimizer for large scale optimization. IEEE Trans. Cybern. 45(2), 191–204 (2015)
15. Li, X., Yao, Y.: Cooperatively coevolving particle swarms for large scale optimization. IEEE Trans. Evol. Comput. 16(2), 1–15 (2011)
16. Yang, Z., Tang, K., Yao, X.: Multilevel cooperative coevolution for large scale optimization. In: Proceedings of IEEE Congress on Evolutionary Computation, pp. 1663–1670. IEEE (2008)
17. Ros, R., Hansen, N.: A simple modification in cma-es achieving linear time and space complexity. In: Parallel Problem Solving from Nature-PPSN X, pp. 296–305 (2008)
18. Hsieh, S.-T., Sun, T.-Y., Liu, C.-C., Tsai, S.-J.: Solving large scale global optimization using improved particle swarm optimizer. In: Proceedings of IEEE Congress on Evolutionary Computation, pp. 1777–1784. IEEE (2008)
19. Zhao, S.-Z., Liang, J.J.: Dynamic multi-swarm particle swarm optimizer with local search for large scale global optimization. In: Proceedings of IEEE Congress on Evolutionary Computation, pp. 3845–3852. IEEE (2008)
20. Mohapatra, P., Das, K.N., Roy, S.: A modified competitive swarm optimizer for large scale optimization problems. Appl. Soft Comput. 59, 340–362 (2017)

Solution of Constrained Optimal Active Power Dispatch Problems Using Exchange Market Algorithm

Abhishek Rajan, T. Malakar and Abhimanyu

Abstract OAPD is basically a Generation Scheduling (GS) problem which is commonly formulated as an Optimal Power Flow (OPF) problem. OPF is a power system optimization tool which aims to optimize certain objective and provide the optimal operating state of power system simultaneously satisfying both physical and operational constraints of power system. The basic aim of OAPD problem is to determine the optimal GS for the committed generators in such a manner that the total fuel cost is optimized. The presence of nonlinear constraints like Valve-Point Loading (VPL), Prohibited Operating Zone (POZ), and Ramp Rate Limits (RRLs) makes the objective function nonlinear, non-convex, and sometimes discontinuous. This paper attempts to investigate the newly developed meta-heuristic algorithm called Exchange Market Algorithm (EMA) in solving highly nonlinear non-convex Optimal Active Power Dispatch (OAPD) problems of power system with VPL, POZ, and RRLs effect. Both continuous and discrete control variables are present in the problem which makes the optimization more complex. The problem is implemented on the standard IEEE-30 bus system. The results are compared with several other meta-heuristic algorithms, and it is found that EMA outperforms many contemporary algorithms in terms of the convergence rate and objective function value.

Keywords Optimal Power Flow · Optimal Active Power Dispatch · POZ Ramp Rate Limits · Exchange Market Algorithm

1 Introduction

Power system economy is one of the crucial issues for both power system practitioners and researchers. The electric power system operation has to be economical to ensure a cost-efficient electric energy supply to the consumer's terminal. The load demand at

A. Rajan (✉) · T. Malakar · Abhimanyu
National Institute of Technology, Silchar, Silchar 788010, Assam, India
e-mail: abhishekrajan099@gmail.com

T. Malakar
e-mail: m_tanmoy1@rediffmail.com

© Springer Nature Singapore Pte Ltd. 2019
J. C. Bansal et al. (eds.), *Soft Computing for Problem Solving*, Advances in Intelligent Systems and Computing 817, https://doi.org/10.1007/978-981-13-1595-4_48

consumer's terminal is continuously varying, and it is required to distribute this load in real time among already running generating units so as to meet the load demand at every interval of time satisfactorily. Active power dispatch problem of power system is basically an optimization problem [1] where scheduling of generators' real power output is performed in a most economic manner, satisfying all physical and operational constraints related to generation and transmission of real power. In such type of optimization problem, the basic aim of power system planner is to gain power system operational economics. So as far as power system economics is concerned, the system operation economics deals with minimum cost of power production. In this case, the problem is called Optimal Active Power Dispatch (OAPD). In realizing any OAPD problems, the optimization is performed by adjusting certain problem variables. These power system variables are usually termed as control variables. These variables are guessed initially, and having initial assumptions, it checks that the given aim is satisfied or not. If it is not, then the value of control variables is adjusted following some optimization techniques and the process is repeated until the objective is satisfied. OAPD problem for a particular power system can be solved as an Optimal Power Flow (OPF) problem. The OPF was first proposed and defined by Dommel and Tinney [2] and developed by Carpentier [3]. Since then, OPF has become most important tool for analyzing power system operation and had been in use for over last few decades. OPF is a power system optimization tool which aims to optimize a given objective function and provide optimal operating state of power system, simultaneously satisfying all physical and operational constraints of the power system [4]. It is a particular case where the power flow in an electrical system occurs optimally [5]. In general, the objectives of OPF problems are nonlinear non-convex and sometimes discontinuous too. Hence, when viewed as an optimization problem, OPF is highly nonlinear non-convex and complex optimization problem.

The most commonly used objective function for OAPD problem is the minimization of generation cost for thermal unit [4]. Some classical optimization techniques [4, 6, 7] were successfully implemented to solve OPF problems. In [4], Lee et al. proposed unified method for real and reactive dispatch for economic operation of power system. Gradient projection method is used as the optimization technique. In [6] of Zehar and Sayah, a multi-objective environmental/economic load dispatch problem, based on an efficient successive linear programming technique, is solved. The problem is solved on Algerian 59-bus power system. The OPF-based real power dispatch problem using linear programming (LP) technique is modeled and discussed in [7]. Though these algorithms have fast convergence speed, their differential calculus-based approach restricts them to solve the non-convex and discontinuous objective functions. Moreover, they have higher tendency to trap into local optimal if the function is multimodal in nature. Hence, as an alternative, since last few decades, researchers and practitioners have begun to show their interest in population-based algorithms instead. In order to overcome the drawbacks of gradient-based optimization techniques, solution based on the behavior of natural evolution and natural objects has been developed and applied to solve many engineering as well as power system problems. In terms of solution methods, these algorithms are termed as meta-heuristic algorithms. Some examples of these types are Genetic Algorithm (GA),

Particle Swarm Optimization (PSO), Ant Colony Optimization (ACO), Differential Evolution (DE), etc. These algorithms are generally population-based and work meticulously toward finding the optimal solution for both constrained and unconstrained optimization problems.

Genetic Algorithm [8], Evolutionary Programming [9], Tabu Search [10], Particle Swarm Optimization [11], Differential Evolution [12], Biogeography-Based Optimization [13], Harmony Search Algorithm [14], Gravitational Search Algorithm [15], Black-Hole-Based Algorithm [16], Teaching–Learning-Based Algorithm [17], etc., have shown promising results when applied to solve power system problems. In [18, 19], the solution methodologies have been improved by researchers to eliminate the drawback associated with the above optimization techniques by either improving its evolution process or by hybridizing it with suitable classical optimization techniques. Multi-objective optimization is also reported in [20], where two or more than two optimization objectives are solved at a time to check the efficiency and capability of the algorithm in finding the global optimal solution. In the above-mentioned works, authors have tried to implement several nonlinearities like VLP, POZ, and RRLs together with the optimization of simple fuel cost.

In this work, relatively new and promising algorithm called Exchange Market Algorithm (EMA) is implemented to solve the OAPD problem with several nonlinearities like VLP, POZ, and RRLs. This algorithm is developed by Ghorbani and Babaei in 2014 which is based on the behavior of shareholders in stock market [21]. In order to prove the efficiency and capability of EMA, many benchmark problems have been solved by authors and results looks promising when compared with other reported literatures. The unique feature of double exploitation and exploration attracts the present authors to use this algorithm in solving complex problems of power system. Till date, EMA has not been applied to solve many power system problems. Therefore, in this paper, authors intend to solve OAPD problems using EMA. The problem is formulated as nonlinear optimization with various objectives associated to OAPD. These problems are implemented on the standard IEEE test systems. The results are compared with other well-established contemporary algorithms.

2 Problem Formulation

OAPD problems are mathematically modeled as Optimal Power Flow (OPF) problems. OPF is a power system optimization tool which aims to optimize a given objective function and provide an optimal operating state through the proper adjustments of various power system controllers while simultaneously satisfying the equality and inequality constraints present in the system. It is expressed as [1]:

$$\text{Minimize} \quad f(x, u) \tag{1}$$

$$\text{Subjected to} \quad \begin{cases} g(x, u) = 0 \\ h_{min} \leq h(x, u) \leq h_{max} \end{cases} \tag{2}$$

where f, x, u, g(x, u), and h(x, u) are the objective function, set of dependent variables, set of independent variables, sets of equality, and inequality constraints, respectively. Slack generators' real (P_{G1}) and reactive power outputs (Q_{G1}), load bus voltage magnitudes ($V_{L1}, \ldots V_{L_{N_{PQ}}}$), reactive power generations ($Q_{G1}, \ldots Q_{G_{N_{PV}}}$), and line loadings $\left(S_{L1}, \ldots S_{L_{N_{TL}}}\right)$ are considered as dependent variables in power system. Hence, the vector of dependent variables 'x' can be expressed as:

$$x^T = \left[P_{G1}, V_{L1}, \ldots V_{L_{N_{PQ}}}, Q_{G1}, \ldots Q_{G_{N_{PV}}}, S_{L1}, \ldots S_{L_{N_{TL}}}\right] \quad (3)$$

The vector of independent/control variables 'u' comprise of all real power generations ($P_{G2}, \ldots P_{G_{N_{PV}}}$) and their generation voltages ($V_{G1}, \ldots V_{G_{N_{PV}}}$), tap-changing transformer's positions ($Tap_1, \ldots Tap_{NT}$), capacitors VAr output ($Q_{C1}, \ldots Q_{C_{NC}}$). Similarly, the vector u is represented mathematically as

$$u^T = [\overbrace{P_{G2}, \ldots P_{G_{N_{PV}}}, V_{G1}, \ldots V_{G_{N_{PV}}}}^{continuous} \overbrace{Tap_1, \ldots Tap_{NT}, Q_{C1}, \ldots Q_{C_{NC}}}^{discrete}] \quad (4)$$

where NT and NC represent number of tap changers and switchable capacitors, respectively.

3 Objective Functions

In this work, the major objective is to find the optimal scheduling of thermal generators to meet the load demand economically by simultaneously maintaining the various physical and operational constraints present in the power system. The nonlinearities present in power generating units such as Valve-Point Loading (VPL), Prohibited Operating Zone (POZ), and Ramp Limits (RLs) are also considered. Inclusion of these nonlinearities makes the cost function non-convex and discontinuous, and hence, the optimization problem becomes a complex one.

(a) *Minimization of fuel cost with Valve-Point Loading effect*

Simple fuel cost expression is an approximated cost expression. In real-time practice, cost expression is not so simple; rather they are complex and nonlinear in nature. Practical cost functions are generally non-convex and contain multiple ripples. This is because, practically, valve is used to control the steam flow to the turbine with the help of nozzle. Nozzles generally achieve high efficiency at full output. To achieve maximum efficiency, sequence operation of nozzle group is required. This results in a rippled efficiency curve which makes the curve non-convex in nature. This effect is called Valve-Point Loading (VPL) effect [18]. Mathematically, the cost function with VPL is expressed as [18]

Fig. 1 Graphical representation of Valve-Point Loading (VPL) effect

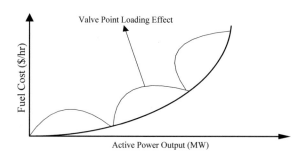

$$F_{cost}^{valvepoint} = \sum_{i=1}^{i=N_g} \left[c_i + b_i P_{Gi} + a_i P_{Gi}^2 + |e_i \sin(f_i(P_{Gi}^{min} - P_{Gi}))| \right] \$/hr. \quad (5)$$

e_i and f_i are the coefficient related to Valve-Point Loading. The numerical values of these coefficients are given in the corresponding result section. From Fig. 1, it can be seen that the simple fuel cost curve which is convex in nature is changed to a non-convex function with multiple ripples.

4 Constraints

There are two types of constraints present in the power system operation, i.e., equality constraints and inequality constraints.

Equality constraints
The mathematical expression of active and reactive power balance equation at each node of the power system network is given below [15]:

$$P_i - P_{Di} - |V_i| \sum_{j=1}^{NB} |V_j| \{ G_{ij} cos(\theta_i - \theta_j) + B_{ij} sin(\theta_i - \theta_j) \} = 0 \quad (6)$$

$$Q_i - Q_{Di} - |V_i| \sum_{j=1}^{NB} |V_j| \{ G_{ij} sin(\theta_i - \theta_j) - B_{ij} cos(\theta_i - \theta_j) \} = 0 \quad (7)$$

where P_i and Q_i are the real power injections at ith node of the network, and P_{Di} and Q_{Di} are the active and reactive load associated with the ith node. NB is the total number of busses. V_i and V_j are the voltage of the ith and jth bus, and θ_i and θ_j are the corresponding angles. G_{ij} and B_{ij} are the conductance and susceptance of the transmission line connected between ith and jth bus.

Inequality constraints

In power system, generally two types of constraints are there: (i) inequality constraints on independent variable side and (ii) inequality constraints on dependent variable side.

(1) Inequality constraints on independent variable side

$$P_{Gi}^{min} \leq P_{Gi} \leq P_{Gi}^{max} \quad i \in NG \tag{8}$$

$$V_{Gi}^{min} \leq V_{Gi} \leq V_{Gi}^{max} \quad i \in NPV \tag{9}$$

$$Tap_i^{min} \leq Tap_i \leq Tap_i^{max} \quad i \in NT \tag{10}$$

$$SC_i^{min} \leq SC_i \leq SC_i^{max} \quad i \in NC \tag{11}$$

POZ constraints

Due to the several steam valve operation and vibration in a shaft bearing of thermal generators, some physical limitations are imposed by the manufacturers. These limitations may result in the non-operation of thermal units within certain range of the power output. These restricted zones are called Prohibited Operating Zone (POZ). The presence of POZs makes the cost function discontinuous and it becomes difficult to determine the exact shape of the cost curve. By using Eq. (8), the feasible operating zones (FOZs) of the ith thermal generating unit are given by

$$P_{Gi}^{min} \leq P_{Gi} \leq P_{Gi,1}^{l} \tag{12}$$

$$P_{Gi,K-1}^{u} \leq P_{Gi} \leq P_{Gi,K}^{l} K = 2, 3 \dots N_{i,PZ} \tag{13}$$

$$P_{Gi,N_{i,PZ}}^{u} \leq P_{Gi} \leq P_{Gi}^{max} \tag{14}$$

where $P_{Gi,K}^{l}$ and $P_{Gi,K}^{u}$ are the lower and upper bonds of the kth POZs of ith unit. $N_{i,PZ}$ is the total number of POZs of ith generating unit.

RRL constraints

The physical limits of thermal generating units restrict the operating range of all units by their Ramp Rate limits (RRLs) [22]. After incorporating the RRLs, Eq. (8) becomes as follows

$$max\{P_{Gi}^{min}, (P_{Gi}^{o} - DR_i)\} \leq P_{Gi} \leq min\{P_{Gi}^{max}, (P_{Gi}^{o} + UR_i)\} \tag{15}$$

where P_{Gi}^{o} is the power output at previous time interval and P_{Gi} is the power output at current time interval. UR_i and DR_i are the up ramp limit and down ramp limit of ith generating unit in (MW/hr), respectively.

(2) Inequality constraints on dependent variable side.

$$P_{G1}^{min} \leq P_{G1} \leq P_{G1}^{max} \tag{16}$$

$$Q_{Gi}^{min} \leq Q_{Gi} \leq Q_{Gi}^{max} \qquad i \in NG \tag{17}$$

$$V_{Li}^{min} \leq |V_{Li}| \leq V_{Li}^{max} \qquad i \in NPQ \tag{18}$$

$$S_{Li} \leq S_{Li}^{max} \tag{19}$$

5 Exchange Market Algorithm

Exchange Market Algorithm is designed by Ghorbani and Babaei after carefully observing the behavior of shareholders of exchange market under different market conditions. The algorithm is designed on the basis of maximizing the profit in the exchange market. Hence, the algorithm is initially developed for maximizing the objective function. However, it can also be used to solve the minimization problem [21]. As a general fact, in the exchange market, shareholders trade different shares in the virtual stock market, under diverse market scenarios. Political and economic policies of country sometimes drag the stock market from non-oscillated to oscillated market condition. If the market is balanced (non-oscillated mode), it is easier to predict the market condition and shareholder can increase their shares as well as profit without taking any unconventional risk. On the contrary, when the market condition is unbalanced (oscillated mode), situation becomes adverse and it involves certain risk in selecting shares for trading. It becomes difficult to predict the behavior of market, and thus, the action of the shareholders can be profitable or disadvantageous. In this algorithm, each shareholder is considered as the potential solution to the problem. Shareholders who are active and experienced are elite stock dealers. Profit of each shareholder is calculated and termed as fitness of the objective function. Based on the fitness values of each individual in both balanced and unbalanced market modes, sorting of the population is done. The individuals with highest, average, and low fitness as first, second, and third groups, respectively. Since first group members are highly experienced and can earn more profit at any market conditions, they remain unaffected in all stages of algorithm.

The steps of EMA in solving any optimization problem are given below:

Step 1: Initialization of the shareholders and their shares

In this step, share quantity (dimension of the problem), initial shareholders desired iterations, and the values of shares (control variable values) $x_{ij}, \{i = 1, 2 \dots m; j = 1, 2 \dots n\}$ where m is the total dimension of the control variables and n is the population size) are initialized. The following formula is used for initialization.

$$x_i = x_i^L + rand \times \left(x_i^U - x_i^L\right) \tag{20}$$

Step 2: Computation of fitness and sorting

The total population is divided into high-, middle-, and low-ranked shareholders. In this step, the fuel costs are calculated and classified into three different groups depending on the effectiveness of their total shares.

Step 3: Updating of the shares of the second group in balanced market condition

The changes in the second group members are carried out in the following manner:

$$pop_j^{group(2)} = r \times pop_{1,i}^{group(1)} + (1-r) \times pop_{2,i}^{group(1)} \tag{21}$$

$i = 1, 2, 3 \ldots n_i$ and $j = 1, 2, 3 \ldots n_j$

Step 4: Updating the shares of the third group in balanced market condition

$$S_k = 2 \times r_1 \times \left(pop_{i,1}^{group(1)} - pop_k^{group(3)}\right) + 2 \times r_2 \times \left(pop_{i,2}^{group(1)} - pop_k^{group(3)}\right) \tag{22}$$

$$pop_k^{group(3),new} = pop_k^{group(3)} + 0.8 \times S_k \tag{23}$$

where S_k is the variation in the share of the kth shareholder of the third group.

Step 5: Computation of shareholders cost (fitness) and ranking

Based on the fitness, the shareholders are sorted and divided into three groups.

Step 6: Adjustment of shares of second group members under market unbalanced market condition

In this step, mean members of shareholders vary some of their shares according to the following equations.

$$\Delta n_{t1} = n_{t1} - \delta + (2 \times r \times \mu \times \eta_1) \tag{24}$$

The detailed process of calculation of these parameters and the meaning associated with it are described in [21].

Step 7: Change of shares of third group members under unbalanced condition

In this step, contrary to the previous step, shareholders exchange some shares according to Eq. (24), irrespective of their total share amount.

$$\Delta n_{t3} = (4 \times r_s \times \mu \times \eta_2) \tag{25}$$

The calculation methods and meanings associated with the parameters are given in [21].

Step 8: End-up criteria

Maximum number of iterations is considered as the terminating criteria in this paper.

Table 1 POZs data for case-1

Prohibited zones
Pg_1 [55–66], [200–230]
Pg_2 [24–30], [45–55]
Pg_5 [10–18]
Pg_8 [10–15]
Pg_{11} [10–15]
Pg_{13} [11–18]

6 Results and Analysis

In this section, EMA is explored in solving OAPD problems of power system. The simulation is performed on IEEE-30 bus system [23]. Both continuous and discrete variables are considered as the control variables. Nonlinearities like VPL, POZ, and RRL are also considered to verify the efficiency of EMA in solving complex optimization problem. A population size of 50 is taken for all case studies. First 20% population is chosen as first, next 60% as second, and rest 20% as third group members. Constraints are handled by well-known penalty function method. Maximum cycle and trial runs are taken as 200 and 100, respectively. The results are compared by implanting the problem on some well-established meta-heuristic algorithms like FA, GSA, ABC, CSA. The results are compared with that of EMA both numerically and graphically. The control variable range and cost coefficients data are taken from [18].

Result Analysis of Case-1 (Fuel cost with VPL and POZ)

In this case, altogether, 25 control variables are used to optimize the fuel cost. The control variable includes active power output of generators, its voltages, transformers tap positions, and shunt capacitors. In addition to VPL, POZs are used with fuel cost in this case which makes the problem more complicated. Along with EMA, problem is also simulated on some promising meta-heuristic algorithms like ABC [24], FA [19], CSA [25], and GSA [15], and results so obtained are compared with that of EMA. The POZs data used in this case are given in Table 1. The combined convergence plots of EMA and other algorithms for the best solution obtained after 100 trials are represented in Fig. 2, whereas the best control settings obtained from these algorithms are compared with EMA in Table 2.

From Fig. 2, it can be seen that the convergence of EMA is very fast in comparison to other algorithms, and it also took very less iterations to converge. Thought convergence plot reflects that the GSA converges little earlier than EMA, but the optimal results obtained from EMA **(830.9393 \$/h)** are lesser than that of GSA (831.44075 \$/h). From Table 2, it can be seen that the optimal control settings of each variable obtained from different algorithms including EMA are within their given range and they also followed the POZs restrictions, which signifies the successful implementation all the above-mentioned algorithms. Both numerical and graphical presentations

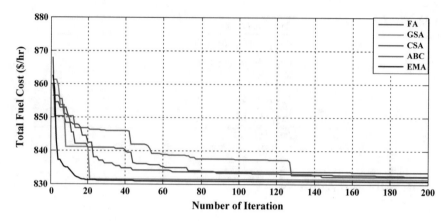

Fig. 2 Convergence plot for case-1

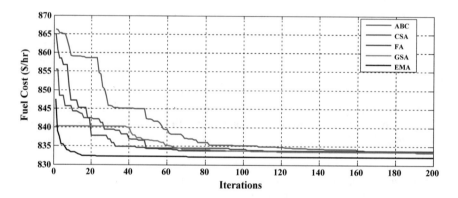

Fig. 3 Convergence plot for case-2

reveal that the EMA outperforms several other promising algorithms in solving such a complex nonlinear and discontinuous objective function.

Results for Case-2 (Fuel cost with VLP, POZ, and RRLs)

As mentioned in Sect. 3, the generators cannot increase or decrease its generation suddenly to any value when the system experiences a load change. Their increment and decrement in generations depend on the up ramping and down ramping limits, respectively. In this case study, same IEEE-30 bus system is taken as previous cases, and the ranges of control variables, POZ data, and cost coefficients are taken from [18]. The UR limit, DR limits, and initial generations (P_{Gi}^o) are given in Table 3 [22].

In this case, the problem is simulated with EMA and the results are compared graphically and numerically with several contemporary algorithms like ABC, FA, CSA, and GSA. The combined convergence plot of all the algorithms mentioned above along with that of EMA is presented in Fig. 3. The optimal control settings found for each method after 100 trials are given in Table 4. From Fig. 2, it can be seen that EMA converges faster than its contemporary algorithms. EMA converges at

Table 2 Comparison of simulation results for case-4 (b)

Variables	ABC	CSA	FA	GSA	EMA
Pg_1	199.98142	199.94782	200.00051	199.741	199.7859
Pg_2	45	43.248285	39.397676	45	45
Pg_5	18.106504	19.844703	18.566638	19.652582	18
Pg_8	10	10	15	10	10
Pg_{11}	10	10	10	10	10
Pg_{13}	11	11	11	11	11
Vg_1	1.09173	1.0885275	1.0451061	0.9549027	1.1
Vg_2	1.07171	1.0656825	1.0120227	1.0143829	1.0516363
Vg_5	1.0380489	1.0415522	0.9838249	1.0215506	1.0542925
Vg_8	1.0333034	1.039999	0.9927971	0.963568	1.0795661
Vg_{11}	0.9842589	0.9630442	1.1876888	0.4184401	1.1
Vg_{13}	1.0473202	1.0330667	3.0346211	1.9638265	1.1
T_{6-9}	1.0258657	1.1	1.7580171	1.6781856	0.9642508
T_{6-10}	0.9610222	0.9	1.7848064	4.5385482	0.9471938
T_{4-12}	0.9786692	0.9816544	2.8781137	2.479595	0.9706705
T_{28-27}	0.9853553	0.9904716	3.4673909	3.6978057	0.9208261
Qc_{10}	1.1379254	2.7591335	2.5386399	4.1823588	4.9976107
Qc_{12}	4.9995384	3.9574064	2.9179724	1.4860359	0.5385628
Qc_{15}	5	2.9890051	2.918599	2.4881503	4.9813202
Qc_{17}	5	5	1.0890878	1.0606953	0.0025519
Qc_{20}	4.7864185	5	1.0674744	1.0345174	4.6090319
Qc_{21}	5	3.6265382	1.0276487	0.9697528	5
Qc_{23}	4.9602783	1.357168	1.0396423	0.9677014	2.6363138
Qc_{24}	3.8667484	5	1.0333033	1.009063	4.919458
Qc_{29}	2.396309	3.8220988	0.9995657	1.0468263	0.4915342
Cost ($/h)	832.10769	832.30725	833.42228	831.44075	**830.9393**
Loss (MW)	10.68792	10.640804	10.564819	11.993581	**10.38593**

Table 3 Ramp Rate Limits of IEEE-30 bus system [22]

Units	P^o_{Gi} (MW/h)	UR_i (MW/h)	DR_i (MW/h)
1	150	60	80
2	35	28	10
3	39	10	20
4	20	10	05
5	18	10	05
6	20	15	06

Table 4 Optimal control settings of different methods for case-2

Control settings	ABC	CSA	FA	GSA	EMA
Pg_1	207.92777	207.92747	208.06596	207.96852	207.5132
Pg_2	25	25	25	25	25
Pg_5	19	19.004729	19	19	19
Pg_8	15	15.00854	15	15	15
Pg_{11}	13.009165	13.004599	13	13	13
Pg_{13}	14	14	14	14	14
Vg_1	1.0946403	1.0904963	1.0852024	1.0917504	1.1
Vg_2	1.0688301	1.065443	1.0605096	1.0683272	1.08
Vg_5	1.0326919	1.0332523	1.0262487	1.036175	1.058
Vg_8	1.0425804	1.0361277	1.0346637	1.0403727	1.066
Vg_{11}	1.0061195	0.9696367	1.006751	1.0539735	1.1
Vg_{13}	1.0204623	1.0496216	1.0422215	1.0316439	1.1
T_{6-9}	0.9795929	1.0521864	1.0179205	1.0807292	1.047
T_{6-10}	1.0634502	0.9130505	1.007955	0.9	0.9016
T_{4-12}	0.9597269	0.9727112	1.0150316	0.9477566	0.9989
T_{28-27}	0.9916507	0.9786864	1.0109243	0.9745052	0.9781
Qc_{10}	1.9739053	2.2729398	2.6973317	0.0009942	1.2
Qc_{12}	3.7964077	0.1875152	2.4849725	3.8797133	5
Qc_{15}	5	2.7080108	2.0377753	4.9013219	5
Qc_{17}	5	5	3.2786065	0.5169132	5
Qc_{20}	4.0458936	4.4030709	3.4339994	0.5886965	5
Qc_{21}	3.9367355	4.4293553	1.3465702	2.6584298	5
Qc_{23}	5	4.8790692	1.8397726	0.4824813	5
Qc_{24}	4.2248185	3.9794151	1.3397933	2.0041334	5
Qc_{29}	2.1774603	2.6063759	3.1384295	2.0907184	2.369
Cost ($/h)	833.51004	833.42519	833.71745	833.42938	**832.0843**
Loss (MW)	10.536939	10.54534	10.665958	10.568525	**10.11**

the lower value while other algorithms prematurely converged at a value higher than that of EMA. The optimal fuel cost obtained from EMA is **832.0843 $/h** where the results obtained from CSA which is closer to that of EMA are found to be 833.4252 $/h. Other algorithms such as ABC, FA, and GSA settles down at 833.51, 833.71, and 833.4294 $/h, respectively. These can be also verified from the optimal control settings given in Table 4. From the graphical as well as numerical results, it can be concluded that EMA performs better than other well-established algorithms in solving such a complex nonlinear discontinuous objective functions.

7 Conclusion

In this work, a newly developed meta-heuristic algorithm called Exchange Market Algorithm is applied to solve the complex, nonlinear, non-convex OAPD problems of power system. The problem is implemented on IEEE-30 bus system. The basic objective is to optimize the fuel cost of thermal generating units by simultaneously satisfying all the physical and operational constraints of the power system. The various nonlinear constraints VLP, POZ, and RRLs are incorporated in order to test the applicability and efficiency of EMA in solving the non-convex and discontinuous objectives. The optimal control settings in both the cases suggest that EMA is successfully implemented to solve such a complex optimization problem. The comparison results with various other methods confirm that EMA has faster convergence and provides better near-optimal solution over other methods. Hence, EMA can be treated as one of the efficient members of evolutionary algorithms and can further be used to solve the other complex power system optimization problems.

References

1. Wood, J., Wollenberg, B.F.: Power Generation Operation and Control, 2nd edn. Wiley, Inc
2. Dommel, H.W., Tinney, W.F.: Optimal power flow solutions. IEEE Trans. Power Appar. Syst. **87**(10), 1866–1876 (1968)
3. Carpentier, J.: Optimal power flows. Electric. Power Energy Syst. **1**(1), 3–15 (1979)
4. Lee, K.Y., Park, Y.M., Ortiz, J.L.: A united approach to optimal real and reactive power dispatch. IEEE Trans. Power Appar. Syst. **104**(5), 1147–1153 (1985)
5. Grainger, J.J., Stevention Jr. W.D.: Power System Analysis, Tata McGraw-Hill Edition (2003)
6. Zehar, K., Sayah, S.: Optimal power flow with environmental constraint using a fast successive linear programming algorithm-application to the Algerian power system. Energy Conv. Manage. **49**, 3361–3365
7. Alsac, O., Bright, J., Prais, M., Stott, B.: Further development in LP-based optimal power flow. IEEE Trans. Power Syst. **5**(3) (1990)
8. Iba, K.: Reactive power optimization by genetic algorithm. IEEE Trans. Power Syst. **9**(2), 685–692 (1994)
9. Yuryevich, J., Wong, K.P.: Evolutionary programming based optimal power flow algorithm. IEEE Trans. Power Syst. **14**(4), 1245–1250 (1999)
10. Abido, M.A.: Optimal power flow using tabu search algorithm. Electr. Power Compon. Syst. **30**(5), 469–483 (2002)
11. Abido, M.A.: Optimal power flow using particle swarm optimization. Electric Power Energy Syst. **24**, 563–571 (2014)
12. Abou ElEla, A.A., Abido, M.A.: Optimal power flow using differential evolution algorithm. Electr. Power Syst. Res. **80**(7), 878–885 (2010)
13. Bhattacharya, A., Chattopadhyay, P.K.: Application of biogeography-based optimisation to solve different optimal power flow problems. IET Gener. Trans. Distrib. **5**(1), 70–80 (2011)
14. Sivasubramani, S., Swarup, K.S.: Multi-objective harmony search algorithm for optimal power flow problem. Electric. Power Energy Syst. **33**, 745–752 (2011)
15. Duman, S., Sonmez, Y., Guvenc, U., Yorukeren, N.: Optimal reactive power dispatch using a gravitational search algorithm. IET Gener. Transm. Distrib. **6**(6), 563–76 (2012)
16. Bouchekara, H.R.E.H.: Optimal power flow using black-hole-based optimization approach. Appl. Soft Comput. (2014)

17. Ghasemi, M., Ghavidel, S., Rahmani, S., Roosta, A., Falah, H.: A novel hybrid algorithm of imperialist competitive algorithm and teaching learning algorithm for optimal power flow problem with non smooth cost functions. Eng. Appl. Artif. Intell. **29**, 54–69 (2014)
18. Niknam, T., Narimani, M.R., Azizipanah-Abarghooee, R.: A new hybrid algorithm for optimal power flow considering prohibited zones and valve point effect. Energy Conv. Manage. **58**, 197–206 (2012)
19. Rajan, A., Malakar, T.: Optimal active power dispatch using hybrid firefly algorithm. In: 2014 Annual IEEE India Conference (INDICON)
20. Basu, M.: Economic environmental dispatch using multio-objective differential evaluation. Appl. Soft Comput. **11**(2) (2011)
21. Ghorbani, N., Babaei, E.: Exchange market algorithm. Appl. Soft Comput. **19**, 177–187 (2014)
22. Arul, R., Ravi, G., Velusami, S.: Non-convex economic dispatch with heuristic load patterns, valve point loading effect, prohibited operating zones, ramp-rate limits and spinning reserve constraints using harmony search algorithm. Electr. Eng. **95**, 53–61 (2013)
23. The IEEE 30-Bus Test System. http://www.ee.washington.edu/research/pstca/pf30/pg_tca30b us.htm
24. Karaboga, D., Basturk, B.: On the performance of artificial bee colony (ABC) algorithm. Appl. Soft Comput. **8**, 687–697 (2008)
25. Xin-She, Y., Deb, S.: Cuckoo search via lévy flights. In: World Congress on Nature & Biologically Inspired Computing (NaBIC 2009), Dec 2009, India, pp. 210–214. IEEE Publications, USA (2009)

Demand Side Management of a Commercial Customer Based on ABC Algorithm

T. Malakar, S. K. Goswami and Abhishek Rajan

Abstract The annual consumption of electrical energy has been increasing through-out the world. Most recently, the optimal utilization of electrical energy has gained more importance. Traditionally, the load has been considered as passive component of the grid but, with the rapid changes in power system operation, the demand side management technologies (DSM) now play an important role to improve the energy efficiency of power grid. DSM technique helps the utility to reshape the electric utility load curve and to reduce the peak demand. This article analyses the demand side management strategy of a commercial building situated in a city of South Assam, India. The purpose of the study is to minimize the electrical energy consumption cost of the building for a given period of time. Physical observation reveals that the energy consumption cannot be reduced. Therefore, cost saving can only be achieved by shifting some loads at cheaper billing periods and by optimizing the operation of self-generation systems. This paper focuses on the modelling aspects of the customers' load and self-generation system for its load management purpose. Further, this article discusses the computational aspects of the optimization model. The load activity strategy is formulated based on the consumption forecast scenarios with significant uncertainty. The customer comfort index during business hours is considered in this work as practical constraints. The algorithm is formulated as mixed integer problem, and the demand side management is obtained under local Time Of Use (TOU) tariff structure. The DSM problem has been implemented using MATLAB and solved using Artificial Bee Colony (ABC) algorithm. The results demonstrate the effectiveness of the load management strategies in reducing the energy consumption cost.

T. Malakar (✉) · A. Rajan
Electrical Engineering Department, National Institute of Technology, Silchar, Silchar 788010, Assam, India
e-mail: m_tanmoy1@rediffmail.com

A. Rajan
e-mail: abhishekrajan099@gmail.com

S. K. Goswami
Electrical Engineering Department, Jadavpur University, Kolkata 700 032, India
e-mail: skgoswami_ju@yahoo.co.in

© Springer Nature Singapore Pte Ltd. 2019
J. C. Bansal et al. (eds.), *Soft Computing for Problem Solving*, Advances in Intelligent Systems and Computing 817, https://doi.org/10.1007/978-981-13-1595-4_49

617

Keywords Demand side management · Time Of Use (TOU) · Load scheduling

1 Introduction

Due to the continuous increase in electrical energy consumption throughout the world, the energy conversion, management and utilization have attained more attention in recent times. In today's power system operation, a portion of load growth has been accommodated locally by using sustainable energy sources. The aim is to restrict additional CO_2 emissions from the Central Generators (CGs) to cope up the same load growth. The increasing penetration of Renewable Energy Sources (RESs) in power system operation has been causing uncertainties in power supply, changing the conventional power flow scenario. Moreover, power system deregulation or restructuring results in enormous energy transaction between utilities and loads. All these have resulted in additional stress on the existing infrastructure. In these circumstances, the power system operation and control has become more challenging than before. Smart grid technology has emerged as a new grid control mechanism in this new era of power system operation. Smart grid is a modern power system grid with improved reliability, efficiency, security, capacity to accommodate RES, automated with modern communication channels, etc. [1]. The present reformed electric structure and new challenges have been encouraging to transform the existing grid into a smart grid. The large-scale power system operation and control starting from centralized bulk generation to distribution can be accommodated within the scope of smart grid. A reliable power system operation largely depends on the real-time information sharing between CG and end-users. The issues related to importance of communication technologies and challenges in smart grid environment have been discussed in [2]. An effective communication system helps reliable, secure and cost-effective decisions in real-time power system operation.

Another dimension of smart grid technology is the participation of the electricity user or consumer to improve the demand response. Demand side management (DSM) is an essential architecture to improve demand response (DR). DSM is more effective than supply side management as it benefits both consumer and utility. Environmental concern, transmission and distribution deficiencies, peak load problems, etc., are the reasons for the development of DSM. Here, the activities at the customer side are designed and planned to improve the load curve. That means demand side management is a program that controls the energy consumption at the customer side. An incentive-based autonomous energy consumption schedule is discussed in [3] as demand side management. The formulation is made to reduce the energy consumption cost. An industrial DSM project for supplying in-line hot water to users is discussed in [4]. The DSM technology not only provides the strategies to be adopted to reschedule the end-user demand for minimizing uses of power during peak hours but also facilitates the optimal operating conditions for the RES. The work in [5] describes how a DSM program can be effective in maximizing the use of RES and can result in simultaneous financial savings for the customer.

One of the major application areas of DSM is thermostatically controlled loads of customers. Thermostatically controlled loads such as air conditioners (ACs) and refrigerators are used by the customers primarily to keep inside temperature within a desirable range. The compressor of the cooler is switched on to cool the inside temperature and switched off when desired temperature is attained. The working schedules of cooling systems are mostly identical in a particular sink area. However, cooling activities can be regulated by pre-cooling or delayed cooling in DSM application for thermostatically controlled loads. The economical effects on the customer are discussed in [6] with DSM application for cooling loads. To achieve an effective cooling system working schedule, accurate mathematical modelling of the thermostatically controlled load [7] is required.

In order to improve the operational efficiency, the utilities energy management program encourages the household owners and building managers to adopt optimal load operation schedule. The operation schedule must be prepared based on requirement, comfort level of the customer and, of course, the energy price. For a household load scheduling problem, the appliance commitment algorithm requires price and consumption forecast scenarios for a particular given objective [8]. The objective may be reduction in payment or maximization of comfort level. For a building manager, considerable amount of energy consumption can be saved through optimized operation and management without changing much of the infrastructure. An energy-efficient building is one which adopts the optimal load scheduling strategy and coordinates the operation of various energy sources [9]. The impact of energy storage in buildings on electricity demand side management is discussed in [10]. Load management is designed to control demands of various consumers of a power utility. Such control and modification enable the utility to meet the demand at all times in most economic manner [11]. Time Of Use (TOU) tariff has been used in several countries as a tool to reduce the energy billing cost [12].

This paper proposes an optimal load and self-generation scheduling strategy of a commercial building in order to minimize the energy billing cost. The main focus of this paper is on modelling of loads and self-generation systems of a commercial customer together with the computational aspects of the proposed optimization model. The load activity strategy is formulated based on the consumption forecast scenarios with significant uncertainties. The customer comfort index during business hours is considered in this work as practical constraints. The algorithm is formulated as mixed integer dynamic optimization problem, and the demand side management is obtained under local TOU tariff structure. The proposed optimization problem is solved using Artificial Bee Colony (ABC) algorithm and implemented in MATLAB. The simulation results show a comprehensive load management and optimized self-generation approach to minimize the energy consumption cost of the commercial building.

2 Problem Formulation

In this section, the mathematical formulation for the proposed algorithm of optimum load management of a commercial building is discussed. The structure of the commercial customer is presented and discussed in Sect. 4. The proposed algorithm is developed based on load modelling and self-generation scheduling. The electrical load of the commercial building has a composite structure, and each component of the load is modelled separately based on its characteristics and constraints. Similarly, the characteristic of self-generation systems is also modelled. The scheduling is planned for one day, and the entire planning period is divided into 24 numbers of 1 h intervals. The purpose of the planning is to minimize the payment for electricity consumption by optimal scheduling of loads and by proper utilization of self-generation system. In the present work, the objective function of minimizing energy billing cost (EBC) can be written mathematically as:

$$\text{Minimize} \quad \sum_{t=1}^{T} \sum_{k=1}^{NL} (EC_t^k R_t) \tag{1}$$

Subject to

$$L_t - RES_t - SS_t = 0 \tag{2}$$

$$\sum_{k=1}^{NL} (X_k^t P_k^t) \leq EC^{\max}, \forall t \in T \tag{3}$$

$$P_k^t \leq P_k^{\max}, \forall t \in (t_s, t_e) \tag{4}$$

where EC_t^k is the energy consumption for load k at t hour. NL represents total number of load, and R_t is the market energy price in Rs./KWh for interval t. Here, T and t are set and index of hourly periods. In this formulation, T = 24 h. Equation (2) represents the power balance equation which is to be satisfied at each hour. L_t represents total load at interval t, and RES_t and SS_t represent the power supplied by self-generation system and substation, respectively. There must be a limit on total energy consumption by all the loads during the entire time of operation. This is expressed mathematically in Eq. (3) that total energy consumption must not exceed the maximum energy consumption $EC^{\max}$. Here, X_k^t and P_k^t represent load activity and quantity for the kth load at hour t, respectively. The energy consumption of each load during a day depends on its activity schedule or energy consumption schedule. Different loads may have different energy consumption schedule. The load activity schedule is designed based on customers' requirement and comfort. Hence, the load activity of a particular load is to be scheduled between t_s, $t_e \in T$ as the beginning and end of the time interval for energy consumption. However, the load must be operated within its maximum capacity $P_k^{\max}$. This fact is expressed mathematically in Eq. (4). The details of the loads and self-generation system modelling aspects considered in this work are mentioned below.

2.1 Load Model

The case study presented in this formulation involves a commercial load; consisting of base load, lighting load, escalator load, lift load, air conditioner load, energy storage and pump load. The composite load structure and the percentage share of its components are mentioned in Table 1. In the present case study, it is observed that some of the loads are fixed and have a fixed time range to be turned on. Some loads are adjustable; i.e. their time of operation can be altered. That means each load has its own activity strategy during the whole 24 h planning period. The load activity strategy is formed by considering users' requirements and comfort level. Therefore, the load activity vector of each load has equal number of elements as the number of intervals. The working hour of the commercial building is considered to be from 8 to 22 h in all weekdays.

2.1.1 Base Load

For the case study performed in this work, the commercial building has a base load of 36 KW. Base loads are considered as must-run load, and hence, it is fixed throughout the scheduling period. The base load of the commercial building composed of ventilation fan, CFL, fire pump and alarm system with Speaker, etc.

Table 1 Composite load structure of the connected load

Sl. No	Type	Percentage (%)	Load (KW)
1	Base load	3	36
2	Lighting load	8.3	99.75
3	Escalator load	12	144
4	AC load	39.5	473.9
5	Lift load	22	262
6	Pump load	11.2	134.4
7	Storage battery	4	50
Total		100	1200 KW

2.1.2   Lighting Load

The use of lighting load of the commercial building has a fixed time range from 8 to 23 h, and its hourly values have been shown in Fig. 4. The lighting load scenario for each interval is picked by the proposed algorithm as input.

2.1.3   Escalator Load

There are ten numbers of escalators present in the building, and the rating of each of them is 14.4 KW. Therefore, the total escalator load for the building comes out to be 144 KW. The load activity strategy for the escalator load is shown in Fig. 4. The hourly demand for the escalator load is based on the number of users. The scenario of predicted hourly demand is taken as input for the proposed formulation.

2.1.4   Lift Load

A major portion of the load is shared by the lift load. The load activity strategy of the lift load is shown in Fig. 4. The hourly demand of the lift load is dependent on the numbers of people movement inside the building. The forecasted hourly lift load demand is shown in Fig. 4 and considered as input for the proposed algorithm for computation of energy consumption.

2.1.5   Pump Load

The commercial building has water pump of different categories such as submersible pump, recirculation pump, centrifugal pump and tullu pump. The equivalent water pump load of total capacity 134.4 KW is considered in this work. Such load is adjustable and its operation can be altered during the day. In this work, the pump load activity is kept null from 24 to 5 h and pumping activity is made active from 6 to 23 h. The required duration of pumping operation is based on the use of water in the building and the reservoir capacity. In this work, minimum 6 h pumping operation per day is considered as pumping load constraint. This is expressed mathematically as follows

$$\sum_{t=1}^{T} xp_t = Ap \tag{5}$$

where xp_j is pump load activity at interval j and Ap is the minimum active time of the pump load. N is the total number of duration. In this case, it is 24.

$$xp = \{xp_1, xp_2, \ldots xp_N\} \tag{6}$$

$$xp_t \in \{0, 1\} \tag{7}$$

Here, $xp_t = 0$ represents pump load is inactive at tth interval and $xp_t = 1$ represents pump load is active at interval t. The total cost of energy consumed by the pump load is determined by the following expression.

$$EC_P = \sum_{t=1}^{T} (Pmp \cdot xp_t \cdot R_t) \tag{8}$$

where Pmp is the pump load in KW and R_t is the market energy price in Rs./KWh for interval t.

2.1.6 AC Load

The air conditioner (AC) is one of the most vital load components for any business house in a commercial building. The primary purpose is to maintain the inside temperature of the business area within an acceptable range. The AC regulates the temperature by using its thermostat and relay actuator. Measuring of AC load requires calculation of room temperature following a compressor operation. The desired temperature set point and acceptable room temperature band regulate the compressor operation. The AC load is active, if compressor is switched on; otherwise, it is inactive. The cost of energy consumption of the AC load can therefore be determined from the following expression.

$$\sum_{t=1}^{T} P_{AC} \cdot x_t^{AC}(\theta) \cdot R_t \tag{9}$$

$$x_t^{AC}(\theta) \in \{0, 1\} \tag{10}$$

where P_{AC} denote the equivalent AC load in KW. In this work, it is considered as 473.9 KW and assumed to be covering a major portion of the total load demand. The AC load activity is represented by $x_t^{AC}(\theta)$ for interval t. The AC load activity vector is function of room temperature. The dynamic behaviour of AC, discussed in [7] for calculating the room temperature at any instant t, is utilized in this work for finding the load activity schedule and is represented in Eq. (11).

$$\frac{d\theta(t)}{dt} = -\frac{1}{CR}[\theta(t) - \theta_a + m(t)RP + w(t)] \tag{11}$$

where θ_a represents the ambient temperature; C and R are the thermal capacitance (KWh/°C) and thermal resistance (°C/KW) of the room being cooled, respectively. P is the thermal power of the AC. $w(t)$ is the random thermal disturbance, i.e. heat loss or gain. The values of R, C and P are referred from [7], and the random thermal disturbance $w(t)$ is considered as decision variable in the present work.

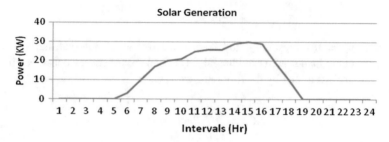

Fig. 1 Hourly solar power generation pattern

2.2 Energy Source Model

In the present study, the business house is energy-efficient one and is equipped with small-scale Renewable Energy Sources (RESs). The RES is in the form of a 50 KW solar power plant with a battery storage capacity of 100 KW. Each PV module composed of 72 PV cells; ten such modules are connected in series to form one string of 400 V, and 18 such strings are connected in parallel. The peak efficiency at 30 °C ambient is 91% at half load and 92% at full load, respectively. The hourly solar power generation was monitored on various days of November with average ambient temperature of 26 °C. The average solar power generation pattern is plotted in Fig. 1 and is used in this work as self-generation system of the commercial customer. The maximum charging and discharging rate for the storage battery is 10 KW with 90% efficiencies.

3 Solution Algorithm and Its Implementation

The Artificial Bee Colony (ABC) algorithm [13], proposed by Karaboga in 2005, is a very well-established swarm optimization method. In this algorithm, the foraging behaviour of bees in terms of their work allocation to optimize the accumulation of nectar has been simulated. The bee colony in ABC algorithm is made up of three groups. Employed artificial bees and artificial onlookers, respectively, constitute the first and second half of the bee colony. The initial exploration is not done by engaging all the bees. Instead, some employed artificial bees are selected for exploring flower patches of the surrounding environment and checking their "profitability". This term "profitability" takes into consideration several parameters like the sugar content present in nectar, the amount of nectar available in concerned flower patches, the distance of flower patches from beehive. One employed bee is allotted to each food source available around the beehive. If a food source allocated to an employed bee is discarded by other bees, then that employed bee becomes a scout bee. Once the employed bee exhausts its food source, then the scout bee begins its search cycle.

In the ABC algorithm, a probable solution to the optimization problem is represented by a position of food source. Also, the fitness of the solution is given by the amount of nectar present in the corresponding food source.

Mathematical formulation based on the behaviour of bees in work allocation among themselves for optimal nectar collection has been given in [14].

ABC algorithm is a comparatively new optimization technique among all other heuristic-based search algorithms found in literature, and the performance and efficiency of this algorithm in solving power system problems have been demonstrated recently [15, 16]. The work in [15] demonstrates the successful application of the ABC algorithm in solving nonlinear power system generation scheduling problem and the comparison of the results obtained using ABC algorithm with that obtained from other methods. In [16], a power dispatch problem has been formulated with limited control movements of the concerned power apparatus. This problem has been designed as a nonlinear dynamic optimization problem and has been solved using ABC algorithm.

The ABC algorithm has been proved to provide acceptable and encouraging performances when aforesaid multidimensional multi-model nonlinear power system problems were attempted to be solved using the said algorithm.

In the present work, the demand side management problem is a mixed integer dynamic optimization. As such, it shall be beneficial to attempt solving the current problem using ABC algorithm. The food sources are assumed to be scattered over the total time span of 24 h, as the position of these sources denotes probable solutions.

Following is a brief description of the steps required for the implementation of the ABC algorithm:

Step 1: Initialization

Initial solutions (food source positions) x_{ij} are generated randomly as $x_{ij} = (LU_1, Pmp_1, AC_1, H_1, SB_1 \ldots LU_n, Pmp_n, AC_n, H_n, SB_n)^T$ where $j = 1, 2, \ldots, D$. Here, D denotes dimension of the vector signifying the number of optimization parameters, and n gives the maximum number of interval, which is 24 in this work. Also, $i = 1, 2, \ldots, SN$, where SN indicates the size of the bee swarm, which is 20 in this case. The solution generated above consists of parameters like percentage of inevitable uncertainty of forecasted loads, status of compressor and pump operation, heat loss/gain, status of battery charging/discharging for every time period t.

Step 2: Evaluation of solution

Considering the initial randomly generated values of problem variables, the net load demands and net self-generation are computed for every time period t. The total power supplied by the utility and the cost of energy purchased from the utility for each hour of 24 h period is calculated. The objective function is the sum of the hourly energy cost.

Step 3: Employed Bee phase

New solutions (problem variables) Y_{ij} are generated for the employed artificial bees in the vicinity of X_{ij} and are given by $Y_{ij} = X_{ij} + \theta_{ij} * (X_{ij} - X_{kj})$. These values are estimated in the same method as described above. The process of greedy selection is performed between X_{ij} and Y_{ij}, where θ_{ij} is generated randomly.

Information regarding the solution corresponds to total energy consumption cost, and these are doled out to the onlookers by the employed artificial bees for added foraging. This guarantees combined intelligence in the entire search process.

Step 4: Onlooker Bee phase

New solutions are generated based on the probability p_i for the onlooker bees for the solution X_i using $p_i = \frac{fitness_i}{\sum_{j=1}^{N} fitness_j}$ where *fitness$_i$* is the fitness value of the solution i, which is proportional to the nectar amount of the food source position or problem variables "*i*". The solution is evaluated in the same manner discussed above, and greedy selection is performed between new and old solutions. In this process, the onlookers exploit for better solutions with higher probability.

Step 5: Scout Bee phase

In case the solution *i* is found to be inferior, then it is affirmed to be discarded. It is substituted with new solution for the scout that is randomly produced as follows $x_i^j = x_{min}^j + rand(x_{max}^j - x_{min}^j)$.

In the current work, the abandonment counter limit is selected to be 100 to make a solution better.

Step 6: Save best solution

The best solution is stored which corresponds to

 (i) the least energy consumption cost for the whole span of 24 h and
(ii) all limits and conditions.

Step 7: Stopping criteria

The cycle is stopped when the Maximum Cycle Number (MCN) is attained. The value of MCN is determined on the basis of experience, and in the present work, the value is fixed at 500.

4   Results and Analysis

The proposed algorithm of demand side management is investigated on a commercial building/business house situated in a city of South Assam, India. The peak electricity demand of the building is 1200 KW and is supplied from a nearby substation owned by Assam Power Distribution Company. The business house/building has different category of loads, and its composite structure is as described in Sect. 2. The building is equipped with a small PV solar power plant with energy storage facility. The illustration of the architecture of the loads and self-generation system is shown in Fig. 2.

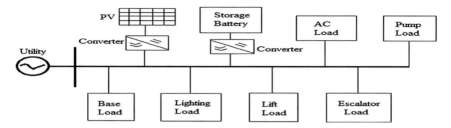

Fig. 2 Illustration of loads and generations

The hourly load variation of different days of November was monitored, and an average hourly loading pattern is developed. The developed loading pattern is discritized in time, and it is assumed that the load demand is fixed for 1 h duration. The hourly variation of the load is shown in Fig. 3. In the present case study, it is observed that the pump load is the only adjustable load, and others, e.g. lighting load, lift load, escalator load and AC load, have a fixed time range of operation. Among these, AC load time of operation is decided based on temperature set point of the business area and the ambient temperature. The power consumption of the AC load is fixed when its compressor is switched on. Therefore, load activity strategy of the AC load is function of prevailing room temperature and ambient temperature. On the other hand, the load activity strategy along with hourly demand for lighting load, escalator load and lift load can be derived and shown in Fig. 4. In this part of the country, the average ambient temperature varies from 25 to 34 °C during first few days of November, and an average hourly temperature plot as shown in Fig. 5 is utilized in this work.

In the present case study, it is observed that total load consumption of the business house cannot be reduced in order to decrease the payment for cost of electricity consumption. However, the saving in cost of electricity can be achieved by shifting some loads to cheap pricing period and by utilizing the available self-generation facility to the best possible way. This is termed as demand side management and is beneficial to both utility and customer. Such type of planning and implementation

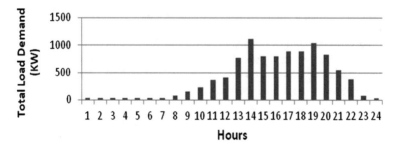

Fig. 3 Hourly variations of total load

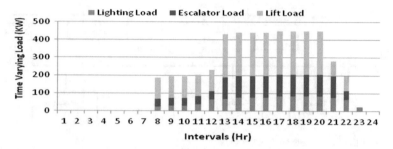

Fig. 4 Load activity of lighting, escalator and lift load

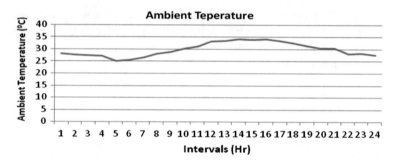

Fig. 5 Variations of average ambient temperature

Table 2 Electricity billing price	Time (h)	Billing price (Rs./KWh)
	23.00–5.00	3.35
	6.00–16.00	4.00
	17.00–22.00	6.25

of the activities for both load and self-generation system requires forecasting of future operating conditions. However, predictions are not always perfect. Actual load variations may differ than what was predicted. In the present case study, 5–20% uncertainty is considered for hourly load variations. The present electricity pricing rate of Assam Power Distribution Company is mentioned in Table 2 and utilized in this work.

The proposed demand side management algorithm is implemented in MATLAB. The solution of the algorithm provides the load management strategies to be followed by the business house together with managing its self-generation systems. The detail loading scenario for each type of load along with total hourly load consumption for the whole 24 h planning period is mentioned in Table 3. It is observed that the solution of the load scheduling algorithm obeys the activity schedule of each type of load. For example, the AC load is active from 10.00 to 22.00 h to maintain the room temperature of the business house. Figure 6 shows the instances of AC load operation to maintain the room temperature between the acceptable temperature

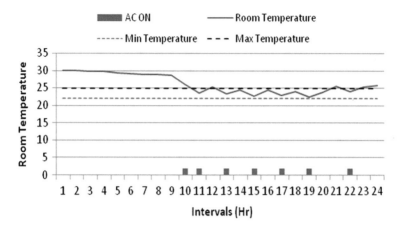

Fig. 6 AC operation schedule and room temperature

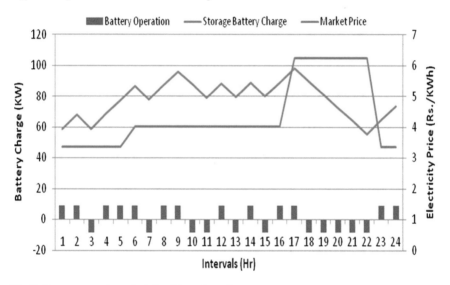

Fig. 7 Battery operation schedule with market price

bands. In this work, a temperature band from 22 to 25 °C is considered as the acceptable temperature. The energy storage facility of the business house is in the form of a 100 KW battery as discussed earlier which acts as load while charging and source while discharging. In this work, the initial battery charge is considered to be 55 KW, as the battery discharge should not be more than 50%. The instances of battery charging and discharging along with corresponding storage battery charge are shown in Fig. 7 and are compared with the hourly market price. It is observed that the battery has charged mostly during low pricing hours; whereas during peak/high pricing hours, it has discharged which obeys the energy management philosophy.

Table 3 Hourly load consumption

Hour	1	2	3	4	5	6	7	8	9	10	11	12
Base load	36	36	36	36	36	36	36	36	36	36	36	36
Lighting load	0	0	0	0	0	0	0	24	30	30	40	67
Escalator load	0	0	0	0	0	0	0	42.75	38.7	47.25	40.16	36.15
Lift load	0	0	0	0	0	0	0	101.7	131.3	111.92	110.32	112.18
Pump load	0	0	0	0	0	134.4	134.4	0	0	0	134.4	134.4
AC load	0	0	0	0	0	0	0	0	0	473.9	473.9	0
Storage battery	9	9	0	9	9	9	0	9	9	0	0	9
Total load (KW)	45	45	36	45	45	179.4	170.4	213.52	245.01	699.07	834.78	394.73

(continued)

Table 3 (continued)

Hour	13	14	15	16	17	18	19	20	21	22	23	24
Base load	36	36	36	36	36	36	36	36	36	36	36	36
Lighting load	70	78	78	78	86	86	86	86	78	65	24	0
Escalator load	114	105.4	96	99.63	138	103.7	110.9	112.3	96.51	56.75	0	0
Lift load	268.7	201.6	206.4	214.47	215	210.7	201.6	196	70.18	85	0	0
Pump load	134.4	0	0	0	0	0	0	0	0	0	134.4	0
AC load	473.9	0	473.9	0	473.9	0	473.9	0	0	473.9	0	0
Storage battery	0	9	0	9	9	0	0	0	0	0	9	9
Total load (KW)	1097	430	890.3	437.1	957.9	436.4	908.43	430.3	280.69	716.65	203.4	45

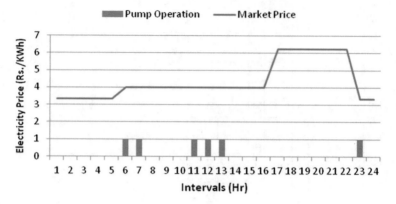

Fig. 8 Pump operation versus market price

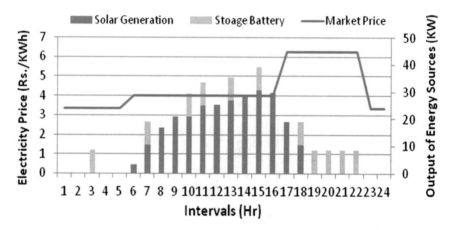

Fig. 9 Self-generation pattern w.r.t market price

The solution of the algorithm provides the required pumping operation during 24 h planning period as shown in Fig. 8. It is observed that the required 6 h of pumping activities is being distributed among the low pricing hours only, in order to reduce the power purchase requirement from the utility during peak pricing periods. The energy dispatch schedule of the self-generation system is shown in Fig. 9. It is observed that during daylight periods, the solar generation has contributed more, while during evening hours, the storage battery has contributed. The total hourly load demand is basically served by both the utility and self-generation system of the business house. The complete dispatch schedule is shown in Fig. 10, and the corresponding energy billing costs are shown in Fig. 11.

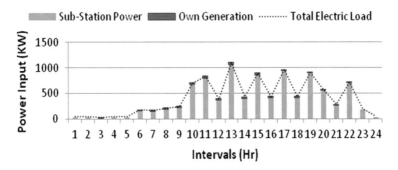

Fig. 10 Complete dispatch schedule

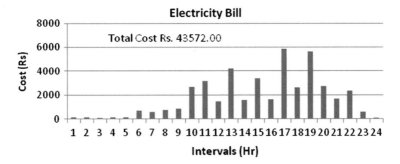

Fig. 11 Variations of electricity consumption cost

5 Conclusion

In this paper, an optimal and automated energy consumption scheduling algorithm for a commercial customer is discussed. This study investigates the effectiveness of demand side management techniques in electric power system. The problem is formulated to reduce the electric energy consumption cost by maximizing the benefits of self-generation system and by reshaping the load curve for a given consumers' comfort constraints. The formulation of the proposed optimization model is based on forecasted demand and self-generation scenarios with significant uncertainties. The implementation of the solution algorithm (ABC) searches for the strategies to be adopted for the controllable loads and for the self-generation system. The result of analysis shows that the proposed load management technique benefits the commercial customer in reducing its electricity consumption cost by effective utilization of energy storage system. The proposed algorithm can be used for other building managers in making an automated load scheduling scheme based on their tariff and comfort setting.

References

1. Santacana, E., Rackliffe, G., Tang, L., Feng, X.: Getting Smart. IEEE Power Energy Mag. **8**(2), 41–48 (2010)
2. Vehbi, C.G., Dilan, S., Taskin, K., Salih, E., Concettina, B., Carlo, C., Gerhard, P.H.: Smart grid technologies: communication technologies and standards. IEEE Trans. Industr. Inf. **7**(4), 529–539 (2011)
3. Mohsenian-Rad, A.-H., Vincent, W.S., Juri, W.J., Robert, S., Leon-Garcia, A.: Autonomous demand side management based on game-theory energy consumption scheduling for the future smart grid. IEEE Trans. Smart Grid **1**(3), 320–331 (2010)
4. Rankin, R., Rousseau, P.G.: Demand side management in South Africa at industrial residence water heating systems using inline water heating methodology. Energy Conver. Manag. **49**(1), 62–74 (2008)
5. Finn, P., OConnell, M., Fitzpatrick, C.: Demand side management of a domestic dishwasher: wind energy gains, financial savings and peak time load reduction. Appl. Energy **101**, 678–685 (2013)
6. Alparslan Zehir, M., Bagriyanik, M.: Demand side management by controlling refrigerators and its effects on consumers. Energy Conv. Manage. **64**, 238–244 (2012)
7. Perfurmo, C., Kofman, E., Braslavsky, J.H., Ward, J.K.: Load management: model-based control of aggregate power for populations of thermostatically controlled loads. Energy Conv. Manage. **55**, 36–48 (2012)
8. Pengwei, D., Ning, L.: Appliance commitment for household load scheduling. IEEE Trans. Smart Grid **2**(2), 411–419 (2011)
9. Guan, X., Xu, Z., Jia, Q.-S.: Energy-efficient buildings facilitated by micro grid. IEEE Trans. Smart Grid **1**(3), 243–252 (2010)
10. Qureshi, W.A., Nirmal-Kumar, C., Nair, M., Farid, M.: Impact of energy storage in buildings on electricity demand side management. Energy Conv. Manage. **52**, 2110–2120 (2011)
11. Kostková, K., Omelina, Ľ., Kyčina, P., Jamrich, P.: An introduction to load management. Electr. Power Syst. Res. **95**, 184–191 (2013)
12. Godoy-Alcantar, J.M., Cruz-Maya, J.A.: Optimal scheduling and self generation for load management in the Mexican power sector. Electric Power Syst. Res. **81**, 1357–1362 (2011)
13. Karaboga, D.: An Idea based on honey bee swarm for numerical optimization. Technical Report TR06, Computer Engineering Department, Erciyes University, Turkey (2005)
14. Karaboga, D., Basturk, B.: On the performance of artificial bee colony (ABC) algorithm. Elsevier Appl. Soft Comput. **8** (2008)
15. Hemamalini, S., Simon, S.P.: Artificial bee colony algorithm for economic load dispatch problem with non-smooth cost functions. Electric Power Comp. Syst. **38**(7), 786–803 (2010)
16. Malakar, T., Goswami, S.K.: Active and reactive dispatch with minimum control movements. Int. J. Electr. Power Energy Syst. **44**(1), 78–87 (2013)

Optimal Energy Sharing Within a Solar-Based DC Microgrid

V. S. K. V. Harish, Naqui Anwer and Amit Kumar

Abstract Solar-based DC microgrids and home-based systems have been a reliable option to solve rural electrification problems in many parts of the world. However, there is a pertaining problem of mismatch between local demand and on-site generation in such microgrids. Proposed work attempts to address this problem by developing an optimal peer-to-peer (P2P) energy sharing among the individual households in a DC microgrid. A nonlinear optimization problem is formulated that aims to minimize the power transmission loss and overall energy cost in a distribution network consisting of a number of households incorporating practical constraints (e.g. power balance and battery's operational constraints). Three different aspects of operation, viz. battery usage, power from grid and P2P sharing, have been considered in order to facilitate maximum utilization of local distributed energy resources, thereby saving the energy bills for all households.

1 Introduction

India has made proficient progress in terms of generation capacity and managing the demand; achieving lowest ever demand–supply gap in terms of both energy (0.7% deficit) and peaking (1.6% deficit) in 2016–17 as compared to 2015–16 [1]. Efforts have been made to alleviate the problems of power shortages, rural electrifications, poor financial health of DISCOMs and non-performing energy assets through various policy interventions.

While electricity generation and capacity building are on track, the challenge lies in making the generated electricity/energy accessible to all (power-to-all). Studies

V. S. K. V. Harish (✉) · N. Anwer
Department of Energy and Environment, TERI School of Advanced Studies,
Vasant Kunj, Delhi 110070, New Delhi, India
e-mail: harishvskv.iitr@gmail.com

A. Kumar
Social Transformation, the Energy and Resources Institute (TERI),
IHC Complex, Delhi 110003, New Delhi, India

© Springer Nature Singapore Pte Ltd. 2019
J. C. Bansal et al. (eds.), *Soft Computing for Problem Solving*, Advances in Intelligent
Systems and Computing 817, https://doi.org/10.1007/978-981-13-1595-4_50

report that more than 200 million people in India are living without reliable electricity, especially large parts of the rural areas. India aims at achieving 100% electrification by 2022 [2], through its principle vehicle of Deen Dayal Upadhyaya Gram Jyoti Yojana (DDUGJY) with an investment plan of around ₹758.93 Billion (~US $12 billion) [3].

As of August 2017, out of the 18,452 un-electrified census villages in India, 14,255 (77%) villages have been electrified and 100% household connectivity has been achieved in 1,161 villages [4] with electrification works under progress in 3,211 villages. With this, India achieves 99.4% of village household or rural electrification [5]. India aims to achieve 100% village electrification by 2019–22 [2]. Despite advances in electrifying rural villages, the quality of electricity service to rural households is dismal, and the problem of electricity "access" has not improved appreciably. States having high electrification rates still have poor household electrification, and there are several other regions (hamlets) which were not covered in Census 2011 and other national surveys. NITI Aayog has identified that "… connection is not the only factor—even duration, quality and reliability are important".

Renewable energy-based decentralized energy systems (DESs), off-grid and on-grid-based microgrids (MGs)/home-based solar products are being considered as an optimistic solution to the rural electrification problems in under developed and developing economies like India [6]. Non-energy storage home-based products, though cheap face the problem of unreliable supply of electricity due to intermittent nature of solar or wind generation. Affordability of purchasing power from the grid- or solar-based products from the market is one prime challenge that has hampered the growth rate of rural electrification in India. Problem of intermittency is solved by implementing a distributed storage unit which increases the cost and suffers energy loss while power transmission due to degradation of battery while charging and discharging processes.

Development of a transactive energy system which shall facilitate peer-to-peer (P2P) sharing of energy can solve the problem of delivering sustainable power at affordable prices for low-income rural people. Designing decentralized P2P MGs which deliver localized generated power to households and businesses in rural areas enables the local villagers to trade their (excess) electricity for profit, thereby creating a sustainable business model. A reliable, economically competitive and environmentally sustainable electric power system can be achieved through an active P2P microgrid, thereby addressing the issues of energy security and environmental strains.

In a P2P energy sharing network, any household (peer) could produce its own energy, using renewable energies like solar energy, and sell its surplus to others (other peers) who need it. For instance, at any time (hour) of the day, if a household has excess electric power beyond its own demand, then under such a condition excess power can be transferred to some other households in need via dedicated transmission lines. This P2P energy can also be scaled up and operate as an energy sharing network within a microgrid and with other MGs. In such a situation, a MG then becomes a peer. By exploiting the diversified energy generation and consumption profiles of the MGs at different geographical locations, the P2P sharing network gains many advantages.

For instance, the energy transmission loss can be reduced for the short distance transmissions between neighbouring MGs. Moreover, with a well-designed trading scheme, each individual MG can benefit from the P2P energy network, e.g. each MG will enjoy a lower purchase price and a higher selling price in the P2P network, in comparison with the external main grid.

The main challenge, however, remains in implementing a P2P sharing strategy for villages with no electrical infrastructure. In such a case, a local entrepreneur can step up and buy local energy generation products with batteries and a P2P sharing strategy could be developed taking into account transportation cost, degradation of the battery while moving it from one peer to the other. Also, such strategies should be flexible enough to be scaled up or down in situations where other households join or unjoin the developed P2P network.

2 System Model

Proposed power sharing algorithm has been developed for a set of rural households, few of which are equipped with solar panels. Conceptual schematic of the proposed power sharing mechanism is shown in Fig. 1.

There are a number of rural households connected together with power lines and to the main grid at the point of common coupling (PCC). Out of the connected RHs,

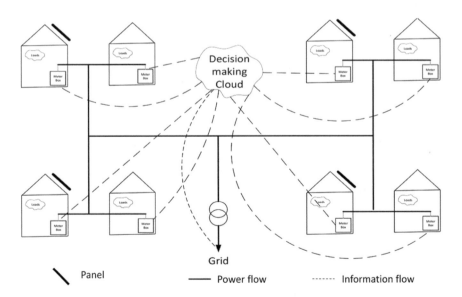

Fig. 1 Peer-to-peer energy sharing model

Fig. 2 Illustrative example of the P2P network

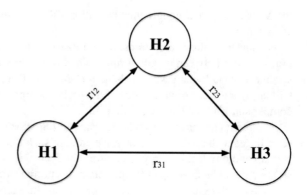

few of them are equipped with roof top solar panels of different capacity. The dashed lines represent the flow of information from one RH to a centralized cloud and to the other RHs. The main purpose of this communication is to enable transparency of information on any transaction and sharing that would happen at any instant of time between RHs. This information can only be READ by the RHs and can be edited only at the centralized decision making cloud with administrator privileges. Communication protocols, rules and strategy of communication are beyond the scope of this paper. Solid lines represent power flow from one RH to the other. Every RH is equipped with loads and a metre box capable of saving the transactions, history of energy sharing, thereby facilitating energy management.

Let us consider a set of R rural households (RHs) in a distribution network, given by Eq. (1).

$$\mathfrak{R} := \{RH_r : r \in Z\} \tag{1}$$

where

$Z := \{1, 2 \ldots R\} =$ index set of rural households.

For illustration, consider an example of three RHs (H1, H2 and H3) connected together with power lines as shown in Fig. 2.

In Fig. 2, H1, H2 and H3 are connected to each other via dedicated distribution lines of resistance r12, r23, r31. Value of resistance of the power lines varies with the distance between each and every RH it is connected to. At every time slot, t, if RH_1 transfers P_{12}^* amount of power from RH_1 to RH_2, then the actual power received by RH_2 can be written as Eq. (2).

$$P_{12} = P_{12}^* - P_{Loss}^{1-2} \tag{2}$$

where

P_{12} Actual power received by RH_2 from RH_1,
P_{12}^* Power transmitted from RH_1, and
P_{Loss}^{1-2} Loss of power during transmission.

Now, P_{Loss}^{1-2} for a power line can be rewritten as Eq. (3).

$$P_{Loss}^{1-2} = I_{1-2}^2 \times r_{12} \tag{3}$$

where,

I_{12}^1 Current carrying capacity of the power line connecting RH_1 and RH_2, (A) and
r_{12} Resistance of power line connecting RH_1 and RH_2.

Thus, Eq. (2) can be simplified as Eq. (4).

$$P_{12} = P_{12}^* - I_{1-2}^2 \times r_{12} \tag{4}$$

Transmission voltage (V) at which power is shared between the RHs is assumed to be constant, and hence, Eq. (4) is modified into Eq. (5).

$$P_{12} = P_{12}^* - \left(\frac{P_{12}^*}{V}\right)^2 r_{12} \tag{5}$$

If there are a number of households participating in P2P energy sharing (Fig. 1), then Eq. (5) can be written in generic terms as Eq. (6).

$$P_{m,n}(t) = P_{m,n}^*(t) - \left(\frac{P_{m,n}^*(t)}{V}\right)^2 r_{m,n} \tag{6}$$

where

$P_{m,n}(t)$ Actual power received by nth RH from mth RH, W,
$P_{m,n2}(t)^*$ Power transmitted from mth RH to nth RH, W and
$r_{m,n}$ Resistance of line connecting mth and nth RH, Ω.

Developed model for energy sharing is applicable to a P2P network of any topology, and the proposed algorithm is majorly focused on the amount of power being shared within the RHs participating in the P2P scenario.

A set of valid assumptions has been considered for simulation of the proposed model and P2P concept. Dedicated power lines are available and operated by a Local Site Operator (LSO) for direct P2P sharing among rural households and are responsible for determining the topology of the P2P network [7]. Voltage at which power sharing occurs is constant throughout the simulation. Scheduling strategies and maintenance of home-based solar systems are decided by the LSO.

3 Simulation Results

Performance of the developed strategy is by considering three different scenarios. Loads in rural households have been considered to be DC appliances (ratings derived from the 2015–16 Global LEAP Award winners) (Figs. 3 and 4).

Power surplus status of each RH is calculated as Eq. (7)

$$S(t) = G(t) - D(t) \tag{7}$$

where,

$S(t)$ Surplus power available at a particular household at time, t, W,
$G(t)$ Power generated at a particular household at time, t, W and
$D(t)$ Power demanded by a particular household at time, t, W.

Decision for P2P sharing of a particular RH at any time, t, is taken as per Eq. (8).

$$S(t) := \begin{cases} > 0 & E \\ < 0 & R \\ = 0 & I \end{cases} \tag{8}$$

E represents the state of excess power where the RH is available to share its excess power, R represents the requirement state where the RH will need to draw power from the P2P network, and I represents the ideal state.

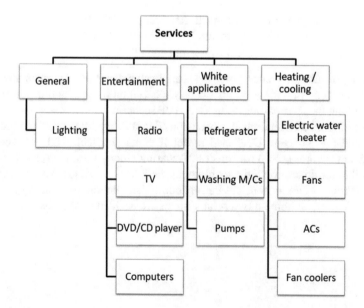

Fig. 3 Electrical services available for off-grid rural households

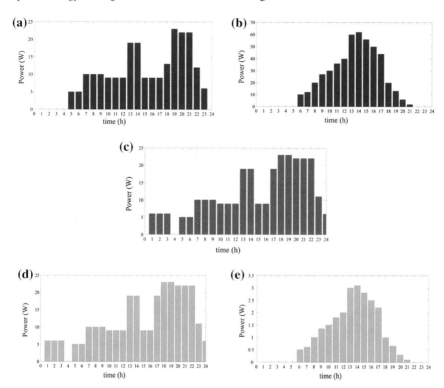

Fig. 4 Hourly demand (**a**), (**c**), (**d**) and generation (**b**), (**e**) of RH_1, RH_2 and RH_3 in P2P system, respectively

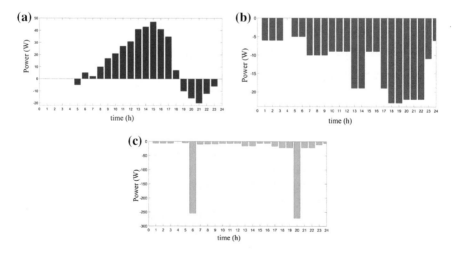

Fig. 5 Surplus power (**a**), (**b**), (**c**) status of RH_1, RH_2 and RH_3, respectively

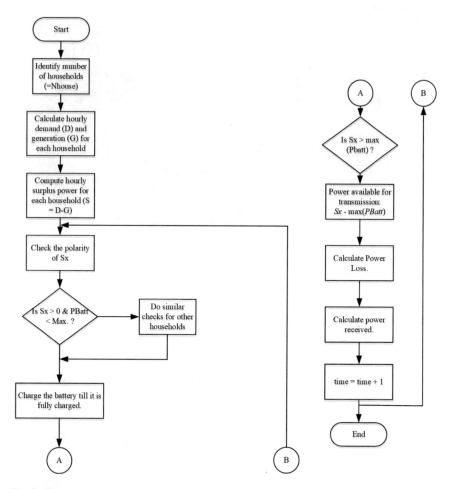

Fig. 6 Flow chart for energy management

Surplus power status of each RH participating in the P2P network is given as Fig. 5.

Numerical values to run the simulations are given as: $r_{12} = 1.5\ \Omega$, $r_{23} = 2.5\ \Omega$ and $r_{31} = 2\ \Omega$; $V = 100$ V; $t = 24$ h; $\delta t = 1$ h; battery: 6000 mAh LiFePO4 3.2 V, 1x USB in/out, 2A, 1x USB out; charging efficiency: 95%; discharging efficiency: 98%, max. and min. power charge of the batteries installed at RH_1 and $RH_3 = 100$ W, 10 W and 10 W, 1 W, respectively. Primary objective is to minimize the power mismatch of every RH on per hour basis, i.e. to make $S(t)$ for each RH minimum by making use of the P2P network. Also, if $S(t)$ exceeds the depth of charge of the battery installed at that RH, then the difference of $S(t)$ and max. power charge of the battery will be available for transmission via P2P network. Flow chart for the optimal strategy is as shown in Fig. 6.

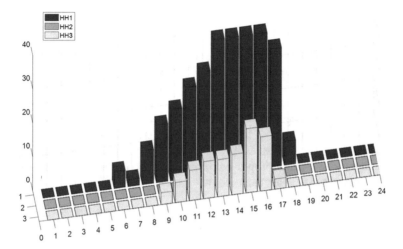

Fig. 7 Power status of battery in RH_1 and RH_3 after the P2P sharing

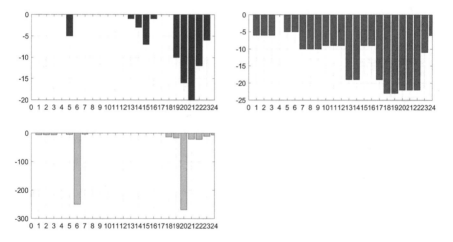

Fig. 8 Surplus power status of RH_1 (blue), RH_2 (red) and RH_3 (green) after one-day simulation of the developed P2P strategy

Power in the battery of RH_1 and RH_2 is shown in Fig. 7.

Surplus power status of every RH after one day of P2P energy sharing is given in Fig. 8.

4 Conclusion

In this paper, a basic peer-to-peer (P2P) energy sharing model for off-grid solar-based rural microgrids is developed. A simple energy management problem is formulated with an objective to minimize the power mismatch of every rural household participating in the P2P system. Simulation results carried out for a basic three rural households connected together with power lines have been discussed. The developed strategy minimizes dependency to manage power during non-solar times of the day on the battery and increases reliability of power to households with no solar panels. The developed model can be improved by incorporating the grid dynamics and work on application of block chain technology.

Acknowledgements Authors wish to acknowledge the contribution of funding agency Netherlands Organization for Scientific Research (NWO) and Eindhoven University of Technology, Netherlands (TU/e).

References

1. Central Electricity Authority (CEA): Load Generation Balance Report 2017–18, May 2017. http://www.cea.nic.in/reports/annual/lgbr/lgbr-2017.pdf
2. Deen Dayal Upadhyaya Gram Jyoti Yojana (DDUGJY) Dashboard, Ministry of Power, Government of India, August 2017. http://www.ddugjy.in/
3. Grameen Vidyutikaran (GARV) Dashboard, Ministry of Power, Government of India, Accessed 30 Aug 2017
4. NITI Aayog, Government of India, Draft National Energy Policy, June 2017. http://niti.gov.in/writereaddata/files/new_initiatives/NEP-ID_27.06.2017.pdf
5. Press Information Bureau (PIB), Government of India, Ministry of Power, May 2017. http://pib.nic.in/newsite/mberel.aspx?relid=161968
6. Harish, V.S.K.V., Kumar, A.: A review on modeling and simulation of building energy systems. Renew. Sustain. Energy Rev. **56**, 1272–1292. ISSN 1364-0321 (2016)
7. Liu, T., Tan, X., Sun, B., Wu, Y., Guan, X., Tsang, D.H.: Energy management of cooperative microgrids with p2p energy sharing in distribution networks. In: 2015 IEEE International Conference on Smart Grid Communications (SmartGridComm), pp. 410–415. IEEE (2015)

Implementation of Hebbian-LMS Learning Algorithm Using Artificial Neural Network

Vartika and Sakshi

Abstract This paper presents the study and analysis of a learning algorithm called as Hebbian-LMS learning rule by the means of an artificial neural network (ANN). Hebbian-LMS itself is a combination of two learning paradigms, which are LMS and Hebb's rule being supervised and unsupervised, respectively. The combined Hebbian-LMS acts as unsupervised learning and has a wide application in the field of engineering. In this paper, Hebbian-LMS is combined with LMS algorithm using ANN, which makes the whole neural network architecture supervised in nature.

Keywords Artificial neural network · Hebbian-LMS · Learning
Supervised learning · Unsupervised learning

1 Introduction

Artificial neural networks usually termed as ANN are the networks that process data and consist processing elements called as neurons; their architecture has been inspired by the biological neuron, but the processing has gone far from the biological inspiration. There are different types of neural networks [1], but the underneath principle remains the same. Multilayer feed-forward neural networks are the prominent neural networks which have numerous applications in the field of chemistry [2] and engineering. In the engineering field, pattern recognition is the main application of neural networks. Machine learning is required to make ANN learn and perform a specific task, and hence, implementing machine learning to the ANN is called as training. There are two types of learning processes: supervised learning and unsupervised learning. In supervised learning (e.g., multilayer neural network), the neural network knows the desired response and weights are adjusted in such a

Vartika (✉) · Sakshi
Thapar University, Patiala, Punjab, India
e-mail: vartika.chauhan.16@gmail.com

Sakshi
e-mail: sakshi.bajaj@thapar.edu

© Springer Nature Singapore Pte Ltd. 2019   645
J. C. Bansal et al. (eds.), *Soft Computing for Problem Solving*, Advances in Intelligent Systems and Computing 817, https://doi.org/10.1007/978-981-13-1595-4_51

manner that calculated output and desired output must be as close as possible. An error signal which is a difference of calculated and desired outputs is generated in order to do so. In unsupervised learning (e.g., Kohonen network [3]), instead of any desired response, the network is supposed to learn itself from the provided pattern of inputs after some iterations. Hebbian-LMS is the learning algorithm obtained by combining the Hebbian and LMS learning paradigms [4]. Widrow along with other authors discussed various applications and future aspects of Hebbian-LMS learning thoroughly in his paper (2015) [4].

2 Hebbian-LMS Learning Rule

The Hebbian-LMS algorithm is the combination of two widely used algorithms, one being supervised and other unsupervised, and the combination itself is an unsupervised learning. Hebbian-LMS is expected to provide some insight into the living neural system learning [4].

To understand Hebbian-LMS, knowledge of LMS and Hebbian individually is essential. For LMS learning algorithm, a number of variants have been introduced [5], but the basic LMS is used in most of the engineering applications. Assume the input vector to be X_k, where k = 1, 2 ..., N with corresponding weight matrix $W_K = [w_{1k}, w_{2k}, \ldots, w_{Nk}]^T$, and their inner product $y_K = X_K^T W_K = (SUM)_K y_k$. The input vector consisted value between -1 and $+1$. d_k is the desired response for the applied input vector. The weights are trained as follows:

$$W_{k+1} = W_k + 2\mu e_k X_K \tag{1}$$

$$e_k = d_k - y_k \tag{2}$$

An error function e_k has been generated according to which the weights are modified. The Hebbian learning rule has no desired response and hence called as unsupervised learning. The error function is calculated as follows:

$$e_k = SGM((SUM)_k) - \gamma (SUM)_k \tag{3}$$

Instead of SGN(.), sigmoid activation function is preferred in case of Hebbian-LMS and represented as SGM(.). Initially, the weights have randomized and the adaptation is performed using Eq. (1), and the error is determined by using Eq. (3) in case of Hebbian-LMS learning rule. The parameter μ [6] controls the speed of convergence and stability and γ must have any value less than the initial slope of the sigmoid function.

The final output of the neuron is given by:

$$(OUTPUT)_k = \begin{cases} SGM(X_k W_k), X_k^T W_k \geq 0 \\ 0, X_k^T W_k \leq 0 \end{cases} \tag{4}$$

Equation (4) shows that the final response of the network which must lie between 0 and +1. In ANN, the output nature is depended on the input nature. If the input vector is not linearly independent, then the output of the first layer will never be purely binary. The output of the second layer will be more close to the binary as compared to the first one. The capacity of the network is equally important in neural networks [7]. A neural network implementing Hebbian-LMS can also be used as element of cognitive memory system [8], and Hebbian-LMS algorithm requires the parameter μ, the learning step to be chosen same as in case of LMS learning [6].

3 Structure and Methodology Used

Here, a system implementing both the learning paradigms, i.e., supervised and unsupervised for different layers of ANN, has been introduced. This section describes the method of constructing and training the multilayer feed-forward neural network used by the proposed algorithm, consisting three layers named input, hidden, and output layers. An ANN can have more than one hidden layer with each layer having different number of neurons, and each neuron has an activation function [9]. In this particular case, the hyperbolic tangent and sigmoid activation functions have been used. The proposed system has hidden layer which was trained using Hebbian-LMS learning rule [4], and the output layer was trained by LMS learning rule [10].

The developed ANN architecture acted as a supervised system as the last layer implemented LMS and therefore had desired response. Figure 1 shows a multilayer neural network with three layers consisting different number of neurons. The network shown is 4-4-3 network; i.e., the input layer (first layer) and hidden layer (second layer) has four neurons, and output layer (third layer) has three neurons. This particular shown network is used for iris dataset's initial analysis as it has four number of attributes (that are sepal length, petal length, sepal width, and petal width) and three number of output classifications (namely Setosa, Versicolor, Verginica). For further analysis of other datasets [11], the number of hidden layer neurons was changed accordingly. The operation performed on these synapses was parallel which means that the weights were adjusted or adapted concurrently, which increases the speed of convergence of the complete network. If these layers were trained until their individual convergences, i.e., the second layer synapse adaptation would have started after the convergence of the first layer and third layer adaptation commenced after the convergence of the second layer, then the time taken for training would be three times of the single neuron training. Parallel operation trains all the layers simultaneously and makes the network faster.

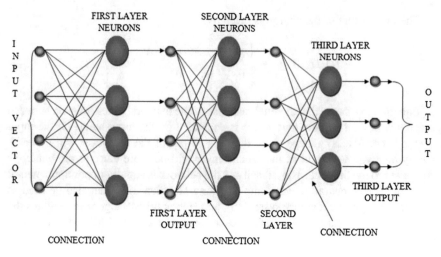

Fig. 1 Multilayer neural network

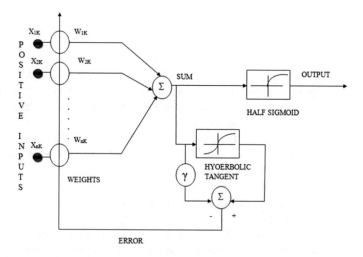

Fig. 2 A neuron trained with Hebbian-LMS learning

Figure 2 shows a diagram of a neuron whose weights were trained with Hebbian-LMS learning. All the synaptic weights were initially set to random values, and then, weights were adjusted by employing Eqs. (1) and (3). The proposed system used the hyperbolic tangent function as its activation function for hidden layer. So the error Eq. (3) can be modified as:

$$error = e_k = \tanh X_k^T W_k - \gamma X_k^T W_k \tag{5}$$

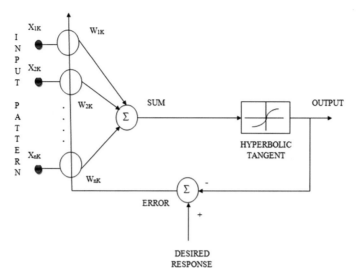

Fig. 3 A neuron trained with LMS algorithm

The final output of this layer is obtained by using Eq. (4); the output layer received its input from the hidden layer neurons. The half-sigmoid function was applied to introduce the nonlinearity to the hidden layer output, and it limited the values between 0 and 1. Figure 3 shows the diagram of a neuron implementing LMS learning algorithm. The notations used are similar to the previous one; i.e., X_k is input vector to output layer which is now the output of the hidden layer, W_k is weight vector, and e_k is the error.

4 Proposed Algorithm Results

The analysis of the proposed algorithm was done for different cases considering different conditions using MATLAB tool. The analysis has been performed:

1. For different standard datasets, i.e., iris, heart disease, gesture phase, and breast cancer datasets. These datasets were obtained from UCI repository [11].
2. For different number of hidden layer neurons for particular datasets, initially starting from number of attributes and gradually increasing to 10, 20, 50, and 100 and for different number of epochs, that is, 100, 1000, and 2000.
3. Mean square error (MSE) comparison with back-propagation learning algorithm [12].

Figures 4, 5, and 6 represent the bar graph of the average success rate of the classification for different datasets considering different number of hidden layer neurons for 100, 1000, and 2000 epochs, respectively. Hidden layer neurons were

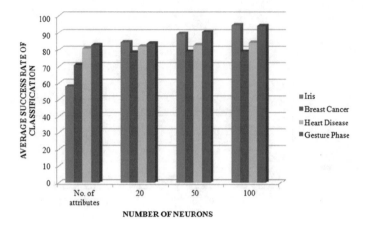

Fig. 4 Average success rate of classification for 100 epochs

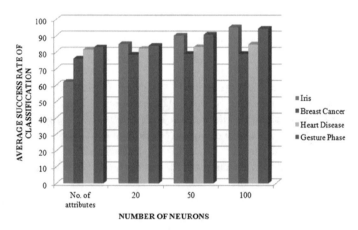

Fig. 5 Average success rate of classification for 1000 epochs

varied starting from the number of attributes in a dataset to 20, 50, and 100. Iris, breast cancer, heart disease, and gesture phase have 4, 9, 13, and 18 number of attributes, respectively. A gradual increase in the success rate can be observed with the increase of the hidden layer neurons. From all three bar graphs, it can also be inferred that the average success rate escalated with increase of epochs too.

Table 1 represents the comparison of the mean square error (MSE) of the proposed algorithm and back-propagation algorithm. The MSE was recorded for different number of neurons in the hidden layers starting from the number of attributes of the datasets for 2000 epochs.

Figures 7, 8, 9, and 10 show the box plots of the mean square error for iris, breast cancer, gesture phase, and heart disease datasets, respectively, using Table 1.

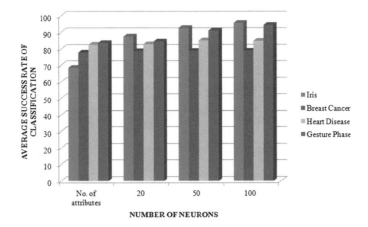

Fig. 6 Average success rate of classification for 2000 epochs

Table 1 Comparison of MSE calculated for proposed and back-propagation algorithm

Algorithm type	Proposed algorithm				Back propagation			
Hidden layer neurons	No. of attributes	20	50	100	No. of attributes	20	50	100
Iris	0.005	0.029	0.013	0.015	0.035	0.103	0.040	0.041
Breast cancer	0.019	0.018	0.017	0.018	0.032	0.034	0.032	0.032
Heart disease	0.015	0.016	0.024	0.047	0.103	0.111	0.117	0.013
Gesture phase	0.004	0.004	0.007	0.007	0.034	0.039	0.030	0.023

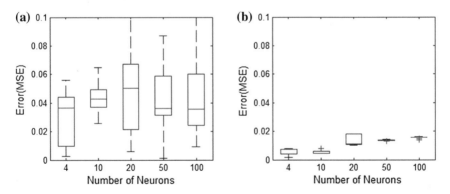

Fig. 7 Box plot of iris dataset MSE versus number of neurons when trained using **a** back-propagation algorithm, **b** proposed algorithm

The (a) in all figures represents the MSE calculated by using back-propagation algorithm [12], and the (b) of all figures shows the MSE computed for the proposed algorithm. It can be seen from Fig. 7a, b that the MSE obtained is more for back-

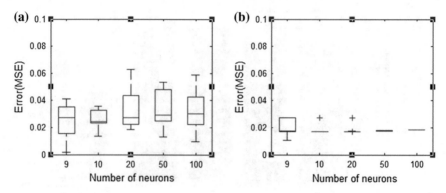

Fig. 8 Box plot of breast cancer dataset MSE versus number of neurons when trained using **a** back-propagation algorithm, **b** proposed algorithm

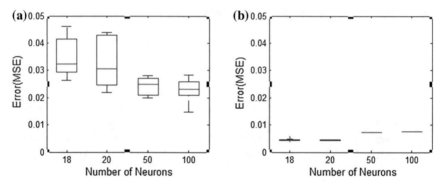

Fig. 9 Box plot of gesture phase dataset MSE versus number of neurons when trained using **a** back-propagation algorithm, **b** proposed algorithm

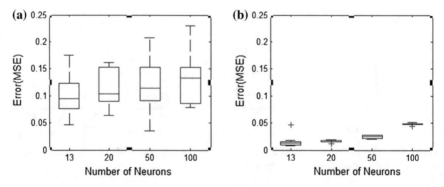

Fig. 10 Box plot of heart disease dataset MSE versus number of neurons when trained using **a** back-propagation algorithm, **b** proposed algorithm

propagation algorithm as compared to the MSE computed by using the proposed algorithm.

The range variation is also more and vast in case of back propagation. Same can be observed in Figs. 8, 9, and 10. The variation between the minimum value of MSE to maximum value for all datasets in case of the proposed algorithm is very less; it can be inferred that the system is more stable as compared to when it is trained with back-propagation algorithm.

5 Conclusion

With the increasing demand of machine learning in technical world and real-time applications, new learning rules and algorithms are being proposed and implemented. These new algorithms must be less time-consuming and should have high efficiency in order to be applicable to new technologies. The research work in this paper presents a neural network system implementing machine learning. The proposed architecture results are comparable to the existing algorithms. Its advantage over Hebbian is that the weights not only keeps increasing, but it can increase or decrease depending upon the neuron state which provides it more stability than Hebb's rule. Hebbian-LMS alone is an unsupervised learning, but when combined with LMS in neural network, the proposed system became supervised. The mean square error (MSE) analyzed for the proposed system is found to be less than the normal back-propagation algorithm MSE, and the range of its maximum value and minimum value was quite less too, making the ANN architecture stable.

6 Future Prospects

Implementing the proposed algorithm to artificial neural network has the tendency to reduce the error and hence making it easy to learn and makes it more viable to numerous practical engineering applications like signal processing, control circuits, pattern recognition and classification, and even in forensic field of science. The neurobiologists are trying to figure out that what kind of learning takes place in human brain or in neural system of living organisms. And by studying and analyzing the Hebbian-LMS learning algorithm in more detail, it is expected that some part of the brain must have been performing this learning. The future scope of Hebbian-LMS using neural network makes scientists and engineers look into its other extensions. And if this is the learning which takes place in human nervous system, the psychobiology field will have better chance to treat patients with psychiatric problems.

References

1. Haykin, S.: Neural Networks—A Comprehensive Foundation. Prentice Hall, New Jersey (1999)
2. Zupan, J., Gasteiger, J.: Neural Networks for Chemists. VCH, New York (1993)
3. Kohonen, T.: Self-organization and Associative Memory. Springer, Berlin (1988)
4. Widrow, B., Kim, Y., Park, D.: The Hebbian-LMS learning. IEEE Comput. Intell. Mag. **10**(4), 37–53 (2015)
5. Narula, V., et al.: Assessment of variants of LMS algorithms for noise cancellation in low and medium frequency signals. In: International Conference Recent Advancements in Electrical, Electronics and Control Engineering (ICONRAEeCE) (2011)
6. Widrow, B., Stearns, S.D.: Adaptive Signal Processing. Prentice Hall, Englewood Cliffs, NJ (1985)
7. Widrow, B., Greenblatt, A., Kim, Y., Park, D.: The No-Prop algorithm: a new learning algorithm for multilayer neural networks. Neural Netw. **37**, 182–188 (2012)
8. Widrow, B., Aragon, J.C.: Cognitive memory. Neural Netw. **41**, 3–14 (2013)
9. Chung, H. et al.: Deep neural network using trainable activation functions. In: Neural Networks (IJCNN), 2016 International Joint Conference on IEEE, Nov (2016)
10. Widrow, B., Hoff Jr., M.E.: Adaptive switching circuits. In: Proceedings IRE WESCON Convention Records, pp. 96–104 (1960)
11. Lichman, M.: UCI machine learning repository. (2013). http://archive.ics.uci.edu/ml
12. Werbos, P.J.: Backpropagation and neural control: a review and prospectus. In: International Joint Conference of Neural Networks, vol. 1, pp. 209–216. Washington (1989)

Modeling Vas Deferens Smooth Muscle Electrophysiology: Role of Ion Channels in Generating Electrical Activity

Chitaranjan Mahapatra and Rohit Manchanda

Abstract The vas deferens smooth muscle (VDSM) cells contract to direct and propel sperms from the epididymis to the urethra. It is well known that membrane electrical activity, particularly the action potential (AP) is an essential prerequisite for the initiation of contraction in all types of muscle cells. As the coordinated activation of a number of ion channels in the VDSM cell membrane causes AP generation, any mutation or dysfunction of any ion channel will modulate the AP generation and hence the contraction. To explore the quantitative contribution of individual active ionic current to the AP generation, a biophysically based single guinea-pig VDSM cell model is presented. The simulated ionic currents and AP show good agreement with the experimental recordings in terms of several parameters. Therefore, this electrophysiological model can be a preliminary platform to investigate the various electrical properties of VDSM cells in both normal and pathological conditions.

Keywords Vas deferens excitability · Ionic currents · Action potential Electrophysiological model

1 Introduction

The vas deferens, also called ductus deferens, a part of male reproductive system is a tubular muscular structure that travels from the epididymis to the urethra. The physiological function of the vas deferens is to contract for the transportation of sperms into the urethra, after the sperm has passed through the epididymis [1, 3, 8]. The coordinated firing of sympathetic nerves causes contraction of the VDSM cells to achieve this transportation task. The elevation of intracellular cytoplasmic calcium $[Ca^{2+}]_i$ is essential for all types of muscle cell contractions. Different experimental studies have demonstrated that membrane electrical activity in the form of APs initiates VDSM

C. Mahapatra (✉) · R. Manchanda
Department of Biosciences & Bioengineering, Indian Institute of Technology
Bombay, Mumbai 400076, India
e-mail: chitaranjan@iitb.ac.in

© Springer Nature Singapore Pte Ltd. 2019
J. C. Bansal et al. (eds.), *Soft Computing for Problem Solving*, Advances in Intelligent Systems and Computing 817, https://doi.org/10.1007/978-981-13-1595-4_52

655

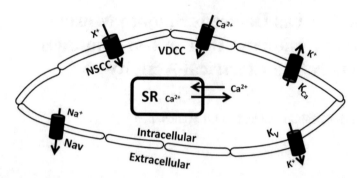

Fig. 1 Illustration of all major ionic currents in the vas deferens smooth muscle cells

cell contraction by permitting extracellular Ca^{2+} influx via the voltage-gated Ca^{2+} channels [1, 2, 13, 14]. Membrane electrical activity also propagates to the neighboring cells as a result of spatial propagation due to the presence of the gap junctions. The gap junctions mediate intracellular connection between the smooth muscle cells [1, 9, 10]. Like other excitable cells, activation of a set of intrinsic active ion channels across the VDSM cell membrane causes AP generation. It is well documented that the voltage-gated Na^+ channels, voltage-gated L-type, and T-type Ca^{2+} channels initiate and mediate the rising phase of the AP, whereas openings of various K^+ channels are essential to mediate the repolarization and after hyperpolarization phases [1–3, 11–13]. Of the large family of mammalian K^+ channel with functional diversity, two types of Ca^{2+}-dependent potassium channels (I_{BK} and I_{SK}) and two types of voltage-dependent potassium channels ($I_{KV1.2}$ and I_{KDR}) are found in VDSM cell [4, 6, 7, 11–13]. Figure 1 illustrates the flow of inward and outward ionic currents via the specific ion channels across the VDSM cell membrane. The NSCC, Na_v, VDCC, K_{Ca}, and K_v are known as the non-specific cation channel, voltage-activated sodium channel, voltage-activated calcium channel, calcium-activated potassium channel, and voltage-activated potassium channel, respectively. SR, also called sarcoplasmic reticulum, is the internal Ca^{2+} store in the VDSM cell. In all excitable cells, active ion channel mutation or dysfunction can lead to pathophysiological conditions. Therefore, a quantitative study of ion channels in modulating the AP generation is essential to evaluate such dysfunctions in the genesis of pathophysiological conditions. Unfortunately, the experimental investigations face inherent difficulties in investigating the modulating roles of all ion channels by pharmaceutical maneuver.

In the present computational era, soft computing provides a significant paradigm shift in modeling information processing for various biological complex systems. Computational models in electrophysiology mimic the experimental results and provide an invaluable platform to aid in developing our understanding of the individual active ionic current contribution to the experimentally observed cellular activities. Although a large number of computational models have been successfully implemented for the neuronal and cardiac cells, there is a dearth of such models in smooth muscle electrophysiology. No computational model of VDSM cell electrophysiology has been developed so far. We present here the first biophysically detailed single VDSM cell model constrained by experimental data which enable us to look at the interactions among various ion channels in mediating the action potential.

2 Methods

A cylindrical structure is considered as the single VDSM cell morphology in our model. The value of the length and diameter of the cylinder are 300 μm and 15 μm, respectively [13]. Like other types of excitable cells, the VDSM cell membrane is also described as a simple resistor–capacitor (RC) circuit with a membrane capacitance C_m in parallel with the variable ion channel conductance g_{ion}. The absolute value of membrane capacitance (C_m) and membrane resistance (R_m) is taken as 14 pF and 6G Ω, respectively [13]. The parallel conductance model of the VDSM cell is illustrated in Fig. 2. The individual ion channel conductances g_{ion} are in series with the respective Nernst potential E_{ion}. In addition to the active ionic currents, one leakage current is also added to compensate for the non-specific currents in the VDSM cell.

Mathematical interpretation of the biological system is critically essential in developing any computational model to visualize the biological system. In this study, the mathematical interpretation is established upon the conventional Hodgkin–Huxley equation. The software platform 'NEURON' is adopted as it is designed to create an environment for simulating both individual and network of excitable tissues. Equation 1 describes the time-dependent characteristics of the membrane potential V_m as a function of total membrane current. The total membrane current is the sum of all ionic currents (I_{Na}, I_{Ca}, I_{KCa}, I_{Kv}, and I_l) and the injected external stimulus I_{stim} to trigger the AP in the model.

$$\frac{dV_m(t)}{dt} = -\frac{1}{C_m}\left(I_{Na} + I_{Ca} + I_{KCa} + I_{Kv} + I_l + I_{stim(t)}\right) \quad (1)$$

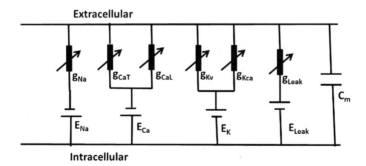

Fig. 2 Illustration of the parallel conductance model for vas deferens smooth muscle cell

The individual ion channel current is calculated according to the Hodgkin–Huxley formulation mechanism which is presented in Eq. 2.

$$I_{ion} = \bar{g}\left[m\left(V_m, t, \left[Ca^{2+}\right]_i\right)\right]^x \left[h\left(V_m, t, \left[Ca^{2+}\right]_i\right)\right]^y (V_m - E_{rev}) \tag{2}$$

where $\bar{g}$ and E_{rev} are the maximum conductance and reversal potential (Nernst potential) of the respective ion channel. The time-dependent dimensionless gating variable 'm' and 'h' are the voltage/calcium-dependent activation and inactivation parameters. The 'x' and 'y' are the power to 'm' and 'h' parameters, respectively.

The instantaneous value of any gating variable is calculated by a first-order differential equation. For example, Eq. 3 calculates the instantaneous value of the activation variable 'm' in the VDSM cell model.

$$\frac{dm(V_m, t)}{dt} = \frac{m_\infty(V_m) - m(V_m, t)}{\tau_m} \tag{3}$$

where m_∞, the steady-state value, and τ_m, the time constant, all being functions of voltage and/or ionic concentrations.

Here, the state parameter dependence on V_m for ion channels is described by the Boltzmann equation.

$$m_\infty = 1/1 + \exp\left(\left(V_m + V_{m\frac{1}{2}}\right)/S_m\right) \tag{4}$$

where $V_{1/2}$ is the half-activation potential, and S is the slope factor. These values are adapted from the published experimental data.

2.1 Intracellular Ion Concentration

As the intracellular/cytoplasmic calcium concentration elevation is essential for the VDSM contraction, calcium dynamics plays an important role in electrophysiological modeling. In order to trigger the calcium-dependent properties of Ca^{2+}-activated K^+ channels and to update the equilibrium potential of the Ca^{2+} ion, it was necessary to calculate the intracellular Ca^{2+} concentration at every time instant during the AP simulation. The whole cell calcium dynamics mechanisms consist of several sub-mechanisms such as diffusion, buffering, SR uptake and release, Na^+–Ca^{2+} exchanger, and pump extrusion to update the intracellular Ca^{2+} concentration. Due to lack of experimental data, we did not incorporate a biophysically detailed realistic intracellular Ca^{2+} dynamics in our model. Instead, we assumed that the Ca^{2+} which enters via Ca^{2+} channels instantaneously diffused within a thin sub-membrane shell and that determining the decay of $[Ca^{2+}]_i$ could be lumped into a single-exponential function.

$$\frac{d[Ca^{2+}]_i}{dt} = \left(-100 * \frac{i_{Ca}}{2 * F * d}\right) + \frac{[Ca^{2+}]_{i\infty} - [Ca^{2+}]_i}{\tau_r} \tag{5}$$

where i_{Ca} is the inward Ca^{2+} flux due to voltage-gated Ca^{2+} channels, d is the depth of the sub-membrane cell, $[Ca^{2+}]_{i\infty}$ is the baseline Ca^{2+} concentration, and τ_r is the time constant.

3 Results

3.1 Inward Currents

It is well known that the AP generation is achieved by the influx of cations (sodium ions or calcium ions) or efflux of anions (chloride ions) through the selective ion channels. In the VDSM cells, the inward current is mainly carried through the voltage-gated Ca^{2+} channels (L-type and T-type) and voltage-dependent Na^+ channel [1–3, 11–13]. The fast component of the total inward current is provided by the voltage-gated sodium channels [2, 14]. Figure 3 represents the voltage-gated Na^+ channel current under the voltage clamp protocol. The voltage is stepped from a holding potential (V_h) of -70 to -10 mV for a duration of 50 ms. The reverse potential E_{Na} is taken as 50 mV in Eq. 6 to calculate the I_{Na}. The half-activation potential, half-inactivation potential, activation slope factor, and inactivation slope factors are -26 mV, -46.3 mV, 7 mV, and 7.5 mV, respectively, to fit Boltzmann function in Eqs. 7 and 8. The τ_m and τ_h in Eqs. 9 and 10 are the activation and inactivation time constants, respectively.

$$I_{Na} = \overline{g_{Na}}m^2h(V - E_{Na}) \tag{6}$$

$$m_\infty = \frac{1}{1 + \exp\left(\frac{-(V + 26)}{7}\right)} \tag{7}$$

Fig. 3 VDSM cells I_{Na} model. The voltage is stepped from a holding potential (V_h) of -70 to -10 mV for a duration of 50 ms with respect to different concentrations of TTX

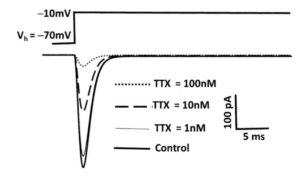

$$h_\infty = \cfrac{1}{1 + \exp\left(\frac{(V+46.3)}{7.5}\right)} \tag{8}$$

$$\tau_m = 0.45 + \cfrac{3.9}{1 + \exp\left(\frac{(V+66)}{26}\right)} \tag{9}$$

$$\tau_h = 150 - \cfrac{150}{\left(1 + \exp\left(\frac{(V-417.43)}{203.18}\right)\right) * \left(1 + \exp\left(-\frac{(V+61.11)}{8.07}\right)\right)} \tag{10}$$

The simulated I_{Na} in Fig. 3 mimics the experimental current (Fig. 3a, [14]) with good agreement for the TTX drug concentration of 1, 10, and 100 nM. The TTX drug model was induced with the Na$^+$ channel after simulating the concentration-response relationship curve (Fig. 3b, [14]).

3.2 Outward Currents

Several types of K$^+$ channels modulate membrane excitability and mediating repolarizing phase of the action potential in VDSM cells (see Introduction). For our model, the electrophysiological data for two types of Ca^{2+}-dependent potassium channels (I_{BK} and I_{SK}) and two types of voltage-dependent potassium channels ($I_{KV1.2}$ and I_{KDR}) are adapted from the VDSM cells of the rat, mouse, guinea pig, and human tissue [4, 6, 13]. All potassium currents have been simulated by their voltage/calcium-dependent activation properties and sensitivity to pharmacological blockers. Figure 4 represents the fast-inactivating K$^+$ current (I_{Kf}) under the voltage clamp protocol. The membrane potential (V_m) is stepped from a holding potential (V_h) of -80 to 60 mV for a duration of 500 ms. The reverse potential E_K is taken as -75 mV in Eq. 11 to calculate the I_K. The half-activation potential, half-inactivation potential, activation slope factor, and inactivation slope factors are -26.1 mV, -56.1 mV, 15.4 mV, and 9.8 mV, respectively, to fit Boltzmann function in Eqs. 12 and 13. The τ_m and τ_h in Eqs. 14 and 15 are the activation and inactivation time constants, respectively.

$$I_{Kf} = \overline{g_{Kf}} mh(V - E_K) \tag{11}$$

Fig. 4 Fast-inactivating voltage-gated K$^+$ channel model in VDSM cell. The membrane potential (V_m) is stepped from a holding potential (V_h) of -80 to 60 mV for a duration of 500 ms

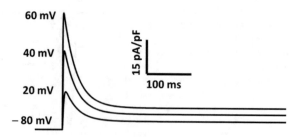

$$m_\infty = \frac{1}{1 + \exp\left(\frac{-(V + 26.1)}{15.4}\right)} \tag{12}$$

$$h_\infty = \frac{0.94}{1 + \exp\left(\frac{(V + 56.1)}{9.8}\right)} \tag{13}$$

$$\tau_m = \frac{1}{1 + \left(\frac{(V + 15)}{20}\right)^2} \tag{14}$$

$$\tau_h = \left(80 - \frac{80}{\left(1 + exp\left(\frac{(V - 800)}{300}\right)^1\right)\left(1 + \exp\left(\frac{-(V + 20)}{8}\right)\right)}\right) \tag{15}$$

The simulated I_{Kf} in Fig. 4 mimics the experimental current (Fig. 3a, [4]) with good agreement.

3.3 AP Simulation

The resting membrane potential V_m is determined mostly by the balance between depolarizing and repolarizing currents through Na_v, K_{Ca}, K_v, K_{leak}, and T-type Ca^{2+} channel. In this model, resting V_m is fine-tuned to around -52 mV [1–3, 7, 11–13] by adjusting the conductances of all ion channels within the physiological range. Resting intracellular calcium concentration is taken as 50 nM. An external stimulus current of 50 pA as a brief square pulse is injected for 50 ms to evoke an AP (solid line in Fig. 5) in our VDSM model. The threshold voltage to trigger the AP is approximately -30 mV. The resting membrane potential, the peak of the AP overshoot, and after-hyperpolarization are -52, 16, and -64 mV, respectively. The adapted experimental data from Imaizumi et al. [7] are plotted (filled triangle in Fig. 4) against the simulated AP.

Fig. 5 Simulated AP (solid line) after inducing current of 50 pA as a brief square pulse for 50 ms. Filled triangles are experimental data adapted from Imaizumi et al. [7]

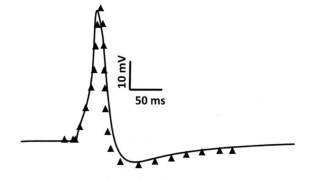

Table 1 Comparison of RMP, peak amplitude, AP duration, and AHP between the simulated AP and experimental AP [7]

Objects	RMP (mV)	Peak (mV)	AHP (mV)	Duration (ms)
Experiment	−52	18	−66	35
Simulation	−52	16	−64	38

This comparison shows a good agreement between the experimental and simulated result. However, at first glance, the major discrepancy from the experimental recording is that the experimental AP exhibits a more prominent positive peak and shifted repolarization phase. The primary explanation for this discrepancy may be attributed to experimental recordings in isolated bundle strip against single cell model in simulation (Table 1).

4 Conclusion

Computational modeling in electrophysiology is established as an essential complement to the experimental research, because it elaborates all active ionic currents quantitatively underlying the electrophysiological phenomena. We developed a biophysically detailed electrophysiological model for the VDSM cell and this is the first model of its kind to the best of our knowledge. The major active ion channel currents were simulated and validated against the data from which they were adapted. After incorporating all ion channels in a single VDSM cell model, we balanced all ionic conductances to maintain the resting membrane potential. We then injected an external current stimulus to evoke the action potential. The simulated action potential showed good agreement with its experimental counterpart. However, several queries can be raised regarding the assumptions and simplifications considered during the development of the model. The major limitation in the model development is that of borrowing experimental data from VDSM cell electrophysiology from various species. A perfect physiologically realistic model must adopt electrophysiological data recorded from a single species. Due to lack of experimental data from a single species, we therefore adopted modeling parameter values from the VDSM tissue in different species (rat, human, mouse). However, despite some limitations, this model was able to generate the guinea-pig VDSM cellular behavior and delivered a fundamental description available to date for the ionic selectivity and activation mechanisms underlying the AP generation in VDSM cell.

In the future, additional components such as Na^+–Ca^{2+} exchanger, plasma membrane Ca^{2+} ATPase pump, and sarcoplasmic reticulum Ca^{2+} ATPase pump can be incorporated into this elementary model toward developing a more comprehensive model. The comprehensive model can simulate the influence of dynamic fluctuations of the intracellular Ca^{2+} concentration and ion channel activation parameters on the membrane action potential. In parallel, this model can be extended into the network level in order to establish a more robust and physiologically realistic computational model for the future investigation at the tissue level.

Acknowledgements This work is supported in part by Department of Biotechnology (DBT), India (grant number BT/PR12973/MED/122/47/2016).

References

1. Beattie, D.T., Cunnane, T.C., Muir, T.C.: Effects of calcium channel antagonists on action potential conduction and transmitter release in the guinea-pig vas deferens. Br. J. Pharmacol. **89**, 235–244 (1986)
2. Belevych, A.E., Zima, A.V., Vladimirova, I.A., Hirata, H., Jurkiewicz, A., Jurkiewicz, N.H., Shuba, M.F.: TTX-sensitive Na^+ and nifedipine-sensitive Ca^{2+} channels in rat vas deferens smooth muscle cells. In: Biochimica et Biophysica Acta (BBA)-Biomembranes, vol. 1419, pp. 343–352 (1999)
3. Bennett, M.R.: The effect of intracellular current pulses in smooth muscle cells of the guinea pig vas deferens at rest and during transmission. J. General Physiol. **50**, 2459–2475 (1967)
4. Harhun, M.I., Jurkiewicz, A., Jurkiewicz, N.H., Kryshtal, D.O., Shuba, M.F., Vladimirova, I.A.: Voltage-gated potassium currents in rat vas deferens smooth muscle cells. Pflügers Archiv **446**, 380–386 (2003)
5. Hines, M.L., Carnevale, N.T.: The NEURON simulation environment. Neural Comput. **9**, 1179–1209 (1997)
6. Huang, Y., Lau, C.W., Ho, I.H.M.: NS 1619 activates Ca^{2+}-activated K^+ currents in rat vas deferens. Eur. J. Pharmacol. **325**, 21–27 (1997)
7. Imaizumi, Y., Torii, Y., Ohi, Y., Nagano, N., Atsuki, K., Yamamura, H., Bolton, T.B.: Ca^{2+} images and K^+ current during depolarization in smooth muscle cells of the guinea-pig vas deferens and urinary bladder. J. Physiol. **510**, 705–719 (1998)
8. Koslov, D.S., Andersson, K.E.: Physiological and pharmacological aspects of the vas deferens—an update. Front. Pharmacol. **4**, 101 (2013)
9. Manchanda, R., Venkateswarlu, K.: Effects of heptanol on electrical activity in the guinea-pig vas deferens. Br. J. Pharmacol. **120**, 367–370 (1997)
10. McLean, M.J., Pelleg, A., Sperelakis, N.: Electrophysiological recordings from spontaneously contracting reaggregates of cultured smooth muscle cells from guinea pig vas deferens. J. Cell Biol. **80**, 539–552 (1979)
11. Ohi, Y., Yamamura, H., Nagano, N., Ohya, S., Muraki, K., Watanabe, M., Imaizumi, Y.: Local Ca^{2+} transients and distribution of BK channels and ryanodine receptors in smooth muscle cells of guinea-pig vas deferens and urinary bladder. J. Physiol. **534**, 313–326 (2001)
12. Ohya, S., Yamamura, H., Muraki, K., Watanabe, M., Imaizumi, Y.: Comparative study of the molecular and functional expression of L-type Ca^{2+} channels and large-conductance, Ca^{2+}-activated K^+ channels in rabbit aorta and vas deferens smooth muscle. Pflügers Archiv. **441**, 611–620 (2001)
13. Park, S.Y., Lee, M.Y., Keum, E.M., Myung, S.C., Kim, S.C.: Ionic currents in single smooth muscle cells of the human vas deferens. J. Urol. **172**, 628–633 (2004)
14. Zhu, H.L., Aishima, M., Morinaga, H., Wassall, R.D., Shibata, A., Iwasa, K., Teramoto, N.: Molecular and biophysical properties of voltage-gated Na^+ channels in murine vas deferens. Biophys. J. **94**, 3340–3351 (2008)

Energy-Efficient Data Aggregation Using Cluster-Based Comb–Needle Model in Wireless Sensor Networks

M. Shanmukhi, Rajesh Eshwarawaka, K. Renuka and K. Durga Preethi

Abstract Nowadays, sensors are being used in a wide spectrum of applications including mission-critical military as well as civil domains. Data dissemination and data aggregation are important aspects of wireless sensor networks. In this paper, the motivation for use of clustering in comb–needle model (CNM) in WSN is presented. Then, the proposed cluster-based CNM (CCNM) is elaborated with suitable illustrations. The outcome of mathematical analysis of the CCNM in terms of communication, computation, and storage overheads is included next. The simulation environment is also explained. This paper is concluded with the performance evaluation of the proposed model in terms of the criteria, namely communication cost, energy consumption, throughput, delay, packet delivery ratio, and packet loss. The communication cost of the proposed model is $O(\sqrt{N/Nc})$. N denotes the total number of sensor nodes in the network, and N_c denotes total number of clusters identified in the sensor network.

Keywords WSN · CNM · Energy

1 Introduction

Wireless sensor network (WSN) is a collection of spatially distributed autonomous sensor nodes deployed in different environments for sensing and gathering data. A sensor node is a tiny special purpose circuit, which senses, computes, and communicates to the sink, and it is specifically manufactured for the particular purpose. A sensor is a transducer which senses physical parameters such as temperature, pressure, sound, and movement of objects to mention few. WSN can be used for

M. Shanmukhi (✉) · R. Eshwarawaka · K. Durga Preethi
BVRIT HYDERABAD College of Engineering for Women, Bachupally, Hyderabad, Telangana State, India
e-mail: shanmukhi.m@gmail.com

K. Renuka
CMREC, Kandlakoya, Medchal, Hyderabad, Telangana State, India

© Springer Nature Singapore Pte Ltd. 2019
J. C. Bansal et al. (eds.), *Soft Computing for Problem Solving*, Advances in Intelligent Systems and Computing 817, https://doi.org/10.1007/978-981-13-1595-4_53

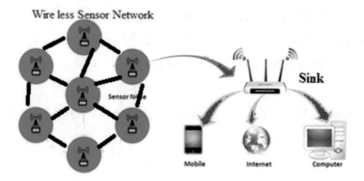

Fig. 1 A typical wireless sensor network scenario

monitoring different physical and environmental conditions. They are best used for surveillance in civil and military environments. The sensor nodes can be deployed in an environment where it is difficult for humans to have manual surveillance. The sensor nodes in WSN can cooperatively pass sensed data to the main location where a special node called sink or base station is deployed. All sensor nodes send data to the base station, where sensed data can be used by users. Sensor networks became ubiquitous as they are used now in many areas such as machine health monitoring (prognosis), monitoring household events, industrial process monitoring besides precision agriculture and consumer applications [1, 2]. The typical WSN is shown in Fig. 1.

Sensor nodes in WSN are resource constrained. They are often deployed to work from hostile environments where they are supposed to sense data for different purposes. An important observation here is that the nodes have a limited lifetime due to energy constraint. It is not possible either to recharge or replenish the battery of a sensor node. Therefore, it is indispensable to have energy-efficient operations of WSN such as data gathering, information dissemination. Optimization of WSN functionality to improve lifespan of the network is an open research problem to be addressed. The sensor nodes in a WSN sense the data and communicate to the sink; in this case, same information may come from the particular region to the base station, and there is redundancy. It is a waste of the energy of sensor node. As it is not possible to recharge a sensor node battery, it should be used conservatively. To avoid the redundancy, data can be aggregated while it is being communicated to the sink. The rationale behind the data aggregation is that it reduces the number of data transmissions. This has an influence on energy efficiency and thus made data aggregation is an open research problem in WSNs. The intermediate nodes also participate in data aggregation [3] to reduce number of transmissions and thereby extend the life of the network. As shown in Fig. 2, intermediate nodes in WSN are acting as aggregators.

Different aggregation techniques came into existence. They are meant for promoting lifespan of WSN besides reducing communication overhead in the network.

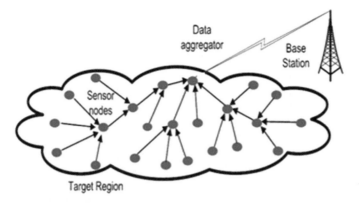

Fig. 2 Data aggregation in WSN

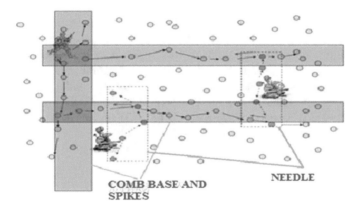

Fig. 3 A comb–needle model example

The research in this paper is exploring CNM in WSN. The comb–needle model (CNM) came into existence which paved way for comb–needle-based method for energy-efficient data aggregation. The CNM is an important research area which can have huge impact on the WSN, which is briefed in Fig. 3. The CNM can be further improved with clustering. The objective of our research is to augment the performance of the CNM with clustering concept in grid wireless sensor networks.

The organization of the paper is as follows: Section 1.1 describes the CNM, the related work is discussed in Sect. 2, Sect. 3 illustrates the proposed work, the simulation results presented in Sect. 4, and the paper concluded in Sect. 5.

1.1   Comb–Needle Model (CNM)

A comb–needle model [4] is based on ancient method, where combing is performed in sand or haystack to find out needles. In the same fashion to get the data from WSN, the querying is done, when the query is generated that is traversed in the network that forms a comb structure that extracts the needles called events. Figure 3 illustrates the comb–needle model.

In the basic of comb–needle model, whenever an event occurs, all the sensor nodes present on the comb (base as well as spikes) participate in transmitting the reply towards the base station. This results in lot of communication cost, which depletes nodes' energy. To lessen this communication cost, cluster-based approach is used for data gathering and subsequently data aggregation before forwarding the result towards the base station. Cluster-based approach groups the sensor nodes in the network into certain number of clusters. For each such cluster, one node is designated as cluster head (CH). All the nodes detecting the events communicate event occurrence to their CH. Then, CH aggregates the received data and forwards it towards the base station. There are few issues involved in this clustering approach. First issue is, how many clusters need to be formed that could optimize the communication cost. Second issue is which node should act as CH. By overcoming, these issues can leverage the energy-efficient data aggregation in WSN. This is the motivation behind the research which is aimed at exploring energy-efficient data aggregation in WSN by implementing different variants of CNM.

2   Related Work

Data aggregation is an important phenomenon in WSN for improving communication efficiency besides reducing power consumption. Invariably, it leads to increased lifetime of the network. As the nodes in such network have limited resources, it is essential to have strategies to utilize them in an optimal way. Data aggregation is one of the approaches that can lead to reduced energy consumption in WSN. In this chapter, a comprehensive review of the literature is made on data aggregation techniques. This review helped to identify the research objectives. Various techniques found in the literature suggest that reduction of data transfer or reducing communication overhead can increase energy efficiency in WSN. The rest of the chapter throws light on different techniques used by researchers for data aggregation. Most of the techniques have optimized the data aggregation in order to achieve energy efficiency.

Cam et al. [5] proposed a protocol named energy-efficient secure pattern-based data aggregation for energy-efficient data aggregation in WSN. The concept of pattern codes reduces the overhead of cluster head (CH) as CH need not know the data of sensors for data aggregation, which improves the performance significantly. Thein et al. [6] employed cluster head selection approach for energy-efficient data aggregation. CH is elected in a manner the CH positions are rotated and energy load is distributed among all sensor nodes. Thereby, the life of the network is expanded. Xiang et al. [7, 8] explored the concept of compressed sensing (CS) while collecting data in WSN. They employed both joint routing and compressed aggregation to achieve energy-efficient data aggregation. They also used greedy heuristics to validate their approach. Girao et al. [9] proposed an approach that conceals sensed data and efficient in-network aggregation of data for reducing energy and achieve reverse multicast traffic. In-network aggregation techniques were reviewed by Fasolo et al. [10]. Chen et al. [11] explored an energy-efficient protocol for aggregator selection in order to optimize data aggregation task and reduce energy consumption. Kumar et al. [12] explored heterogeneity of nodes in WSN for energy efficiency. They proposed energy-efficient heterogeneous cluster scheme (EEHC) for the same and found that it could prolong network lifetime. Rajendran et al. [13] proposed an energy-efficient cluster-based routing protocol and compared it with leach and achieved optimized results.

Al-karaki et al. [14] focused on different exact and approximation algorithms for efficient data aggregation. Several factors may affect data aggregation such as the aggregation function, placement of aggregation points, and the density of the sensors in the network. The determination of an aggregation point is extremely important. They presented an exact and as well as approximate algorithms to find the lowest number of aggregation points in order to maximize the network lifetime. Kalpakis et al. [15] focused on data gathering problem and addressed it by using solutions with polynomial running time to improve the lifetime of network. Ciancio et al. [16] also explored data aggregation problem in WSN by using distributed wavelet compression algorithm. They made use of routing strategies and data representation algorithms in order to achieve energy-efficient data aggregation. Luo et al. [17] also explored compressive data gathering, where it will reduce global communication cost without introducing complicated transmission control. Li et al. [18] proposed a scheme known as energy-efficient high accuracy (EEHA). This scheme for secure data aggregation and found its accurate aggregation abilities. Chatterjea et al. [19] proposed a distributed and self-organizing scheduling algorithm for eliminating dropped messages besides achieving energy-efficient data aggregation. Dimokas et al. [20] explored distributed clustering for energy efficiency and network longevity. Liu et al. [21] exploited the spatiotemporal correlation in WSN energy-efficient data collection. Based on the spatiotemporal correlation of sensor data, the sensor nodes are clustered for achieving this.

3 Proposed Work

3.1 Assumptions

In the research pertaining to CCNM, there are certain assumptions. They are as follows:

- An event, against which sensor nodes have to respond, may occur anywhere in the sensing area and any time.
- Queries come to nodes from time to time. Any node may get a query at any given time.
- The nodes are aware of their locations in the sensing area.
- Regular network and reliable links are assumed.

3.2 Proposed Cluster-Based Cnm

This section describes the proposed cluster-based CNM. The basic CNM is enhanced with clustering. In this model, it is assumed that sensor nodes are installed in fixed locations. The WSN is divided into clusters. The nodes in a cluster provide sensed data to their CH and in turn CH sends data to BS after aggregation.

In CNM, the query dissemination resembles comb and event notification appears as a needle [4] that looks like extracting needles from sand using a comb. Figure 4 shows the basic CNM with regular network topology.

Nodes are deployed in the form of a grid (square) size $n \times n$, i.e., there are n^2 nodes in the WSN. The position of a sensor node is (x, y) where $0 \leq x$, $y < n$. The number of clusters and the size of each cluster are fixed. If the n value is 9, then the total number of nodes in the network is 81. There are three clusters; each cluster has 27 nodes including CH.

Each CH has communication with 26 nodes. Those sensor nodes send sensed data to their CH when an event occurs. Stated differently, a node located at (x, y) founds

Fig. 4 Basic comb–needle model [3]

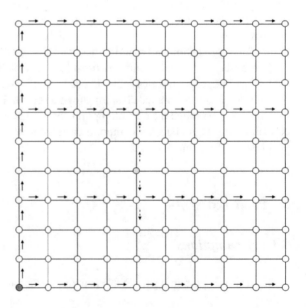

an event and sends data to a node located at (x, y + 1) which is denoted as d1. Then, the data is sent to (x, y + s/2) which is denoted as d-d1 horizontally and vertically to (x ± 1, y), (x ± 2, y), ..., which is denoted as *l*. Here, *s* denotes distance between comb spikes called as degree of the comb. Comb is formed by query propagation while the needle is formed by event notification. In this fashion, multiple combs (multiple queries) and multiple needles (multiple replies) may exist.

In the real-world example, multiple soldiers might request the location of enemy tanks at a time, which results in having several combs and corresponding needles. In this research, a single comb and single event notification are considered to prove the hypothesis.

3.3 Analysis of the CCNM

This section provides information regarding the formation of comb–needle structure in the proposed cluster-based CNM. The number of nodes and the number of clusters in WSN are fixed. The clusters are shown in Fig. 5. The nodes in WSN are in a grid. The number of nodes in the network is denoted as N.

$$N = n \times n \qquad (1)$$

For each query in the WSN, the communication cost is computed. Every event generates a spike which has length *l* (number of hops) from the corresponding comb. Therefore, the communication cost is

$$C_l = l - 1 \qquad (2)$$

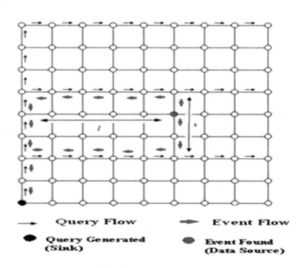

Fig. 5 Basic comb–needle model, where l = 6 and s = 3

→ Query Flow   ━● Event Flow

● Query Generated (Sink)   ● Event Found (Data Source)

Any node in the network can generate an event. There are many parameters considered regarding events and queries. They are known as query frequency, event frequency, and frequency of events per sensor node.

f_q = Query frequency
f_e = Event frequency
$f_d = f_e/n^2$, Frequency of events per sensor node.

The communication cost is also influenced by the fact that broadcast or unicast is used. For proof of the concept, unicast is used in this research. The cost of having a single spike with length l is computed as $l - 1$. The number of events generated by a query in this model is represented as $l - 1(f_e/f_q)$. The event generation cost per needle is

$$C_l\left(\frac{f_e}{f_q}\right) \tag{3}$$

The process of query dissemination is same for both CNM and CCNM models. Hence, the communication cost for query dissemination for both the models is denoted as C_{qd}.

The query dissemination cost is computed as follows:

The query dissemination over the base of the comb is $(n - 1)$
The number of spikes per comb = floor $(n - 1/s) + 1$.

Hence, the total query dissemination cost = query dissemination cost over the base + (the dissemination cost per spike * number of spikes)

$$Cqd = n - 1 + (n - 1)\left(\left\lfloor \frac{n-1}{s} \right\rfloor + 1\right) \tag{4}$$

If $s = l$, then the query dissemination is maximum. The CCNM determines the values for both s and l. However, the cost of query response depends on the underlying scheme employed for data aggregation. Clustering concept is applied in CNM, called as CCNM. In case of CCNM, every node pushes sensed data to CH. The data is pushed either vertically upward or downward based on the location of queried node. Figure 5 shows data traversal in downward direction against an event encountered in the WSN.

When an event is generated, the data is pushed vertically and horizontally upwards. This is denoted as follows.

$$\text{The cost for upward direction} = l + n + d1 \tag{5}$$
$$\text{The cost for downward direction} = l + (n - d) + (d - d1) \tag{6}$$

By combining (5) and (6), the query reply cost can be obtained for CCNM:

$$C_{qr} = 2(n + l) \tag{7}$$

Fig. 6 Proposed
cluster-based comb–needle
model

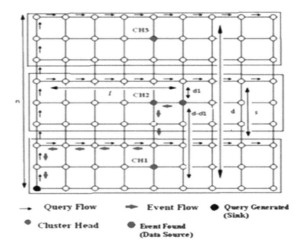

When needle is found in a single hop, the query reply cost is reduced by 50% in case of CCNM as shown in Fig. 6 (whereas Fig. 5 shows the data traversal in both directions so query reply cost is double in CNM). The rationale behind this is that the data traversal is made in one direction only. Thus, it is considered $n+l$. Therefore, approximately it is computed as follows.

$$C_{qr} = n + l \tag{8}$$

The aim of this is to reduce the cost of communication. The total cost per query involves $C_l + C_{qd} + C_{qr}$. By combining (3), (4), and (8), and simplifying, total communication cost C is

$$= (n-1)\left(\frac{n-1+2s}{s}\right) + n + l + (l-1)\left(\frac{f_e}{f_q}\right) \tag{9}$$

As $s = l - 1$

$$C = (n-1)\left(\frac{n-1+2s}{s}\right) + n + l + S\left(\frac{f_e}{f_q}\right) \tag{10}$$

Here, the parameters s and l are known as control parameters of comb–needle structure. Further details can be found in [3]

$$S_{optimal} \approx (n-1)\left(\frac{\sqrt{f_q}}{\sqrt{f_e}}\right) \approx \left(\frac{\sqrt{f_q}}{\sqrt{f_d}}\right) \tag{11}$$

Here, $f_q < f_e$. The minimum cost for communication in the CCNM is denoted as $C_{optimal}$. By substituting s value as shown in (10) and (11), the following equation computes $C_{optimal}$.

$$Coptimal = (n-1)\left(\sqrt{fe/fq}\right)\left(2/2\left(\sqrt{fq/fe}\right)\right)|n|l = O(n) \qquad (12)$$

Since $n^2 = N$

Assuming that N_c denotes number of clusters, $C_{optimal}$ is computed as follows.

$$C_{optimal} = O\left(\frac{\sqrt{N}}{N_c}\right) \qquad (13)$$

When there are less number of events and more number of queries, the needle length (l) and comb width (s) need to be more for optimal communication cost.

4 Simulation

Linux OS-based platform is used for experiments. NS2 is the simulator used for simulation study of the proposed CCNM for energy-efficient data aggregation in WSN. The CCNM model, as expressed earlier, is based on CNM. The open-source simulator NS2 is used for the proof of concept. It supports wireless sensor network for proving the hypothesis. Six performance metrics are used for evaluating the efficiency of the proposed system and comparing it with that of CNM. They are communication cost, energy consumption, delay, packet loss, packet delivery ratio, and throughput.

4.1 Simulation Environment

The simulation with NS2 is carried out with a WSN having 81 nodes that are positioned regularly in the network area covering 10×10 square area. Two-ray ground reflection model is used as radio propagation model.

Table 1 shows the simulation environment. By considering the various parameters, namely number of nodes, MAC protocol used, node deployment model, radio

Table 1 Simulation environment

Parameter	Value
Mac protocol	802.11
Number of nodes	81
Node deployment	Regular
Radio propagation model	Two-ray ground
Radio transmission range	200 m

Table 2 Performance metrics

Metric	Description
Packet delivery ratio (PDR)	It is the performance measure used to know the ratio between the number of packets received and the number of packets sent
Throughput	The rate of messages transferred successfully in network
Average delay	The time difference between packets received and packet sent
Energy consumption	It is the measure used to know how much energy is consumed by the network in data transmission, sensing, data processing, and idle sleep state
Communication cost	It is the measure used to determine the number of packets received and transmitted in order to have event and query notifications
Packet loss	It is the measure used to determine the number of packets lost while transmission in order to disseminate query and notification

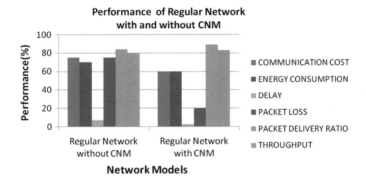

Fig. 7 Performance of regular network with and without CNM

transmission model, and radio propagation model, the simulation is carried out. Simulation parameters and their values are given in the next section.

4.2 Performance Metrics

Performance metrics are used to evaluate the simulation results which are shown in Table 2. The purpose of each metric is described here.

4.3 Performance Evaluation

The simulation results show the comparison of cluster-based CNM for regular networks and basic CNM. The evaluation is based on the different parameters such as communication cost, energy consumption, delay, packet loss, packet delivery ratio, and throughput. The outcome is presented in Figs. 7 and 8.

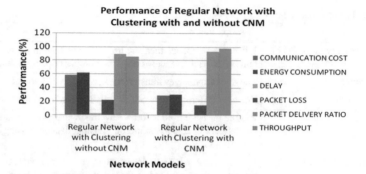

Fig. 8 Performance of cluster-based regular network with and without CNM

Table 3 Performance of regular network with and without CNM

Parameter (Units)	Regular network without CNM	Regular network with CNM	Improvement (%)
Communication cost (packets)	75	60	20
Energy consumption (J)	70	60	14.28
Delay (ms)	7.3	2.8	61.64
Packet loss (packets)	75	20	73.33
Packet delivery ratio (packets)	84	89	5.95
Throughput (packets)	80	83	3.75

The simulation results shown in Table 3 reveal the performance of regular networks with and without CNM employed. The regular network with CNM has comparable performance improvement over its counterpart without CNM. It is true with all the parameters considered. With respect to communication cost, regular network with CNM shows 60 packets which are 20% less communication overhead. Energy consumption with CNM is reduced by 14.28%. It is also to be noted that delay is reduced by 61.64% in regular network with CNM. Packet loss is reduced by 73.33%, and packet delivery ratio is increased by 5.95% while throughput is increased by 3.75%. Thus, the results reveal that CNM with regular network improves the performance of the WSN, which is shown in Fig. 7.

The performance results of two network models are compared and presented in Table 4. The network models are cluster-based regular network with CNM and the cluster-based regular network without CNM. The parameters considered for comparison are same as used earlier. All parameters showed improved performance with respect to cluster-based regular network with CNM.

Energy consumption is reduced by 51.61%, communication cost by 50%, delay by 50%, packet loss by 31.81%; packet delivery ratio is increased by 4.49%, and

Table 4 Performance of cluster-based regular network with and without CNM

Parameter (Units)	Regular network with clustering and without CNM	Regular network with clustering and with CNM	Improvement (%)
Communication cost (packets)	58	29	50
Energy consumption (J)	62	30	51.61
Delay (ms)	3	1.5	50
Packet loss (packets)	22	15	31.81
Packet delivery ratio (packets)	89	93	4.49
Throughput (packets)	85	98	15.29

throughput is increased by 15.29%. The results thus reveal that the addition of clustering mechanism to CNM has improved the performance of the WSN substantially in terms of reducing overheads and increasing PDR and throughput.

5 Conclusion

This paper proposed the concept of proposed cluster-based CNM for efficient and energy-aware data aggregation in WSN. The information discovery and dissemination are based on the hybrid model of CNM that combines the features of both push and pull models. The synergic effect of push and pull models is realized in CNM. However, it is further improved with CCNM and consumes energy optimally besides being efficient in computational cost as well. The simulation results revealed that proposed CCNM shows significant performance improvement over basic CNM model in terms of communication cost, energy consumption, delay, packet loss, packet delivery ratio, and throughput. Further, it can be implemented in random wireless sensor networks.

References

1. Krishnamachari, B.: An introduction to wireless sensor networks. In: Proceedings of ICISIP, Tutorial Presented at the Information Processing (ICISIP), Chennai, India (2005)
2. Heinzelman, W.R., Chandrakasan, A., Balakrishnan, H.: Energy-efficient communication protocol for wireless microsensor networks. In: Proceedings of the 33rd Hawaii International Conference on System Sciences, pp. 1–10. IEEE (2000)
3. Ozdemir, S., Xiao, Y.: Secure data aggregation in wireless sensor networks: a comprehensive overview. Comput. Netw. **23**, 2022–2037 (2009)
4. Liu, X., Huang, Q., Zhang, Y.: Balancing push and pull for efficient information discovery in large-scale sensor networks. IEEE Trans. Mob. Comput. **6**(3), 241–251 (2007)

5. Cam, H., Ozdemir, S., Nair, P., Muthuavinashiappana, D., Ozgur Sanli, H.: Energy-efficient secure pattern based data aggregation for wireless sensor networks. In: Computer Communications, vol. 29, pp. 446–455. Elsevier (2006). https://doi.org/10.1016/j.comcom.20 04.12.029

6. Thein, M.C.M., Thein, T.: An energy efficient cluster-head selection for wireless sensor networks. In: 2010 International Conference on Intelligent Systems, Modelling and Simulation, pp. 1–5. IEEE (2010)

7. Luo, L.X.J., Vasilakos, A.: Compressed data aggregation for energy efficient wireless sensor networks. In: 8th annual IEEE communications society conference on sensor, mesh and Ad Hoc communications and networks, pp. 1–9. IEEE (2011)

8. Xiang, L., Luo, J., Rosenberg, C.: Compressed data aggregation: energy-efficient and high-fidelity data collection. IEEE/ACM Trans. Netw. 21(6), 1–14. IEEE (2013)

9. Girao, J., Westhoff, D., Schneider, M.: CDA: concealed data aggregation for reverse multicast traffic in wireless sensor networks. In: International Conference on Communications, pp. 1–6. IEEE (2005)

10. Fasolo, E., Rossi, M., Zorzi, W.A.: In-network aggregation techniques for wireless sensor networks: a survey. IEEE Wireless Commun. 1–18 (2007)

11. Chen, Y.P., Liestman, A.L., Liu, J.: A hierarchical energy-efficient framework for data aggregation in wireless sensor networks. IEEE Trans. Veh. Technol. 55(3), 1–8. IEEE (2006)

12. Dilip Kumar, A., Aseri, T.C., Patel, R.B.: EEHC: energy efficient heterogeneous clustered scheme for wireless sensor networks. In: Elsevier, Computer Communications, vol. 32, pp. 662–667 (2009)

13. Rajendran, V., Obraczka, K., Garcia-Luna-Aceves, J.J.: Energy-efficient, collision-free medium access control for wireless sensor networks. In: Wireless Networks, vol. 12, pp. 63–78. Springer Science+Business Media (2006)

14. AI-Karaki, J.N., UI-Mustafa, R., Kamal, A.E.: Data aggregation in wireless sensor networks—exact and approximate algorithms. In: Workshop on High Performance Switching and Routing, pp. 1–5. IEEE (2004)

15. Kalpakis, K., Dasgupta, K., Namjoshi, P.: Efficient algorithms for maximum lifetime data gathering and aggregation in wireless sensor networks. In: Elsevier, Computer Networks, vol. 42, pp. 697–716 (2003)

16. Ciancio, A., Pattem, S., Ortega, A., Krishnamachari, B.: Energy-efficient data representation and routing for wireless sensor networks based on a distributed wavelet compression algorithm. In: The International Conference on Information Processing in Sensor Networks, pp. 1–8. ACM (2006)

17. Luo, C., Wu, F., Sun, J., Chen, C.W.: Compressive data gathering for large-scale wireless sensor networks. In: MobiCom'09, Proceedings of the 15th Annual International Conference on Mobile Computing and Networking, pp. 145–156. ACM (2009)

18. Li, H., Lin, K., Li, K.: Energy-efficient and high-accuracy secure data aggregation in wireless sensor networks. In: Computer Communications, vol. 34, pp. 591–597. Elsevier (2011)

19. Chatterjea, S., Nieberg, T., Meratnia, N., Havinga, P.: A distributed and self-organizing scheduling algorithm for energy-efficient data aggregation in wireless sensor networks. Trans. Sensor Netw. 4(4), 1–41. ACM (2008)

20. Dimokasa, N., Katsaros, D., Manolopoulosa, Y.: Energy-efficient distributed clustering in wireless sensor networks. Elsevier-J. Parallel Distrib. Comput. 70, 371–383 (2010)

21. Liu, C., Wu, K., Pei, J.: An energy-efficient data collection framework for wireless sensor networks by exploiting spatiotemporal correlation. Trans. Parallel Distrib. Syst. 18(7), 1–14. IEEE (2007)

Fourth-, Fifth-, Sixth-Order Linear Differential Equations (LDEs) via Homotopy Perturbation Method Using Laplace Transform

Rajnee Tripathi and Hradyesh Kumar Mishra

Abstract In this article, we construct the solution of fourth-, fifth-, and sixth-order boundary value problems. To solve these boundary value problems, we apply the homotopy perturbation method (HPM) using Laplace transform (LT). The proposed method is easy, effective, and the accuracy of this method has been proved by comparing the results with homotopy perturbation method (HPM), variational iterative method (VIM), Adomian decomposition method (ADM), and exact solutions by using Mathematica package. The results obtained by this LT-HPM have been shown in tables and graphs.

Keywords Fourth-, fifth-, sixth-order differential equations · Homotopy perturbation method using Laplace transform (LT-HPM) · Boundary value problems (BVPs)

AMS Classification 34B05

1 Introduction

The boundary value problems are mostly used in various scientific and technical fields. Such problems arise in physical and mathematical modeling. The main significance of higher-order boundary value problems (BVPs) can be determined from their various existences such as hydrodynamic stability problems, non-Newtonian fluids, viscoelastic flows, and convection of heat by Chandrasekhar [1]. These problems also occur in various areas of engineering which are applicable in fluid flow of water, oil, and gas through ground layers, introduced in theory of elastic stability by

R. Tripathi · H. K. Mishra (✉)
Department of Mathematics, Jaypee University of Engineering and Technology, Guna 473226, Madhya Pradesh, India
e-mail: hk.mishra@juet.ac.in; hkm1975@yahoo.co.in

R. Tripathi
e-mail: rajneetripathi@hotmail.com

© Springer Nature Singapore Pte Ltd. 2019
J. C. Bansal et al. (eds.), *Soft Computing for Problem Solving*, Advances in Intelligent Systems and Computing 817, https://doi.org/10.1007/978-981-13-1595-4_54

Timoshenko [2]. Generally, the nth-order boundary value problems can be defined by [3] as follows:

$$\Gamma^N(\tau) = \varsigma(\Gamma, \Gamma, \ldots, \Gamma^{(N-1)}) + \phi(\tau), \quad a < \tau < b, \tag{1}$$

Boundary conditions are:

$$\Gamma^n(a) = \mu_i \text{ and } \Gamma^n(b) = \eta_i. \tag{2}$$

where n and N are said to be non-negative integers (n < N), real constants are denoted by μ_i and η_i, and $\phi(\tau)$ is defined by a continuous function on $[a, b]$.

Recently, many authors introduced the brief description of higher-order boundary value problems and corresponding their applications, for example fluid flow through porous media by Hajji [4], optimal bridge design by Geng and Cui [5], a new meshless local Petrov–Galerkin (MLPG) approach in computational mechanics by Atluri and Zhu [6], the comparison of variational iterative method with finite element analysis for analysis of convective straight and redial fins with temperature-dependent thermal conductivity by Coskun and Atay [7]. Some authors also describe various methods, explicit analytical solutions of the generalized Burger and Burger–Fisher equations by homotopy perturbation method by Rashidi et al. [8], fifth- and sixth-order results via homotopy asymptotic approach by Ali et al. [9], optimal homotopy perturbation approach to thin film flow of a fourth-grade fluid by Marinca and Herisanu [10], a numerical method based on polynomials sextic spline functions for the solution of special fifth-order boundary value problems by Islam and Khan [11], variational iteration technique for solving higher-order boundary value problems by Noor and Mohamud-Din [12], numerical solution of sixth-order boundary value problems by Wazwaz [13], an efficient method for fourth-order boundary value problems by Noor and Mohyud-din [14], various perturbation techniques [15–25]. The existence and uniqueness conditions for nonlocal boundary value problems for nth-order differential equations are solved by Eloe and Henderson [26] and convergence analysis of HPM by Hussein [27]. Mishra and his co-workers solved two-point boundary value problems by [28, 29].

The proposed method has an objective to examine the convergence of series solution numerically and compared it by various methods. Here, our methodology is more reliable by applying LT-HPM. In some cases, a solution of the differential equation is very difficult in the time domain; therefore, we first convert differential equation into frequency domain by using LT and after solving the problem again transposed into time domain via inverse Laplace transform (ILT). This article is originated as follows: The explanation of the method is given in Sect. 2. Numerical examples are illustrated in Sect. 3. Finally, the conclusion is explained in Sect. 4.

2 Explanation of the Method

The nth-order nonlinear differential equation (BVP) is considered as follows:

$$\Gamma^n + p_1\left[\Gamma^{n-1}\right] + p_2\left[\Gamma^{n-2}\right] + \ldots p_n[\Gamma] + p_{n+1}[\zeta(\Gamma)] = \zeta(\tau) \tag{3}$$

$$\Gamma^n(a) = \alpha_i \text{ and } \Gamma^n(b) = \beta_i. \tag{4}$$

where $p_1, \ldots p_n, p_{n+1}, \alpha_i, \beta_i$ are constant. ζ (Γ) is a nonlinear function and $\zeta(\tau)$ is the source term.

Implementing Laplace transformation (denoted by L) on both sides of Eq. (3), we get

$$L\left\{\left[\Gamma^n\right]\right\} + L\left\{p_1\left[\Gamma^{n-1}\right]\right\} + \cdots + L\{p_n[\Gamma]\} + L\{p_{n+1}\zeta(\Gamma)\} = L[\zeta(\tau)] \tag{5}$$

By linear property of Laplace transformation,

$$L\left(\Gamma^n\right) + p_1 L\left(\Gamma^{n-1}\right) + \cdots + p_n L(\Gamma) + p_{n+1} L(\zeta(\Gamma)) = L(\zeta(\tau)) \tag{6}$$

Using the formula of Laplace transform, we obtain

$$\left[s^n L(\Gamma) - s^{n-1}\Gamma(0) - \cdots - \Gamma^{n-1}(0)\right] + p_1\left[s^{n-1}L(\Gamma) - s^{n-2}\Gamma(0) - \cdots - \Gamma^{n-2}(0)\right] + \cdots$$
$$+ p_n L(\Gamma) + p_{n+1}L(\zeta(\Gamma)) = L(\zeta(\tau)) \tag{7}$$

Applying the initial conditions in Eq. (7), we have

$$(s^n + p_1 s^{n-1} + \cdots + s p_{n-1})L(\Gamma) = \alpha_1 s^{n-1} + (\alpha_1 + p_1\alpha_2 + \cdots + p_{n-1}\alpha_n)s^{n-2} + \cdots$$
$$+ (\alpha_1 p_{n-1} + \alpha_2 p_{n-2} + \ldots Ap_n + Bp_{n-1} + C)$$
$$- p_n L(\Gamma) - p_{n+1}L(\zeta(\Gamma)) + L(\zeta(\tau)) \tag{8}$$

Or

$$L(\Gamma) = \frac{\alpha_1 s^{n-1} + (\alpha_1 + \cdots + p_1\alpha_2)s^{n-2} + \cdots + (\alpha_1 p_{n-1} + \alpha_2 p_{n-2} + \ldots Ap_2 + Bp_1 + C) + L(\zeta(\tau))}{(s^n + p_1 s^{n-1} + \cdots + s p_{n-1})}$$
$$- \frac{p_n L(\Gamma) - p_{n+1}L(\zeta(\Gamma))}{(s^n + p_1 s^{n-1} + \cdots + s p_{n-1})} \tag{9}$$

After inverse Laplace transform of Eq. (9), we get

$$\Gamma(\tau) = G(\tau) - L^{-1}\left(\frac{p_n L(\Gamma)}{(s^n + p_1 s^{n-1} + \cdots + s p_{n-1})}\right) - L^{-1}\left(\frac{p_{n+1}L(\zeta(\Gamma))}{(s^n + p_1 s^{n-1} + \cdots + s p_{n-1})}\right) \tag{10}$$

where $G(\tau)$ is combination of source term and the given initial conditions. The homotopy perturbation method [30] of Eq. (10) is as follows

$$\Gamma(\tau) = \sum_{n=0}^{\infty}\left(p^n\Gamma_n(\tau)\right) \tag{11}$$

where $\Gamma_n(\tau)$ is a term which must be step by step calculated and the nonlinear term $\zeta(\Gamma)$ can be defined as

$$\zeta(\Gamma) = \sum_{n=0}^{\infty} \left(p^n H_n(\Gamma) \right) \tag{12}$$

where $H_n(\Gamma)$'s are He's polynomials (see [28, 31 and their reference in]) which are given by

$$H_n(\Gamma_0, \Gamma_1, \Gamma_2 \ldots \Gamma_n) = \frac{1}{n!} \frac{\partial^n}{\partial p^n} \left[\zeta \left(\sum_{i=0}^{\infty} p^i \Gamma_i \right) \right]_{p=0} \quad n = 0, 1, 2, \ldots. \tag{13}$$

Substituting the value of (21) and (22) in (20), we get

$$\sum_{n=0}^{\infty} \left(p^n \Gamma_n(\tau) \right) = G(\tau) - p \left\{ \begin{array}{l} L^{-1} \left(\frac{p_n}{(s^n + p_1 s^{n-1} + \cdots + s p_{n-1})} L \left(\sum_{n=0}^{\infty} (p^n \Gamma_n(\tau)) \right) \right) \\[2ex] + L^{-1} \left(\frac{p_{n+1}}{(s^n + p_1 s^{n-1} + \cdots + s p_{n-1})} L \left(\sum_{n=0}^{\infty} (p^n H_n(\Gamma)) \right) \right) \end{array} \right\} \tag{14}$$

The above expression contains Laplace transformation (LT), homotopy perturbation method, He's polynomials. In the next step, the coefficients of same powers of p are compared step by step:

$$p^0 : \Gamma_0(\tau) = G(\tau)$$

$$p^1 : \Gamma_1(\tau) = - \left\{ \begin{array}{l} L^{-1} \left(\frac{p_n}{(s^n + p_1 s^{n-1} + \cdots + s p_{n-1})} L(\Gamma_0) \right) \\[2ex] + L^{-1} \left(\frac{p_{n+1}}{(s^n + p_1 s^{n-1} + \cdots + s p_{n-1})} L(H_0(\Gamma)) \right) \end{array} \right\}$$

$$p^2 : \Gamma_2(\tau) = - \left\{ \begin{array}{l} L^{-1} \left(\frac{p_n}{(s^n + p_1 s^{n-1} + \cdots + s p_{n-1})} L(\Gamma_1) \right) \\[2ex] + L^{-1} \left(\frac{p_{n+1}}{(s^n + p_1 s^{n-1} + \cdots + s p_{n-1})} L(H_1(\Gamma)) \right) \end{array} \right\} \tag{15}$$

$$p^3 : \Gamma_3(\tau) = - \left\{ \begin{array}{l} L^{-1} \left(\frac{p_n}{(s^n + p_1 s^{n-1} + \cdots + s p_{n-1})} L(\Gamma_2) \right) \\[2ex] + L^{-1} \left(\frac{p_{n+1}}{(s^n + p_1 s^{n-1} + \cdots + s p_{n-1})} L(H_2(\Gamma)) \right) \end{array} \right\}$$

$$\vdots$$

3 Numerical Examples

Here, five examples are considered for the applicability of the method.

Example 3.1 The fourth-order boundary value problem [32] is given as

$$\Gamma^4(\tau) = \tau(1 + e^\tau) + 3e^\tau + \Gamma(\tau) - \int_0^\tau \Gamma(\tau)d\tau, \quad 0 < \tau < 1, \quad (16)$$

Boundary conditions are:

$$\Gamma(0) = 1, \quad \Gamma''(0) = 2,$$
$$\Gamma(1) = 1 + e, \quad \Gamma''(1) = 3e. \quad (17)$$

The exact solution is

$$\Gamma = 1 + \tau e^\tau \quad (18)$$

Taking Laplace transform (LT) on both sides of Eq. (16), we obtain

$$L\left\{\Gamma^4(\tau)\right\} = L\left\{\tau(1 + e^\tau)\right\} + 3L\left\{e^\tau\right\} + L\left\{\Gamma(\tau)\right\} - L\left\{\int_0^\tau \Gamma(\tau)d\tau\right\},$$

$$\Gamma(s) = \frac{1}{s} + \frac{A}{s^2} + \frac{2}{s^3} + \frac{B}{s^4} + \frac{1}{s^6} + \frac{1}{s^4}\left[\frac{1}{(s-1)^2}\right] + \frac{3}{s^4}\left[\frac{1}{(s-1)}\right] + \left\{\frac{L}{s^4}\left[\int_0^\tau \Gamma(\tau)d\tau\right]\right\}.$$

By applying inverse Laplace transform (ILT) on both sides of above equation, we have

$$\Gamma(\tau) = 1 + A\tau + \tau^2 + \frac{B\tau^3}{3!} + \frac{\tau^5}{5!} + L^{-1}\left\{\frac{1}{s^4}\left[\frac{1}{(s-1)^2}\right]\right\} + L^{-1}\left\{\frac{3}{s^4}\left[\frac{1}{(s-1)}\right]\right\}$$

$$+ L^{-1}\left\{\frac{L}{s^4}\left[\int_0^\tau \Gamma(\tau)d\tau\right]\right\},$$

$$= 2 + A\tau + \frac{\tau^2}{2} + \frac{(B-2)\tau^3}{3!} + \frac{\tau^5}{5!} - (1 - \tau)e^\tau + L^{-1}\left\{\frac{L}{s^4}\left[\int_0^\tau \Gamma(\tau)d\tau\right]\right\}.$$

Using HPM, the above expression has the following form

$$\sum_{n=0}^\infty \left(p^n \Gamma_n(\tau)\right) = 2 + A\tau + \frac{\tau^2}{2} + \frac{(B-2)\tau^3}{3!} + \frac{\tau^5}{5!} - (1 - \tau)e^\tau$$

$$+ pL^{-1}\left\{\frac{L}{s^4}\left(\sum_{n=0}^\infty \left(p^n\left[\int_0^\tau \Gamma(\tau)d\tau\right]\right)\right)\right\}. \quad (19)$$

By comparing the coefficient of same powers of p, we obtain

$$
\left.
\begin{aligned}
& p^0 : \Gamma_0(\tau) = 2 + A\tau + \frac{\tau^2}{2} + \frac{(B-2)\tau^3}{3!} + \frac{\tau^5}{5!} - (1-\tau)e^\tau, \\
& p^1 : \Gamma_1(\tau) = -1 - \tau - \frac{\tau^2}{2} - \frac{\tau^3}{3!} + \frac{(A-2)\tau^5}{5!} + \frac{(1-A)\tau^6}{6!} + \frac{(B-3)\tau^7}{7!} \\
& \qquad - \frac{(B-2)\tau^8}{8!} + \frac{\tau^9}{9!} - \frac{\tau^{10}}{10!},
\end{aligned}
\right\}
\tag{20}
$$

where A and B are the constants.

Thus, the solution of the given problem (16) in terms of constants A and B is given as:

$$
\begin{aligned}
\Gamma(\tau) &= 2 + A\tau + \frac{\tau^2}{2} + \frac{(B-2)\tau^3}{3!} + \frac{\tau^5}{5!} - (1-\tau)e^\tau \\
&\quad -1 - \tau - \frac{\tau^2}{2} - \frac{\tau^3}{3!} + \frac{(A-2)\tau^5}{5!} + \frac{(1-A)\tau^6}{6!} + \frac{(B-3)\tau^7}{7!} \\
&\quad - \frac{(B-2)\tau^8}{8!} + \frac{\tau^9}{9!} - \frac{\tau^{10}}{10!} + e^\tau \cdots \\
&= 1 + (A-1)\tau + \frac{(B-3)\tau^3}{3!} + \tau e^\tau + \frac{(A-1)\tau^5}{5!} + \frac{(1-A)\tau^6}{6!} \\
&\quad + \frac{(B-3)\tau^7}{7!} - \frac{(B-2)\tau^8}{8!} + \frac{\tau^9}{9!} - \frac{\tau^{10}}{10!} + \cdots
\end{aligned}
$$

After applying the boundary conditions (17) in above equation, we conclude the values of the constants for $\tau = 0, 1$.

$$A = 0.999818465, \quad B = 3.001229433;$$

Therefore, the approximate solution of the problem (16) is given as follows:

$$
\begin{aligned}
\Gamma(\tau) &= 2 + A\tau + \frac{\tau^2}{2} + \frac{(B-2)\tau^3}{3!} + \frac{\tau^5}{5!} - (1-\tau)e^\tau - 1 - \tau - \frac{\tau^2}{2} - \frac{\tau^3}{3!} + \frac{(A-2)\tau^5}{5!} + \frac{(1-A)\tau^6}{6!} + \frac{(B-3)\tau^7}{7!} \\
&\quad - \frac{(B-2)\tau^8}{8!} + \frac{\tau^9}{9!} - \frac{\tau^{10}}{10!} + e^\tau \cdots \\
&= 1 + (A-1)\tau + \frac{(B-3)\tau^3}{3!} + \tau e^\tau + \frac{(A-1)\tau^5}{5!} + \frac{(1-A)\tau^6}{6!} + \frac{(B-3)\tau^7}{7!} - \frac{(B-2)\tau^8}{8!} + \frac{\tau^9}{9!} - \frac{\tau^{10}}{10!} + \cdots \\
\Gamma(\tau) &= 1 - 0.000181535\tau + 0.0002049055\tau^3 - 0.000001512791667\tau^5 + 0.0000002521319444\tau^6 \\
&\quad + 0.000000243935119\tau^7 + 0.00002483207919\tau^8 + 0.000002755731922\tau^9 \\
&\quad - 0.0000002755731922\tau^{10} + \tau e^\tau + \cdots
\end{aligned}
\tag{21}
$$

Here, Table 1 shows that the result is obtained by the LT-HPM with the help of Mathematica package. Figure 1 shows the comparison of proposed method with the exact solution. Furthermore, this method shows the absolute maximum error $E(\tau) = |(1 + \tau e^\tau) - \Gamma(\tau)|$.

Table 1 Comparison of LT-HPM with VIM and exact

τ	Approximate solution (LT-HPM)	Approximate solution (VIM)	Exact solution	Max. absolute error (LT-HPM)	Max. absolute error (VIM) [30]
0.0	1.0	1.0	1.0	0.0	0.0
0.2	1.24424	1.24411	1.24428	3.4668E−05	1.65737E−04
0.4	1.59667	1.59645	1.59673	5.9497E−05	2.84804E−04
0.6	2.09320	2.09296	2.09327	6.43182E−05	3.12185E−04
0.8	2.71828	2.78022	2.78043	3.6194E−05	2.13844E−04
1.0	3.71828	3.71828	3.71828	0.0	0.0

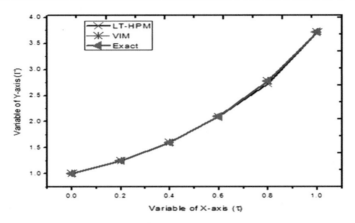

Fig. 1 Comparison of proposed method with VIM and exact solution

Example 3.2 The fourth-order boundary value problem [33] is considered as:

$$\Gamma^4(\tau) = \Gamma(\tau) + \Gamma''(\tau) + e^\tau(\tau - 3), \tag{22}$$

Boundary conditions are

$$\Gamma(0) = 1, \quad \Gamma'(0) = 0,$$
$$\Gamma(1) = 0, \quad \Gamma'(1) = -e. \tag{23}$$

The exact solution of the problem (22) is

$$\Gamma(\tau) = (1 - \tau)e^\tau \tag{24}$$

By taking Laplace transform (LT) on both sides, we obtain

$$L\left(\Gamma^4(\tau)\right) = L[\Gamma(\tau)] + L[\Gamma''(\tau)] + L[e^\tau(\tau - 3)],$$

$$s^4 L(\Gamma(\tau)) - s^3 \Gamma(0) - s^2 \Gamma'(0) - s\Gamma''(0) - \Gamma'''(0) = L(\Gamma(\tau)) + L\left(\Gamma''(\tau)\right) + \frac{1}{(s-1)^2} - 3\frac{1}{(s-1)},$$

$$\Gamma(s) = \frac{1}{s} + \frac{A}{s^3} + \frac{B}{s^4} + \frac{L(\Gamma(\tau))}{s^4} + \frac{L(\Gamma''(\tau))}{s^4} + \frac{1}{s^4}\left[\frac{1}{(s-1)^2}\right] - \frac{3}{s^4}\left[\frac{1}{(s-1)}\right].$$

After applying the inverse Laplace transform (ILT) on both sides, we obtain

$$\Gamma(\tau) = 1 + \frac{A\tau^2}{2} + \frac{B\tau^3}{3!} + L^{-1}\left\{\frac{L}{s^4}\left((\Gamma(\tau)) + \Gamma''(\tau)\right)\right\} + L^{-1}\left\{\frac{1}{s^4}\left[\frac{1}{(s-1)^2}\right]\right\}$$

$$- 3L^{-1}\left\{\frac{1}{s^4}\left[\frac{1}{(s-1)}\right]\right\},$$

$$= 8 + 6\tau + \frac{(A+5)\tau^2}{2} + \frac{(B+4)\tau^3}{3!} + \tau e^\tau - 7e^\tau + L^{-1}\left\{\frac{L}{s^4}\left((\Gamma(\tau)) + \Gamma''(\tau)\right)\right\}.$$

Now, taking HPM on both sides of above equation, we obtain

$$\sum_{n=0}^{\infty}\left(p^n\Gamma_n(\tau)\right) = 8 + 6\tau + \frac{(A+5)\tau^2}{2} + \frac{(B+4)\tau^3}{3!} + \tau e^\tau - 7e^\tau$$

$$+ pL^{-1}\left\{\frac{L}{s^4}\left(\sum_{n=0}^{\infty}\left(p^n((\Gamma(\tau)) + \Gamma''(\tau))\right)\right)\right\}. \qquad (25)$$

By comparing the coefficient of same powers of p, we have

$$p^0 : \Gamma_0(\tau) = 8 + 6\tau + \frac{(A+5)\tau^2}{2} + \frac{(B+4)\tau^3}{3!} + \tau e^\tau - 7e^\tau,$$

$$p^1 : \Gamma_1(\tau) = 20 + 2\tau e^\tau - 20e^\tau + 18\tau + 8\tau^2 + \frac{7}{2}\tau^3 + \frac{(A+13)\tau^4}{4!} + \frac{(B+10)\tau^5}{5!} + \cdots,$$

$$p^2 : \Gamma_1(\tau) = 52 + 4\tau e^\tau - 52e^\tau + 48\tau + 22\tau^2 + \frac{20}{3}\tau^3 + \frac{3}{2}\tau^4 + \frac{4\tau^5}{15} + \cdots,$$

$$\vdots$$

$$(26)$$

where A and B are the constants.

Thus, the solution of the problem (22) in constant terms is given as follows:

$$\Gamma(\tau) = 512 + 480\tau + \frac{(A+449)\tau^2}{2} + \frac{(B+418)\tau^3}{3!} + \frac{(385+A)\tau^4}{4!} + \frac{(B+354)\tau^5}{5!}$$

$$+ \frac{(2A+322)}{6!}\tau^6 + \frac{(B+146)\tau^7}{2520} + \frac{(254+3A)\tau^8}{8!} + \cdots. \qquad (27)$$

By applying the boundary conditions (23) in Eq. (27), we conclude the values of the constants for $\tau = 0, 1$

$$A = -0.999916277, \quad B = -2.00299259;$$

Table 2 Comparison of LT-HPM with exact

τ	HPM via LT	Exact	Error
0	1	1	0.0E+0
0.1	0.994654749	0.9946538264	9.228E−07
0.2	0.977127897	0.9771222072	5.690884E−06
0.3	0.944918488	0.9449011666	1.7322704E−05
0.4	0.895133584	0.8950948205	3.876597E−05
0.5	0.824433424	0.8243606378	7.2788749E−05
0.6	0.728969053	0.7288475228	1.21530761E−05
0.7	0.604311571	0.6041258147	1.85759376E−04
0.8	0.445371527	0.4451081875	2.63341665E−04
0.9	0.246307288	0.2459603111	3.469769E−04
1.0	0.000000000	0.000000000	0.0000E+0

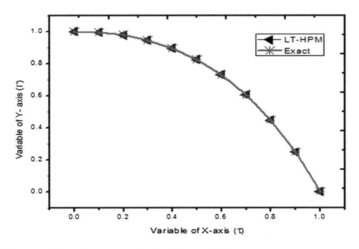

Fig. 2 The comparison of propose method and exact solution

Now, the approximate series solution of the problem (22) is given as follows:

$$\Gamma(\tau) = 512 + 480\tau + 24.0000418615\tau^2 + 69.332834568333\tau^3 + 16.000003488458\tau^4 + 2.9333083950833\tau^5$$
$$+ 0.44444467700833\tau^6 + 0.057141669607143\tau^7 + 0.0062252046420883\tau^8 + \ldots, \tag{28}$$

Here, Table 2 shows that the result is obtained by the LT-HPM with the help of Mathematica package. The proposed method is compared with the exact solution. Figure 2 shows the comparison of proposed method with the exact solution. Furthermore, this method shows the absolute maximum error $E(\tau) = |(1-\tau)e^\tau - \Gamma(\tau)|$.

Example 3.3 The fourth-order linear nonhomogeneous boundary value problem [34] is given as:

$$\Gamma^4(\tau) - 2\Gamma''(\tau) + \Gamma(\tau) = -8e^\tau, \quad \tau \in [0, 1], \tag{29}$$

Boundary conditions are

$$\begin{aligned} \Gamma(0) = 0, \quad \Gamma''(0) = 0, \\ \Gamma'(1) = -e, \ \Gamma''(1) = -4e. \end{aligned} \tag{30}$$

The exact solution of the given problem (29) is

$$\Gamma(\tau) = \tau(1 - \tau)e^\tau \tag{31}$$

By applying Laplace transform (LT) on both sides of Eq. (29), we get

$$L\big(\Gamma^4(\tau)\big) = 2L\big[\Gamma''(\tau)\big] - L[\Gamma(\tau)] - 8L\big[e^\tau\big],$$

$$\Gamma(\tau) = \frac{A}{s^2} + \frac{B}{s^4} + \frac{L}{s^4}\big(2[\Gamma''(\tau)] - [\Gamma(\tau)]\big) - \frac{8}{s^4}\left[\frac{1}{(s-1)}\right].$$

Taking inverse Laplace transform on both sides of above equation, we obtain

$$\Gamma(\tau) = A\tau + \frac{B\tau^3}{6} + L^{-1}\left\{\frac{L}{s^4}\big(2[\Gamma''(\tau)] - [\Gamma(\tau)]\big)\right\} - L^{-1}\left\{\frac{8}{s^4}\left[\frac{1}{(s-1)}\right]\right\},$$

$$= 8 + (A + 8)\tau + 4\tau^2 + + \frac{(B+8)\tau^3}{6} - 8e^\tau + L^{-1}\left\{\frac{L}{s^4}\big(2[\Gamma''(\tau)] - [\Gamma(\tau)]\big)\right\}$$

After using HPM on both sides of above equation, we obtain

$$\sum_{n=0}^{\infty}\big(p^n\Gamma_n(\tau)\big) = 8 + (A + 8)\tau + 4\tau^2 + + \frac{(B+8)\tau^3}{6} - 8e^\tau$$

$$+ pL^{-1}\left\{\frac{L}{s^4}\left(\sum_{n=0}^{\infty}\big(p^n\big(2(\Gamma_n''(\tau)) - (\Gamma_n(\tau))\big)\big)\right)\right\} \tag{32}$$

Now, comparing the coefficient of same powers of p, we get

$$p^0 : \Gamma_0(\tau) = 8 + (A+8)\tau + 4\tau^2 + + \frac{(B+8)\tau^3}{6} - 8e^\tau,$$

$$p^1 : \Gamma_1(\tau) = L^{-1}\left\{\frac{L}{s^4}\left(2\left(\Gamma_0''(\tau)\right) - \left(\Gamma_0(\tau)\right)\right)\right\},$$

$$= 8 + 8\tau + 4\tau^2 + \frac{4}{3}\tau^3 + \frac{\tau^4}{3}$$
$$- \frac{(A-2B-8)\tau^5}{120} - \frac{\tau^6}{90} - \frac{(B+8)\tau^7}{7!} - 8e^\tau,$$

$$p^2 : \Gamma_2(\tau) = L^{-1}\left\{\frac{L}{s^4}\left(2\left(\Gamma_1''(\tau)\right) - \left(\Gamma_1(\tau)\right)\right)\right\},$$

$$= 8 + 8\tau + 4\tau^2 + \frac{4}{3}\tau^3 + \frac{\tau^4}{3} + \frac{\tau^5}{15}$$
$$+ \frac{\tau^6}{90} - \frac{(A-2B-4)\tau^7}{2520} - \frac{\tau^8}{5040}$$
$$+ \frac{(A-4B-24)\tau^9}{9!} + \frac{\tau^{10}}{453600} + \frac{(B+8)\tau^{11}}{11!} - 8e^\tau,$$

$$p^3 : \Gamma_3(\tau) = L^{-1}\left\{\frac{L}{s^4}\left(2\left(\Gamma_2''(\tau)\right) - \left(\Gamma_2(\tau)\right)\right)\right\},$$

$$= 8 + 8\tau + 4\tau^2 + \frac{4}{3}\tau^3 + \frac{\tau^4}{3} + \frac{\tau^5}{15} + \frac{\tau^6}{90}$$
$$+ \frac{\tau^7}{630} + \frac{\tau^8}{1680} - \frac{(A-2B-2)\tau^9}{90720}$$
$$- \frac{\tau^{10}}{64800} + \frac{(A-3B-14)\tau^{11}}{9979200} + \frac{\tau^{12}}{11975040}$$
$$- \frac{(A-6B-40)\tau^{13}}{6227020800} - \frac{\tau^{14}}{10897284000}$$
$$- \frac{(B+8)\tau^{15}}{13067674368000} - 8e^\tau,$$

$$p^4 : \Gamma_4(\tau) = L^{-1}\left\{\frac{L}{s^4}\left(2\left(\Gamma_3''(\tau)\right) - \left(\Gamma_3(\tau)\right)\right)\right\},$$

$$= 8 + 8\tau + 4\tau^2 + \frac{4}{3}\tau^3 + \frac{\tau^4}{3} + \frac{\tau^5}{15} + \frac{\tau^6}{90}$$
$$+ \frac{\tau^7}{630} + \frac{\tau^8}{5040} + \frac{\tau^9}{45360} + \frac{\tau^{10}}{453600}$$
$$- \frac{(A-2B-1)\tau^{11}}{4989600} - \frac{\tau^{12}}{373621248000}$$
$$+ \frac{(3A-8B-30)\tau^{13}}{1556755200} + \frac{17\tau^{14}}{10897286400}$$
$$- \frac{(3A-12B-68)\tau^{15}}{653873784000} - \frac{\tau^{16}}{373621248000}$$
$$+ \frac{(A-8B-56)\tau^{17}}{355687428096000} + \frac{\tau^{18}}{800296713216000}$$
$$+ \frac{\tau^{19}}{121645100408832000} - 8e^\tau,$$

$$\vdots$$

$$(33)$$

Thus, the approximate solution of the problem (29) including constant terms is given as follows:

$$\Gamma(\tau) = \Gamma_0 + \Gamma_1 + \Gamma_2 + \Gamma_3 + \dots$$

$$= 40 + (40 + A)\tau + 20\tau^2 + \frac{1}{6}(40 + B)\tau^3 + \frac{4}{3}\tau^4$$

$$- \frac{1}{120}(A - 2B - 32)\tau^5 + \frac{\tau^6}{45} - \frac{(2A - 3B - 16)\tau^7}{5040}$$

$$- \frac{\tau^8}{5040} - \frac{(3A - 4B + 8)\tau^9}{362880} - \frac{\tau^{10}}{90720} - \frac{(4A - 5B + 40)\tau^{11}}{39916800}$$

$$- \frac{\tau^{12}}{5987520} + \frac{(11A - 26B - 80)\tau^{13}}{6227020800}$$

$$+ \frac{\tau^{14}}{681080400} - \frac{(6A - 23B - 128)\tau^{15}}{1307674368000} - \frac{\tau^{16}}{373621248000}$$

$$+ \frac{(A - 8B + 56)\tau^{17}}{35568742809600} - \frac{\tau^{18}}{800296713216000}$$

$$+ \frac{(B + 8)\tau^{19}}{121645100408832000} + \dots \tag{34}$$

By applying the boundary conditions (29) in Eq. (34), we conclude the values of the constants for $\tau = 0, 1$

$$A = 1.0000000 \quad B = -3.0000000;$$

Now, after substituting the values of A and B we get the approximate solution of problem (29) as follows:

$$= 40 - 40e^\tau + 41\tau + 20\tau^2 + 6.166666667\tau^3 + 1.333333333\tau^4 + 0.208333333\tau^5 + 0.022222222\tau^6$$

$$- 0.0009920634921\tau^7 - 0.000198412698\tau^8 - 0.000000275573192\tau^9 - 0.00001102292769\tau^{10}$$

$$- 0.0.000001478074395\tau^{11} - 0.0000001670140559\tau^{12} + 0.00000002713978408\tau^{13}$$

$$+ 0.000000001468255437\tau^{14} + O(\tau^{15}) + \dots$$

Here, Table 3 shows that the result is obtained by the LT-HPM with the help of Mathematica package, which is compared with the exact solution. Figure 3 shows the comparison of proposed method with the exact solution. Furthermore, this method shows the absolute maximum error $E(\tau) = |\tau(1 - \tau)e^\tau - \Gamma(\tau)|$.

Example 3.4 The fifth-order boundary value problem [35] is considered as:

$$\Gamma^5(\tau) - \Gamma(\tau) + 15e^\tau + 10\tau e^\tau = 0, \quad 0 < \tau < 1, \tag{35}$$

Table 3 Comparison of proposed method with the exact solution

τ	LT-HPM	Exact	Absolute error
0.0	0	0	0
0.1	0.0994653	0.0994653	3.7E-10
0.2	0.1954244	0.1954244	1.3E-09
0.3	0.28347035	0.2834703	4.0E-10
0.4	0.35803793	0.3580379	2.6E-09
0.5	0.4121803	0.4121803	2.0E-11
0.6	0.4373085	0.4373085	9.E-11
0.7	0.42288807	0.42288806	1.4E-09
0.8	0.3560865	0.3560865	2.56E-09
0.9	0.2213642	0.2213642	1.0E-08
1.0	−4.0E-8	0.000000	−4.0E-8

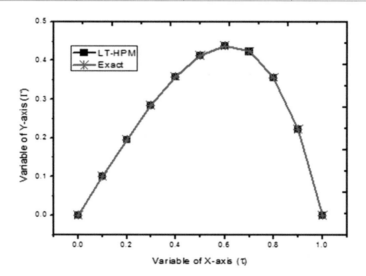

Fig. 3 The comparison of proposed method and exact

Boundary conditions are

$$\Gamma(0) = 0, \ \Gamma'(0) = 1, \quad \Gamma''(0) = 0$$
$$\Gamma(1) = 0, \ \Gamma'(1) = -e,$$

(36)

The exact solution of the given problem (35) is

$$\Gamma(\tau) = \tau e^\tau (1 - \tau)$$

(37)

By taking Laplace transform (LT) on both sides of above Eq. (35), we have

$$L[\Gamma^5(\tau)] = L[\Gamma(\tau)] - 15[e^\tau] - 10L[\tau e^\tau],$$

$$L[\Gamma(\tau)] = \frac{1}{s^2} + \frac{A}{s^4} + \frac{B}{s^5} + \frac{L}{s^5}[\Gamma(\tau)] - 15\left[\frac{1}{s^5(s-1)}\right] - 10\left[\frac{1}{s^5(s-1)^2}\right].$$

By taking inverse Laplace transform on both sides of the above equation, we have

$$\Gamma(\tau) = \tau + \frac{A\tau^3}{6} + \frac{B\tau^4}{24} + L^{-1}\left\{\frac{L}{s^5}(\Gamma(\tau))\right\} - L^{-1}\left\{15\left[\frac{1}{s^5(s-1)}\right]\right\} - L^{-1}\left\{10\left[\frac{1}{s^5(s-1)}\right]\right\},$$

$$= -35e^\tau - 35 - 24\tau - \frac{15}{2}\tau^2 + \frac{(A-5)}{6}\tau^3 + \frac{(B+5)}{24}\tau^4 - 10\tau e^\tau + L^{-1}\left\{\frac{L}{s^5}(\Gamma(\tau))\right\}.$$

By taking HPM on both sides of above equation, we obtain

$$\sum_{n=0}^{\infty}[p^n(\Gamma_n(\tau))] = 35e^\tau - 35 - 24\tau - \frac{15}{2}\tau^2 - \frac{(A-5)}{6}\tau^3 + \frac{(B+5)}{24}\tau^4$$

$$- 10\tau e^\tau + pL^{-1}\left\{\frac{L}{s^5}\left(\sum_{n=0}^{\infty}[p^n(\Gamma_n(\tau))]\right)\right\}. \tag{38}$$

By comparing the coefficient of same powers of p, we get

$$p^0 : \Gamma_0(\tau) = 35e^\tau - 35 - 24\tau - \frac{15}{2}\tau^2 - \frac{(A-5)}{6}\tau^3 + \frac{(B+5)}{24}\tau^4 - 10\tau e^\tau,$$

$$p^1 : \Gamma_1(\tau) = L^{-1}\left\{\frac{L}{s^5}(\Gamma_0(\tau))\right\},$$

$$= 85e^\tau - 85 - 75\tau - \frac{65}{2}\tau^2 - \frac{55}{6}\tau^3 - \frac{45}{24}\tau^4 - \frac{35}{5!}\tau^5 - \frac{24\tau^6}{6!} - \frac{15\tau^7}{7!}$$

$$+ \frac{(A-5)}{8!}\tau^8 + \frac{(B+5)}{9!}\tau^9 - 10\tau e^\tau,$$

$$p^2 : \Gamma_2(\tau) = L^{-1}\left\{\frac{L}{s^5}(\Gamma_1(\tau))\right\},$$

$$= 135e^\tau - 135 - 24\tau - \frac{115}{2}\tau^2 - \frac{105}{6}\tau^3 - \frac{95}{24}\tau^4 - \frac{85\tau^5}{5!} - \frac{75\tau^6}{6!} - \frac{65\tau^7}{7!} - \frac{55\tau^8}{8!}$$

$$- \frac{45\tau^9}{9!} - \frac{35\tau^{10}}{10!} - \frac{24\tau^{11}}{11!} - \frac{15\tau^{12}}{12!} + \frac{(A-5)}{13!}\tau^{13} + \frac{(B+5)}{14!}\tau^{14} - 10\tau e^\tau,$$

$$\vdots \tag{39}$$

Thus, the series solution of the problem is given as follows:

$$\Gamma(\tau) = \Gamma_0 + \Gamma_1 + \Gamma_2 + \ldots$$

$$= -255 - 224\tau + 255e^\tau - \frac{195\tau^2}{2} + \left(\frac{1}{6}A - \frac{55}{2}\right)\tau^3 + \left(-\frac{135}{24} + \frac{1}{24}B\right)\tau^4 - \tau^5$$

$$- \frac{99}{6!}\tau^6 - \frac{80}{7!}\tau^7 + \left(\frac{1}{40320}A - \frac{1}{672}\right)\tau^8 + \left(\frac{1}{362880}B + \frac{1}{9072}\right)\tau^9$$

$$-\frac{1}{103680}\tau^{10} - \frac{1}{1663200}\tau^{11} - \frac{1}{31933440}\tau^{12} + \left(\frac{1}{6227020800}A - \frac{1}{1245404160}\right)\tau^{13}$$

$$+ \left(\frac{1}{17435658240} + \frac{1}{871782941200}B\right)\tau^{14} - 30\tau e^{\tau} + \ldots$$

$$\Gamma(\tau) = \tau + \frac{A\tau^3}{6} + \frac{B\tau^4}{24} + L^{-1}\left\{\frac{L}{s^5}(\Gamma(\tau))\right\} - L^{-1}\left\{15\left[\frac{1}{s^5(s-1)}\right]\right\} - L^{-1}\left\{10\left[\frac{1}{s^5(s-1)}\right]\right\},$$

$$= -35e^{\tau} - 35 - 24\tau - \frac{15}{2}\tau^2 + \frac{(A-5)}{6}\tau^3 + \frac{(B+5)}{24}\tau^4 - 10\tau e^{\tau} + L^{-1}\left\{\frac{L}{s^5}(\Gamma(\tau))\right\}.$$

By taking HPM on both sides of above equation, we obtain

$$\sum_{n=0}^{\infty}\left[p^n(\Gamma_n(\tau))\right] = 35e^{\tau} - 35 - 24\tau - \frac{15}{2}\tau^2 + \frac{(A-5)}{6}\tau^3 + \frac{(B+5)}{24}\tau^4$$

$$- 10\tau e^{\tau} + pL^{-1}\left\{\frac{L}{s^5}\left(\sum_{n=0}^{\infty}\left[p^n(\Gamma_n(\tau))\right]\right)\right\}.$$

By comparing the coefficient of same powers of p, we get

$$p^0 : \Gamma_0(\tau) = 35e^{\tau} - 35 - 24\tau - \frac{15}{2}\tau^2 + \frac{(A-5)}{6}\tau^3 + \frac{(B+5)}{24}\tau^4 - 10\tau e^{\tau},\Bigg\}$$

Thus the series solution of the problem is given as follows:

$$\Gamma(\tau) = \Gamma_0 + \Gamma_1 + \Gamma_2 + \ldots\ldots$$

$$= -255 - 224\tau + 255e^{\tau} - \frac{195\tau^2}{2} + \left(\frac{1}{6}A - \frac{55}{2}\right)\tau^3 + \left(-\frac{135}{24} + \frac{1}{24}B\right)\tau^4 - \tau^5$$

$$- \frac{99}{6!}\tau^6 - \frac{80}{7!}\tau^7 + \left(\frac{1}{40320}A - \frac{1}{672}\right)\tau^8 + \left(\frac{1}{362880}B + \frac{1}{9072}\right)\tau^9$$

$$- \frac{1}{103680}\tau^{10} - \frac{1}{1663200}\tau^{11} - \frac{1}{31933440}\tau^{12} + \left(\frac{1}{6227020800}A - \frac{1}{1245404160}\right)\tau^{13}$$

$$+ \left(\frac{1}{17435658240} + \frac{1}{871782941200}B\right)\tau^{14} - 30\tau e^{\tau} + \ldots\ldots$$

By applying the boundary conditions (37), we conclude the values of the constants for $\tau = 0, 1$

$$A = -3.000001074, \quad B = -7.999995199;$$

Now, we obtain the series solution of problem (35) as:

$$\Gamma(\tau) = \Gamma_0 + \Gamma_1 + \Gamma_2 + \ldots$$

$$= -255 - 224\tau + 255e^{\tau} - \frac{195\tau^2}{2} - 3.000244826\tau^3 - 5.95808503\tau^4 - \tau^5$$

$$- 0.1375\tau^6 - 0.015873015\tau^7 - 0.002083624793\tau^8 + 0.00008819984368\tau^9$$

$$- 0.0000096450617728\tau^{10} - 0.0000006012506013\tau^{11} - 0.0000000313513548\tau^{12}$$
$$+ 0.000000001284959407\tau^{13} + 0.0000000005746835597\tau^{14} - 30\tau e^{\tau} + \ldots$$

Example 3.5 The sixth-order linear boundary value problem [3, 36] is considered as:

$$\Gamma^6(\tau) - (\Gamma(\tau) - 6e^{\tau}) = 0, \quad 0 < \tau < 1, \tag{40}$$

Boundary conditions are

$$\Gamma(0) = 1, \ \Gamma''(0) = -1, \ \Gamma''''(0) = -3,$$
$$\Gamma(1) = 0, \ \Gamma''(1) = -2e, \ \Gamma''''(1) = -4e. \tag{41}$$

The exact solution of the given problem (40) is

$$\Gamma(\tau) = (1 - \tau)e^{\tau} \tag{42}$$

By applying Laplace transform on both sides of above problem (40), we get

$$L\big(\Gamma^6(\tau)\big) - L(\Gamma(\tau) - 6e^{\tau}) = 0,$$

$$\Gamma(s) = \frac{1}{s} + \frac{A}{s^2} - \frac{1}{s^3} + \frac{B}{s^4} - \frac{3}{s^5} + \frac{C}{s^6} + \frac{L(\Gamma(\tau))}{s^6} - \frac{6}{s^6(s-1)}.$$

By applying inverse Laplace transform on both sides of the above equation, we get

$$\Gamma(\tau) = 7 + (A+6)\tau + \frac{5}{2}\tau^2 + \frac{(B+6)\tau^3}{6} + \frac{3\tau^4}{4!} + \frac{(C+6)\tau^5}{5!} - 6e^{\tau} + L^{-1}\left\{\frac{L}{s^6}(\Gamma(\tau))\right\}.$$

Now, applying HPM on both sides, we get

$$\sum_{n=0}^{\infty} p^n \Gamma_n(\tau) = 7 + (A+6)\tau + \frac{5}{2}\tau^2 + \frac{(B+6)\tau^3}{6} + \frac{3\tau^4}{4!} + \frac{(C+6)\tau^5}{5!} - 6e^{\tau}$$
$$+ pL^{-1}\left\{\frac{L}{s^6}\left(\sum_{n=0}^{\infty} p^n \Gamma_n(\tau)\right)\right\}. \tag{43}$$

By comparing the coefficient of same powers of p, we obtain

$$p^0 : \Gamma_0(\tau) = 7 + (A + 6)\tau + \frac{5}{2}\tau^2 + \frac{(B + 6)\tau^3}{6} + \frac{3\tau^4}{4!} + \frac{(C + 6)\tau^5}{5!} - 6e^\tau,$$

$$p^1 : \Gamma_1(\tau) = L^{-1}\left\{\frac{L}{s^6}(\Gamma_0(\tau))\right\}$$

$$= 6 + 6\tau + 3\tau^2 + \tau^3 + \frac{1}{4}\tau^4 + \frac{1}{20}\tau^5$$

$$+ \frac{7}{6!}\tau^6 + \frac{(A + 6)\tau^7}{7!} + \frac{5\tau^8}{8!} + \frac{(B + 6)\tau^9}{9!}$$

$$+ \frac{3\tau^{10}}{10!} + \frac{(C + 6)\tau^{11}}{11!} - 6e^\tau,$$

$$p^2 : \Gamma_2(\tau) = L^{-1}\left\{\frac{L}{s^6}(\Gamma_1(\tau))\right\}$$

$$= 6 + 6\tau + 3\tau^2 + \tau^3 + \frac{1}{4}\tau^4 + \frac{1}{20}\tau^5 + \frac{1}{5!}\tau^6$$

$$+ \frac{6\tau^7}{7!} + \frac{6\tau^8}{8!} + \frac{6\tau^9}{9!} + \frac{6\tau^{10}}{10!} + \frac{6\tau^{11}}{11!} + \frac{7\tau^{12}}{12!}$$

$$+ \frac{(A + 6)\tau^{13}}{13!} + \frac{5\tau^{14}}{14!} + \frac{(B + 6)\tau^{15}}{15!} + \frac{3\tau^{16}}{16!} + \frac{(C + 6)\tau^{17}}{17!} - 6e^\tau,$$

$$\vdots$$

$$(44)$$

Thus the series solution of the problem is given as follows:

$$\Gamma(\tau) = \Gamma_0 + \Gamma_1 + \ldots\ldots$$

$$= 13 + (A + 12)\tau + \frac{11}{2}\tau^2 + \frac{(B + 12)}{6}\tau^3 + \frac{9}{7!}\tau^4$$

$$+ \frac{(C + 12)}{5!}\tau^5 + \frac{7}{6!}\tau^6 + \frac{(A + 6)}{7!}\tau^7 + \frac{5}{8!}\tau^8 + \frac{(B + 6)}{9!}\tau^9$$

$$+ \frac{(C + 6)}{11!}\tau^{11} - 12e^\tau.$$

By applying the boundary conditions (41) in Eq. (45), we conclude the values of the constants for $\tau = 0, 1$.

$$A = -0.000002082634658 \quad B = -2.0000137, \quad C = -3.999975769;$$

Thus, the approximate solution of the problem (40) is given as follows:

Table 4 Comparision of LT-HPM with ADM, HPM VIM and exact

τ	Approximate solution (LT-HPM)	Exact solution	Absolute error (LT-HPM)	Absolute error (ADM) [3]	Absolute error (VIM) [3]	Absolute error (HPM) [3]
0.0	1	1	0	0	0	0
0.1	0.994653616	0.994653826	3.8E–07	4.1E–04	4.1E–04	4.1E–04
0.2	0.977121772	0.977122206	4.34E–07	7.8E–04	7.8E–04	7.8E–04
0.3	0.94490048	0.944901165	6.85E–07	1.1E–03	1.1E–03	1.1E–03
0.4	0.895093833	0.895094818	9.84E–07	1.3E–03	1.3E–03	1.3E–03
0.5	0.82435931	0.824360635	1.32E–06	1.3E–03	1.3E–03	1.3E–03
0.6	0.728845791	0.72884752	1.72E–06	1.3E–03	1.3E–03	1.3E–03
0.7	0.604123662	0.604125812	2.21E–06	1.1E–03	1.1E–03	1.1E–03
0.8	0.44510541	0.445108185	2.77E–06	4.1E–04	4.1E–04	4.1E–04
0.9	0.24585688	0.245960311	3.43E–06	7.8E–04	7.8E–04	7.8E–04
1.0	0	0	0.0	0.0	0.0	0.0

$$\Gamma(\tau) = \Gamma_0 + \Gamma_1 + \ldots$$
$$= 13 + 11.99999792\tau + 5.5\tau^2 + 1.666664383\tau^3 + 0.375\tau^4$$
$$+ 0.066666868\tau^5 + .0097222222\tau^6 + 0.001190475777\tau^7$$
$$+ 0.0001240079365\tau^8 + 0.00001102288994\tau^9 + \ldots \qquad (45)$$

Here, Table 4 shows that the result is obtained by the LT-HPM with the help of Mathematica package, which is compared with the exact, Adomian decomposition method, variational iterative method, and homotopy perturbation method. Figure 4 shows the comparison of proposed method with the exact solution. Furthermore, this method shows the absolute maximum error $E(\tau) = |(1 - \tau)e^\tau - \Gamma(\tau)|$.

4 Conclusion

In this paper, the homotopy perturbation method with Laplace transform has been implemented to solve complex linear higher-order boundary value problems. The present method is very effective, simple, easy to implement. Here, the approximate solutions have been also compared with variational iterative method, Adomian decomposition method, and homotopy perturbation method by the sum of at least three iterations and by using Mathematica package. The numerical results found by using this approach are superior to conventional approach and very close to exact solutions. The attractive property of the method is that it is implemented directly in a straightforward manner without using restrictive assumptions, linearization, and discretization.

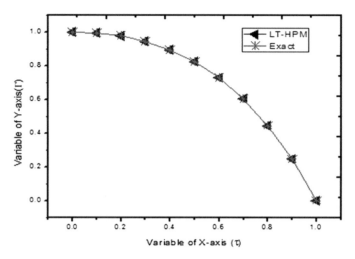

Fig. 4 The comparison of the proposed method with exact solution

Acknowledgements The second author is very much thankful for grant no. 1013/CST/R&D/Phy&EnggSc/2015 given by Madhya Pradesh Council of Science and Technology (MPCST), Bhopal, India.

References

1. Chandrasekhar, S.: Hydrodynamic and hydro-magnetic stability. Dover Press, New York (1981)
2. Timoshenko, S.: Theory of Elastic Stability. McGraw-Hill, New York (1961)
3. Ghani, F., Islam, S., Ozel, C., Liaqat Ali, M., Rashidi, M.: Application of modified optimal homotopy perturbation method to higher order boundary value problems in a finite domain. Chaos Solit. Fract. **41**, 1905–1909 (2009)
4. Hajji, M.A.: Multi-point special boundary-value problems and applications to fluid flow through porous media. In: Proceedings of the International Multi-Conference of Engineers and Computer Scientists II IMECS 2009, pp. 18–20 (2009)
5. Geng, F., Cui, M.: Multi-point boundary value problem for optimal bridge design. Int. J. Comput. Math. **87**, 1051–1056 (2010)
6. Atluri, S.N., Zhu, T.: A new meshless local Petrov-Galerkin (MLPG) approach in computational mechanics. Comput. Mech. **22**, 117–127 (1998)
7. Coskun, S.B., Atay, M.T.: Analysis of convective straight and radial fins with temperature-dependent thermal conductivity using variational iteration method with comparison with respect to finite element analysis. Math. Prob. Eng. **200**, 1–15 (2007)
8. Rashidi, M.M., Ganji, D.D., Dinarvand, S.: explicit analytical solutions of the generalized Burger and Burger-Fisher equations by homotopy perturbation method. Numer. Methods Partial Diff. Equ. **25**, 409–417 (2009)
9. Ali, J., Islam, S., Shah, S., Khan, H.: The optimal homotopy asymptotic method for the solution of fifth and sixth order boundary value problems. World Appl. Sci. J. **15**, 1120–1126 (2011)
10. Marinca, V., Herisanu, N.: Optimal homotopy perturbation approach to thin film flow of a fourth-grade fluid. AIP Conf. Proc. **1479**, 2383–2386 (2012)

11. Islam, S.U., Khan, M.A.: A numerical method based on polynomials sextic spline functions for the solution of special fifth-order boundary value problems. Appl. Math. Comput. **181**, 356–361 (2006)
12. Noor, M.A., Mohyud-Din, S.T.: Variational iteration technique for solving higher order boundary value problems. Appl. Math. Comput. **189**, 1929–1942 (2007)
13. Wazwaz, A.M.: The numerical solution of sixth-order boundary value problems by the modified decomposition method. Appl. Math. Comput. **118**, 311–325 (2001)
14. Noor, M.A., Mohyud-Din, S.T.: An efficient method for fourth-order boundary value problems. Comput. Math Appl. **54**, 1101–1111 (2007)
15. Liao, S.: Beyond Perturbation: Introduction to the Homotopy Analysis Method. Chapman & Hall/CRC Press, Florida (2004)
16. He, J.H.: Homotopy perturbation method for solving boundary value problems. Phys. Lett. A **350**, 87–88 (2006)
17. Chun, C., Sakthivel, R.: Homotopy perturbation technique for solving two-point boundary value problems—comparison with other methods. Comput. Phy. Commu. **181**, 1021–1024 (2010)
18. Hesameddini, E., Latifizadeh, H.: A new vision of the He's homotopy perturbation method. Int. J. Nonlinear Sci. Numer. Simul. **10**, 1415–1424 (2009)
19. Wu, B.Y., Li, X.Y.: A new algorithm for a class of linear nonlocal boundary value problems based on the reproducing kernel method. Appl. Math. Lett. **24**, 156–159 (2011)
20. Siddiqui, S.S., Akram, G.: Solutions of sixth order boundary-value problems using nonpolynomial spline technique. Appl. Math. Comput. **181**, 708–720 (2006)
21. Siddiqui, S.S., Akram, G.: Solutions of fifth order boundary-value problems using nonpolynomial spline technique. Appl. Math. Comput. **175**, 1574–1581 (2006)
22. Tatari, M., Dehghan, M.: The use of the Adomian decomposition method for solving multipoint boundary value problems. Phys. Scripta. **73**, 672–676 (2006)
23. Ali, J., Islam, S., Islam, S., Zaman, G.: The solution of multipoint boundary value problems by the optimal homotopy asymptotic method. Comput. Math Appl. **59**, 2000–2006 (2010)
24. Biazar, J., Ghazvini, H.: Convergence of the homotopy perturbation method for partial differential equations. Nonlinear Anal. Real World Appl. **10**, 2633–2640 (2009)
25. He, J.H.: Homotopy perturbation technique. Comput. Methods Appl. Mech. Eng. **178**, 257–262 (1999)
26. Eloe, P.W., Henderson, J.: Uniqueness implies existence and uniqueness conditions for nonlocal boundary value problems for nth order differential equations. J. Math. Anal. Appl **331**, 240–247 (2007)
27. Hussein, A.J.: Study of error and convergence of homotopy perturbation method for two and three dimensions linear Schrödinger equation. J. College Educ. **1**, 21–43 (2011)
28. Mishra, H.K.: He-Laplace method for the solution of two-point boundary value problems. Amer. J. Math. Anal. **2**, 45–49 (2014)
29. Tripathi, R., Mishra, H.K.: Homotopy perturbation method with Laplace transform (LT-HPM) for solving Lane-Emden type differential equations (LETDEs), Springer Plus **5**(1859), 1–21 (2016)
30. Wazwaz, A.M.: A composition between Adomian's decomposition method and Taylor series method in the series solution. Appl. Math. Comput. **79**, 37–44 (1998)
31. Mishra, H.K.: He-Laplace method for special nonlinear partial differential equations. Math. Theory Model. **3**, 113–117 (2013)
32. Sweilam, N.H.: Fourth-order integro-differential equations using variational iteration method. Int. J. Modern Phys. B **20**, 1086–1091 (2006)
33. Momani, S., Moadi, K.: A reliable algorithm for solving fourth-order boundary value problems. J. Appl. Math. Comput. **22**, 185–197 (2006)
34. Kelesoglu, Omer: The solution of fourth-order boundary value problem arising out of the beam-column theory using Adomain decomposition method. Math. Prob. Eng. **2014**, 1–6 (2014)
35. Noor, M.A., Mohyud-Din, S.T.: Variational iterative method for fifth-order boundary value problems using He's polynomials. Math. Prob. Eng. **2008**, 1–12 (2008)
36. Fazal-I-Haq, A.A., Hussain, I.: Solution of sixth-order boundary-value problems by collocation method using Haar wavelets. Int. J. Phys. Sci. **7**, 5729–5735 (2012)

Lung Cancer Detection: A Deep Learning Approach

Siddharth Bhatia, Yash Sinha and Lavika Goel

Abstract We present an approach to detect lung cancer from CT scans using deep residual learning. We delineate a pipeline of preprocessing techniques to highlight lung regions vulnerable to cancer and extract features using UNet and ResNet models. The feature set is fed into multiple classifiers, viz. XGBoost and Random Forest, and the individual predictions are ensembled to predict the likelihood of a CT scan being cancerous. The accuracy achieved is 84% on LIDC-IRDI outperforming previous attempts.

Keywords Lung cancer detection · Deep residual networks · XGBoost
Random Forests · Ensemble · Deep learning

1 Introduction

Lung cancer is the leading cause of cancer-related deaths all around the world. One of the important steps in detecting early stage cancer is to find out whether there are any pulmonary nodules in the lungs which may grow to a tumor in recent future. This work aims to determine the likelihood of a given CT scan of lungs to be cancerous. In a nutshell, we employ deep residual networks to extract features from preprocessed images which are fed to classifiers, the predictions of which are ensembled for the final output. We explain in this paper the proposed methodology, evaluation, and results using the LIDC-IDRI dataset.

Rest of the paper is organized as follows. Earlier studies on lung cancer detection have been delineated in Sect. 2. Section 3 explains the dataset used. We explain the various background techniques employed in Sect. 4. Section 5 elaborates on the proposed methodology, preprocessing steps, feature extraction, and classification. The results are further described in Sect. 6. We conclude with future directions in Sect. 7.

S. Bhatia · Y. Sinha (✉) · L. Goel
Department of Computer Science and Information Systems, BITS,
Pilani, Pilani Campus, Pilani, Rajasthan, India
e-mail: mail.yash.sinha@gmail.com

© Springer Nature Singapore Pte Ltd. 2019
J. C. Bansal et al. (eds.), *Soft Computing for Problem Solving*, Advances in Intelligent
Systems and Computing 817, https://doi.org/10.1007/978-981-13-1595-4_55

2 Related Work

Lung cancer detection has earlier been studied using image processing techniques [1–3]. With the advent of neural networks and deep learning techniques, these have recently been used in the medical imaging domain [4–6]. Various researchers [7–12] have tried to classify, detect lung cancer using machine learning and neural networks. Not many deep learning techniques have been applied to detect lung cancer. This is because of the lack of a large dataset for medical images especially lung cancer. Shimizu et al. [13] use urine samples to detect lung cancer.

The technique proposed by Hua et al. [14] simplifies the image analysis pipeline of conventional computer-aided diagnosis of lung cancer. Sun et al. [15] experimented using convolutional neural networks (CNN), deep belief networks (DBNs), and stat denoising autoencoder (SDAE) on Lung Image Database Consortium image collection (LIDC-IDRI) [16]. Their accuracies were 79%, 81%, and 79%, respectively.

The National Lung Screening Trial (NLST), a randomized control trial in the USA, including more than 50,000 high-risk subjects, showed that lung cancer screening using annual low-dose computed tomography (CT) reduces lung cancer mortality by 20% in comparison to annual screening with chest radiography [23].

In 2013, the US Preventive Services Task Force (USPSTF) has given low-dose CT screening a grade B recommendation for high-risk individuals [24], and in early 2015, the US Centers for Medicare and Medicaid Services (CMS) has approved CT lung cancer screening for Medicare recipients [25].

The recent challenge launched "LUNA16" [26] aims to predict the position of nodule in the given lung region. Zatloukal et al. [27] present a study of localization of non-small lung cancer cell with chemotherapy techniques. Zhou et al. [27] present a cancer cell identification technique based on neural network ensembles.

3 Dataset

The Lung Image Database Consortium image collection (LIDC-IDRI) [16] contains diagnostic and lung cancer screening thoracic computed tomography (CT) scans with marked-up annotated lesions. It consists of more than thousand scans from high-risk patients in the DICOM image format. Each scan contains a series of images with multiple axial slices of the chest cavity. Each scan has a variable number of 2D slices, which can vary based on the machine taking the scan and patient. The DICOM files have a header that contains the details about the patient id, as well as other scan parameters such as the slice thickness. The images are of size (z, 512, 512), where z is the number of slices in the CT scan and varies depending on the resolution of the scanner.

4 Background of Technology Used

4.1 Deep Residual Networks

Deep residual networks [17] have emerged as a family of extremely deep architectures showing compelling accuracy and nice convergence behaviors. Deep residual networks (ResNets) consist of many stacked "Residual Units." The central idea of ResNets is to learn the additive residual function F with respect to $h(x_1)$, with a key choice of using an identity mapping

$$h(x_1) = x_1. \tag{1}$$

Each subsequent layer in a deep neural network is only responsible for, in effect, fine tuning the output from a previous layer by just adding a learned "residual" to the input. This differs from a more traditional approach where each layer had to generate the whole desired output (Fig. 1).

What's happening is that the $F(x)+x$ layer is adding in, element-wise, the input to the F(x) layer. Here, F(x) is the residual.

4.2 XGBoost Regressor

Extreme Gradient Boosting [18] builds on the premise of "boosting" many weak predictive models into a strong one, in the form of ensemble of weak models well known as Gradient Tree Boosting [19]. There are many gradient tree boosting algorithms, but specifically XGBoost uses the second-order method by Friedman et al. [20, 21] and employs a more regularized model formalization to control over-fitting, which gives it better performance.

Fig. 1 Residual learning: a building block

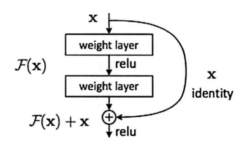

4.3   Random Forest Classifier

It is a meta-estimator [22] based on subsampling over many decision trees which controls over-fitting well. The basis of random forest is that randomization over many decision trees can improve the accuracy of the overall classification by boosting the selection rates of features that contribute more toward the classification among others.

5   Proposed Methodology

In a nutshell, we preprocess the CT scan images which are in the DICOM image format to extract the central region of interest of the lungs, which is more likely to have pulmonary nodule. Features are extracted using deep residual networks that are fed into classifiers for supervised learning. For the final output, we ensemble the predictions of various classifiers.

5.1   Preprocessing

The preprocessing step consists of a series of applications of region growing and morphological operations. It identifies and separates the lung structures and nodules to aid the feature extraction. Segmenting lungs from the CT scan aims to identify distinguishing features to aid the classifier and classify the candidates better. This is also important because the CT scan is too huge to be fed into the classifier directly. It will take a lot of time for the classifier to identify differentiating featured from the huge DICOM images.

The segmentation of lung structures is very challenging problem primarily because there is no homogeneity in the lung region. There are similar densities in the pulmonary structures, different scanners, and scanning protocols.

Lung segmentation is followed by normalization and zero centering.

5.2   Feature Extraction

We feed the preprocessed images to ResNet-50 imagenet11k+Places365 feature extractor [17] (Fig. 2).

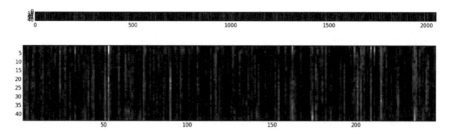

Fig. 2 Visualizing preprocessed features

Fig. 3 % accuracy of various approaches

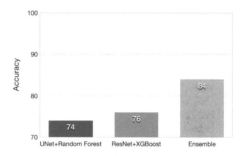

5.3 Classification and Ensemble

The feature dataset created at the feature extraction stage is fed into a number of classifiers like XGBoost and Random Forest. The results are outlined in Fig. 3. The predictions are ensembled by vote for the final output. The hyperparameters for the classifiers are determined using Grid Search, and the model is tested using 10-fold cross-validation.

6 Results and Inferences

We compare our proposed methodologies in Fig. 3.

We are able to get an accuracy of 84% using an ensemble of UNet+RandomForest and ResNet+XGBoost which individually have accuracies 74% and 76%, respectively.

7 Conclusion

In this paper, we propose an approach to lung cancer detection employing feature extraction using deep residual networks. We compare performance of tree-based classifiers like Random Forest and XGBoost. The highest accuracy we get is 84% using ensemble of Random Forest and XGBoost classifier.

References

1. Palcic, B., et al.: Detection and localization of early lung cancer by imaging techniques. CHEST J. **99**(3) 742–743 (1991)
2. Yamomoto, S., et al.: Image processing for computer-aided diagnosis of lung cancer by CT (LSCT). In: Proceedings 3rd IEEE Workshop on Applications of Computer Vision, 1996. WACV'96. IEEE (1996)
3. Gurcan, M.N., et al.: Lung nodule detection on thoracic computed tomography images: preliminary evaluation of a computer-aided diagnosis system. Med. Phys. **29**(11) 2552–2558 (2002)
4. Fakoor, R., et al.: Using deep learning to enhance cancer diagnosis and classification. In: Proceedings of the International Conference on Machine Learning (2013)
5. Greenspan, H., van Ginneken, B., Summers, R.M.: Guest editorial deep learning in medical imaging: overview and future promise of an exciting new technique. IEEE Trans. Med. Imag. **35**(5) 1153–1159 (2016)
6. Shen, D., Wu, G., Suk H.-I.: Deep learning in medical image analysis. Ann. Rev. Biomed. Eng. (2017)
7. Cai, Z., et al.: Classification of lung cancer using ensemble-based feature selection and machine learning methods. Molec. BioSyst. **11**(3) 791–800 (2015)
8. Al-Absi Hamada R.H., Belhaouari Samir B., Sulaiman, S.: A computer aided diagnosis system for lung cancer based on statistical and machine learning techniques. JCP **9**(2) 425–431 (2014)
9. Gupta, B., Tiwari, S.: Lung cancer detection using curvelet transform and neural network. Int. J. Comput. Appl. **86**(1) (2014)
10. Penedo, M.G., et al.: Computer-aided diagnosis: a neural-network-based approach to lung nodule detection. IEEE Trans. Med. Imag. **17**(6) 872–880 (1998)
11. Taher, F., Sammouda, R.: Lung cancer detection by using artificial neural network and fuzzy clustering methods. In: GCC Conference and Exhibition (GCC), 2011 IEEE. IEEE (2011)
12. Kuruvilla, J., Gunavathi, K.: Lung cancer classification using neural networks for CT images. Comput. Methods Program. Biomed. **113**(1), 202–209 (2014)
13. Shimizu, R., et al.: Deep learning application trial to lung cancer diagnosis for medical sensor systems. In: SoC Design Conference (ISOCC), 2016 International. IEEE (2016)
14. Hua, K.-L., et al.: Computer-aided classification of lung nodules on computed tomography images via deep learning technique. Onco Targets Therapy **8** 2015–2022 (2014)
15. Sun, W., Zheng, B., Qian, W.: Computer aided lung cancer diagnosis with deep learning algorithms. In: SPIE Medical Imaging. International Society for Optics and Photonics (2016)
16. Armato, S.G., et al.: The lung image database consortium (LIDC) and image data-base resource initiative (IDRI): a completed reference database of lung nodules on CT scans. Med. Phys. **38**(2) 915–931 (2011)
17. He, K., Zhang, X., Ren, S., Deep, S.J.: residual learning for image recognition. In: Proceedings of the IEEE Conference on Computer Vision and Pattern Recognition, pp. 770–778 (2016)
18. Chen, T., Guestrin, C.: Xgboost: a scalable tree boosting system. In: Proceedings of the 22nd ACM SIGKDD International Conference on Knowledge Discovery and Data Mining, ACM (2016)

19. Friedman, J.: Greedy function approximation: a gradient boosting machine. Ann. Stat. **29**(5), 1189–1232 (2001)
20. Friedman, J., Hastie, T., Tibshirani, R., et al.: Additive logistic regression: a statistical view of boosting (with discussion and a rejoinder by the authors). Ann. Stat. **28**(2), 337–407 (2000)
21. Friedman, J.H.: Greedy function approximation: a gradient boosting machine. Ann. Stat. 1189–1232 (2001)
22. Liaw, Andy, Wiener, Matthew: Classification and regression by random forest. R news **2**(3), 18–22 (2002)
23. Aberle, D.R., Adams, A.M., Berg, C.D., Black, W.C., Clapp, J.D., Fagerstrom, R.M., Ga-reen, I.F., Gatsonis, C., Marcus, P.M., Sicks, J.D.: Reduced lung-cancer mortality with low-dose computed tomographic screening. N. Engl. J. Med. **365**, 395–409 (2011)
24. Moyer, V.A.: U.S. preventive services task force. Screening for lung cancer: U.S. Preventive services task force recommendation statement. Ann. Int. Med. **160**, 330–338 (2014)
25. Armato, S.G., McLennan, G., Bidaut, L., McNitt-Gray, M.F., Meyer, C.R., Reeves, A.P., et al.: The lung image database consortium (LIDC) and image database resource initiative (IDRI): a completed reference database of lung nodules on CT scans. Med. Phys. **38**, 915–931 (2011)
26. LUng Nodule Analysis (LUNA) Challenge. https://luna16.grand-challenge.org/description/
27. Zatloukal, P., et al.: Concurrent versus sequential chemoradiotherapy with cisplatin and vinorel-bine in locally advanced non-small cell lung cancer: a randomized study. Lung Cancer **46**(1) 87–98 (2004)

Convolutional Neural Network-Based Human Identification Using Outer Ear Images

Harsh Sinha, Raunak Manekar, Yash Sinha and Pawan K. Ajmera

Abstract This paper presents a deep learning approach for ear localization and recognition. The comparable complexity between human outer ear and face in terms of its uniqueness and permanence has increased interest in the use of ear as a biometric. But similar to face recognition, it poses challenges such as illumination, contrast, rotation, scale, and pose variation. Most of the techniques used for ear biometric authentication are based on traditional image processing techniques or handcrafted ensemble features. Owing to extensive work in the field of computer vision using convolutional neural networks (CNNs) and histogram of oriented gradients (HOG), the feasibility of deep neural networks (DNNs) in the field of ear biometrics has been explored in this research paper. A framework for ear localization and recognition is proposed that aims to reduce the pipeline for a biometric recognition system. The proposed framework uses HOG with support vector machines (SVMs) for ear localization and CNN for ear recognition. CNNs combine feature extraction and ear recognition tasks into one network with an aim to resolve issues such as variations in illumination, contrast, rotation, scale, and pose. The feasibility of the proposed technique has been evaluated on USTB III database. This work demonstrates 97.9% average recognition accuracy using CNNs without any image preprocessing, which shows that the proposed approach is promising in the field of biometric recognition.

H. Sinha · R. Manekar · P. K. Ajmera (✉)
Department of Electrical and Electronics Engineering, BITS, Pilani, Pilani Campus, Pilani, Rajasthan, India
e-mail: pawan.ajmera@pilani.bits-pilani.ac.in

Y. Sinha
Department of Computer Science and Information Systems, BITS Pilani, Pilani Campus, Pilani, Rajasthan, India
e-mail: mail.yash.sinha@gmail.com

© Springer Nature Singapore Pte Ltd. 2019
J. C. Bansal et al. (eds.), *Soft Computing for Problem Solving*, Advances in Intelligent Systems and Computing 817, https://doi.org/10.1007/978-981-13-1595-4_56

707

1   Introduction

As a form of security enhancement, human traits are being used extensively for bio-metric authentication. The most prominent biometric used today are face, iris, and fingerprint. However, there are some major drawbacks with them such as illumi-nation, intrusiveness, facial makeup, expressions, surgical alterations, birth defects, and physical changes with aging. Hence, researchers are finding different recognition techniques for better security. In recent years, using human ear as a biometric trait has become a popular choice. Descartes Biometrics developed an app for Android that uses ear as a password [5]. An iOS application was developed that allows a medical practitioner to search the medical data on using patient's ear as the key [3].

The motivation behind the present study derives from extensive work done in the field of image recognition using convolutional neural networks (CNNs) and histogram of oriented gradients (HOG) feature extraction technique [9, 23, 24]. This paper demonstrates the use of CNNs for the task of ear recognition and achieves very high scale-invariant ear localization accuracy using a simple (HOG+SVM) framework on a multi-scale image pyramid.

It is stated that most of recognition techniques give very poor accuracy when applied directly on acquired images [12]. Reported works have suggested addition of new image enhancement and feature extraction stages in the image preprocessing pipeline for making recognition invariant to scale, rotation, illumination, etc. [2, 27]. Hence, LeCun et al. [23] have suggested a traditional recognition framework which is preceded by a feature extraction step to extract relevant information from images (training samples). These features (or vectors) are then fed into a generic classifier, which classifies them into one-hot-encoded classes. Extracting features on the basis of performance of classifiers is found to be a much better approach [19, 23]. Hence, CNNs are suitable for such a scenario as they extract relevant features and train the end classifier pivoting on backpropagation algorithm. As a result, any explicit feature extraction pipeline was eliminated.

1.1   Detection

Ear detection is an important preliminary step for ear biometrics. In the literature, researchers achieve detection of ear in a given profile face using image processing pipeline similar to skin detection [4], skin segmentation [29], edge computation [41], ear-candidates generation by frequent subgraph mining [28], and template matching [6]. However, some researchers preferred to achieve it manually.

Prakash and Gupta [31] achieved a ear localization accuracy of 94% over 150 individuals which involved skin segmentation followed by template matching. Yan and Bowyer [41] used contour information while Yuan and Mu [43] used skin color and contour information to localize the ear and edges to detect ear pit. Chen and Bhanu [4] presented a template matching approach which used canny edge detector for segregating convex and concave images. Researchers have also focused on assuming a shape for the ear, e.g., elliptical [4], wavelet based [32] or average of known ear images [31].

Clearly, the detection pipeline in recent literature depends upon texture and edge information for localization of ear. However, the proposed HOG+SVM technique involves only two steps, calculation of gradients on the image and classification using pre-trained SVM on given dataset. There are no pre-assumptions regarding the background or shape of the ear. Moreover, HOG feature extraction has advantage of lower complexity in terms of computational time and greater accuracy as compared to popular feature extraction like Gabor filters and Haar features [9, 34, 37].

1.2 Recognition

In the past decade, most of the work in ear recognition have focused on identifying new features and representations from ear images.

Apart from holistic techniques such as linear discriminant analysis (LDA), wavelet transforms [32, 39], Gabor wavelets [22], log-Gabor filters [18, 21], 2-D quadrature filters [7] which are applied on the entire image, another common way of feature extraction is extracting local image information by dividing an image into many small windows and applying transforms on them. Sparse representation of localized Radon transform created a robust ear representation [20] while an extension of principal components analysis (PCA) was applied to create new features from the features calculated from input image to make recognition less sensitive to environmental variations [15]. Calculating local information for forming new feature set was also suggested [42]. Force-field transformations proved efficient as they achieved 99.2% accuracy on XM2VTS dataset [16]. Moreover, a modular neural network achieved a recognition accuracy of 75.44% using USTB dataset [33]. Image contrast enhancement techniques such as scale-invariant feature transform (SIFT) with artificial bee colony (ABC) algorithm, histogram equalization, and contrast-limited adaptive histogram equalization (CLAHE) have also been explored [12].

The performance of approaches based on 2D images rely on pose and illumination variation. Hence, 3D ear images have also been proposed in order to overcome these problems [25, 28, 30] as 3D models contain depth information which further enhances the accuracy of a biometric system. However, using 3D models as a biometric would lead to bulky databases and a slow verification system.

In this paper, a HOG-based descriptor for localization of ear and CNN for recognition is proposed. HOG computes gradients to classify images. Localized ear is cropped and fed into CNN for recognition which combine the feature extraction and ear recognition. CNNs have the ability to identify which features are most suitable for recognition on the particular training set and learn specific features resulting in most optimized recognition. Hence, it eliminates the need for handcrafted features. The proposed methodology is summarized in Fig. 1.

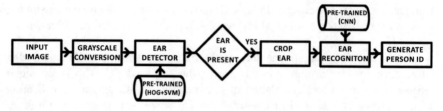

Fig. 1 Proposed pipeline

The paper is organized as follows, the proposed methodology along with HOG and CNN is described in Sect. 2. The dataset used and performance evaluation is presented in Sect. 3. Conclusions based on the experimental results are presented in Sect. 4.

2 Proposed Method

A biometric system mainly consists of of image preprocessing, detection, and recognition. In subsequent sections, detection and recognition methodology used is discussed. Preprocessing is minimal; that is, the acquired images were normalized and grayscaled.

2.1 Ear Localization

Object localization consists of three basic steps, viz. choosing a feature descriptor, training a classifier on positive and negative samples of the object, and classification of pixels inside scanning window using the trained classifier. Different approaches varies in types of feature descriptor and the generic classifier chosen.

Two major approaches which are used for ear localization are Haar feature-based cascade classifier [37] and HOG with SVM [8, 26]. Both techniques have been applied on a multi-scale image pyramid representation of the side profile image for scale-invariant ear detection.

In HOG, an object appearance and shape is characterized by intensity gradients or edge directions. The image window is divided into many small regions. The gradients of an image is calculated. Moreover, the computed gradient is converted to polar coordinates to accommodate gradient directions.

The flowchart in Fig. 2 describes the procedure for ear detection. The first step is to generate an image pyramid for each image in the training dataset which ensures that the object detector is scale invariant. For each region, a histogram of gradient directions is calculated over the pixels. Then, the cells are grouped to perform nor-

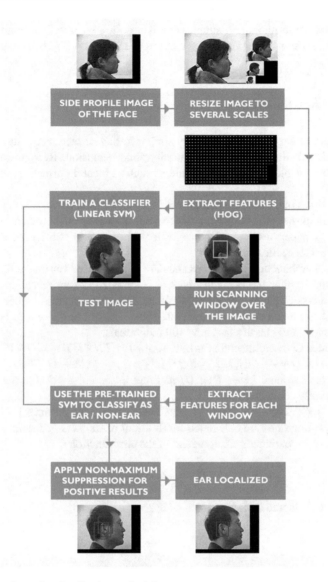

Fig. 2 Flowchart of ear localization methodology

malization across 2×2 cells. The normalized features are concatenated as a single HOG feature descriptor.

A classifier (linear SVM) is trained on extracted features using large number of positive and negative samples of the object (ear). When a test image is fed as input, pixels inside a scanning window are classified using the trained classifier. Generally, it is found that the detector generates several overlapping bounding boxes for a detected image. Hence, non-maximum suppression is used to generate final bounding box. Figures 4 and 5 shows the performance of (HOG+SVM) on USTB III dataset.

2.2 Ear Recognition

The problem of ear recognition is very similar to that of face recognition. It faces problem of bad illumination, pose variation, and occlusion. Recognition models are generally unable to overcome problems such as local deformation, rotation, or translation.

CNNs possess certain characteristic concepts such as local receptive fields, shared weights, and subsampling which makes it capable to overcome aforementioned problems. In this paper, a CNN structure depicted in Fig. 3 is considered for feature extraction and classification.

The CNN architecture chosen is a standard network used for image recognition. However, the architecture is much smaller as compared to very recent architectures used for large-scale general image recognition. As in our case, the number of objects to be learned is much smaller, a moderate-sized network turned out to be fast and sufficient enough to classify less than 100 individuals.

The applied CNN architecture can be described as $[INPUT - CONV - RELU - POOL - CONV - RELU - POOL - FC1 - FC2 - OUTPUT]$ where $CONV$: convolutional layer, $RELU$: rectified linear unit, $POOL$: max pooling layer, FC: fully connected.

The parameters of CNN were optimized using Adagrad optimizer [10] in which different learning rates are calculated for different parameters θ. Assuming, for tth time step and ith parameter, the gradient of the cost function is

$$g_{t,i} = \nabla_\theta J(\theta_i) \tag{1}$$

According to Adagrad update,

$$\theta_{t+1,i} = \theta_{t,i} - \frac{\eta}{\sqrt{G_{t,ii} + \epsilon}} g_{t,i} \tag{2}$$

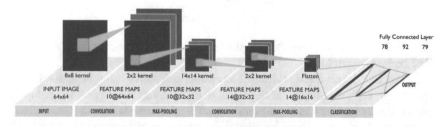

Fig. 3 Proposed convolutional neural network for ear recognition

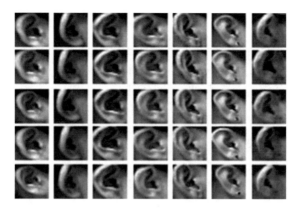

Fig. 4 Cropped ears: localization performance of HOG + SVM

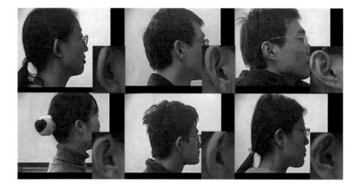

Fig. 5 Test image and cropped ear after localization

Here, G_t is a $d \times d$ diagonal matrix. Every element ii is summation of squares of gradients w.r.t θ_i upto time step t.

However, in stochastic gradient descent (SGD), the algorithm has a fixed learning rate.

$$\theta_{t+1,i} = \theta_{t,i} - \eta g_{t,i} \tag{3}$$

Hence, Adagrad optimizer automatically manipulates the learning rate to achieve a better accuracy.

Hyperparameters such as kernel size or number of layers are used to tune the performance of a machine learning model for unseen data. SigOpt was used to identify best hyperparameters [35, 38]. SigOpt creates a feedback loop between model performance and different values for hyperparameters.

To further improve the performance on the validation set, dropout and L2 regularization were employed [36]. Dropout prevents co-adaptation by nullifying nodes in the network. L2 regularization or weight decay selects the smallest values for different parameters by penalizing for large weights. The weights or parameters were initialized with a defined variance, which ensured that the problem of vanishing gradients is eliminated. Moreover, weight initialization improved consistency of results [1, 13]. Batch normalization was employed which resolves the problem of internal covariate shift [17]. Applying these techniques together during training had a compounding effect on performance of model.

3 Performance Evaluation

In this section, the proposed algorithm has been analyzed and evaluated by performing various experiments on the ear recognition task using the USTB III dataset. A comparative performance has been carried out against some of the existing ear recognition schemes such as wavelet transforms [14], log-Gabor filters [21] and image enhancement techniques [12].

The recognition rates in all the experiments are computed for the correct number of matches out of the total ears used for testing. The various performance measures used are defined as follows:

$$Precision_{micro} = \frac{tp_1 + ... + tp_k}{tp_1 + ... + tp_k + fp_1 + ... + fp_k} \tag{4}$$

$$Recall_{micro} = \frac{tp_1 + ... + tp_k}{tp_1 + ... + tp_k + fn_1 + ... + fn_k} \tag{5}$$

$$Precision_{macro} = \frac{Precision_1 + ... + Precision_k}{k} \tag{6}$$

$$Recall_{macro} = \frac{Recall_1 + ... + Recall_k}{k} \tag{7}$$

where, tp = true positives, fp = false positives, fn = false negatives, tn = true negatives.

3.1 Dataset

The University of Science and Technology in Beijing (USTB) provides three databases for public use. The USTB III ear dataset contains side profile images of face. The database contains images from 79 subjects, each image having a resolution of 768×576. Each subject has approximately 10 uncropped images (a total of 785 images).

3.2 Detection

For ear detection, the dataset was divided into 60:40 as training data and test data. For Haar cascade classifier, the standard pre-trained model of ear was used. But, the classifier was unable to detect ears in images with pose variations. The detector could detect ears with 74.5% accuracy. For HOG+SVM detector, while training the detector achieved an accuracy of 97.02% (457 out of 471 images). The trained detector when used over all images was able to locate an ear in 775 images (Out of a total of 785 images), i.e., an accuracy of 98.72%.

Haar features or other wavelets are good for texture-based characterization as it relies on difference in intensities in their vicinity. But, in object detection, edge orientation is very important to identify objects. HOG descriptors depend on the shape of the object using gradient orientations. Hence, HOG has a better performance than Haar for ear localization.

3.3 Recognition

The dataset contains $8 - 10$ uncropped images per person. However, the dataset was unbalanced (as the number of images per person was unequal). So, gamma correction was used for data augmentation using following equation. Gamma correction also simulates the effects of uneven lighting. After data augmentation, there were 14 images per person. These images were then divided in the ratio of 8:4:2 as training, validation, and test dataset, respectively.

$$I_2 = A \times I_1^{\gamma} \tag{8}$$

In our case, $A = 1$, $\gamma \epsilon (0.5, 1.5)$ was used.

The convolutional neural network optimized using Adagrad optimizer (incorporating adaptive learning rate) obtained an average recognition accuracy of 97.9% on the test dataset (Fig. 6). The CNN also used dropout with a keep probability of 0.5 for incorporating regularization into the network.

For recognition, several variations in parameters for CNN were used before finally arriving at the aforementioned architecture. It was found that applying image enhancement techniques just before CNN allows it to converge much quickly. However, the recognition accuracy saturates around 75%.

To asses the performance of the trained recognition network, confusion matrix was computed which is visualized as a heat map in Fig. 7. The different columns represent the predicted label by trained network, while the rows represent the ground truth. The diagonal shows number of correct predictions, which makes it is easy to identify mis-classified classes.

Fig. 6 Precision and recall
of proposed CNN

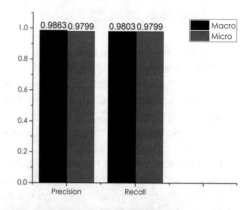

Fig. 7 Confusion matrix of
CNN

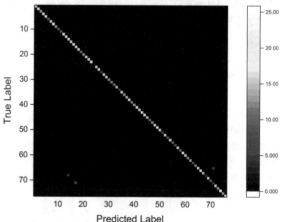

3.4   Comparative Performance

This section compares performance of proposed method with some of the other
prominent techniques which have used the similar dataset.

Yuan and Mu [43] achieved an accuracy of 90% used active shape model for detecting ears. Further, they used full-space linear discriminant Analysis (FSLDA) for recognizing ears. Xie and Mu [40] used local linear embedding (LLE) for nonlinear
dimensionality reduction. Initially, they obtained an accuracy of 60.75% accounting
for all poses. However, they developed a better LLE model which had a recognition rate of 80% for ear poses in the range [−10, 20] and 90% for ear poses in the
range [0, 10]. Hai-Long and Zhi-Chun [14] used orthogonal centroid algorithm over
low-frequency sub images and achieved an accuracy of 97.2%.

Galdmez et al. [11] used CNN over Avila's police school dataset and Bisite videos
dataset and reported an accuracy of 94.79% and 40.13%, respectively. Kumar and Wu
[21] achieved a recognition rate of 96.27% over IITD dataset while Ghoualmi et al.

Fig. 8 Comparative performance of the proposed methodology (numbers in parenthesis denote no. of subjects)

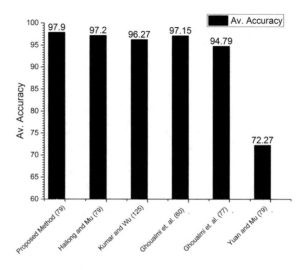

[12] achieved 97.15% and 94.79% over USTB I and USTB II datasets, respectively. In Fig. 8, recent methodology is compared with a comparable subject count. Proposed method achieves highest recognition rate of 97.9% on USTB III dataset.

4 Conclusions

In this paper, a methodology for ear localization and recognition is proposed which reduces the pipeline for a biometric recognition system. The proposed framework uses HOG with SVM for ear localization and CNN for ear recognition. HOG descriptors depend on image gradient orientations which describe the shape of an object. Hence, HOG features provide accurate and faster ear localization. Localization accuracy of 98.72% was achieved using HOG with SVM framework. CNN combines feature extraction and ear recognition into one network. It has the ability to learn most suitable features on a given training set resulting in optimized recognition. The feasibility of the proposed technique has been evaluated on USTB III dataset. This work demonstrates 97.9% average recognition accuracy using CNN without any image preprocessing, which shows that the proposed approach is a promising in the field of biometric recognition.

References

1. Abadi, M., Agarwal, A., Barham, P., Brevdo, E., Chen, Z., Citro, C., Corrado, G.S., Davis, A., Dean, J., Devin, M., Ghemawat, S., Goodfellow, I., Harp, A., Irving, G., Isard, M., Jia, Y., Jozefowicz, R., Kaiser, L., Kudlur, M., Levenberg, J., Mané, D., Monga, R., Moore, S., Murray, D., Olah, C., Schuster, M., Shlens, J., Steiner, B., Sutskever, I., Talwar, K., Tucker, P., Vanhoucke, V., Vasudevan, V., Viégas, F., Vinyals, O., Warden, P., Wattenberg, M., Wicke, M., Yu, Y., Zheng, X.: TensorFlow: large-scale machine learning on heterogeneous systems (2015). http://tensorflow.org/. Software available from tensorflow.org
2. Anwar, A.S., Ghany, K.K.A., Elmahdy, H.: Human ear recognition using geometrical features extraction. Procedia Comput. Sci. **65**, 529–537 (2015)
3. Bargal, S.A., Welles, A., Chan, C.R., Howes, S., Sclaroff, S., Ragan, E., Johnson, C., Gill, C.: Image-based ear biometric smartphone app for patient identification in field settings. VISAPP **3**, 171–179 (2015)
4. Bhanu, B., Chen, H.: Human ear recognition by computer. Springer Science & Business Media (2008)
5. Biometrics, D.: Ergo Ear Biometric App: Unlock Your Phone with Your Ear. http://www.descartesbiometrics.com/ergo-app/ (2017). Accessed February 24, 2017
6. Burge, M., Burger, W.: Ear biometrics in computer vision. In: 15th International Conference of Pattern Recognition (ICPR) pp. 826–830 (2000)
7. Chan, T.S., Kumar, A.: Reliable ear identification using 2-d quadrature filters. Patt. Recogn. Lett. **33**(14), 1870–1881 (2012)
8. Cortes, C., Vapnik, V.: Support-vector networks. Mach. Learn. **20**(3), 273–297 (1995)
9. Dalal, N., Triggs, B.: Histograms of oriented gradients for human detection. In: IEEE Computer Society Conference on Computer Vision and Pattern Recognition, 2005. CVPR 2005, vol. 1, pp. 886–893. IEEE (2005)
10. Duchi, J., Hazan, E., Singer, Y.: Adaptive subgradient methods for online learning and stochastic optimization. J. Mach. Learn. Res. **12**, 2121–2159 (2011)
11. Galdmez, P.L., Raveane, W., Arrieta, A.G.: A brief review of the ear recognition process using deep neural networks. J. Appl. Logic (2016). https://doi.org/10.1016/j.jal.2016.11.014. http://www.sciencedirect.com/science/article/pii/S1570868316300684
12. Ghoualmi, L., Draa, A., Chikhi, S.: An ear biometric system based on artificial bees and the scale invariant feature transform. Exp. Syst. Appl. **57**, 49–61 (2016)
13. Glorot, X., Bengio, Y.: Understanding the difficulty of training deep feedforward neural networks. Aistats **9**, 249–256 (2010)
14. Hai-Long, Z., Zhi-Chun, M.: Combining wavelet transform and orthogonal centroid algorithm for ear recognition. In: 2nd IEEE International Conference on Computer Science and Information Technology, 2009. ICCSIT 2009, pp. 228–231. IEEE (2009)
15. Hanmandlu, M., et al.: Robust ear based authentication using local principal independent components. Exp. Syst. Appl. **40**(16), 6478–6490 (2013)
16. Hurley, D.J., Nixon, M.S., Carter, J.N.: Force field feature extraction for ear biometrics. Comput. Vis. Image Understand. **98**(3), 491–512 (2005)
17. Ioffe, S., Szegedy, C.: Batch Normalization: Accelerating Deep Network Training by Reducing Internal Covariate Shift. arXiv:1502.03167 (2015)
18. Jamil, N., AlMisreb, A., Halin, A.A.: Illumination-invariant ear authentication. Procedia Comput. Sci. **42**, 271–278 (2014)
19. Krizhevsky, A., Sutskever, I., Hinton, G.E.: Imagenet classification with deep convolutional neural networks. In: Advances in Neural Information Processing Systems, pp. 1097–1105 (2012)
20. Kumar, A., Chan, T.S.T.: Robust ear identification using sparse representation of local texture descriptors. Pattern Recogn. **46**(1), 73–85 (2013)
21. Kumar, A., Wu, C.: Automated human identification using ear imaging. Pattern Recogn. **45**(3), 956–968 (2012)

22. Kumar, A., Zhang, D.: Ear authentication using log-gabor wavelets. In: Defense and Security Symposium on International Society for Optics and Photonics, p. 65,390A (2007)
23. LeCun, Y., Bengio, Y., et al.: Convolutional networks for images, speech, and time series. Handbook Brain Theor. Neural Netw. **3361**(10), 1995 (1995)
24. LeCun, Y., Boser, B., Denker, J.S., Henderson, D., Howard, R.E., Hubbard, W., Jackel, L.D.: Backpropagation applied to handwritten zip code recognition. Neural Computat. **1**(4), 541–551 (1989)
25. Liu, Y., Zhang, B., Zhang, D.: Ear-parotic face angle: a unique feature for 3d ear recognition. Pattern Recogn. Lett. **53**, 9–15 (2015)
26. Malisiewicz, T., Gupta, A., Efros, A.A.: Ensemble of exemplar-svms for object detection and beyond. In: 2011 IEEE International Conference on Computer Vision (ICCV), pp. 89–96. IEEE (2011)
27. Omara, I., Li, F., Zhang, H., Zuo, W.: A novel geometric feature extraction method for ear recognition. Exp. Syst. Appl. **65**, 127–135 (2016)
28. Prakash, S., Gupta, P.: A rotation and scale invariant technique for ear detection in 3d. Pattern Recogn. Lett. **33**(14), 1924–1931 (2012)
29. Prakash, S., Gupta, P.: An efficient ear recognition technique invariant to illumination and pose. Telecommun. Syst. 1–14 (2013)
30. Prakash, S., Gupta, P.: Human recognition using 3d ear images. Neurocomputing **140**, 317–325 (2014)
31. Prakash, S., Jayaraman, U., Gupta, P.: A skin-color and template based technique for automatic ear detection. In: Seventh International Conference on Advances in Pattern Recognition, 2009. ICAPR'09, pp. 213–216. IEEE (2009)
32. Sana, A., Gupta, P., Purkait, R.: Ear biometrics: A new approach. In: Advances in Pattern Recognition, pp. 46–50. World Scientific (2007)
33. Sánchez, D., Melin, P.: Modular neural network with fuzzy integration and its optimization using genetic algorithms for human recognition based on iris, ear and voice biometrics. In: Soft Computing for Recognition Based on Biometrics, pp. 85–102. Springer (2010)
34. Shu, C., Ding, X., Fang, C.: Histogram of the oriented gradient for face recognition. Tsinghua Sci. Technol. **16**(2), 216–224 (2011)
35. SigOpt: Sigopt—Amplifies your Research. https://www.sigopt.com/ (2017)
36. Srivastava, N., Hinton, G.E., Krizhevsky, A., Sutskever, I., Salakhutdinov, R.: Dropout: a simple way to prevent neural networks from overfitting. J. Mach. Learn. Res. **15**(1), 1929–1958 (2014)
37. Viola, P., Jones, M.: Rapid object detection using a boosted cascade of simple features. In: Proceedings of the 2001 IEEE Computer Society Conference on Computer Vision and Pattern Recognition, 2001. CVPR 2001, vol. 1, p. I. IEEE (2001)
38. Wang, J., Clark, S.C., Liu, E., Frazier, P.I.: Parallel Bayesian Global Optimization of Expensive Functions. arXiv:1602.05149 (2016)
39. Wang, Y., Mu, Z.c., Zeng, H.: Block-based and multi-resolution methods for ear recognition using wavelet transform and uniform local binary patterns. In: 19th International Conference on Pattern Recognition, 2008. ICPR 2008, pp. 1–4. IEEE (2008)
40. Xie, Z., Mu, Z.: Ear recognition using lle and idlle algorithm. In: 19th International Conference on Pattern Recognition, 2008. ICPR 2008. pp. 1–4. IEEE (2008)
41. Yan, P., Bowyer, K.: Empirical evaluation of advanced ear biometrics. In: IEEE Computer Society Conference on Computer Vision and Pattern Recognition-Workshops, 2005. CVPR Workshops, pp. 41–41. IEEE (2005)
42. Yuan, L., Chun Mu, Z.: Ear recognition based on local information fusion. Pattern Recogn. Lett. **33**(2), 182–190 (2012)
43. Yuan, L., Chun Mu, Z.: Ear recognition based on 2d images. In: First IEEE International Conference on Biometrics: Theory, Applications, and Systems, 2007. BTAS 2007, pp. 1–5. IEEE (2007)

Experimental Study and Optimization of Process Parameters During WEDM Taper Cutting

Anshuman Kumar, K. Abhishek, K. Vivekananda and Chandramani Upadhyay

Abstract An experimental investigation has been performed on wire-EDM of Inconel 718 by using zinc-coated brass wire as tool electrode. Based on L_{27} orthogonal array, the studies have been conducted by varying five process parameters (such as taper angle, pulse on-time, wire speed, wire tension, and discharge current), within the selected experimental domain. The following machining performance yields (viz. angular error and surface roughness) have been investigated during this study. The angular error (AE) and surface roughness (Ra) have been critical responses in die making industry. Simultaneously, the mathematical model has been developed by using nonlinear regression analysis for correlating with various process parameters and performance characteristics. Finally, JAYA algorithm has been applied individually to find out the satisfactory machining performances. Application and potential of JAYA algorithm have been compared with Teaching-Learning based algorithm and Genetic Algorithm (GA) and observed by the JAYA algorithm, which does not require any specific parameter settings and hence easy to implement.

Keywords WEDM · Taper cutting · Surface roughness · JAYA · TLBO · GA

A. Kumar (✉)
Department of Mechanical Engineering, Koneru Lakshmaiah Education Foundation,
Vaddeswaram, Andhra Pradesh, India
e-mail: anshu.mit06@gmail.com

K. Abhishek
Department of Mechanical Engineering, Institute of Infrastructure Research and Management
(IITRAM), Ahmedabad, Gujarat, India

K. Vivekananda
Faculty of Science and Technology, Department of Mechanical Engineering, ICFAI Foundation
for Higher Education, Hyderabad, Telangana state, India

C. Upadhyay
Department of Mechanical Engineering, National Institute of Technology, Rourkela, Rourkela,
Odisha, India

© Springer Nature Singapore Pte Ltd. 2019
J. C. Bansal et al. (eds.), *Soft Computing for Problem Solving*, Advances in Intelligent
Systems and Computing 817, https://doi.org/10.1007/978-981-13-1595-4_57

1 Background

Wire electrical discharge machining (WEDM) is the process of removing electrically conductive material by means of discrete electrical discharges in the presence of dielectric. It is used to manufacture conductive hard metal components with intricate shape with precise and good tolerance limit of about 1 μm. WEDM has become a most advanced machining process for manufacturing for precise geometries and is the best choice for superhard material with intricate shape which is used in the tooling industry. Taper cutting involves the generation of inclined ruled surfaces, and it is especially important in the manufacturing of tooling requiring draft angles. This cutting can be achieved by applying a relative displacement between the upper guide and lower guide of the wire. If the wire had no stiffness, it would exactly adapt the geometry of the guide [1]. In this ideal case, α would be the programmed angle; this angle is the actual cutting angle in the machined part as shown in Fig. 1. However, the deviation of the wire with respect to actual shape happens due to change in stiffness of wire. Angle β represents the AE induced by this effect, and it is shown in minutes. The errors may occur in different aspects such as the stiffness of the wire, distance between the upper and lower guide, and the force exerted during the cutting process. As a result, the machined part is unable to find its accuracy.

In recent times, many research works have been carried out on accuracy in WEDM, since the requirements of tolerances and roughness imposed by industries such as tool making are growing continuously [2, 3]. Tosun et al. [4] studied the effects of process parameter on precision cutting like kerf width and MRR using WEDM operation, and the multi-objective optimization had been used to optimize the process parameter for optimum machining condition. Plaza et al. [1] proposed a design of experiment methodology to study the influence of WEDM process parameters on the angular accuracy of taper cutting operations and influence of response variable related to geometrical problem and part thickness angle. Sanchez et al. [5] discussed

Fig. 1 Theoretical and actual locations of the deformed wire

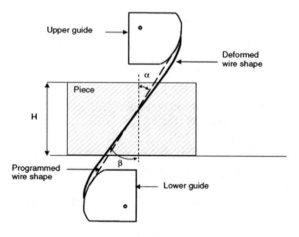

an original contribution to the analysis of the factors that influenced AE in taper cutting and developed experimental and numerical models to predict error. Finally, finite element simulation of mechanical behavior of a wire typically used for taper cutting was presented. Hsue et al. [6] proposed a theoretical model to identify the material removal mechanics and also developed strategy for controlling discharge power and wire tension for precision WEDM taper cutting. Selvakumar et al. [7] reported that the Ra has not affected by taper angle, wire tension, and peak current. High value of Wire tension was the most affected process parameter for better taper accuracy. Sanchez et al. [8] observed that actual contact angle is not equal to the programmed angle and also observed that error has been found out by trial-and-error method and compared with computer simulation. Nayak and Mahapatra [9] presented a mathematical model using artificial neural network (ANN) to determine the relationship between response variable and output response. Finally, Bat algorithm is used to suggest the optimum parametric combination to minimize the AE during taper cutting operation using WEDM operation. Sanchez et al. [10] presented a computer simulation software for analysis of AE in taper during WEDM operation. The work material used was hard materials, such as those used in the tooling industry. The simulated result has been compared with the experimental result by the classical trial-and-error method, yielding good results. Nayak and Mahapatra [11] highlighted a quantum-behaved particle swarm optimization (PSO) approach combined with maximum deviation theory to determine the optimum parameter setting during taper cutting on AISI D2 tool steel as work material. Jangra et al. [12] studied effects of response parameters on MRR and Ra on carbide cobalt (WC-Co) composite as work material. The optimization of response output (MRR and Ra) was carried out using Gray relation analysis (GRA). Results of ANOVA showed that pulse on-time and taper angle were the most significant parameters for the multiple machining characteristics. Nayak and Mahapatra [13] reported an experimental investigation and optimization of different process parameters (part thickness, taper angle, pulse duration, discharge current, wire speed, and wire tension) during taper cutting of deep cryogenic-treated Inconel 718 using WEDM operation. Finally, process model was optimized using the Bat algorithm which yielded good result of performance characteristics during taper cutting.

From the literature survey, it has been observed that limited investigations have been reported and with few literatures reported to optimization of response parameters of taper cutting operations. It has been noted that few or none of the literatures focuses on the performance of zinc-coated brass wire in taper cutting on Inconel 718 using WEDM. Therefore, in this investigation, an attempt has been made to discuss the effect of zinc-coated brass wire as electrode with various process parameters during taper cutting on Inconel 718. The JAYA algorithm has been utilized for assessing optimal machining parameter for achieving the high-quality productivity while varying process parameter of WEDM. Because, it is observed that, during the machining, the compatible balance should be maintained between the quality and productivity, which include high production rate with effective machining cost to maintain the good surface finish and dimensional accuracy.

2 Experimentation

In the present study, experiments have been conducted on a CNC WEDM machine, [AGIE, SWITZERLAND]. Inconel 718 has been selected as work material with dimension (50 mm × 10 mm × 5 mm). Initially, the zinc-coated brass wire electrode material (coated BRONCO CUT) with diameter of 0.20 mm has been used for performing experiments. Deionized water has been used as dielectric fluid.

2.1 Selection of Process Parameters

The selection process parameter is given in Table 1. The experiments have been conducted to investigate the significant contribution of process parameters on machinability (in terms of AE and Ra) during WEDM process. The process parameters and their levels have been chosen based on the literature review, significance, and machining limitations. Researcher informed that taper angle is the most influencing variable in taper cutting using WEDM [1]. Therefore, taper angle, pulse on-time, wire speed, wire tension, and discharge current are considered as process parameters. The pulse off-time and flushing pressure have been kept constant throughout the experiments as the value of 50 μs and 8 kg/mm^2, respectively.

The cutting speed (Cs) in mm/min has been directly obtained and digitally displayed on the computer screen during machining. The taper angles have been measured by using coordinate measuring machine (CMM) (SPECTRA ACCURATE). The surface roughness (Ra) of WEDMed surface of Inconel 718 has been measured using Talysurf (Talysurf 50 Taylor Hobson Ltd., UK). The surface texture has been captured after WEDMed process by using SEM (Nova Nano SEM 450).

AE is calculated by using the following formula in minute [13]

$$AE = \theta - \phi \tag{1}$$

where θ is programmed angle in the machined part and ϕ is the programmed angle obtained from machining. The geometry of the test part is shown in Fig. 2. Taper accuracy may be affected by wire tension, hydraulic forces, and discharge spark forces.

Table 1 Process parameters and domain of variation

Control factors	Code	Unit	Level I	Level II	Level III
Taper angle	A	Degree	10	12	14
Pulse on-time	B	(μs)	28	30	32
Wire speed	C	(mm/s)	95	115	135
Wire tension	D	(N)	8	10	12
Discharge current	E	(A)	13	14	15

Fig. 2 Specimen for measurement of angular accuracy

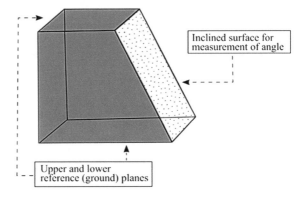

Inclined surface for measurement of angle

Upper and lower reference (ground) planes

3 Methodologies Explored

The present work proposes nonlinear regression with JAYA algorithm in order to assess the optimal parametric combination for the angular error and surface roughness during WEDM of Inconel 718.

Nonlinear regression analysis has been used in order to develop the relationship between the dependent variable and a set of independent variables.

$$Y = \texttt{constant} \times A^{(x)} \times B^{(y)} \times C^{(z)} \times D^{(l)} \times E^{(m)} \tag{2}$$

The derived model for aforementioned output characteristics has been treated as fitness function and finally optimized (maximized) by JAYA algorithm.

Detailed flow of experiment stepping and section describes the chronology of computations steps involved in JAYA algorithm; refer to Fig. 3.

4 Results and Discussions

A 5-factor, 3-level, L_{27} orthogonal array has been selected for conducting the experiments by using MINITAB 16 and collection response data as shown in Table 2. The sample cutting during the experiment using WEDM and angle measurement after machining using CMM as shown in Figs. 4 and 5. The main effects of the controllable process parameters on AE are shown in Fig. 6. Figure 6a shows that AE first increases trend and then decreases. The minimum AE has been observed in 10° of taper angle. Similarly, Fig. 7a indicates that higher taper angle gives maximum Ra. This may also be attributed due to the fact that, longer length of the wire electrode in a thicker workpiece provide enough opportunities for the spark to occur and enough space for movement in U-V axes using upper guide and lower guide [13]. From Fig. 6b, it has been observed that wire speed has a major contribution in minimizing AE. In addition, Fig. 7b indicates that higher wire speed results in higher Ra in

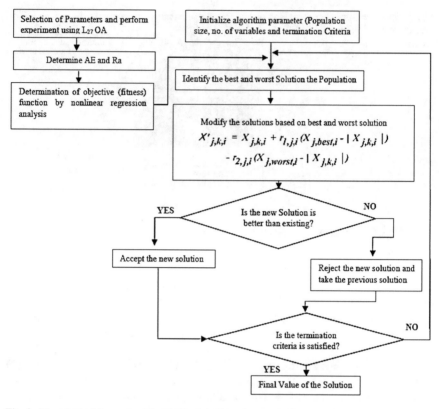

Fig. 3 Flowchart of the proposed optimization route

the range of 95–135 mm/s. These behaviors are probably due to higher wire speed mainly because of stable spark generated by the fresh wire on the workpiece and thus creating minimum AE and Ra. Figure 6c shows that, as wire tension increases, the AE tends to increase and then decrease. This may be due to stronger forces acting on the wire, responsible for wire stability during the discharge process. Figure 7c shows that Ra increases with increase in the discharge current. Increasing the discharge current also increases the discharge energy, which leads to deteriorate surface finish.

The analysis of variance (ANOVA) is used to analyze the effect of various process parameters on AE and Ra. Table 3 reveals the fact that the percentage of contribution were found to affect AE by taper angle (10%), pulse on-time (7%), wire speed (40%), wire tension (17%), discharge current (22%), and experimental error (2%). Similarly, the results for Ra are shown in Table 4. The percentage of contribution affecting Ra were found by taper angle (6%), pulse on-time (13%), wire speed (4%), wire tension (32%), discharge current (38%), and experimental error (6%). Wire speed and discharge current for AE and wire tension and discharge current are found to the major affected factors.

Table 2 Design of experiment and collection response data

Sl. no.	A (°)	B (μs)	C (mm/s)	D (N)	E (A)	AE (min)	Ra (μm)
1	1	1	1	1	1	13.80	2.84
2	1	1	1	1	2	17.40	3.48
3	1	1	1	1	3	21.24	4.20
4	1	2	2	2	1	13.02	3.02
5	1	2	2	2	2	21.00	3.14
6	1	2	2	2	3	25.20	3.20
7	1	3	3	3	1	12.60	3.00
8	1	3	3	3	2	17.76	3.10
9	1	3	3	3	3	22.20	3.40
10	2	1	2	3	1	20.40	2.56
11	2	1	2	3	2	23.40	3.10
12	2	1	2	3	3	24.60	4.10
13	2	2	3	1	1	12.60	3.20
14	2	2	3	1	2	15.00	3.55
15	2	2	3	1	3	19.80	3.80
16	2	3	1	2	1	37.20	3.04
17	2	3	1	2	2	40.80	3.50
18	2	3	1	2	3	43.20	3.73
19	3	1	3	2	1	13.80	3.00
20	3	1	3	2	2	18.00	2.94
21	3	1	3	2	3	21.00	2.80
22	3	2	1	3	1	35.40	2.88
23	3	2	1	3	2	31.80	3.20
24	3	2	1	3	3	28.80	3.71
25	3	3	2	1	1	16.80	3.85
26	3	3	2	1	2	17.52	4.10
27	3	3	2	1	3	27.60	4.40

Fig. 4 Sample cutting process during WEDM

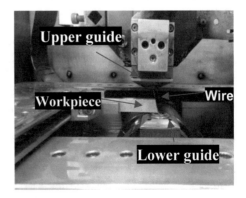

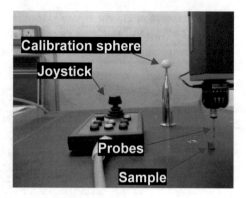

Fig. 5 Sample measurement by CMM

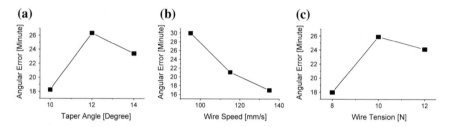

Fig. 6 Main effect of control parameters on AE

Figure 8 shows the surface textures have been captured by scanning electron microscope (SEM) (Nova Nano SEM 450) of Inconel 718 specimen. Three samples were selected for surface texture observation. In this analysis, it has been observed that less number of craters and spherical shape particles were formed due to lower value of taper angle (10°), pulse on-time (28 μs), wire speed (95 mm/s), wire tension (8 N), and discharge current (13 A) (Fig. 8a). This may due to the fact that the location of impingement of spark energy may become stable.

Increasing the discharge current and pulse on-time (Fig. 8b) results in increased deposited materials and uneven surface as observed on the WEDMed surface. This is due to low discharge energy which not completely erodes and evaporates material, and some melted materials are deposited on the WEDMed surface.

From Fig. 8c, it has been found that, the higher value of discharge current (14 A) increases the energy of the spark which unevenly distributed along the machining surface. The uneven spark which exhibits uneven erosion leads to the increase in large number of melted material deposited on the WEDMed surface.

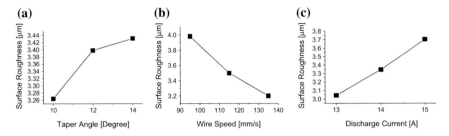

Fig. 7 Main effect of control parameters on Ra

Table 3 ANOVA for AE

Source	DF	Seq SS	Adj SS	Adj MS	% Contribution
A	2	22.948	22.948	11.4742	10
B	2	16.098	16.098	8.0491	7
C	2	96.373	96.373	48.1866	**40**
D	2	40.276	40.276	20.1378	17
E	2	53.830	53.830	26.9150	22
D * E	4	5.902	5.902	1.4755	3
Error	12	5.248	5.248	0.4373	2
Total	26	240.675			100.0

DF degrees of freedom; *SS* sum of square; *MS* mean square

Table 4 ANOVA for Ra

Source	DF	Seq SS	Adj SS	Adj MS	% Contribution
A	2	1.6161	1.6161	0.8080	6
B	2	3.3713	3.3713	1.6856	13
C	2	0.9319	0.9319	0.4659	4
D	2	8.5407	8.5407	4.2703	32
E	2	10.1697	10.1697	5.0848	**38**
D * E	4	0.4349	0.4349	0.1087	2
Error	12	1.6057	1.6057	0.1338	6
Total	26	26.6701			100.0

DF degrees of freedom; *SS* sum of square; *MS* mean square; *A* melted material deposited

4.1 Optimization Results

To obtain the optimum parameter setting for minimizing Ra and AE to satisfy the production needs during taper cutting. The JAYA algorithm was applied in MATLAB 12.0 in the present study for simultaneous optimization of the machining responses.

(a)   (b)   (c)

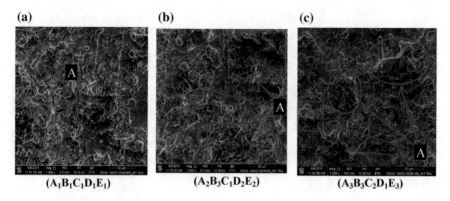

(A₁B₁C₁D₁E₁)   (A₂B₃C₁D₂E₂)   (A₃B₃C₂D₁E₃)

Fig. 8 FESEM images of the machined surface of WEDMed Inconel 718

4.2 Data Analysis

The present study highlights the application of JAYA algorithm in order to assess the
optimal machining condition of WEDMed process. Initially, nonlinear regression
model which is used to establish the relationship between the dependent variable
and a set of independent variables is developed for aforementioned performance
characteristics (AE and Ra).

For angular error:

$$AE = 0.025 \times A^{(0.479)} \times B^{(2.318)} \times C^{(-1.505)} \times D^{(0.586)} \times E^{(1.333)} \qquad (3)$$

For surface roughness

$$Ra = 0.028 \times A^{(0.146)} \times B^{(0.739)} \times C^{(-0.173)} \times D^{(-0.384)} \times E^{(1.366)} \qquad (4)$$

The value of the coefficient of determination (R^2) has been used to check the
adequacy of a mathematical model. The R^2 value for the AE is 96.6%, whereas Ra
is 97.6%.

Hence, it can be concluded that aforesaid models are adequate enough and can
be used as the objective functions (fitness functions) for JAYA, TLBO, and GA
optimization algorithms. Figures 9, 10, 11, 12, 13, and 14 show the convergence
history for the result of the JAYA, TLBO, and GA optimization methods. The global
optimal value is obtained to minimize the Ra and AE along with the fitness value as
shown in Table 5.

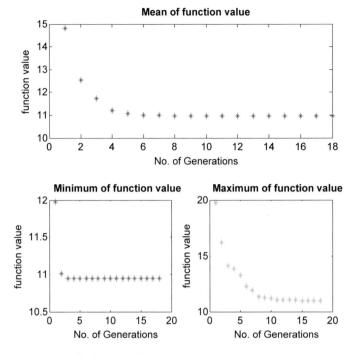

Fig. 9 Convergence plot for AE by JAYA

Table 5 Optimal parametric combination (obtained by TLBO, JAYA, and GA) along with fitness value(s) of the objective function(s)

Algorithm	Responses	Optimal parametric combination					Fitness value
		A	B	C	D	E	
TLBO	AE	10	28	135	8	13	10.95
	Ra	12	28	120	10	13	2.510
JAYA	AE	**10**	**28**	**135**	**8**	**13**	**10.949**
	Ra	**10**	**28**	**135**	**8**	**13**	**2.490**
GA	AE	11	28	132	8	13	11.758
	Ra	10	28	126	12	13	2.556

5 Conclusions

The aforementioned work, highlighted on influence of process parameters during taper cutting to obtain most favorable process environment. The mathematical models have been developed by using nonlinear regression analysis for analyzing the effect of process parameters such as taper Angle, pulse on-time, wire speed, wire tension, and discharge current. AE and Ra have been considered for appraisal of machining

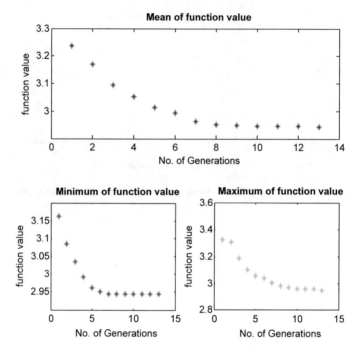

Fig. 10 Convergence plot for Ra by JAYA

performance characteristics during taper cutting process. Based on the aforesaid research, the following conclusions have been drawn.

- Wire speed has been the most influenced factor to minimize the AE and Ra during taper cutting.
- The taper angle was found to be the most important parameter that influence the AE.
- Wire tension has been the insignificant factor for AE during taper cutting.
- Surface finish is predominantly influenced by taper angle, pulse on-time, and discharge current.
- SEM analysis of WEDMed surface at experimental conditions reveals that high spark energy parameters exhibit rougher surface and with lot of melted material deposition, whereas the low energy parameter observed quality surface during the taper cutting.

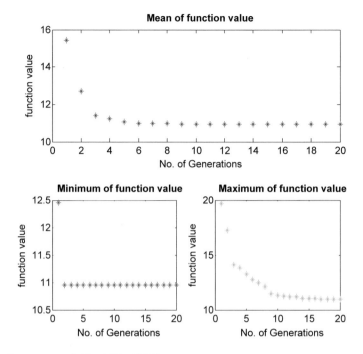

Fig. 11 Convergence plot for AE by TLBO

- From the present investigation, it has been observed that the JAYA algorithm gives reliable result with minimum computation work and optimization time. The optimum results of the machining process parameters are obtained by fitness value. JAYA and TLBO algorithms have the same parameter values for minimizing the angular error (i.e., taper angle $= 10°$, pulse on-time $= 28$ µs, wire speed $= 135$ mm/s, wire tension $= 8$ N, discharge current $= 13$ A).

This work may be extended further by considering the variation of different wire materials or coated materials to find the effects on the angular error and cutting speed. This study also may apply different types of optimization techniques, such as Whale Optimization Algorithm and Grey Wolf Optimizer to obtain results which may be compared with some other optimization techniques to find the better machinability parameters.

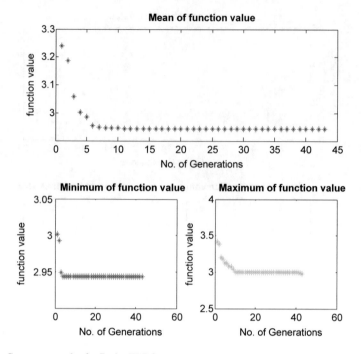

Fig. 12 Convergence plot for Ra by TLBO

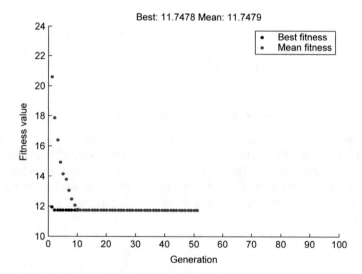

Fig. 13 Convergence plot for AE by GA

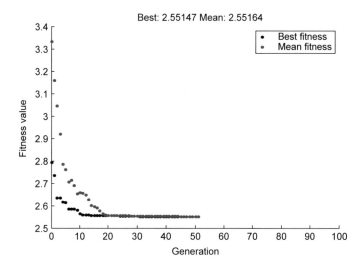

Fig. 14 Convergence plot for Ra by GA

References

1. Plaza, S., Ortega, N., Sanchez, J., Pombo, I., Mendikute, A.: Original models for the prediction of angular error in wire-EDM taper-cutting. Int. J. Advanc. Manufact. Technol. **44**(5–6), 529–538 (2009)
2. Dauw, D., Beltrami, I.: High-precision wire-EDM by online wire positioning control. CIRP Ann. Manufact. Technol. **43**(1), 193–197 (1994)
3. Han, F., Kunieda, M., Sendai, T., Imai, Y.: Simulation on influence of electrostatic force on machining characteristics in WEDM. Int. J. Electric. Mach. **7**, 31–36 (2002)
4. Tosun, N., Cogun, C., Tosun, G.: A study on kerf and material removal rate in wire electrical discharge machining based on Taguchi method. J. Mater. Process. Technol. **152**(3), 316–322 (2004)
5. Sanchez, J., Plaza, S., Ortega, N., Marcos, M., Albizuri, J.: Experimental and numerical study of angular error in wire-EDM taper-cutting. Int. J. Mach. Tools Manufact. **48**(12), 1420–1428 (2008)
6. Hsue, A.W.-J., Su, H.-C.: Removal analysis of WEDM's tapering process and its application to generation of precise conjugate surface. J. Mater. Process. Technol. **149**(1), 117–123 (2004)
7. Selvakumar, G., Jiju, K.B., Veerajothi, R.: Experimental study on wire electrical discharge machining of tapered parts. Arabian J. Sci. Eng. 1–9 (2016)
8. Sanchez, J.A., Plaza, S., Lopez De Lacalle, L., Lamikiz, A.: Computer simulation of wire-EDM taper-cutting. Int. J. Comput. Integr. Manuf. **19**(7), 727–735 (2006)
9. Nayak, B.B., Mahapatra, S.S.: A hybrid approach for process optimization in taper cutting operation using wire electrical discharge machining. Appl. Mech. Mater. Trans Technical Publications, 83–88 (2014)
10. Sanchez, J., Rodil, J., Herrero, A., De Lacalle, L.L., Lamikiz, A.: On the influence of cutting speed limitation on the accuracy of wire-EDM corner-cutting. J. Mater. Proc. Technol. **182**(1), 574–579 (2007)

11. Nayak, B.B., Mahapatra, S.S.: A quantum behaved particle swarm approach for multi-response optimization of WEDM process. In: International Conference on Swarm, Evolutionary, and Memetic Computing, pp. 62–73. Springer (2014)
12. Jangra, K., Grover, S., Aggarwal, A.: Simultaneous optimization of material removal rate and surface roughness for WEDM of WC-Co composite using grey relational analysis along with Taguchi method. Int. J. Ind. Eng. Comput. **2**(3), 479–490 (2011)
13. Nayak, B.B., Mahapatra, S.S.: Optimization of WEDM process parameters using deep cryo-treated Inconel 718 as work material. Eng. Sci. Technol. Int. J. **19**(1), 161–170 (2016)

Reinforcement Learning-Based Controller for Field-Oriented Control of Induction Machine

Ashish Kushwaha and Madan Gopal

Abstract This paper presents the concept of reinforcement learning-based field-oriented control (FOC) of induction motor. Conventional controllers such as PID used for FOC of induction machines are model-based controllers and face the issue of parameter tuning. Periodic retuning of PID controllers is required to take care of model approximations, parameter variations of the system during operation and external disturbances which are random in character, magnitudes, and place of occurrences in the system. Reinforcement learning is a model-free and online learning technique which can take care of parameter variations. These properties make reinforcement learning a potential candidate, to act as an adaptive controller which can replace conventional controllers. In this study, reinforcement learning-based controller has been designed to control the speed of induction machine using filed-oriented control. The controller performance has been verified for various operating conditions by computer simulation in MATLAB/SIMULINK.

Keywords Reinforcement learning · Q-learning · Induction motor
Field-oriented control · MATLAB/SIMULINK

1 Introduction

Induction machines are very popular in industrial applications due to their rugged construction, lesser maintenance, and high reliability. Speed and position control of induction machines are two major requirements in industrial applications. Due to nonlinear and multivariable behavior of machine dynamics, it is difficult to control the machine. However, field-oriented control [1] is very widely used method because of its ease of implementation and good transient response [2, 3]. Field-oriented control (FOC) method decouples the rotor flux and rotor speed. It makes the control of induction machine similar to separately excited DC machine.

A. Kushwaha (✉) · M. Gopal
Shiv Nadar University, GB Nagar, Greater Noida 201314, India
e-mail: ak999@snu.edu.in

© Springer Nature Singapore Pte Ltd. 2019   737
J. C. Bansal et al. (eds.), *Soft Computing for Problem Solving*, Advances in Intelligent
Systems and Computing 817, https://doi.org/10.1007/978-981-13-1595-4_58

The design of field-oriented controller is based on presumption that there will be no variation and uncertainty in machine or environment parameters. Due to parameter change, decoupling may get affected and controller performance deflects from desired nature [4, 5]. In FOC, normally conventional PID controllers are being used because of their simplicity. PID controllers are system parameter dependent and face the tuning issues. If system parameters change, PID controllers are required to be tuned again.

Due to overheating, rotor resistance of induction machine changes which effects the decoupling of rotor flux and rotor speed [6]. To compensate the effect of change in rotor resistance, so many methods have been proposed. Model reference adaptive controllers (MRACs) are quiet often used but this method depends on parameters which are taken to create reference models [7]. Some other techniques as observer-based methods are very complicated to implement [8].

Reinforcement learning is a model-free algorithm which learns with experience achieved from interaction with the environment. Reinforcement learning has been used several in applications as robotics, retail markets, and others [9, 10]. In this work, reinforcement learning-based controller is proposed to implement field-oriented control of induction machine. Q-learning has been used to implement reinforcement learning. The learning method adopted is online learning by taking actions for the given state to reduce the error between reference and rotor speed. No offline training is required for the controller.

The paper is organized as follows: Sect. 2 presents the basic details of field-oriented control. In Sect. 3, reinforcement learning is described. In Sect. 4, design of reinforcement learning-based controller is explained. Section 5 shows the simulation results of controller for various cases. Section 5 contains the conclusion of proposed algorithm.

2   FOC of Three-Phase Induction Motor

The indirect field-oriented control of three-phase induction motor has been a well-established technique. The block diagram of field-oriented control is presented in Fig. 1. Three-phase currents are decoupled into d-axis and q-axis currents in rotor flux orientation frame. On the q-axis, speed control loop is performed and flux control loop operated in d-axis. The outer speed loop generates reference torque which generates q-axis reference current i_q and flux control loop generates d-axis reference current i_d. The reference d-axis and q-axis currents are transformed again in three-phase currents. The inner current control loop compares the reference and actual three-phase currents and generates the switching pulses for three-phase voltage source inverter (VSI) switches by hysteresis control.

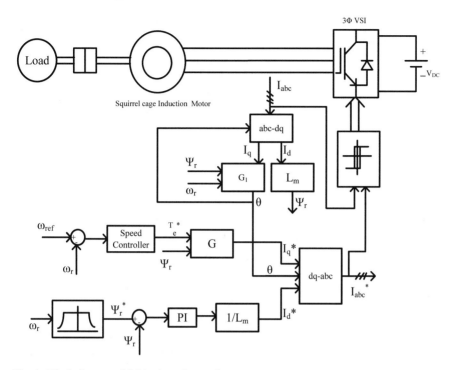

Fig. 1 Block diagram of field-oriented control

3 Reinforcement Learning

Due to unavailability of precise models and uncertain environment, model-free techniques are proved to be quite effective and advantageous. The reinforcement learning (RL) is such an intelligent scheme which is free from model parameters and environmental disturbances. In the reinforcement learning, the learning approach is similar to human learning tendencies. The individual (agent) interacts with the environment through actions and as it gets experience with time, it becomes expert for the individual task [11]. Reinforcement learning emphasizes the action which is followed by improved outcomes. Each action taken by the agent leads to another state. The reward received by agent for an action depends on the favorability of next state. The basic block of RL is shown in Fig. 2.

Markov decision process (MDP) forms the groundwork for reinforcement learning structure. Mathematical expression of MDP has a set of four variables (S, A, P, r) where S is set of possible states of the environment, A is set of actions, P is state transition probability, and r is reward function. $V(s)$ corresponds to the value of state s. For action a at state s, $P(s, a, s')$ is the probability of transition from state s to s' and the value of state s is presented as,

Fig. 2 Block diagram of
reinforcement learning

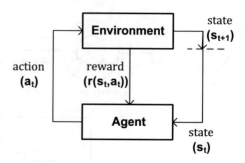

$$V(s) = r(s) + \max_a \Upsilon \sum_{s'} P(s, a, s')V(s') \qquad (1)$$

$\Upsilon \in [0, 1]$ is discount factor that determines the effect of future reward on current reward. Υ close to 1 shows provident approach while Υ close 0 shows myopic approach.

The assignment of action from action set A in a state to lead from initial state to final state is called policy represented as $\pi(s)$. The aim is to find the optimal policy $(\pi^*(s))$ to achieve final state without getting affected by initial states, decisions, and the final state. To obtain optimal policy, value iteration and policy iteration are used. Action with the maximum discounted reward for a state forms optimum policy and shown as

$$\pi^*(s) = \arg\max_a \sum_{s'} P(s, a, s')V^*(s') \qquad (2)$$

Involvement of probability in MDP makes it parameter dependent. Temporal difference (TD) approach is a model-free method where the agent interacts with the environment in episodes and updates the value estimates by one-step real-time transition and learned estimate of successor state.

A milestone in reinforcement learning has been made by Watkins [12] who has proposed an algorithm named Q-learning. The convergence of Q-learning to optimal policy has also been verified for decision problems. Q-learning has the Q-value $Q(s, a)$, which represents the value of taking action a in state s. Updation of Q-values is based on value iteration to get the optimal Q-values.

Q-value update method for TD-based Q-learning is given in (3) which gives the updated Q-value whenever action a is taken in state s leading to state s',

$$Q(s, a) = Q(s, a) + \eta \left(r(s) + \Upsilon \left[\max_{a'} Q(s', a') \right] - Q(s, a) \right) \qquad (3)$$

TD-based Q-learning has the potential to be applied in real-time problems. In the real-time Q-learning, all the evaluated Q-values are reserved in a look-up table with one entry for each state-action pair. Only one Q-value is amended in one-time step, others remain unmodified. In real-time application, if the agent identifies state s_t at time t and takes action a_t, which drives the system to state s_{t+1}, the agent earns the immediate reward r_t. Q-value for state-action pair (s_t, a_t) is updated as given in (4).

$$Q_{t+1}(s_t, a_t) = Q_t(s_t, a_t) + \eta_t \left(r_t + \Upsilon \left[\begin{array}{c} max \\ a \end{array} Q_t(s_{t+1}, a) \right] - Q_t(s_t, a_t) \right) \quad (4)$$

The look-table is updated as follows,

$$Q_{t+1}(s, a) = Q_t(s, a) \forall (s, a) \neq (s_t, a_t) \quad (5)$$

$\eta \in (0, 1)$ is learning parameter. For efficient learning, higher learning rate is desired for initial interactions and as agent gains experience, learning rate should be decreased. The selection of action a_t in state s_t is done by greedy policy for better exploitation shown in (6) which leads to maximization of $Q_t(s_t, a)$.

$$a_t = \begin{array}{c} arg\ max \\ a \end{array} Q_t(s_t, a) \quad (6)$$

However, better exploitation leads to pay coast in terms of exploration. Convergence condition demands visiting each state-action pair theoretically infinite number of times. But, greedy policy reinforces the actions which were taken in early stages. Therefore, agent should follow the policy by maintaining the balance between exploitation and exploration. ε-greedy policy is such a scheme which makes the agent becomes greedy in nature, but it also selects an action randomly with small probability ε. The selection of random action is done uniformly and is independent of current state. Need of explorations decreases with experience. The probability ε is kept high when agent is unlearned, and with time as it gains experience, probability ε is decreased for effective learning.

4 Reinforcement Learning-Based Controller

As described in Sect. 1, the speed controller output is torque reference $(Te*)$ for inner current control loops. Q-learning-based speed controller takes the speed error e and change in speed error ce as inputs and reference torque $Te*$ is the output. The block diagram of Q-learning controller is shown in Fig. 3.

The controller works in two modes–coarse mode and fine mode. The selection of mode depends on magnitude of error. If magnitude of error is greater than specific value, controller works in coarse mode otherwise it works in fine mode. In the coarse mode, the output of the controller is proportional to error with constant coefficient. In the fine mode, the Q-learning is activated and the output of the controller is decided by the action selected by Q-learning controller.

The major components of Q-learning-based controllers are space of states, space of actions, and rewards. Factors affecting the performance of Q-learning are learning

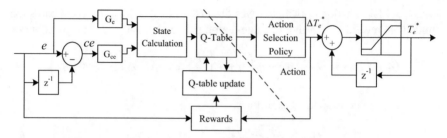

Fig. 3 Block diagram of Q-learning-based controller

rate, discount factor, and probability (ε) of selection of random action to create evenness between exploitation and exploration.

4.1 Space of States (S)

Increment on reference torque depends on nature of current speed error and current change in speed error. As Q-learning is model-free algorithm, no system parameter is given as input to controller. Only inputs are e and ce which together represent the operating state of the system with respect to controller. So, the space of states has been defined by these two variables.

The error e and change in error ce divided into equal intervals for the fine control. The number of intervals is m and n for e and ce. The controller can recognize the state by receiving the feedback of e and ce. The total number of states is $m \times n$.

During operation, observed state at ant step will be (e_i, ce_j) where e_i and ce_j belong to ith interval in error range and jth interval in change in error range, respectively. The space of states can be defined as,

$$S = \left\{ s_{ij} \mid s_{ij} = n * (i-1) + j, i \in [1, 2, \ldots, m], j \in [1, 2, \ldots, n] \right\} \qquad (7)$$

where

$$i = \arg_i (e_i, ce_j) \, \& \, j = \arg_j (e_i, ce_j) \qquad (8)$$

4.2 Space of Actions (A) and Rewards

Once the controller observes the state, it selects an action so that current speed tries to track the reference speed by reducing the error. The action in this controller is rise, drop, or no change in reference torque. When the error is large, the change in reference torque (ΔTe) should be large in right direction in order to make the tracking of reference speed fast and similar logic for small errors. The action set is given in (9). The actions are coefficients that should be multiplied with error to get the output ΔTe. There are three negative actions, five positive actions and one zero action to maintain ΔTe zero in steady state. When the error magnitude is greater than eth, the ΔTe does not rely on action selected by Q-learning as shown in (10). The current reference torque $Te * (t)$ is given in (11).

$$A = \{a \mid a = a_t, a_t \in \{-K_1, -K_2, -K_3, 0, K_4, K_5, K_6, K_7, K_8\}\} \quad (9)$$

$$\Delta T_e = \begin{cases} a_t e_t & for\ |e_t| < e_{th};\ finemode \\ K_c e_t & for\ |e_t| > e_{th};\ coarsemode \end{cases} \quad (10)$$

$$T_t^* = T_{t-1}^* + \Delta T_t \quad (11)$$

$$r_t = \begin{cases} 1 & for\ a = K_4 \\ 0.8 & for\ a = K_5 \\ 0.6 & for\ a = -K_1, K_6 \\ 0.4 & for\ a = -K_2, K_7 \\ 0.2 & for\ a = -K_3, K_8 \\ 0.1 & for\ a = 0 \end{cases} \quad (12)$$

The controller receives reward for each action. The values of the rewards associated with each action are shown in (12). The reward will be positive or negative depends on whether next state is desired or not. Positive reward is given if error magnitude is decreased in next state else negative reward is given for the taken action.

4.3 Learning Factors

The learning rate of Q-learning decides pace of learning. When system is unexplored, learning rate should be high and as the system interacts with environment, it should decrease with time. For the online learning, learning of each state depends on environment. Theoretically, learning is complete when each state is being visited infinite times. Controller cannot wait for all the states to be visited. So, each state has been assigned its own learning rate. Each state has high initial learning rate which decreases as the number of visits to the state increases. Learning rate of other states remains same in each iteration. The learning rate for state s_{ij} is η_{ij}, and its value depends on number of visit to the state k_{ij} as shown in (13).

$$\eta_{ij} = \frac{1}{\left(1 + k_{ij} * 0.08\right)^{0.8}} \quad (13)$$

The performance of Q-learning depends on balance between exploration and exploitation. Less explored system has high Q-values for frequently occurring states

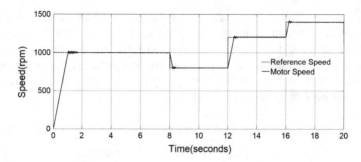

Fig. 4 Speed tracking at no load

which will leave the other states unlearned. ε-greedy policy takes care of this issue. For the exploration, controller selects the random action with ε probability. The random action selection is done uniformly. The value of probability ε decreases with time for better exploitation. The discount factor has been chosen as 0.6. Discount factor maintains the check in increase in Q-value of undesirable action of less explored state.

5 Simulation Results

The reinforcement learning-based controller has been tested by computer simulation in MATLAB/SIMULINK. The induction machine is of 5HP, 415 V. The machine parameters are—$R_s = 4.5$ Ω, $R_r = 4.5$ Ω, $L_{ls} = L_{lr} = 8.3$ mH, $L_m = 592.6$ mH, $J = 0.29$ Kg-m^2. The motor has been simulated with reinforcement learning controller and performance of the controller has been verified with different operating conditions. The step size of simulation is 2 μs and step size of speed control loop is 400 μs.

The threshold value of speed error e_{th} for the controller to work in coarse mode and fine mode is 0.01 pu. For coarse mode, the constant K_c value is 0.2. For fine mode, number of states are 400. The value of actions K_1, K_2, K_3, K_4, K_5, K_6, K_7 and K_8 are 1, 0.4, 0.2, 1, 0.6, 0.4, 0.3, and 0.2, respectively.

5.1 Speed Tracking

Controller has been started with zero initial Q-values for each state. Initially, motor has been run at no load with different reference speed changing with time. The controller speed tracking performance is shown in Fig. 4.

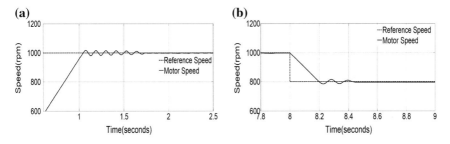

(a)

(b)

Fig. 5 Settling time with learning for speed tracking

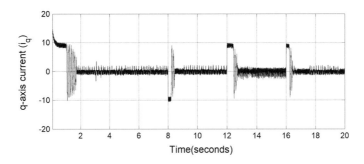

Fig. 6 Stator q-axis current

The machine is started with initial reference speed of 1000 rpm. The controller learns to track the reference speed online and as the controller explores the operating conditions; its performance improves with time.

It can be observed from the speed tracking waveform that initially controller takes more time to reach steady state. When controller is enough learned; it takes less time to track the next reference speed. Reference speed has been changed to 800 rpm at $t = 8$ s. Once the speed reaches the reaches to reference speed, it takes 0.6 s to settle down as shown in Fig. 5a. At next change, system is learned and it takes 0.2 s to settle down as shown in Fig. 5b. Figure 6 shows the q-axis current i_q which generates the electromagnetic torque T_e. It can be observed that transient oscillations of torque decrease with time as learning improves.

5.2 Load Disturbance Rejection

The controller performance has been tested while external disturbance is applied. In this test also, the controller has been started with zero initial Q-values. Load torque disturbances applied to the induction machine in the form of stator q-axis current. The speed tracking performance with the load disturbances is shown in Fig. 7. The machine is being started at no load with reference speed of 1000 rpm. At t = 5 s, load

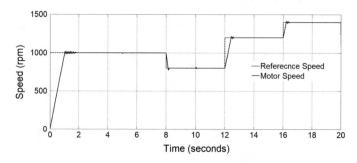

Fig. 7 Motor speed and reference speed

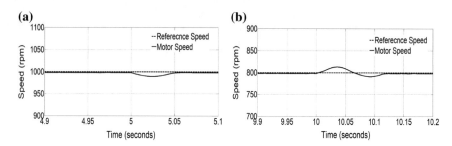

Fig. 8 Transient at load disturbance

torque of 15 Nm has been applied. The controller increases the torque and tracks the speed back in 0.05 s with 1% undershoot as shown in Fig. 8a. At $t = 8$ s, the reference speed has been changed to 800 rpm with no change in load torque. The controller tracks the speed at $t = 10$ s, load torque is being removed and machine runs at no load. The controller decreases the torque and tracks the speed back in 0.12 s with 1.3% undershoot as shown in Fig. 8b. The results show controller successfully tracks the speed with load disturbances. Also, it can be observed from Fig. 7 that when machines run for more time, transient performance improves as controller learns more.

5.3 Effect of Parameter Variation

In this section, controller performance has been tested for parameter variations of induction machine and compared with the performance of PI-based controller. First, the machine has been started with normal inertia value and the machine is started with increasing the inertia value by a factor of 1.5. Figure 9a and 9b shows the performance of PI controller and reinforcement learning-based controller, respectively.

For reinforcement learning-based controller, already learned Q-table from the previous section has been used. As the inertia increases, the rate of change of speed is

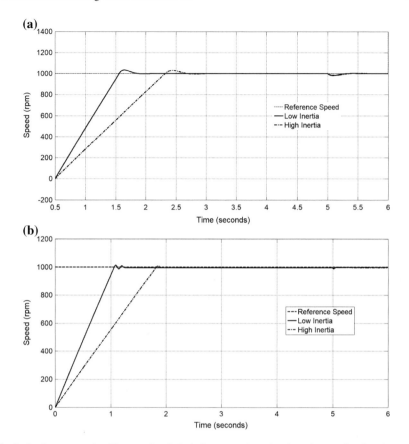

Fig. 9 Performance of **a** PI controller. **b** Reinforcement learning-based controller for change of inertia

reduced in both the cases. Rise time for both the controllers is same; however, settling time for reinforcement learning-based controller is lesser than that of PI controller. Also, at t = 5 s, load torque disturbance of 15 Nm is applied. Reinforcement learning-based controller has less disturbance in speed compared to PI controller. Figure 10 shows the performance of controller with rotor resistance variation. Rotor resistance has been changed by factor of 1.5 at t = 3.5 s. It can be observed that controller is able to track the speed even if rotor resistance is changed.

6 Conclusion

Reinforcement learning-based controller has been designed and implemented for field-oriented control of induction motor. Q-learning-based controller performance

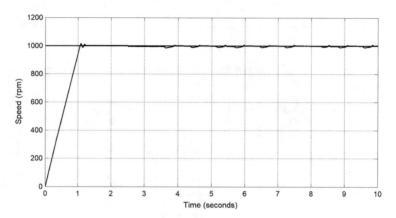

Fig. 10 Controller performance with rotor resistance variation

has been tested for speed tracking, load disturbance rejection, and machine parameter variations. The controller performance improves with interaction with environment. Also, controller is capable to adapt the parameter variations of the system. The controller is tested with parameter variations of induction machine, and it is able to track the speed.

References

1. Bose, B.K.: Modern Power Electronics and AC Drives. Prentice-Hall, New Jersey (2001)
2. Sung, W., Shin, J., Jeong, Y.: Energy-efficient and robust control for high-performance induction motor drive with an application in electric vehicles. IEEE Trans. Vehicul. Technol. **61**(8), 3394–3405 (2012)
3. Amezquita, B.L., Liceaga, C.J., Liceaga, C.E.: Speed and position controllers using indirect field-oriented control: a classical control approach. IEEE Trans. Industr. Electron. **61**(4), 1928–1943 (2014)
4. Nordin, K.B., Novotny, D.W., Zinger, D.S.: The influence of motor parameter deviations in feed forward field orientation drive systems. IEEE Trans. Industr. Appl. **IA-21**(4), 1009–1015 (1985)
5. Krishnan, R., Doran, F.C.: Study of parameter sensitivity in high performance inverter fed induction motor drive systems. IEEE Trans. Industr. Appl. **IA-23**(4), 623–635 (1987)
6. Qiang, D., Jingbo, K., Kai, Z.: Research on rotor resistance adaptation algorithm for indirect vector control of induction machines. In: IEEE Advanced Information Technology, Electronic and Automation Control Conference (IAEAC), pp. 831–836. Chongqing (2015)
7. Kumar, R., Das, S., Syam, P., Chattopadhyay, K.A.: Review on model reference adaptive system for sensorless vector control of induction motor drives. IET Electr. Power Appl. **9**(7), 496–511 (2015)
8. Akatsu, K., Kawamura, A.: Online rotor resistance estimation using the transient state under the speed sensorless control of induction motor. IEEE Trans. Power Electron. **15**(3), 553–560 (2000)

9. Raju, C.V.L., Narahari Y., Ravikumar K.: Reinforcement learning applications in dynamic pricing of retail markets. In: EEE International Conference on E-Commerce, pp. 339–346. USA (2003)
10. Gu, S., Holly, E., Lillicrap, T., Levine, S.: Deep reinforcement learning for robotic manipulation with asynchronous off-policy updates. In: IEEE International Conference on Robotics and Automation (ICRA), pp. 3389–3396. Singapore (2017)
11. Sutton, R.S., Barto, A.G.: Reinforcement Learning: An Introduction. MIT Press, Cambridge (1998)
12. Watkins, C.J.: Learning from Delayed Rewards. PhD Dissertation, Cambridge University, England (1989)

Simulation Modeling for Manufacturing System Application Using Simulink/SimEvents

Om Ji Shukla, Gunjan Soni and Rajesh Kumar

Abstract This paper refers to the utilization of simulation modeling for a manufacturing system by evaluation of the performance measures of the system. The application of manufacturing system model created and simulated by SimEvents toolbox of Simulink library in MATLAB is illustrated. The present study mainly focuses on exploration and evaluation of SimEvents toolbox in case of three jobs and six machines job-shop problem.

1 Introduction

In the last years, the technological development has resulted in strong market competition. Therefore, the manufacturing industries have to set their manufacturing policy as per consumer-oriented due to diversified demands and increasing mass customization of products. In order to remain in the global competition, the industries are in great need of manufacturing systems with agile nature and more flexible to handle uncertain situations [1]. The research shows that new manufacturing systems have been emerged in the last years such as networked manufacturing system, flexible manufacturing system, cloud manufacturing system, agile manufacturing system, distributed manufacturing system [2–4]. All these advanced manufacturing systems are more adaptable to rapidly changing environment. The discrete event simulation places an important role to model and simulate these manufacturing systems. Discrete event simulation is one of the most flexible tools and techniques which analyzes the dynamic nature of manufacturing systems and evaluates various strategies in operation for decision support in the manufacturing problems.

O. J. Shukla (✉) · G. Soni · R. Kumar
Malaviya National Institute of Technology Jaipur, Jaipur, Rajasthan, India
e-mail: om.mechanical@gmail.com

G. Soni
e-mail: gunjan1980@gmail.com

R. Kumar
e-mail: rkumar.ee@gmail.com

© Springer Nature Singapore Pte Ltd. 2019
J. C. Bansal et al. (eds.), *Soft Computing for Problem Solving*, Advances in Intelligent Systems and Computing 817, https://doi.org/10.1007/978-981-13-1595-4_59

751

Numerous researchers have adopted simulation in designing and operating the manufacturing systems which has increased the research interest in this topic. For example, Wang et al. [5] provided a discrete event simulation analysis of dynamic flexible manufacturing system (FMS) in their article by examining the two decisions: the part launching into the system for production and finding the order sequence for the collection of the completed parts. Imran et al. [6] have proposed an integration approach of discrete event simulation and genetic algorithm to solve mathematical model for minimizing value-added work-in-process. Mousavi et al. [7] presented a hierarchical framework for the assessment of energy and water consumption in a manufacturing system, and a simulation analysis was done on a pharmaceutical company as a case industry. Thus, these examples have shown successful applications of simulation in different manufacturing systems.

The main objective of this paper is to explore SimEvents, a discrete event simulation toolbox in Simulink library of MATLAB, by creating simulation model for a manufacturing system. A large number of industrial applications of SimEvents motivate its selection in this research study. The paper is structured as follows: Sect. 2 discusses the related literature on Simulink/SimEvents for manufacturing system applications. Section 3 describes the structure of manufacturing system. Section 4 explains the simulation modeling of the manufacturing system and its components (manufacturing operations). The simulation results are analyzed in Sect. 5, followed by conclusion in Sect. 6.

2   Related Work

SimEvents toolbox of MATLAB has servers, queues, switches, and some predefined blocks. SimEvents solves event-driven system models through a model created by these blocks. The models are created using drag and drop method. This discrete event simulation has discrete items in SimEvents, called as entities. Omar et al. [8] proposed a hybrid simulation model of continuous and discrete modes for predicting energy consumption and flows in automotive production lines using Simulink and SimEvents. Mushiri et al. [9] developed model reference adaptive fuzzy controller (MRAFC) and MRAC with PID controllers for controlling the valve operation under disturbances and nonlinearity. Simulations were carried out in MATLAB–Simulink environment. Ortiz et al. [10] presented dynamic simulation system (DSS) based on dynamic mathematical model and described a simulation analysis of a pilot-scale rotary kiln for activated carbon manufacture. DSS is developed by Simulink simulation framework. He et al. [11] proposed a modeling method which is solved in Simulink environment and implemented to optimize task-oriented energy consumption in machining manufacturing system. Sachdeva et al. [12] described multi-criteria optimization framework for determining optimal preventive maintenance schedules. In this framework, the authors have used Simulink simulation tool box with genetic algorithm for optimization and while availability, maintenance cost and life cycle costs are considered as the criteria for optimization. In recent years, the use of

SimEvents is growing in research publications because it has wide-range analytical tools and built-in modeling capabilities which make them more suitable than other simulation tools for discrete event modeling and simulation [13]. After reviewing the related literature, it can be mentioned that lack of evidence is available in the recent literature to suggest that Simulink/SimEvents simulation tool is yet to be used for addressing job-shop problem in the manufacturing system. Therefore, this paper analyzes the SimEvents discrete event simulation tool in the manufacturing system.

3 Manufacturing System Structure

The overview of a manufacturing system (MS) is shown in Fig. 1. The MS is producing three jobs (products) which are processed under six manufacturing operations, i.e., casting, planing, turning, shaping, drilling, and polishing. Job-1 (J1) follows the processing sequence of casting–planing–turning–polishing, whereas Job-2 (J2) follows shaping–drilling–turning, and the processing sequence of casting–shaping–drilling–planing–polishing is followed by Job-3 (J3). The assumptions made in the present paper are given in the following subsection.

3.1 Assumptions

The present paper focuses on scheduling a dynamic job-shop manufacturing system under the following assumptions:

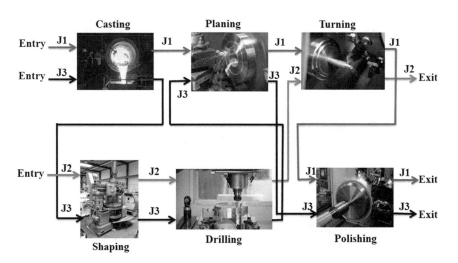

Fig. 1 Schematic diagram of the manufacturing system

- Each machine type consists only one machine in the shop.
- Only one operation occurs at a time on any job.
- The preceding operations of a job must be completed before performing its next operation.
- The manufacturing system can handle three job types at a time.
- Jobs arrive stochastically (Poisson distribution).
- Jobs get processed stochastically (exponential distribution).

4   Simulation Modeling of the Manufacturing System

The MS consists of six machines: casting, planing, turning, shaping, drilling, and polishing. The three different jobs (products) were being produced in this job-shop manufacturing system. Due to the complex nature of the production process in MS, the MS model is developed and simulated using SimEvents (a Simulink toolbox of MATLAB). SimEvents model of the manufacturing system is shown in Fig. 2. The description of different manufacturing operations with respective SimEvents models is given in the following subsections.

4.1   Casting Operation

The casting units facilitate this operation in the MS. It is the first process for jobs J1 and J3, while J2 is not processed under this operation. The SimEvents model of the casting process is shown in Fig. 3, which comprises of 'Set attribute' block for set processing time on each generated entity, 'Event-based random number' block for generating process time distribution, 'Priority queue' block for dispatching the job

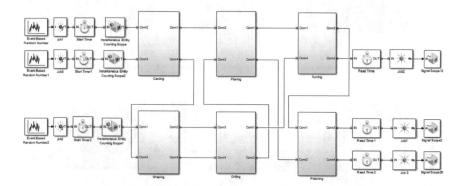

Fig. 2 SimEvents model of the manufacturing system

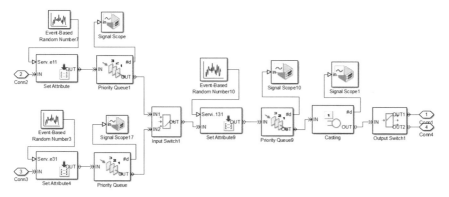

Fig. 3 SimEvents model of casting operation

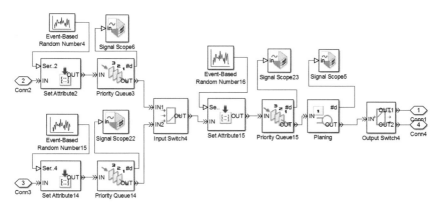

Fig. 4 SimEvents model of planing operation

entities, 'Input switch' and 'Output switch' blocks for scheduling two jobs, and 'N server' block for a number casting units.

4.2 Planing Operation

The planers provide the facility for planing operation in the MS. For job J1, this is the second task, while for job J3, planing occurs in the fourth processing sequence. Its SimEvents model is shown in Fig. 4, which consists of set processing time distribution using 'Set attribute' and 'Event-based random number' blocks, job dispatching rule using 'Priority queue' block, and production scheduling of both jobs J1 and J3 using 'Input switch' and 'Output switch' blocks.

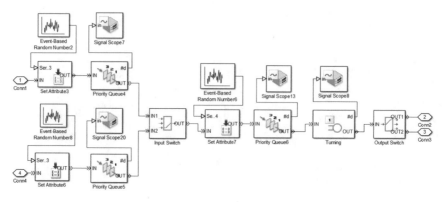

Fig. 5 SimEvents model of turning operation

4.3 Turning Operation

The manufacturing system consists of lathe machines which facilitate turning operation for the jobs. Both jobs J1 and J2 have been processed under turning operation in their third processing sequence. The model of turning operation is shown in Fig. 5, which comprises of different SimEvents blocks such as 'Set attribute,' 'Event-based random number,' 'Priority queue,' 'N server,' and 'Input switch' and 'Output switch' for part production of both jobs J1 and J2.

4.4 Shaping Operation

The shapers provide shaping operation in the MS. Jobs J2 & J3 are being processed in shaping operation in their first and second tasks, respectively. The model of shaping operation is shown in Fig. 6 that consists of different SimEvents blocks for part production of both jobs through set the processing time distribution, job priority rule, and suitable switching criterion for scheduling of both jobs.

4.5 Drilling Operation

A number of drills facilitate drilling operation in the MS. Jobs J2 and J3 are being processed through this drilling process in their second and third processing sequences, respectively. The model of drilling operation is shown in Fig. 7 that comprises of various SimEvents blocks which are 'Set attribute' block, 'Event-based random number'

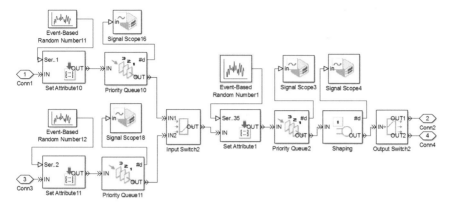

Fig. 6 SimEvents model of shaping operation

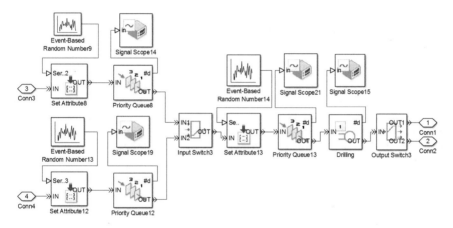

Fig. 7 SimEvents model of drilling operation

block, 'Input switch' and 'Output switch' block, 'Job priority queue' block and 'N server' block. These blocks are combined into one model and give output as drilled parts.

4.6 Polishing Operation

The polishing machines provide polishing operation to the jobs. Jobs J1 and J3 are being processed under this operation in their fourth and fifth tasks. The SimEvents model of polishing operation is shown in Fig. 8, which is the combination of various SimEvents blocks that facilitate part production of the jobs.

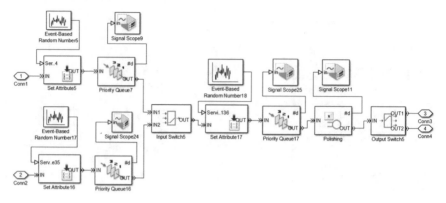

Fig. 8 SimEvents model of polishing operation

5 Simulation Results and Analysis

The previous section has described the simulation modeling of each manufacturing operation in the MS. In this study, the jobs are scheduled equiprobable for production in each operation. In each operation, two jobs have to be scheduled for processing. All the three jobs are dispatched using shortest processing time (SPT) rule. The simulation experiments of this study run for the whole day (24 hours). The present study conducts the performance measurement of the MS. The performance measures are mean flow time, total completed jobs, and work-in-process. After the simulation run, the number of entities for three jobs J1, J2, and J3 is 452, 415, and 466, respectively, that enter into the MS. The number of completed entities of three jobs J1, J2, and J3 is 245, 265, and 265, respectively, which are departed from the MS. The simulation results of arrival and departure for the entities of three jobs are shown in Figs. 9, 10 and 11. Through the simulation results, the work-in-process for all three jobs may be found as 207, 150, and 201 for J1, J2, and J3, respectively. Mean flow time can be calculated as follows.

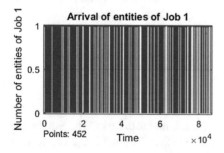
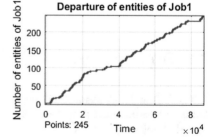

Fig. 9 Arrival and departure of entities of Job 1

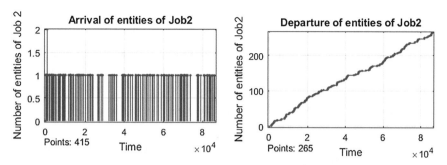

Fig. 10 Arrival and departure of entities of Job 2

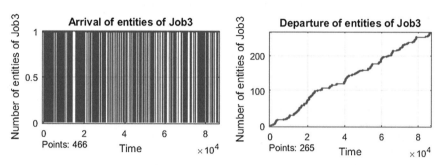

Fig. 11 Arrival and departure of entities of Job 3

Flow time of job i, $F_i = C_i - A_i$

Mean flow time $\bar{F} = \frac{1}{n}[\sum_{i=1}^{n} F_i]$

where C_i is the completion time of job i, and A_i is the arrival time of job i.

6 Conclusion

The modeling and analysis of the MS represent the behavior of the MS. With this created model, it can easily find the answer to those questions which might be hard without simulation. This study has used SimEvents (a simulation toolbox of MATLAB) for simulation model creation and analysis of the performance measures of MS. Hence, this research study assumes significant importance in this context. The created simulation model of the MS in this research study may be the basic framework for developing agent-based manufacturing system. This research work can be further extended with the discussion of different scheduling decision rules, and more performance measures might be considered for analysis such as mean tardiness, absolute lateness, standard deviation of flow time, and standard deviation of tardiness.

References

1. Dominici, G.: Holonic production system to obtain flexibility for customer satisfaction. J. Service Sci. Manag. **1**(3), 251–254 (2008)
2. Peklenik, J., Jerele, A.: Some basic relationships for identification of the machining processes. CIRP Ann. Manufact. Technol. **41**(1), 155–159 (1992)
3. Zhang, L., Luo, Y., Tao, F., Li, B.H., Ren, L., Zhang, X., Guo, H., Cheng, Y., Hu, A., Liu, Y.: Cloud manufacturing: a new manufacturing paradigm. Enterpise Informat. Syst. **8**(2), 167–187 (2014)
4. Manupati, V.K., Putnik, G.D., Tiwari, M.K., Vila, P., Cruz-Cunha, M.M.: Integration of process planning and scheduling using mobile-agent based approach in a networked manufacturing environment. Comput. Indust. Eng. **94**, 63–73 (2016)
5. Wang, Y.C., Chen, T., Chiang, H., Pan, H.C.: A simulation analysis of part launching and order collection decisions for a flexible manufacturing system. Simulat. Modell. Pract. Theor. **69**, 80–91 (2016)
6. Imran, M., Kang, C., Lee, Y.H., Jahanzaib, M., Aziz, H.: Cell formation in a cellular manufacturing system using simulation integrated hybrid genetic algorithm. Comput. Indust. Eng. **105**, 123–135 (2017)
7. Mousavi, S., Kara, S., Kornfeld, B.: A hierarchical framework for concurrent assessment of energy and water efficiency in manufacturing systems. J. Clean. Product. **133**, 88–98 (2016)
8. Omar, M.A., Qilun, Z., Lujia, F., Ali, A.A., Lahjouji, D., Khraisheh, M.: A hybrid simulation approach for predicting energy flows in production lines. Int. J. Sustain. Eng. **9**(1), 25–34 (2016)
9. Mushiri, T., Mahachi, A., Mbohwa, C.: A model reference adaptive control (MRAC) system for the pneumatic valve of the bottle washer in beverages using simulink. Procedia Manufact. **7**, 364–373 (2016)
10. Ortiz, O.A., Surez, G.I., Nelson, A.: Dynamic simulation of a pilot rotary kiln for charcoal activation. Comput. Chem. Eng. **29**(8), 1837–1848 (2005)
11. He, Y., Liu, B., Zhang, X., Gao, H., Liu, X.: A modeling method of task-oriented energy consumption for machining manufacturing system. J. Clean. Product. **23**(1), 167–174 (2012)
12. Sachdeva, A., Kumar, D., Kumar, P.: Planning and optimizing the maintenance of paper production systems in a paper plant. Comput. Indust. Eng. **55**(4), 817–829 (2008)
13. Ravitz, A., Mazzuchi, T.A., Sarkani, S.: Multi-scale modeling and simulation of complex systems: opportunities and challenges. IIE Trans. Healthcare Syst. Eng. **6**(2), 79–95 (2016)

Performance Assessment of Thirteen Crossover Operators Using GA

Ashish Jain, Tripti Mishra, Jyoti Grover, Vivek Verma and Sumit Srivastava

Abstract Performance of genetic algorithms depends on evolutionary operators, i.e., selection, crossover, and mutation, in general, and on the type of crossover operators, in particular. With constant research going on in the field of evolutionary computation, many crossover operators have come into the light, thus making the systematic comparison of these operators necessary. This paper presents comparison of 13 crossover operators on 20 benchmark problems using genetic algorithm. An exhaustive statistical study shows the supremacy of uniform, reduced surrogate, and single-point crossover operators among others.

Keywords Crossover operators · Davis order · Discrete · Flat
Nonparametric statistical test

1 Introduction

The idea to use genetic algorithm (GA) is to use the power of evolution to solve optimization problems. They simulate the survival of the fittest among individuals over consecutive generation for solving a problem. John Holland, also known as father of genetic algorithms, invented GAs in the 1960s to mimic some of the processes

A. Jain · T. Mishra (✉) · J. Grover · V. Verma · S. Srivastava
School of Computing and Information Technology, Manipal University
Jaipur, Jaipur, Rajasthan, India
e-mail: mishra14tripti@gmail.com

A. Jain
e-mail: ashish.jain@jaipur.manipal.edu

J. Grover
e-mail: jyoti.grover@jaipur.manipal.edu

V. Verma
e-mail: vivek.verma@jaipur.manipal.edu

S. Srivastava
e-mail: sumit.srivastava@jaipur.manipal.edu

© Springer Nature Singapore Pte Ltd. 2019 761
J. C. Bansal et al. (eds.), *Soft Computing for Problem Solving*, Advances in Intelligent
Systems and Computing 817, https://doi.org/10.1007/978-981-13-1595-4_60

involved in natural evaluation [1–5]. The performance of a GA is reliant on crossover type [2, 6]. During the evolution process by a GA, more than one parent is selected and one or more offspring are produced through crossover using genetic material of the parent [1–5]. To evaluate the performance of crossover operators, mere comparison of standard deviation values is not enough; rather, we need to perform a thorough statistical analysis [7]. This can be achieved in two ways: parametric statistical analysis and nonparametric statistical analysis [8]. The difference between the two statistical analyses is that the number of parameters is fixed in the case of first, while in the case of second the number of parameters increases with the amount of training data. The nonparametric analysis uses data that is often ordinal; i.e., it does not rely on numbers, but rather a ranking [8]. Here, we use nonparametric statistical analysis to evaluate the efficiency of the crossover operators. Many researches have been performed on this concept, the most notable being by Picek et al. [7] in which the performance of 10 crossover operators were tested on a set of 15 benchmark problems. This paper extends the scope of study by taking 3 more crossover operators, namely Davis order [9], flat [9], and discrete [9]. We have tested totally 13 crossover operators on 20 set of benchmark problems (or functions). Instead of evolutionary computation framework used in [7], the simple Java code has been written and executed for testing and analyzing considered crossover operators. Here, we also used tournament selection instead of roulette wheel selection as it is modern and more accurate than roulette wheel [2]. In this paper, Sect. 2 summarizes the applicable theory, Sect. 3 informs about the parameters used and results obtained, and finally Sect. 4 helps us to draw a conclusion.

2 Overview: Crossover Operators and Test Functions

2.1 Crossover Operators

Crossover is the process in which individuals are generated from the genes of the parents. This paper compares the performance of the following crossover operators:

Crossover operator	Description	Ref.
Single-point crossover	Using this crossover, a new offspring is produced by exchanging the genes of individuals from a point. The point is selected randomly, and tails of individuals are exchanged	[10]
Two-point crossover	It is a modified version of single-point crossover, wherein alternating divisions are merged to produce new offspring	[10]
Half-uniform crossover	It is another form of single point where mixing ratio must be 0.5 to produce new offspring	[11]

(continued)

(continued)

Crossover operator	Description	Ref.
Uniform crossover	In this crossover, each bit/gene is selected randomly, from either parent; then, the bits/gene are swapped to produce new offspring	[10]
Shuffle crossover	Shuffle crossover helps in creation of offspring which have independent crossover points in the parents. This crossover technique first shuffles the gene in both the parents, then selects a random point, and combines both the parents to produce two offspring	[12]
Reduced surrogate crossover	Reduced surrogate crossover minimizes the unwanted crossover operations in case of parents having same gene. In these cases, reduced surrogate crossover first checks for the individual gene in the parents. It creates list of all possible crossover points where genes of both the parent are different. After performing this check, if no crossover point is there, then no action is taken. But in case, if the parents are differing in more than one gene, it keeps the list of all crossover points. It then randomly selects one crossover point from the list and performs single-point crossover to create offspring	[12]
Segmented crossover	The segmented crossover operator processes the logical segment pairs of parent chromosomes one by one. For each segment pair, single-point crossover is performed	[12]
Non-geometric crossover	In this crossover, offspring are generated outside the segment between parents. The process involves: (i) First individual needs to be chosen randomly or based on fitness (ii)Offspring is created as a copy of that individual (iii) The bits of the offspring are reversed with specific possibility when the values of the bits are the same in both parents	[13]
Orthogonal array crossover	They are created based on orthogonal design making a methodical and balanced search in the region of parents	[14]
Hybrid Taguchi crossover	This method is used to exploit the optimum offspring	[15]
Davis order crossover	This permutation-based crossover operator transmits information about the relative ordering of the offspring	[9]
Discrete crossover	Discrete crossover uses random real numbers to create one child from two parents	[9]
Flat crossover	Flat crossover uses random real numbers to create one child from two parents	[9]

2.2 Benchmark Problems

The benchmark problems that have presented in Table 1 are well-known problems in the literature [16–21] that are considered in this paper for measuring the performance of crossover operators in GA procedures.

Table 1 Benchmark problems, their ranges, and global optimum

S. no.	Function	[Range]^n	Optimum
1	$f_1(x) = 0.5 + \frac{sin^2\sqrt{x^2+y^2}-0.5}{\left(1+0.001(x^2+y^2)\right)^2}$	$[-100, 100]^n$	0
2	$f_2(x) = 100\left(x_1^2 - x_2^2\right)^2 + \left(1 - x_1^2\right)^2$	$[-2.048, 2.048]^n$	0
3	$\frac{1}{f_3(x)} = \frac{1}{500} + \sum\limits_{j=1}^{25} \frac{1}{j+\sum_{i=1}^2(x_i-a_{ij})^6}$	$[-65.536, 65.536]^n$	1
4	$f_4(x) = \sum\limits_{i=1}^{n} x_i^2$	$[-100, 100]^n$	0
5	$f_5(x) = \sum\limits_{i=0}^{n} int(x_i)$	$[-100, 100]^n$	0
6	$f_6(x) = \sum\limits_{i=1}^{n} i.x_i^4 + N(0, 1)$	$[-1.28, 1.28]^n$	0
7	$f_7(x) = \sum\limits_{i=1}^{n} -x_i \sin\left(\sqrt{|x_i|}\right)$	$[-500, 500]^n$	0
8	$f_8(x) = \sum\limits_{i=0}^{n} |x_i| + \prod\limits_{i=0}^{n} |x_i|$	$[-10, 10]^n$	0
9	$f_9(x) = \sum\limits_{i=1}^{n} \left(\sum\limits_{j=1}^{i} x_j\right)^2$	$[-100, 100]^n$	0
10	$f_{10}(x) = max_i(|x_i|, 1 \leq i \leq n|)$	$[-100, 100]^n$	0
11	$f_{11}(x) = -20\exp\left(-0.2\sqrt{\frac{1}{n}\sum\limits_{i=1}^{n}x_i^2}\right) - \exp\left(\frac{1}{n}\sum\limits_{i=1}^{n}\cos(2\pi x_i)\right) + 20 + e$	$[-32, 32]^n$	0
12	$f_{12}(x) = 10n + \sum\limits_{i=1}^{n}\left(x_i^2 - 10\cos(2\pi x_i)\right)$	$[-5.12, 5.12]^n$	0
13	$f_{13}(x) = \sum\limits_{i=1}^{n}\left(100\left(x_{i+1}^2 - x_i^2\right)^2 + (1 - x_i)^2\right)$	$[-2.048, 2.048]^n$	0
14	$f_{14}(x) = \frac{1}{4000}\sum\limits_{i=1}^{n}x_i^2 - \prod\limits_{i=1}^{n}\cos\left(\frac{x_i}{\sqrt{i}}\right) + 1$	$[-300, 300]^n$	0
15	$f_{15}(x) = \frac{1}{n}\sum\limits_{i=1}^{n}\sin(x_i).(sin^{20}(i.x_i^2/\pi))$	$[0, \pi]^n$	0

(continued)

Table 1 (continued)

S. no.	Function	[Range]^n	Optimum
16	$f_{16}(x) = \sum_{i=1}^{n} i.x_i^2$	$[-100, 100]^n$	0
17	$f_{17}(x) = \sum_{i=1}^{n} 5.i.x_i^2$	$[-100, 100]^n$	0
18	$f_{18}(x) = -x_i . \sin x_i^2$	$[-100, 100]^n$	0
19	$f_{19}(x) = \sum_{j=1}^{n} \sum_{i=0}^{j} x_i^2$	$[-65.35, 65.35]^n$	0
20	$f_{20}(x) = \sum_{i=1}^{n} [x_i + 0.5]^2$	$[-100, 100]^n$	0

3 Results

In the experiment, binary-coded GA with tournament selection is used with following settings:

- The parents which are binary vectors denote real values.
- Precision is limited to 3 places after decimal point.
- Benchmark problems dimensionality $D = 30$.
- GA population size $= 40$.
- Number of fitness evaluation $= 400000$.
- Simple bit mutation with probability P_m is set to 0.01 per bit.
- Independent runs of algorithm for each crossover operator $= 40$.

The aim of the experiment is to achieve global minimum for considered benchmark functions. The performance measure is the error rate for each crossover operator tested. Experiments are performed in two rounds. In the first round, the best P_c value is determined for the crossover operators mentioned in Sect. 2.1. In the second round, for all the functions in Table 1, the best crossover operator is determined. The steps involved in the first round are as follows:

i. Each operator given in Sect. 2.1 is run 40 times on the functions mentioned in Table 1 with altered values of P_c.
ii. The value of P_c is selected from 0.1 to 1 with step size of 0.1.

In many combinations, there was no notable statistical variance between performances for different values of P_c. However, in 40 runs, the value of P_c that gives us least average error was chosen. The steps involved in the second round are:

i. Each operator was compared using the average error of the best individuals with the best probability value.
ii. Then, a sequence of nonparametric statistical analysis is organized in the given data.

Table 2 Average rankings of crossover operators (Friedman)

Crossover operator	Average rankings	Crossover operator	Average rankings
Single point (SP)	3.5873	Non-geometric (NG)	5.9714
Two point (TP)	4.4011	Orthogonal array (OA)	9.3123
Half-uniform (HU)	4.8121	Hybrid Taguchi (HT)	8.9809
Uniform (UN)	3.7924	Davis order (DO)	9.1011
Segmented (SG)	4.5012	Discrete (DI)	9.0322
Shuffle (SH)	5.4822	Flat crossover (FL)	8.2452
Reduced surrogate (RS)	4.2801		

Table 3 Post hoc comparison (control crossover operator: SP)

Operator	Unadjusted p	P_{Bonf}	$P_{Hochberg}$	P_{Finner}	P_{Li}
OA	0	0.000001	0.000001	0.000001	0.000001
HT	0.000001	0.000005	0.000004	0.000002	0.000004
NG	0.032299	0.291762	0.226231	0.093798	0.212304
SH	0.091331	0.822121	0.548121	0.193867	0.432502
HU	0.277743	2.499559	0.880112	0.443284	0.698790
SG	0.433192	3.897821	0.880203	0.573179	0.783381
TP	0.472187	4.223713	0.880203	0.573179	0.796615
RS	0.507832	4.564511	0.880203	0.573179	0.808938
UN	0.880486	7.923151	0.880203	0.880203	0.880203
DO	0.000001	0.000002	0.000001	0.000001	0.000001
DI	0.000000	0.000001	0.000001	0.000001	0.000002
FL	0.000001	0.000002	0.000003	0.000002	0.000003

Initially, Friedman two-way analysis of variances by ranks is used for testing the differences between two or more related samples. The aim of this analysis is to show statistical difference between performances of given operators. Since statistical difference exists now, post hoc statistical analysis can be investigated to determine those differences. Table 2 encapsulates the ranks obtained by Friedman analysis. The results indicate the single-point crossover operator as the finest crossover operator; therefore, post hoc analysis is done with single point as control method.

Given the level of significance (α) of 0.05, Friedman statistic shows significant differences in operators. In post hoc analysis, Bonferroni–Dunn, Hochberg, Finner, and Li tests were applied on the outcomes of Friedman test [22] which gives 'α' with which one control operator is better than other operators (i.e., for which the null hypothesis is rejected). The adjusted p values are shown in Table 3.

From Table 3, orthogonal array, hybrid Taguchi, Davis order, discrete, and flat crossover operators are notably worse than single-point crossover with $\alpha = 0.05$. For other operators, null hypothesis cannot be rejected with $\alpha = 0.05$. Considering

Table 4 Contrast pairwise comparison

	SP	TP	HU	UN	SG	SH	RS	NG	OA	HT	DO	DI	FL
SP	0	-0.003	-0.004	0.009	-0.022	-0.056	0.001	-0.095	-3.924	-3.684	-3.891	-3.687	-3.512
TP	0.003	0	-0.001	0.015	-0.019	-0.053	0.004	-0.092	-3.920	-3.682	-3.911	-3.685	-3.509
HU	0.004	0.001	0	0.015	-0.018	-0.052	0.005	-0.092	-3.919	-3.682	-3.912	-3.686	-3.510
UN	-0.009	-0.015	-0.015	0	-0.032	-0.065	-0.009	-0.105	-3.933	-3.694	-3.929	-3.697	-3.512
SG	0.022	0.019	0.018	0.032	0	-0.035	0.021	-0.073	-3.901	-3.661	-3.887	-3.664	-3.994
SH	0.056	0.053	0.052	0.065	0.033	0	0.057	-0.042	-3.869	-3.631	-3.861	-3.634	-3.965
RS	-0.001	-0.004	-0.005	0.009	-0.021	-0.057	0	-0.097	-3.924	-3.685	-3.910	-3.688	-3.513
NG	0.095	0.092	0.092	0.105	0.073	0.042	0.097	0	-3.831	-3.591	-3.828	-3.594	-3.373
OA	3.924	3.920	3.919	3.933	3.901	3.869	3.924	3.831	0	0.241	0.208	0.244	0.411
HT	3.684	3.682	3.682	3.694	3.661	3.631	3.685	3.591	-0.241	0	-0.173	-0.142	0.331
DO	3.891	3.911	3.912	3.929	3.887	3.861	3.910	3.828	-0.208	0.173	0	0.101	0.382
DI	3.687	3.685	3.686	3.697	3.664	3.634	3.688	3.594	-0.244	0.142	-0.101	0	0.353
FL	3.512	3.509	3.510	3.512	3.994	3.965	3.513	3.373	-0.411	-0.331	-0.382	-0.353	0

Finner analysis with $\alpha = 0.1$, it rejects another hypothesis; i.e., single-point crossover operator is better than non-geometric crossover operator.

Since there are no significant differences for the majority of the operators, contrast estimation which is based on medians can be used to estimate the differences between the performances of two crossover operators. In this estimation, the performance of the procedures is verified by the differences in error rates as shown in Table 4. A negative value for the crossover operator in each row shows that it performed better than the crossover operator in each column. Results obtained in Table 4 indicate that uniform, reduced surrogate, and single point are the best crossover operators.

4 Conclusions

In this paper, we investigated for the best crossover operators on a certain set of benchmark functions. The results pointed out notable statistical difference between certain crossover operators. A prominent difference can be seen in case of orthogonal array, hybrid Taguchi, Davis order, discrete, and flat, as they perform poorer than the other crossover operators. The results also show that single point, uniform, and reduced surrogate are the best crossover operators. With increasing research in field of evolutionary computation, new crossover operators are coming into light. Thus, comparison between the new crossover operator and the old ones may also prove beneficial for future research purposes.

References

1. Goldberg, D.: Genetic Algorithm in Search Optimization and Machine Learning, Fourth Impression. Pearson Education (2009)
2. Zbigniew, M.: Genetic algorithms + data structures = evolution programs. Springer Science & Business Media (2013)
3. Gonzalez, T.F.: Handbook of Approximation Algorithms and Metaheuristics. CRC Press (2007)
4. Mitchell, M.: An introduction to genetic algorithms. MIT Press (1998)
5. Sivanandam, S.N., Deepa, S.N.: Introduction to Genetic Algorithm. Springer Science & Business Media (2007)
6. Srinivas, M., Patnaik, L.M.: Adaptive probabilities of crossover and mutation in genetic algorithms. IEEE Trans. Syst. Man Cybernet. **24**, 656–667 (1994)
7. Picek, S., Golub, M., Jakobovic, D.: Evaluation of crossover operator performance in genetic algorithms with binary representation. Bio-Insp. Comput. Appl. 223–230 (2012)
8. Sheskin, D.: Handbook of Parametric and Nonparametric Statistical Procedures, 4th ed. Chapman and Hall/CRC (2007)
9. Spears, W., Vic, A.: A study of crossover operators in genetic programming. Methodol. Intell. Syst. 409–418 (1991)
10. Poon, P.W., Carter, J.N.: Genetic algorithm crossover operators for ordering applications. Comput. Oper. Res. **22**, 135–147 (1995)
11. Eshelman, L.J.: The CHC adaptive search algorithm: how to have safe search when engaging in nontraditional genetic recombination. In: Foundations of Genetic Algorithms, pp. 265–283. Morgan Kaufmann, San Francisco, CA, USA (1991)

12. Dumitrescu, D., Lazzerini, B., Jain, L.C., Dumitrescu, A.: Evolutionary Computation. CRC Press, Florida, USA (2000)
13. Ishibuchi, H., Tsukamoto, N., Nojima, Y.: Maintaining the diversity of solutions by non-geometric binary crossover: a worst one-max solver competition case study. In: Proceedings of the Genetic and Evolutionary Computation Conference GECCO'08. pp. 1111–1112 (2008)
14. Chan, K.Y., Kwong, C.K., Jiang, H., Aydin, M.E., Fogarty, T.C.: A new orthogonal array based crossover, with analysis of gene interactions, for evolutionary algorithms and its application to car door design. Expert Syst. Appl. **37**, 3853–3862 (2010)
15. Tsai, J.T., Liu, T.K., Chou, J.H.: Hybrid taguchi-genetic algorithm for global numerical optimization. IEEE Trans. Evol. Comput. **8**(4), 365–377 (2004)
16. Leung, Y.W., Yuping, W.: An orthogonal genetic algorithm with quantization for global numerical optimization. IEEE Trans. Evol. Comput. **5**, 41–53 (2001)
17. Digalakis, J.G., Konstantinos, G.M.: An experimental study of benchmarking functions for genetic algorithms. Int. J. Comput. Math. **79**, 403–416 (2002)
18. Pohlheim, H.: Geatbx Examples of Objective Functions (2006). http://www.geatbx.com/download/GEATbx_ObjFunExpl_v37.pdf
19. Liang, J.J., Qu, B.Y., Suganthan, P.N.: Problem definitions and evaluation criteria for the CEC 2014 special session and competition on single objective real-parameter numerical optimization. Zhengzhou University, Zhengzhou China and Technical Report, Nanyang Technological University, Singapore, Computational Intelligence Laboratory (2013)
20. Liang, J.J., Qu B.Y., Suganthan P.N., Chen, Q.: Problem definitions and evaluation criteria for the CEC 2015 competition on learning-based real-parameter single objective optimization. Technical Report 201411A, Computational Intelligence Laboratory, Zhengzhou University, Zhengzhou China and Technical Report, Nanyang Technological University, Singapore (2014)
21. Beheshti, Z., Shamsuddin, S.M., Shafaatunnur, H.: Memetic binary particle swarm optimization for discrete optimization problems. Inf. Sci. **299**, 58–84 (2015)
22. Derrac, J., Garcia, S., Molina, D., Herrera, F.: A practical tutorial on the use of nonparametric statistical tests as a methodology for comparing evolutionary and swarm intelligence algorithms. Swarm Evolution. Computat. **1**, 3–18 (2011)

Enhancing Saliency of a Target Object Through Color Modification of Every Object Using Genetic Algorithm

Dipanjan Roy and Rajarshi Pal

Abstract Ability to draw an observer's gaze toward an object in an image depends on the visual saliency of the object. Contrast and/or uniqueness of low-level features of an object (e.g., intensity, color, etc.) drives visual saliency in the absence of external influence. Hence, an increased difference between a user-defined target object (whose saliency is being enhanced) and its surrounding objects in terms of these low-level features enhances the visual saliency of the target object. In this paper, enhancement of saliency of the target object has been posed as a maximization problem. Feature values of every segment in an image have been modified within a limit to maximize the feature difference between the target segment and all other segments. The imposed limit restricts the amount of change in any segment in order to preserve the naturalness of the image. In the framework of the proposed maximization problem, a genetic algorithm-based approach has been devised to find suitable features values for every segment. Effectiveness of the proposed method has been demonstrated through suitable objective measures.

1 Introduction

Gaze-based interaction with the system leads to several novel applications [1]. Effective development of these gaze-based systems demands smooth guidance of a viewer's attention to an intended target. A viewer's gaze can be drawn to the intended region through changing the visual saliency of the region.

Literature in computer vision refers to several such techniques to direct a viewer's attention to intended region in an image. For example, field of view of an intelligent camera (ICam) [2] is guided by visual saliency of the objects in the scene to influence

D. Roy
Indian Institute of Technology Indore, Indore, India
e-mail: dipanjanroy13@gmail.com

R. Pal (✉)
Institute of Development & Research in Banking Technology, Hyderabad, India
e-mail: iamrajarshi@yahoo.co.in

© Springer Nature Singapore Pte Ltd. 2019
J. C. Bansal et al. (eds.), *Soft Computing for Problem Solving*, Advances in Intelligent Systems and Computing 817, https://doi.org/10.1007/978-981-13-1595-4_61

the viewer's attention. Image modification-based techniques also attempt to alter the saliency of certain components in the image. Such techniques alter the luminance and chromatic components of the image segments to draw viewer's attention to intended portions [3–5]. In [6], saliency is influenced by altering texture feature. Distracting regions are deemphasized by reducing spatial variation of texture using power maps which capture local frequency of the contents in the image. This leads to enhancement of relative saliency for the target region. A gradual blurring of portions of the image dynamically controls visual attention in [7]. Selected region was not blurred to draw attention to this region. An eye-tracking device is used in [8] to capture an observer's gaze. As the observer's gaze shifts toward the intended region, the region is not modulated further. Similarly, approaches for influencing human attention have been proposed for spherical images [9] and infographics images [10].

Recently, enhancement of saliency of an intended object has been posed as a maximization problem [11]. But a relatively small solution space in [11], where a suitable gray value of the target segment was searched for, made an exhaustive search practical. Inspired by the work in [11], there has been an attempt in [12] to extend the optimization framework for color images. A genetic algorithm-based method has been developed to cause the intended object to be more salient. Formulation of this problem as a constrained maximization problem (considering multiple features) and subsequent application of genetic algorithm to search for suitable color value for the target segment has been suggested in [12]. Contrary to [12], the problem formulation has been revised in the proposed work to search for suitable color values of every segment. Genetic algorithm is used to find suitable feature values for every segment in the image so that the result maximizes the feature dissimilarity (which is weighted by positional proximity) between the target object and all other objects. In the context of this work, the term 'feature' indicates the achromatic and chromatic channels of CIE L* a* b* representation of the image. Experimental analysis involving objective assessment of the method using saliency map demonstrates that the target region has become more salient.

Organization of the remaining parts of the paper is as follows: The maximization framework for enhancing the saliency of a target segment has been presented in Sect. 2. Section 3 depicts the proposed genetic algorithm-based approach for having the intended region as more salient. Section 4 presents the experimental results. Finally, Sect. 5 draws the concluding remarks.

2 Saliency Enhancement as Maximization Problem

This paper attempts to maximize the relative saliency of a target segment. A slight variation of degree centrality-based saliency model [13] is considered here to formulate the problem. In this variation, segments of an image, as obtained using a mean shift-based segmentation [14], represent the nodes in the graph in the graph-based model. Feature difference (modulated by positional proximity) between a pair of segments i and j is considered as the edge weight $E_{f_{ij}}$ between concerned nodes.

Hence, edge weight between nodes i and j ($E_{f_{ij}}$) can be represented as

$$E_{f_{ij}} = D_{f_{ij}} * e^{-D_{s_{ij}}^2 / 2\sigma^2} \tag{1}$$

where $D_{f_{ij}}$ is the difference in terms of a low-level feature f between the concerned segments. The absolute difference between mean feature values of the segments, as represented by nodes i and j, determines the feature difference in Eq. (1). CIE L* a* b* color model is used to represent an image. Hence, the term 'feature' implies the dimension L* for luminance, a* for red–green opponent-color, and b* for blue–yellow opponent-color channels. The spatial distance $D_{s_{ij}}$ between the concerned segments is computed as the Cartesian distance between the center points of these two segments. A complete description of the degree centrality-based saliency model can be found in [13]. Saliency of a segment is estimated by counting the degree of the corresponding node. This is due to the fact that the degree of a node, in this context, represents with how many other segments the feature differences (modulated by positional proximity) are really high. Hence, according to feature f (where $f \in \{L*, a*, b*\}$), the saliency of a segment, which is represented by a node i, can be expressed as

$$S_{f_i} = \sum_{j=1}^{n} E_{f_{ij}} = \sum_{j=1}^{n} D_{f_{ij}} * e^{-D_{s_{ij}}^2 / 2\sigma^2} \tag{2}$$

where n indicates the number of segments of the image. Feature dissimilarity $D_{f_{ij}}$ can be expressed as absolute difference of mean feature values between the segments i and j. Therefore, Eq. (2) can be rewritten as

$$S_{f_i} = \sum_{j=1}^{n} |F_{f_i} - F_{f_j}| * e^{-D_{s_{ij}}^2 / 2\sigma^2} \tag{3}$$

where F_{f_i} and F_{f_j} represent the mean feature values of the segments i and j, respectively, for a particular feature f (where $f \in \{L*, a*, b*\}$). Moreover, $|\cdot|$ represents the absolute value of its argument.

Thus, estimation of the saliency of all segments for a particular feature generates a saliency map for the feature. Hence, saliency maps are generated for all three features in parallel and those are combined using a map normalization and combination strategy as depicted in [13].

Finally, the saliency rank R_t of a target segment t is estimated by counting the number of other segments whose saliency values are less than that of the target segment.

A genetic algorithm-based technique is presented in this paper to enhance the saliency rank R_t of a target segment t by changing the mean feature values of every segment in the image. The mean feature values of each segment i are changed from F_{f_i} to F'_{f_i} for all $f \in \{L*, a*, b*\}$. The objective, here, is to search for suitable values F'_{f_i} for every segment i in each of the three feature channels ($f \in \{L*, a*, b*\}$), which maximize the saliency rank R_t of the target segment t. Therefore, the objective can be formulated as

$$\text{maximize } R_t \tag{4}$$

The spatial distance $D_{s_{ij}}$ between two nodes i and j is always constant. Therefore, the unknown variables are the modified features values for all segments F'_{f_i} in all three feature channels ($L*$, $a*$, and $b*$). Therefore, the purpose is to select the values of different features F'_{f_i} (where $f \in \{L*, a*, b*\}$) for every segment i (including both target and non-target segments) in such a way that they maximize the saliency rank R_t of the target segment t.

To maintain the naturalness of the image, changes in feature values for any segment are restricted within threshold limits T_f. These limits may be different for different features, i.e., $f \in \{L*, a*, b*\}$. Hence, the constraints for maintaining the naturalness can be expressed as

$$|F'_{f_i} - F_{f_i}| \leq T_f \tag{5}$$

where T_f is the maximum allowable change for feature f (where $f \in \{L*, a*, b*\}$). More changes will cause distortion or unwanted alterations in the image. The new value F'_{f_i} must be an integer in the range from 0 to 255 as all the features values are represented as an 8-bit binary representation.

$$0 \leq F'_{f_i} \leq 255 \tag{6}$$

3 Proposed Approach to Enhance Saliency

3.1 RGB to CIE L* a* b* Color Conversion of Image

The input image is converted from RGB color space to CIE L* a* b* color space. Then, the values in each channel are converted into an integer in the range from 0 to 255, so that each feature value can be represented using a 8-bit binary string.

3.2 Segmentation of the Image

A mean shift-based image segmentation technique [14] is followed here to divide the image into a set of homogeneous regions. These segments render a better representation of underlying objects in comparison with the rectangular blocks in [13].

3.3 Obtaining the Values for Known Variables

In order to compute new mean feature values F'_{f_i} for each segment i, the current mean feature values of the segment F_{f_i} are required. These values can be obtained from the segmented image. The spatial distance $D_{s_{ij}}$ between any two segments i and j is computed as the Cartesian distance between the center points of the concerned segments. The threshold values T_f in Eq. (5) are set as 20% of the dynamic range of that particular feature in the image.

3.4 Estimating New Feature Values for All Segments Using Genetic Algorithm

A genetic algorithm-based approach has been developed to solve the optimization problem as specified using Eqs. (4), (5), and (6).

3.4.1 Representation of a Valid Chromosome

Each gene in the chromosome contains a triplet of values representing the three mean feature values for a particular segment (corresponding to feature channels L*, a*, and b*). So, length of the chromosome is n, where n is the number of segments in the image. It is to be noted that any value for a mean feature value F'_{f_i} for any segment i may not be an acceptable solution. This is because the solution space is restricted by the constraint as specified in Eq. (5). Equation (5) suggests that the value of F'_{f_i} for a valid gene must satisfy the following condition:

$$F_{f_i} - T_f \leq F'_{f_i} \leq F_{f_i} + T_f \tag{7}$$

3.4.2 Fitness Criteria

An updated image is constructed by replacing the old feature values F_{f_i} of each segment by the new values F'_{f_i} as represented by a chromosome. Saliency map of an updated image is computed. A saliency rank R_t for the target segment t is estimated as the number of other segments whose saliency values are less than the saliency of the target segment. This saliency rank is considered as fitness score. Higher saliency rank indicates a better solution.

3.4.3 Initialization of Population

An initial pool of 11 chromosomes is considered at the start of the process. Population size is considered to be 11.

3.4.4 Reproduction

Reproduction of chromosomes is carried out using the saliency rank as the fitness score. Fitness of each chromosome in the population is evaluated. Then, a roulette wheel-based implementation of the reproduction stage selects better chromosomes more number of times for next population.

3.4.5 Crossover

The best chromosome in the population is not allowed to change due to crossover. It does not take part in the crossover process as crossover may modify the chromosome. Five non-overlapping pairs are constituted from remaining ten chromosomes. Crossover is performed between the chromosomes in each of these pairs. For a pair of parent chromosomes, three crossover points are randomly selected—each one for the features L*, a*, and b*. Crossover is performed by interchanging the values appearing after the crossover point between corresponding subparts of the parent chromosomes.

3.4.6 Mutation

Similar to crossover stage, the best chromosome of a population does not participate in mutation because it is desired to retain the best solution at every iteration. Mutation is carried out with a small mutation probability (e.g., 0.02) using a procedure as described here. Let at an particular iteration the value of a gene for feature $f \in \{L*, a*, b*\}$ be G_{f_i}. Moreover, the lower and upper bounds for the values of this gene for this particular feature be l_{f_i} and u_{f_i}, respectively. Hence, after mutation, the new gene value for this particular feature will be G'_{f_i} as

$$G'_{f_i} = u_{f_i} - r * (G_{f_i} - l_{f_i}) \tag{8}$$

where r is a random number in the range 0 to 1.

3.4.7 Stopping Criteria

The stages of the proposed genetic algorithm are repeated for N number of iterations (N is ideally a large number). As the crossover and mutation do not modify the best chromosome, the solution does not deteriorate at any point of time. Therefore, the best chromosome as obtained after N iterations provides the result for the optimization problem as discussed above.

3.5 Restoring the Texture

Same feature values (as derived by solving the optimization problem using the proposed genetic algorithm) should not be assigned to every pixel in a segment. This ensures that the texture of the segment can be preserved. Hence, the original look-and-feel of the image is maintained. Therefore, the difference between an estimated mean feature value and the original mean feature value, i.e., $(F'_{f_i} - F_{f_i})$ for each feature $f \in \{L*, a*, b*\}$, is added with existing feature value of the pixel in each region i. As the difference between the new mean and the original mean feature values for each segment i is added to every pixel in the segment, the new mean value of the segment is F'_{f_i} (which is computed using the proposed genetic algorithm-based approached).

4 Experimental Validation

The proposed method is tested on a set of 25 images which was randomly selected from MSRA salient object database [15]. Few of these results have been presented in Fig. 1.

Each row in Fig. 1 presents an input image, its saliency map, the modified image, and the saliency map of the modified image in a left to right arrangement. Saliency maps are generated using a slight variation (as discussed previously) of the degree centrality-based model [13] for both input image and modified image. To carry out the experiment, it was attempted to enhance the relative saliencies of the stem of the tree (top row), the branch of the tree (middle row), and the house (bottom row) in the images in Fig. 1. Visual inspection of these saliency maps (at the portion containing the target segment) indicates enhancement of saliency for the target segment. Higher (brighter) saliency value reflects that the underlying object is more salient. Saliency rank R_t of the target segment t has improved from 135 to 153 (out of 154 segments), from 24 to 71 (out of 72 segments), and from 36 to 142 (out of 175 segments) for these three examples (top to bottom, respectively).

Based on saliency map, two more evaluation metrics—average discrimination ratio (ADR) and normalized saliency (NS)—are used to demonstrate the effectiveness of the proposed approach. ADR is expressed as

Fig. 1 Experimental results (left to right column-wise): Input image, saliency map of the input image, modified image, saliency map of the modified image

$$ADR = \frac{\sum_{(i,j)\in A} M(i,j)/|A|}{\sum_{(i,j)\in A} M(i,j)/|A| + \sum_{(i,j)\in B} M(i,j)/|B|} \qquad (9)$$

where A and B are the set of pixels in the target and all other non-target segments, respectively. M is the saliency map. The average ADR value for 25 experimental cases has improved from 0.43 (for input images) to 0.64 (for modified images).

The mean saliency value of all pixels at the target segment is considered as normalized saliency (NS) using the zero mean and unit standard deviation representation of the saliency map. If the mean and standard deviation of the saliency map M are μ_M and σ_M, respectively, then NS can be mathematically expressed as

$$NS = \frac{1}{\sigma_M |A|} \sum (M(i,j) - \mu_M) \qquad (10)$$

where A is the set of pixels in the target segment. The average value of NS for 25 experimental cases has improved from -0.20 (for input images) to 1.21 (for modified images). Hence, the saliency for the target segment is enhanced as it is showcased using these two objective metrics.

5 Conclusion

A genetic algorithm-based approach to enhance saliency of a target segment has been proposed in this paper. Modification of feature values for every segment in the image is allowed here. Genetic algorithm searches for suitable feature values for every segment within the solution space which is restricted by a set of constraints. Visual inspection of the results as well as objective assessment using saliency maps establishes the success of the proposed method.

The limit in Eq. (5) influences the amount of change in color triplets of a segment. For this current experiment, the limit has been abruptly set to 20% of the dynamic range of the concerned feature map. But with this limit, the change in color values may appear unnatural for some of the images (e.g., the bottom row in Fig. 1). Finding an appropriate permissible range of such changes in feature values is still a research issue.

References

1. Sridharan, S., John, B., Pollard, D., Bailey, R.: Gaze guidance for improved password selection. In: Proceedings of the Ninth Biennial ACM Symposium on Eye Tracking Research and Applications (2016)
2. Pal, R., Mitra, P., Mukhopadhyay, J.: Icam: maximizes viewers attention on intended objects. In: Proceedings of Pacific-Rim Conference on Multimedia (2008)
3. Kim, Y., Varshney, A.: Saliency-guided enhancement for volume visualization. IEEE Trans. Vis. Comput. Graph. **12**(5) (2006)
4. Kokui, T., Takimoto, H., Mitsukura, Y., Kishihara, M., Okubo, K.: Color image modification based on visual saliency for guiding visual attention. In: Proceeding of the 22nd IEEE Symposium on Robot and Human Interactive Communication (2013)
5. Mateescu, V.A., Bajic, I.V.: Attention retargeting by color manipulation in images. In: Proceedings of the 1st International Workshop on Perception Inspired Video Processing (2014)
6. Su, S.L., Durand, F., Agarwala, M.: De-emphasis of distracting image regions using texture power maps. In: Proceedings of the 4th IEEE International Workshop on Texture Analysis and Synthesis (2005)
7. Hata, H., Koike, H., Sato, Y.: Visual guidance with unnoticed blur effect. In: Proceedings of the International Working Conference on Advanced Visual Interfaces (2016)
8. Bailey, R., McNamara, A., Sudarsanam, N., Grimm, C.: Subtle gaze direction. ACM Trans. Graph. **28**(4) (2009)
9. Tanaka, R., Narumi, T., Tanikawa, T., Hirose, M.: Attracting users attention in spherical image by angular shift of virtual camera direction. In: Proceedings of 3rd ACM Symposium on Spatial User Interaction (2015)
10. Yasuda, K., Takahashi, S., Wu, H. Y.: Enhancing inforgraphics based on symmetry saliency. In: Proceeding of the 9th International Conference on Visual Information Communication and Interaction (2016)
11. Kumar, R.K., Pal, R.: Constrained maximization of saliency of intended objects for guiding attention. In: Proceedings of 12th IEEE India International Conference (2015)
12. Pal, R., Roy, D.: Enahncing saliency of an object using genetic algorithm. In: Proceedings of the Conference on Computer and Robot Vision (2017)
13. Pal, R., Mukherjee, A., Mitra, P., Mukherjee, J.: Modelling visual saliency using degree centrality. IET Comput. Vis. **4**(3), 218–229 (2010)
14. Comaniciu, D., Meer, P.: Mean shift: a robust approach towards feature-space analysis. IEEE Trans. Pattern Anal. Mach. Intell. **24**(5) (2002)
15. Liu, T., Sun, J., Zheng, N.N., Tang, X., Shum, H.Y.: Learning to detect a salient object. In: Proceedings of the Conference on Computer Vision and Pattern Recognition, pp. 1–8 (2007)

Temperature Resolution and Spatial Resolution Improvement of BOCDR-Based DTS System Using Particle Swarm Optimization Algorithm

Tangudu Ramji and Prasant Kumar Sahu

Abstract Temperature resolution and spatial resolution are major performance metrics in any distributed temperature sensing (DTS) system. In this paper, we have presented a detailed analysis on the performance of a Brillouin optical correlation domain reflectometry (BOCDR)-based DTS (BOCDR-DTS) system. Particle swarm optimization (PSO) evolutionary algorithm is being used in this paper to improve the performance of the proposed BOCDR-DTS system. Using this optimization algorithm, we minimized the Brillouin frequency shift (BFS) error in sensing system. As a consequence of this, the achieved temperature and spatial resolution are ~0.839 °C and ~43 cm, respectively. The results were simulated using MATLAB version 15.0.

Keywords Spontaneous Brillouin scattering · Distributed temperature sensing
Optical correlation domain reflectometry (OCDR) · Particle swarm optimization
(PSO) · Brillouin frequency shift · Temperature resolution and spatial resolution

1 Introduction

Distributed fiber optic sensors (DFOS) based on Brillouin scattering are widely developed in engineering sciences. One of the widely used versions of DFOS is Brillouin distributed temperature sensing (B-DTS) system useful for distributed temperature measurement. This system has more immunity to electromagnetic interference, small in size, less complexity, and more survivability under harsh environmental conditions. Due to these advantages, Brillouin-DTS system is able to detect fire locations, leakage positions in oil and gas pipelines and able to monitor the power cables and structural health conditions [1]. This sensing system is highly applicable in medical and military areas. Changes of temperature on the optical fiber produce variations

T. Ramji (✉) · P. K. Sahu
Indian Institute of Technology Bhubaneswar, Argul, Bhubaneswar, Odisha, India
e-mail: rt10@iitbbs.ac.in

P. K. Sahu
e-mail: pks@iitbbs.ac.in

© Springer Nature Singapore Pte Ltd. 2019
J. C. Bansal et al. (eds.), *Soft Computing for Problem Solving*, Advances in Intelligent
Systems and Computing 817, https://doi.org/10.1007/978-981-13-1595-4_62

in acoustic properties of medium and generate changes in its characteristic Brillouin frequency shift (BFS) parameter. These changes are used to estimate actual temperature values on the optical fiber. Typically, BFS is ~11 GHz for 1550 nm pumping wavelength under room temperature along the fiber [2, 3].

As reported in the literature, the B-DTS system can be broadly classified into two categories based on two types of scatterings, namely the spontaneous and stimulated Brillouin scattering. Under spontaneous Brillouin scattering (SpBS), Brillouin optical time domain reflectometry (BOTDR) [4–10], and Brillouin optical correlation domain reflectometry (BOCDR) [11–16] techniques will be come. Similarly, the stimulated Brillouin scattering (SBS) systems comprises Brillouin optical time domain analysis (BOTDA) [17–21], Brillouin optical frequency domain analysis (BOFDA) [22, 23], and Brillouin optical correlation domain analysis (BOCDA) [24, 25] techniques. Compared to analysis-based techniques, reflectometry-based techniques are having less complexity. Because the analysis-based techniques contain double ended accesses (pump and probe lights). Other important drawbacks of BOTDA, BOFDA, and BOCDA techniques are high nonlinearity effects and cost (usage of electro-optic modulators and vector network analyzers) [17–25].

In reflectometry-based techniques, BOTDR is used for temperature sensing up to tens of kilometers. But it has limitation in spatial and temperature resolution. The typical spatial resolution as reported in the literature for a BOTDR-based DTS system is in terms of meters. Because of the theoretical limit, it is not possible to select the pulse width of input optical signal lesser than 10 ns, as nonlinearity effects will be generated in an optical fiber [4–10]. For avoiding this problem, BOCDR is used in DTS system [11–16]. Mostly, the spatial resolution by BOCDR technique is in terms of tens of centimeters. But the major drawback is frequently occurs in BOCDR-based DTS system which is BFS error. This error occurs for either room or non-room temperature sections along the fiber. This BFS error value is in terms of MHz for temperature and/or strain [12–16].

In this paper, we have proposed and analyzed the performance of a BOCDR-DTS system. The proposed system is having the potential application in the areas of defense applications preferably in fighter jet, unmanned aerial vehicles (UAV), aircraft carrier naval ships, etc. The paper reports the analysis on BFS error in BOCDR-based DTS system. Along with this BFS error, we also focused on spatial resolution improvement. For minimization of BFS error and enhancement in spatial resolution, soft and evolutionary computing algorithm, such as particle swarm optimization (PSO) [7, 9] is used. The reasons behind the usage of PSO algorithm are high efficiency result and fast convergence in both linear and nonlinear problems and less complexity for analysis and implementation point of views [26, 27]. Compared to PSO algorithm, genetic algorithm (GA) gives more complexity and takes more time to produce convergence [28, 26], whereas ant colony optimization (ACO) gives uncertainty in time to give convergence and differential evolution (DE) algorithm gives more complexity and takes more time to make convergence [29, 30]. By minimization of this BFS error, we extracted or captured almost true values of temperatures and their locations on the fiber.

2 Principle and Theory

2.1 BOCDR Technique

When an input optical signal is launched into an optical fiber, it interacts with acoustic phonons, generating backscattered Stokes and anti-Stokes lights. This phenomenon is known as spontaneous Brillouin scattering (SpBS). Its spectrum is called as Brillouin gain spectrum (BGS). The frequency difference between the pump or input optical frequency and central frequency of BGS spectrum is called as the Brillouin frequency shift (BFS). When the pump wavelength is ~1550 nm, the BFS in silica single-mode fiber (SMF) is ~11 GHz, and it slightly varies depending on the fiber fabrication process. If different temperatures are applied on silica SMF, the BFS shifts toward a higher frequency with 0.9876 MHz/°C at ~1550 nm. So, if we measures BFS distribution along a fiber, the information of applied temperature distribution can be estimated BOCDR is one of the Brillouin-based distributed sensing techniques. Conceptual BOCDR system is shown in Fig. 1. In Fig. 1, laser is narrow-band distributed feedback (DFB) laser. This laser source generates an optical signal with a frequency f_0. After launching process, this optical signal's intensity is divided into two equal parts. One part is directed in the forward direction and other part is directed in the downward direction as shown in the figure. These two intensities contain same frequency (f_0). In straightforward direction, three-port clockwise circulator is arranged. When the light signal comes to first port of circulator, circulator sends this signal into its second port. Light from the second port is coupled into the fiber under test (FUT). FUT is subjected to temperature variation. When the input optical signal with frequency (f_0) comes in FUT, the optical signal's frequency will be changed to ($f_0 \pm f_B$). This frequency change (f_B) is known as BFS. Then, the optical signal will be backscattered with this new frequency. This backscattered light passes through the third port of circulator and will be collected by a balanced photodetector (PD) device. And at the same time, the reference light with frequency (f_0) comes from coupler to this balanced PD device. The backscattered (Stokes or anti-Stokes) light and reference light are mixed at an input side of balanced PD device. The beat signal between these two signals is converted into an electrical signal with this balanced PD device and is detected as a BGS with an electrical spectrum analyser (ESA). By modulating the laser frequency with a sinusoidal waveform, correlation peaks are periodically synthesized along the FUT. Typically, FUT length is double to measurement range (distance between two consecutive correlation peaks). It means FUT is completely depending on measurement range. FUT will be decided; in such a way that only one correlation peak is in the FUT. By controlling the optical path difference (e.g., by changing the reference path length or delay), a correlation peak of any order can be located in the FUT.

The measurement range d_m of this setup is expressed as [13]

$$d_m = c/(2nf_m) \tag{1}$$

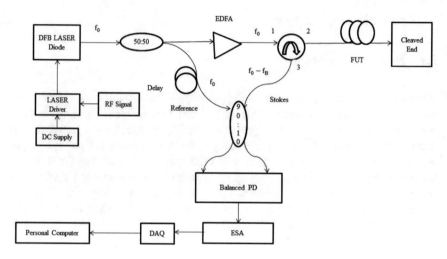

Fig. 1 Schematic setup of BOCDR. *EDFA* Erbium-doped fiber amplifier; *FUT* fiber under test; *PD* photodetector; *ESA* electrical spectrum analyzer; *RF* radio frequency; *DC* direct current; *DAQ* data acquisition

where c indicates velocity of light in vacuum and n indicates refractive index of fiber core. By tuning the modulating frequency (f_m), sensing point or location will be scanned along the fiber to extract the BGS and BFS distributions.

According to theory, when f_m is lower than the Brillouin bandwidth $\Delta\vartheta_B$, the spatial resolution Δz is written by [13]

$$\Delta z = c\Delta\vartheta_B/(2\pi n f_m \Delta f) \tag{2}$$

Here, Δf is the modulation amplitude of the optical frequency. Assume f_m is higher than $\Delta\vartheta_B$ does not contribute to the enhancement of Δz, and that Δf is practically limited to a half of BFS (ϑ_B) of the fiber because of the Rayleigh noise, the limitation of the spatial resolution Δz_{min} is given by [13]

$$\Delta z = c/(\pi n \vartheta_B) \tag{3}$$

The number of effective sensing points N_R, this is an evaluation parameter of this system, and it is calculated by [13]

$$N_R = d_m/\Delta z \tag{4}$$

For achieve a maximum value of N_R, Δf needs to be increased but it must be lower than $\vartheta_B/2$. So that [13],

$$N_{Rmax} = \pi\vartheta_B/(2\Delta\vartheta_B) \tag{5}$$

Brillouin frequency shift (ϑ_B) is linearly dependent on the temperature (T °C). It can be written as [2, 31]:

$$\vartheta_B = \vartheta_{B0} + C_T(T_{unknown} - T_{ref}) \tag{6}$$

Here, ϑ_{B0} indicates reference Brillouin frequency shift (under free temperature change). C_T is thermal coefficient of SMF. Typically, C_T has 0.9876 MHz/°C and ϑ_{B0} has ~11 GHz for ~1550 nm for silica SMF [31]. $T_{unknown}$ and T_{ref} are non-room and room temperatures.

Reference Brillouin frequency shift (ϑ_{B0}) is expressed as [2–3]:

$$\vartheta_{B0} = 2n_{eff}V_a/\lambda_0 \tag{7}$$

where n_{eff} is effective refractive index of SMF fiber, V_a is the sound (acoustic) velocity of phonons in SMF fiber, and λ_0 is the reference or pumping (input) optical light's wavelength.

2.2 Particle Swarm Optimization Algorithm

An optimization algorithm is highly necessary to achieve the best global solution. These algorithms give either global minima or global maxima for particular function or signal. Depends upon our requirements, we can focus on maxima or minima value. This algorithm is called as an evolutionary algorithm. Because its current result fully depends upon its previous result. It helps to produce convergence in advance manner.

In PSO evolutionary optimization algorithm, a group of candidate solutions referred as particles and these are trying to reach an original solution. According to two basic mathematical equations, the optimum solution will be decided. These equations are called as velocity update and position update. In an every iteration or generation, these two basic expressions must be taken. P_{Best} and G_{Best} are called as individual and global best solutions of swarm. These solutions must be calculated in an each generation or iteration.

The update equation for velocity is given by [7, 9]:

$$V_i^{g+1} = WV_i^g + C_1 rand_1\left(P_{Besti} - X_i^g\right) + C_2 rand_2\left(G_{Best} - X_i^g\right) \tag{8}$$

where V_i^g is the velocity of ith particle at gth generation, W is the inertial constant, C_1 and C_2 are the acceleration constants, $rand_1$ and $rand_2$ are an uniformly distributed random number between 0 and 1, X_i^g is the current position of ith particle at gth generation, P_{Besti} is an individual best fitness value of the ith particle, and G_{Besti} is the global best fitness value among all particles.

Here, $1 < C_1$ and $C_2 < 2$ and $W > \left(\frac{C_1+C_2}{2}\right) - 1$

The location of an every particle is modified by the following formulae [7, 9]

$$X_i^{g+1} = X_i^g + V_i^{g+1} \qquad (9)$$

The procedure of PSO is explained as follows:

Step 1: First consider the position X_i^1 and the velocity V_i^1 of an each particle in a random manner. Create a loop with $g_{min} = 1$ and the maximum number g_{max} of generations in an entire search. In this paper, we have taken 10 initial particles and g_{max} as 300.

Step 2: Find out the fitness value $BFS_{Error}(X_i^g)$ of each particle X_i^g for an estimate the local best position P_{Besti} and the global best position G_{Best} up to the current instant.

Step 3: Estimate V_i^{g+1} and X_i^{g+1} values using (8) and (9), respectively, for obtaining the new position and velocity of an each particle.

Step 4: If the current fitness value of P_{Besti} is lower than the previous fitness value of P_{Besti}, replace P_{Besti} with the present one. Else previous value of P_{Besti} will be continued again. Similarly, if the current fitness value of G_{Best} is lower than the previous G_{Best}, replace G_{Best} with the present one. Else previous value of G_{Best} will be continued again.

Step 5: If $g < g_{max}$ then $g = g+1$ and jump to **step 2**, else halt the searching process.

3 Simulated Results and Discussions

In this paper, we present analysis of BOCDR-based DTS system through MATLAB software. These simulation results are obtained through PSO technique.

The basic parameters of this system are optical power with 10 dBm, delay in reference optical path with 2 km, velocity of light with 3×10^8 m/s, optical wavelength with $\sim 1.55\,\mu$m, and FUT (sensing range) with 500 m. Here, some of the parameters in BOCDR-based DTS systems are optimized through PSO evolutionary algorithm. The required simulation parameters of our proposed system are listed in Table 1.

In this proposed system, FUT is used as standard SMF with a length of 500 m. In this FUT length, 6 m section of FUT was heated from 24.8 to 72.5 °C. This 6-m-heated section in this FUT starts at 255 m and ends at 261 m. This structure is shown in Fig. 2. The fiber considered is having a propagation loss of 0.2 dB/km. Temperature and spatial resolutions are performance metrics in this proposed system. Improvements of these metrics are presented in the following sections.

When the temperature variation is applied at specific region of FUT, reference Brillouin frequency shift is modified at that specific region. By the BFS distribution of BOCDR technique, we are extracted or estimated this temperature variation values. In Fig. 3, we have shown the applied temperature variation, and it has been varied from 24.8 to 72.5 °C. All these temperature values have been extracted through respective BFS values. We observed that the temperature is varying over the range from 255 to 261 m of FUT, which is shown in Fig. 3. In this BOCDR-based DTS system, the BFS value has an error of ± 2.3 MHz for an either room temperature or temperature variations. Due to this BFS error, we got ± 2.328 °C fluctuation in the

Table 1 Simulation parameters of proposed (BOCDR-PSO-based DTS) system

Simulated parameters	Symbol	Values
Measurement range	d_m	250 m
Effective refractive index of fiber core	n_{eff}	1.46–1.50
Modulating frequency	f_m	410.96 kHz
Reference BFS	ϑ_{B0}	10806 ± 2.3 MHz
Reference (room) temperature	T_{ref}	27 °C
Thermal coefficient of SMF	C_T	0.9876 MHz/°c
Brillouin bandwidth	$\Delta\vartheta_B$	30 MHz
Modulation amplitude of the optical frequency	Δf	5403 ± 1.15 MHz
Acoustic wave velocity	V_a	5300–5900 m/s
Pumping wavelength	λ_0	1.5–1.55 μm

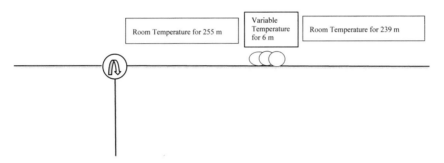

Fig. 2 Structure of SMF under test

extracted temperature over the entire FUT. It can be observed by Fig. 3. This is one of the limitation of the BOCDR-based DTS system. To minimize this BFS error, we have done modifications in BOCDR-based DTS system.

In BOCDR-based DTS system, effective refractive index of fiber, acoustic velocity, and pumping wavelength parameters are responsible for BFS. In order to avoid this BFS fluctuation, we made a link between PSO evolutionary algorithm and BOCDR-based DTS system. By this PSO evolutionary computing algorithm, we optimized above-specified parameters values. In this PSO algorithm, the obtained optimum values of these parameters (n_{eff}, V_a and λ_0) are 1.479, 5482.1 m/s, and 1.523 μm. These optimum values came at tenth-generation process of PSO algorithm. By these optimum values, we got BFS error as $\sim \pm 0.8303$ MHz. It means less temperature error reduced over the length of FUT. It can be observed by Fig. 4. With this BFS error fluctuation, we can say that our proposed system produces ~±0.839 °C temperature resolution. In this temperature extraction process, thermal coefficient of SMF is 0.9876 °C/MHz, which is fixed for the systems.

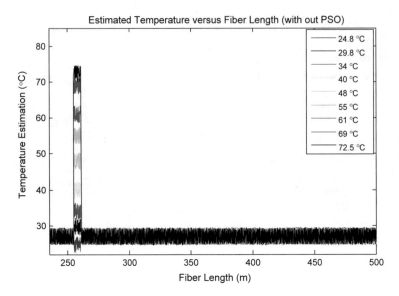

Fig. 3 Estimated temperatures through BFS in BOCDR-based DTS system

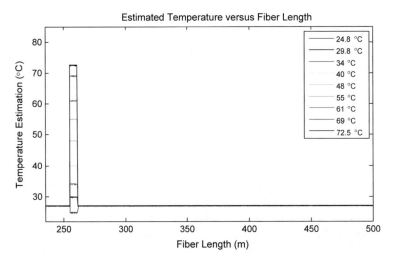

Fig. 4 Estimated temperatures through BFS in proposed system

When the temperature varied from 24.8 to 72.5 °C, the BFS also varied. The relation between BFS and temperature has been shown in Fig. 5. From Fig. 5, we have two linear lines. Among these linear lines, blue color line indicates BOCDR-based DTS system and red color line indicates proposed (PSO-BOCDR)-based system. Over the given range of temperature the proposed system go offers less BFS error as compared with BOCDR based system.

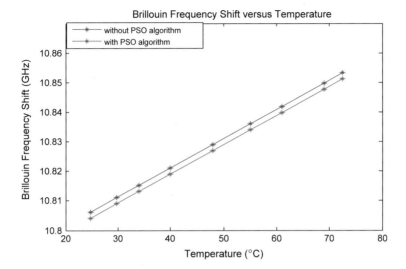

Fig. 5 Estimated BFS for different non-room temperatures in existed and proposed systems

Figure 6 shows spatial resolution of BOCDR-based DTS system. This calculation was carried out by assuming a modulation frequency f_m of 410.96 kHz, optical fiber core refractive index of (n) is 1.46, corresponding to the measurement range d_m of 250 m according to Eq. (1). When the modulation amplitude of the optical frequency Δf was kept to 5403 ± 1.15 MHz and Brillouin bandwidth $\Delta \vartheta_B$ is 30 MHz, according to Eq. (2) resulting spatial resolution is reported as 44.19 cm. According to theory, spatial resolution is calculated by considering the 10 to 90% value of estimated temperature [7]. By this concept, we have shown the spatial resolution of BOCDR-based DTS system in Fig. 6.

By the optimization of effective refractive index of fiber (n_{eff}) and minimization of BFS (ϑ_B) error, we have achieved a spatial resolution of 43.42 cm. Here, the improvement in spatial resolution is ~0.8 cm. This improvement is shown in Fig. 7.

4 Conclusion

To conclude, BOCDR-PSO-based distributed temperature sensing system is presented and analyzed for improving the spatial and temperature resolution. By performing multi-parameter optimization on a few parameters, such as effective refractive index of fiber, pumping wavelength, and acoustic velocity in fiber, we have achieved better temperature resolution and spatial resolution than existed BOCDR-based DTS systems. Here, we have used 500 m length SMF for sensing range. With our proposed DTS system, ~0.839 °C temperature resolution and ~43 cm spatial resolution are achieved and shown through MATLAB simulations. Thus, we believe that

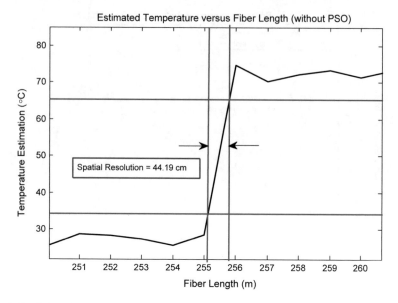

Fig. 6 Spatial resolution in BOCDR-based DTS system

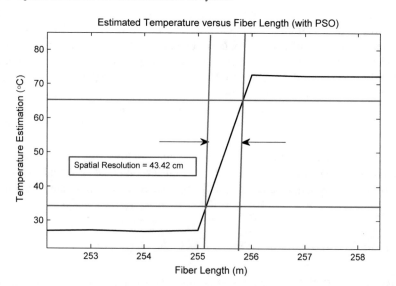

Fig. 7 Spatial resolution in proposed system

our proposed sensing system will be applicable in real-time sensing areas like structural health monitoring in defense applications, in space applications, fire detection.

References

1. Lee, B.: Review of the present status of optical fiber sensors. J. Opt. Fiber Technol. **9**(2), 57–79 (2003)
2. Abdurrahman, G., Gunes, Y., Karlik, S. E.: Spontaneous Raman power and Brillouin frequency shift method based distributed temperature and strain detection in power cables. J. Opt. Fiber Technol. (2008)
3. Yu, J.W., Park Y., Oh, K.: Brillouin frequency shifts in silica optical fiber with the double cladding structure. J. Optic Exp. **10**(19), 996–1002. OSA (2002)
4. Song, M.P., Zhao, B.: Accuracy enhancement in Brillouin scattering distributed temperature sensor based on Hilbert transform. J. Opt. Commun. 252–257. Elsevier (2005)
5. Bernini, R., Minardo, A., Zeni, L.: An accurate high-resolution technique for distributed sensing based on frequency-domain Brillouin scattering. Lett. Photon. Technol. **18**(1), 280–282. IEEE (2006)
6. Soto, M.A., Sahu, P.K., Bolognini, G., Pasquale, F.D.: Brillouin-based distributed temperature sensor employing pulse coding. J. Opt. Sens. **8**(3), 225–226. IEEE (2008)
7. Himansu, S.P., Sahu, P.K.: High-performance Brillouin distributed temperature sensor using fourier wavelet regularized deconvolution algorithm. J. Opto Elect. **8**(6), 203–209. IET (2014)
8. Yanjun, Z., Yu, C., Xinghu, F., Wenzhe, L., Weihong, B.: Spectrum parameter estimation in Brillouin scattering distributed temperature sensor based on Cuckoo search algorithm combined with the improved differential evolution algorithm. J. Opt. Commun. 15–20. Elsevier (2015)
9. Himansu, S.P., Sahu, P.K.: Brillouin distributed temperature sensor employing phase modulation and optimization techniques. J. Opt. Comm. 788–794. Elsevier (2015)
10. Xia, L., Zhao, Q., Chen, J., Wu, P., Zhang, X.: A distributed Brillouin temperature sensor using a single-photon detector. J. Opt. Sens. **16**(7), 2180–2185. IEEE (2016)
11. Mizuno, Y., Zou, W., He, Z., Hotate, K.: Operation of Brillouin optical correlation-domain reflectometry: theoretical analysis and experimental validation. J. Lightw. Technol. **28**(22), 3300–3306. IEEE (2010)
12. Mizuno, Y., Zou, W., He, Z., Hotate, K.: Proposal of Brillouin optical correlation-domain reflectometry (BOCDR). J. Opt. Exp. **16**(16), 12148–12153. OSA (2008)
13. Hayashi, N., Mizuno, Y., Nakamura, K.: Distributed Brillouin sensing with centimeter-order spatial resolution in polymer optical fibers. J. Lightw. Technol. **32**(21), pp. 3999–4003. IEEE (2014)
14. Hayashi, N., Mizuno, Y., Nakamura, K.: First demonstration of distributed Brillouin measurement with centimeter-order resolution based on plastic optical fibers. In: International Conference, pp. 377–379. IEEE (2014)
15. Hayashi, N., Mizuno, Y., Nakamura, K.: Simplified configuration of Brillouin optical correlation-domain reflectometry. J. Photon. Technol. **6**(5). IEEE (2014)
16. Hayashi, N., Mizuno, Y., Nakamura, K.: Alternative implementation of simplified Brillouin optical correlation-domain reflectometry. J. Photon. Technol. **6**(5). IEEE (2014)
17. Taki, M., Soto, M.A., Pasquale, F.D., Bolognini, G.: Long-range BOTDA sensing using optical pulse coding and single source bi-directional distributed raman amplification. J. Optic Sensors IEEE (2011)
18. Cui, Q., Pamukcu, S., Lin, A., Xiao, W., Herr, D., Toulouse, J., Pervizpour, M.: Distributed temperature sensing system based on Rayleigh scattering BOTDA. J. Optic Sensors **11**(2), 399–403. IEEE (2011)
19. Zornoza, A., Sagues, M., Loayssa, A.: Self-heterodyne detection for SNR improvement and distributed phase-shift measurements in BOTDA. J. Lightw. Technol. **30**(8), 1066–1072. IEEE (2012)

20. Lopez-Gil, A., Angulo-Vinuesa, X., Dominguez-Lopez, A., Martin-Lopez, S., Gonzalez-Herraez, M.: Simple baseband method for the distributed analysis of Brillouin phase-shift spectra. Lett. Photon. Technol. **28**(13), 1379–1382. IEEE (2016)
21. Li, Z., Yan, L., Shao, L., Pan, W., Luo, B., Liang, J., He, H.: Coherent BOTDA sensor with single-sideband modulated probe. Light. J. Photon. Technol. **8**(1). IEEE (2016)
22. Garus, D., Krebber, K., Schliep, F.: Distributed sensing technique based on Brillouin optical-fiber frequency domain analysis. Opt. Lett. **21**(17), 1402–1404 (1996)
23. Minardo, A., Bernini, R., Zeni, R.: Distributed temperature sensing in polymer optical fiber by BOFDA. IEEE Photon. Technol. Lett. **24**(4), 387–390. IEEE (2014)
24. Hotate, K., Hasegawa, T.: Measurement of Brillouin gain spectrum distribution along an optical fiber using correlation-based technique-proposal experiment and simulation. Trans. Electron., **E83-C**(3), 405–412. IEICE (2000)
25. Song, K.Y., Hotate, K.: Brillouin optical correlation domain analysis in linear configuration. Lett. Photon. Technol. **20**(24), 2150–2152. IEEE (2008)
26. Ramachandra Murthy, K.V.S., Ramalinga Raju M., Govinda Rao, G.: Comparison of three evolutionary algorithms: GA, PSO, and DE. In: International Conference of INDICON. IEEE (2010)
27. Voratas, K.: Comparison of three evolutionary algorithms: GA, PSO, and DE. J. Ind. Eng. Manag. Syst. **11**(3), 215–223. IEEE (2012)
28. McCall, J.: Genetic algorithms for modelling and optimization. J. Comp. Appl. Math. 205–222. Elsevier (2005)
29. Jiang, T., Su, X., Han, W.: Optimization of support scheduling on deck of carrier aircraft based on improved differential evolution algorithm. In: 3rd International Conference on Control Science and System Engineering, pp. 136–140. IEEE (2017)
30. Blum, C.: Ant colony optimization: introduction and recent trends. J. Phys. 353–373. Elsevier (2005)
31. Horiguchi, T., Shimizu, K., Kurashima, T., Tateda, M., Koyamada, Y.: Development of a Distributed Sensing Technique Using Brillouin Scattering. J. Lightw. Technol. **13**(7), 1296–1302. IEEE (1995)

Multi-agent Navigation and Coordination Using GA-Fuzzy Approach

Buddhadeb Pradhan, Diptendu Sinha Roy and Nirmal Baran Hui

Abstract This paper presents the coordinated navigation problem of multiple wheeled robots. Two motion planners such as PFM and GA-tuned FLC are combined with a coordination strategy block to solve such problems. Performances of the developed approaches are tested through computer simulations. A total hundred numbers of scenarios are taken to show the efficacy of the proposed navigation schemes. GA-tuned FLC has wholly outperformed the PFM in most of the situations. Also, with the increase in some robots, coordination count has increased, and the need for the strategies was prominent. It is experienced that proposed coordination strategy along with the developed motion planners results in a time-optimal and realistic solution to the discussed problem.

Keywords Multi-agent system (MAS) · Mobile robot navigation
Potential field method (PFM) · Fuzzy logic control (FLC)
Genetic algorithm (GA) · Coordination strategies

1 Introduction

The motion planning problem of multiple mobile robots is one of the most exciting research areas in robotics. The primary challenge here is to sense neighbouring information and control the movement of robots. In a dynamic environment, a mobile robot must have fundamental ability to find a collision-free path from a starting point

B. Pradhan · N. B. Hui (✉)
Mechanical Engineering, National Institute of Technology, Durgapur, Durgapur, India
e-mail: nirmalhui@gmail.com

B. Pradhan
e-mail: buddhadebpradhan@gmail.com

D. S. Roy
Computer Science & Engineering, National Institute of Technology, Meghalaya, Shillong, Meghalaya, India
e-mail: diptendu.sr@gmail.com

© Springer Nature Singapore Pte Ltd. 2019
J. C. Bansal et al. (eds.), *Soft Computing for Problem Solving*, Advances in Intelligent Systems and Computing 817, https://doi.org/10.1007/978-981-13-1595-4_63

to a planned goal location. Some researchers are working in this field, and at least one researcher in each robotics laboratory across the world is trying to solve the mobile robot navigation problem. Some of them are discussed here. Padhy et al. [1] proposed a PID-based controlled rule to obtain consistent linear and rotational velocities of the mobile robot. Peng and Akella [2] considered it as a mixed integer nonlinear programming problem. But this approach is not at all suitable for real-time systems, as the environment is highly dynamic. The soft computing methods such as neural network [3], fuzzy logic, genetic algorithm [4] and their different combinations [5] are also available in the literature. Quite a few researchers have successfully applied the neural network [3] to develop the motion planning algorithm of mobile robots. Authors in [6] considered GA-FLC and GA-NN approach for solving path planning problem in a static environment. Very few researchers as like [7] did real experiments for solving navigation problems. However, they tested in a static environment. The important literature related to authors' research field is surveyed according to their capabilities and presented in Table 1.

In the Table 1, tick ($\sqrt{}$) mark indicates the compliance of the corresponding capability and cross (x) marks are meant for non-compliance of the same ability. The table shows that lots of cross (x) marks are there at "coordination issues" columns. It indicates that the coordination problem among the multiple mobile robots is still a challenging research issue for robotics researchers. The primary challenge is to convert the cross (x) marks into tick ($\sqrt{}$) mark. In the present research, different approaches have been tried to solve the motion planning problem as well as coordination of multi-agent systems in a typical dynamic environment. The remaining part of the paper is structured as follows: in Sect. 2, developed cooperative multi-agent systems have been studied. Results and discussions, as well as comparison with previous research works, are presented in Sect. 3. Conclusions and the scopes for future work are discussed in Sect. 4.

2 Problem Formulation of Multi-agent Systems

In MAS, all the robots are trying to find their collision-free paths while traversing from a predefined position to a goal point. Depending upon the situations, a particular robot may see many collision-free paths in the environment. Keeping that in mind, the authors plan to identify the time-optimal path. Figure 1 shows a typical problem scenario, in which four two-wheeled differential drive robots are navigating in a common dynamic environment.

Figure 2 shows the sequence of solution methodology applied in this paper. Starting and goal positions of any planning robot (say, R1) are denoted by S and G. Initially, all the robots are moving with the same velocity, but in a different direction. In case, a robot finds any other robot as a critical robot (which has a chance of making collision with the planning robot), the motion planner is activated or else the robot moves towards the goal in a straight path with the maximum velocity. Critical robots are identified based on two inputs: distance from the planning robot and angle from

Table 1 Brief description of important literature

References	A	B	C	D	E	F	G
Hui et al. [8]	√	×	C1	√	×	×	FLC, NN, PFM
Abadi and Khooban [9]		×		×	×	×	FLC, PSO
Chakraborty and Ghosal [10]	√	×	C1	×	×	×	
Wang and Yang [11]	√	×	C1	√	×	×	FLC, NN, NN-FLC
Ren et al. [12]	√	×	C1	√	√	√	FLC
Yousfi et al. [13]	√	√	C1	√	×	×	FLC
Qing et al. [14]	√	√	C2	×	×	×	FLC
Boubertakh et al. [15]	√	√	C1	√	×	×	FLC, Reinforcement learning
Pandey et al. [16]	×	×	–	×	×	×	×
Fujimori et al. [17]	√	√	C1	√	√	×	Adaptive navigation technique
Pratihar et al. [18]	√	√	C1, C2	√	×	×	GA-fuzzy
Wang et al. [19]	×	√	C2	√	√	×	Inverse optimal control strategy
Maa et al. [20]	√	√	C2	√	×	×	Time-varying nonlinear programming, SA-PSO

A Kinematic constraints of the robot
B Dynamic constraints of the robot
C Environmental condition (C1 for dynamic and C2 for static)
D Navigation capability
E Coordination capability
F Real experiments
G Techniques developed/applied

the goal line of the planning robot. Less distance and angle indicate the robot is more critical to the planning robot. At a time instance, only one robot is treated as the most critical one, and the motion of the planning robot is obtained based on that particular robot. The developed motion planner calculates the acceleration (a) and deviation (θ) of the planning robot required to avoid collision with the critical robot. This is repeated for all the robots. So, two robots mutually can treat each other as the critical robot. Accordingly, the motion of both the two robots may be hampered. Therefore, there is a need for a coordination strategy to identify the robot which will follow the planned motion and other will sacrifice the same.

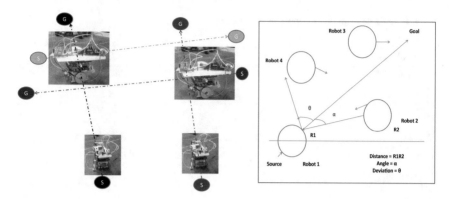

Fig. 1 Motion planning scenario of four navigating robots

2.1 Developed Motion Planning Methods

Two different approaches are developed for this purpose. These are briefed below.

2.1.1 Approach 1: Potential Field Method

It is a well-known and prevalent method [9]. The robot moves along the resultant direction of the attractive and repulsive force with a velocity proportional to the resultant force. Interested readers may go through [9] for detailed information of this method.

2.1.2 Approach 2: Manually Constructed Fuzzy Logic Controller

In the field of mobile robot navigation, this technique generated additional significance among researchers and is also a popular technique. In this paper, inputs and outputs are represented using linguistic terms as shown below.

- *Distance*: Very Near, Near, Far and Very Far.
- *Angle* and *deviation*: Left, Ahead Left, Ahead, Ahead Right and Right.
- *Acceleration*: Very Low, Low, High and Very High.

Therefore, there will be 20 different rules. One such rule may look like the following.

If *the distance* is Very Near and *angle* is Left, then *acceleration* is Very Low and *deviation* is Right.

In this approach, both database and rule base are manually constructed.

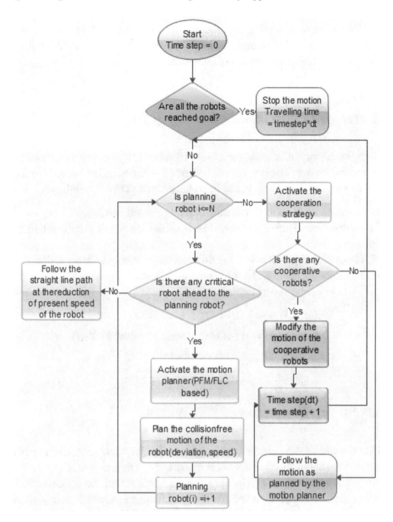

Fig. 2 Flowchart of motion planning scheme

2.1.3 Approach 3: GA-Tuned Fuzzy Logic Controller

A manually constructed knowledge base is not an optimal one. Therefore, a binary-coded GA is used to evolve the knowledge base in this approach. A GA string consisting of 140 bits is considered to represent the KB of the FLC as shown below:

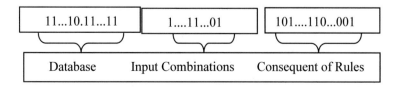

First 40 bits are representing the database (10 bits for the half-base width of the two inputs and two outputs), and the next 100 bits are used to identify the output combination for a particular input combination.

2.2 Developed Coordination Strategy (CS)

There may be a possibility that the planning robot might become critical robot to another planning robot. The present navigation problem might be solved from general cooperative attitude, and there could be at least three possible solutions.

- **No coordination**: No robot will sacrifice its planned motion.
- **Full coordination**: Both the planned and critical robot will sacrifice their planned motion.
- **Partial coordination**: One will sacrifice its motion but the other will not. Identification of the cooperative robot is made as follows.

$$\text{if(crit(i)} == j \;\&\&\; crit(j) \;==\; i)$$
$$\text{coopcount} \;=\; 1; \;\text{if (deviation(i)} > \text{deviation(j))}$$
$$\text{cooprobot} = j; \text{else cooprobot} = i$$

3 Results and Discussions

Two different cases are considered with eight and twelve robots navigating in a grid of $40 \times 40 \, \text{m}^2$. Maximum speed of a robot is restricted to 4 m/s, and acceleration is varied between 0.005 and 0.05 m/s^2. The time interval is 2 s, and the maximum distance that a robot can travel in a step is 8 m. The performance is tested through computer simulations against one hundred random scenarios. All the scenarios are completely different by starting and goal positions of the robots. All three approaches have been applied; however, to get the best result, it is necessary to conduct GA parametric study. With the optimal GA parameters, tuning of FLC is carried out. GA parametric study has been done on a step-by-step method.

- **Step 1**: Crossover probability (p_c), population size (N) and generation number (G_{max}) are fixed to 0.9, 100 and 100, respectively, and mutation probability (p_m) has been varied from 0.001 to 0.011 in steps of 0.001, and optimal mutation probability (p_m^*) is noted corresponding to the lowest fitness.
- **Step 2**: The values of p_c, $p_m = p_m^*$ and G_{max} have been taken as 0.9, 0.002 and 100, respectively, and the population size is varied from 50 to 150 in steps of 10, and optimal population size (N*) is noted.
- **Step 3**: In this stage, p_c, $p_m = p_m^*$ and N=N* are set equal to 0.9, 0.002 and 100, respectively, and the experiments are carried out after setting the number of

Table 2 Average travelling time of eight robots for different approaches

Robots	Approach 1		Approach 2		Approach 3	
	Without CS	With CS	Without CS	With CS	Without CS	With CS
R1	59.77	59.38	62.05	61.68	61.59	61.32
R2	68.48	68.45	71.68	71.68	70.00	69.69
R3	62.54	61.72	64.96	64.18	64.14	63.83
R4	54.36	53.94	55.76	55.34	55.88	55.77
R5	103.30	102.86	104.87	104.43	105.31	104.86
R6	82.55	82.44	84.69	84.60	84.16	83.98
R7	115.02	114.91	115.99	115.89	116.34	116.12
R8	82.34	82.21	84.99	84.86	84.06	83.54
Average	**78.55**	**78.24**	**80.62**	**80.33**	**80.18**	**79.89**
Std. dev.	**21.61**	**21.70**	**21.26**	**21.35**	**21.60**	**21.56**

maximum generations at different values (starting from 50 to 150 in steps of 10), and optimal generation number $(G_{max}*)$ is obtained.

- After successful completion of parametric study, the optimal parameter values are found to be: population size $(N*) = 70$, generation $(G_{max}*) = 60$ and mutation probability $(p_m*) = 0.006$.

3.1 Motion Planning of Eight Robots

Eight robots are sharing a common environment. Average travelling time taken by all the eight robots following three different approaches for one hundred different scenarios is presented in Table 2. Average travel time using PFM for all the robots in 100 situations was found to be 78.55 s for non-cooperation case and that of cooperation case it is 78.24 s. Similarly, the average travelling time following Approach 2 is recorded as 80.62 s for non-cooperation and 80.18 s for cooperation case and Approach 3, and those are 80.18 s for non-cooperation and 79.89 s for cooperation. Standard deviation (**Std. Dev.** denoted as in Table 2) also has been calculated to show the efficacy of the simulation results.

Finally, the movement of eight robots for a particular scenario is presented in Fig. 3. It has been observed that cooperation scheme has been fired so many times out of those 100 scenarios indicating the need for such a strategy. It is true that the need for such a strategy is not there for all the scenarios. However, the frequency is very high. For example, Robot 1 has cooperated in 16 scenarios following Approach 1 (refer to Table 3). It is also noted that the average cooperation count of the robot for Approach 3 is the highest and Approach 1 is the lowest. Therefore, Approach 3 will provide a more realistic solution.

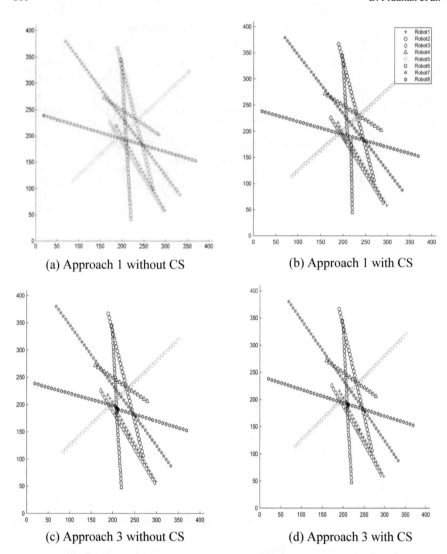

(a) Approach 1 without CS

(b) Approach 1 with CS

(c) Approach 3 without CS

(d) Approach 3 with CS

Fig. 3 Movement of eight robots for a particular best scenario

3.2 Motion Planning of Twelve Robots

In this case, twelve different robots are allowed to move in a common environment. Average travelling time for all the twelve robots in 100 different scenarios is presented in Table 4. These are found to be 69.06, 77.03 and 72.77 s for the Approaches 1, 2 and 3, respectively, without any cooperative strategies compared to 68.35, 75.17, 71.11 s with cooperative strategies. Average travelling time following FLC-based approaches has come out to be more, because most of the time majority of the robots

Table 3 Cooperative count of a robot

	R1	R2	R3	R4	R5	R6	R7	R8
PFM	12	6	6	7	11	7	7	18
Fuzzy	13	5	2	7	14	5	7	17
GA-fuzzy	12	12	13	11	23	20	17	29

Table 4 Average travelling time of twelve robots for different approaches

Robots	Approach 1		Approach 2		Approach 3	
	Without CS	With CS	Without CS	With CS	Without CS	With CS
R1	58.27	56.58	67.79	64.62	63.63	61.57
R2	68.49	68.46	76.74	75.34	71.14	70.17
R3	60.47	58.07	70.33	67.65	65.53	64.08
R4	53.91	52.52	64.25	61.16	57.94	56.06
R5	101.97	101.92	106.92	106.10	106.61	105.21
R6	82.66	82.14	89.42	87.76	85.99	84.29
R7	114.68	114.23	116.11	115.54	117.96	116.39
R8	82.36	82.15	90.20	88.67	85.57	83.83
R9	53.86	54.29	61.37	60.56	56.10	54.37
R10	43.05	43.55	52.18	51.22	44.74	43.17
R11	57.22	54.62	65.42	62.45	63.34	61.29
R12	51.72	51.72	63.64	60.96	54.73	52.90
Average	**69.06**	**68.35**	**77.03**	**75.17**	**72.77**	**71.11**
Std. dev.	**21.938**	**22.12**	**19.58**	**20.09**	**22.07**	**22.16**

Table 5 Cooperative count of a robot

	R1	R2	R3	R4	R5	R6	R7	R8	R9	R10	R11	R12
PFM	13	9	11	10	16	9	12	18	8	8	8	10
Fuzzy	13	10	7	10	18	8	10	19	5	4	8	9
GA-fuzzy	14	24	14	18	29	17	19	37	11	15	25	12

has taken more turning compared to the Approach 1. Also, numbers of the firing of the cooperative strategies are more for Approach 3 compared to the Approach 1.

Movement of the twelve robots for a particular scenario is shown in Fig. 4. As the robot number increases, a number of cooperation strategies also have been increased. A number of times the cooperation strategies have been fired are presented in Table 5.

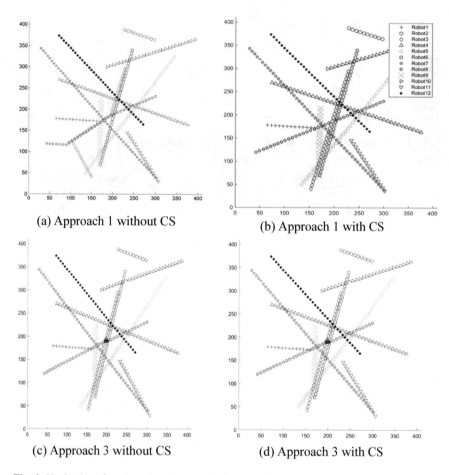

(a) Approach 1 without CS

(b) Approach 1 with CS

(c) Approach 3 without CS

(d) Approach 3 with CS

Fig. 4 Navigation of twelve robots for a particular scenario

3.3 Computational Complexity

Simulation time (in seconds) for the three approaches is presented in Table 6. Approach 3 is found to be computationally expensive one among all the three approaches, and it increases with the increase in a number of robots.

3.4 Comparison with Others' Work

In the present work, GA-fuzzy system has been developed to solve the navigation problem of multiple mobile robots. Hui et al. [8] developed some SC-based approaches for the same. But they have considered motion planning of single robot,

Table 6 Computation time (in seconds) calculated for simulation

Approaches	Without CS		With CS	
	Robots		Robots	
	8	12	8	12
Approach 1	2037.7	2251.3	2096.8	2245.3
Approach 2	2163.2	2278.9	2218.1	2527
Approach 3	4478.18	12,119	8388.25	10,968

not the multiple robots. Abadi et al. [9] used non-optimal FLC and PSO, but they have not made any real experiments. Chakraborty et al. [10] used traditional methods to solve the MAS problem without any coordination. Researchers of [12–14] have considered FLC as their motion planning algorithm. Fujimori et al. [17] suggested adaptive navigation techniques for planning the motion of a robot in the presence of static obstacles. Boubertakh et al. [15] made a comparison between the FLC-based approach to reinforcement learning technique.

In the present work, GA-fuzzy approach has been developed with PFM too. By using the GA-fuzzy technique, the multiple robots can reach their path using collision-free and optimal time way as coordination issue among the moving robots has been resolved.

4 Conclusion

By applying potential field method and fuzzy logic technique as well as GA-tuned FLC approach, multiple mobile robots' navigation is demonstrated in the present work. Cooperation scheme has also been deployed to cooperate between robots. With the incorporation of cooperation strategy, one particular may be inferior in terms of travelling time but gives rise to the overall improvement in the average travelling time of all the robots. FLC has been constructed manually according to the author's concept of the particular problem. So it is evident that depending on the rule base performance of FLC has also been measured. But it may not be the best possible one. The performance of FLC approach is worst on increasing the number of robots. Though it is no way optimal, average travelling time will not be improved. But cooperation scheme has been successfully implemented on non-optimal FLC also. The primary challenge was to solve coordination problem as quite a few researchers have solved this. The computational complexity of all the approaches has also been measured. In some cases, it may be high depending on CPU performance. In such cases to get better performance, a cloud-based framework can be considered for more numbers of robots. The game-theoretic technique may be applied in future for strategy building to obtain the best cooperation between robots. Authors are presently working in this direction.

References

1. Padhy, P.K., Sasaki, T., Nakamura, S., Hashimoto, H.: Modeling and position control of mobile robot. In: Proceedings of 11th IEEE International Workshop on Advanced Motion Control, Nagaoka, Japan, pp. 100–105 (2010)
2. Peng, J., Akella, S.: Coordinating the motion of multiple robots with kinodynamic constraints. In; Proceedings of IEEE International Conference on Robotics and automation, Taipei, Taiwan, pp. 4066–4073 (2003)
3. Low, K.H., Leow, W.K., Ang, M.H., Jr. (2002) Integrated planning and control of mobile robot with self-organizing neural network. In: Proceedings of IEEE International Conference on Robotics and Automation (ICRA'02), Washington, DC, pp. 3870–3875
4. Thomaz, C.E., Pacheco, M.A.C., Vellasco, M.M.B.: Mobile robot path planning using genetic algorithms. In: Mira, J., Snchez-Andrés, J.V. (eds.) Foundations and Tools for Neural Modeling, pp. 671–679. Berlin: Springer
5. Hassanzadeh, I., Sadigh, S.M.: Path planning for a mobile robot using fuzzy logic controller tuned by GA. In: Proceedings of 6th International Symposium on Mechatronics and its Applications (ISMA'09), Sharjah, UAE, pp. 1–5 (2009)
6. Khelchandra, T., Huang, J., Debnath, S.: Path planning of mobile robot with neuro-genetic-fuzzy technique in static environment. Int. J. Hybrid Intelligen. Syst. 11(2), 71–80 (2014)
7. Liu, Q., Lu, Y.G., Xie, C.X.: Optimal genetic fuzzy obstacle avoidance controller of autonomous mobile robot based on ultrasonic sensors. In: Proceedings of IEEE Int. Conference on Robotics and Biomimetics (ROBIO'06), Kunming, China, pp. 125–129 (2006)
8. Hui, N.B., Mahendar, V., Pratihar, D.K.: Time-optimal, and collision-free navigation of a car-like mobile robot using neuro-fuzzy approaches. Elsevier Fuzzy Sets Syst. 157(16), 2171–2204 (2006)
9. Abadi, D.N.M., Khooban, M.H.: Design of Optimal Mamdani Type Fuzzy Controller for Nonholonomic Wheeled Mobile Robots. Journal of King Saud University-Engineering Sciences 27(1), 92–100 (2015)
10. Chakraborty, N., Ghosal, A.: Kinematics of Wheeled Mobile Robots on Uneven Terrain. ELSEVIER Mechanism and Machine Theory 39(12), 1273–1287 (2004)
11. Wang, X., Yang, S.X.A.: Neuro-Fuzzy approach to obstacle avoidance of a nonholonomic mobile robot. In: IEEE/ASME International Conference on Advanced Intelligent Mechatronics, Japan, pp. 29–34 (2003)
12. Ren, L., Wang, W., Du, Z.: (2012) A New Fuzzy Intelligent Obstacle Avoidance Control Strategy for Wheeled Mobile Robot. IEEE Int. Conference on Mechatronics and Automation (ICMA), China, pp. 1732–1737
13. Yousfi, N., Rekik, C., Jallouli, M., Derbel, N.: Optimized Fuzzy controller for mobile robot navigation in a cluttered environment. In: IEEE 7th International Multi-Conference on Systems, Signals and Devices, pp. 1–7
14. Qing-yong, B., Shun-ming, L., Wei-yan, S., Mu-jin, A.: A Fuzzy behavior-based architecture for mobile robot navigation in unknown environments. In: IEEE International Conference on Artificial Intelligence and Computational Intelligence, China, pp. 257–261
15. Boubertakh, H., Tadjine, M., Glorennec, P., Labiod, S.: A simple goal seeking navigation method for a mobile robot using human sense. Fuzzy Logic Reinforce Learn. J. Automat. Control 18(1), 23–27 (2008)
16. Pandey, A., Pandey, S., Parhi, D.R.: Mobile robot navigation and obstacle avoidance techniques: a review. Int. Robot. Automat. J. 2(3), 00022 (2017). https://doi.org/10.15406/iratj.2017.02.00023

17. Fujimori, A., Teramoto, M., Nikiforuk, P., Gupta, M.: Cooperative collision avoidance between multiple mobile robots. J. Robot. Syst. **17**(7), 347–363 (2000)
18. Pratihar, D.K., Deb, K., Ghosh, A.: A genetic-fuzzy approach for mobile robot navigation among moving obstacles. Int. J. Approx. Reason. **20**, 145–172 (1999)
19. Wang, J., Xin, M.: Distributed optimal cooperative tracking control of multiple autonomous robots. Robot. Autonom. Syst. **60**(4), 572–583 (2012)
20. Ma, Y., Hongwei, W., Xie, Y., Guo, M.: Path planning for multiple mobile robots under the double—warehouse. Informat. Sci. **278**, 357–379 (2014)

A Survey on Pareto-Based EAs to Solve Multi-objective Optimization Problems

Saykat Dutta and Kedar Nath Das

Abstract Most of the real-world optimization problems have multiple objectives to deal with. Satisfying one objective at a time may lead to the huge deviation in other. Therefore, an efficient tool is required which can handle multiple objectives simultaneously in order to provide a set of desired solutions. In view of this, multi-objective optimization (MOO) attracts the attention of the researchers since last few decades. Many classical optimization techniques have been proposed by the researchers to solve the multi-objective optimization problems. However mostly, the gradient-based approaches fail to handle complex MOO problems. Hence, as an alternative, researchers have shown their interest toward population-based optimization approaches to solve the MOO problems and come up with convincing results even in the complex environment. Evolutionary algorithms (EAs), which are the first in the group of population-based approach, enjoy almost a decade in providing the solutions to MOO problems. The real challenge is to achieve the set of solutions called Pareto-optimal set. The smooth landing on such set is only possible if there exists diversified solution in the population. Due to the continuous effort, there is a gradual development in the proposition of various efficient Pareto-based approaches in the literature to solve MOEAs. A critical review of those approaches is being carried out in this present study.

Keywords Multi-objective optimization · Pareto optimality · Genetic algorithms
Evolutionary algorithms · Archive-based algorithms

S. Dutta (✉) · K. N. Das
Department of Mathematics, National Institute of Technology,
Silchar, Silchar 788010, Assam, India
e-mail: saykatdutta@gmail.com

K. N. Das
e-mail: kedar.iitr@gmail.com

© Springer Nature Singapore Pte Ltd. 2019
J. C. Bansal et al. (eds.), *Soft Computing for Problem Solving*, Advances in Intelligent
Systems and Computing 817, https://doi.org/10.1007/978-981-13-1595-4_64

1   Introduction

Handling single-objective optimization problem is rather easier than solving MOO problems. However, many real-world decision-making problems engage with multiple objectives simultaneously [1]. Instead of getting a single solution, a set of solutions (namely Pareto-optimal solution) is expected in MOO scenario. Over the time, researchers successfully proposed several classical techniques to handle MOPs [2]. All those techniques work with the help of some utility function (sometimes called weight or preference). Those preferences are provided by the decision makers. According to the given preferences, researchers solve the optimization dilemma and handed over the result to the decision makers. Difficulty arises while the preferences are unknown and the objectives are conflicting in nature. The difficulty level increases when the search space is non-convex and the objective functions are highly nonlinear. In such cases, the classical techniques become handicapped due to the lack of auxiliary knowledge of the functions and sometimes provide worst solution to the problems in hand. Therefore, quite a good number of evolutionary algorithm (EA)-based frameworks have been cited in the existing literature.

Initially, EAs mainly suffer from a difficulty to compare two solutions, as finding fitness of a probable solution is itself a difficult task due to the involvement of multiple objectives. This issue is probably addressed first during 1984 in a dissertation [3]. Later in [4], Schaffer proposed an optimization technique to solve multiple objectives by evolutionary algorithm which is termed as vector evaluated genetic algorithm (VEGA) and provides a new avenue on research community. Further, Schaffer pointed out that after some generations it suffers from speciation. To overcome this problem, Goldberg came with a new idea of non-dominated sorting procedure and niching techniques for maintaining the diversity. Sometimes, due to the crossover and mutation operators of genetic algorithm (GA), inefficient solutions may occur in the population. In order to avoid such situation, Rudolph introduced the mechanism of 'elitism,' which has the inherent mechanism of retaining the better solutions obtained so far.

Solving the multi-objective evolutionary algorithm (MOEA) has three objectives such as (i) convergence (to find the set of solutions close to true Pareto front), (ii) diversity (to find the well-spread solutions on the search space), and (iii) coverage (to cover the maximum Pareto front). Due to their contradictory behavior, it is quite a tough job to care for all the objectives simultaneously. In spite of that, various MOEAs have already been proposed in this literature. However, those MOEAs can be broadly categorized into three parts as shown in Fig. 1.

In this study, different types of Pareto-based MOEAs have been discussed in Sect. 2. Section 3 introduces various computational tools and software used in the literature for solving MOEAs. The conclusion of the paper is drawn in Sect. 4, along with highlighting a few of the future scopes of research.

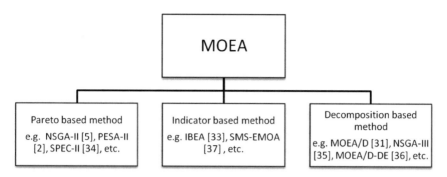

Fig. 1 MOEAs and its varieties

2   Survey of Literature

In the literature, various non-elitist MOEAs are proposed in [5–7]. Among them, non-dominated sorting genetic algorithm (NSGA) [7] has received much popularity due to its niching technique to maintaining the diversity in the population resulting in better solutions. However, these methods are being dominated by other methods, where the 'elitism' is used. During the year 1999, an elite-preservation mechanism is introduced in MOEAs, which is being treated as the improved version of MOEAs. Similarly, researchers suggested the technique of maintaining an external population to accumulate the non-dominated solutions, namely archive-based approach. This aims to create Pareto-based (or dominance-based) MOEAs, which helps in providing an easy platform for coding and simulation for achieving improved solutions.

Therefore, keeping in view the above, a survey of the literature of such techniques (those use elitism, archive-based technique, and Pareto-based approaches) proposed since 1999 is being acknowledged in the following subsections. The brief outlines of these methods are presented in Table 1. In this table, d represents the user-defined value, M represents the number of objectives, and N represents the number of variables.

2.1   PAES

Pareto archived evolutionary strategy (PAES) was grown up when Knowles et al. [8] investigated a network design problem telecommunication (specifically, on routing calls through a sparse network). They observed local search strategy has more potential than population-based strategy in solving single-objective model of network design problem. The same idea is being implemented later to solve the same problem in multi-objective environment. Gradually, this approach attracted the attention of the research community for a longer period. The methodology, which differs PAES from any existing algorithms, is the involvement of the local search strategy

Table 1 Survey of archive-based MOEAs

Year	Proposed algorithm	Key contribution	Beats the algorithms	Computational complexity	Concluding remark
1999	PAES [8]	PAES uses $(1+1)$ evolutionary strategy and maintains the diversity using the grid-based techniques	NPGA [28, 5]	$O(MN^2d)$	The functions having non-uniform search space can be solved by PAES But computationally inefficient if the d is large
1999	SPEA [10]	By using two interacting populations, fitness of an individual is assigned Instead of distance-based, Pareto-based pruning method is introduced	VEGA [4], NPGA, HLGA [29]	$O(MN^2)$	More diverse set of solution achieved because of the clustering approach for pruning the solution If the large-sized archive is used, SPEA is unable to reach the Pareto-optimal front. Therefore, if any individual from archive dominates more solutions, gets a worse fitness
2000	PESA [13]	It borrows several concepts from both PAES and SPEA and used hyper-grid-based techniques for both the selection and diversity schemes	SPEA, PAES	NA[a]	SPEA outperforms other state-of-the-art problems in the test suit
2000	NSGA-II [16]	It introduces a new non-dominated sorting algorithm which needs very less computation. It uses elitism in a different fashion. Also, it proposes a new technique for pruning the extra solutions named crowding distance sorting	SPEA, PAES	$O(MN^2)$	In this proposed crowding distance technique, no extra parameter needs to be fixed. This makes the algorithm more flexible
2001	PESA-II [14]	Region-based selection scheme is proposed, where the fitness is assigned to the hyperboxes in the objective space	SPEA, PAES, PESA	NA[a]	With the help of region-based selection, one can choose three times more likely isolated individuals

(continued)

Table 1 (continued)

Year	Proposed algorithm	Key contribution	Beats the algorithms	Computational complexity	Concluding remark
2001	SPEA2 [11]	A new superior fitness scheme is introduced Environment selection and kth nearest neighbor (for diversity) is proposed here	SPEA, PESA, NSGA-II (TIE)	$O(MN^2)$	SPEA2 and NSGA-II show almost same performance in the test suit. But in a higher dimensional space, SPEA2 provides better convergence
2004	SPEA2+ [12]	For maintaining the search ability and better diversity neighborhood, crossover mechanism and archive mechanism are introduced over SPEA2, respectively	SPEA2, SPEA2	$O(MN^2)$	It provides a better variety of individuals in the genotype and maintains a widespread solution with better precision
2006	The two-archive algorithm [20]	Introduces the concept of two types of archive such as (i) a non-dominated solution with domination and (ii) a non-dominated solution without domination in a MOEA	PESA	NA[a]	It provides more diverse solution than PESA and has potential to solve many-objective (four or more objective) problems
2008	AMGA [17]	It works with a small population and maintains an archive which stores the non-dominated solutions. The crowding distance operator is modified for this particular problem. It also uses the SBX crossover operator	FastPGA [30], NSGA-II (for two objective problems)	$O(TMA^2 \log A/N)$	Self-governing tuning of the parameters is engaged, which are independent of initial population size, archive size, working population size, and the probability of mutation rate. Hence, it is free of fine-tuning of parameters

(continued)

Table 1 (continued)

Year	Proposed algorithm	Key contribution	Beats the algorithms	Computational complexity	Concluding remark
2011	AMGA2 [18]	Parent population is chosen based on the domination and rank. Mating pool is selected from both of archive and parent	NSGA-II, FastPGA, GDE3 [31], AMGA	$O\left(T N_a^2 \log N_a / N_p\right)$	AMGA2 uses a large external population and small working population for less function evaluations
2013	IPESA-II [15]	After entering all the entity in the internal population, the archive is maintained. Some special cure has been taken on the boundary solution. Eliminates the worst performed entity from the crowded hyperbox	NSGA-II, SPEA2, IBEA[], ε-MOEA [16], TDEA [32], PESA-II	NA[a]	The three major changes done in the algorithm help toward more uniformity, extensity, and convergence in the solution space
2015	ASMiGA [19]	A new MOEA has been proposed with new environmental and mating strategy. Moreover, a new crossover mechanism (DE-3) has also been introduced in the algorithm	AMGA2, MOEA/D-DE [33], AbYSS, NSGA-II, SPEA2	$O(M N^2)$	Exploration is reduced in less probable search region for the environmental selection and enhanced the exploration in more probable search region for the mating election Comparison in [19] concludes that ASMiGA is one of the best algorithms in the literature. It has the potential to handle constrained high-dimensional MOOPs too

[a]NA: Not applicable

and the archive strategy. PAES kept all the non-dominated solutions in an external population called 'archive.' PAES is recognized as a $(1 + 1)$ evolutionary strategy. This means that mating is executed between two parents to produce the single offspring where the first one is from EA population and the second one is from the archive. So, the fitness of the parent and the offspring is compared for the process of domination. If the parent dominates the offspring, it needs to find the new muted solution. If the parent is dominated by the offspring, then the offspring is a part of the archive. But the difficulties arise when both of them are non-dominated. In such case, the fitness of both offspring and archive population is compared. Again, if the member of archive dominates the offspring, then the offsprings are rejected. If the member of archive is dominated by the offspring, then the offspring is tolerated as an archive member and the archive member is eliminated. In the last case, when non-domination is the only relation among them, then the offspring added to the archive depends on the availability of free slots and the density of the solution. Here, hyper-grid division of objective space maintains the diversity of archive. In this way, convergence toward the true Pareto-optimal front and diversity of the solution in the front are both maintained.

After almost a year, a GA that uses the non-dominated sorting approach, namely NSGA-II, is proposed. It seems to be more robust and efficient in solving MOEAs. The detailed behavior of NSGA-II is illustrated in the following subsection.

2.2 NSGA-II

The initial proposition of non-dominated sorting genetic algorithm (NSGA) is proven to be inefficient in the later stage, due to the existence of the three major factors such as (i) high computational complexity, (ii) dependency upon the sharing parameter, and (iii) the use of non-elitist approach. Therefore, Deb et al. [9] designed the second phase of NSGA (namely NSGA-II) was first proposed by. So, the author proposed the NSGA-II to avoid the above-mentioned difficulties. Interestingly, NSGA-II employs a crowded comparison operator (which is independent of any parameter), in order to maintain the diversity. The methodology is very easy to understand. During the process of elitism, the parent and the offspring solutions combined to provide a population of double strength. The candidates for the Pareto front will be selected one after another on the basis of fast non-dominated sorting technique. However, the complicacy arises while choosing individuals from a particular front does not require all solutions to be filled. In such scenario, the crowded comparison is used to choose the best-diversified individuals. Some of the other algorithms discussed below are proposed during the same period of NSGA-II.

2.3 SPEA

Strength Pareto evolutionary algorithm (SPEA) was first introduced by Zitzler et al. [10]. SPEA implements elitism by maintaining an archive (external population) which contains the best non-dominated (i.e., first front) solutions. The archive also participates in creating the mating pool with EA population. SPEA has two tire fitness (i.e., strength) assignment processes. For each individual $i \in \overline{P}$ (archive), a strength $S_i = n/(N+1)$ is assigned. Here, N is the size of the original EA population (P) and n is the number of individuals in the EA population dominated by a particular individual i. For each $j \in N$, the fitness is calculated by $f_j = 1 + \sum_{i, i \geq j} s_i$. The best individuals are selected from $P \cup \overline{P}$ to create the mating pool and perform a genetic operation for the next generation. In case the archive size overflows to its pre-defined size during upgradation, clustering technique based on average linkage method for pruning is used. Moreover, SPEA employs Pareto-based clustering technique rather than distance-based approaches. Over the time, improved versions of SPEA have been proposed as discussed below.

2.4 SPEA2

Strength Pareto evolutionary algorithm 2 (SPEA2) is an enhanced version of the SPEA, which is again proposed by Zitzler et al. [11]. SPEA2 uses the archive of a fixed size. In SPEA, solutions those are dominated by the identical archive members have similar fitness values, which make the scheme inefficient when the external population contains only one member. However in SPEA2, both the dominated and dominating individuals have been considered. In order to reequip the archive, SPEA uses a clustering technique for maintaining the diversity, which discards the outer solutions, but in SPEA2 the diversity deals with the kth nearest neighbor method.

2.5 SPEA2+

Strength Pareto evolutionary algorithm 2+ (SPEA2+) is suggested by Kim et al. [12] where two new mechanisms (crossover and archive) were introduced to maintaining the diversity in the decision variable space and objective space. Unlike SPEA2, SPEA2+ uses the new operators such as neighborhood crossover, mating selection. During simulation, this algorithm allows archive to catch hold of well-spread solution in both the decision variable (genotypic) space and the objective (phenotypic) space, which makes the algorithm more efficient than the SPEA. Very few years later, the concepts of hybridization of some of the above-discussed methods are developed. They are discussed below.

2.6 PESA

Pareto envelope-based selection algorithm (PESA) is developed by Corne et al. [13]. This algorithm borrows several concepts from SPEA and PAES. Similar to SPEA, it uses an undersized internal population (PI) (EA population) and a large archive (i.e., external population, abbreviated as PE). Similarly like PAES, hyper-grid crowding strategy is used for achieving the more diverse solution. After evaluating the population, an individual takes place in PE if it is non-dominated within the PI and not dominating by any existing members of the archive. The selection mechanism (binary tournament selection) of the parent population depends on the degree of crowding (squeeze factor) in a different region of the archive. During the selection process, two random individuals are taken from the archive, the squeeze factor is calculated, and the lowest squeeze factor is chosen. Then, genetic operations are performed to produce the child solution (called offspring). During upgradation of archive, hyper-grid crowding strategy is used while maximum allowable size of archive is achieved.

2.7 PESA-II

The author Corne et al. [14] proposed the modified and improved version of PESA, namely PESA-II. The working mechanism of PESA-II remains same as that of PESA, except the selection of parent population. For this, authors proposed a new selection scheme, which is known as region-based selection. A selective fitness is assigned to the all the hyperboxes, instead of individuals in phenotypic space. Binary tournament selection is used for selecting the hyperbox. The hyperbox which contains fewer individuals is the winner. Randomly, individuals are selected from that hyperbox to undergo with genetic operations to create the child solutions. As a whole, PESA-II provides better diversity in solution space.

2.8 IPESA-II

After over a one decade, Li et al. [15] pointed out some massive shortcomings of PESA-II and proposed an improved Pareto envelope-based selection algorithm II (IPESA-II). In IPESA-II, environmental selection scheme of PESA-II is improved in three aspects as explained below.

(a) Unlike PESA-II, IPESA-II maintains the archive after entering all the individuals (i.e., uniformity improvement) into it.
(b) In fact, maintenances of diversity in the distribution range are a major expect while solving MOO problem. Larger the boundary solutions better the diversity in the Pareto front. Hence in IPESA-II, an attempt is made to enlarge the

boundary solution points for achieving a diversified distribution range, which was not there in case of PESA-II.

(c) PESA-II randomly eliminates the non-dominated solutions from the archive whenever the archive is getting full. But IPESA-II has implemented a distance-based elimination strategy which is inspired by ϵ-MOEA [16].

Due to the continuous achievement of well-spread solutions over the search space, authors highly recommended the performance of IPESA-II over that of PESA-II.

2.9 AMGA

The author Tiwari et al. [17] suggested as well as implemented a multi-objective evolutionary algorithm which is supportive for solving the large-scale problem. The prototype of the archive-based micro-genetic algorithm (AMGA) has been influenced by the key actuality that in most of the optimization framework, the total time span for simulation is very high with respect to performing the selection, crossover, mutation, and diversity preservation due to lack of computational resources. AMGA works through an undersized population and maintains an external archive of good solutions. The initial population generated by the Latin hypercube sampling coupled with unbiased Knuth shuffling. AMGA works with a bilayer fitness mechanism, the domination level, and the diversity measure of the solution in the population. The diversity-based selection helps us to achieve the less crowded solutions.

After a couple of years, an appropriate modification has been in limelight which has been described below.

2.10 AMGA2

Archive-based micro-genetic algorithm 2 (AMGA2) [18] is a modified version of AMGA, which is claimed to be better in the following manner.

(a) Unlike AMGA, AMGA2 selects the parent population, depending on the domination rank and the diversity.

(b) AMGA2 maintains the diversity in objective space only, whereas in AMGA diversity is being maintained in both objective and decision variable spaces.

(c) AMGA2 employs differential evolution (DE) operator as an alternative of simulated binary crossover (SBX) operator (which is used in AMGA).

(d) The mutation probability is user defined in case of AMGA, but it is based on the rank of the parent solution to AMGA2.

Afterward, researchers started working on steady-state micro-GA with an aim to decrease the computational time by reducing the number of function evaluations. One such algorithm is discussed below.

2.11   ASMiGA

An archive-based steady-state micro-genetic algorithm is proposed by Nag et al. [19]. Author proposed a new archive-maintaining scheme, where the concept of minimum threshold value of the archive is additionally introduced. As a result, the archive can never be empty even in the worst case that makes the archive adaptive and dynamic. Hence, it preserves the boundary solution and also helps to maintain diversity in the objective space. The beauty of the steady-state micro-GA is that the algorithm can maximize the number of generation whenever the number of function evolution is fixed. A new crossover mechanism DE-3 has been proposed. Moreover, depending upon the size of archive, DE-3 and SBX crossover operator have been used. For the diversity measure, it uses crowding distance for bi-objective problem and Euclidean nearest neighbor for more than two objectives.

Nevertheless, archive-based algorithms have crossed many successful milestones over the last decades in solving MOEAs. Recently, algorithms based on two archives are becoming alternate paradigm to handle MOEAs and hence treated as the immediate competitors of single-archive-based algorithms. An algorithm based on such concept is illustrated below.

2.12   The Two-Archive Algorithm

Praditwong and Yao [20] first introduced the two-archive algorithm to solve MOEAs. The non-dominated solution is inserted into two different archives as follows

(a) Convergence archive (CA)
(b) Diversity archive (DA).

The working mechanism of this algorithm is as same as its near competitors, except the implementation of archive mechanism and the pruning of the archive. In this strategy, all the 'non-dominated solutions with domination' credit to the account of CA and all the 'non-dominated solutions without domination' to DA. Whenever the size of the archive exceeds the maximum threshold, pruning techniques is applied on DA (as it maintains the diversity). However, this technique is not applicable to CA. Especially, it has an additional ability to solve the many-objective problems and is proven to be robust in the literature.

3   Existing Tools and Software for MOEAs

Various software and tools are available for the evolutionary multi-objective optimization. Out of these, OPT4J [21] is the most popular open-source framework that uses evolutionary computation. [21] cites numerous algorithms and test suits.

Among this, Jmetal [22] and MOEA [23] are very successful. A MATLAB-based platform PlatEMO [24] is also available for evolutionary multi-objective optimization. Apart from that, ParadisEO [25] is an ANSI-C++ based tool, which provides the meta-heuristics multi-objective optimization algorithms. Again, Python [26] also provides a package for solving MOEAs through meta-heuristics algorithms as well. Yarpiz [22] and KanGAL [27] are two research communities, who provide easy implementation of MOEAs codes in C, C++, and MATLAB platforms.

4 Conclusion and Future Scope

In this paper, the archive-based approaches proposed in the literature to solve MOOPs are being surveyed. Archive-based method became more popular as it takes less function evaluations, preserves all non-dominated solutions, and deserves ease of implementation. Since the inception of archive-based approaches by Knowles et al., quite a good number of research works have been carried out during last few decades. However, these approaches basically make modifications in the fitness function, selection criterion, and crossover operators. Some of the researchers also introduced the adaptive mutation in order to make the proposition more robust. It is also interesting here to note that both SPEA2 and NSGA-II perform equally good in all respect. Moreover, it can be concluded that the two-archive-based method performs better than single-archive-based method. Anyway, no much work has been done on two-archive-based method yet. Some of the hassle-free approaches are also proposed like AMGA, which do not require any fine-tune of parameters. Such approaches help to reduce the computation time. Overall speaking, due to the conflicting behavior of the MOOPs, arises in the real life, it is very difficult to rank the available algorithms. One is better than other in one or some other sense. However, it may finally be concluded that in order to handle MOOPs, employment of two archives, using micro-population concept, parameter-free approaches, and self-adaptive operators makes sense and increases the robustness of the algorithm.

As a future scope of research, one can work on multiple archive-based algorithms. Numerous works can also be done to solve constrained MOOPs.

References

1. Guliashki, V., Toshev, H., Korsemov, C.: Survey of evolutionary algorithms used in multiobjective optimization. Prob. Eng. Cybernet. Robot. 42–54 (2009)
2. Elarbi, M., Bechikh, S., Ben Said, L., Datta, R.: Multi-objective optimization: classical and evolutionary approaches. In: Bechikh, S., Datta, R., Gupta, A. (eds.) Recent Advances in Evolutionary Multi-objective Optimization. Adaptation, Learning, and Optimization, p. 20. Springer, Cham (2017)
3. Schder, J.D.: Some experiments in machine learning using vector evaluated genetic algorithms, Unpublished doctoral dissertation, Vanderbilt University (1984)

4. Schaffer, J.D.: Multiple objective optimization with vector evaluated genetic algorithms. In: Grefenstette, J. (ed.), Proceedings of an International Conference on Genetic Algorithms and their Applications, pp. 93–100 (1985)
5. Horn. J., Nafpliotis, N., Goldberg, D.E.: A niched pareto genetic algorithm for multiobjective optimization. In: Proceedings of the First IEEE Conference on Evolutionary Computation, IEEE World Congress on Computational Intelligence, vol. 1, pp. 67–72. IEEE Service Centre, Piscataway, NJ (1994)
6. Fonseca, C.M., Fleming, P.J.: Genetic algorithms for multiobjective optimization: Formulation, discussion, and generalization. In: Proceedings of the Fifth International Conference on Genetic Algorithms, pp. 416–423 (1993)
7. Srinivas, N., Deb, K.: Multi-objective function optimization using non-dominated sorting genetic algorithms. Evol. Comput. **2**, 221–248 (1994)
8. Knowles, J.D., Corne, D.W.: The Pareto archived evolution strategy: a new baseline algorithm for pareto multiobjective optimization. In: CEC'99: Proceedings of the 1999 Congress on Evolutionary Computation, IEEE Service Center, Piscataway, New Jersey (1999)
9. Deb, K., Pratap, A., Agarwal, S., Meyarivan, T.: A fast and elitist multiobjective genetic algorithm: NSGA-II. IEEE Trans. Evol. Comput. **6**, 182–197 (2002)
10. Zitzler, E., Thiele, L.: Multiobjective evolutionary algorithms: a comparative case study and the strength pareto approach. IEEE Trans. Evol. Comput. **3**, 257–271 (1999)
11. Zitzler, E., Laumanns, M., Thiele, L.: SPEA2: improving the strength pareto evolutionary algorithm. Comput. Eng. Netw. Lab. (TIK), Swiss Federal Institute of Technology (ETH), Zurich, Switzerland, Tech. Rep. 103 (2001)
12. Kim, M., Hiroyasu, T., Miki, M., Watanabe, S.: SPEA2+: improving the performance of the strength pareto evolutionary algorithm 2 parallel problem solving from nature—PPSN VIII, pp. 742–751. Springer (2004)
13. Corne, D.W., Knowles, J.D., Oates, M.J.: The Pareto envelope-based selection algorithm for multiobjective optimization. In: Deb, K., Rudolph, G., Lutton, E., Merelo, J.J., Schoenauer, M., Schwefel, H.-P., Yao, X. (eds.) PPSN VI. LNCS. Springer, Heidelberg, 1917, pp. 839–848 (2000)
14. Corne, D.W., Jerram, N.R., Knowles, J.D., Oates, M.J.: PESA-II: Region-based selection in evolutionary multiobjective optimization. In: Proceedings of the Genetic and Evolutionary Computation Conference (GECCO 2001). Morgan Kaufmann, San Francisco, 283–290 (2001)
15. Li, M., Yang, S., Liu, X., Wang, K.: IPESA-II: improved pareto envelope-based selection algorithm II. In: Purshouse R.C., Fleming P.J., Fonseca C.M., Greco S., Shaw J. (eds.) Evolutionary Multi-Criterion Optimization. EMO 2013. Lecture Notes in Computer Science, vol. 7811. Springer, Berlin, Heidelberg (2013)
16. Deb, K., Mohan, M., Mishra, S.: Evaluating the ε-dominated based multi-objective evolutionary algorithm for a quick computation of pareto-optimal solutions. Evol. Comput. **13**(4), 501–525 (2005)
17. Tiwari, S., Koch, P., Fadel, G., Deb, K.: AMGA: An archive-based micro genetic algorithm for multi-objective optimization. In: Genetic and Evolutionary Computation Conference (GECCO 2008), pp. 729–736. ACM (2008)
18. Tiwari, S., Fadel, G., Deb, K.: AMGA2: improving the performance of the archive-based micro-genetic algorithm for multi-objective optimization. Eng. Opt. 371–401, Taylor and Francis (2011)
19. Nag, K., Pal, T., Pal, N.: ASMiGA: an archive-based steady state micro genetic algorithm. IEEE Trans. Cybern. **45**(1), 40–52 (2015)
20. Praditwong, K., Yao, X.: A new multi-objective evolutionary optimisation algorithm: The two-archive algorithm. In: International Conference on Computational Intelligence and Security, pp. 286–291. IEEE Press (2006)
21. Lukasiewycz, M.: Opt4J: a modular framework for metaheuristic optimization. In: Proceedings of the 13th Annual Conference on Genetic and Evolutionary Computation. ACM (2011)
22. YARPIZ: http://yarpiz.com/category/multiobjective-optimization
23. MOEA https://sourceforge.net/projects/moeaframework/

24. Tian, Y., Cheng, R., Zhang, X., Jin, Y.: PlatEMO: a MATLAB platform for evolutionary multi-objective optimization. IEEE Comput. Intelligen. Mag.
25. http://paradiseo.gforge.inria.fr/
26. PYTHON https://pypi.python.org/pypi
27. KANGAL LAB: http://www.iitk.ac.in/kangal/codes.shtml
28. Horn, J., Nafpliotis, N.: Multiobjective Optimisation Using The Niched Pareto Genetic Algorithm. IlliGAL Report 93005, Illinois Genetic Algorithms Laboratory, University of Illinois, Urbana, Champaign (1994)
29. Hajela, P., Lin, C.Y.: Genetic search strategies in multicriterion optimal design. In: Structural Optimization, New York, Springer, vol. 4, pp. 99–107 (1992)
30. Eskandari, H., Geiger, C.D., Lamont, G.B.: Fastpga a dynamic population sizing approach for solving expensive multiobjective optimization problems. In: Evolutionary Multiobjective Optimization Conference on EMO, pp. 141–155 (2007)
31. Kukkonnen, S., Lampinen, J.: GDE3: the third evolution step of generalized differential evolution. In: Proceedings of the IEEE Congress on Evolutionary Computation, pp. 443–450 (2005)
32. Karahan, I., Köksalan, M. (2010). A territory defining multiobjective evolutionary algorithm and preference incorporation. IEEE Trans. Evol. Comput. 14(4), 636–664
33. Li, H., Zhang, Q.: Multiobjective optimization problems with complicated Pareto sets MOEA/D and NSGA-II. IEEE Trans. Evolution. Comput. 12(2), 284–302 (2009)
34. JMETAL: http://sourceforge.net/projects/jmetal/
35. Nebro, A.J., Luna, F., Alba, E., Dorronsoro, B., Durillo, J.J., Beham, A.: AbYSS: adapting scatter search to multiobjective optimization. IEEE Tans. Evol. Comput. 12(4), 439–453 (2008)
36. Zhang, Q., Li, Hui: MOEA/D: a multiobjective evolutionary algorithm based on decomposition. IEEE Trans. Evol. Comput. 11(6), 712–731 (2007)
37. Zitzler, E., Künzli, S.: Indicator-based selection in multiobjective search. In: Yao, X., Burke, E.K., Lozano, J.A., Smith, J., Merelo-Guerv´os, J.J., Bullinaria, J.A., Rowe, J.E., Tiˇno, P., Kab´an, A., Schwefel, H.-P. (eds.) PPSN VIII. LNCS, vol. 3242, pp. 832–842. Springer, Heidelberg (2004)
38. Deb, K., Jain, H.: An evolutionary many-objective optimization algorithm using reference-point-based nondominated sorting approach, part i: solving problems with box constraints. IEEE Trans. Evolution. Comput. 18(4), 577–601 (2014)
39. Emmerich, M., Beume, N., Naujoks, B.: An EMO algorithm using the hypervolume measure as selection criterion. European J. Operation. Res. 3, 1653–16 (2007)

A Novel Optimal Gabor Algorithm for Face Classification

Lingraj Dora, Sanjay Agrawal, Rutuparna Panda and Ajith Abraham

Abstract Over the past decade, most of the research in the area of pattern classification has emphasized the use of Gabor filter (GF) banks for extracting features. Typically, the design and the choice of GF banks are done on experimentation basis. In this paper, an attempt is made on obtaining an optimized set of GFs for improving the performance of face classification. The bacteria foraging optimization (BFO) is utilized to get optimized parameters of GF. The proposed BFO-Gabor technique is utilized to derive the feature vectors from the face images based on Gabor energy. These feature vectors are then used by probabilistic reasoning model (PRM) to perform the classification task. The ORL and UMIST datasets are utilized to investigate the superiority of the proposed approach. In addition, the experimental results of the proposed approach and the classical methods are compared. It is observed that the proposed BFO-Gabor method is superior than the classical methods.

Keywords Face classification · Gabor filter · Bacteria foraging optimization
Probabilistic reasoning model

L. Dora
Department of EEE, VSSUT, Burla 768018, India
e-mail: lingraj02uce157ster@gmail.com

S. Agrawal (✉) · R. Panda
Department of Electronics and Telecommunication Engineering, VSSUT, Burla 768018, India
e-mail: agrawals_72@yahoo.com

R. Panda
e-mail: r_ppanda@yahoo.co.in

A. Abraham
MIR Labs, Auburn, Washington, USA
e-mail: ajith.abraham@ieee.org

© Springer Nature Singapore Pte Ltd. 2019
J. C. Bansal et al. (eds.), *Soft Computing for Problem Solving*, Advances in Intelligent
Systems and Computing 817, https://doi.org/10.1007/978-981-13-1595-4_65

1   Introduction

Face classification (FC) is a process of identifying or verifying the class of an unknown subject (person) from labeled trained facial images. A group of facial images of the same subject is called a class. The FC task is carried out by extracting the facial features from the labeled training samples. The derived facial features are then compared with that of the unknown test image. FC methods and algorithms for extracting the features are mainly divided into two categories: (1) appearance based and (2) feature based. The appearance-based methods make use of the whole face image for extracting the facial features and seem to perform better than the feature-based methods [1, 2]. For instance, eigenfaces and fisherfaces do well on several challenging situations and occlusions such as facial hair, low light, and sunglasses [2]. However, the main drawback with these methods is the requirement of prepro-cessing steps. Usually, these techniques give better results for identifying images with frontal face. Nevertheless, the performance of such methods is degraded at the time of classifying subjects having face images containing multiple rotation as well as inclination. In order to deal with the above-stated shortcomings, GF bank method could be an appropriate choice. GFs have the capability to capture different salient visual properties such as spatial frequency, spatial localization, and angle selectivity. These abilities of GF make it suitable to be utilized in the applications like face classification, facial expression, gender classification, texture classification, and segmentation [3–10].

Liu and Wechsler [6] used the bank containing 40 GFs comprising of five scales and eight angles for feature extraction from the face images followed by principal component analysis, independent component analysis, and PRM for face recogni-tion. Kruizinga and Petkov [7] used a bank of filters emphasizing on Gabor energy operator for texture classification and segmentation. Thus, the filter bank approach is an efficient technique for feature extraction. However, a limitation in the approach is the selection of filter parameters, which is typically done on a trial-and-error basis. To overcome this problem, the authors in [8] have investigated an evolutionary Gabor filter optimization (EGFO) technique. It uses genetic algorithm (GA) to get the opti-mal set of parameters for filter design. An incremental clustering method is then utilized for selecting the filters. However, the authors applied the algorithm for vehi-cle detection. To the best of our knowledge, optimization of GF parameters using BFO has not been reported in the literature and this has motivated us to use it for optimizing filter parameters and extend the idea to FC.

In this paper, a BFO-Gabor method is proposed for FC. The method overcomes the limitations of classical filter banks approach by employing BFO [11, 12], a global optimization algorithm for optimizing the parameter set of filters. In this framework, an incremental clustering algorithm is integrated to group the similar filters in the parameter space and allowing us to design an optimal and selecting a concise set of filters. The response of the optimal filters is utilized to derive the feature vectors from the facial samples based on Gabor energy. The PRM is then employed for classifying the test face images. The remaining part of the paper proceeds as follows:

Part 2 contains a concise description about the GF approach. Part 3 is concerned with the proposed BFO-Gabor method used in this study. The experimental findings and discussions are in Part 4. Finally, Part 5 presents the conclusion.

2 GF Method

The GFs exhibits strong properties of spatial locality and angle selectivity. Additionally, they are optimally localized in space and frequency domains. The GFs can be defined [3–10]:

$$G(x, y) = exp\left(-\frac{\widetilde{x}^2 + \sigma^2\widetilde{y}^2}{2\mu^2}\right) cos\left(2\pi F\widetilde{x} + \phi\right) \qquad (1)$$

where $\widetilde{x} = (x - \xi)cos\theta - (y - \eta)sin\theta$, $\widetilde{y} = (x - \xi)sin\theta + (y - \eta)cos\theta$, (x, y) represents the 2D spatial coordinates, and ξ and η are the center of the receptive field in image coordinate. θ specifies the GF orientation. The parameter μ represents the standard deviation of Gaussian envelope and defines the effective size. The eccentricity of μ is determined by spatial aspect ratio σ, which is limited in the range $0.23 < \sigma < 0.92$ [7]. The parameter F is called the spatial frequency of the cosine factor, and $\lambda = 1/F$ is the spatial wavelength. The ratio μ/λ is given as follows:

$$\frac{\mu}{\lambda} = \frac{1}{\pi}\sqrt{\frac{ln2}{2}} \times \frac{2^B + 1}{2^B - 1} \qquad (2)$$

where B is the bandwidth of the filter. Finally, ϕ is the phase-offset parameter in the argument of the cosine function, which defines the symmetry of the function $G(x, y)$. The GFs as given in (1) are designed by selecting appropiate values of the four parameters in $\zeta = \{\theta, F, \sigma, B\}$. Figure 1a, b shows response of the GF in spatial domain for two different parameter sets. Figure 1c, d shows their corresponding responses in frequency domain.

(a)   (b)   (c)   (d)

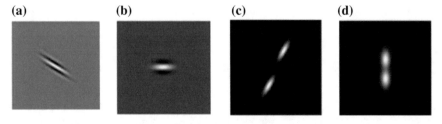

Fig. 1 Sample GF responses. Spatial response **a** for $\zeta = \{30°, 0.2, 0.5, 1\}$, **b** for $\zeta = \{0°, 0.1, 0.5, 1\}$, **c** response of (a) in frequency domain, and **d** response of **b** in frequency domain

Hence, random parameter sets result in different filter responses, which makes filter design a challenging task for applications like FC. One way to eliminate this problem is the optimal selection of the parameter set, i.e., $\{\theta, F, \sigma, B\}$. For instance, if a problem involves N GFs, then $4N$ parameters need to be optimized. In this paper, BFO is used to solve such a high-dimensional multivariate problem.

2.1 Gabor Feature Extraction

The output of a GF is the convolution of the filter with a face image. It is given as:

$$\widehat{R}(x, y) = I(x, y) \otimes G(x, y) \tag{3}$$

where $I(x, y)$ is a face image and $\otimes$ indicates convolution operation. For a specific $\zeta = \{\theta, F, \sigma, B\}$, the output $\widehat{R}$ extracts facial features only in that particular scale and angle. Thus, a group of GFs containing several parameter sets can obtain the facial features in all possible scales and angles. In this way, it can be utilized to extract the local and the distinguishing features for face classification. In our work, we have proposed a Gabor energy feature vector for calculation of features from a sample face using a group of GFs, which is given as below:

$$\widehat{H}(x, y) = \sum_{i=1}^{N} \|R_i(x, y)\| \tag{4}$$

After getting the feature vector, it is normalized as follows:

$$\widehat{G}(x, y) = \frac{\widehat{H}(x, y) - min(\widehat{H})}{max(\widehat{H}) - min(\widehat{H})} \tag{5}$$

However, the design of filter bank by choosing different values of the parameter set ζ is quite a difficult task and is application oriented. Therefore, BFO technique is utilized here to determine the optimal values of parameter set ζ and the objective function used here is the correct classification rate as calculated from PRM method.

3 Proposed BFO-Gabor Method

In this paper, a BFO-Gabor method is proposed for FC by employing an optimized set of GFs to derive the feature vectors from the face samples. These vectors are then utilized by PRM for classification. BFO is applied here for optimizing the GF bank parameters. To design N number of filters, BFO uses a swarm of $4N$-dimensional bacteria, which undergoes through three necessary steps: chemotaxis, reproduction and elimination and dispersal to find the solution of such high-dimensional optimization problem. During optimization, the above method uses an incremental clustering algorithm, which eliminates the redundant filters and clusters. The filters having similar properties are grouped in a cluster. The optimal GFs are then utilized to extract the features from the face samples.

3.1 Parameter Estimation

Initially, the BFO technique creates a group of random parameters for GFs, limited to be specified within the bounds. The orientation variable (θ) is bounded within $[0, \pi)$. The value of the variable F (spatial frequency) is application based. Here, its value is bounded within $[0, 0.5]$. The parameter σ (spatial aspect ratio) is in the range of $[0.23, 0.92]$. As per the literature, the value of B is bounded within 1.0–1.8 octaves [7]. All the above-mentioned bounds are utilized to obtain the parameter set of a GF.

3.2 Incremental Clustering Algorithm

To generate N number of filters, BFO uses $4N$-dimensional bacteria. Some of the filters in the bacteria may have similar property. An incremental clustering algorithm [13] is used here to cluster the filters having similar properties in the parameter space, by comparing the distance between two filters, with a threshold value. The steps of the algorithm are as follows:

1. A set of N GFs is generated randomly.
2. A filter is represented as one cluster.
3. Then, the distance between the centroid of the filter and the rest filters are calculated.
4. The smallest distance is chosen.
5. If the smallest distance is less than a threshold value, it is assigned to that corresponding cluster; otherwise, it is represented as a new cluster. Let $\{\theta, F, \sigma, B\}$ be the parameters of a filter and $\{\theta^i, F^i, \sigma^i, B^i\}$ be the parameters of ith filter for i = 1 to N. This filter is assigned to the ith cluster, if the following conditions are satisfied:

$$\theta^i - 0.5 \times \theta_{th} \leq \theta \leq \theta^i + 0.5 \times \theta_{th} \tag{6}$$

$$F^i - 0.5 \times F_{th} \leq F \leq F^i + 0.5 \times F_{th} \tag{7}$$

$$\sigma^i - 0.5 \times \sigma_{th} \leq \sigma \leq \sigma^i + 0.5 \times \sigma_{th} \tag{8}$$

$$B^i - 0.5 \times B_{th} \leq B \leq B^i + 0.5 \times B_{th} \tag{9}$$

where $\{\theta_{th}, F_{th}, \sigma_{th}, B_{th}\}$ are the threshold values. The following threshold values are used here: $\theta_{th} = \pi/K$, $F_{th} = (F_{max} - F_{min})/K$, $\sigma_{th} = (\sigma_{max} - \sigma_{min})/K$, $B_{th} = (B_{max} - B_{min})/K$. The value of K is application dependent. In our experiments, we have taken K as 3.

6. Steps 2–4 are repeated, for all the randomly generated filters.
7. At last, each cluster is represented as one filter whose parameters are the average value of the corresponding parameter values in that cluster.

In this way, the above algorithm discards the redundant filters by selecting the similar filters in the parameter space. Then, the selected optimal filters are utilized to derive the feature vectors from the face samples.

3.3 Objective Function for BFO-Gabor Approach

The objective function used in BFO technique for choosing an optimal bank of GFs is the performance parameter of PRM method. The PRM technique [14] uses Gabor energy features derived from both training and test dataset for classifying the test images. As the features derived from filter bank are discriminant, it seems logical to assume that the PRM technique describes a Bayes linear classifier having identical and diagonal covariance matrices [6, 14]. The maximum aposteriori (MAP) classification rule categorizes Z (test image feature vector) into class C_q as given below:

$$\sum_{i=1}^{T} \frac{(z_i - m_{qi})^2}{\omega_i^2} = \min_{j} \left\{ \sum_{i=1}^{T} \frac{(z_i - m_{ji})^2}{\omega_i^2} \right\} \rightarrow Z \in C_q \qquad (10)$$

The symbols represent the usual meaning as explained in [14].

Let T_s be the total number of test images and T_c be the number of correctly classified test images Z_t according to (10) for $t = 1, 2, \ldots, T_s$. Then, the objective function used in BFO is given as:

$$F_{opt} = arg\max_{S}\{O\} \qquad (11)$$

$$O = \frac{T_C}{T_S} \times 100 \qquad (12)$$

The optimization technique determines the filter bank ζ_k for which the function O attains a maximum value.

3.4 Algorithm of the Proposed Method

The following steps are followed by BFO-Gabor method for feature extraction and classification of face images:

1. Let X be a training dataset of T_n face images of size $w \times h$.
2. At first, the four parameters, $\zeta = \{\theta, F, \sigma, B\}$, are initialized for each of the N GFs as described in Sect. 3.1.
3. A swarm of S bacteria is created, each having dimension $4N$.
4. Incremental clustering algorithm as described in Sect. 3.2 is used to group similar filters into distinct clusters for each bacteria. Each cluster is represented as one filter, whose parameters are the average value of the corresponding parameter values in that cluster.
5. The above filters are utilized to obtain the feature vectors from T_n training images.
6. The same set of filters is also utilized to obtain the feature vectors from T_s test images.
7. Each vector in the test dataset is compared with all vectors of training dataset using (10) to classify the test face images.
8. Classification rate is calculated using (12).
9. BFO is used to update the swarm S such that the objective function given by (11) is maximized.
10. Repeat step 4 until the termination condition is not met.
11. BFO generates an optimal GF bank for which the objective function in (11) attains a maximum value.

4 Results and Discussions

Two face image database, namely ORL [15] and UMIST [16], are utilized to investigate the effectiveness of the proposed BFO-Gabor approach. The ORL dataset contains 400 different face samples of 40 subjects (or classes). Each class contains ten face images. Similarly, the UMIST dataset consists of 1012 different face samples of 20 subjects. At first, all these face samples are normalized to a dimension 70×90. Then, randomly some sample face images are taken from both the databases to form the training dataset and rests are used as test images. Figure 2 shows an example of training face images from ORL database. In this figure, five images of the same subject are represented as one class. A sample face image from the UMIST database is shown in Fig. 3. Here, eight images of the same person represent one class.

Figures 4 and 5 illustrate the classification of face samples in 3D space to validate the class separability measure.

Figure 4a, b shows 25 face samples of five subjects from ORL training dataset in 3D space, by considering three most important features derived from GF technique and BFO-Gabor technique. Similarly, Fig. 5a, b shows the 3D plot of face samples by considering three most important features from 40 face samples of five distinct

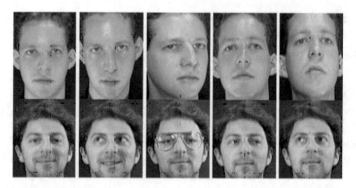

Fig. 2 Sample training face images from the ORL

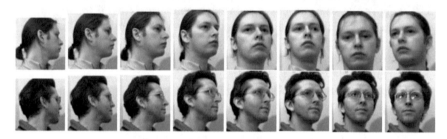

Fig. 3 Sample training face images from the UMIST

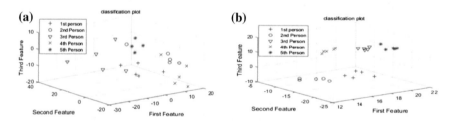

Fig. 4 3D classification plots of 25 training face samples of five distinct persons taken from ORL. **a** Classification done by GF method and **b** classification done by BFO-Gabor method

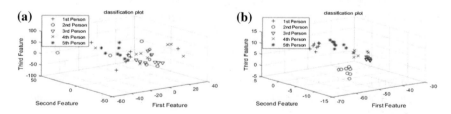

Fig. 5 3D classification plots of 40 training face samples of five distinct persons taken from UMIST. **a** Classification done by GF method and **b** classification done by BFO-Gabor method

Table 1 Parameters of bacteria foraging algorithm

Parameters	Value
Number of bacteria, S	50
Number of chemotaxis steps, N_c	5
Swimming length, N_s	4
Number of reproduction steps, N_{re}	10
Number of elimination-dispersal events, N_{ed}	2
Size of the step taken, C	0.05
$d_{attract}$	0.01
$W_{attract}$	0.2
$h_{repellent}$	0.01
$W_{repellent}$	10

persons from UMIST training dataset. All the plots in Figs. 4 and 5 use five different symbols to represent five different persons. From Figs. 4a and 5a, it is evident that the features derived from GF technique are such that face samples of different subjects are not clearly separated and some face samples are also overlapped, whereas the plot in Figs. 4b and 5b shows that BFO-Gabor method separates the different subjects and clusters the same subject. This suggests that BFO-Gabor technique has improved class separability index over the simple GF technique [17, 18].

In this paper, we have also compared the error rates of GF method, EGFO method, particle swarm optimized (PSO) Gabor method [19], and BFO-Gabor method. At first, GF with three scales and five angles was employed with PRM for classification. Secondly, GF with five scales and eight angles was utilized for FC using PRM. Then, EGFO method, which uses GA, is used to select a group of 32 filters and then used with PRM for FC. Additionally, PSO-Gabor which uses PSO to select 32 optimal filters is also included in the comparison. However, the proposed method can also be explored using artificial bee colony (ABC) or social media optimization (SMO). At last, our proposed BFO-Gabor approach is used which selects a group of 20 filters and then used with PRM for FC. The parameter setting for BFO-Gabor is given in Table 1.

The error rate of all the above methods are listed in Table 2. From Table 2, it is seen that the BFO-Gabor approach gives better results in comparison with other techniques with a lowest error rate of 2% for ORL database and 1.44% for UMIST database. In all the above experiments, the value of L (no. of images in a class) is set to 5 for ORL database and 8 for UMIST database. The error rate of BFO-Gabor for different values of L is listed in Table 3. and it is seen from the table that the proposed method performs better with a lowest error rate of 2% for $L = 5$ in ORL database and 1.44% for $L = 8$ in UMIST database.

We have also carried out an experiment to calculate the error rate for BFO-Gabor method for $K = 2$ and 3, and the corresponding plots are shown in Fig. 6.

Table 2 Error rate for face classification

Face database	Methods				
	GF method (3 × 5) (%)	GF method (5 × 8) (%)	EGFO method (%)	PSO-Gabor method (%)	BFO-Gabor method (%)
ORL	9.5	8.5	4.5	3	2
UMIST	10.91	8.45	2.11	1.99	1.44

Table 3 Error rate for different values of L

Face database	Methods					
	L	GF method (3 × 5) (%)	GF method (5 × 8) (%)	EGFO method (%)	PSO-Gabor method (%)	BFO-Gabor method (%)
ORL	2	12.5	34	16.5	11.88	10
	3	11	19.5	10	7.14	6
	4	11	10.5	6.5	4.59	3.5
	5	9.5	8.5	4.5	3	2
UMIST	2	34.03	24.53	6.22	5.04	4.68
	4	21.36	15.02	4.22	4.08	3.53
	6	15.02	8.68	2.23	2.01	2
	8	10.91	8.45	2.11	1.99	1.44

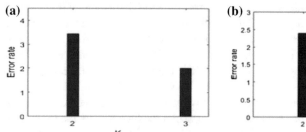 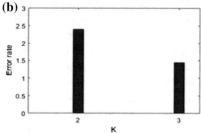

Fig. 6 Error rate plot for BFO-Gabor method using different values of K. **a** Plot for $K = 2$ and 3 in ORL database. **b** Plot for $K = 2$ and 3 in UMIST database

The value of $K = 2$ results in a group of 12 filters and $K = 3$ results in a group of 20 filters. The results show that we can customize the group of filters in a cluster by changing the value of K. Thus, the value of K affects both clustering and error rates.

5   Conclusion

In this work, we have investigated a BFO-Gabor approach for FC. It uses BFO to select an optimal set of GFs. These filters are used to extract discriminant features from the Gabor energy representation of face images and then used by PRM for classification. The performance of the suggested technique is validated by utilizing the ORL and the UMIST databases. From the results, it is seen that the proposed method performs better than classical GF method, EGFO method, and PSO-Gabor method under changing facial expression and occlusions. The proposed technique can be validated with more complex database, like FERET. The concept can also be investigated to texture classification and segmentation.

References

1. Chellappa, R., Wilson, C.L., Sirohey, S.: Human & machine recognition of faces: a survey. Proc. IEEE **83**, 705–740 (1995)
2. Turk, M.: A random walk through Eigenspace. IEICE Trans. Inform. Syst. **E84-D**, 1586–1695 (2001)
3. Daugman, J.G.: Uncertainty relation for resolution in space, Spatial frequency and orientation optimized by two-dimensional visual cortical filters. J. Opt. Soc. Amer. A. **2**, 1160–1169 (1985)
4. Daugman, J.G.: Complete discrete 2-D Gabor transforms by neural networks for image analysis and compression. IEEE Trans. Pattern Anal. Mach. Intell. **36**, 1169–1179 (1988)
5. Tong, L., Wong, W.K., Kwong, C.K.: Differential evolution-based optimal Gabor filter model for fabric inspection. Neurocomputing **173**, 1386–1401 (2016)
6. Liu, C., Wechsler, H.: Independent component analysis of Gabor features for face recognition. IEEE Trans. Neural Netw. **14**, 919–928 (2003)
7. Kruizinga, P., Petkov, N.: Nonlinear operator for oriented texture. IEEE Trans. Image Process. **8**(10), 1395–1407 (1999)
8. Sun, Z., Bebis, G., Miller, R.: On-road vehicle detection using evolutionary gabor filter optimization. IEEE Trans. Intell. Trans. Syst. **6**(2), 125–137 (2005)
9. Khan, S., Hussain, M., Aboalsamh, H., Mathkour, H., Bebis, G., Zakariah, M.: Optimized Gabor features for mass classification in mammography. Appl. Soft Comput. **44**, 267–280 (2016)
10. Singh, V.K., Mathai, K.J., Singh, S.: Fingerprint segmentation: optimization of a filtering technique with Gabor filter. In: IEEE International Conference on CSNT, pp. 823–827 (2014)
11. Passino, K.M.: Biomimicry of bacteria foraging for distributed optimization & control. IEEE Control Syst. Mag. 52–67 (2002)
12. Dasgupta, S., Das, S., Abraham, A., Biswas, A.: Adaptive computational chemotaxis in bacterial foraging optimization: an analysis. IEEE Trans. Evol. Comput. **13**(4), 919–941 (2009)
13. Jain, A., Murty, M., Flynn, P.: Data clusering: a review. ACM Comput. Surrv. **31**(3), 265–323 (1999)
14. Liu, C., Wechsler, H.: Probabilistic reasoning models for face recognition. In: Proceedings of IEEE Computer Society Conference on Computer Vision & Pattern Recognition, Santa Barbara, California (1998)
15. ORL Face Database. http://www.com-orl.co.uk/facedatabase.html
16. UMIST Face Database. http://images.ee.umist.ac.uk/danny/database.html
17. Serrano, Á., de Diego, I.M., Conde, C., Cabello, E.: Analysis of variance of Gabor filter banks parameters for optimal face recognition. Pattern Recognit. Lett. **32**(15), 1998–2008 (2011)

18. Naik, M.K., Panda, R.: A novel adaptive cuckoo search algorithm for intrinsic discriminant analysis based face recognition. Appl. Soft Comput. **38**, 661–675 (2016)
19. Abhishree, T.M., Latha, J., Manikantan, K., Ramachandran, S.: Face recognition using Gabor filter based feature extraction with anisotropic diffusion as a pre-processing technique. Proc. Comput. Sci. **45**, 312–321 (2015)

Trapezoidal Intuitionistic Fuzzy Fractional Transportation Problem

Shailendra Kumar Bharati

Abstract Naturally decision makers try to maximize ratio of return/risk, return/cost, and time/cost of transportation problem under uncertain environment. In such cases fractional transportation problem (FTP) play perfect role in instead of ordinary transportation problem. Some limited methods for the solution of fractional transportation problem using exact or fuzzy parameters are available in literature, and existing methods cannot deal uncertain FTP properly having hesitation factors due to inexact information. Here, I introduce the concept of intuitionistic fuzzy expectation of trapezoidal intuitionistic fuzzy numbers, and some related theorems are presented. Further, I present the concept of trapezoidal intuitionistic fuzzy fractional transportation (TIFFTP) problem, and a methodology for the solution of TIFFTP problem based on expectation of trapezoidal intuitionistic fuzzy numbers. Finally, I illustrate the presented TIFFTP problem by using an example.

Keywords Non-membership function · Transportation problem
Fractional objective function · Optimization method

1 Introduction

Fractional transportation problem is a generalization of transportation problem of [1], and it occurs in real life optimization problems where we maximize the ratio of return and risk, estimated cost and time etc. which occurs in resource allocation, agriculture management, the analysis of financial enterprises and undertaking, finance sector, industrial maintenance, transportation. Swarup [2] introduced the concept of fractional transportation with fixed numbers and presented a method for its solutions. In conventional fractional transportation problem fixed values of parameters were used. It is very to note that cost and time of transportation problem depend on

S. K. Bharati (✉)
Department of Mathematics, Kamala Nehru College, University of Delhi,
New Delhi 110049, India
e-mail: skmaths.bhu@gmail.com

© Springer Nature Singapore Pte Ltd. 2019   833
J. C. Bansal et al. (eds.), *Soft Computing for Problem Solving*, Advances in Intelligent
Systems and Computing 817, https://doi.org/10.1007/978-981-13-1595-4_66

many uncertain situations like speed breakers, traffic management, uncertain change in diesel price, driver health, mental condition of driver, uncertain weather conditions. Therefore due to such situations conventional fractional transportation models cannot deal properly problems.

Zadeh [3] contribution in the form of fuzzy set (FS) became more helpful for dealing real life problems under uncertainty. Further, Bellman and Zadeh [4] applied FS in modeling and decision making of real life optimization problems. Recently Bharati et al. [5] have studied fully multiobjective programming with trapezoidal fuzzy numbers, Liu [6] studied fractional transportation problem with fuzzy parameters. It is observed that a transportation problem cannot be restricted to a single objective thus multiple objective transportation problems are naturally occurred, and Verma et al. [7] have presented methodology for the solution of multiple objective transportation (MOTP) problems using some well-known non-linear membership functions using fuzzy programming, Das et al. [8] have introduced a new solution methods for the MOTP where cost, source and destinations are interval parameters, and Wahed [9] has found optimal compromise solution of MOTP problems, and also its quality is tested by using degree of closeness of the compromise solution to the ideal solution using a class of distance functions. Wahed and Lee [10] have studied fuzzy goal programming approach and presented interactive approach for the efficient solution of MOTP problem, Liu and Kao [11] have studied fuzzy transportation problems, where cost, supply and demand are fuzzy numbers and gave the solution approach using extension principle. Liu [12] has studied the total cost bounds of the transportation problem with varying demand and supply. It very interesting to point out that approaches based fuzzy set do not deal transportation problem having imprecision, uncertainty and hesitation properly.

Atanassov [13] has presented the extension of the fuzzy set term as intuitionistic fuzzy sets (IFS) by using non-membership and hesitation degrees. It was Angelov [14] who applied IFS in optimization problem particularly in TP, Later Bharati et al. [15–23] studied MOLP problem and presented computational algorithms for it based on IFS, and further it is implemented in various agricultural and industrial production planning problems. Jana and Roy [24] have presented a solution methodology for MOLP, and compared the results obtained using linear and non-linear membership functions. Antony et al. [25] have presented a solution method for triangular intuitionistic fuzzy version of transportation problem, and further, Singh and Yadav [26] investigated more powerful method for the solution of intuitionistic fuzzy transportation problem of type-2. Pandian [27] have studied fully intuitionistic fuzzy transportation problem, and Kumar and Hussain [28, 29] have proposed computational method to minimize the computational complexity of fully intuitionistic fuzzy balanced and unbalanced transportation problem by using existing ranking function. Recently, Bharati and Malhotra [30] have studied time minimizing transportation problem using IFS, and presented a computational algorithm for its solutions.

Fractional transportation problem has one special feature that it reduces the number of objective functions of ordinary transportation problem. Pradhan [31] presented a method for the solution of fractional transportation problem, Porchelvi and Sheela [32] have presented computational method for the solution of interval-valued frac-

Table 1 Comparison among different fractional transportation problems

S. no.	Article	Approach
1	Porchelvi and Sheela [32]	Fractional transportation with interval parameters
2	Liu Shiang-Tai [6]	Fractional transportation with fuzzy parameters
3	Pradhan and Biswal [31]	Fractional transportation with fixed numbers
4	Kocken et al. [33]	Fractional transportation with interval parameters
5	Proposed method	Fractional transportation with intuitionistic fuzzy parameters

tional transportation problem. Here, we present comparison of some famous methods in Table 1. There are several articles are presented but there is no any article related to fractional transportation problems with intuitionistic fuzzy parameters is available in literature. Here, we investigate the concept of fractional transportation problem, where cost, demand, supply are full of imprecise and hesitation, and all these imprecise parameters are presented by trapezoidal intuitionistic fuzzy numbers, and for its solutions, we developed a method based on expectation of trapezoidal intuitionistic fuzzy numbers.

Definition 1 The mathematical model of the fractional transportation problem can formulated as:

$$\text{Minimize } \frac{\sum_{i=1}^{m} \sum_{j=1}^{n} c_{ij} x_{ij} + \alpha}{\sum_{i=1}^{m} \sum_{j=1}^{n} d_{ij} x_{ij} + \beta}$$

$$\text{Such that } \sum_{j=1}^{n} x_{ij} = a_i, i = 1, 2, ..., m \tag{1}$$

$$\sum_{i=1}^{n} x_{ij} = b_j, j = 1, 2, ..., n,$$

$$x_{ij} \geq 0.$$

Definition 2 If atleast one parameter of the problem (1) is a intuitionistic fuzzy number, called intuitionistic fuzzy fractional transportation problem.

Definition 3 An intuitionistic fuzzy set A in X is expressed as $\tilde{A} = \{(x, \mu_{\tilde{A}}(x), \nu_{\tilde{A}}(x))\}$, where $\mu_{\tilde{A}}(x) : X \rightarrow [0, 1]$ and $\nu_{\tilde{A}}(x) : X \rightarrow [0, 1]$ define the degree of membership and degree of non-membership of the element $x \in X$, respectively and for every $x \in X$, satisfy $0 \leq \mu_{\tilde{A}}(x) + \nu_{\tilde{A}}(x) \leq 1$.
The term defined by $\pi_{\tilde{A}}(x) = 1 - \mu_{\tilde{A}}(x) - \nu_{\tilde{A}}(x)$, is called the degree of hesitation of the element $x \in X$ to the intuitionistic fuzzy set A.
If $\pi_{\tilde{A}}(x) = 0$, an intuitionistic fuzzy set reduce to fuzzy set and takes the form $\tilde{A} = \{(x, \mu_{\tilde{A}}(x), 1 - \mu_{\tilde{A}}(x))\}$.

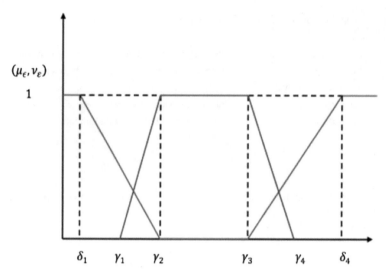

Fig. 1 Graphical representation of trapezoidal intuitionistic fuzzy numbers

Definition 4 $\tilde{A} \cap \tilde{B} = \{(x, \min(\mu_{\tilde{A}}(x), \mu_{\tilde{B}}(x)), \max(\nu_{\tilde{A}}(x), \nu_{\tilde{B}}(x)) : x \in X\}.$
$\tilde{A} \cup \tilde{B} = \{(x, \max(\mu_{\tilde{A}}(x), \mu_{\tilde{B}}(x)), \min(\nu_{\tilde{A}}(x), \nu_{\tilde{B}}(x)) : x \in X\}.$

Definition 5 An intuitionistic fuzzy set can be defined on any non-empty set but intuitionistic fuzzy number can always be defined on set of real numbers. An IFS $\tilde{A} = \{(x, \mu_{\tilde{A}}(x), \nu_{\tilde{A}}(x))\}$ is defined on set of real numbers is called intuitionistic fuzzy numbers if it satisfies following conditions:
(i) $\mu_{\tilde{A}}$ and $\nu_{\tilde{A}}$ both form fuzzy numbers.
(ii) $\mu_{\tilde{A}}(x) \leq \nu_{\tilde{A}}^{c}(x)$ for all x.

Definition 6 Any intuitionistic fuzzy number $\{(\gamma_1, \gamma_2, \gamma_3, \gamma_4); (\delta_1, \gamma_2, \gamma_3, \delta_4)\}$, where $\gamma_1, \gamma_2, \gamma_3, \gamma_4, \delta_1, \gamma_2, \gamma_3, \delta_4 \in \mathbb{R}$ such that $\delta_1 \leq \gamma_1 \leq \gamma_2 \leq \gamma_3 \leq \gamma_3 \leq \delta_4$ is said to be a trapezoidal intuitionistic fuzzy number if its membership and non-membership functions have the form

$$\mu_{\tilde{A}}(x) = \begin{cases} 1, & x \in [\gamma_2, \gamma_3] \\ \frac{x-\gamma_1}{\gamma_2-\gamma_1}, & \gamma_1 < x < \gamma_2 \\ \frac{\gamma_4-x}{\gamma_4-\gamma_3}, & \gamma_3 < x < \gamma_4 \\ 0, & x \geq \gamma_3, x \leq \gamma_1 \end{cases} \quad \nu_{\tilde{\varepsilon}}(x) = \begin{cases} 1, & x > \delta_4, x < \delta_1 \\ \frac{\gamma_2-x}{\gamma_2-\delta_1}, & \delta_1 < x < \gamma_2 \\ \frac{x-\gamma_3}{\delta_4-\gamma_3}, & \gamma_3 < x < \delta_4 \\ 0, & x \in [\gamma_2, \gamma_3] \end{cases} \quad (2)$$

In above, if $\gamma_2 = \gamma_3$ trapezoidal intuitionistic fuzzy number reduce to triangular intuitionistic fuzzy number. The graphical representation of trapezoidal intuitionistic fuzzy number is presented in Fig. 1.

Definition 7 Trapezoidal intuitionistic fuzzy number cannot be used directly in computational method so here first we convert it into its parametric form by introducing a parameter $t, 0 \leq t \leq 1$. Now suppose $\{(\gamma_1, \gamma_2, \gamma_3, \gamma_4); (\delta_1, \gamma_2, \gamma_3, \delta_4)\}$, where $\gamma_1, \gamma_2, \gamma_3, \gamma_4, \delta_1, \gamma_2, \gamma_3, \delta_4 \in \mathbb{R}$ such that $\delta_1 \leq \gamma_1 \leq \gamma_2 \leq \gamma_3 \leq \gamma_3 \leq \delta_4$ is a representation of trapezoidal intuitionistic fuzzy number. Then its parametric equations are: $u(t) = (\overline{u(t)}, \underline{u(t)})$ and $u(t) = (\overline{v(t)}, \underline{v(t)})$ where $u(t)$ and $v(t)$, where $\overline{u(t)} = \gamma_4 - t(\gamma_4 - \gamma_3)$, $\underline{u(t)} = \gamma_1 + t(\gamma_2 - \gamma_1)$ and $\overline{v(t)} = \gamma_3 - (1-t)(\delta_4 - \gamma_3)$, $\underline{v(t)} = \gamma_2 + (1-t)(\gamma_2 - \delta_2)$.

Definition 8 Often we use non-negative intuitionistic numbers for dealing real life problems because negative intuitionistic fuzzy numbers have no importance in real life problems. Here we define positive intuitionistic fuzzy numbers only for the application point of view in real life problems. Let $\{(\gamma_1, \gamma_2, \gamma_3, \gamma_4); (\delta_1, \gamma_2, \gamma_3, \delta_4)\}$ be a representation of trapezoidal intuitionistic fuzzy number $\widetilde{\varepsilon}$. If $\delta_1 > 0$, it results all $\gamma_1, \gamma_2, \gamma_3, \gamma_4, \delta_4$ are positive real numbers. Such intuitionistic fuzzy numbers are called positive.

Property 1. Suppose $\{(\gamma_1, \gamma_2, \gamma_3, \gamma_4); (\delta_1, \gamma_2, \gamma_3, \delta_4)\}$, and $\{(\eta_1, \eta_2, \eta_3, \eta_4), (\xi_1, \eta_2, \eta_3, \xi_4)\}$ are representation of two trapezoidal intuitionistic fuzzy numbers $\widetilde{\varepsilon}_1$ and $\widetilde{\varepsilon}_2$ respectively. Then

$$\widetilde{\varepsilon}_1 \oplus \widetilde{\varepsilon}_2 = \{(\gamma_1 + \eta_1, \gamma_2 + \eta_2, \gamma_3 + \eta_3, \gamma_4 + \eta_4); (\delta_1 + \xi_1, \gamma_2 + \eta_2, \gamma_3 + \eta_3, \delta_4 + \xi_4)\} \tag{3}$$

Property 2. Suppose $\{(\gamma_1, \gamma_2, \gamma_3, \gamma_4); (\delta_1, \gamma_2, \gamma_3, \delta_4)\}$ is a representation of trapezoidal intuitionistic fuzzy number $\widetilde{\varepsilon}$. Then multiplication of $\widetilde{\varepsilon}$ with any constant real number c is defined as:

$$k\widetilde{A} = \begin{cases} (c\gamma_1, c\gamma_2, c\gamma_3, c\gamma_4; c\delta_1, c\gamma_2, c\gamma_3, c\delta_4), & c > 0 \\ (c\gamma_4, c\gamma_3, c\gamma_2, c\gamma_1; c\delta_4, c\delta_3, c\delta_2, c\delta_1), & c < 0 \end{cases} \tag{4}$$

Property 3. Two trapezoidal intuitionistic fuzzy numbers are said to be equal if and only if their representations are same, i.e. if $\widetilde{\varepsilon}_1 = \{(\gamma_1, \gamma_2, \gamma_3, \gamma_4); (\delta_1, \gamma_2, \gamma_3, \delta_4)\}$ and $\widetilde{\varepsilon}_2 = \{(\eta_1, \eta_2, \eta_3, \eta_4), (\xi_1, \eta_2, \eta_3, \xi_4)\}$ are said to be equal iff $\gamma_1 = \eta_1, \gamma_2 = \eta_2, \gamma_3 = \eta_3, \gamma_4 = \eta_4; \delta_1 = \xi_1, \gamma_2 = \eta_2, \gamma_3 = \eta_3, \delta_4 = \xi_4$.

2 Intuitionistic Fuzzy Expectation

Suppose $IF(\mathbb{R})$ represents a collection of all trapezoidal intuitionistic fuzzy numbers on $\mathbb{R}$, $\widetilde{\varepsilon} = \{(\gamma_1, \gamma_2, \gamma_3, \gamma_4); (\delta_1, \gamma_2, \gamma_3, \delta_4)\}$ we denote and define intuitionistic fuzzy expectation of triangular intuitionistic fuzzy number as:

$$e(\widetilde{\varepsilon}) = \tfrac{1}{4}\{\int_0^1 \underline{u(t)}dt + \int_0^1 \overline{u(t)}dt + \int_0^1 \underline{v(t)}dt + \int_0^1 \overline{v(t)}dt\}, \text{ where } t \text{ is a parameters such}$$

that $0 \le t \le 1$

$$= \frac{1}{4}\{\int_0^1 \{\gamma_1 + t(\gamma_2 - \gamma_1)\}dt + \int_0^1 \{\gamma_4 - t(\gamma_4 - \gamma_3)\}dt + \int_0^1 \{\gamma_2 + (1 - t)(\gamma_2 - \delta_2)\}dt$$

$$+ \int_0^1 \{\gamma_3 - (1 - t)(\delta_4 - \gamma_3)\}dt\},$$

$$= \frac{\gamma_1 + 2\gamma_2 + 2\gamma_3 + \gamma_4 + \delta_1 + \delta_4}{8},$$

$$\therefore e(\tilde{\varepsilon}) = \frac{\gamma_1 + 2\gamma_2 + 2\gamma_3 + \gamma_4 + \delta_1 + \delta_4}{8}.$$

Remark If $\tilde{\varepsilon} = \{(\gamma_1, \gamma_2, \gamma_3, \gamma_4); (\delta_1, \gamma_2, \gamma_3, \delta_4)\}$ be trapezoidal intuitionistic fuzzy number is positive TIFNs, Then $e(\tilde{\varepsilon}) > 0$.

3   Theorems Based on Intuitionistic Fuzzy Expectation

Theorem 9 $e(c\tilde{\varepsilon}) = ce(\tilde{\varepsilon})$.

Proof Case (i): If $c = 0$, prove is obvious.
Case (ii): If $c > 0$
Let $\tilde{\varepsilon} = \{(\gamma_1, \gamma_2, \gamma_3, \gamma_4); (\delta_1, \gamma_2, \gamma_3, \delta_4)\}$ be trapezoidal intuitionistic fuzzy number and c be a real number.
$c\tilde{\varepsilon} = c\{(\gamma_1, \gamma_2, \gamma_3, \gamma_4); (\delta_1, \gamma_2, \gamma_3, \delta_4)\} = \{(c\gamma_1, c\gamma_2, c\gamma_3, c\gamma_4); (c\delta_1, c\gamma_2, c\gamma_3, c\delta_4)\}$
by using property 2, Now taking IF expectation of $c\tilde{\varepsilon}$, we get
$e(c\tilde{\varepsilon}) = e(c\tilde{\varepsilon}) = e((c\gamma_1, c\gamma_2, c\gamma_3, c\gamma_4); (c\delta_1, c\gamma_2, c\gamma_3, c\delta_4))$

$$= \frac{c\gamma_1 + 2c\gamma_2 + 2c\gamma_3 + c\delta_1 + c\gamma_4 + c\delta_4}{8}$$

$$= k.\frac{\gamma_1 + 2\gamma_2 + \gamma_3 + \delta_1 + \gamma_4 + \delta_4}{8}$$

$$= c.e(\tilde{\varepsilon}).$$

Case (iii): If $c < 0$
Suppose $\tilde{\varepsilon} = \{(\gamma_1, \gamma_2, \gamma_3, \gamma_4); (\delta_1, \gamma_2, \gamma_3, \delta_4)\}$ is a representation of trapezoidal intuitionistic fuzzy number
$c\tilde{\varepsilon} = c\{(\gamma_1, \gamma_2, \gamma_3, \gamma_4); (\delta_1, \gamma_2, \gamma_3, \delta_4)\} = \{((\gamma_4, \gamma_3, \gamma_2, \gamma_1); (\delta_4, \gamma_3, \gamma_2, \delta_1))\}$, where c is a real number, after using property 2 and taking IF expectation both sides, we have

$$e(c\tilde{\varepsilon}) = \frac{c\gamma_4 + 2c\gamma_3 + 2c\gamma_2 + c\delta_4 + c\delta_1}{8}$$

$$= c.\frac{\gamma_4 + 2\gamma_3 + 2\gamma_2 + \delta_1 + \delta_4}{8}$$

$$= c.e(\tilde{\varepsilon}).$$

Theorem 10 $e(\tilde{\varepsilon}_1 + \tilde{\varepsilon}_2) = e(\tilde{\varepsilon}_1) + e(\tilde{\varepsilon}_2)$.

Proof Suppose $\tilde{\varepsilon}_1 = \{(\gamma_1, \gamma_2, \gamma_3, \gamma_4); (\delta_1, \gamma_2, \gamma_3, \delta_4)\}$ and $\tilde{\varepsilon}_2 = \{(\eta_1, \eta_2, \eta_3, \eta_4), (\xi_1, \eta_2, \eta_3, \xi_4)\}$ are representations of two trapezoidal intuitionistic fuzzy numbers,

$\tilde{\varepsilon}_1 + \tilde{\varepsilon}_2 = \{(\gamma_1 + \eta_1, \gamma_2 + \eta_2, \gamma_3 + \eta_3, \gamma_4 + \eta_4); (\delta_1 + \xi_1, \gamma_2 + \eta_2, \gamma_3 + \eta_3, \delta_4 + \xi_4)\}$, After using property 2 and taking IF expectation, we obtain

$e(\tilde{\varepsilon}_1 + \tilde{\varepsilon}_2) = \dfrac{(\gamma_1 + \eta_1) + 2(\gamma_2 + \eta_2) + 2(\gamma_3 + \eta_3) + (\gamma_4 + \eta_4) + (\delta_1 + \xi_1) + 2(\gamma_2 + \eta_2) + 2(\gamma_3 + \eta_3) + (\delta_4 + \xi_4)}{8}$

$= \dfrac{(\gamma_1 + \eta_1) + (2\gamma_2 + 2\eta_2) + (\gamma_3 + \eta_3) + (\gamma_4 + \eta_4) + (\delta_1 + \xi_1) + (2\gamma_2 + 2\eta_2) + (\gamma_3 + \eta_3) + (\delta_4 + \xi_4)}{8}$

$= \dfrac{(\gamma_1 + \delta_1 + 2\gamma_2 + 2\gamma_2 + \gamma_3 + \gamma_3 + \gamma_4 + \delta_4) + (\eta_1 + \xi_1 + 2\eta_2 + 2\eta_2 + \eta_3 + \eta_3 + \eta_4 + \xi_4)}{8}$

$= \dfrac{\gamma_1 + \delta_1 + 2\gamma_2 + 2\gamma_3 + \gamma_4 + \delta_4}{8} + \dfrac{\eta_1 + \xi_1 + 2\eta_2 + 2\eta_3 + \eta_4 + \xi_4}{8}$

$= e(\tilde{\varepsilon}_1) + e(\tilde{\varepsilon}_2)$.

Theorem 11 *For constant trapezoidal intuitionistic fuzzy number* $\tilde{\varepsilon}$, $e(\tilde{\varepsilon}) = \varepsilon$.

Proof Constant trapezoidal intuitionistic fuzzy number $\tilde{\varepsilon}$ can be written as: $\tilde{\varepsilon} = \{(\varepsilon, \varepsilon, \varepsilon, \varepsilon); (\varepsilon, \varepsilon, \varepsilon, \varepsilon)\}$ be trapezoidal intuitionistic fuzzy number. By definition expected value, we get

$e\{\{(\varepsilon, \varepsilon, \varepsilon, \varepsilon); (\varepsilon, \varepsilon, \varepsilon, \varepsilon)\}\}$

$= \dfrac{\varepsilon + \varepsilon + 2.\varepsilon + 2.\varepsilon + \varepsilon + \varepsilon}{8} = \dfrac{8\varepsilon}{8}$

$= \varepsilon$.

Remark In particular, if $\tilde{\varepsilon} = 0$

$e(\tilde{\varepsilon}) = e(0) = 0$.

Theorem 12 $e\{\frac{\tilde{c}x + \tilde{\alpha}}{\tilde{d}x + \tilde{\beta}}\} = \frac{e\{\tilde{c}x + \tilde{\alpha}\}}{e\{\tilde{d}x + \tilde{\beta}\}}$, *provided* $e\{\tilde{d}x + \tilde{\beta}\} > 0$.

Proof Let $\frac{\tilde{c}x + \tilde{\alpha}}{\tilde{d}x + \tilde{\beta}} = \lambda$

$\tilde{c}x + \tilde{\alpha} = \lambda(\tilde{d}x + \tilde{\beta})$

Taking IF expectation both sides we, get

$e(\tilde{c}x + \tilde{\alpha}) = \lambda e(\tilde{d}x + \tilde{\beta})$

or $\frac{e\{\tilde{c}x + \tilde{\alpha}\}}{e\{\tilde{d}x + \tilde{\beta}\}} = \lambda$, ... (i)

$e(\frac{\tilde{c}x + \tilde{\alpha}}{\tilde{d}x + \tilde{\beta}}) = e(\lambda) = \lambda$

or

$e(\frac{\tilde{c}x + \tilde{\alpha}}{\tilde{d}x + \tilde{\beta}}) = \lambda$, ... (ii)

From (i) and (ii), we get

$e(\frac{\tilde{c}x + \tilde{\alpha}}{\tilde{d}x + \tilde{\beta}}) = \frac{e(\tilde{c}x + \tilde{\alpha})}{e(\tilde{d}x + \tilde{\beta})}$.

4 Solution Method

Solution method of fractional transportation problem consists the following steps:

Step 1: First we construct trapezoidal intuitionistic fuzzy number for each cost of fractional transportation problem.

Step 2: In this step we transform intuitionistic fuzzy fractional TP into crisp fractional

TP using Theorem 12.

Step 3: In this step we convert crisp fractional transportation problem into linear programming problem by using Charnes and Cooper transformation method [34].

Step 4: We solve linear programming problem that is obtained from previous step and we get optimal solution and stop the process.

5 Numerical Verification

Let us consider an trapezoidal intuitionistic fuzzy fractional transportation problem Minimize $z = \dfrac{\{(2,5,5.5,7),(4,5,5.5,6)\}x_{11} + 6x_{12} + 2x_{13} + 3x_{14} + 3x_{21} + 6x_{22} + 8x_{23} + 2x_{24} + 4x_{31} + 3x_{32} + 9x_{33} + 11x_{34} + 55}{\{(1,3,4,6),(2,3,4,8)\}x_{11} + 4x_{12} + x_{13} + 2x_{14} + 2x_{21} + 4x_{22} + 7x_{23} + x_{24} + 2x_{31} + x_{32} + 6x_{33} + 9x_{34} + 45}$

Such that

$x_{11} + x_{12} + x_{13} + x_{14} \leq 50,$

$x_{21} + x_{22} + x_{23} + x_{24} \leq \{(50, 60, 65, 70), (40, 60, 65, 80)\},$

$x_{31} + x_{32} + x_{33} + x_{34} \leq 80,$

$x_{11} + x_{21} + x_{31} \geq 30,$

$x_{12} + x_{22} + x_{32} \geq 20,$

$x_{13} + x_{23} + x_{33} \geq \{(30, 40, 45, 50), (20, 40, 45, 60)\},$

$x_{14} + x_{24} + x_{34} \geq 30,$

$x_{ij} \geq 0, \quad i = 1, 2, 3, \quad j = 1, 2, 3, 4.$

Using solution method, we get the following optimal solutions:

Minimum $\quad z = 281.25, x_{11} = 0, x_{12} = 0, x_{13} = 20, x_{14} = 30, x_{21} = 40, x_{22} = 0,$ $x_{23} = 21.25, x_{24} = 0, x_{31} = 60, x_{32} = 20, x_{33} = 0, x_{34} = 0.$

6 Conclusions

In this paper, we investigate the concept of trapezoidal intuitionistic fuzzy fractional transportation problem, concept of expectation of trapezoidal intuitionistic fuzzy numbers. We also propose and prove some theorems which are necessary for method for the optimal solution of trapezoidal intuitionistic fuzzy fractional transportation problem. Here, we have considered all cost coefficients, coefficient matrix are trapezoidal intuitionistic fuzzy numbers. Proposed method is very easy to handle the intuitionistic fuzzy fractional transportation problem generally occurs in real life transportation problem. Presented idea can be extended for the case of multiobjective fractional transportation problems.

References

1. Hitchcock, F.L.: The distribution of a product from several sources to numerous localities. J. Math. Phys. **20**, 224–230 (1941)

2. Swarup, K.: Transportation technique in linear fractional programming. J. R. Naval Sci. Serv. **21**(5), 256–260 (1996)
3. Zadeh, L.A.: Fuzzy Sets. Informat. Control **8**, 338–353 (1965)
4. Bellman, R.E., Zadeh, L.A.: Decision making in a fuzzy environment. Manag Sci. **17**, 141–164 (1970)
5. Bharati, S.K.: Abhishekh, Singh, S.R. Int. J Dynam. Control (2017). https://doi.org/10.1007/s40435-017-0355-1
6. Shiang-Tai, Liu: Fractional transporation problem with fuzzy parameters. Soft Comput. **20**, 3629–3636 (2016)
7. Verma, R., Biswal, M.P., Biswal, A.: Fuzzy programming technique to solve multi-objective transportation problems with some non-linear membership functions. Fuzzy Sets Syst. **91**, 37–43 (1997)
8. Das, S.K., Goswami, A., Alam, S.S.: Multiobjective transportation problem with interval cost, source and destination parameters. Eur. J. Operation. Res. **117**, 100–112 (1997)
9. Wahed Abd, F., Waiel, : A multi-objective transportation problem under fuzziness. Fuzzy Sets Syst. **177**, 27–33 (2001)
10. Wahed Abd, F., Waiel, Lee S.M.: A multi-objective transportation problem under fuzziness. Fuzzy Sets Syst. **34**, 158–166 (2006)
11. Shiang-Tai, Liu: The total cost bounds of the transportation problem with varying demand and supply. Omega **31**, 247–251 (2003)
12. Shiang-Tai, Liu, Kao, Chiang: Solving fuzzy transportation problems based on extension principle. Eur. J. Operatio. Res. **153**, 661–674 (2004)
13. Atanassov, K.T.: Intuitionistic fuzzy sets. Fuzzy Sets Syst. **20**, 87–96 (1986)
14. Angelov, P.P.: Optimization in an intuitionistic fuzzy environment. Fuzzy Sets Syst. **86**, 299–306 (1997)
15. Bharati, S.K., Singh, S.R.: Soft Comput (2018). https://doi.org/10.1007/s00500-018-3100-6
16. Bharati, S.K.: Hesitant fuzzy computational algorithm for multiobjective optimization problems. International Journal of Dynamics and Control (2018). https://doi.org/10.1007/s40435-018-0417-z
17. Bharati, S.K., Singh, S.R.: Int. J. Fuzzy Syst. (2018). https://doi.org/10.1007/s40815-018-0470-y
18. Malhotra, R., Bharati, S.K.: Intuitionistic fuzzy two stage multiobjective transportation problems. Advanc. Theor. Appl. Math. **11**, 305–316 (2016)
19. Bharati, S.K., Singh, S.R.: A note on solving a fully intuitionistic fuzzy linear programming problem based on sign distance. Int. J. Comput. Appl. **119**(23), 30–35 (2015)
20. Nishad, A.K., Bharati, S.K., Singh, S.R.: A new centroid method of ranking for intuitionistic fuzzy numbers. In: Proceedings of the Second International Conference on Soft Computing for Problem Solving (SocProS 2012), 28–30 Dec, 2012, pp. 151–159. Springer India (2014)
21. Bharati, S.K., Singh, S.R.: Intuitionistic fuzzy optimization technique in agricultural production planning: a small farm holder perspective. Int. J. Comput. Appl. **89**(6), 17–23 (2014)
22. Bharati, S.K., Nishad, A.K., Singh, S.R.: Solution of multi-objective linear programming problems in intuitionistic fuzzy environment. In: Proceedings of the Second International Conference on Soft Computing for Problem Solving (SocProS 2012), December 28–30, 2012, pp. 161–171. Springer India
23. Bharati, S.K., Singh, S.R.: Solving multi objective linear programming problems using intuitionistic fuzzy optimization method: a comparative study. Int. J. Model. Optimizat. **4**(1), 10–16 (2014)
24. Jana, B., Roy, T.K.: Multi-objective intuitionistic fuzzy linear programming and its application in transportation model. Notes Intuition. Fuzzy Sets **13**, 34–51 (2007)
25. Atony, R.J.P., Savarimuthu, S.J., Pathinathan, T.: Method for solving the transporation problem using triangular intutitionistic fuzzy number. Int. J. Comput. Algorithm **03**, 590–605 (2014)
26. Singh, S.K., Yadav, S.P.: A new approach for solving intutionistic fuzzy transporation problem of type-2. Ann. Oper. Res. (2014). https://doi.org/10.1007/s10479-014-1724-1

27. Pandian, P.: Realistic method for solving fully intuitionistic fuzzy transportation problems. Appl. Math. Sci. **8**(113), 5633–5639 (2014)
28. Kumar, P.S., Hussain, R.J.: A method for solving unbalanced intuitionistic fuzzy transportation problems. Notes Intuition. Fuzzy Sets **21**(3), 54–65 (2015)
29. Kumar, P.S., Hussain, R.J.: Computationally simple approach for solving fully intuitionistic fuzzy real life transportation problems. Int. J. Syst. Assur. Eng. Manag. **1**, 1–12 (2015)
30. Bharati, S.K., Malhotra, R.: Int. J. Syst. Assur. Eng. Manag. **8**(Suppl 2), 1442 (2017). https://doi.org/10.1007/s13198-017-0613-9
31. Pradhan, A., Biswal, M.P.: Computational methodology for linear fractional transportation problem. In: Proceedings of the 2015 Winter Simulation Conference, pp. 3158–3159. IEEE Press (2015)
32. Porchelvi, R.S., Sheela, A.: A linear fractional interval transportation problem with and without budgetary constraints. Intern. J. Fuzzy Math. Arch. **9**(2), 165–170 (2015)
33. Kocken, H.G., Emiroglu, I., Guler, C., Tasci, F., Sivri, M.: The fractional transportation problem with interval demand, supply and costs. In: AIP Conference Proceedings, vol. 1557, No. 1, pp. 339–344. AIP (2013)
34. Charnes, A., Cooper, W.W.: Programming with linear fractional functionals. Naval Res. Logist. Quart. **9**(3–4), 181–186 (1962)

Chemo-Inspired GA for Non-convex Economic Load Dispatch

Rajashree Mishra and Kedar Nath Das

Abstract Powered by the chemotactic step of bacterial foraging optimization (BFO), a new hybrid genetic algorithm is proposed in this paper for solving non-linear constrained optimization problems. In the recent past, researchers attempted to hybridize the GA and BFO for improving the quality of the solution. However, this hybridization unnecessarily increases the computational burden as some of the mechanisms/steps are seem to be technically repeated. It is due to the fact that the internal mechanism of *selection* in GA and the *reproduction* in BFO; and the *elitism* in GA and *elimination-dispersal* step in BFO is almost similar. Undoubtedly, *chemotactic step* plays the vital role in the better performance of BFO. Therefore in this present study, only the chemotactic step of BFO is considered for hybridization with GA. Further, it is designed to tackle constrained optimization problems and is named as chemo-inspired genetic algorithm for constrained optimization (CGAC). Here in this paper, it is applied to solve economic load dispatch (ELD) problem, and finally, the result comparison has been done with other state-of-the-art algorithms to validate the superiority of CGAC.

Keywords Economic load dispatch · Valve-point loading effect
Ramp rate limit · Constrained optimization

1 Introduction

Economic load dispatch (ELD) problem is a powerful and challenging optimization problem in present-day scenario in small-, medium- and large-scale power systems. The said problem has the objective of determining the power output taking into

R. Mishra (✉)
School of Applied Sciences, KIIT University, Bhubaneswar 751024, India
e-mail: rajashreemishra011@gmail.com

K. N. Das
Department of Mathematics, National Institute of Technology, Silchar 788010,
Assam, India
e-mail: kedar.iitr@gmail.com

© Springer Nature Singapore Pte Ltd. 2019
J. C. Bansal et al. (eds.), *Soft Computing for Problem Solving*, Advances in Intelligent
Systems and Computing 817, https://doi.org/10.1007/978-981-13-1595-4_67

account various operational constraints. Various constraints are included in the above problem such as power balance constraints, ramp rate limits, prohibited operating zones and valve-point loading effect. There exist various deterministic methods to tackle the above problem. They are (i) dynamic programming [1], (ii) Lagrangian relaxation algorithm [2], (iii) lambda iteration method [3], (iv) quadratic programming [4], (v) gradient method [5], (vi) linear programming [6] and many other methods. But all the conventional methods are mostly handicapped to handle the higher order nonlinearities arising in large-scale ELD problem. It requires huge computational steps to reach global optimal solution. Many times, the method is trapped in local optima instead of global one. To overcome the deficiencies in the above methods, in the recent past, many popular evolutionary algorithms came into existence. To list a few, they are GA [7], PSO [8], interactive honey bee mating optimization (IHBMO) [9], iteration particle swarm optimization (IPSO) [10], firefly algorithm [11], quick group search optimizer (QGSO) [12], krill herd algorithm (KHA) [13], chaotic PSO [14], gravitational search algorithm (GSA) [15], artificial immune system (AIS) [16], θ-PSO [17], pattern search method [18].

Later, quite a good number of hybridized algorithms have been developed to handle medium- and large-scale ELD problems. Some of the efficient algorithms are differential harmony search algorithm (DHS) [19], shuffled differential algorithm (SDE) [20], hybrid differential evolution (HDE) [21], DE based on PSO (DEPSO) [22], hybrid PSO and gravitational search (HPSO-GSA) algorithm [23], hybrid (PSO-SQP) [24], etc.

Inspired by the novel concept of hybridization of several algorithms, in this paper GA has been hybridized with chemotactic step of BFO. It is named as chemo-inspired GA for constrained optimization (CGAC). The proposed algorithm is applied to solve the following problem where three power system problems of economic load dispatch (ELD) of different generator constraints have been taken into account.

The rest of the paper is organized as follows. Design of CGAC for ELD problem is discussed in Sect. 2. Section 3 describes the problem definition of economic load dispatch (ELD). Three test cases are considered in Sect. 4, and the experimental set-up and the corresponding result analysis for them are given in Sects. 5 and 6, respectively. The final conclusion of this paper is drawn in Sect. 7.

2 Design of CGAC for ELD Problem

Initially, the proposed algorithm has been designed to solve unconstrained optimization problems [25, 26]. The algorithm is verified over several benchmark problems available in the literature [26]. To test its efficacy, it is also applied to various real-life applications on unconstrained optimization [27]. Later, the algorithm is developed to handle the constrained optimization problems [28, 29]. The pseudocode of CGAC is discussed as follows.

Pseudocode of CGAC:

The **algorithm *CGAC* ( )** is given below. The parameters used in the CGAC are taken from our previous paper [25, 26].

Algoithm CGAC ( )

{

Step1: **//Initialize the CGAC parameters**.

Set $i = j = 0$. $S = 4$, and generate the initial population $P(t)$

Step 2: Evaluate the objective function value *fit* $X^i(j)$

Step 3: Apply selection operator.

Step 4: Create offspring $C(t)$ applying crossover

Step 5: Create offspring $C'(t)$ applying mutation

Step 6: Apply complete elitism, evaluate the objective function value, and update F_{best} and X_{best}

Step 7: **If** $(Gen > 10)$ Go to Step 8

Else Go to Step 3

Step 8: *//Initialize the Chemo tactic Loop*;

$j = j + 1$

   (a) *//Initialize the Bacteria loop*;

      $i = i + 1$

   (b) *//Tumble and Move*:

      (i) Reinitialize the SearchSpace within *Min(j) and Max(j)* as [25, 26] Create population and evaluate objective function value. Update F_{best} and X_{best}

      (ii) Evaluate the new position in $j + 1$th chemotactic step as Eq. (1)

$$X^i(j + 1) = X^i(j) + C_{step} \frac{\Delta(i)}{\sqrt{\Delta(i)\Delta^T(i)}} \qquad (1)$$

      (iii) Update F_{best} and X_{best}

   (c) *//Swim loop:* Set m=0

        While ($m < N_s$) **do**

           $m = m + 1$

           If ($Fit\_Fun\left(X^i\left(j + 1\right)\right) < F_{best}$)

              {

              Update F_{best} and X_{best}

              Evaluate the modification of new position in $j + 1^{th}$ step   as Eq. (2)

$$X^i(j+1) = X^i(j+1) + C_{step} \frac{\Delta(i)}{\sqrt{\Delta(i)\Delta^\mathsf{T}(i)}} \qquad (2)$$

Evaluate $Fit\_Fun\left(X^i(j+1)\right)$

}

Else

$X^i(j+1) = X^i(j)$ and Set $m = N_S$

End While

(d)   If $(i < S)$ Go to step 8(a)

Else Go to Step 8(e)

(e)   If $(j < N_c)$ Go to Step 8

Else Go to Step 9

Step 9: If $(gen < gen\_max)$ Go to Step 3

Else Exit

}

3   Problem Definition of ELD Problem

The ELD problem is mathematically modelled by Eq. (3)

$$MinF = \sum_{j=1}^{n} F_j(P_j) \qquad (3)$$

The objective function is taken as F. The power output of jth unit is P_j. The generation cost $F_j(P_j)$ is a quadratic cost function defined as Eq. (4)

$$F_i = a_j P_j^2 + b_j P_j + c_j \qquad (4)$$

The generator coefficients are a_j, b_j and c_j, respectively.

The sinusoidal terms are added to the quadratic cost function if the valve-point loading effect is taken into consideration, and modified generation cost is defined as follows by Eq. (5)

$$F_j(P_j) = a_j P_j^2 + b_j P_j + c_j + \left| e_j \sin\left(f_j\left(P_j^{min} - P_j\right)\right)\right| \qquad (5)$$

The coefficients e_j and f_j of unit j reflect the valve-point effects. Here, P_j is the jth generator, s power output. The minimum power output of the jth generator is P_j^{min}.

The ELD problem is solved under the following constraints.

(I) Generator Capacity Constraints:

The variation of power output is within minimum power output $\left(P_j^{\min}\right)$ and maximum power output $\left(P_j^{\max}\right)$. It is defined as Eq. (6)

$$P_j^{\min} \leq P_j \leq P_j^{\max} \tag{6}$$

(II) Power Balance Constraint:

The power balance constraint is the sum of system load demand and total transmission loss which is defined as Eq. (7)

$$\sum_{j=1}^{n} P_j = P_D + P_L, \tag{7}$$

The power loss P_L is calculated by means of the B-coefficient matrix, which can be expressed as the quadratic function of unit' s power output defined as Eq. (8).

$$P_L = \sum_{j=1}^{n}\sum_{k=1}^{n} P_j B_{jk} P_k + \sum_{j=1}^{n} B_{0j} P_j + B_{00} \tag{8}$$

where B_{jk} is the jkth component of loss coefficient square matrix of size n.

(III) Ramp Rate Limits:

Let P_j is the present power output of jth unit and P_j^0 is the previous power output of the jth unit. Then, the ramp rate limits are described by Eqs. (9) and (10).

$$P_j - P_j^0 \leq UR_j \tag{9}$$

$$P_j^0 - P_j \leq DR_j \tag{10}$$

UR_j and DR_j are, respectively, the up-ramp and down-ramp limits of the jth generator (in units of MW/time period). The inclusion of the ramp rate limit is included in the generator constraints as follows.

$$Max\left(P_j^{\min}, P_j^0 - DR_j\right) \leq P_j \leq Min\left(P_j^{\max}, P_j^0 + UR_j\right) \tag{11}$$

(IV) Prohibited Operating Zones:

The limitation of machine components and instability concerns of the operation zones impose certain restrictions on the generating units. So, the prohibited operating zones can be mathematically formulated as follows.

$$\begin{cases} P_j^{\min} \leq P_j \leq P_{j,1}^{lb} \\ P_{j,k-1}^{ub} \leq P_j \leq P_{j,k}^{lb}, \ k = 2, 3, \ldots, nP_j \\ P_{j,k}^{ub} \leq P_j \leq P_j^{\max}, \ k = nP_j \end{cases} \quad (12)$$

Here, $P_{j,k}^{lb}$ is the lower bound and $P_{j,k}^{ub}$ is the upper bound of prohibited operating zone k of the generator j, respectively. The total number of POZs of generator j is nP_j.

4   Test Cases Under Consideration

Three test cases have been considered comprising 3, 13 and 40 generating units. The expected power demands to be met by the generating units are 850 MW, 2520 MW and 10,500 MW, respectively. The system data can be found from [30, 31]. The test cases consisting of 3, 13 and 40 generators are solved by taking into account the valve-point loading effect with the power balance constraint and generator capacity constraint. Methodology adapted for evaluating the slack generator and evaluating the power output of the dependent unit violating the capacity constraints are discussed as follows.

(I) Calculation of Slack Generator [32]
Let P_n be the power output of the dependent generator.

$$P_n = P_D + P_L - \sum_{j=1}^{n-1} P_j \quad (13)$$

where the transmission loss is a function of all the generator outputs including the slack generator and it is given by

$$P_L = \sum_{j=1}^{n-1}\sum_{k=1}^{n-1} P_j B_{jk} P_k + 2P_n \left(\sum_{j=1}^{n-1} B_{nj} P_j \right) + B_{nn} P_n^2 + \sum_{j=1}^{n-1} B_{0j} P_j + B_{0n} P_n + B_{00} \quad (14)$$

Substituting the power loss term P_L into Eq. (13), the quadratic equation obtained is given in Eq. (15)

$$B_{nn} P_n^2 + 2\left(\sum_{j=1}^{n-1} B_{nj} P_j + B_{0n} - 1 \right) P_n + \left(P_D + \sum_{j=1}^{n-1}\sum_{k=1}^{n-1} P_j B_{jk} P_k + \sum_{j=1}^{n-1} B_{0j} P_j - \sum_{j=1}^{n-1} P_j + B_{00} \right) = 0 \quad (15)$$

It is a quadratic equation of the form

$$xP_n^2 + yP_n + z = 0, \ \text{where } x = B_{nn} \quad (16)$$

$$y = \left(2 \sum_{j=1}^{n-1} B_{nj} P_j + B_{0n} - 1 \right) \tag{17}$$

$$z = \left(P_D + \sum_{j=1}^{n-1} \sum_{k=1}^{n-1} P_j B_{jk} P_k + \sum_{j=1}^{n-1} B_{0j} P_j - \sum_{j=1}^{n-1} P_j + B_{00} \right) \tag{18}$$

$$P_n = \frac{-y \pm \sqrt{y^2 - 4xz}}{2x} \tag{19}$$

where

$$y^2 - 4xz \geq 0 \tag{20}$$

If the inequality constraint Eq. (20) is violated, then the power output of the generator is again reinitialized until the constraint equation is satisfied including all other constraints of the problem concerned. If the equation is still violated at later stage of the algorithm, the same procedure is repeated.

(II) Calculation of Power Output of Dependent Unit Violating the Capacity Constraints

If the power generation of the dependent unit P_n exceeds the maximum limit, then

$$Difference = P_n - P_n^{\max} \tag{21}$$

Let

$$E = \frac{Differnce}{n - 1} \tag{22}$$

$$P_n = \begin{cases} P_n^{\min}, & \text{if } P_n < P_n^{\min} \\ P_n^{\max} \text{ and } P_j = P_j + E \text{ for } j = 1, 2, \ldots, n - 1, & \text{if } P_n > P_n^{\max} \end{cases} \tag{23}$$

If for any of P_j^s, the generator capacity constraint Eq. (23) is violated, then

$$P_j = \begin{cases} P_j^{\min}, & \text{if } P_j \leq P_j^{\min} \text{ for } j = 1, 2, \ldots, n - 1 \\ P_j^{\max}, & \text{if } P_j \geq P_j^{\max} \text{ for } j = 1, 2, \ldots, n - 1 \end{cases} \tag{24}$$

5 Experimental Set-Up

Three test cases are considered with valve-point loading effect, generator capacity constraint, and power balance constraint. It has been compared with firefly algorithm and many other algorithms. The comparison of result for three-unit system is done in terms of total generation cost and CPU time which is shown in Table 1. Comparison is made in terms of best, average, worst value, SD, number of function evaluations and shown in Table 2. Table 3 represents the best solutions obtained for 13-unit system. Tables 4 and 5 represent the result comparison for 40-unit system in terms of mean cost, best cost, mean time, SD and number of function evaluations. Keeping the parameters fixed, the comparison graphs have been plotted for GA-BFO and CGAC.

Table 1 Best solutions of system 1 (3 unit) [30]

Method	P_1 (MW)	P_2 (MW)	P_3 (MW)	Total generation cost ($/h)	CPU time (s)
EP	300.264	400	149.736	**8234.07**	6.78
EP-SQP	300.267	400	149.733	**8234.07**	5.12
PSO	300.268	400	149.732	**8234.07**	4.37
PSO-SQP	300.267	400	149.733	**8234.07**	3.37
CPSO	300.267	400	149.733	**8234.07**	2.25
CPSO-SQP	300.266	400	149.734	**8234.07**	2.06
GA-BFO	300.264	400	149.736	**8234.07**	2.0472
CGAC	300.266	400	149.733	**8234.07**	**1.7866**

Table 2 Best, average and worst results of different ELD solution methods for the three-unit test system in terms of function evaluation [11]

Methods	Generation cost ($/h)				
	Best	Average	Worst	S.D	No. of function evaluations
GAB	8234.08	NA	NA	NA	10,000
GAF	**8234.07**	NA	NA	NA	10,000
CEP	**8234.07**	8235.97	8241.83	NA	**1000**
FEP	**8234.07**	8234.24	8241.78	NA	**1000**
MFEB	**8234.07**	8234.71	8241.8	NA	**1000**
IFEP	8234.08	8234.16	8234.54	NA	**1000**
FA	**8234.07**	**8234.08**	8241.23	3.63	5000
GA-BFO	**8234.07**	8249	8310.1	15.5337	1.09e+006
CGAC	**8234.07**	8235.25	**8240.91**	**2.5328**	22,139

Table 3 Best solution of system 2 (13 unit) [30]

Generator	Unit generation in MW								
	GA	SA	GA-SA	EP-SQP	PSO-SQP	CPSO	CPSO-SPQ	GA-BFO	CGAC
P_1	628.32	668.40	628.23	628.3136	628.3205	628.32	628.31	628.152	628.315
P_2	356.49	359.78	299.22	299.1715	299.0524	299.83	299.83	297.713	299.198
P_3	359.43	358.20	299.17	299.0474	298.9681	299.17	299.16	297.713	299.196
P_4	159.73	104.28	159.12	159.6399	159.4680	159.70	159.73	165.103	159.723
P_5	109.86	60.36	159.95	159.6560	159.1429	159.64	159.73	159.941	159.715
P_6	159.73	110.64	158.85	158.4831	159.2724	159.67	159.73	160.645	159.725
P_7	159.63	162.12	157.26	159.6749	159.5371	159.64	159.73	167.449	159.727
P_8	159.73	163.03	159.93	159.7265	158.8522	159.65	159.73	159.824	159.728
P_9	159.73	161.52	159.86	159.6653	159.7845	159.78	159.73	156.657	159.725
P_{10}	77.31	117.09	110.78	114.0334	110.9618	112.46	109.07	40.0782	77.3798
P_{11}	75.00	75.00	75.00	75.00	75.00	74.00	77.40	77.3021	77.3873
P_{12}	60.00	60.00	60.00	60.00	60.00	56.50	55.00	91.7889	92.39119
P_{13}	55.00	119.58	92.62	87.5884	91.6401	91.64	92.85	117.634	87.7637
Total generation cost in ($/h)	24,398.23	24,970.91	24,275.71	24,266.44	24,261.05	24,211.56	24,190.97	24,417.6	**24,170.03929**

Table 4 Comparison of the total generation cost in system 3 (40 unit) [30]

Method	Mean time (s)	Best cost ($/h)	Mean cost ($/h)
EP	1167.35	122,624.35	123,382.00
EP-SQP	997.73	122,323.97	122,379.63
PSO	933.39	123,930.45	124,154.49
PSO-SQP	733.97	122,094.67	122,245.25
GA-PS-SQP	46.98	121,458	122,039
CPSO	114.65	121,865.23	122,100.87
CPSO-SQP	98.49	121,458.54	122,028.16
GA-BFO	**25.1906**	121,598	121,869
CGAC	51.5447	**120,816.413**	**121,309**

One hundred independent runs are considered for the above problem. The size of the population is considered to be 40. In the chemotactic loop, four bacteria are initialized and are allowed to search the optimum intensively. Total number of chemotactic steps used is 40. Four swim steps are allowed for the bacteria. The length of the string is taken 20. The program stops at 500 generations or the optimum is achieved within 200 generations.

6 Result and Discussion

It can be visualized from the Tables 1, 2, 3, 4 and 5 that the best cost, mean cost and mean time are better in comparison with other algorithms mentioned in the table. It can also be observed that CGAC also performs better in comparison with genetic algorithm hybridized bacterial foraging optimization (GA-BFO) in terms of success rate, best, average, worst function values, standard deviation, time and total number of function evaluations as can be observed from Tables 1, 2, 3, 4 and 5, respectively. It can also be visualized from Figs. 1, 2 and 3 that CGAC outperforms genetic algorithm hybridized bacterial foraging optimization (GA-BFO). Though the function evaluation is more, CGAC is performing better in terms of best, average, worst function value, mean time as compared to many popular continuous methods and hybridized algorithms listed in the table.

7 Conclusion

In this paper, chemo-inspired GA for constrained optimization (CGAC) has been successfully applied to handle economic load dispatch (ELD) problem. Three test cases of economic load dispatch (ELD) problems have been considered and solved

Table 5 Best, average and worst results of different ED solution in terms of function evaluation for the 40-unit test system) [11]

Methods	Generation cost ($/h)				
	Best	Average	Worst	S.D	No. of function evaluation
HGPSO	124,797.13	126,855.70	NA	1160.91	NA
SPSO	124,350.40	126,074.40	NA	1153.11	NA
PSO	123,930.45	124,154.49	NA	NA	10,000
CEP	123,488.29	124,793.48	126,902.89	NA	NA
HGAPSO	122,780.00	124,575.70	NA	906.04	NA
FEP	122,679.71	124,119.37	127,245.59	NA	NA
MFEP	122,647.57	123,489.74	124,356.47	NA	NA
IFEP	122,624.35	123,382.00	125,740.63	NA	NA
TM	122,477.78	123,078.21	124,693.81	NA	4050
EP-SQP	122,323.97	122,379.63	NA	NA	10,000
MPSO	122,252.26	NA	NA	NA	NA
ESO	122,122.16	122,558.45	123,143.07	NA	75,000
HPSOM	122,112.40	124,350.87	NA	978.75	NA
PSO-SQP	122,094.67	122,245.25	NA	NA	10,000
PSO-LRS	122,035.79	122,558.45	123,461.67	NA	20,000
IMPROVED GA	121,915.93	122,811.41	123,334.00	NA	100,000
HPSOWM	121,915.30	122,844.40	NA	497.44	NA
IGAMU	121,819.25	NA	NA	NA	NA
HDE	121,813.26	122,705.66	NA	NA	100
DEC(2)SQP(1)	121,741.97	122,295.12	122,839.29	386.181	18,000
PSO	121,735.47	122,513.91	123,467.40	NA	20,000
APSO(i)	121,704.73	122,221.36	122,995.09	NA	20,000
ST-HDE	121,698.51	122,304.30	NA	NA	100
NPSO-LRS	121,664.43	122,209.31	122,981.59	NA	20,000
APSO(ii)	121,663.52	122,153.67	122,912.39	NA	20,000
SOHPSO	121,501.14	121,853.57	122,446.30	NA	62,500
BBO	121,479.50	121,512.06	121,688.66	NA	50,000
BF	121,423.63	121,814.94	NA	124.876	10,000
GA-PS-SQP	121,458.00	122,039.00	NA	NA	1000
PS	121,415.14	122,332.65	125,486.29	NA	1000
FA	121,415.05	121,416.57	121,424.56	1.784	25,000
GA-BFO	121,598	121,869	122,077	156.148	31,600
CGAC	**120,816.413**	**121,309**	**121,810.14929**	109.675	73,292

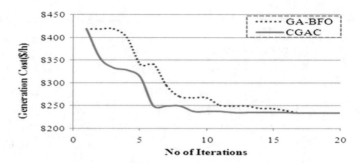

Fig. 1 Convergence graph of three generators

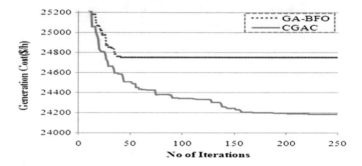

Fig. 2 Convergence graph of 13 generators

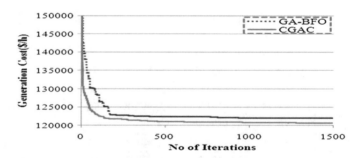

Fig. 3 Convergence graph of 40 generators

by CGAC in which the parallel result is achieved with a conclusion that CGAC performs equally or better than hybrid chaotic PSO and sequential quadratic programming (CPSO-SQP) algorithm, evolutionary programming (EP) and other popular and continuous methods available in the literature. It is further realized that instead of hybridization of the entire BFO with GA, only the hybridization of chemotactic step with GA has a greater contribution towards the effectiveness and robustness of the CGAC algorithm in solving the real-life problems, especially in solving economic load dispatch (ELD) problem. This has been validated through the statistical and graphical results illustrated in the paper. Therefore, the proposed hybrid algorithm CGAC is more reliable, more efficient and more accurate as compared to recent state-of-the-art constrained evolutionary algorithms for solving non-convex economic load dispatch (ELD) problems.

References

1. Lowery, G.: Generating unit commitment by dynamic programming. IEEE Trans. Power Appar. Syst. PAS **85**(5), 422–426 (1996)
2. Bard, J.F.: Short term scheduling of the thermal electric generators using Lagrangian relaxation. Oper. Res. **36**(5), 756–766 (1988)
3. Chen, C.L., Wang, S.C.: Branch and bound scheduling for thermal generating units. IEEE Trans. Energy Convers. **8**(2), 184–189 (1993)
4. Fan, J.I.Y., Zhang, L.: Real time economic dispatch with line flow and emission constraints using quadratic programming. IEEE Trans. power Syst. **13**, 320–325 (1998)
5. Dodu, J.C., Martin, P., Merlin, A., Pouget, J.: An optimal formulation and solution of Short range operating problems for a power system with flow constraints. Proc. IEEE **60**(1), 54–63 (1972)
6. Parikh, J., Chattopadhyay, D.: A multi area linear programming approach for analysis of economic operation of the Indian power system. IEEE Trans. Power Syst. **11**, 52–58 (1996)
7. Chiang, C.L.: Genetic based algorithm for power economic load dispatch. IEEE Proc. Gener. Trans. Distrib. **1**(2), 261–269 (2007)
8. Chaturvedi, K.T., Pandit, M., Srivastava, L.: PSO with crazy particles for non-convex economic dispatch. Appl. Soft Comput. **9**, 962–969 (2009)
9. Ghasemi, A.: A fuzzified multi objective interactive honey bee mating optimization for environmental/economic power dispatch with valve point effect. Electr. Power Syst. Res. **49**, 308–321 (2013)
10. Safari, A., Shayeghi, H.: Iteration PSO procedure for economic load dispatch with generator constraints. Expert Syst. Appl. **38**, 6043–6048 (2011)
11. Yang, X.S., Hosseini, S.S.S., Gandomi, A.H.: Firefly algorithm for solving non-convex economic dispatch problems with valve loading effect. Appl. Soft Comput. **12**, 1180–1186 (2012)
12. Moradi-Dalvand, M., Mohammadi-Ivatloo, B., Najafi, A., Rabiee, A.: Continuous quick group search optimizer for solving non-convex economic dispatch problems. Electr. Power Syst. Res. **93**, 93–105 (2012)
13. Mandal, B., Ray, P.K., Mandal, S.: Economic load dispatch using kill-herd algorithm. Electr. Power Syst. Res. **57**, 1–10 (2014)
14. Jiejin, C., Xiaoqian, M., Lixiang, L., Haipeng, P.: Chaotic PSO for economic dispatch considering the generator constraints. Energy Convers. Manag. **48**, 645–653 (2007)
15. Swain, R.K., Sahu, N.C., Hota, P.K.: Gravitational search algorithm for optimal economic dispatch. In: 2nd International Conference on Communication, Computing and Security (ICCCS-2012); Proc. Technol. **6**, 411–419. Elsevier (2012)
16. Panigrahi, B.K., Yadav, S.R., Agarwal, S., Tiwari, M.K.: A clonal algorithm to solve economic load dispatch. Electr. Power Syst. Res. **77**, 1381–1389 (2007)
17. Hosseinnezhad, V., Babaei, E.: Economic load dispatch using θ $-$PSO. Electr. Power Syst. Res. **49**, 160–169 (2013)
18. Al-Summait, J.S., Al-Othman, A.K., Sykulski, J.K.: Application of pattern search method to power system valve point ELD. Electr. Power Syst. Res. **29**, 720–730 (2007)
19. Wang, L., Li, L.P.: An effective differential harmony search algorithm for solving non-convex economic load dispatch problems. Electr. Power Syst. Res. **44**, 832–843 (2013)
20. Reddy, A.S., Vaisakh, K.: Shuffled differential evolution for large scale economic dispatch. Electr. Power Syst. Res. **96**, 237–245 (2013)

21. Bhattacharya, A., Chattopadhyay, P.K.: Solving economic emission load dispatch problems using hybrid differential evolution. Appl. Soft Comput. **11**, 2526–2537 (2011)
22. Sayah, S., Hamouda, A.: A hybrid differential evolution algorithm based on PSO for non convex economic dispatch problems. Appl. Soft Comput. **13**, 1608–1619 (2013)
23. Jiang, S., Ji, Z., Shen, Y.: A novel hybrid PSO and gravitational search algorithm for solving economic emission load dispatch problems with various practical constraints. Electr. Power Syst. Res. **55**, 628–644 (2014)
24. Victoire, T.A.A., Jeyakumar, A.E.: Hybrid PSO-SQP for economic dispatch with valve-point effect. Electr. Power Syst. Res. **71**, 51–59 (2004)
25. Das, K.N., Mishra, R.: Chemo-inspired genetic algorithm for function optimization. Appl. Math. Comput. **220**, 394–404 (2013)
26. Das, K.N., Mishra, R.: A performance study of chemo inspired genetic algorithm on benchmark functions. In: Proceedings of 7th International Conference on Bio-inspired Computing: Theories and Applications (BICTA-2012); Adv. Intell. Syst. Comput. **2** 489–501 (2013)
27. Mishra, R., Das, K.N.: Chemo-inspired genetic algorithm and application to model order reduction problem. In: Proceedings of 5th International Conference on Soft Computing for Problem solving (SocProS-2015), IIT Roorkee, Saharanpur Campus, AISC series. Springer (2015)
28. Mishra, R., Das, K.N: A novel chemo-inspired GA for solving constrained optimization problem. In: International Conference on Computing, Communication and Automation (ICCCA2015), ISBN:978-1-4799-8890-7/15/$31.00 ©2015 IEEE 156, Galgotias University, Greater Noida, U.P
29. Mishra, R., Das, K.N.: A Novel Hybrid Genetic Algorithm for Constrained and Unconstrained Function Optimization, Bio-inspired Computing for Information Retrieval Applications. IGI Global Publisher (2015). ISBN 10:1522523758
30. Cai, J., Li, Q., Li, L., Peng, H., Yang, Y.: A hybrid CPSO-SQP method for economic dispatch considering the valve-point effects. Energy Convers. Manag. **53**, 175–181 (2012)
31. Chaturvedi, K.T., Pandit, M., Srivastava, L.: Self-organizing hierarchical PSO for non-convex economic dispatch. IEEE Trans. Power Syst. **23**(3) (2008)
32. Bhattacharya, A., Chattopadhyay, P.K.: Hybrid differential evolution with biogeography based optimization for solution of economic load dispatch. IEEE Trans. Power Syst. **25**(4) (2010)

Test Data Generation for Mutation Testing Using Genetic Algorithm

Deepti Bala Mishra, Rajashree Mishra, Arup Abhinna Acharya
and Kedar Nath Das

Abstract Mutation testing is a fault-based unit testing in which faults are detected by executing certain test data designed by any white box testing technique. This paper presents a hybridized method for path testing as well as mutation testing by generating the test data automatically using genetic algorithm. In the proposed approach, first path coverage-based test data is generated and further this data is exercised to cover all mutants present in the specific program under test. The proposed method can improve the testing efficiency by deleting the redundant test data obtained from the path testing in terms of better mutation score, and fault detection matrix is used to delete the duplicate data covering same mutants.

Keywords Software test case · Genetic algorithm (GA) · Path coverage
Mutation testing

1 Introduction

Quality and reliability of software can be maintained by successful testing. So, software testing is the most important phase of software development life cycle (SDLC)[1]. There are mainly two types of software testing techniques such as black

D. B. Mishra · A. A. Acharya
School of Computer Engineering, KIIT University, Bhubaneswar 751024, India
e-mail: dbm2980@gmail.com

A. A. Acharya
e-mail: aacharyafcs@kiit.ac.in

R. Mishra (✉)
School of Applied Sciences, KIIT University, Bhubaneswar 751024, India
e-mail: rajashreemishra011@gmail.com

K. N. Das
Department of Mathematics, National Institute of Technology, Silchar, Silchar 788010, Assam,
India
e-mail: kedar.iitr@gmail.com

© Springer Nature Singapore Pte Ltd. 2019
J. C. Bansal et al. (eds.), *Soft Computing for Problem Solving*, Advances in Intelligent
Systems and Computing 817, https://doi.org/10.1007/978-981-13-1595-4_68

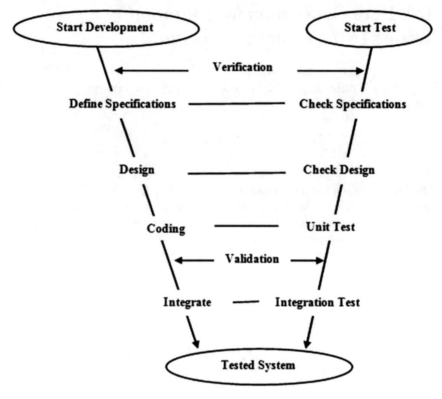

Fig. 1 V-testing model

box testing and white box testing. In black box testing, desired output of the software is checked only on the basis of inputs but in white box testing the functional behavior of the software is checked through the source code [1, 2].

Testing process consists of different steps as shown in Fig. 1. In testing phase, several kinds of testing are done for specific software under test (SUT), including unit testing, integration testing, and system testing. Among all testing, technique unit testing is the basis of all other testing techniques [3, 4].

Manual testing suffers from the drawbacks such as operation speed, high investment of cost and time, limited availability of resources, redundancy of test cases, inefficient and inaccurate test checking. These drawbacks can be overcome by automated testing [4, 5]. It is found that 50% of the resources of the software development are consumed for software testing. So, development cost and time can be decreased by automatic testing [6].

During testing process, many optimization problems related to time, cost, and huge number of test data for execution arise and these problems can be reached to a global optimum solution by using GA [7, 8].

This paper presents a new approach to find path coverage-based test data by using real coded GA, and further they can be used for mutation testing to kill mutants present in SUT. The rest of the paper is organized as follows. Section 2 describes about mutation testing technique. Section 3 explains about the various GA operators used in the proposed algorithm. Section 4 is the discussion about the related work on mutation testing using GA. Section 5 deals with the proposed method for path coverage-based test data generation and further optimizing test data for mutation testing. The implementation of the proposed approach with a case study has been done in Sect. 6. The conclusion and future work of the paper is drawn in Sect. 7.

2   Mutation Testing

Unit testing is the basis of all other types of testing techniques, and if this is not done properly other testing might require more resources. So, unit testing plays a vital role in achieving qualitative and reliable software [9]. Mutation testing is a fault injection testing technique in which user has to insert some syntactic errors to test the SUT. The faulty SUT is called as mutated software [6]. This testing technique is first developed by Offutt and Untch [10], which evaluates and improves the quality of a test data by killing the existing mutants present in the program. The effectiveness of test data can be calculated in terms of higher mutation scores [11–13]. In the proposed algorithm, path coverage-based test data is taken for mutation testing and GA is used to achieve a representative suite that covers 100% path.

3   Genetic Algorithm

Genetic algorithm is an evolutionary search technique used to solve many optimization problems. It is inspired by the biological concept of evolution and based on "surviving the fittest" [7]. In software engineering field, it is used to solve many complex and real-life problems by producing high-quality test data automatically during testing phase. There are four different operators used in GA such as selection, crossover, mutation, and elitism [8]. Real encoding [14] with average crossover [15, 16] and insertion mutation [17] operators are used in this paper to find the new offspring chromosome.

3.1   Average Crossover

Average crossover takes two parents to perform crossover and creates only one offspring by finding the average of two parents [15, 16]. New offspring can be found by averaging the genes of two different chromosomes shown in Fig. 2.

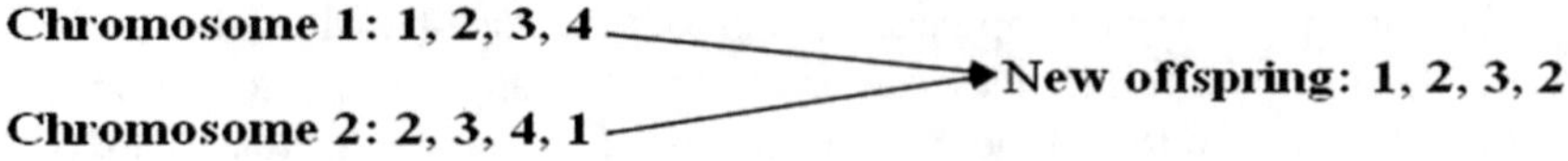

Chromosome 1: 1, 2, 3, 4

New offspring: 1, 2, 3, 2

Chromosome 2: 2, 3, 4, 1

Fig. 2 New offspring after crossover

Fig. 3 Insertion mutation   **Initial Chromosome: 1, 2, 3, 4**

Mutated Chromosome: 1, 2, 3, 4

New Chromosome after Mutation: 1, 3, 2, 4

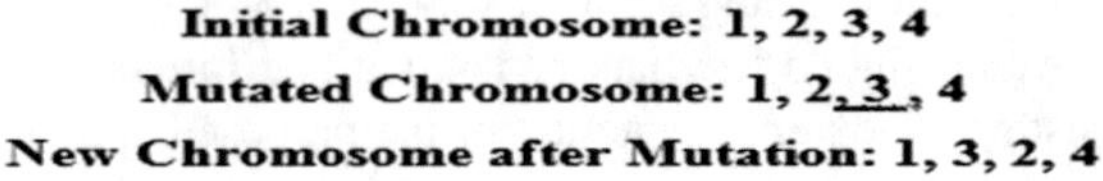

3.2 Insertion Mutation

The position of a particular gene in a specific chromosome is arbitrarily changed by reinserting in a new position which results in a new chromosome [17] as shown in Fig. 3.

3.3 Elitism

The redundant genes are removed (if present) from the chromosome and checked for the optimum solution by using the fitness given in Eq. (1).

3.4 Fitness Function

The fitness function evaluates individual's performance. Based on the fitness value, the individuals with higher fitness are selected to the next generation for better optimum solution [8]. The proposed algorithm is exercised with the fitness defined in Eq. (1), which is based on total path covered by individual chromosome.

4 Literature Survey

Last and Eyal [18] developed a fuzzy-based age extension of GA known as (FAexGA) approach for automatic test data generation and minimization to expose faults using mutated version of the original program. Fuzzy logic controller (FLC) is used to determine the probability of crossover. Quyen et al. [19] proposed a GA-based technique to minimize the mutants of a dynamic model. They have used the Simulink tool to design the system. Their experimental results confirm that the time and cost of mutation testing are reduced without any loss of information. In their proposed approach, GA is used not only to create a subset of mutants but also to delete all active mutants present in the system. Khan and Amjad [11] proposed an optimizing technique for automatic test case generation using GA and mutation analysis. Their proposed technique can find 100% path, and boundary coverage and test case adequacy are measured by the mutation score. The method improves the mutation score by covering maximum number of mutants present in the software. In [20], authors have proposed a method to optimize the efficiency of software test data by using GA. They have used a mutating function for measuring the adequacy of test data. Again in [21], authors proposed a hybrid GA for automatic test data generation and used data flow testing technique for mutation testing. Masud et al. [22] proposed a GA-based model to reveal faults and killed mutants. Their approach first creates the source and mutant programs and divides the whole program into small unit to check the mutants. Rani and Suri [10] proposed a novel approach to generate test data by using GA and delete mutant analysis. The test data is optimized by deleting the redundant test data, and they found that the delete operator saves huge time and effort. Haga and Suehiro [13] developed a technique for automatic test case generation using GA and mutation analysis. They have used only three mutation operators such as COR, CSR, and BVI to kill the mutants. The test cases are reduced by using detection matrix, and then GA is applied to the minimal set of test cases for generating new test data. Mishra et al. [23] developed a new strategy for generating efficient test data for mutation testing. They have extended the work of Bybro and Arnborg [24] and Masud et al. [22] by using elitist version of GA to find a faulty unit in the code.

5 Proposed Algorithm for Mutation Testing

Mutation testing is a structural testing technique and is also known as fault-based testing. Test cases are generated to detect faults present in the SUT. In this testing technique, test cases are designed to detect specific types of faults in a program. Initially, the SUT is tested by the test suite prepared in any white box testing process and then mutation testing can be carried out by executing the same data [12, 25].

In the proposed method, the SUT is tested by path testing technique as total path coverage-based test data results in 100% statement coverage and 100% branch coverage [9]. In path testing, test cases are designed to execute all linearly independent paths, obtained from the corresponding control flow graph (CFG) of the program [2].

The proposed algorithm aims to find a representative test suite which achieves 100% path coverage, and then the test suite is exercised for mutation testing. GA is used to generate total path coverage-based test data, and then the algorithm finds the minimum test data which covers all mutants by eliminating the redundant test data from the fault detection matrix [13]. The flow of the algorithm is shown in Fig. 4.

Fig. 4 Flow of proposed
algorithm

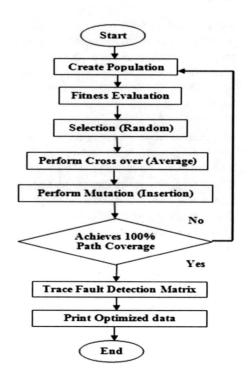

5.1   Fault Detection Matrix (FDM)

The FDM detects which mutant is detected by which test case and can be traced by
using Eq. (1). Table 1 represents an example of FDM in which three mutants M1,
M2, and M3 are detected by different test cases as T1, T2, T3, and T4.

$$FDM\,(i,\,j) = \begin{cases} 1, & \textit{if the ith mutant is detected by jth testcase} \\ 0, & \textit{otherwise} \end{cases} \tag{1}$$

where $i = (1, 2, 3, \ldots, m)$ and $j = (1, 2, 3 \ldots, n)$. m and n represent the number
of mutants generated and the number of test cases, respectively.

Table 1 shows that test case T1 detects only one mutant M1 and similarly test case
T2 detects mutants M2, M3, and so on.

Table 1 Example of fault
detection matrix

	T1	T2	T3	T4
M1	1	0	0	1
M2	0	1	1	1
M3	0	1	0	1

Fig. 5 Function for GCD of
two numbers

```
1.  int GCD(int m, int n)
2.  {
3.  int r;
4.  if(n>m)
5.  { r=m;
6.  m=n;;
7.  n=r;}
8.  r=m%on;
9.  while(r!=0){
10. m=n;
11. n=r;
12. r=m%on;}
13. return n;
14. }
```

Fig. 6 CFG for GCD
function

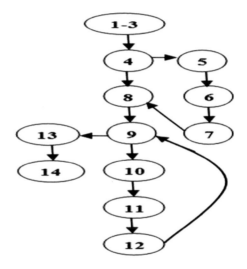

6 Experimental Setup

The proposed algorithm is implemented on a case study as finding the greatest common divisor of any two inputted numbers, and the function for GCD (int *m*, int *n*) is shown in Fig. 5. The CFG for the program is drawn in Fig. 6. To perform mutation testing, some faults are inserted in the given program code shown in Fig. 5 as it aims to cover maximum number of mutants by the path coverage-based test data. Table 2 shows the details of mutants inserted to the program shown in Fig. 5.

The corresponding paths can be obtained from CFG shown in Fig. 6 as:

PATH 1: 1–3, 4, 5, 6, 7, 8, 9, 10, 11, 12, 13, 14
PATH 2: 1–3, 4, 8, 9, 10, 11, 12, 13, 14

Table 2 Mutant details

Line no.	Operator present	Replace with operator	Mutant no.
4	>	!=	M1
6	=	/=	M2
8	%	/	M3
9	!=	= =	M4
12	%	/	M5

Table 3 Parameters used for GA operation

S. no.	Parameters	Value	
1	Population size	10	
2	Input range	1–50	
3	Encoding type	Real encoding	
4	Crossover type and rate	Average crossover (AX), 0.8	
5	Mutation type and rate	Insertion mutation 0.02	
6	Generation	5	

PATH 3: 1–3, 4, 8, 9, 13, 14
PATH 4: 1–3, 4, 5, 6, 7, 8, 9, 13, 14

The fitness function is designed by considering the total number of path covered by a chromosome, and it is defined in Eq. (2). The obtained result found from GA operations is further exercised for mutation testing where the test data adequacy can be represented by the mutation score defined in Eq. (3). The minimum number of test cases required for total mutant coverage can be obtained from the fault detection matrix. Table 3 shows the parameters taken for GA operation.

$$f(x) = \left(\frac{Number\ of\ path\ Covered\ by\ a\ Chromose}{Total\ Number\ of\ path} \right) \times 100 \qquad (2)$$

$$Mutation\ Score = \frac{Number\ of\ mutants\ killed\ by\ a\ testcase}{Total\ Number\ of\ mutants\ inserted\ in\ a\ program} \qquad (3)$$

6.1 Result Analysis

The various steps of the proposed algorithm and the results obtained in GA operations are shown in Tables 4, 5, and 6. The resultant test suite that covers 100% path is shown in Table 7. The FDM is traced out form the optimized test suite, which indicates the mutant coverage by individual test case shown in Table 8. The mutation score is

Table 4 Initial population

Test suite no.	Test data	Target path	Fitness
1	(12, 4), (8, 27), (45, 8), (9, 44)	3, 1, 3, 1	0.5
2	(14, 9), (23, 8), (33, 45), (14, 5)	2, 2, 2, 3	0.5
3	(49, 9), (7, 33), (28, 5), (39, 8)	2, 1, 2, 2	0.5
4	(7, 12), (6, 18), (16, 4), (9, 42)	1, 4, 3, 1	0.75
5	(32, 6), (44, 16), (20, 7), (17, 12)	2, 2, 2, 2	0.25

Table 5 Best population

Test suite no.	Fitness	
1	0.5	
4	0.75	

Table 6 New trait (after crossover)

New trait	Test data	Target path	Fitness
1	(9, 8), (7, 22), (30, 6), (19, 43)	2, 1, 3, 1	0.75

Table 7 Path coverage test data

Test suite	Test data	Target path	Fitness
(T1, T2, T3, T4)	(21, 7), (14, 9), (7, 22), (6, 18)	1, 2, 3, 4	100%

calculated using Eq. (3) and shown in Table 9, and redundant test cases are eliminated to find the optimum test data which is shown in Fig. 7.

Table 8 Fault detection matrix

Test case/mutant	T1	T2	T3	T4
M1	1	0	1	0
M2	0	0	1	1
M3	0	1	1	0
M4	1	1	1	1
M5	0	1	1	0

Table 9 Mutation score

Test case	Mutation score
T1	0.4
T2	0.6
T3	1
T4	0.4

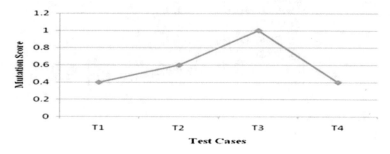

Fig. 7 Mutation score of test cases

7 Conclusion and Future Work

Unit testing plays an important role in achieving qualitative and reliable software. Although mutation testing is computationally very expensive, it improves the quality of test data in terms of higher mutation score and enhances the reliability of a program by killing the active mutants. The proposed method can generate test data not only for path testing but also for mutation testing. GA is applied to generate test data for total path coverage, and further the same data is executed to kill the mutants present in the SUT.

In future, it is planned to use a new hybridized algorithm which can generate test data for mutation testing. It is also planned to develop an automatic tool to generate mutants for a specific program under test.

References

1. Mathur, A.P.: Foundations of Software Testing, 2/e. Pearson Education India (2013)
2. Ahmed, M.A., Hermadi, I.: GA-based multiple paths test data generator. Comput. Oper. Res. **35**(10), 3107–3124 (2008)
3. Sharma, A., Rishon, P., Aggarwal, A.: Software testing using genetic algorithms. Int. J. Comput. Sci. Eng. Surv. (IJCSES) **7**(2), 21–33 (2016)
4. Mishra, D.B., Mishra, R., Das, K.N., Acharya, A.A.: A systematic review of software testing using evolutionary techniques. In: Proceedings of Sixth International Conference on Soft Computing for Problem Solving, pp. 174–184. Springer, Singapore (2017)
5. Srivastava, P.R., Kim, T.H.: Application of genetic algorithm in software testing. Int. J. Softw. Eng. Appl. **3**(4), 87–96 (2009)

6. Silva, R.A., de Souza, S.D.R.S., de Souza, P.S.L.: A systematic review on search based mutation testing. Inf. Softw. Technol. **81**, 19–35 (2017)
7. Deb, K.: Optimization for Engineering Design: Algorithms and Examples. PHI Learning Pvt, Ltd (2012)
8. Mishra, D.B., Bilgaiyan, S., Mishra, R., Acharya, A.A., Mishra, S.: A review of random test case generation using genetic algorithm. Indian J. Sci. Technol. **10**(30) (2017)
9. Ghiduk, A.S.: Automatic generation of basis test paths using variable length genetic algorithm. Inf. Process. Lett. **114**(6), 304–316 (2014)
10. Rani, S., Suri, B.: An approach for test data generation based on genetic algorithm and delete mutation operators. In: 2015 Second International Conference on Advances in Computing and Communication Engineering (ICACCE), pp. 714–718. IEEE, May 2015
11. Khan, R., Amjad, M.: Automatic test case generation for unit software testing using genetic algorithm and mutation analysis. In: 2015 IEEE UP Section Conference on Electrical Computer and Electronics (UPCON), pp. 1–5. IEEE, Dec 2015
12. Bashir, M.B., Nadeem, A.: Improved genetic algorithm to reduce mutation testing cost. IEEE Access **5**, 3657–3674 (2017)
13. Haga, H., Suehiro, A.: Automatic test case generation based on genetic algorithm and mutation analysis. In: 2012 IEEE International Conference on Control System, Computing and Engineering (ICCSCE), pp. 119–123. IEEE, Nov 2012
14. Goldberg, D.E.: Real-coded genetic algorithms, virtual alphabets, and blocking. Complex Syst. **5**(2), 139–167 (1991)
15. Umbarkar, A.J., Sheth, P.D.: Crossover operators in genetic algorithms: a review. ICTACT J. Soft Comput. **6**(1) (2015)
16. Singh, G., Gupta, N., Khosravy, M.: New crossover operators for real coded genetic algorithm (RCGA). In: 2015 International Conference on Intelligent Informatics and Biomedical Sciences (ICIIBMS), pp. 135–140. IEEE, Nov 2015
17. Boopathi, M., Sujatha, R., Kumar, C.S., Narasimman, S.: The mathematics of software testing using genetic algorithm. In: 2014 3rd International Conference on Reliability, Infocom Technologies and Optimization (ICRITO) (Trends and Future Directions), pp. 1–6. IEEE, Oct 2014
18. Last, M., Eyal, S.: A fuzzy-based lifetime extension of genetic algorithms. Fuzzy Sets Syst. **149**(1), 131–147 (2005)
19. Quyen, N.T.H., Tung, K.T., Binh, N.T.: Improving mutant generation for Simulink models using genetic algorithm. In: 2016 International Conference on Electronics, Information, and Communications (ICEIC), pp. 1–4. IEEE, Jan 2016
20. Khan, R., Amjad, M.: Optimize the software testing efficiency using genetic algorithm and mutation analysis. In: 2016 3rd International Conference on Computing for Sustainable Global Development (INDIACom), pp. 1174–1176. IEEE, Mar 2016
21. Khan, R., Amjad, M., Srivastava, A.K.: Generation of automatic test cases with mutation analysis and hybrid genetic algorithm. In: 2017 3rd International Conference on Computational Intelligence & Communication Technology (CICT), pp. 1–4. IEEE, Feb 2017
22. Masud, M., Nayak, A., Zaman, M., Bansal, N.: Strategy for mutation testing using genetic algorithms. In: 2005 Canadian Conference on Electrical and Computer Engineering, pp. 1049–1052. IEEE, May 2005
23. Mishra, K.K., Tiwari, S., Kumar, A., Misra, A.K.: An approach for mutation testing using elitist genetic algorithm. In: 2010 3rd IEEE International Conference on Computer Science and Information Technology (ICCSIT), vol. 5, pp. 426–429. IEEE, July 2010
24. Bybro, M., Arnborg, S.: A mutation testing tool for java programs. Master's thesis, Stockholm University, Stockholm, Sweden (2003)
25. Dave, M., Agrawal, R.: Search based techniques and mutation analysis in automatic test case generation: a survey. In: 2015 IEEE International Advance Computing Conference (IACC), pp. 795–799. IEEE, June 2015

Local Invariant Feature-Based Gender Recognition from Facial Images

Vivek Kumar Verma, Sumit Srivastava, Tarun Jain and Ashish Jain

Abstract Human gender is an important demographic characteristic in the society. Recognizing demography characteristics of individuals, for example, age and gender using automatic image recognition taken much consideration in last few years. This paper proposes the extraction of geometric and appearance feature of face automatically from the front view. For extracting the feature, cumulative benchmark approach is used. Two basic categories as supervised as well as unsupervised methodology may be applied for gender grouping. In this paper, we used supervised machine learning approach. We have used three diverse classifiers, for this approach as SVM, neural network, and adobos. We have trained all the classifiers by means of identical training dataset and similar feature. We have done a comparative study of the performance of these classifiers and which classifier is best for our primary dataset over face images.

Keywords Benchmark test · Neural network · Support vector machine · AdaBoost

1 Introduction

During the past few years, gender recognition plays an essential role in machine learning. Gender recognition system is likely to contain a human face, which is the most important biometric features and holds more key biometric information out of which, pose helps in providing the usable information about gender. Automatic facial

V. K. Verma (✉) · S. Srivastava · T. Jain · A. Jain
School of Computing and Information Technology, Manipal University Jaipur, Jaipur, Rajasthan, India
e-mail: vermavivek123@gmail.com

S. Srivastava
e-mail: sumit.srivastava@jaipur.manipal.edu

T. Jain
e-mail: tarunjainjain02@gmail.com

A. Jain
e-mail: ashish.jain@jaipur.manipal.edu

© Springer Nature Singapore Pte Ltd. 2019
J. C. Bansal et al. (eds.), *Soft Computing for Problem Solving*, Advances in Intelligent Systems and Computing 817, https://doi.org/10.1007/978-981-13-1595-4_69

gender recognition has become a challenging research problem in current year. Facial elements used geometric features for recognition of human faces, texture color of the skin, and histograms also used for feature extraction. These are four major phases: pre-processing, feature extraction, classification, and evaluation together develop a complete automatic gender recognition system.

This paper is devoted to the development of an automatic system capable to distinguish people's gender by analyzing their faces on computerized images [1]. Gender can be classified into two classes, i.e., (Man and Woman) in which one form is allocated to given face image (M/F).

Contribution: In this paper, we compare classifier models with a standard face image dataset of face images and it is performed over the entire system module. The extraction of feature vector and histogram features of face automatically. Geometric feature is extracted using all the values of feature vectors and histogram features, color moment, and normalized values are used for feature matching.

1.1 Application of Gender Recognition System

In the auto-tracking zone, it may be useful for limited control of a particular kind, such as a specific bus or subway train hostel. An automated surveillance zone may also indicate paying more attention or assigning a higher level of threat to a defined gender.

Content and search retrieval with extensive use of electronic devices such as cameras and a lot of photos and videos are created. Indexing or annotating information such as the number of people in the photo or video, age, and gender will be easier with automated systems that use computer vision. On the other hand, for content-based search and finding a person's photo, identifying gender as a pre-processing step will reduce the amount of research needed in the database.

Sophisticated man–machine interaction systems can be built if they are able to identify a human attribute such as sex and age. The system can do very human, how and respond appropriately. A simple situation would be a mechanism for interacting with a human being; they require the knowledge of mankind to deal adequately (like Mr. or Ms.) [1]. In the area where biometric systems that use facial recognition, time to exploration in the face database can be reduced and separate face recognition can be trained for each gender and improve the corresponding precision [2].

An electronic billboard is used to display ads on flat screens. Targeted ad is used to view the relevant ad for the person who sees compatible attributes as a gender sign. We can say by way of example, the billboard can choose to display wallet ads when a man is caught in the hand of another woman in the case of bags. In Japan, vending machines using customer age and sex information to recommend beverages have seen a rise in sales [3]. Demographic study systems are designed to bring together consumer statistics and demographic data such as sex, for example, getting into a store, or looking at a billboard. A system that uses computer vision can be used to automate the task.

1.2 Challenges

Facial images are perhaps the best common biometric feature used by individuals to make their own recognition [1]. The face area, which may include peripheral features, such as a hair and neck region, is used for gender identification. A person's face image shows many differences that can affect the ability of a vision computer to identify the exact identification of the genre. We can classify these changes due to humans or the method of capturing images. Human identification factors are due to individual characteristics such as age, cultural characteristics, and facial expressions. Main features: The cause of the processing is the position of the person's head, light or illumination, and corresponding image quality (blur, noise, resolution). The head position corresponds to the head orientation with respect to the viewing angle of the capturing devices. Generally, a human head can be limited to three degrees of common freedom called pitch, roll, and yaw [4].

2 Related Work

This section will focus on studies carried out on gender recognition and the related face parts information analysis.

Over the past few years, there has been increase research on gender recognition due to its ever-increasing applications in texture classification, pattern recognition facial analysis, human detection, and many other tasks. Therefore, overcome all the limitation of gender recognition technologies.

Ardakany and Joula [5] proposed a novel technique of feature vector extraction for the purpose of gender classification, which includes geometric and shape-based features as well they used few appearance-based features in the same classification parameters. These feature vector extraction process achieved by calculating the local derivative in the entire pixels of the face images and after that by constructing a common histogram of the on the edges their magnitudes with directions. For experiment purpose, they were used FERET database. Classifiers such as LDA, AdaBoost, and SVM with linear kernel. Therefore, increase the performance of SVM first linear kernel used than they found that using SVM with linear kernel give best result and accuracy by proposed method. In this work, histogram based on magnitude and direction of the edge while an existing algorithm defines histogram based on intensity.

Dong and Woodard [6] investigate gender recognition method based on the shape of eye and eyebrow, because the shape of eyes and eyebrow large difference in the structure and variety that makes each human face very much different. Actually, they have extracted nineteen different geometrical shape-based features from each eyebrow. Collectively, entire feature vectors were normalized to binary scale [0 1] using min–max normalization method and then compared with different classification methods such as minimum distance classifier (MD), linear discriminate analysis classifier (LDA), and support vector machine classifier (SVM) where two image datasets demonstrate the feasibility of utilizing geometrical shape-based features extracted from noisy images for biometric recognition and same for gender classification.

Sasikala and Niirosha [7] proposed a new gender recognition system, based on neural networks. Gaussian filter and Gabor filter used for the feature extraction of face image. For the classification, process author used the AdaBoost (an adaptive algorithm to boost a sequence of classifier) to combine a collection of weak classification function. AdaBoost and SVM classifiers provide fast result and reliable classification process. In this paper for recognition, they used artificial neural networks. This method applied on 400 images and 100 * 100 resolutions. Experiment performed on Web and BioID database, 502 images collected from the Web with 80.9% for male and 84.6% for Female and (968 + 543) image collected from the BioID database with accuracy 89.2% and 81.4%, respectively.

Khan et al. [8] have made widespread comparison of state-of-the-art research techniques that are gait based, body based, and face based for gender classification. Their analyses have been presented along with their strengths and weaknesses. They have also discussed standard datasets like face database, gait database. Most of the researchers focused on face image for gender classification but there are some significant problems, which are facing by the researchers, they are facial variation like occlusion, change in pose, illumination effect, blur, noisy, high data dimensions. It is also stated that SVM classifier may work better than other classifiers and enhances classification accuracy rate.

By going through the papers in detail, we have found that most of the experiments performed in gray-scale and colored images reveal color information improves on face recognition and gender retrieval performance. In, geometric-based methods, number of fiducially points considered to determine different gender. However, it was found that the geometric-based methods are more complicated than appearance-based methods in terms of implementing for gender recognition system because in geometric-based methods feature points are need to be set manually. Most of the classifiers used are SVM, AdaBoost, and neural network.

3   Proposed Model

Survey has revealed that various model and combination of these can be applied for the implementation and testing of gender recognition system. In proposed work, we decided to use combination of geometric and appearance features at the same time. The identified problem of gender recognition needs attention on robust feature extraction as well as efficient method. Here, the success of solution to this problem crucially depends on the stability and scalability of the facial features, which used the characteristics of features vector. Also, we examined different structure and color-based features for the identification of human face gender.

In this paper, we have a set of images and firstly used a dataset designed for the face recognition IITK [9]. Secondly, we use R language Rattle and compare the

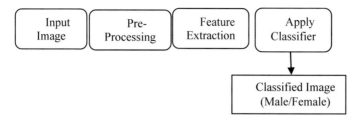

Fig. 1 System model for gender recognition

entire predefined classifier model and compare which model gives us best result for the classification (Fig. 1).

Face Image: The face image samples are taken from standard human face database and created corresponding indexing for each images. Entire images are stored in the form of 24-bit depth with RGB as only JPEG format for the standard.

Proposed Algorithm:

In this work, firstly facial components were detected from selected image by using skin color segmentation and morphological operations. Secondly, the extracted facial components were subjected to proposed algorithm.

Input: Apply testing for individual selected image

Output: Classify gender of selected image, number of face images in database, number of male and female in the database individually

Step 1: Load training and testing database 70%, 30%, respectively, for feature extraction.
Step 2: Pre-processing: It consists of noise reduction, color conversion, select skin area, and resizing the image.
Step 3: Crop face image to convert it into standard face image.
Step 4: Feature extraction total number of white pixels, and distance and direction calculation, color moment.
Step 5: Feature matching: Euclidean distance of each class of the images in the database is calculated. The class having the minimum distance, the test image belongs to that class.

A. *Pre-processing*

Once the face of the entire image is detected, apply a few pre-processing on the image. It helps to improve understanding of classifier differences such as lighting, head posture detection, and inaccuracies. Ding et al. [10] such signals as brightness and dimensions can be learned from the SVM classifier example can produce a better afectly performance.

Pre-processing can be applied to the face image including normalizing the contrast of the image and the corresponding brightness. You can also include the removal of non-contextual and external content features as applied here to the hair and neck region.

More important for the accuracy of the output picture is the geometric alignment of the region of interest (manually or by automatic techniques).

B. *Facial feature extraction*

Usually, cataloging feature extraction strategies for targeting gender targeting (based on geometric characteristic) and subjective (based on appearance) strategies [1, 5]. The first is based on the measurement of facial features. Geometric relationships between these points remain, but other useful information can be discarded and extraction point positions must be accurate. Look-based approaches are based on an operation or transformation made in pixels in an image. This can be done globally or locally. Locally, the face is divided into distinct regions such as eyes, nose, and mouth or windows spaced frequently. In aspect-based methods, geometric relationships of course [5] taken, which is advantageous when gender discriminatory features are not exactly known. However, they are sensitive to appearance variations (due to sight, illumination, expression, etc.) and the large number of features [11].

C. *Classification*

A wide range of classifiers is used for classification in gender recognition system. Some of them are Euclidean distance, support vector machine (SVM), neural networks, linear discriminative analysis (LDA), AdaBoost classifier are widely used in classification process. In classification, training images are trained and put into different classes. Then, the test images are tested with the training images and their respective classes and output is found to which class the test image belongs to classifiers is presented in the following subsection.

Support vector machine is first time proposed and developed by Vapnik and Boser, the support vector machine (SVM) and its variants are the most commonly used supervised approaches for classification. In general, SVM is a discriminative binary classifier that attempts to find a separating hyperplane that has the widest margin to the closest training data sample of any class. This hyperplane acts as a decision function to predict the category to which a new query sample belongs. Generally speaking, the wider the margin, the lower generalization error of the SVM classifier. Hence, SVM can be modeled as a maximization problem that finds a maximum margin hyperplane with the largest possible distance to the nearest data points so-called support vectors of each class.

AdaBoost is a sort of high margin classifier. It is an adaptation in the sense that weak students are subsequently adjusted for these cases incorrectly classified by the previous classifier. AdaBoost has been widely used to improve the accuracy of any existing algorithm. Increasing is a method to combine a collection of weak classification function to form a strong classifier. Some authors have combination of AdaBoost with SVM as weak component classifiers to be used in gender classification task. For the better classification, AdaBoost-based classifier demonstrates better performance as compared to SVM on imbalance classification problems.

Fig. 2 Facial test images [9]

4 Result Analysis

When an algorithm is compared it is always advisable to use a standard image dataset so that researchers can compare results with existing systems. While there are many databases in use today, choosing a database suitable to be used should be based on the task taken and for that we have chosen face images of Indian environment only. The image database is the selected as a subset of the face image database collected for the research on face Database (Fig. 2).

This face image database is a collection of male and female faces with a standard environment to evaluate the result. This database only used for academic and research purpose and cannot be used in commercial products. We have selected about 100 images from the database as our training or testing. In this we have different ratio with their different results such as Ratio 1 is left eye to right eye distance divided by eye to nose distance, Ratio 2 is left eye to right eye distance divided by eye to lip distance, Ratio 3 is eye to nose distance divided by eye to chin, Ratio 4 is eye to nose distance divided by eye to lip distance (Fig. 3 and Table 2).

Most of the images are considered from the taken standard databases and trained them. We have applied our proposed technique over both male and female face images. The above necessary steps are performed on the trained images and ratios are calculated and are compared with the threshold as given in Table 1. The testing of face images is performed by using only Indian face database. It is examined that the gender recognition rate is overall 97.6%.

The complete result of this comparison study is carried out in tabular form as shown in Table 1, and corresponding graph is shown in Fig. 4.

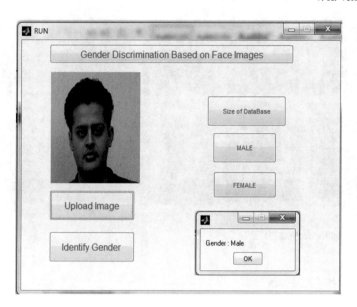

Fig. 3 Identify face image gender

Table 1 Recognition parameter ratio

Image	Ratio 1	Ratio 2	Ratio 3	Ratio 4
1.jpg	1.5574	0.9096	0.5841	2.5062
2.jpg	1.396	0.9423	0.6585	0.8452
3.jpg	1.5089	0.9516	0.6359	1.8448
4.jpg	1.8963	0.9065	0.4756	1.5498
7.jpg	1.2441	0.8849	0.7052	2.5448
9.jpg	1.6259	0.8956	0.5539	1.8745
8.jpg	1.5828	0.9958	0.8456	0.4949
10.jpg	1.3249	0.9934	0.7546	0.9553
12.jpg	1.6641	0.9092	0.5468	1.4949
13.jpg	1.2656	0.9915	0.8546	1.4945
14.jpg	1.1734	0.9351	0.745	1.9455
Normalized value	1.45475	0.93874167	0.66775	1.536808
Variance	0.048528534	0.0015321	0.014025	0.423103

Table 2 Result comparison

Author	Feature extraction	Classifier	Dataset	Accuracy
Thai Hoang Le, Len Bui	2DPCA	MLP, k-N, SVM	FERET and AT&T	95.7, 96.2, 97.3%
Liao, Pin, et al.	PCA	SVM	Random taken	86.67%
A. R. Ardakany	Edges; histogram	LDA, AdaBoost, and SVM	FERET	97.61% male and 91.61% female
Miss. P. Sasikala et al.	Gaussian fitted based on Neural Networks	AdaBoost, SVM	Web and BioID database	<90%
Ramsha K, KB Raja, et al.	FEBFRGAC system	ANN	Frontal face images	~95%
Proposed work	*Benchmark test*	*SVM*	*Indian faces*	*98.70%*

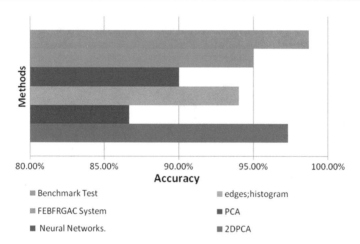

Fig. 4 Accuracy comparison with proposed algorithm

5 Conclusion

In this work, we have proposed a proficient gender discrimination method based on feature vectors of face image. To improve the discrimination accuracy of shape indexing techniques, we encode the nominal quantity of 2-D spatial information in the index by extracting properties like texture, shape, intensity distribution feature vectors with another hybrid image features. This paper demonstrates the shape as well as color-based feature extraction. The performance of the system compares with different classifiers with dataset. Therefore, future work can be done on developing algorithms and code for real-time analysis.

References

1. Ng, C.B., Tay, Y.H., Goi, B.M.: Recognizing human gender in computer vision: a survey. In: Pacific Rim International Conference on Artificial Intelligence. Springer Berlin Heidelberg (2012)
2. Jeganlal, R., Gopi, V., Rajeswari, S.: Robust automatic face, gender and age recognition using ABIFGAR algorithm. Int. J. Emerg. Trends Electr. Electron. IJETEE—ISSN: 2320-9569
3. Le, T.H., Bui, L.: Face recognition based on SVM and 2DPCA. Int. J. Signal Process. Image Process. Pattern Recogn. **4**(3) (2011)
4. Chaichulee, S., et al.: Multi-task Convolutional Neural Network for Patient Detection and Skin Segmentation in Continuous Non-contact Vital Sign Monitoring (2017)
5. Ardakany, A.R., Joula, A.M.: Gender recognition based on edge histogram. Int. J. Comput. Theory Eng. **4**(2) (2012)
6. Dong, Y., Woodard, D.L.: IEEE-Eyebrow Shape-Based Features for biometric recognition and gender classification, 978-1-4577-1359 (2011)
7. Sasikala, P., Niirosha, M.N., et al.: Identification of gender and face recognition using recognition using adaboost and SVM **3**(11), 9305–9312 (2014)
8. Khan, S.A., et al.: A comparative analysis of gender classification techniques. Middle-East J. Sci. Res. **20**(1), 1–13 (2014)
9. Amitabh Mukherjee of CSE, IIT Kanpur, Facial Image database. http://vis-www.cs.umass.edu/~vidit/AI/dbase.html
10. Ding, C., et al.: Multi-directional multi-level dual-cross patterns for robust face recognition. IEEE Trans. Pattern Anal. Mach. Intell. **38**(3), 518–531 (2016)
11. Liao, P., et al.: Nesting differential evolution to optimize the parameters of support vector machine for gender classification of facial images. In: 3rd International Conference on Technological Advances in Electrical, Electronics and Computer Engineering (TAEECE) (2015)
12. Jeganlal, R., Gopi, V., Rajeswari, S.: Robust automatic face, gender and age recognition using ABIFGAR algorithm. Int. J. Emerg. Trends Electr. Electron. (IJETEE—ISSN: 2320-9569)
13. Kalam, S., et al.: Gender classification using geometric facial features. Int. J. Comput. Appl. **85**(7), 0975–8887 (2014)
14. Nagdeve, A.K., et al.: Automated facial features points extraction. Int. J. Comput. Electron. Res. **1**(3) (2012)
15. El Manhraby, A., et al.: Detect and analyse face parts information using viola jone and geometric approces. IJCA **101**(3) (2014)

Performance Comparison of SMO-Based Fuzzy PID Controller for Load Frequency Control

Debasis Tripathy, Amar Kumar Barik, Nalin Behari Dev Choudhury
and Binod Kumar Sahu

Abstract This paper presents dynamic performance comparison of a fuzzy logic-based proportional, integral, and derivative controller (FPID) with different membership functions such as triangular, trapezoidal, and Gaussian for load frequency control (LFC) in an interconnected two-area thermal power system. The parameters of controller are optimized by using spider monkey optimization (SMO) algorithm. The superiority of the proposed algorithm is established by comparing the results with popularly used algorithms like particle swarm optimization (PSO) and teaching–learning-based optimization (TLBO). Initially, the linearized model of the system is considered with reheat turbine; then, the study is extended by imposing nonlinearity such as generation rate constraints (GRC) and governor dead band (GDB). The result comparison is analyzed using various time domain specifications like peak undershoot, peak overshoot, and settling time of different area frequencies and tie-line power deviation between them applying a step load perturbation (SLP) of 1%.

Keywords Fuzzy PID controller · Fuzzy membership function · Load frequency control · Spider monkey optimization

D. Tripathy (✉) · A. K. Barik · N. B. D. Choudhury
Department of Electrical Engineering,
NIT Silchar, Silchar 788010, India
e-mail: debasis.2404@gmail.com

A. K. Barik
e-mail: akbeee@gmail.com

N. B. D. Choudhury
e-mail: nalinbdc@yahoo.com

B. K. Sahu
Siksha 'O' Anusandhan University, Bhubaneswar 751030, India
e-mail: binodsahu@soauniversity.ac.in

© Springer Nature Singapore Pte Ltd. 2019   879
J. C. Bansal et al. (eds.), *Soft Computing for Problem Solving*, Advances in Intelligent
Systems and Computing 817, https://doi.org/10.1007/978-981-13-1595-4_70

1   Introduction

The deviation in system frequency and tie-line power flow in an interconnected power system depends upon mismatch of active power generation and load demand. Issues of controlling the active power output of generating units in response to retain frequencies and tie-line powers within precise limits are known as load frequency control (LFC), which is very crucial for quality power generation [1, 2]. It is found from the literature over the world many researchers have proposed several techniques to maintain the system frequency and tie-line power exchange within specified limits during normal operation and also due to small step perturbations.

Initially in 1956, Chon [3] proposed the concept of AGC. Elgard and Fosha [4] in 1970 introduced the concept of modern optimal control for AGC of an interconnected power system. Ghosal [5] used PSO algorithm to optimize the gains of PID controller for AGC. Kocaarslan and Ertuğrul [6] proposed fuzzy gain schedule PI controller to damp out oscillations resulted from load perturbation, in comparison with PI and fuzzy PI controller. Panda et al. [7] proposed a hybrid neuro-fuzzy approach for AGC of two-area non-reheat thermal system. Saikia et al. [8] estimated that the integral double derivative controller performs better over other classical controllers for AGC. Sahu et al. [9] compared the performance of hybrid firefly pattern search (PS) algorithm with bacteria foraging optimization (BFO) and genetic algorithm (GA) for AGC for two-area and three-area systems. The parameter's optimization of fuzzy PID controller using hybrid PSO-PS algorithm is proposed by Sahu et al. [10], for two-area thermal system with and without nonlinearity. Sahu et al. [11] introduced TLBO fuzzy PID controller for AGC. Hybrid DE-PSO fuzzy PID controller for two- and three-area reheat thermal systems was proposed by Sahu et al. [12]. Sahu et al. [13] projected hybrid gravitational search algorithm (GSA) and PS-based PI controller for two-area thermal system. Sahu et al. [14] proposed fuzzy PID controller optimized by hybrid firefly PS initially for two-area thermal system and then extended to three-area system with nonlinearity. The gain of fuzzy PID controller optimized by hybrid local uni-modal sampling (LUS)–TLBO is introduced by Sahu et al. [15] for LFC of two-area multi-source interconnected system with and without HVDC link.

This work is focused to introduce SMO in optimizing the fuzzy PID controller parameters for LFC of a two-area thermal power system, and then, the study extended by considering GRC and GDB. The performance of system is compared for different membership functions of fuzzy PID controller. Finally, the superiority of SMO algorithm is proved over other popular algorithms like PSO and TLBO.

The paper is structured as follows: Sects. 2 and 3 describe the proposed system model and controller design. Section 4 presents a brief discussion of the SMO algorithm. Section 5 provides the analysis of simulation results, and finally, Sect. 6 concludes the paper.

2 System Model

The linear transfer function model of proposed two-area interconnected power system including GRC of 3% [7, 8, 10, 14] and GDB of 0.06% (0.036 Hz) [9, 13] nonlinearity is shown in Fig. 1. Rating of each reheat thermal unit is 2000 MW having nominal load of 1000 MW. The fuzzy PID controller is used as a secondary controller in both areas for controlling the frequency and tie-line power flow. Inputs to the controller are area control errors (ACEs) of the respective area, which is the linear combination of frequency and tie-line error and U_1, U_2 are the output. When a sudden small disturbance is subjected to the system, ACEs are the actuating signal for respective controllers. In Fig. 1, B_1 and B_2 are the frequency bias factor, R_1 and R_2 are speed regulation of governors, Δf_1 and Δf_2 are the deviations in frequency, ΔP_{12} and ΔP_{21} are tie-line power flow changes between the area, T_{G1} and T_{G2} are time constants of governors, T_{T1} and T_{T2} are time constants of turbines, K_{P1} and K_{P2} are gains of power systems, T_{P1} and T_{P2} are power systems time constants, K_{r1} and K_{r2} are coefficients of reheat unit, and T_{r1} and T_{r2} are time constants of reheat unit of area 1 and area 2, respectively. T_{12} is the synchronizing torque coefficient. The values of these parameters used in the proposed model study are provided in Appendix.

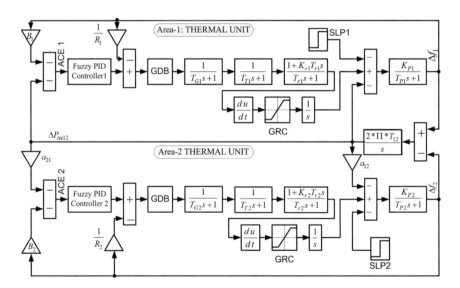

Fig. 1 Transfer function model of two-area thermal system with GRC and GDB nonlinearity

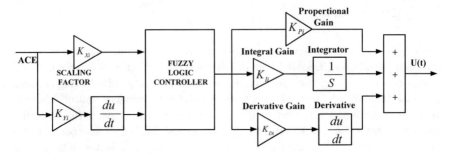

Fig. 2 Block diagram of proposed fuzzy PID controller

3   Controller Structure

The fuzzy PID controller structure for both the area is shown in Fig. 2 [9, 14]. It is a cascade connection of fuzzy PD and PID controllers. The input gains are K_{Xi}, K_{Yi}, and the output gains are K_{Pi}, K_{Ii}, K_{Di} with suffix $i = 1, 2$ representing area 1 and area 2, respectively. These gains of the fuzzy PID controller must be chosen properly to improve the performance of the system. ACE and derivative of ACE (ΔACE) are the two inputs for the fuzzy controller.

The triangular, trapezoidal, and Gaussian membership functions (MF) are used with five fuzzy linguistic variables such as: large negative (LN), small negative (SN), zero error (ZE), small positive (SP), and large positive (LP) for both the input and the output. The purpose of this study is to give a performance comparison in terms of undershoot, overshoot, and settling time of the system using fuzzy PID controller with different membership functions. Structure of the membership function for both the inputs and outputs for triangular [15], trapezoidal, and Gaussian is shown in Fig. 3a, b, respectively.

Mamdani fuzzy inference engine and center of gravity method of defuzzification are selected for this paper for simplicity. The FLC output is determined by the two-dimensional rule base for error, derivative of error, and FLC output is shown in Table 1 [10, 12]. The objective or cost function considered for the present study is integral time absolute error (ITAE), as its performance is better than others for AGC [9, 10, 12].

4   Spider Monkey Optimization Algorithm

The spider monkey optimization (SMO) algorithm is a meta-heuristic nature-inspired population-based algorithm which is inspired from fission–fusion social structure foraging behavior of spider monkey. This algorithm was proposed and introduced by Bansal et al. [16] in 2014. Spider monkeys dwell socially in a group of 40–50 individuals. They forage at different places by dividing themselves into subgroups

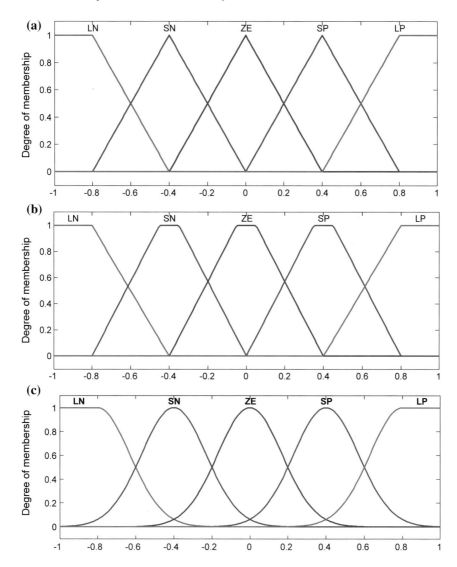

Fig. 3 **a** Structure of triangular fuzzy membership function, **b** structure of trapezoidal fuzzy membership function, and **c** structure of Gaussian fuzzy membership function

and then recombine to share the collected food. The food availability for group members is led by a female monkey. She divides the group into subgroups, if she does not find enough food for the group. These subgroups forage in different directions during day and gather at a common place to share the collected food.

Table 1 Rule base for error, change in error, and fuzzy logic controller output

ACE	ΔACE				
	LN	SN	ZE	SP	LP
LN	LN	LN	SN	SN	ZE
SN	LN	SN	SN	ZE	SP
ZE	SN	SN	ZE	SP	SP
SP	SN	ZE	SP	SP	LP
LP	ZE	SP	SP	LP	LP

The proposed foraging behavior of spider monkeys is different from their natural foraging behavior. The strategy for leader selection depends upon the ability of leader to search for food but not permanent, and the type of communication tactics is not simulated by the proposed strategy. The main advantages of this algorithm are balancing both exploration and exploitation of search space and taking care of premature convergence and stagnation issue. There are four control parameters of SMO, namely local leader limit (LL_{lim}), global leader limit (GL_{lim}), maximum number of groups (MG), and perturbation rate (pr). The pseudocode for SMO algorithm is shown in Fig. 4, and the working phases are discussed below.

begin:
initialize the swarm using equation (1)
initialize Local leader limit(LL_{lim}), Global leader limit(GL_{lim}) Maximum
    number of group(MG). Perturbation rate(pr)
Iteration=0
Calculate fitness value of the position of each spider monkey in the swarm
Select Global leader and local leaders by applying greedy selection
While*(termination criterion is not satisfied)* **do**
    // Local Leader phase(refer section 4.2)
    // Calculate Probabilities using equation (4)
    // Global Leader Phase(refer section 4.3)
    // Global Leader Learning Phase(refer section 4.4)
    // Local Leader Learning phase(refer section 4.5)
    // Local Leader Decision phase (refer section 4.6)
    // global Leader Decision phase(refer section 4.7)
    Iteration=iteration+1
    end while
  end

Fig. 4 Pseudocode for SMO [17]

4.1 Initialization of the Population

SMO initially generates the population with a dimension of $(n \times d)$, where 'n' is number of monkeys or population size and 'd' is the number of variables which has to be optimized. Each M_x (xth monkey) in the population is initialized by Eq. (1).

$$M_{xy} = M_{min_y} + \left(M_{max_y} - M_{min_y}\right) \times U(0, 1) \tag{1}$$

where M_{min_y} and M_{max_y} are the lower and upper bounds of xth *monkey* in yth direction and $U(0, 1)$ is a random number distributed uniformly in the range $[0, 1]$.

4.2 Local Leader Phase (LLP)

The main purpose of this phase is to explore the search space, where individual monkey updates their current position based on the experience of local leader and group members. Equation (2) is to calculate fitness value of xth monkey (which is a member of pth local group) corresponding to new position.

$$M_{newxy} = M_{xy} + \left(LL_{py} - M_{xy}\right) \times U(0, 1) + \left(M_{qy} - M_{xy}\right) \times U(-1, 1) \tag{2}$$

where M_{xy} is the yth dimension of the xth monkey, and LL_{py} represents the yth dimension of pth local group leader position. M_{qy} is the yth dimension of qth monkey which is chosen randomly within pth group such that $q \neq x$, and $U(0, 1)$ is a random number in the range $[0, 1]$.

4.3 Global Leader Phase (GLP)

Global leader phase is mainly used for exploitation, where all the monkeys update their position using the experience of local group members and global leader. Eq. (3) is for position update.

$$M_{newxy} = M_{xy} + \left(GL_y - M_{xy}\right) \times U(0, 1) + \left(M_{qy} - M_{xy}\right) \times U(-1, 1) \tag{3}$$

where GL_y represents the yth dimension of the global leader position. With Eq. (3), the position of xth monkey gets updated based on a factor known as probabilities (pr_x) and it can be calculated using Eq. (4).

$$pr_x = 0.1 + \left(\frac{fit_x}{max\_fit}\right) \times 0.9 \tag{4}$$

where fit_x is the fitness value of xth monkey and max_fit is the maximum fitness in the group. In this way, better monkey gets more chance for update.

4.4 Local Leader Learning (LLL) Phase

In this phase, local leader position gets updated using greedy selection (i.e., position of monkey with best fitness in that group). If the position of the local leader is not updated, then increment local limit count by unity.

4.5 Global Leader Learning (GLL) Phase

In this phase, greedy selection in the population is used for updating the position of global leader. Further, compare new position of global leader with old one; if it is not updated, then increment global limit count by unity.

4.6 Local Leader Decision (LLD) Phase

In this phase, the position of local leader if not updated in a pre-specified number of trials (local leader limit) then each and every member of that group must be reinitialized. Initialization of individual member of that group can either randomly or by collecting information from local leader and global leader based on a probability factor 'pr', using Eq. (5).

$$M_{newxy} = M_{xy} + \left(GL_y - M_{xy}\right) \times U(0, \ 1) + \left(M_{xy} - LL_{py}\right) \times U(0, \ 1) \qquad (5)$$

The above equation shows that the updated dimension of the monkey gets attracted toward global leader and repelled from local leader.

4.7 Global Leader Decision (GLD) Phase

In this phase, if global leader position remains unchanged for predefined number of trials (global leader limit) then whole population is divided into a number of small groups.

Table 2 Optimum values of the controller gains with and without nonlinearity

Gain		Without GRC and GDB			With GRC and GDB		
		Triangular	Trapezoidal	Gaussian	Triangular	Trapezoidal	Gaussian
Area-1	K_{X1}	1.9996	1.4058	1.9993	1.9998	2.0000	1.5948
	K_{Y1}	0.5345	0.4760	0.8236	0.0171	0.4605	1.2934
	K_{P1}	1.9777	1.9465	1.9998	1.7394	1.3030	0.3912
	K_{I1}	1.9998	2.0000	1.9954	1.4844	1.1867	1.5937
	K_{D1}	0.1124	0.0010	0.0217	0.8031	0.4047	0.0010
Area-2	K_{X2}	1.9684	0.7444	1.6863	1.9669	2.0000	0.8617
	K_{Y2}	0.1512	0.0010	0.1949	0.2686	0.3872	1.1304
	K_{P2}	1.9433	1.3217	1.4321	0.7024	1.1196	0.6111
	K_{I2}	0.9518	1.0536	1.4376	1.3013	1.7686	2.0000
	K_{D2}	0.7940	0.9857	0.4826	1.7035	0.5012	1.4274

5 Result Analysis

The usefulness of proposed controller with triangular membership function was demonstrated by considering two cases, i.e., with and without GRC and GDB nonlinearity in the system. SMO algorithm has been used to optimize the controller parameters. The best value of solution is chosen after running the optimization for 20 times for proposed controller parameters. The optimum gain values of the controller with different membership functions are depicted in Table 2, with application of 1% step load perturbation (SLP) in area 1.

The dynamic performance of the system without and with GRC and GDB nonlinearity is shown in Figs. 5a–c and 6a, b, respectively. The corresponding values of peak undershoot (U_{SH}), peak overshoot (O_{SH}), and settling times (T_S) (with 0.0002 band) in frequency deviations and tie-line power deviations are depicted in Tables 3 and 4. Further to demonstrate the effectiveness of SMO algorithm over PSO and TLBO, a step load perturbation (SLP) of 1% is applied to area 1 for the same system without GRC and GDB.

It is observed from the dynamic performance of the system with different algorithm as shown in Fig. 7a, b, that proposed optimization technique is performing better in terms of peak undershoot (U_{SH}), peak overshoot (O_{SH}), and settling times (T_S) in comparison with PSO and TLBO optimizations.

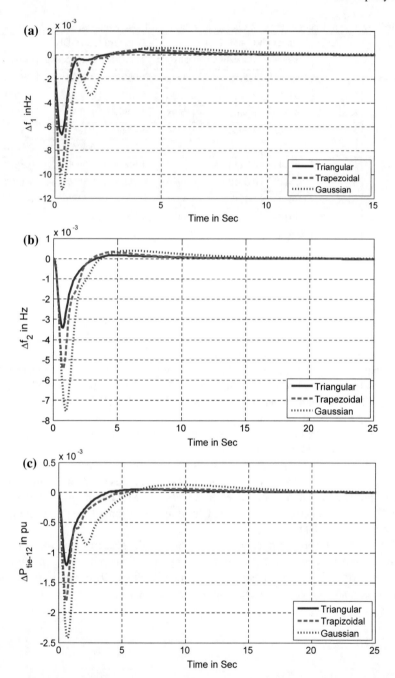

Fig. 5 **a** Frequency deviation in area 1 for different membership functions, **b** frequency deviation in area 2 for different membership functions, and **c** tie-line power deviation for different membership functions

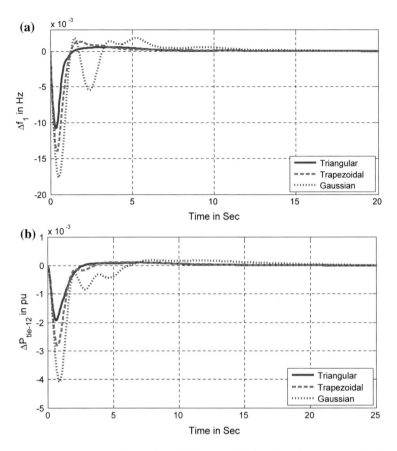

Fig. 6 **a** Frequency deviation in area 1 for different membership functions with nonlinearity and **b** tie-line power deviation for different membership functions with nonlinearity

6 Conclusion

In this paper, spider monkey optimization (SMO) is introduced to tune the controller parameters for load frequency control (LFC) issue. Initially, a two-area reheat thermal power system without nonlinearity and then with nonlinearity is considered for the same system to demonstrate the performances of fuzzy PID controller with different membership functions. It is observed that the performance of controller with triangular membership function is better in terms of peak undershoot, peak overshoot, and settling time for the system in both the cases, i.e., without and with GRC and GDB. Further, the proposed algorithm is compared with other optimization algorithms like PSO and TLBO and found to be performing better in terms of time domain specifications. The study could be further extended for multi-area systems.

Table 3 U_{sh}, O_{sh}, and T_s of controller for different membership functions without nonlinearity

Membership function		Triangular	Trapezoidal	Gaussian
Δf_1	$U_{SH} \times 10^{-3}$ in Hz	−6.693	−9.704	−11.302
	$O_{SH} \times 10^{-3}$ in Hz	0.271	0.506	0.588
	T_S in s	5.014	7.078	10.493
Δf_2	$U_{SH} \times 10^{-3}$ in Hz	−3.425	−5.436	−7.542
	$O_{SH} \times 10^{-3}$ in Hz	0.164	0.351	0.398
	T_S in s	2.718	6.734	10.830
$\Delta P_{tie\text{-}12}$	$U_{SH} \times 10^{-3}$ in Hz	−1.208	−1.810	−2.421
	$O_{SH} \times 10^{-3}$ in Hz	0.049	0.061	0.131
	T_S in s	2.374	2.850	4.292

Table 4 U_{sh}, O_{sh}, and T_s of controller for different membership functions with nonlinearity

Membership function		Triangular	Trapezoidal	Gaussian
Δf_1	$U_{SH} \times 10^{-3}$ in Hz	−10.77	−13.96	−17.59
	$O_{SH} \times 10^{-3}$ in Hz	0.586	1.336	1.827
	T_S in s	6.348	6.907	12.150
Δf_2	$U_{SH} \times 10^{-3}$ in Hz	−4.851	−8.151	−9.875
	$O_{SH} \times 10^{-3}$ in Hz	0.560	1.215	1.438
	T_S in s	6.169	7.710	12.827
$\Delta P_{tie\text{-}12}$	$U_{SH} \times 10^{-3}$ in Hz	1.925	−2.802	−4.077
	$O_{SH} \times 10^{-3}$ in Hz	0.107	0.121	0.192
	T_S in s	2.119	1.928	5.443

Appendix [8, 12]

$f = 60$ Hz; $R_1 = R_2 = 2.4$ Hz/pu.; $T_{G1} = T_{G2} = 0.08$ s; $T_{T1} = T_{T2} = 0.3$ s; $B_1 = B_2 = 0.425$ p.u. MW/Hz; $T_{r1} = T_{r2} = 10$ s; $K_{r1} = K_{r2} = 0.5$ $T_{P1} = T_{P2} = 20$ s; $K_{P1} = K_{P2} = 120$ Hz/p.u. MW $T_{12} = 0.545$, $a_{12} = -1$.

Appendix

Population size (n) = 50, number of iteration = 100, and number of runs = 20.

Control parameters for SMO: $LL_{lim} = (n/2) = 25$, $GL_{lim} = ((n * d)/2) = 250$. MG. = 10, pr = 0.1–0.9 *increasing linearly*.

Control parameters for PSO: Inertia weight (w), decreases linearly from 0.9–0.1; acceleration coefficients (c1 = 2, c2 = 2).

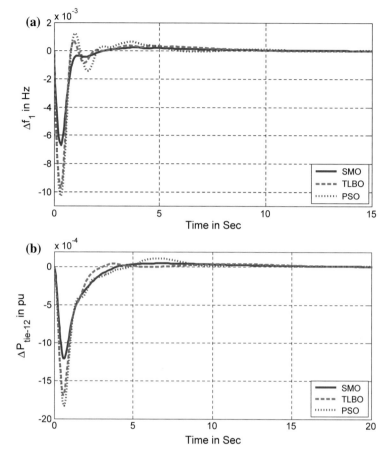

Fig. 7 **a** Frequency deviation in area 1 for different algorithms and **b** tie-line power deviation for different algorithms

References

1. Elgerd, O.I.: Electric Energy Systems Theory: An Introduction, 2nd edn. Tata McGraw-Hill, New Delhi (2007)
2. Kundur, P.: Power System Stability and Control. Tata McGraw-Hill, New Delhi (2009) (8th reprint)
3. Cohn, N.: Some aspects of tie-line bias control on interconnected power systems. Trans. Am. Inst. Electr. Eng. Part III Power Appar. Syst. **75**(3), 1415–1436 (1956)
4. Elgerd, O.I., Fosha, C.E.: Optimum megawatt-frequency control of multiarea electric energy systems. IEEE Trans. Power Appar. Syst. **4**, 556–563 (1970)
5. Ghoshal, S.P.: Optimizations of PID gains by particle swarm optimizations in fuzzy based automatic generation control. Electr. Power Syst. Res. **72**(3), 203–212 (2004)
6. Kocaarslan, I., Ertuğrul, Ç.: Fuzzy logic controller in interconnected electrical power systems for load-frequency control. Int. J. Electr. Power Energy Syst. **27**(8), 542–549 (2005)

7. Panda, G., Panda, S., Cemal, A.: Automatic generation control of interconnected power system with generation rate constraints by hybrid neuro fuzzy approach. Int. J. Electr. Electron. Eng. (2009)

8. Saikia, L.C., Nanda, J., Mishra, S.: Performance comparison of several classical controllers in AGC for multi-area interconnected thermal system. Int. J. Electr. Power Energy Syst. **33**(3), 394–401 (2011)

9. Sahu, R.K., Panda, S., Pradhan, S.: A hybrid firefly algorithm and pattern search technique for automatic generation control of multi area power systems. Int. J. Electr. Power Energy Syst. **64**, 9–23 (2015)

10. Sahu, R.K., Panda, S., Sekhar, G.T.C.: A novel hybrid PSO-PS optimized fuzzy PI controller for AGC in multi area interconnected power systems. Int. J. Electr. Power Energy Syst. **64**, 880–893 (2015)

11. Sahu, B.K., et al.: Teaching–learning based optimization algorithm based fuzzy-PID controller for automatic generation control of multi-area power system. Appl. Soft Comput. **27**, 240–249 (2015)

12. Sahu, B.K., Pati, S., Panda, S.: Hybrid differential evolution particle swarm optimization optimized fuzzy proportional–integral derivative controller for automatic generation control of interconnected power system. IET Gener. Transm. Distrib. **8**(11), 1789–1800 (2014)

13. Sahu, R.K., Panda, S., Padhan, S.: A novel hybrid gravitational search and pattern search algorithm for load frequency control of nonlinear power system. Appl. Soft Comput. **29**, 310–327 (2015)

14. Sahu, R.K., Panda, S., Pradhan, P.C.: Design and analysis of hybrid firefly algorithm-pattern search based fuzzy PID controller for LFC of multi area power systems. Int. J. Electr. Power Energy Syst. **69**, 200–212 (2015)

15. Sahu, B.K., et al.: A novel hybrid LUS–TLBO optimized fuzzy-PID controller for load frequency control of multi-source power system. Int. J. Electr. Power Energy Syst. **74**, 58–69 (2016)

16. Bansal, J.C., et al.: Spider monkey optimization algorithm for numerical optimization. Memetic Comput. **6**(1), 31–47 (2014)

17. Gupta, K., Deep, K., Bansal, J.C.: Spider monkey optimization algorithm for constrained optimization problems. Soft. Comput. **21**(23), 6933–6962 (2017)

A Third-Order Accurate Finite Difference Method and Compact Operator Approach for Mildly Nonlinear Two Spatial Dimensions Elliptic BVPs with Integral Form of Source Term

Navnit Jha, Venu Gopal and Bhagat Singh

Abstract The use of finite difference approximations on geometric mesh network has been reported for the numerical estimation of two spatial dimensions mildly nonlinear elliptic boundary value problems with integral form of the source term. The new scheme applied the special type of unequal mesh spacing for differentials and integrals and obtained compact formulation for the mildly nonlinear equations. The estimated order of the scheme is three if the meshes are unequally spaced and achieves an accuracy of order four for evenly spaced mesh points. The necessary condition for the scheme to converge has been obtained with the help of monotonicity and irreducibility of the iteration matrix. The accuracy and efficiency of the proposed compact scheme have been illustrated using Helmholtz equation, stationary heat equation and steady-state mass transfer equation. The root mean squared error on the test problems with known true solutions justifies the third (four)-order accuracies of the new scheme. It is demonstrated that the third-order compact scheme on geometric meshes improves the efficiency and accuracy while comparing with fourth-order scheme on uniform mesh network.

Keywords Lagrange's interpolation · Finite difference approximations · Compact operators · Geometric meshes · Elliptic equations · Root mean squared error

Mathematics Subject Classification 35J25 · 35J60 · 65N06

N. Jha (✉) · B. Singh
Faculty of Mathematics and Computer Science, South Asian University, Chanakyapuri
110021, New Delhi, India
e-mail: navnitjha@sau.ac.in

V. Gopal
The School of Mathematical Sciences, University of Science and Technology of China,
Hefei 230026, Anhui, Republic of China
e-mail: vgopal@ustc.edu.cn

© Springer Nature Singapore Pte Ltd. 2019
J. C. Bansal et al. (eds.), *Soft Computing for Problem Solving*, Advances in Intelligent
Systems and Computing 817, https://doi.org/10.1007/978-981-13-1595-4_71

1   Introduction

Consider the two spatial dimensions elliptic boundary value problems (EBVPs) involving nonlinear partial derivatives of first order and integral form of source term

$$U^{xx} + U^{yy} = F\left(x,\ y,\ U,\ U^x,\ U^y\right) + \int_{s=0}^{1} \int_{r=0}^{1} K(r,\ s,\ x,\ y) drds,\ (x,\ y) \in \mathcal{D}.$$

$$(1.1)$$

Here, the solution domain is defined as $\mathcal{D} = \{(x,\ y)\ :\ a < x < b,\ c < y < d\}$ and the steady-state process $U(x,\ y)$ is known on the boundary $\partial \mathcal{D}$ of the fixed finite region $\mathcal{D}$ and is specified by the Dirichlet's conditions:

$$U(x,\ c) = S_1(x),\ U(x,\ d) = S_2(x),\ U(a,\ y) = S_3(y),\ U(b,\ y) = S_4(y).\quad (1.2)$$

Let us denote

$$R(x,\ y) = \int_{s=0}^{1} \int_{r=0}^{1} K(r,\ s,\ x,\ y) drds \qquad (1.3)$$

and

$$G\left(x,\ y,\ U,\ U^x,\ U^y\right) = F\left(x,\ y,\ U,\ U^x,\ U^y\right) + R(x,\ y) \qquad (1.4)$$

We assume that $U \in C^4(\Omega)$, the partial derivative G_U is non-negative, and the derivatives G_{U_x}, G_{U_y} are uniformly bounded, Bers [1].

Many physical phenomena are modelled using integro-differential equations and are one of the principal tools in the application areas of mathematics and physics. For example, in the context of fluid flow, the Navier–Stokes equation describing blood flow in heart incurs force density and is the integral of intensity to the boundary forces, Peskin [2]. The source term appearing in the interface problems is often expressed as integrals of source strength on the interface, Zhilin et al. [3]. Such kind of two spatial dimensions EBVPs with nonlinear first-order derivatives and source term in the form of integral may not possess analytical solution in general. Therefore, we require an efficient numerical technique, which provides an approximate solution to the differential integral problems. In the past, the numerical solution of EBVPs was treated with the help of finite difference, finite element, Haar wavelets, spline collocation and cubic spline by Kim and Lee [4], Mohanty and Singh [5], Tian and Dai [6], Fairweather et al. [7] and Aziz and Šarler [8]. The defect correction method for the singular perturbation problems of the elliptic type exhibiting parallel boundary layers and/or vertical boundary layers has been discussed by Segal [9].

Among various numerical methods, the finite difference approximation (FDA) is one of the elegant tools for obtaining the solution of ordinary and partial differential equations. The beauty of the solutions obtained by FDA is mainly measured by the convergence order, computational efficiencies and solution accuracies. The lower-order method developed on less number of stencils or higher-order method developed on more number of stencils often lands up with instability or extended processing time and thus impractical for the numerical solution. As an example, in the context of one space dimension, first- and second-order derivatives can be represented using n mesh points with $(n-1)$ and $(n-2)$-order accurate finite difference equations $(n \geq 3)$, Mullges and Uhlig [10]. If the second-order derivative replaces five mesh point discretization formula, the corresponding system of the difference equations yields five-band matrix and their numerical solution requires massive computing time. Thus, we restrict the design of scheme with a minimum number of mesh points needed to discretize the highest order derivative appearing in the given problems. Since our aim is to solve second-order EBVPs in two spatial dimensions, the least number of mesh points required to discretize the second-order derivative is three in each dimension. Thus, we formulate a new scheme on a 3×3 mesh point network, known as compact scheme. Such a scheme is convergent and placed among the most stable one. On the same line, we evaluate the double integral by constructing an interpolating polynomial with basis $\{1, r, r^2\} \times \{1, s, s^2\}$. The integrals will be calculated on two panels of the nonuniformly spaced mesh points in both r- and s-directions. If the meshes are uniformly spaced, the integral formula reduces to Simpson's one-third rule. The higher-order terms of differential and integral parts are jointly treated to describe the order of numerical method.

This work is organized in the following manner. In Sect. 2, we develop some compact operators on a new mesh network. In Sect. 3, the integral formula by means of bivariate Lagrange's interpolation on nonuniformly spaced mesh points has been obtained and examined the error of double integral. In Sect. 4, the compact scheme of order three and four is presented in algorithmic form. Section 5 describes the mathematical derivations of the numerical scheme. The detailed error analysis is presented in Sect. 6. In Sect. 7, some linear and nonlinear EBVPs are solved and presented the results in tabular form. In the last, the paper is concluded with the advantage of using the proposed scheme.

2 Geometric Meshes and Compact Operators

Let the two spatial dimensions mesh points $\{(x_i, y_j) : i = 0(1)I + 1, j = 0(1)J + 1\}$ are given by $a = x_0 < x_1 < \cdots < x_I < x_{I+1} = b$, $\Delta x_i = x_i - x_{i-1}$, $i = 1(1)I + 1$, $c = y_0 < y_1 < \cdots < y_J < y_{J+1} = d$, $\Delta y_j = y_j - y_{j-1}$, $j = 1(1)J + 1$ and the next mesh spacing are $\Delta x_{i+1} = \rho \Delta x_i$ and $\Delta y_{j+1} = \sigma \Delta y_j$, $i = 1(1)I, j = 1(1)J$, where ρ and σ are positive geometric mesh parameters in x- and y-directions, respectively. If $\rho = \sigma = 1$, then $\Delta x_i = (b-a)/(I+1)$, $\Delta y_j = (d-c)/(J+1)$, $\forall i = 1(1)I + 1$, $j = 1(1)J + 1$. If $\rho \neq 1$ and $\sigma \neq 1$, then the first step size is given by $\Delta x_1 = (b-a)(\rho-1)/(\rho^{I+1}-1)$ and $\Delta y_1 = (d-c)(\sigma-1)/(\sigma^{J+1}-1)$. Such a kind of mesh network has been considered in the application areas of electrochemistry by Britz [11] and two-point singularly perturbed boundary value problems by Jain et al. [12], Mohanty [13], Kadalbajoo and Kumar [14] and Jha and Bieniasz [15]. The numerical solutions $U(x, y)$ of Eqs. (1.1) and (1.2) are obtained by means of finite difference replacement of partial derivatives on a nine-point stencil $\Omega_{i,j}^{xy} = \{(x_p, y_q) : (p, q) \in \mathcal{T}\}$, $\mathcal{T} = \{i - 1, i, i + 1\} \times \{j - 1, j, j + 1\}$. The symbols $U_{i,j}$ and $u_{i,j}$ refer to the exact and approximate solution values, respectively, at the mesh point (x_i, y_j).

Following the definition of the average and central differencing operators, it is feasible to define compact operators on geometric meshes, which approximate first- and second-order partial derivatives on $\Omega_{i,j}^{xy}$. These operators are defined as

$$
\begin{bmatrix} \mathcal{F}_x U_{i,j} \\ \mathcal{S}_x U_{i,j} \end{bmatrix} = \begin{bmatrix} 1/[\rho(\rho + 1)] & 1 - 1/\rho & -\rho/(\rho + 1) \\ 2/[\rho(\rho + 1)] & -2/\rho & 2/(\rho + 1) \end{bmatrix} \begin{bmatrix} U_{i+1,j} \\ U_{i,j} \\ U_{i-1,j} \end{bmatrix}, \quad (2.1)
$$

$$
\begin{bmatrix} \mathcal{F}_y U_{i,j} \\ \mathcal{S}_y U_{i,j} \end{bmatrix} = \begin{bmatrix} 1/[\sigma(\sigma + 1)] & 1 - 1/\sigma & -\sigma/(\sigma + 1) \\ 2/[\sigma(\sigma + 1)] & -2/\sigma & 2/(\sigma + 1) \end{bmatrix} \begin{bmatrix} U_{i,j+1} \\ U_{i,j} \\ U_{i,j-1} \end{bmatrix}. \quad (2.2)
$$

With the application of Taylor's series, one obtains

$$
\begin{cases} \mathcal{F}_x U_{i,j} = \Delta x_i U_{i,j}^x + \rho \Delta x_i^3 U_{i,j}^{xxx}/6 + O(\Delta x_i^4), \\ \mathcal{F}_y U_{i,j} = \Delta y_j U_{i,j}^y + \sigma \Delta y_j^3 U_{i,j}^{yyy}/6 + O(\Delta y_j^4), \end{cases} \quad (2.3)
$$

$$
\begin{cases} \mathcal{S}_x U_{i,j} = \Delta x_i^2 U_{i,j}^{xx} + (\rho - 1)\Delta x_i^3 U_{i,j}^{xxx}/3 + O(\Delta x_i^4), \\ \mathcal{S}_y U_{i,j} = \Delta y_j^2 U_{i,j}^{yy} + (\sigma - 1)\Delta y_j^3 U_{i,j}^{yyy}/3 + O(\Delta y_j^4). \end{cases} \quad (2.4)
$$

Thus, the first- and second-order derivatives in x- and y-directions can be written with the help of operators $\mathcal{F}_x$, $\mathcal{S}_x$ and $\mathcal{F}_y$, $\mathcal{S}_y$, respectively. The composite of these operators yields the approximations to mixed partial derivatives.

3   Bivariate Lagrange's Interpolation

Given a mesh point (x_i, y_j) in xy plane, let $\mathcal{P}^{i,j}(r, s)$ denote the interpolating polynomial of $K(r, s, x_i, y_j)$ on the tabular points $\{(r_p, s_q, \mathcal{P}_{p,q}^{i,j}) : (p, q) \in \mathcal{H}\}$, $\mathcal{H} = \{l - 1, l, l + 1\} \times \{m - 1, m, m + 1\}$, $\mathcal{P}_{p,q}^{i,j} = K(r_p, s_q, x_i, y_j)$, where $0 = r_0 < r_1 < \cdots < r_{L+1} = 1$, $\Delta r_l = r_l - r_{l-1}$, $l = 1(1)L + 1$, $0 = s_0 < s_1 < \cdots < s_{M+1} = 1$, $\Delta s_m = s_m - s_{m-1}$, $m = 1(1)M + 1$ and $\Delta r_{l+1} = \rho \Delta r_l$, $\Delta s_{m+1} = \sigma \Delta s_m$, $l = 1(1)L$, $m = 1(1)M$. The bivariate Lagrange's

Table 1 Values of coefficients appearing in Eq. (3.2)

$A_{l,m}$	$(\rho + 1)^3(\sigma + 1)^3/(\rho\sigma)$
$A_{l+1,m}$	$(2\rho - 1)(\rho + 1)(\sigma + 1)^3/(\rho\sigma)$
$A_{l-1,m}$	$(\rho + 1)(2 - \rho)(\sigma + 1)^3/\sigma$
$A_{l,m+1}$	$(\rho + 1)^3(\sigma + 1)(2\sigma - 1)/(\rho\sigma)$
$A_{l+1,m+1}$	$(\rho + 1)(2\rho - 1)(\sigma + 1)(2\sigma - 1)/(\rho\sigma)$
$A_{l-1,m+1}$	$(\rho + 1)(2 - \rho)(\sigma + 1)(2\sigma - 1)/\sigma$
$A_{l,m-1}$	$(\rho + 1)^3(\sigma + 1)(2 - \sigma)/\rho$
$A_{l+1,m-1}$	$(\rho + 1)(2\rho - 1)(\sigma + 1)(2 - \sigma)/\rho$
$A_{l-1,m-1}$	$(\rho + 1)(\rho - 2)(\sigma + 1)(\sigma - 2)$

interpolating polynomial on the nine-point stencil $\Omega^{rs}_{i,j} = \{(r_p, s_q) : (p, q) \in \mathcal{H}\}$ is therefore expressed as:

$$
\mathcal{P}^{i,j}(r, s)
$$

$$
= \sum_{(p,q)\in\mathcal{H}} \prod_{\substack{e = l - 1, l, l + 1 \\ e \neq p}} \frac{r - r_e}{r_p - r_e} \prod_{\substack{t = m - 1, m, m + 1 \\ t \neq q}} \frac{s - s_t}{s_q - s_t} \mathcal{P}^{i,j}_{p,q}.
$$

$$(3.1)$$

Integrating Eq. (3.1) on two panels, we obtain

$$
\int_{s=s_{m-1}}^{s_{m+1}} \int_{r=r_{l-1}}^{r_{l+1}} \mathcal{P}^{i,j}(r, s)drds = \frac{1}{36}\Delta r_l \Delta s_m \sum_{(p,q)\in\mathcal{H}} A_{p,q}\mathcal{P}^{i,j}_{p,q},
$$

$$(3.2)$$

where the expressions $A_{p,q}$, $(p, q) \in \mathcal{H}$ are enumerated in Table 1.

Thus, the value of double integral at the mesh point (x_i, y_j) based on geometric meshes is given by

$$
R(x_i, y_j) = \int_{s=0}^{1} \int_{r=0}^{1} K(r, s, x_i, y_j)drds
$$

$$
= \sum_{l=1(2)L} \sum_{m=1(2)M} \int_{s=s_{m-1}}^{s_{m+1}} \int_{r=r_{l-1}}^{r_{l+1}} \mathcal{P}^{i,j}(r, s)drds
$$

$$(3.3)$$

With the help of Taylor's expansion, it is possible to write

$$
K(r, s, x_i, y_j) = \sum_{e=0}^{\infty} \sum_{t=0}^{\infty} \frac{1}{e!\,t!}(r - r_l)^e (s - s_m)^t \left(\frac{\partial^{e+t}K}{\partial r^e \partial s^t}\right)_{(r_l,s_m,x_i,y_j)}
$$

$$(3.4)$$

Upon integrating (3.4) and neglecting higher-order terms, we get

Table 2 Values of coefficients appearing in Eq. (3.5)

$B_{0,0}$	1
$B_{1,0}$	$(\rho - 1)/2$
$B_{0,1}$	$(\sigma - 1)/2$
$B_{2,0}$	$(\rho^2 - \rho + 1)/6$
$B_{1,1}$	$(\rho - 1)(\sigma - 1)/4$
$B_{0,2}$	$(\sigma^2 - \sigma + 1)/6$
$B_{3,0}$	$(\rho - 1)(\rho^2 + 1)/24$
$B_{1,2}$	$(\rho - 1)(\sigma^2 - \sigma + 1)/12$
$B_{2,1}$	$(\sigma - 1)(\rho^2 - \rho + 1)/12$
$B_{0,3}$	$(\sigma - 1)(\sigma^2 + 1)/24$

$$\int_{s=s_{m-1}}^{s_{m+1}} \int_{r=r_{l-1}}^{r_{l+1}} K(r, s, x_i, y_j)drds$$

$$= (\rho + 1)(\sigma + 1)\Delta r_l \Delta s_m \sum_{\substack{e,t = 0(1)3 \\ e+t \leq 3}} B_{e,t}\Delta r_l^e \Delta s_m^t \left(\frac{\partial^{e+t}K}{\partial r^e \partial s^t}\right)_{(r_l, s_m, x_i, y_j)} \quad (3.5)$$

where the expressions of $B_{e,t}$ are enumerated in Table 2.

Subtracting (3.5) from (3.2), error on the nine-point network $\Omega_{i,j}^{rs}$ is obtained as follows:

$$\mathcal{E}_{l,m} = \frac{1}{72}(\sigma + 1)(\rho - 1)(\rho + 1)^3 \Delta r_l^4 \Delta s_m \left(\frac{\partial^3 K}{\partial r^3}\right)_{(r_l, s_m, x_i, y_j)}$$

$$+ \frac{1}{72}(\sigma - 1)(\rho + 1)(\sigma + 1)^3 \Delta r_l \Delta s_m^4 \left(\frac{\partial^3 K}{\partial s^3}\right)_{(r_l, s_m, x_i, y_j)}$$

$$+ O(\Delta r_l \Delta s_m(\Delta r_l^4 + \Delta r_l^2 \Delta s_m^2 + \Delta s_m^4)). \quad (3.6)$$

Let $\Delta r = \max_{l=1(1)L+1} \Delta r_l$, $\Delta s = \max_{m=1(1)M+1} \Delta s_m$, $Z = \max_{\substack{0 \leq r, s \leq 1 \\ a \leq x, y \leq b}} \left\{\left|\frac{\partial^3 K}{\partial r^3}\right|, \left|\frac{\partial^3 K}{\partial s^3}\right|\right\}$.

Then, from Eq. (3.6), bounds of error can be estimated as:

$$\mathcal{E}_{l,m} \leq \frac{Z}{72}(\sigma + 1)(\rho - 1)(\rho + 1)^3 \Delta r^4 \Delta s$$

$$+ \frac{Z}{72}(\sigma - 1)(\rho + 1)(\sigma + 1)^3 \Delta r \Delta s^4 + O(\Delta r \Delta s(\Delta r^4 + \Delta r^2 \Delta s^2 + \Delta s^4)).$$

Then, the total error $(\mathcal{E})$ in the composite integral (3.3) is obtained as:

$$\mathcal{E} = \sum_{l=1(2)L,m=1(2)M} \left| \mathcal{E}_{l,m} \right|$$

$$\leq \begin{cases} O\big(\Delta r \Delta s \big(\Delta r^4 + \Delta r^2 \Delta s^2 + \Delta s^4\big)\big)(L+1)(M+1), \ \rho = 1, \ \sigma = 1 \\ O\big(\Delta r \Delta s \big(\Delta r^3 + \Delta s^3\big)\big)(L+1)(M+1), \qquad \rho \neq 1, \ \sigma \neq 1 \end{cases}$$

$$\approx \begin{cases} O\big(\Delta r^4 + \Delta r^2 \Delta s^2 + \Delta s^4\big), \ \rho = 1, \ \sigma = 1 \\ O\big(\Delta r^3 + \Delta s^3\big), \qquad\qquad \rho \neq 1, \ \sigma \neq 1 \end{cases} \tag{3.7}$$

Since the quantities $\Delta r(L+1)$ and $\Delta s(M+1)$ exactly (approximately) equal the length of the intervals in r- and s-directions, respectively, both of them are one if the meshes are uniformly (nonuniformly) spaced. Moreover, if $\rho = \sigma = 1$, the integral formula (3.3) reduces to Simpson's one-third rule in two spatial dimensions, Gerald et al. [16]. In order to implement the formulation (3.3), we must take L and M to be odd integers, since the integrals in Eq. (3.2) are performed on two panels.

4 Compact Scheme of Order Three and Four

We begin with the two spatial dimensions Poisson's equation $\nabla^2 U(x, y) = G(x, y)$, and their finite difference replacement on the nine-point geometric meshes is expressed by the recursive equation

$$\mathcal{L}_{i,j} U_{i,j} = 2\Delta x_i^2 \Delta y_j^2 \sum_{(p,q)\in \mathcal{T}} Q_{p,q} G_{p,q} + T_{i,j}, \ i = 1(1)I, \ j = 1(1)J, \tag{4.1}$$

where

$$\mathcal{L}_{i,j} = 3\Delta x_i^2 \big[12 + 4(\rho - 1)\mathcal{F}_x + \big(\rho^2 - \rho + 1\big)\mathcal{S}_x\big]\mathcal{S}_y$$
$$+ 3\Delta y_j^2 \big[12 + 4(\sigma - 1)\mathcal{F}_y + \big(\sigma^2 - \sigma + 1\big)\mathcal{S}_y\big]\mathcal{S}_x,$$

is the geometric mesh discretization of the Laplace operator $\nabla^2 = \partial_{xx}^2 + \partial_{yy}^2$, and the scheme has local truncation error (LTE) as $T_{i,j} = \big(\Delta x_i^2 \Delta y_j^2 \big(\Delta x_i^3 + \Delta x_i^2 \Delta y_j + \Delta x_i \Delta y_j^2 + \Delta y_j^3\big)\big)$, for the arbitrary choice of ρ and σ. However, if the meshes are uniformly spaced ($\rho = \sigma = 1$), the LTE turns out to be $T_{i,j} = \big(\Delta x_i^2 \Delta y_j^2 \big(\Delta x_i^4 + \Delta x_i^2 \Delta y_j^2 + \Delta y_j^4\big)\big)$, Jha [17]. The values of coefficients $Q_{p,q}$, $(p, q) \in \mathcal{T}$ are listed in Table 3.

Extending the compact scheme (4.1) to the elliptic Eqs. (1.1) and (1.2), which contains $U(x, y)$ along with their partial derivatives of first order as nonlinear terms and following the technique by Shaw et al. [18], the algorithmic formulation in the compact operators form is mentioned below:

For $i = 1(1)I, j = 1(1)J$:

Table 3 Values of coefficients appearing in Eq. (4.1)

$Q_{l,m}$	$[(\rho^2+1)(2\sigma^2-\sigma+2)-\rho(\sigma^2-8\sigma+1)]/(\sigma\rho)$
$Q_{l+1,m}$	$[(\rho-1)(2\sigma^2-\sigma+2)+3\rho^2\sigma]/[(\rho+1)\sigma\rho]$
$Q_{l-1,m}$	$[\rho(1-\rho)(2\sigma^2-\sigma+2)+3\sigma]/[(\rho+1)\sigma]$
$Q_{l,m+1}$	$[2(\rho^2+1)(\sigma-1)+\rho(3\sigma^2-\sigma+1)]/[\sigma\rho(\sigma+1)]$
$Q_{l+1,m+1}$	$2(\sigma-1)(\rho-1)/[\sigma\rho(\sigma+1)(\rho+1)]$
$Q_{l-1,m+1}$	$-2\rho(\sigma-1)(\rho-1)/[\sigma(\sigma+1)(\rho+1)]$
$Q_{l,m-1}$	$[(\rho^2+1)2\sigma(1-\sigma)+\rho(\sigma^2-\sigma+3)]/[\rho(\sigma+1)]$
$Q_{l+1,m-1}$	$-2\sigma(\sigma-1)(\rho-1)/[\rho(\sigma+1)(\rho+1)]$
$Q_{l-1,m-1}$	$2\sigma\rho(\sigma-1)(\rho-1)/[(\sigma+1)(\rho+1)]$

Compute $R(x_i, y_j)$ using Eq. (3.3).

$$\tilde{U}^x_{i,j+\tau} = [U_{i+1,j+\tau} + (\rho^2-1)U_{i,j+\tau} - \rho^2 U_{i-1,j+\tau}]/[\rho(\rho+1)\Delta x_i], \quad \tau = 0, \pm 1, \tag{4.2}$$

$$\tilde{U}^y_{i+\tau,j} = [U_{i+\tau,j+1} + (\sigma^2-1)U_{i+\tau,j} - \sigma^2 U_{i+\tau,j-1}]/[\sigma(\sigma+1)\Delta y_j], \tag{4.3}$$

$$\tilde{U}^x_{i+1,j+\tau} = [(1+2\rho)U_{i+1,j+\tau} - (\rho+1)^2 U_{i,j+\tau} + \rho^2 U_{i-1,j+\tau}]/[\rho(\rho+1)\Delta x_i], \tag{4.4}$$

$$\tilde{U}^y_{i+\tau,j+1} = [(1+2\sigma)U_{i+\tau,j+1} - (\sigma+1)^2 U_{i+\tau,j} + \sigma^2 U_{i+\tau,j-1}]/[\sigma(\sigma+1)\Delta y_j], \tag{4.5}$$

$$\tilde{U}^x_{i-1,j+\tau} = [-U_{i+1,j+\tau} + (\rho+1)^2 U_{i,j+\tau} - \rho(\rho+2)U_{i-1,j+\tau}]/[\rho(\rho+1)\Delta x_i], \tag{4.6}$$

$$\tilde{U}^y_{i+\tau,j-1} = [-U_{i+\tau,j+1} + (\sigma+1)^2 U_{i+\tau,j} - \sigma(\sigma+2)U_{i+\tau,j-1}]/[\sigma(\sigma+1)\Delta y_j], \tag{4.7}$$

$$\tilde{G}_{p,q} = F\left(x_p, y_q, U_{p,q}, \tilde{U}^x_{p,q}, \tilde{U}^y_{p,q}\right) + R(x_p, y_q), \ (p, q) \in \mathcal{T} \sim \{(i, j)\}, \tag{4.8}$$

$$\tilde{U}^{xx}_{i,j\pm1} = 2[U_{i+1,j\pm1} - (\rho+1)U_{i,j\pm1} + \rho U_{i-1,j\pm1}]/[\rho(\rho+1)\Delta x_i^2], \tag{4.9}$$

$$\tilde{U}^{yy}_{i\pm1,j} = 2[U_{i\pm1,j+1} - (\sigma+1)U_{i\pm1,j} + \sigma U_{i\pm1,j-1}]/[\sigma(\sigma+1)\Delta y_j^2], \tag{4.10}$$

$$\hat{U}^x_{i,j} = \tilde{U}^x_{i,j} + \upsilon\Delta x_i\left(\tilde{G}_{i+1,j} - \tilde{G}_{i-1,j} - \tilde{U}^{yy}_{i+1,j} + \tilde{U}^{yy}_{i-1,j}\right), \tag{4.11}$$

$$\hat{U}^y_{i,j} = \tilde{U}^y_{i,j} + \zeta\Delta y_j\left(\tilde{G}_{i,j+1} - \tilde{G}_{i,j-1} - \tilde{U}^{xx}_{i,j+1} + \tilde{U}^{xx}_{i,j-1}\right), \tag{4.12}$$

$$\tilde{G}_{i,j} = F\left(x_i, y_j, U_{i,j}, \hat{U}_{i,j}^x, \hat{U}_{i,j}^y\right) + R(x_i, y_j), \tag{4.13}$$

$$\mathcal{L}_{i,j} U_{i,j} = 2\Delta x_i^2 \Delta y_j^2 \sum_{(p,q)\in\mathcal{T}} Q_{p,q} \tilde{G}_{p,q} + \tilde{T}_{i,j}, \ \tilde{T}_{i,j} = T_{i,j} + \Delta x_i^2 \Delta y_j^2 \mathcal{E}. \tag{4.14}$$

The difference Eq. (4.14) approximates the EBVPs (1.1) and (1.2) on the nine-point stencils $\Omega_{i,j}^{xy}$, and thus, we arrive at a compact formulation of the scheme. The order of magnitude to $\tilde{T}_{i,j}$ is eight (seven) on uniformly (nonuniform) spaced meshes. For the solution of a system of difference equations (4.14), it is essential to incorporate the boundary values (1.2) and omit the higher-order term $\tilde{T}_{i,j}$. The resulting system of equation forms a block tri-diagonal matrix and can be solved by the standard solution technique, such as Gauss–Seidel iterative method or Newton–Raphson method, depending upon the linear or nonlinear characters of the elliptic problems; see Kelley [19].

5   Mathematical Description of the Difference Scheme

In this section, we discuss the derivation of a compact scheme (4.14) and obtain the values of free parameters υ and ζ appearing in Eqs. (4.11) and (4.12). In the following simplifications, we denote $G_{p,q} = G\left(x_p, y_q, U_{p,q}, U_{p,q}^x, U_{p,q}^y\right)$, $(p, q) \in \mathcal{T}$ and $C(x, y) = \partial G/\partial U^x$, $D(x, y) = \partial G/\partial U^y$. Applying the approximation of first-order partial derivatives (4.2)–(4.7) defined on the stencils $\Omega_{i,j}^{xy}$ to Eq. (4.8) and using two-dimensional finite Taylor's expansion, we obtain

$$\tilde{G}_{i-1,j-1} = G_{i-1,j-1} - (\rho+1)\Delta x_i^2 C_{i,j} U_{i,j}^{xxx}/6 - (\sigma+1)\Delta y_j^2 D_{i,j} U_{i,j}^{yyy}/6 + O\left(\Delta x_i^3\right), \tag{5.1}$$

$$\tilde{G}_{i-1,j} = G_{i-1,j} - (\rho+1)\Delta x_i^2 C_{i,j} U_{i,j}^{xxx}/6 - \sigma\Delta y_j^2 D_{i,j} U_{i,j}^{yyy}/6 + O\left(\Delta x_i^3 + \Delta y_j^3\right), \tag{5.2}$$

$$\tilde{G}_{i-1,j+1} = G_{i-1,j+1} - (\rho+1)\Delta x_i^2 C_{i,j} U_{i,j}^{xxx}/6 - \sigma(\sigma+1)\Delta y_j^2 D_{i,j} U_{i,j}^{yyy}/6 + O\left(\Delta x_i^3\right), \tag{5.3}$$

$$\tilde{G}_{i,j-1} = G_{i,j-1} + \rho\Delta x_i^2 C_{i,j} U_{i,j}^{xxx}/6 - (\sigma+1)\Delta y_j^2 D_{i,j} U_{i,j}^{yyy}/6 + O\left(\Delta x_i^3 + \Delta y_j^3\right), \tag{5.4}$$

$$\tilde{G}_{i,j+1} = G_{i,j+1} + \rho\Delta x_i^2 C_{i,j} U_{i,j}^{xxx}/6 - \sigma(\sigma+1)\Delta y_j^2 D_{i,j} U_{i,j}^{yyy}/6 + O\left(\Delta x_i^3 + \Delta y_j^3\right), \tag{5.5}$$

$$\tilde{G}_{i+1,j-1} = G_{i+1,j-1} - \rho(\rho+1)\Delta x_i^2 C_{i,j} U_{i,j}^{xxx}/6 - (\sigma+1)\Delta y_j^2 D_{i,j} U_{i,j}^{yyy}/6 + O\left(\Delta x_i^3\right), \tag{5.6}$$

$$\tilde{G}_{i+1,j} = G_{i+1,j} - \rho(\rho+1)\Delta x_i^2 C_{i,j} U_{i,j}^{xxx}/6 + \sigma\Delta y_j^2 D_{i,j} U_{i,j}^{yyy}/6 + O\left(\Delta x_i^3 + \Delta y_j^3\right), \tag{5.7}$$

$$\tilde{G}_{i+1,j+1} = G_{i+1,j+1} - \rho(\rho+1)\Delta x_i^2 C_{i,j} U_{i,j}^{xxx}/6 - \sigma(\sigma+1)\Delta y_j^2 D_{i,j} U_{i,j}^{yyy}/6 + O\left(\Delta x_i^3\right) \tag{5.8}$$

Eqs. (4.11) and (4.12) are formulated by considering the linear combination of a minimum number of various functional approximations (5.1)–(5.8) and approximated second-order partial derivatives (4.9) and (4.10). The free parameters υ and ζ in Eqs. (4.11) and (4.12) are obtained to achieve the magnitude of the LTE to the compact scheme (4.14) same as $T_{i,j}$. With the help of Eqs. (4.11) and (4.12), the new approximation (4.13) at the central mesh point (x_i, y_j) can be evaluated as

$$\tilde{G}_{i,j} = G_{i,j} + \Delta x_i^2 [\upsilon(\rho + 1) + \rho/6] C_{i,j} U_{i,j}^{xxx}$$
$$- \Delta y_j^2 [\zeta(\sigma + 1) + \sigma/6] D_{i,j} U_{i,j}^{yyy} + O\left(\Delta x_i^3 + \Delta y_j^3\right). \qquad (5.9)$$

Refer to the compact scheme (4.1); the functional approximations $G_{p,q}$ and $\tilde{G}_{p,q}$, $(p, q) \in T$ defined, respectively, for the linear Poisson's equation and mildly nonlinear elliptic Eq. (1.1) satisfies the relation

$$2\rho\sigma \sum_{(p,q)\in T} Q_{p,q}(\tilde{G}_{p,q} - G_{p,q})$$
$$= [\rho\sigma(\rho^2 + \rho + 1) - 2\upsilon(\rho + 1)\{\rho(\sigma^2 - 8\sigma + 1)$$
$$- (\rho^2 + 1)(2\sigma^2 - \sigma + 2)\}]\Delta x_i^2 C_{i,j} U_{i,j}^{xxx} \qquad (5.10)$$
$$+ [\rho\sigma(\sigma^2 + \sigma + 1) - 2\zeta(\sigma + 1)\{\rho(\sigma^2 - 8\sigma + 1)$$
$$- (\rho^2 + 1)(2\sigma^2 - \sigma + 2)\}]\Delta y_j^2 D_{i,j} U_{i,j}^{yyy} + O\left(\Delta x_i^3 + \Delta y_j^3\right).$$

Eq. (5.10) yields an accuracy of the third order provided the coefficient to the second-order terms Δx_i^2 and Δy_j^2 vanishes. Therefore, we see that the following values:

$$\upsilon = \rho\sigma(\rho^2 + \rho + 1)/[2(\rho + 1)\{\rho(\sigma^2 - 8\sigma + 1) - (2\sigma^2 - \sigma + 2)(\rho^2 + 1)\}],$$

$$\zeta = \rho\sigma(\sigma^2 + \sigma + 1)/[2(\sigma + 1)\{\rho(\sigma^2 - 8\sigma + 1) - (2\sigma^2 - \sigma + 2)(\rho^2 + 1)\}],$$

furnish the scheme (4.14) to the accuracy of third order. In particular, the detailed simplification shows that $O(\Delta x_i^3) \approx O\left(\Delta x_i^3 + \Delta x_i^2 \Delta y_j + \Delta x_i \Delta y_j^2 + \Delta y_j^3\right)$ terms in (5.10) vanish for $\rho = \sigma = 1$, and consequently, the scheme (4.14) proves to be fourth-order accurate.

6 Error Analysis

In this section, we determine the necessary conditions on the geometric mesh parameters (ρ, σ) and the function G so that the proposed high-order compact scheme converges when the mesh net unit is sufficiently small. Eq. (1.1) at the mesh point (x_i, y_j), $i = 1(1)I$, $j = 1(1)J$ is expressed as

$$U_{i,j}^{xx} + U_{i,j}^{yy} = G\left(x_i, y_j, U_{i,j}, U_{i,j}^x, U_{i,j}^y\right), \tag{6.1}$$

and their corresponding third-order geometric mesh discretization (4.14) can be represented by the equation

$$\Phi_{i,j} + W_{i,j} = 0, \; i = 1(1)I, \; j = 1(1)J, \tag{6.2}$$

where $W_{i,j} = \Delta x_i^{-2} \Delta y_j^{-2} \tilde{T}_{i,j}$ and

$$\begin{aligned} \Phi_{i,j} = &-3\Delta y_j^{-2}\left[12 + 4(\rho - 1)\mathcal{F}_x + \left(\rho^2 - \rho + 1\right)\mathcal{S}_x\right]\mathcal{S}_y U_{i,j} \\ &- 3\Delta x_i^{-2}\left[12 + 4(\sigma - 1)\mathcal{F}_y + \left(\sigma^2 - \sigma + 1\right)\mathcal{S}_y\right]\mathcal{S}_x U_{i,j} \\ &+ 2\sum_{(p,q)\in\mathcal{T}} Q_{p,q} \tilde{G}_{p,q}. \end{aligned}$$

Eq. (6.2) in its vector representations leads to

$$\boldsymbol{\Phi}(\boldsymbol{U}) + \boldsymbol{W} = 0, \tag{6.3}$$

where $\boldsymbol{U} = [U_{11}, U_{21}, \ldots, U_{I1}, \ldots, U_{1J}, U_{2J}, \ldots, U_{IJ}]^T$ is the vector of exact solution, $\boldsymbol{W} = [W_{11}, W_{21}, \ldots, W_{I1}, \ldots, W_{1J}, W_{2J}, \ldots, W_{IJ}]^T$ denote the third-order accurate truncation error vector and $\boldsymbol{\Phi}(\boldsymbol{U}) = [\Phi_{11}, \Phi_{21}, \ldots, \Phi_{I1}, \ldots, \Phi_{1J}, \Phi_{2J}, \ldots, \Phi_{IJ}]^T$.

In addition, the vector $\boldsymbol{u} = [u_{11}, u_{21}, \ldots, u_{I1}, \ldots, u_{1J}, u_{2J}, \ldots, u_{IJ}]^T$ which closely approximates the solution values satisfies the matrix equation

$$\boldsymbol{\Phi}(\boldsymbol{u}) = \boldsymbol{0}_{IJ \times IJ}. \tag{6.4}$$

Therefore, from Eqs. (6.3) and (6.4), we get

$$\boldsymbol{\Phi}(\boldsymbol{u}) - \boldsymbol{\Phi}(\boldsymbol{U}) = \boldsymbol{W}. \tag{6.5}$$

Let $\varepsilon_{i,j} = u_{i,j} - U_{i,j}$ be the point-wise error due to the discretizations and $\boldsymbol{\varepsilon} = [\varepsilon_{11}, \varepsilon_{21}, \ldots, \varepsilon_{I1}, \ldots, \varepsilon_{1J}, \epsilon_{2J}, \ldots, \varepsilon_{IJ}]^T$ be the error vector.

Let us construct the new functional on approximate solution values and their derivatives as follows:

$$\begin{cases} \tilde{g}_{p,q} = F\left(x_p, y_q, u_{p,q}, \tilde{u}^x_{p,q}, \tilde{u}^y_{p,q}\right) + R\left(x_p, y_q\right) \approx \tilde{G}_{p,q}, \ (p, q) \in \mathcal{T} \sim \{(i, j)\} \\ \tilde{g}_{i,j} = F\left(x_i, y_j, u_{i,j}, \hat{u}^x_{i,j}, \hat{u}^y_{i,j}\right) + R\left(x_i, y_j\right) \approx \tilde{G}_{i,j}. \end{cases}$$

$$(6.6)$$

Now, the application of mean value theorem results in

$$\begin{cases} \Upsilon_{p,q} = \tilde{g}_{p,q} - \tilde{G}_{p,q} = \alpha_{p,q}\tilde{\varepsilon}^x_{p,q} + \beta_{p,q}\tilde{\varepsilon}^y_{p,q} + \lambda_{p,q}\varepsilon_{p,q}, \ (p, q) \in \mathcal{T} \sim \{(i, j)\}, \\ \Upsilon_{i,j} = \tilde{g}_{i,j} - \tilde{G}_{i,j} = \alpha_{i,j}\hat{\varepsilon}^x_{i,j} + \beta_{i,j}\hat{\varepsilon}^y_{i,j} + \lambda_{i,j}\varepsilon_{i,j}, \end{cases}$$

$$(6.7)$$

where $\alpha_{p,q}$, $\beta_{p,q}$ and $\lambda_{p,q}$ are constants appearing due to the application of mean value theorem, Mohanty et al. [20], Jha et al. [21]. Also, the formulation of $\tilde{\varepsilon}^x_{p,q}$, $\tilde{\varepsilon}^y_{p,q}$, $\hat{\varepsilon}^x_{i,j}$ and $\hat{\varepsilon}^y_{i,j}$ can be established from Eqs. (4.2)–(4.7) and (4.11), (4.12), respectively, upon replacing U by ε.

Therefore, one obtains the difference equation to the discretization errors as follows:

$$\boldsymbol{\Phi}(u) - \boldsymbol{\Phi}(U) = \left[-\Delta x_i^{-2}\Delta y_j^{-2}\mathcal{L}_{i,j}\varepsilon_{i,j} + 2\sum\nolimits_{(p,q)\in\mathcal{T}}\mathcal{Q}_{p,q}\Upsilon_{p,q}\right]_{i=1(1)I, j=1(1)J} \quad (6.8)$$

The right-hand side of Eq. (6.8) in the matrix form is expressed by

$$\boldsymbol{\Phi}(u) - \boldsymbol{\Phi}(U) = \mathcal{P}\boldsymbol{\varepsilon} \quad (6.9)$$

where $\mathcal{P} = \left[\mathcal{P}_{l,m}\right]$, l, $m = 1(1)IJ$ is the block tri-diagonal matrix and mentioning the most significant expressions in terms of mesh step sizes to the nonzero elements of the matrix $\mathcal{P}$ are calculated as follows:

$$\mathcal{P}_{(j-1)I+i,(j-2)I+i-1} = \frac{12(\rho^2 - \rho - 1)}{(\rho+1)(\sigma+1)\Delta y_j^2} + \frac{12(\sigma^2 - \sigma - 1)}{(\rho+1)(\sigma+1)\Delta x_i^2}, \ i = 2(1)I, \ j = 2(1)J,$$

$$\mathcal{P}_{(j-1)I+i,(j-2)I+i} = -\frac{12(\rho^2 + 3\rho + 1)}{\rho(\sigma+1)\Delta y_j^2} - \frac{12(\sigma^2 - \sigma - 1)}{\rho(\sigma+1)\Delta x_i^2}, \ i = 1(1)I, \ j = 2(1)J,$$

$$\mathcal{P}_{(j-1)I+i,(j-2)I+i+1} = -\frac{12(\rho^2 + \rho - 1)}{\rho(\rho+1)(\sigma+1)\Delta y_j^2} + \frac{12(\sigma^2 - \sigma - 1)}{\rho(\rho+1)(\sigma+1)\Delta x_i^2}, \ i = 1(1)I - 1, \ j = 2(1)J,$$

$$\mathcal{P}_{(j-1)I+i,(j-1)I+i-1} = -\frac{12\left(\rho^2 - \rho - 1\right)}{(\rho+1)\sigma\Delta y_j^2} - \frac{12\left(\sigma^2 + 3\sigma + 1\right)}{(\rho+1)\sigma\Delta x_i^2}, \ i = 2(1)I, \ j = 1(1)J,$$

$$\mathcal{P}_{(i-1)I+i,(j-1)I+i} = \frac{12\left(\rho^2 + 3\rho + 1\right)}{\rho\sigma\Delta y_j^2} + \frac{12\left(\sigma^2 + 3\sigma + 1\right)}{\rho\sigma\Delta x_i^2}, \ i = 1(1)I, \ j = 1(1)J,$$

$$P_{(j-1)I+i,(j-1)I+i+1} = \frac{12(\rho^2 + \rho - 1)}{\rho\sigma(\rho+1)\Delta y_j^2} - \frac{12(\sigma^2 + 3\sigma + 1)}{\rho\sigma(\rho+1)\Delta x_i^2}, \ i = 1(1)I - 1, j = 1(1)J,$$

$$P_{(j-1)I+i,jI+i-1} = \frac{12(\rho^2 - \rho - 1)}{\sigma(\rho+1)(\sigma+1)\Delta y_j^2} - \frac{12(\sigma^2 + \sigma - 1)}{\sigma(\rho+1)(\sigma+1)\Delta x_i^2}, \ i = 2(1)I, j = 1(1)J - 1,$$

$$P_{(j-1)I+i,jI+i} = -\frac{12(\rho^2 + 3\rho + 1)}{\rho\sigma(\sigma+1)\Delta y_j^2} + \frac{12(\sigma^2 + \sigma - 1)}{\rho\sigma(\sigma+1)\Delta x_i^2}, \ i = 1(1)I, j = 1(1)J - 1,$$

$$P_{(j-1)I+i,jI+i+1} = -\frac{12(\rho^2 + \rho - 1)}{\rho\sigma(\rho+1)(\sigma+1)\Delta y_j^2} - \frac{12(\sigma^2 + \sigma - 1)}{\rho\sigma(\rho+1)(\sigma+1)\Delta x_i^2}, \ i = 1(1)I - 1, j = 1(1)J - 1.$$

Note that whenever $\rho, \sigma \neq (\sqrt{5} \pm 1)/2$, the lower, upper and main tri-diagonal blocks to the matrix $\mathcal{P}$ have nonzero elements at sub-diagonal, diagonal and sup-diagonal. Moreover, the directed graph of the matrix $\mathcal{P}$ is strongly connected, and hence, the matrix $\mathcal{P}$ is irreducible, Varga [22], Young [23].

Combining Eqs. (6.5) and (6.9), one obtains

$$\mathcal{P}\varepsilon = W \tag{6.10}$$

In order that the scheme (4.14) to be convergent, it is necessary to prove $\|\varepsilon\| \to 0$, whenever $\Delta x = \max_{i=1(1)I+1} \Delta x_i$ and $\Delta y = \max_{j=1(1)J+1} \Delta y_j$ are sufficiently small. Thus, we need to show that the matrix $\mathcal{P}$ is invertible in order to proceed further.

Let
$$\alpha = \min_{l,m} \alpha_{l,m}, \ \beta = \min_{l,m} \beta_{l,m}, \ \lambda = \min_{l,m} \lambda_{l,m}, \ l, \ m = 1(1)IJ$$
and
$$\pi_l = \sum_{m=1(1)IJ} \mathcal{P}_{l,m}, l = 1(1)IJ.$$

If $\left|\rho - \sqrt{5}/2\right| < 1/2$ and $\left|\sigma - \sqrt{5}/2\right| < 1/2$, then the lth weak row sum π_l to the matrix $\mathcal{P}$ is strictly positive except corresponding to the main diagonal of the matrix $\mathcal{P}$, as evident from the following lower bounds

$$\pi_1 \geq 12[(\rho^2 + 5\rho + 5)\Delta y^{-2} + (\sigma^2 + 5\sigma + 5)\Delta x^{-2}]/[(\rho+1)(\sigma+1)] > 0,$$

$$\pi_q \geq 72\Delta y^{-2}/(\sigma+1) > 0, q = 2(1)I - 1,$$

$$\pi_l \geq 12[(5\rho^2 + 5\rho + 1)\Delta y^{-2} + (\sigma^2 + 5\sigma + 5)\Delta x^{-2}]/[\rho(\rho+1)(\sigma+1)] > 0,$$

$$\pi_{(r-1)I+1} \geq 72\Delta x^{-2}/(\rho+1) > 0, \ r = 2(1)J - 1,$$

$$\pi_{(r-1)I+q} = 36\lambda \geq 0, \ r = 2(1)J - 1, \ q = 2(1)I - 1, \ \lambda \geq 0,$$

$$\pi_{(r-1)I+I} \geq 72\Delta x^{-2}/[\rho(\rho+1)] > 0, \ r = 2(1)J - 1,$$

$$\pi_{(J-1)I+1} \geq 12[(\rho^2 + 5\rho + 5)\Delta y^{-2} + (5\sigma^2 + 5\sigma + 1)\Delta x^{-2}]/[\sigma(\rho+1)(\sigma+1)] > 0,$$

$$\pi_{(J-1)I+q} \geq 72\Delta y^{-2}/[\sigma(\sigma+1)] > 0, \ q = 2(1)I - 1,$$

$$\pi_{JI} \geq 12[(5\rho^2 + 5\rho + 1)\Delta y^{-2} + (5\sigma^2 + 5\sigma + 1)\Delta x^{-2}]/[\rho\sigma(\rho+1)(\sigma+1)] > 0.$$

Thus, we find that the matrix $\mathcal{P}$ is monotone, Henrici [24]. The matrix properties irreducibility and monotonicity to $\mathcal{P}$ show that $\mathcal{P}^{-1}$ exists and their elements have strictly positive values.

Let us denote $\mathcal{P}^{-1} = [\mathcal{P}_{l,m}^{-1}], \ l, m = 1(1)IJ$ and define the matrix vector norm as

$$\|\mathcal{P}^{-1}\|_\infty =$$

$$\max_{l=1(1)IJ} \left[\begin{array}{c} |\mathcal{P}_{l,1}^{-1}| + \sum_{m=2}^{I-1} |\mathcal{P}_{l,m}^{-1}| + |\mathcal{P}_{l,I}^{-1}| + \left|\mathcal{P}_{l,(J-1)I+1}^{-1}\right| + \sum_{m=2}^{I-1} |\mathcal{P}_{l,(J-1)I+m}^{-1}| \\ + |\mathcal{P}_{l,IJ}^{-1}| + \sum_{e=2}^{J-1} \left(\left|\mathcal{P}_{l,(e-1)I+1}^{-1}\right| + \sum_{m=2}^{I-1} \left|\mathcal{P}_{l,(e-1)I+m}^{-1}\right| + |\mathcal{P}_{l,eI}^{-1}| \right) \end{array} \right]$$

and $\|W\|_\infty = \max_{i=1(1)I} \sum_{j=1(1)J} W_{i,j}$.

With the help of matrix property $\mathcal{P}^{-1}(\mathbf{P}I) = I$, where I is the $IJ \times 1$ matrix with all of its elements as one, we get the following identity

$$\sum_{q=1(1)IJ} \mathcal{P}_{p,q}^{-1}\pi_q = 1, \ p = 1(1)IJ. \tag{6.11}$$

Using relation (6.11) and series expansion in two variables, it is easy to determine the bounds on the nonzero elements of matrix $\mathcal{P}^{-1}$ as follows

For $p = 1(1)IJ$:

$$\mathcal{P}_{p,1}^{-1} \leq 1/\pi_1 \leq (\rho+1)(\sigma+1)\Delta x^2/[12(\sigma^2 + 5\sigma + 5)] + O(\Delta y),$$

$$\sum_{q=2}^{I-1} \mathcal{P}_{p,q}^{-1} \leq 1/\min_{q=2(1)I-1} \pi_q \leq (\sigma+1)\Delta y^2/72 + O(\Delta y^3),$$

$$\mathcal{P}^{-1}_{p,I} \le 1/\pi_I \le \rho(\rho + 1)(\sigma + 1)\Delta x^2/\left[12\big(\sigma^2 + 5\sigma + 5\big)\right] + O(\Delta y),$$

$$\sum_{r=2(1)J-1} \mathcal{P}^{-1}_{p,(r-1)I+1} \le 1/\min_{r=2(1)J-1} \pi_{(r-1)I+1} \le \Delta x^2(\rho + 1)/72 + O\big(\Delta x^3\big),$$

$$\sum_{r=2}^{J-1}\sum_{q=2}^{I-1} \mathcal{P}^{-1}_{p,(r-1)I+q} \le \sum_{q=1}^{IJ} \mathcal{P}^{-1}_{p,q}\pi_q = 1,\ \lambda = 0,$$

$$\sum_{r=2}^{J-1}\sum_{q=2}^{I-1} \mathcal{P}^{-1}_{p,(r-1)I+q} \le 1/\min_{\substack{q=2(1)I-1 \\ r=2(1)J-1}} \pi_{(r-1)I+q} \le 1/(36\lambda),\ \lambda > 0,$$

$$\sum_{r=2}^{J-1} \mathcal{P}^{-1}_{p,rI} \le 1/\min_{r=2(1)J-1} \pi_{rI} \le \Delta x^2\rho(\rho + 1)/72 + O\big(\Delta x^3\big),$$

$$\mathcal{P}^{-1}_{p,(J-1)I+1} \le 1/\pi_{(J-1)I+1} \le \Delta x^2\sigma(\rho + 1)(\sigma + 1)/\left[12\big(5\sigma^2 + 5\sigma + 1\big)\right] + O(\Delta y),$$

$$\sum_{q=2}^{I-1} \mathcal{P}^{-1}_{p,(J-1)I+q} \le 1/\min_{q=2(1)I-1} \pi_{(J-1)I+q} \le \Delta y^2\sigma(\sigma + 1)/72 + O\big(\Delta y^3\big),$$

$$\mathcal{P}^{-1}_{p,IJ} \le 1/\pi_{IJ} \le \Delta x^2\rho\sigma(\rho + 1)(\sigma + 1)/\left[12\big(5\sigma^2 + 5\sigma + 1\big)\right] + O(\Delta y).$$

With the use of above inequalities, the bounds on the matrix norm are computed as

$$\left\|\mathcal{P}^{-1}\right\|_\infty$$

$$= \begin{cases} 1 + \frac{\Delta y^2}{72}(\sigma + 1)^2 + \frac{\Delta x^2}{72}\frac{(11\sigma^4 + 96\sigma^3 + 171\sigma^2 + 96\sigma + 11)}{(\sigma^2 + 5\sigma + 5)(5\sigma^2 + 5\sigma + 1)(\rho + 1)^{-2}} + O\big(\Delta x^3 + \Delta y^3\big),\ \lambda = 0, \\[2ex] \frac{1}{36\lambda} + \frac{\Delta y^2}{72}(\sigma + 1)^2 + \frac{\Delta x^2}{72}\frac{(11\sigma^4 + 96\sigma^3 + 171\sigma^2 + 96\sigma + 11)}{(\sigma^2 + 5\sigma + 5)(5\sigma^2 + 5\sigma + 1)(\rho + 1)^{-2}} + O\big(\Delta x^3 + \Delta y^3\big),\ \lambda > 0, \end{cases}$$

Since $\|W\|_\infty = O\big(\Delta x^3 + \Delta y^3 + \Delta r^3 + \Delta s^3\big)$, as a result, we get

$$\|\epsilon\|_\infty \le \left\|\mathcal{P}^{-1}\right\|_\infty \cdot \|W\|_\infty \le O\big(\Delta x^3 + \Delta y^3 + \Delta r^3 + \Delta s^3\big) \approx O\big(\Delta x^3\big),\ \lambda \ge 0. \tag{6.12}$$

Summing up this section, we state that the compact scheme (4.14) converges to the true solution of EBVPs (1.1) and (1.2) provided $\left(\sqrt{5}-1\right)/2 < \rho, \sigma < \left(\sqrt{5}+1\right)/2$ and $\partial G/\partial U \geq 0$, since λ refers to the coefficients of $\varepsilon_{p,q}$, $(p, q) \in T$ in Eq. (6.7), which corresponds to $\partial G/\partial U$.

7 Numerical Validations

In this section, we compute the root mean squared error of approximate and true solution values for various mesh spacing by means of the proposed third (fourth)-order schemes. The root mean squared errors used to compute the accuracies in the solution are given by the formula, $\|\epsilon\|_2 = \left((1/IJ) \sum_{j=1}^{J} \sum_{i=1}^{I} \left|U_{i,j} - u_{i,j}\right|^2\right)^{1/2}$, and the numerical convergence order is computed using $\Theta_2 = \log_2\left(\|\epsilon\|_2^{(I,J)}/\|\epsilon\|_2^{(2I+1,2J+1)}\right)$. In all the test problems, Dirichlet's boundary values along the four boundary lines of the rectangular domain $\mathcal{D}$ are received through the true solutions. The Gauss—Seidel iterative method solves the system of linear difference equations, while Newton–Raphson method applies to the nonlinear difference equations, in each case zero vector is taken to be initial iterate with the tolerance of error as 10^{-10}, Hageman and Young [25]. To ease the computations and compare the convergence order, the numerical values of L and M are taken to be same, in particular, odd integers due to the integral formula. Computer programs are coded in C, and symbolic computations were performed in Maple software.

Example 1 We consider generalized heterogeneous Helmholtz equation

$$U^{xx} + U^{yy} = -\left[4 + (e-1)^2/\{(x+1)(\cos(2y) + \sin(2y))\}\right]U + \int_{s=0}^{1}\int_{r=0}^{1} e^{r+s}drds,$$

The true solution is given by $U(x, y) = (x+1)(\cos(2y) + \sin(2y))$, Tikhonov et al. [26]. Table 4 presents the root mean squared error and numerical order of convergence for various values of L.

Example 2 Consider the steady-state convective heat and mass transfer equations

Table 4 Numerical order and root mean squared errors in Example 1

L	ρ	σ	$\|\epsilon\|_2$	Θ_2	ρ	σ	$\|\epsilon\|_2$	Θ_2
3	1.0	1.0	1.41e−04	–	0.75	1.0	2.44e−05	–
7	1.0	1.0	7.53e−06	4.2	0.87	1.0	1.42e−06	4.0
15	1.0	1.0	4.37e−07	4.1	0.93	1.0	8.29e−08	4.1
31	1.0	1.0	2.04e−08	4.4	0.97	1.0	3.92e−09	4.4

$$U^{xx} + U^{yy} = (1-y)^2 U^x + (1-x)^2 U^y + \chi(x,y) + \int_{s=0}^{1} \int_{r=0}^{1} \frac{1}{1+r^2 e^s} \, dr \, ds,$$

The function $\chi(x, y)$ is obtained so as to have true solution as $U(x, y) = e^x \cosh(y)$, Polyanin [27]. Table 5 presents the root mean squared error and numerical order of convergence for various values of L.

Example 3 Consider the stationary heat equation with nonlinear source

$$U^{xx} + U^{yy} = \left[x + \log(U)\right]U + \int_{s=0}^{1} \int_{r=0}^{1} K(r, s, x, y) \, dr \, ds,$$

where

$$K(r, s, x, y) = \frac{1}{8}\left[10\pi e^{-1} - 3\pi \operatorname{erf}(1)\right]\operatorname{erf}(1) + \frac{r^4 + s^4}{e^{r^2 + s^2}}$$

The true solution is given by $U(x, y) = e^{3/2 - x + (y+1)^2/4}$, Polyanin and Zaitsev [28]. Due to logarithmic nonlinear term, the initial guess is taken as one for the iterative procedure of solutions. Table 6 presents the root mean squared error and numerical order of convergence for various values of L.

Example 4 Consider the nonlinear equation appearing in combustion theory

$$U^{xx} + U^{yy} = 2e^{\omega U} + \int_{s=0}^{1} \int_{r=0}^{1} K(r, s, x, y) \, dr \, ds$$

where

Table 5 Numerical order and root mean squared errors in Example 2

L	ρ	σ	$\|\epsilon\|_2$	Θ_2	ρ	σ	$\|\epsilon\|_2$	Θ_2
3	1.0	1.0	2.69e−05	–	0.98	0.77	5.30e−06	–
7	1.0	1.0	1.23e−06	4.4	0.98	0.91	1.36e−07	5.3
15	1.0	1.0	7.09e−08	4.1	0.98	0.95	8.30e−09	4.0
31	1.0	1.0	1.30e−09	5.8	0.98	0.98	7.06e−10	3.6

Table 6 Numerical order and root mean squared errors in Example 3

L	ρ	σ	$\|\epsilon\|_2$	Θ_2	ρ	σ	$\|\epsilon\|_2$	Θ_2
3	1.0	1.0	1.20e−05	–	0.950	1.0	2.62e−06	–
7	1.0	1.0	7.44e−07	4.0	0.970	1.0	1.19e−07	4.5
15	1.0	1.0	4.54e−08	4.0	0.984	1.0	6.31e−09	4.2
31	1.0	1.0	6.27e−09	4.0	0.984	1.0	1.14e−09	2.5

Table 7 Numerical order and root mean squared errors in Example 4

L	ρ	σ	$\|\epsilon\|_2$	Θ_2	ρ	σ	$\|\epsilon\|_2$	Θ_2
3	1.0	1.0	9.81e−04	–	1.06	0.960	1.13e−04	–
7	1.0	1.0	5.28e−05	4.1	1.03	0.980	6.70e−06	4.1
15	1.0	1.0	3.06e−06	4.1	1.02	0.990	4.58e−07	3.9
31	1.0	1.0	1.81e−07	4.1	1.01	0.996	1.92e−08	4.6

Table 8 Numerical order and root mean squared errors in Example 4

L	ρ	σ	$\|\epsilon\|_2$	Θ_2	ρ	σ	$\|\epsilon\|_2$	Θ_2
3	1.0	1.0	4.91e−03	–	0.80	0.980	8.67e−04	–
7	1.0	1.0	2.64e−04	4.2	0.91	0.990	4.86e−05	4.2
15	1.0	1.0	1.53e−05	4.1	1.00	0.990	3.20e−06	4.0
31	1.0	1.0	9.19e−07	4.1	1.00	0.995	1.93e−07	4.1

$$K(r, s, x, y) = \frac{4e^r \cos(s+1)}{\omega(e-1)[\sin(1) - \sin(2)][\cosh^2(x+y+1) - 1]}.$$

The true solution is given by $U(x, y) = \log\left(4/\left[\omega\sinh^2(x+y+1)\right]\right)/\omega$, Kamenet-skii [29]. For large values of ω, the solution exhibits regular behaviour; however, as the value of ω decreases the solution behaviour changes sharply. Tables 7 and 8 present the root mean squared error and numerical order of convergence for $\omega = 0.01$ and $\omega = 0.002$, respectively, for various values of L.

8 Conclusion

We summarize our numerical method as follows: the two spatial dimensions mildly nonlinear elliptic equation with integral form of source term is discretized by the finite differences on geometric meshes and obtained a new scheme in compact operator form. The accuracy of order three and four is accomplished in a single formulation of the compact scheme depending on the equal and unequal spacing of meshes. Using geometric meshes, the root mean squared error of the test problems exhibits improved results compared with the fourth-order scheme on equal mesh spacing. This happens due to the formation of parallel boundary layers and/or vertical boundary layers, and the geometric mesh network takes care of such layers. Moreover, the compact character of the proposed scheme can be easily programmed and will take small computing time. The development of such types of a scheme for quasi-linear elliptic equations in an arbitrary shape domain by means of conformal mapping will help to understand solution behaviour of elliptic partial differential equations.

Acknowledgements Science and Engineering Research Board, Department of Science and Technology, Government of India (No. SR/FTP/MS-020/2011) has supported the present research work.

References

1. Bers, L.: On mildly nonlinear partial differential equations of elliptic type. J. Res. Nat. Bur. Stand. **51**, 229–236 (1953)
2. Peskin, C.S.: Numerical analysis of blood flow in the heart. J. Comput. Phys. **25**, 220–252 (1977)
3. Zhilin, L., Lin, T., Wu, X.: New Cartesian grid methods for interface problems using the finite element formulation. Numer. Math. **96**, 61–98 (2003)
4. Kim, S.D., Lee, Y.H.: Analysis on eigenvalues for preconditioning cubic spline collocation method of elliptic equations. Linear Algebra Appl. **327**, 1–15 (2001)
5. Mohanty, R.K., Singh, S.: A new fourth order discretization for singularly perturbed two dimensional non-linear elliptic boundary value problems. Appl. Math. Comput. **175**, 1400–1414 (2006)
6. Tian, Z.F., Dai, S.Q.: High-order compact exponential finite difference methods for convection–diffusion type problems. J. Comput. Phy. **220**, 952–974 (2007)
7. Fairweather, G., Karageorghis, A., Maack, J.: Compact optimal quadratic spline collocation methods for the Helmholtz equation. J. Comput. Phys. **230**, 2880–2895 (2011)
8. Aziz, I., Šarler, B.: Wavelets collocation methods for the numerical solution of elliptic BV problems. Appl. Math. Model. **37**, 676–694 (2013)
9. Segal, A.: Aspects of numerical methods for elliptic singular perturbation problems. SIAM J. Sci. Comput. **3**, 327–349 (1982)
10. Mullges, G.E., Uhlig, F.: Numerical Algorithms with C. Springer, Berlin, Heidelberg (1996)
11. Britz, D.: Digital Simulation in Electrochemistry. Springer, Berlin Heidelberg (2005)
12. Jain, M.K., Iyengar, S.R.K., Subramanyam, G.S.: Variable mesh method for the numerical solution of two point singular perturbation problems. Comput. Meth. Appl. Mech. Eng. **42**, 273–286 (1984)
13. Mohanty, R.K.: A family of variable mesh methods for the estimates of (du/dr) and solution of non-linear two point boundary value problems with singularity. J. Comput. Appl. Math. **182**, 173–187 (2005)
14. Kadalbajoo, M.K., Kumar, D.: Geometric mesh FDM for self-adjoint singular perturbation boundary value problems. Appl. Math. Comput. **190**, 1646–1656 (2007)
15. Jha, N., Bieniasz, L.K.: A fifth(six) order accurate, three-point compact finite difference scheme for the numerical solution of sixth order boundary value problems on geometric meshes. J. Sci. Comput. **64**, 898–913 (2015)
16. Gerald, C.F., Wheatley, P.O.: Applied Numerical Analysis. Pearson Education, Inc. (1999)
17. Jha, N.: Convergence analysis of exponential expanding meshes compact-FDM for Poisson equation in polar coordinate system. Neural, Parallel Sci. Comput. **23**, 293–306 (2015)
18. Shaw, R.E., Garey, L.E., Lizotte, D.J.: A parallel numerical algorithm for Fredholm integro-differential two-point boundary value problems. Int. J. Comput. Math. **77**, 305–318 (2001)
19. Kelley, C.T.: Solving Nonlinear Equations with Newton's Method. SIAM (2003)
20. Mohanty, R.K., Jain, M.K., Dhall, D.: High accuracy cubic spline approximation for two dimensional quasi-linear elliptic boundary value problems. Appl. Math. Model. **37**, 155–171 (2013)
21. Jha, N., Kumar, N., Sharma, K.K.: A third (four) order accurate nine-point compact scheme for mildly-nonlinear elliptic equations in two space variables. Differ. Equ. Dyn. Syst. (2015). https://doi.org/10.1007/s12591-015-0263-9
22. Varga, R.S.: Matrix Iterative Analysis. Springer Series in Computational Mathematics. Springer, Berlin (2000)
23. Young, D.M.: Iterative Solution of Large Linear Systems. Elsevier (2014)
24. Henrici, P.: Discrete Variable Methods in Ordinary Differential Equations. Wiley, New York (1962)

25. Hageman, L.A., Young, D.M.: Applied Iterative Methods. Dover Publication, New York (2004)
26. Tikhonov, A.N., Samarskii, A.A.: Equations of Mathematical Physics. Dover Publication, New York (1990)
27. Polyanin, A.D.: Handbook of Linear Partial Differential Equations for Engineers and Scientists. CRC Press (2010)
28. Polyanin, A.D., Zaitsev, V.F.: Handbook of Nonlinear Partial Differential Equations. CRC press (2004)
29. Kamenetskii, D.A.F.: Diffusion and Heat Transfer in Chemical Kinetics. Nauka, Moscow (1987)

How Delay Can Affect the Survival of Species in Polluted Environment

Saroj Kumar Sahani

Abstract In this article, a four-state delayed differential model of biomass–species interaction in a polluted environment has been proposed. It is well-established phenomenon that the effects of any external changes in the environment have some kind of effect on species dynamics and they are in fact not instantaneous. In particular, when a species live in the polluted environment, its growth is hampered but it is an instantaneous process. It takes time to go inside species then it affects the growth as a whole. The situation can be more severe if their food resources are also subjects to toxicity present in the environment. Therefore, it is very much necessary to model the effects of a toxicant on this species–biomass interaction. Well-established stability and bifurcation have been used to study and predict the long-term dynamics of these interactions. The result is worth to note that a survival of species is in fact guaranteed if this model is assumed to govern the dynamics. However, the model needs to be modified to take into account of other parameters and interaction to fully justify the underlying phenomena.

Keywords Bifurcation · Delay · Equilibrium point · Stability · Toxicant

1 Introduction

Almost all living species live in atmosphere which are nowadays affected by pollutants/toxicants emitted into the environment. The main cause of toxicant present in our surrounding is the because of mankind revolution which have led to massive industrialisation, and therefore, various kinds of toxicants are discharged into both aquatic and terrestrial environments. These toxicants not only have adverse effect on the species but also their food resources [10, 17, 19, 21, 22]. There are varieties of pollutants which are abundant in our surrounding such as airborne pollutants CO, CO_2, SO_2, NO_x, fluorides, heavy metals. These pollutants directly affect not only

S. K. Sahani (✉)
South Asian University, Akbar Bhawan, Chanakyapuri, New Delhi, India
e-mail: sarojkumar@sau.ac.in

© Springer Nature Singapore Pte Ltd. 2019
J. C. Bansal et al. (eds.), *Soft Computing for Problem Solving*, Advances in Intelligent Systems and Computing 817, https://doi.org/10.1007/978-981-13-1595-4_72

the human population but also other living organisms. Most of them have also very severe effects on plant population such as SO_2, NO_x. The presence of these pollutants/toxicants in the environment decreases not only the overall growth rates of species but also their carrying capacities [14, 23, 27, 28]. The gradual degradation of the environment due to the presence of toxicant causes both physiological and bio-spherical changes in the population, and sometimes it can lead to very catastrophic changes in their life cycle. Due to the uncontrolled industrialisation and their discharge of the pollutant into the environment, many species heve become extinct, and many species are on the verse of their extinction.

There has been many studies on experimental basis to ascertain the global effect of pollutant present in environment, but it is very tedious process. The main difficulties in these experimental studies is the data collection and their analysis based on some statistical tools available to us. The beauty of mathematical studies through proper modelling of these biological interaction lies in their simplicity and robustness of the mathematical tools. Through model, one can also simulate the long-term dynamics of the toxicant effects on species and their survival. There has been many mathematical studies applying the simple ordinary differential equation model to efficiently model these interactions [1, 3, 4, 7, 16, 24]. Some authors have studied as to how the presence of toxicant affects the food chain model and survival of species [1, 7, 13, 26]. All of such types of model considers the instantaneous effect of toxicant on the species and their biomass. Although it is well-established result that almost every phenomena in their world involves some amount of delay [5, 15, 25], but to make model simple, it is developed as ODE model. Many of the authors [8, 9, 11, 20] do have considered the delay effects when toxicants are affecting a species but this delay are not one. The delayed model are of natural choices in these scenarios because of its effectiveness in dealing the instability in thesis kind of models [6, 18]. They are present in almost every compartment of the model being considered. This way model leads to a very complicated one, but it gives more insight into the dynamics of the interactions being considered.

In this present model, a situation where the toxicant inside the body of the species affects the species growth but the toxicant in the environment affects the growth of the biomass with a delay has been considered. The toxicant from the environment does not directly affect the species unless it enters inside the body. Here it is assumed that the toxicant takes another delay to enter inside the body where it actually starts affecting the growth of species. With these two delays, the model analysis becomes almost impossible to proceed, but with the help of computer algebra system such as Mathematica, it is possible to determine the long-term dynamics of the system.

2 The Proposed Model

Suppose, u_1, u_2, u_3 and u_4 denote the densities of biomass, species, toxicant in species and toxicant in environment. It has been assumed that biomass density and species grow with logistic growth rate, and whereas the toxicant present in environment has negative impact on growth of biomass instantly, the growth of species is diminished in same way by the presence of toxicant inside the body of species but with a delay τ_1. Again the toxicant enters inside the body from environment with a certain lag which is denoted by τ_2. All interactions among the species, biomass and toxicant are assumed to follow the law of mass action type interaction with interaction rates α, β and σ. The total influx of the toxicant into the environment is assumed to be at constant rate ϕ. Further, the toxicants are flushed out from the body of the species by the internal mechanisms such as excretion, precipitation, urination. At a constant per capita rate δ_0 and in the same way the toxicant in the environment also decays at the rate δ_1. The constants γ and ν denote the efficiency of conversion from one compartment to another taking the value between 0 and 1. Therefore, the following model is proposed

$$
\begin{aligned}
\frac{du_1}{dt} &= u_1\left(-r_1 u_1 - r_2 u_4 + r_0\right) - \alpha u_1 u_2 \\
\frac{du_2}{dt} &= u_2\left(-s_1 u_2 - s_2 u_3(t - \tau_1) + s_0\right) + \alpha \gamma u_1 u_2, \\
\frac{du_3}{dt} &= \sigma u_2(t - \tau_2) u_4(t - \tau_2) - \beta u_2 u_3 - \delta_0 u_3 \\
\frac{du_4}{dt} &= \phi + \beta \nu u_2 u_3 - \delta_1 u_4 - \sigma u_2 u_4.
\end{aligned}
\tag{1}
$$

In the model (1), it is assumed that all parameters are positive and $\tau_1 \geq 0$, $\tau_2 \geq 0$. Again because the model is a system of delayed differential equation, for the solution to exists and unique, the following initial conditions so-called the history functions [15] are assumed.

$$
\left.
\begin{aligned}
u_1(\theta) &= \psi_1(\theta) \\
u_2(\theta) &= \psi_2(\theta) \\
u_3(\theta) &= \psi_3(\theta) \\
u_4(\theta) &= \psi_4(\theta)
\end{aligned}
\right\}
\quad \theta \in [-\tau, 0], \ \tau = \max\{\tau_1, \tau_2\},
\tag{2}
$$

with $\psi_i(\theta) \in C\left([-\tau, 0], \mathbb{R}^4_{0+}\right)$, $i = 1, 2, 3, 4$, the Banach space of all continuous functions defined from interval $[-\tau, 0]$ to $\mathbb{R}^4_{0+}$ where

$$
\mathbb{R}^4_{0+} = \{(u_1, u_2, u_3, u_4) : u_i \geq 0, \ i = 1, 2, 3, 4\}.
$$

In the next section, the existence of positive roots of the underlying system (1) has been proved.

3 Preliminary Results

The following theorem gives the positive invariance of the positive orthant.

Theorem 1 *If initial functions (2) are positive, then all solutions of the system* (1) *remain positive for all time* $t > 0$.

Proof The first equation of system (1) gives

$$\frac{du_1}{u_1} = (-r_1 u_1 - r_2 u_4 + r_0 - \alpha u_2)\, dt, \tag{3}$$

which on integration produces

$$u_1(t) = u_1(0)\exp\left\{\int_0^t (-r_1 u_1 - r_2 u_4 + r_0 - \alpha u_2)\right\}.$$

The form of $u_1(t)$ shows that $u_1(t)$ remains positive as long as $u_1(0) > 0$. Similarly, the second equation of system (1) on integration gives

$$u_2(t) = u_2(0)\exp\left\{\int_0^t (-s_1 u_2 - s_2 u_3(t - \tau_1) + s_0 + \alpha\gamma u_1)\right\}.$$

Similar arguments prove that $u_2(t)$ remains positive as long as $u_2(0) > 0$. Now to prove positivity of $u_3(t)$, consider the interval $[0, \tau_2]$. In this interval, $\sigma u_2(t - \tau_2)u_4(t - \tau_2) = \sigma\psi_2(\theta)\psi_4(\theta) \ge 0$. Therefore, the third equation simplifies to

$$\frac{du_3}{dt} \ge -\beta u_2 u_3 - \delta_0 u_3. \tag{4}$$

The direct integration now yields

$$u_3(t) \ge u_3(0)\exp\left\{\int_0^{\tau_2} (-\beta u_2 - \delta_0)\right\},$$

which clearly shows the positivity of $u_3(t)$ in $[0, \tau_2]$. Now in the same interval $[0, \tau_2]$, the fourth equation gives

$$\frac{du_4}{dt} \ge -\delta_1 u_4 - \sigma u_2 u_4, \tag{5}$$

which on integration gives

$$u_4(t) \ge u_4(0)\exp\left\{\int_0^{\tau_2} (-\delta_1 - \sigma u_2)\right\},$$

proving the positivity of u_4 in $[0, \tau_2]$. Now consider the interval $[\tau_2, 2\tau_2]$ in which again $\sigma u_2(t - \tau_2)u_4(t - \tau_2) = \sigma \psi_2(\theta)\psi_4(\theta) \geq 0$ for $\theta \in [0, \tau_2]$. So the third equation of system (1) gives

$$\frac{du_3}{dt} \geq -\beta u_2 u_3 - \delta_0 u_3. \tag{6}$$

Similar integration yields

$$u_3(t) \geq u_3(\tau_2)\exp\left\{\int_{\tau_2}^{2\tau_2} (-\beta u_2 - \delta_0)\right\}.$$

Similarly, the fourth equation gives

$$u_4(t) \geq u_4(\tau_2)\exp\left\{\int_{\tau_2}^{2\tau_2} (-\delta_1 - \sigma u_2)\right\}.$$

These two integral therefore gives the positivity of $u_3(t)$ and $u_4(t)$ in $[\tau_2, 2\tau_2]$. Continuing in this way, one can easily prove the positivity of $u_3(t)$ and $u_4(t)$ in any finite interval $[0, t]$. This completes the proof.

4 Stability of Boundary Equilibrium Points

To determine the local stability behaviour of the system, it is necessary to find out the equilibrium points it has. The system has three boundary equilibrium points, namely

1. $E_0(0, 0, 0, \frac{\phi}{\delta_1})$ which always exists,
2. $E_1\left(\frac{\delta_1 r_0 - r_2\phi}{\delta_1 r_1}, 0, 0, \frac{\phi}{\delta_1}\right)$ exists when $\phi < \frac{\delta_1 r_0}{r_2}$
3. $E_2(0, \bar{u}_2, \bar{u}_3, \bar{u}_4)$,
4. $E_i(\tilde{u}_1, \tilde{u}_2, \tilde{u}_3, \tilde{u}_4)$.

The conditions on existence of E_2 and E_i are difficult to obtain because of highly nonlinear underlying equation. However, it can be easily verified that there exists equilibrium point of type E_2 and E_i in addition to other boundary equilibrium points.

In order to discuss the local stability of the equilibrium point, it is necessary to linearise the system around equilibrium point. Let $\hat{u} = (\hat{u}_1, \hat{u}_2, \hat{u}_3, \hat{u}_4)$ denotes any equilibrium coordinates, then the linearisation [12] of system (1) is given by

$$\frac{du}{dt} = Au(t) + Bu(t - \tau_1) + Cu(t - \tau_2) \tag{7}$$

for $u = (u_1, u_2, u_3, u_4)$ and the matrix A, B and C are given as

$$A = \begin{pmatrix} \Delta_1(\hat{u}) & -\alpha\hat{u}_1 & 0 & -r_2\hat{u}_1 \\ \alpha\gamma\hat{u}_2 & \Delta_2(\hat{u}) & 0 & 0 \\ 0 & -\beta\hat{u}_3 & -\beta\hat{u}_2 - \delta_0 & 0 \\ 0 & \beta v\hat{u}_3 - \sigma\hat{u}_4 & \beta v\hat{u}_2 & -\sigma\hat{u}_2 - \delta_1 \end{pmatrix}$$

$$B = \begin{pmatrix} 0 & 0 & 0 & 0 \\ 0 & 0 & -\hat{u}_2 s_2 & 0 \\ 0 & 0 & 0 & 0 \\ 0 & 0 & 0 & 0 \end{pmatrix}$$

$$C = \begin{pmatrix} 0 & 0 & 0 & 0 \\ 0 & 0 & 0 & 0 \\ 0 & \sigma\hat{u}_4 & 0 & \sigma\hat{u}_2 \\ 0 & 0 & 0 & 0 \end{pmatrix}$$

The functions Δ_1 and Δ_2 take the form

$$\Delta_1(\hat{u}) = \begin{cases} -r_1\hat{u}_1, & \text{for } \hat{u}_1 \neq 0 \\ r_0 - \alpha\hat{u}_2 - r_2\hat{u}_4, & \text{for } \hat{u}_1 = 0 \end{cases}$$

$$\Delta_2(\hat{u}) = \begin{cases} -s_1\hat{u}_2, & \text{for } \hat{u}_2 \neq 0 \\ s_0 - \hat{u}_3 s_2 + \alpha\gamma\hat{u}_1, & \text{for } \hat{u}_2 = 0 \end{cases}$$

In such case, the characteristic equation with two delays takes the form

$$\det\left(\lambda I - A - Be^{-\lambda\tau_1} - Ce^{-\lambda\tau_2}\right) = 0. \tag{8}$$

It should be noted that the $(\hat{u}_1, \hat{u}_2, \hat{u}_3, \hat{u}_4)$ is a general equilibrium point. In order to decide the local stability of any equilibrium point, it is necessary to determine the nature of equilibrium point; therefore, the following theorems give the condition s under which each of the equilibrium point is stable or unstable.

Theorem 2 *The equilibrium point E_0 is always unstable for any values of system parameters.*

Proof It is obvious that at E_0, the characteristic equation (8) reduces to

$$(\delta_0 + \lambda)(\delta_1 + \lambda)(\lambda - s_0)\left(\lambda + \frac{r_2\phi}{\delta_1} - r_0\right) = 0$$

which has one eigenvalues $s_0 > 0$, two eigenvalues $-\delta_0, -\delta_1$ and one eigenvalue $r_0 - \frac{r_2\phi}{\delta_1}$ which can be positive or negative, hence E_0 will be a saddle and hence unstable.

Theorem 3 *The equilibrium point E_1 if it exists then it is also always unstable for any values of system parameters.*

Proof In this case, the characteristic equation (8) simplifies to

$$(\delta_0 + \lambda)(\delta_1 + \lambda)(\delta_1 \lambda + \delta_1 r_0 + r_2(-\phi))\left(\lambda - \frac{\alpha\gamma(\delta_1 r_0 - r_2\phi)}{\delta_1 r_1} - s_0\right) = 0.$$

Clearly, the eigenvalues are given by $-\delta_0, -\delta_1, \frac{r_2\phi - \delta_1 r_0}{\delta_1}$ and $\frac{\alpha\gamma\delta_1 r_0 - \alpha\gamma r_2\phi + \delta_1 r_1 s_0}{\delta_1 r_1}$. It is obvious that if E_1 exists, then one eigenvalue becomes positive and hence it is unstable.

Theorem 4 *The equilibrium point E_2 if it exists is stable if and only if all the eigenvalues of the characteristic equation*

$$(r_2\bar{u}_4 + \alpha\bar{u}_2 + \lambda - r_0)\left(Q_0 + Q_1 e^{-\lambda\tau_1} + Q_2 e^{-\lambda\tau_2} + Q_3 e^{-\lambda\tau_1 - \lambda\tau_2}\right) = 0 \quad (9)$$

The form of characteristic equation at E_2 is very complex, and therefore, it is impossible to determine the require condition for stability; however, E_2 is unstable if

$$r_0 > r_2\bar{u}_4 + \alpha\bar{u}_2.$$

The main aim in this paper is to establish the stability bifurcation of E_i, so to discuss the stability and stability switches, the parameter τ_2 is taken as bifurcation parameter. It is left as future scope to determine the stability and bifurcation of E_2.

5 Stability and Stability Switch of E_i

The characteristic equation at E_i takes the form

$$P_0(\lambda) + P_1(\lambda)e^{-\lambda\tau_1} + P_2(\lambda)e^{-\lambda\tau_2} + P_3(\lambda)e^{-\lambda\tau_1 - \lambda\tau_2} = 0 \quad (10)$$

where each polynomial takes the form

$$\begin{aligned}
P_0(\lambda) &= a_0 + a_1\lambda + a_2\lambda^2 + a_3\lambda^3 + \lambda^4 \\
P_1(\lambda) &= b_0 + b_1\lambda + b_2\lambda^2 \\
P_2(\lambda) &= c_0 + c_1\lambda + c_2\lambda^2 \\
P_3(\lambda) &= d_0 + d_1\lambda + d_2\lambda^2.
\end{aligned} \quad (11)$$

Moreover, the coefficients a_i, b_i, c_i and d_i depend on system parameters other than delay parameters. Because of complexity of the Eq. (10), it is impossible to state the particular stability switch criteria for the underlying system. Therefore in next section, it is verified that the stability switch condition leading to Hopf bifurcation [2, 15] occurs in such type of equation for some specific set of parameter.

6 Numerical Simulations and Illustrations

To have in-depth idea about the dynamics specifically the long-term dynamics, the following parametric values with proper units have been taken:

$$r_0 = 2, r_2 = 1.5, r_1 = 0.005,$$
$$s_0 = 4, s_2 = 2.00225, s_1 = 0.0006,$$
$$\gamma = 0.6, \alpha = 0.000075, \delta_0 = 12, \sigma = 0.0072, \beta = 0.0036,$$
$$\phi = 86.4, \delta_1 = 72, \nu = 0.8, \tau_1 = 0.01. \tag{12}$$

For this set of parameters, the equilibrium points of the system which exists are:

- $E_0(0, 0, 0, 1.2)$,
- $E_1(40, 0, 0, 1.2)$,
- $E_2(0, 3330.63, 0.999685, 1.00009)$,
- $E_i(50, 3333.33, 1, 1)$,

i.e. system possesses all kind of equilibrium points. To establish the stability of E_i or the range of τ_2 in which the interior point is stable can be determined by plotting the $\text{Re}(\lambda)$ with respect to parameter τ_2 which is shown in Fig. 1. Clearly, for $\tau_2 \in [0, 2.035)$, the system is stable and beyond that it is unstable. This also gives the critical point of Hopf bifurcation which is $\tau_2^{cr} \approx 2.04$. The system is therefore locally asymptotically stable at $\tau_2 = 1$ as shown in Fig. 2 and unstable for $\tau_2 = 10$ as shown Fig. 3.

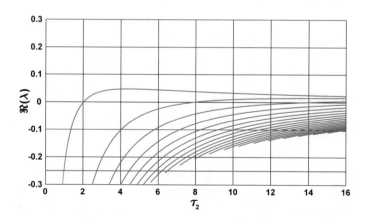

Fig. 1 Plot of variation of $\Re(\lambda)$ with τ_2

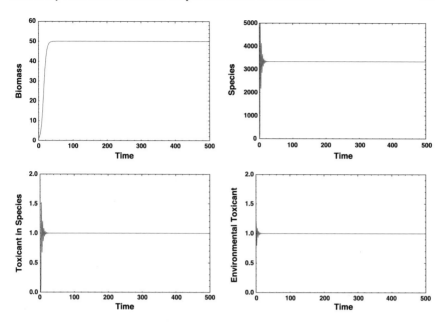

Fig. 2 Locally asymptotically stability of the interior equilibrium E_i when $\tau_2 = 1$

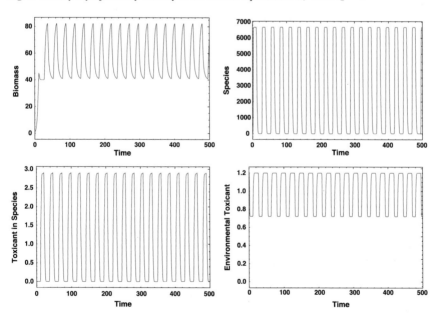

Fig. 3 Presence of periodic oscillation around interior equilibrium E_i $\tau_2 = 10$

7 Conclusions and Future Scope

In the present article, here a simultaneous effect of two delays which are due to toxicant effects on species and their resources is proposed and analysed. As it has been shown that the analytical system analysis of the model is very complicated and so it was difficult to do some rigorous analysis on the model. However, the model has been simulated with the given parametric values and has been found that the system indeed is stable for certain values of the parameter. This leads to conclusion that in spite of toxicant which not only affects the growth of species but also their biomass can still survive in the system. This may be attributed to the fact that almost every living creatures in this universe have evolved just to adopt themselves to the new environment. This unique feature of the species actually equipped themselves to find another source of their livelihood from environment itself, and hence, they can still survive in this world in spite of limited or no main resources at all. The oscillatory behaviour in this model also depicts that the species can also survive from intermittent ups and downs in their growth from time to time.

Although the present model is quite simple in the sense that it takes only the one resource and their species in the environment, it will be interesting to see as to how the system will behave if the two or more species with their resources are present in the system. In the present model too, there is further scope to study for the future analysis of the system through global bifurcation phenomena.

References

1. De Luna, J., Hallam, T.: Effects of toxicants on populations: a qualitative approach iv. resource-consumer-toxicant models. Ecol. Model. **35**(3–4), 249–273 (1987)
2. Diekmann, O., Van Gils, S.A., Lunel, S.M., Walther, H.O.: Delay Equations: functional-, complex-, and nonlinear analysis, vol. 110. Springer Science & Business Media (2012)
3. Dubey, B., Hussain, J.: A model for the allelopathic effect on two competing species. Ecol. Model. **129**(2), 195–207 (2000)
4. Dubey, B., Hussain, J.: Models for the effect of environmental pollution on forestry resources with time delay. Nonlinear Anal. Real World Appl. **5**(3), 549–570 (2004)
5. Erneux, T.: Applied Delay Differential Equations, vol. 3. Springer Science & Business Media (2009)
6. Foit, K., Kaske, O., Liess, M.: Competition increases toxicant sensitivity and delays the recovery of two interacting populations. Aquat. Toxicol. **106**, 25–31 (2012)
7. Freedman, H., Shukla, J.: Models for the effect of toxicant in single-species and predator-prey systems. J. Math. Biol. **30**(1), 15–30 (1991)
8. Gakkhar, S., Sahani, S.K.: A delay model for the effects of environmental toxicant on biological species. J. Biol. Syst. **15**(04), 525–537 (2007)
9. Gakkhar, S., Sahani, S.K.: A model for delayed effect of toxicant on resource-biomass system. Chaos, Solitons & Fractals **40**(2), 912–922 (2009)
10. Garcia-Montiel, D., Scatena, F.N.: The effect of human activity on the structure and composition of a tropical forest in puerto rico. For. Ecol. Manag. **63**(1), 57–78 (1994)
11. Gopalsamy, K.: Delayed responses and stability in two-species systems. ANZIAM J. **25**(4), 473–500 (1984)

12. Gu, K., Chen, J., Kharitonov, V.L.: Stability of Time-Delay Systems. Springer Science & Business Media (2003)
13. Hallam, T., De Luna, J.: Effects of toxicants on populations: a qualitative: approach iii. environmental and food chain pathways. J. Theor. Biol. **109**(3), 411–429 (1984)
14. Hamilton, S.J.: Review of selenium toxicity in the aquatic food chain. Sci. Total Environ. **326**(1), 1–31 (2004)
15. Kuang, Y.: Delay Differential Equations: with applications in population dynamics, vol. 191. Academic Press (1993)
16. Lata, K., Misra, A., Upadhyay, R.: A mathematical model for the conservation of forestry resources with two discrete time delays. Model. Earth Syst. Environ. **3**(3), 1011–1027 (2017)
17. McLaughlin, S.: Effects of air pollution on forests: a critical review. Technical Report, Oak Ridge National Lab., TN (USA) (1985)
18. Moe, S., Stenseth, N., Smith, R.: Effects of a toxicant on population growth rates: sublethal and delayed responses in blowfly populations. Funct. Ecol. **15**(6), 712–721 (2001)
19. Mudd, J.B.: Responses of Plants to Air Pollution. Elsevier (2012)
20. Mukherjee, D.: Persistence and global stability of a population in a polluted environment with delay. J. Biol. Syst. **10**(03), 225–232 (2002)
21. Padoch, C., Vayda, A.P.: Patterns of resource use and human settlement in tropical forests. Ecosyst. World (1983)
22. Patin, S.A.: Pollution and the Biological Resources of the Oceans. Elsevier (2016)
23. Schulze, E.D.: Air Pollution and Forest Decline in a spruce (Picea Abies) forest. Science 776–783 (1989)
24. Shukla, J., Sharma, S., Dubey, B., Sinha, P.: Modeling the survival of a resource-dependent population: effects of toxicants (pollutants) emitted from external sources as well as formed by its precursors. Nonlinear Anal. Real World Appl. **10**(1), 54–70 (2009)
25. Smith, H.: An Introduction to Delay Differential Equations with Applications to the Life Sciences, vol. 57. Springer Science & Business Media (2010)
26. Thomann, R.V., Connolly, J.P.: Model of pcb in the lake michigan lake trout food chain. Environ. Sci. Technol. **18**(2), 65–71 (1984)
27. Woodman, J.N., Cowling, E.B.: Airborne chemicals and forest health. Environ. Sci. Technol. **21**(2), 120–126 (1987)
28. Woodwell, G.M.: Effects of pollution on the structure and physiology of ecosystems. Science **168**(3930), 429–433 (1970)

A Review on Scale Factor Strategies in Differential Evolution Algorithm

Prashant Sharma, Harish Sharma, Sandeep Kumar
and Jagdish Chand Bansal

Abstract Differential evolution (DE) algorithm is a well-known and straightforward population-based optimization approach to deal with nonlinear and composite problems. The scale factor (F) and crossover rate (CR) are two control parameters which play a crucial role to keep up the proper equilibrium between exploration and exploitation processes. The perturbation in the new solutions is controlled by CR, and the step size is managed by F during the solution search process. The step size of an individual is tuned to explore or exploit the search region of the solving problem. Large step size is used to explore while small step size is used to exploit the search region. Therefore, a fine-tuned step size can avoid the situation of skipping the true optima while maintaining the proper convergence speed. Researchers are working hard to adjust the step size as per the search progress. Therefore, this paper presents descriptive details of DE and a review on the various scale factor strategies in DE with their comparative impact on the solution search process.

Keywords Evolutionary optimization · Differential evolution · Meta-heuristics

P. Sharma
Career Point University, Kota, Rajasthan, India
e-mail: prashant.tiwari555@gmail.com

H. Sharma (✉)
Rajasthan Technical University, Kota, Rajasthan, India
e-mail: hsharma@rtu.ac.in; harish.sharma0107@gmail.com

S. Kumar
Amity University, Jaipur, Rajasthan, India
e-mail: sandpoonia@gmail.com

J. C. Bansal
South Asian University, New Delhi, India
e-mail: jcbansal@sau.ac.in

© Springer Nature Singapore Pte Ltd. 2019
J. C. Bansal et al. (eds.), *Soft Computing for Problem Solving*, Advances in Intelligent
Systems and Computing 817, https://doi.org/10.1007/978-981-13-1595-4_73

1 Introduction

Storn and Price developed a new evolutionary algorithm (EA) to solve the complex optimization problems in 1995 named as differential evolution (DE) [14, 20] algorithm. The algorithm is adapted to find out the true solution (optimum solution) of the multi-model, nonlinear, and non-convex functions.

Researchers are regularly working to boost the competence and efficacy of the DE algorithm and have developed many competitive variants in the field of EAs [5, 14]. The developed DE algorithm and its variants have been tested well on the numerous real-world optimization problems that are very complex.

In DE, the true solution is identified while going through the iterative search process among the feasible solutions like in the other EA, even though the DE is conceptually different from the EAs and evolutionary programming (EP). These algorithms are based on the principal of evolution. The population evolution is based on the fitness of individuals. In this process, first the crossover operation is performed among the selected solutions and then mutation operation is applied to get a new solution. Based on the fitness of the recently generated solution and the existing one, the better solution fixed its place in the next iteration or generation population. While in DE, first mutation operation takes place which is a vector difference of randomly selected solutions and the next step is crossover operation that is carried out between the mutated solution and the parent solution. The vector difference decides the step size of the solution that is required to diversify the solutions in the search space. The crossover and mutation operations automatically adopt the step size for the solutions to search the global optima in the feasible region of the given problem. In this process, the DE has need of a single relative crossover rate CR and scale factor F for all the solutions. So, it is clear that both F and CR engaged in deciding the convergence and diversification of the population in the feasible region. The DE and its variants have exhibited impressive performances in various competitions among the nature-inspired algorithms (EAs) like IEEE CEC Conferences, which make it more reliable compared to the other existing EAs. To describe the competitiveness of the DE and its variants to solve the single objective functions, Neri et al. [14] presented a survey in 2010 with the practical performance evaluations and recommendations. In that survey, the considered DE variants were tested over standard unbiased benchmark functions. However, the survey was not able to present the multi-objective DE, self-adaptive DE variants and complexity in solving high-dimensional problems. Further, a detailed review on the DE and its variants was presented by Das et al. [5] in 2011 which explained the solution search principal of DE and the practical application on the complex engineering optimization problems. Further in 2016, [6] descriptive analyses on the two aspects of DE: adaptive parameters in DE and combination of the DE with other meta-heuristics were presented.

Since the inception of the DE, it has improved significantly. It is observed through the literature that the crossover rate (CR) and scale factor (F) have a considerable effect on the performance of the DE algorithm. As the scale factor F plays a crucial role in the solution search process of DE (controls the individuals step size while searching for solution), a number of DE variants are only developed while modifying this parameter.

Therefore, in this article, an exhaustive review of the important scale factor strategies of the DE variants with design principals, structures, and comparative analyses is presented.

2 Overview of the Algorithm

Initialization, mutation, crossover, and selection are the four basic steps in standard DE algorithm. In the successive DE iterations, only the last three steps are repeated out of these four steps. The iterations prolong while the termination condition (such as exhaustion of utmost functional evaluations) is fulfilled. Generally, DE strategy used notation: $DE/x/y/z$; in this notation, DE is a symbol of differential evolution, where 'x' denotes the way in which target solution selected, 'y' stands for the quantity of differential vectors used for perturbation of x, and the crossover type being used is denoted by 'z'. There are a number of strategies available in DE based on the choice of x, y, and z [15]. For instance, in the strategy $DE/rand/1/bin$, 'rand' indicates that it selects the target solution arbitrarily, the number of differential vector is one, and 'bin' denotes that *binomial* crossover is used.

The DE has a population of individuals that investigates the optimum solution, similar to other population-based search strategies. The solutions are denoted by a vector $(x_{i_1}, x_{i_2}, \ldots, x_{i_D})$ where $i = 1, 2, \ldots, NP$, (NP represents the size of population).

The three operators, namely mutation, crossover, and selection, play an important role in DE. In the first step, a uniformly distributed population engendered; then, these three operators are used to engender a fresh population. In DE process, trial solution development is a pivotal step. These three DE operators are discussed in detail in the next subsections.

2.1 Mutation

In biology, an unexpected transformation in the gene characteristics of a chromosome is known as 'mutation.' Here in EAs, an alteration or perturbation with an arbitrary component is known as mutation. In DE, generally three solutions are used: *target* solution, *doner* solution, and *trial* solution. *Target* solution represents a parent solution from the current iteration, *doner* solution denotes a mutant solution generated by the differential mutation operation, and *trial* solution is an offspring generated by the combination of the target with the donor solution. In the crossover operation, an offspring is generated using the recently generated trial solution. In order to produce a trial solution t_i from the parent solution x_i, the mutation operator can be defined in three simple steps as follows:

- Choose three solutions x_{i_1}, x_{i_2}, and x_{i_3} arbitrarily but should not be equivalent to each other.
- Then, the trial solution is calculated through mutation operator as shown in Eq. 1

$$t_i = x_{i_1} + F \times (x_{i_2} - x_{i_3}) \tag{1}$$

where augmentation of the differential variation [8] is controlled by mutation scale factor $F \in [0, 1]$.

2.2 Crossover

DE applies uniform crossover to generate offspring O'_i by combining the parent solution (a solution going to update), x_i, and the trial solution, t_i, as described below:

$$O'_{ij} = \begin{cases} t_{ij}, & \text{if } j \in S \\ x_{ij}, & \text{otherwise.} \end{cases}$$

where S represents the set of crossover points, and x_{ij} is the jth element of the solution x_i. It is presumed that if $j \in S$, the trial parameter is selected from the mutant t_{ij}; contrarily, the parameter is copied from solution x_{ij}.

There are various methods to select the set S. Exponential crossover and binomial crossover are the most popular used [8] crossover techniques to decide the set S. The binomial crossover is described in Algorithm 1. In this crossover operator, the points are selected from the set of available dimensions (problem space), $\{1, 2, ..., D\}$. Here, D presents the number of dimensions of the problem. The important steps of binomial crossover is described in Algorithm 1. In this algorithm, CR represents the probability that the measured crossover point will be incorporated. More crossover points get selected if the value of CR is high.

In Algorithm 1, S denotes the set of points on which crossover operation has to be applied, CR represents the probability of crossover, and $U(1, D)$ denotes an evenly dispersed arbitrarily integer between 1 and D.

Algorithm 1 Pseudo code of Crossover (Binomial):

$S = \phi$
$d^* \sim U(1, D)$;
$S \leftarrow S \cup d^*$;
for every $d \in 1...D$ **do**
    if $U(0, 1) < CR$ and $d \neq d^*$ **then**
        $S \leftarrow S \cup d$;
    end if
end for

2.3 Selection

A better individual between trial solution and parent solution is selected with the help of selection operator. Mainly, selection operator performs a couple of functions: First, it determines an individual to engender the trial solution for the mutation, and second, it carefully selects the most suitable individual for the next iteration, between the trial solutions and their predecessors depending on their fitness value as shown in the following equation.

$$x_i = \begin{cases} O_i', & \text{if objective Value of } O_i' \text{ is more than } x_i. \\ x_i, & \text{otherwise.} \end{cases}$$

This process makes sure that the average fitness of population does not depreciate. The major steps of DE strategy [8] are encapsulated in Algorithm 3.

Algorithm 2 DE Algorithm:

Initialize the initial population P randomly;
Initialize the F, CR, and NP as per parameter setting;
while Required conditions not meet **do**
  for Every solution, $x_i \in P$ **do**
    Assess the objective value of (x_i) of ith solution;
    Generate the offspring O_i' through the trial solution t_i and parent solution by using mutation and crossover operators;
    if offspring O_i' is better than x_i **then**
      Include offspring O_i' to P;
    else
      Include parent x_i to P;
    end if
  end for
end while
The best ever found solution is declared the global optima and solution of the problem;

Here, CR (crossover probability), P is the population, and F (scale factor) are two important control parameters. The scale factor controls the diversity of the population, and the crossover rate controls the number of components inherited from the mutant solution.

3 Cauchy Distribution-Based Scale Factor

In the prior knowledge-guided DE (PKDE) algorithm, Fan et al. [9] proposed a Cauchy distribution-based scale factor F as follows:

$$F_i = Cauchy(1 - 0.6 \times t/T, \sigma_1), i = 1, \ldots, Np \tag{2}$$

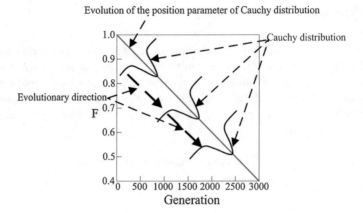

Fig. 1 Representation of scale factor evolving process [9]

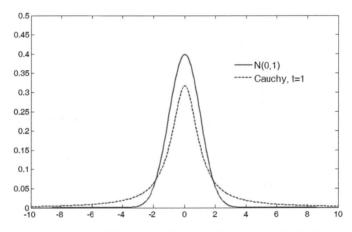

Fig. 2 Representation of probability density of the normal and Cauchy distribution functions [9]

where $Cauchy(x, y)$ represents the Cauchy distribution, x denotes the location parameter, y is the scaling parameter, and $\sigma_1 = 0.7 - 0.4(1 - (t/T)^2)$. Here, t and T are the current iteration and the maximum number of iterations, respectively. With the help of Eq. 2, every iteration has new value of F for target solution. From Eq. 2, it can be easily identified that the value of location parameter and the scaling parameter gradually decreases and increases, respectively, during the evolution (as shown in Fig. 1). During the early stage of evolution, the PKDE concentrates on the global exploration, and in the later stage of evolution, it improves the local exploitation. A large F value is generated by the Cauchy distribution with a large value of location parameter, and it produces small F values when the Cauchy distribution has a small value of location parameter.

Note that if F_i not lies in 0.0 to 1.0, then it is regenerated with the help of Eq. 3.

$$F_i = \begin{cases} 1, & \text{if } F_i > 1 \\ |N(0, 0.2)|, & \text{if } F_i < 0. \end{cases} \tag{3}$$

where $N(m, n)$ stands for the normal distribution, and m and n denote mean value and the standard deviation, respectively. Here, normal distribution function is selected as it has narrower tails in comparison with Cauchy distribution (as depicted in Fig. 2). It can be observed from Eq. 3 that it has good global search potential due to the high value of the F as well as it has good local search ability due to small value of F, and hence, a balanced search process may take place.

4 Fitness-Based Self-adaptive Scale Factor

Sharma et al. proposed a fitness-based self-adaptive scale factor (FSADE) F [19] in DE and enhanced diversification and convergence potential of DE. A small value of F leads to small step size (as shown in Eq. 1) in solution search and motivates an individual to search a new solution that results in better exploitation potential. While solution search process takes a large step if the value of F is large, it provides better exploration capability. In FSADE, value of F is different for each individual i, as it is set adaptively and is based on the fitness-based probability fp_i of the individual in order to balance diversification and convergence capability of DE, which may be computed as using Eq. 4.

$$fp_i = 0.9 \times \frac{fit_i}{maximumfit} + 0.1, \tag{4}$$

Here, fit_i and $maximumfit$ denote the fitness value of the ith solution and the maximum fitness in the population, respectively. The value of the probability fp_i always lies in $\in [0.1, 1]$ as per Eq. 4.

Additionally, according to fp_i, the F is adaptively changed as shown in Eq. 5.

$$F_i = (2 \times Const - fp_i) \times Rand(-0.5, 0.5), \tag{5}$$

Here, $Const = 1.1$ is a constant and $F_i \in [-1.05, 1.05]$.

It is clear from Eq. 5 that highly fitted solution enhances exploitation capability of DE. The large value of fp_i leads to a small value of F_i and vice versa. As a result, highly fitted solution moves with small step sizes, and thus, it could select feasible solutions from the solution search space in its proximity while the individuals with low fitness value can explore the search area more efficiently with large step size to find out new solutions.

5 Randomized Scale Factor

This scale factor is presented by Li et al. [12] in the modified DE with self-adaptive parameter method (MDE) algorithm. The intensification of the differential variation between two individuals is controlled by the scale factor, so it should be decided sensitively. The small values of F lead to the premature convergence, and the large values of F reduce the speed of the solution search process. But due to the versatility in the problems, there could not be a single optimal value of the F [12].

$$x'_{i,j} = x_{i_1,j} + rand(0, 1) \times (x_{i_2,j} - x_{i_3,j}) + F \times (x_{i_4,j} - x_{i_5,j}) \tag{6}$$

In Eq. 6, x_{i_1} is the parent solution while x_{i_2}, x_{i_3}, x_{i_4}, and x_{i_5} are the randomly selected solutions from the population. Equation 6 indicates that the value of F always remains same that leads to loss of the diversity of the population as the same scale factor [12] is used to compute all the individuals. In order to solve this problem, Gaussian distribution is used to generate the value of F. Due to short tail, it generates almost all values within unity. These randomly generated value of F minimizes chances of trap in local optima. The selection of the scale factor may be described as given below:

> **if** rand()$\leq$ rand() **then**
>   F $= |Gauss(0.3, 0.3)|$
> **else**
>   F $= |Gauss(0.7, 0.3)|$
> **end if**

$|Gauss(a, b)|$ represents the Gaussian distribution in the range (a, b). In order to ensure exploitation and exploration capability of DE, F=$|Gauss(0.3, 0.3)|$ and F=$|Gauss(0.7, 0.3)|$ are selected randomly with the range (0.6, 1.2) and (0.2, 1.6), respectively. However, there are more chances of stagnation and slow down the search process due to these parameter settings especially first one.

6 Self-adaptive Scale Factor (*jDE*)

In [1] *jDE* algorithm, Brest et al. presented a self-adapting control parameter system for '*rand*/1/*bin*' strategy. The self-adaptive control mechanism was used to change the control parameters scale factor F and crossover probability *CR* during the simulation. The scale factor F is self-adaptively updated through iterations as follows:

$$F_{i+1} = \begin{cases} F_l + rand_1.F_u, & \text{if } rand_2 < \tau_1 \\ F_i, & \text{otherwise} \end{cases} \tag{7}$$

The above equation engenders control parameter F in a new parent solution. Here, $rand_j, j \in (1, 2)$ denotes uniform arbitrary values in the range [0, 1]. τ_1 probability is used to alter value of F. τ_1, F_l, F_u constant and their values are fixed as 0.1, 0.1, 0.9, respectively. A random value from [0.1, 1.0] is assigned to F_i. The control parameter value of F_i is computed before implementation of mutation operator. That is the reason that it has effect on the mutation, crossover, and selection operations of the new solution x_i.

7 Adaptive Scale Factor (*JADE*)

In algorithm JADE, Zhang et al. [23] suggested an adaptive mechanism f or the scale factor F. In JADE, the scale factor F_i of each individual x_i is independently computed using a Cauchy distribution with location parameter μ_F and scale parameter 0.1, in every iteration as shown in Eq. 8, and then truncated to be 1 if $F_i = 1$ or recomputed if $F_i = 0$.

$$F_i = randc_i(\mu_F, 0.1) \tag{8}$$

Let us consider S_F as the set of all feasible scale factors in a single iteration. The Cauchy distribution's location parameter μ_F is initially set as 0.5 and then modernized at the end of each iteration as follows:

$$\mu_F = (1 - c).\mu_F + c.mean_L(S_F) \tag{9}$$

where $mean_L(.)$ is the Lehmer mean and c is a nonnegative constant between 0 and 1.

$$Mean_L(S_F) = \frac{\sum_{F \in S_F} F^2}{\sum_{F \in S_F} F} \tag{10}$$

8 Stochastic Scale Factor

A stochastic mutation scale factor is introduced in DE in [2] to enhance exploitation potential of the DE algorithm. The newly introduced stochastic mutation scale factor is motivated from levy flight random walk. Levy flight is a random walk used to determine step size randomly that is explained in Eq. 11.

$$L(st) \sim |st|^{-1-\alpha}, \text{ where } \alpha \ (0 < \alpha \leq 2) \text{ is an index and } st \text{ is the size of step.} \tag{11}$$

The Mantegna algorithm (Yang) [21] is used to generated random step size for a levy stable distribution in either positive or negative direction. The st (step size) is calculated as shown in Eq. 12:

$$step\_size(st) = \frac{u}{|v|^{\frac{1}{\alpha}}} \tag{12}$$

where u and v are generated using normal distribution, i.e., $u \sim N(0, \sigma_u{}^2)$, $v \sim N(0, \sigma_v{}^2)$ here,

$$\sigma_u = \left\{ \frac{\Gamma(1+\alpha)\sin(\pi\alpha/2)}{\alpha\Gamma[(1+\alpha)/2]2^{(\alpha-1)/2}} \right\}^{1/\alpha}, \quad \sigma_v = 1. \tag{13}$$

The st represents the l'evy distribution where $|st| \geq |s_0|$; here, s_0 denote the size of smallest step. In Eq. 13, $\Gamma(.)$ function is calculated using Eq. 14.

$$\Gamma(1+\alpha) = \int_0^\infty t^\alpha e^{-t} dt. \tag{14}$$

Particularly when α is an integer, then we have $\Gamma(1+\alpha) = \alpha!$.

The scale factor F is computed as follows:

$$F = step\_size \times rand\,(0, 1) \tag{15}$$

Here, levy flight random walk with the help of Eq. 12 decides the value of *step*. The range of u and v is modified from [0, 1] to [0.5, 1] in order to maintain diversification and to avoid premature convergence in the computation of step as displayed in Eq. 12.

9 Self-balanced Scale Factor

The self-balanced scale factor is introduced by Sharma et al. [18] to reduce the possibilities of premature convergence of the population. The problems of premature convergence and stagnation are a serious issue for the performance of DE. Therefore, to enhance the search potential of the DE, Sharma et al. [18] presented cognitive learning factor and dynamic scale factor as follows.

The mutation Eq. 1 of DE may be presented as follows:

$$x' = A \times x_1 + B \times \overbrace{(x_2 - x_3)}^{\text{Variation}}$$

i.e., the trial solution x' is generated through weighted sum of target solution x_1 and the difference vector of two randomly selected solutions $(x_2 - x_3)$. As we know that the weight to the target solution (A) is 1 in DE while B is a constant value in the range $(0, 1)$ named as scale factor. Various research and study has been done for varying scaling factor F [1] for better exploration and exploitation mechanisms. The weight A is named as cognitive learning factor (CLF) and denoted by 'C'. CLF denotes the weight of the current position of an individual, or it may be considered as the weight to self-confidence and that is why it was named so.

The improved mutation operation of DE is shown in Eq. 16.

$$x' = C \times x_1 + F \times (x_2 - x_3) \tag{16}$$

The small value of C and large value of F enhance the possibility of exploration as the weight for individual's current position is low and the weight of variation component is high that can be easily identified from Eq. (16). Moreover, exploitation potential enhanced with large value of C and small value of F, as the weight to variation component is low and weight to individual's current position is high in this case. Therefore, a proper managed C and F can balance the diversity in the population. The proposed strategy dynamically controls the diversification and convergence abilities of DE algorithm and hence named as self-balanced scale factor. Here, $C \in [0.11]$ is proposed as mentioned in Eq. (17):

$$C_i = \begin{cases} C_i + prob_i & \text{if } fit_i(t) > fit_i(t-1) \text{ and } C_i(t) < 1 \\ 1 & \text{if } fit_i(t) > fit_i(t-1) \text{ and } C_i(t) \geq 1 \\ C_i & \text{if } fit_i(t) < fit_i(t-1) \text{ and } tri_i \leq limit \\ 0.1 & \text{if } fit_i(t) < fit_i(t-1) \text{ and } tri_i > limit \end{cases} \tag{17}$$

In Eq. (17), the CLF of the ith solution is represented by C_i while the fit_i represents the fitness of the ith solution. The tri_i shows the number of iterations for which the ith solution was not updated. The *maxlimit* is the maximum counts upto which the solution is allowed not to update. At last, the probability $prob_i(t)$ which is a function of fitness is calculated through Eq. 18. In Eq. 18, *maximumfit* shows the fitness of the fittest solution

$$prob_i(t) = \frac{0.9 \times fit_i(t)}{maximumfit} + 0.1 \tag{18}$$

Further, the proposed dynamic scale factor (F) is calculated using Eq. (19).

$$F_i = (rand(0, 1) - 0.5) \times (1.5 - prob_i) \tag{19}$$

In the proposed scheme, in any iteration, the individuals of the population work twofolded: exploration and exploitation. Usually, large step size produces exploration while small step size produces exploitation. It is clear from Eqs. (18) and (19) that the value of CLF increases while F_i decreases with the fitness fit_i. Hence, the step size is controlled through this weighted combination of F_i and CLF. It is observed from Eqs. (17) and (19) that the high fit solutions will exploit while the low fit solutions will explore the solution search space.

10 Adaptive Mutation Factor

The adaptive mutation/scale factor is proposed by Cui et al. in [3]. In the proposed scheme, at each iteration t, the scaling factor F_i of each individual x_i is generated individually based on a Cauchy distribution, which is formulated by:

$$F_i = cauchy(F_m, 0.1) \tag{20}$$

where $Cauchy = (F_m, 0.1)$ is an arbitrary real number computed with the help of Cauchy distribution with the location parameter F_m and scaling parameter (0.1). F_i is truncated to be 1 when $F_i > 1$, and F_i is reestablished when $F_i < 0$. The S_F contains all the successful scaling factors at current iteration. The location parameter F_m is initialized to be 0.5, and then update at the end of each iteration as defined by:

$$F_M = w_F \times F_m + (1 - w_F) \times Mean_L(S_F) \tag{21}$$

where $mean_L(.)$ is the Lehmer mean [22] as depicted in Eq. 22 and w_F is a random weight factor in [0.8; 1.0] as proposed in [10], which is displayed in Eq. 23.

$$Mean_L(S_F) = \sum_{F_i \in S_F} F_i^2 / \sum_{F_i \in S_F} F_i \tag{22}$$

$$w_F = 0.8 + 0.2 \times rand(0, 1) \tag{23}$$

The empty set S_F indicates that in the current iteration there is no successful scaling factor and the parameter F_m at this iteration t is an unacceptable parameter to a particular point. In this case, F_m is modified as follows [3]:

$$F_m = C_F \times F_m + (1 - C_F) \times rand(0, 1) \tag{24}$$

11 Dynamic Scaling Factor

A dynamic scale factor was developed by Sharma et al. in [17]. In that paper, a dynamic scale factor DSF was proposed which controls the perturbation rate in the mutation process. The proposed DSF adjusts the step size of the solutions during the solution search process dynamically and provides a balanced diversification and convergence equilibrium among the solutions. Initially, the range of DSF is set to be $[-0.8, 0.8]$, and then, iteratively the range is squeezed to $[-0.4, 0.4]$.

$$x_i' = x_i + F \times (x_{i_1} - x_{i_2}) \tag{25}$$

Through Eq. 25, it was claimed that the use of DSF provides an appropriate balance of diversity and convergence in DE [17]. The perturbation on the position x_i decreases with decreasing difference between the parameters of the x_{i_1} and x_{i_2}, as displayed in Eq. 25.

12 Self-adapting Scale Factor

An easy and efficient approach has been anticipated in [1] to overcome the problem of parameter setting of F. This approach is known as self-adapting control parameters in differential evolution. The variation operations are headed by the parameter update to synchronize with a self-adaptive logic, e.g., [7]. Particularly when, at each iteration, the ith individual x_i is considered and additional three individuals are taken out pseudo-randomly, its parameter F_i is modified with the help of below-mentioned process [13].

$$F_i = \begin{cases} F_l + rand_1.F_u, & \text{if } rand_2 < \tau_1 \\ F_i, & \text{otherwise} \end{cases} \tag{26}$$

where $rand_1$ and $rand_2$ are uniformly generated pseudo-random values in the range 0 and 1, τ_1 denotes the probability of parameter update that is a constant value, F_l and F_u are lower and upper bounds for F_i, respectively. The offspring is generated using newly computed values of F_i.

13 Local Search-Based Scale Factor

In [13], Neri et al. developed a self-adaptive scale factor by incorporation of two cooperative/competitive local search algorithms, namely golden section search (GSS) and hill-climb (HC) local search algorithms. Here, at each iteration, best performing individuals are considered; scale factor is modified using below-mentioned rules:

$$F_i = \begin{cases} \text{GSS}, & \text{if } rand_3 < \tau_2 \\ \text{HC}, & \text{if } \tau_2 \leq rand_3 < \tau_3 \\ \begin{cases} F_l + F_u rand_1, & \text{if } rand_2 < \tau_1 \\ F_i, & \text{otherwise} \end{cases} & , \text{ if } rand_3 > \tau_3 \end{cases} \tag{27}$$

where $rand_j$, $j \epsilon 1, 2, 3$ are uniform pseudo-random values between 0 and 1; τ_k, $k \epsilon 1, 2, 3$ are constant threshold values. The probability of the updating parameters is denoted by the value of τ_1, while τ_2 and τ_3 are associated with the establishment of local search. F_l and F_u are lower and upper bounds for F_i, respectively. The details about the GSS scale factor and HC scale factor are described in the subsequent Sects. 13.1 and 13.2, respectively.

13.1 Scale Factor Golden Section Search (SFGSS)

Kiefer [11] proposed a classical local search algorithm for non-differentiable fitness functions, namely golden section search. It is used in DE with scale factor to get a superior offspring. The SFGSS makes use of the interval $[a = 0.1, b = 1]$ and produces two intermediary points:

$$F_i^1 = b - \frac{b-a}{\Phi} \tag{28}$$

$$F_i^2 = a + \frac{b-a}{\Phi} \tag{29}$$

The pseudo-code of the SFGSS algorithm is shown in Algorithm 3.

Algorithm 3 SFGSS Algorithm
---
insert a and b;
while Termination condition **do**
  compute $F_i^1 = b - \frac{b-a}{\Phi}$ and $F_i^2 = a + \frac{b-a}{\Phi}$;
  compute $f(F_i^1)$ and $f(F_i^2)$;
  if $f(F_i^1) < f(F_i^2)$ **then**
    $b = F_i^2$;
  else
    $a = F_i^1$;
  end if
end while

In Algorithm 3, fitness function is denoted by f. Here, $\Phi = \frac{1+\sqrt{5}}{2}$ is the golden ratio. It is clear from the Algorithm 3 that the range of the scale factor generation is iteratively reduced to provide the good step size to get the global optima.

13.2 Hill-Climb-Based Scale Factor

One of the most accepted optimization algorithms that is very simple and also available in any book of optimization, e.g., [16] is the unidimensional hill-climb local search. The current value of F is considered as an initial point and is composed of an investigative and a decisional move. The investigative move samples $F - h$

and $F + h$ with step size h. The minimum of $f(F - h), f(F), f(F + h)$ is computed by decisional move and also decides the analogous point as the center of the subsequently investigative move. The step size h is halved if the center for the next exploratory move is unchanged. The local search is stopped with termination criteria fulfilled. The pseudo-code of the scale factor hill-climb is displayed in Algorithm 4.

Algorithm 4 Hill-climb based Scale Factor

Initialize F and h;
while Termination condition **do**
    Calculate function values $f(F - h), f(F)$, and $f(F + h)$;
    Identify the best performer F^*;
    if $F^* == F$ **then**
        $h = h/2$;
    end if
    $F = F^*$
end while

The Algorithm 4 presents the step size of the solutions, is updated as per the feedback received from the previous solutions, and hence helps the next-generation solutions to move accordingly to find the global optima.

14 Dimension-Based Scale Factor

Zou et al. introduced a dimension-based scale factor and a new mutation strategy in [25] while proposing an effective modified differential evolution (EMDE) algorithm. As we know that in the classical scale factor F, it has fixed value for one problem in case of all dimensions. Specifically, every dimension $x_{i,j}$, has the same intensification factor of difference vector $(x_{i_1,j} - x_{i_2,j})$ as shown in Eq. 1. But, here, Zou et al. proposed dynamically balanced D scale factors each for every dimensions. Generally, F_j is decided by using the following steps:

1. Generate D scale exponential scale factors $F = (F_1, F_2, ..., F_D)$. Here, F_j is calculated using Eq. 30.

$$F_j = a \times exp(b \times j) \tag{30}$$

    where $b = ln(UB/LB)/(T - 1)$ and $a = exp(-b) \times LB$. UB and LB are the maximum and minimum values of the scale factors, respectively. T denotes the maximum number of iterations.

2. In every iteration, arbitrarily select an dimension d ($1 \leq d \leq D$), and unshuffle d dimensions of F. $F = (F_{d+1}, ..., F_D, F_1, ..., F_d)$ is the newly engendered sequence of scale factors. As per these settings of F, exploitation and exploration of search space is carried out by dimension with small-scale factors and large-scale factors, respectively.

15 Normal Distribution-Based Scale Factor

Zhao et al. [24] introduced a normal distribution-based scale factor. They proposed an iteration-based mutation scale factor F_i for each target solution X_i separately with the help of normal distribution with mean μ_F and a standard deviation of 0.1. The scale factor F_i is computed as follows:

$$F_i = Normalrand(\mu_F, 0.1) \tag{31}$$

The value of F_i is fixed at 1 if it exceeds 1. Every target solution X_i and its corresponding F_i are used to engender a new trial solution U_i. If U_i is better than its target solution X_i, then the associated F_i will enter a winning set w_F. The mean μ_F of the normal distribution is initially fixed at 0.5. After each iteration, the mean μ_F is modified using following equation:

$$\mu_F = a \times \mu_F (1 - a) \times \frac{\sum_{F \in wF} F}{NP} \tag{32}$$

where a is a nonnegative constant within the range (0, 1), and NP denotes the size of population. The set wF are reset to empty after updating μ_F.

16 Random and Time-Varying Scale Factor

In [4], Das et al. introduced two scale factor, namely random and time-varying scale factors. The proposed scale factors are explained as follows:

16.1 Random Scale Factor

In basic DE [20], a number between 0.4 and 1 is most appropriate choice for this control parameter. Das et al. [4] proposed a random varying scale factor in the range (0.5, 1) by using the relation shown in Eq. 33.

$$F = 0.5 \times (1 + U(0, 1)); \tag{33}$$

where $U(0, 1)$ is an evenly scattered arbitrary value in the interval [0, 1]. The mean value of F is 0.75 that leads for stochastic variations in the intensification of the difference vector and maintains diversity of population with progress in search. Hence, the newly generated trial solution has good probability of indicating toward feasible location on the multi-modal functional surface, even if the majority of population

Fig. 3 Pictorial representation of the time-varying scale factor in a 2D parameter space [4]

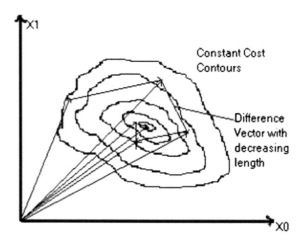

indicates toward locations concentrated near a local optimum by the reason of the arbitrarily scaled difference vector.

16.2 Time-Varying Scale Factor

Almost all population-based optimization methods maintain large step size to get a better exploration of search space during early iterations while it is essential to fine-tune the movements of trial solution very gently in order to get better exploitation of the solution search space. To achieve these goals, Das et al. reduced the value of the scale factor from a maximum to a minimum value linearly with time as shown in Eq. 34.

$$F = (UB - LB) \times (T - t)/T; \tag{34}$$

Here, UB and LB denote the highest and lowest values of scale factor F, respectively. t represents the current iteration, while maximum iterations are represented by T. Figure 3 pictorially shows the locus of the tip of the solution found by the proposed strategy.

17 Conclusion

This paper provides a detailed scenario of the differential evolution (DE) algorithm and the research carried out on the scale factor strategies of DE. Initially in this paper, an importance of the scale factor is described in the mutation process of DE which further affects the solution search process. As the primary role of the scale factor is to balance the step size of an individual during the solution search process, a large step size may increase the computation cost while a small step size may converge the population prematurely. Hence, a step size should be fine-tuned according to the search process. This paper presents a detailed survey of the various scale factor strategies and shows the versatility of scale factors in terms of efficiency and reliability.

References

1. Brest, J., Greiner, S., Boskovic, B., Mernik, M., Zumer, V.: Self-adapting control parameters in differential evolution: a comparative study on numerical benchmark problems. IEEE Trans. Evol. Comput. **10**(6), 646–657 (2006)
2. Choudhary, N., Sharma, H., Sharma, N.: Differential evolution algorithm using stochastic mutation. In: 2016 International Conference on Computing, Communication and Automation (ICCCA), pp. 315–320. IEEE (2016)
3. Cui, L., Li, G., Lin, Q., Chen, J., Nan, L.: Adaptive differential evolution algorithm with novel mutation strategies in multiple sub-populations. Comput. Oper. Res. **67**, 155–173 (2016)
4. Das, S., Konar, A., Chakraborty, U.K.: Two improved differential evolution schemes for faster global search. In: Proceedings of the 7th Annual Conference on Genetic and Evolutionary Computation, pp. 991–998. ACM (2005)
5. Das, S., Suganthan, P.N.: Differential evolution: a survey of the state-of-the-art. IEEE Trans. Evol. Comput. **15**(1), 4–31 (2011)
6. Dragoi, E.N., Dafinescu, V.: Parameter control and hybridization techniques in differential evolution: a survey. Artif. Intell. Rev. **45**(4), 447–470 (2016)
7. Eigen, M.: Ingo Rechenberg Evolutionsstrategie Optimierung technischer Systeme nach Prinzipien der biologishen Evolution. mit einem Nachwort von Manfred Eigen, Friedrich Frommann Verlag, Struttgart-Bad Cannstatt (1973)
8. Engelbrecht, A.P.: Computational Intelligence: An Introduction. Wiley (2007)
9. Fan, Q., Yan, X., Xue, Y,: Prior knowledge guided differential evolution. Soft Comput. 1–18 (2016)
10. Islam, S.M., Das, S., Ghosh, S., Roy, S., Suganthan, P.N.: An adaptive differential evolution algorithm with novel mutation and crossover strategies for global numerical optimization. IEEE Trans. Syst. Man Cybern., Part B (Cybernetics) **42**(2):482–500 (2012)
11. Kiefer, J.: Sequential minimax search for a maximum. Proc. Am. Math. Soc. **4**(3), 502–506 (1953)
12. Li, X., Yin, M.: Modified differential evolution with self-adaptive parameters method. J. Comb. Optim. **31**(2), 546–576 (2016)
13. Neri, F., Tirronen, V.: Scale factor local search in differential evolution. Memetic Comput. **1**(2), 153–171 (2009)
14. Neri, F., Tirronen, V.: Recent advances in differential evolution: a survey and experimental analysis. Artif. Intell. Rev. **33**(1–2), 61–106 (2010)
15. Price, K.V.: Differential evolution: a fast and simple numerical optimizer. In: 1996 Biennial Conference of the North American Fuzzy Information Processing Society, NAFIPS, pp. 524–527. IEEE (1996)
16. Russell, S., Norvig, P.: Artificial Intelligence: A Modern Approach. Prentice-Hall, Egnlewood Cliffs, 25–27 (1995)
17. Sharma, H., Bansal, J.C., Arya, K.V.: Dynamic scaling factor based differential evolution algorithm. In: Proceedings of the International Conference on Soft Computing for Problem Solving (SocProS 2011), pp. 73–85. Springer (2012), 20–22 Dec 2011

18. Sharma, H., Bansal, J.C., Arya, K.V.: Self balanced differential evolution. J. Comput. Sci. **5**(2), 312–323 (2014)
19. Sharma, H., Shrivastava, P., Bansal, J.C., Tiwari, R.: Fitness based self adaptive differential evolution. In: Nature Inspired Cooperative Strategies for Optimization (NICSO 2013), pp. 71–84. Springer (2014)
20. Storn, R., Price, K.: Differential evolution-a simple and efficient heuristic for global optimization over continuous spaces. J. Global Optim. **11**(4), 341–359 (1997)
21. Viswanathan, G.M., Afanasyev, V., Buldyrev, S.V., Murphy, E.J., et al.: Lévy flight search patterns of wandering albatrosses. Nature **381**(6581), 413 (1996)
22. Wang, Y., Cai, Zixing, Zhang, Qingfu: Differential evolution with composite trial vector generation strategies and control parameters. IEEE Trans. Evol. Comput. **15**(1), 55–66 (2011)
23. Zhang, J., Sanderson, A.C.: Jade: adaptive differential evolution with optional external archive. IEEE Trans. Evol. Comput. **13**(5), 945–958 (2009)
24. Zhao, Z., Yang, J., Ziyu, H., Che, H.: A differential evolution algorithm with self-adaptive strategy and control parameters based on symmetric latin hypercube design for unconstrained optimization problems. Eur. J. Oper. Res. **250**(1), 30–45 (2016)
25. Zou, D.X., Pan, G., Qi, H.W., Li, Y.P.: An effective modified differential evolution algorithm for reliability problems. In: 2016 4th International Symposium on Computational and Business Intelligence (ISCBI), pp. 132–136. IEEE (2016)

Troop Search Optimization Algorithm for Unconstrained Problems

Biplab Chaudhuri and Kedar Nath Das

Abstract In defence sector, the combined effort of both commander and captain helps to prepare an efficient troop with optimized strength, before each combat operation. A continuous attempt is being maintained in order to maximize both protection of the combatants in the troop and rate of killing enemies. Hence during the process of selecting the best troop, there could be a number of readjustments, shuffling and exchanging mechanisms applied over the combatants. These mechanisms or the course of actions taken by a commander/captain has been modelled to an optimization algorithm, proposed as 'Troop Search Optimization (TSO) algorithm' in this paper. TSO takes utmost care to balance both exploration and exploitation in the population during simulation. In order to realize the better strength of TSO together with proposed crossover, a set of 11 unconstrained benchmark problems have been solved by TSO and the results are compared. The numerical results and statistical analysis confirm the better strength of TSO over many of the recently established bio-inspired algorithms including ABC, GABC, TLBO, ITLBO, HS and IBA.

Keywords Unconstrained optimization · Local search · Simplex search method

1 Introduction

The need for optimality is intrinsic to human nature. The search for extremities motivates climbers, researchers, mathematicians, and the rest of the human race. A handsome and useful mathematical model of optimization (i.e. search-for-optimum strategies) is established since the year 1966 when computers became available. The objective is to form consistent methods in order to catch the extrema of a function by

B. Chaudhuri (✉) · K. N. Das
Department of Mathematics, National Institute of Technology, Silchar, Silchar 788010, Assam, India
e-mail: biplabs2008@gmail.com

K. N. Das
e-mail: kedar.iitr@gmail.com

© Springer Nature Singapore Pte Ltd. 2019
J. C. Bansal et al. (eds.), *Soft Computing for Problem Solving*, Advances in Intelligent Systems and Computing 817, https://doi.org/10.1007/978-981-13-1595-4_74

an intelligent procedure of its evaluations (measurements). Modelling is essentially important for modern engineering and planning that incorporate optimization at every step of the complicated decision-making process. Over the increasing complexity of the optimization problems, the classical optimization methods become handicapped to solve them. However, the evolutionary algorithms (EAs) help as an alternate prototype. Such methods have received much popularity nowadays as they do not need the auxiliary information (like differentiability, continuity) of the optimization problem in hand.

During the last 15 years, a number of EAs have been proposed in the literature. In the challenging competition of balancing the exploration and exploitation, one algorithm often performs better than the other, but not always. Hence, continuous attempts are made by the researchers in order to establish an 'optimization algorithm' that can solve wide varieties of problems, if not to all of them. A few recent works, which are very close competitors of the proposed algorithm, are cited below.

Rao et al. [1] proposed an algorithm, known as TLBO, which simulates the traditional teaching–learning process of a classroom. Rao et al. [2] apply their optimization on continuous nonlinear large-scale problems. Rao and Patel [3] solve complex constraint optimization problems using elitism technique. Gradually, they improved the basic TLBO and named it as Improved TLBO (ITLBO) [4, 5]. Applying TLBO, many researchers [6–9] have shown that this optimization algorithm performs better than other EAs. It is realized that proper balance of exploration and exploitation over the search space is still missing in recent methods in the literature survey. Therefore, an entirely new algorithm of similar taste is proposed in this paper, namely 'Troop Search Optimization (TSO)' algorithm based on the Artificial Bee Colony (ABC) as described in this paper. The TSO is inspired by the searching mechanism of a best troop by the commander in the defence sector, just before/during each combat operation. The repeated phenomena of rejection, selection, shuffling and swapping of combatants in preparing best troop are being modelled to design the proposed algorithm. The major components of the proposed algorithm are defined in the next section.

2 Major Components of the Proposed Algorithm

For better understanding the proposed method (Sect. 3), it is essential to know the working principle of few of its components separately, which is discussed as follows.

2.1 Swapping Crossover

During the process of simulation, it is often observed that the value/s of one/more variables in one string may be a better fit for the corresponding variables of another string. Thus, a swapping mechanism is likely to help in generating a better individual

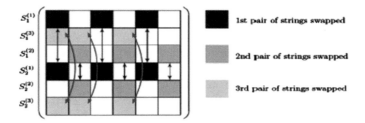

Fig. 1 Illustration of swapping crossover

in the current population. Keeping in view, a new crossover operator, namely 'swapping crossover', is proposed in this paper where the variable values are swapped arbitrarily across the variables. The working principle of swapping crossover is briefed below.

Let D be dimension of the problem. $\lceil x \rceil$: Ceiling function of x.

1. Repeat steps 1–4 for $2D$ number of times.
2. Select two arbitrary strings S_1 and S_2 from the population.
3. In S_1, randomly select $\lceil D/2 \rceil$ number of distinct variables x_1, $i \in \{1, 2, \ldots, D\}$. Store the values of '$i$' chosen.
4. Select the corresponding variables y_i in S_2, for same values of 'i', chosen in step 2.
5. Swap $x_i's$ and $y_i's$ in S_1 and S_2.
6. Exit.

The concept of swapping crossover is diagrammatically represented in Fig. 1, where three different pairs of strings participated in swapping. In figure, $s_1^{(i)}$ and $s_2^{(i)}$ are the two strings swapped in ith iteration.

2.2   Cut and Fill Method

The basic idea behind 'cut and fill' mechanism is to replace worst 20% of the population by better ones from rest 80% of the population by using Eq. (1).

$$x_{i,j} = x_{best,j} + rand(0, 1)(x_{r1,j} - x_{r2,j}) \tag{1}$$

where $x_{i,j}$ is the jth variable of ith string and x_{best} is the best-fit string in the population. r_1 and r_2 are two random strings from the top 80% of population, and 'j' represents the variable number. $rand(0, 1)$ is a random number between 0 and 1. If the value of any variable obtained by Eq. (1) is lying outside the range, then that will be automatically replaced by the corresponding value (x_{best}) of the best string.

2.3 Simplex Search Method

The fitness function the simplex search method (SSM) [10] was originally proposed by Spendley et al. in 1962. SSM starts with an initial simplex that should not form a zero volume N-dimensional hypercube. For N variable problems, only $N+1$ points are used in the initial simplex. The basic idea of SSM is to reflect the worst point in the population through the centroid of the initial simplex, in order to find possibly a better point over the worst.

Simplex Search Algorithm

Let D be the dimension of the problem and $D+1$ be the vertices or points in the initial simplex. Let

x_{bp}: best point, x_{wp}: worst point, x_{nwp}: next to worst point
x_{cp}: centroid of the simplex, x_{rp}: reflected point.
Contraction factor $\beta = 0.5$ and expansion factor $\gamma = 1.5$ are fixed in this paper, as reported in [10].

Steps:

1 Generate the initial simplex by picking $(D + 1)$ strings randomly.
2 Repeat the steps 2–4 for $(D+1)$ number of times, which is the termination criterion of simplex search method.
3 Calculate $x_{cp} = \frac{1}{D} \sum_{i=1, i \neq h}^{D+1} x_i$, $x_{rp} = 2x_{cp} - x_{wp}$ and set $x_{new} = x_{rp}$.
4 If $f(x_{rp}) < f(x_{bp})$, then set $x_{new} = (1 + \gamma)x_{cp} - \gamma x_{wp}$ for expansion case.

Else If $f(x_{rp}) \geq f(x_{wp})$, then set $x_{new} = (1 - \beta)x_{cp} + \beta x_{wp}$ for contraction case.
Else If $f(x_{nwp}) \leq f(x_{rp}) \leq f(x_{wp})$, then set $x_{new} = (1 + \beta)x_{cp} - \beta x_{wp}$ for contraction case.

5 Replace worst point x_{wp} by improved point x_{new}.
6 Exit.

2.4 Modified Quadratic Approximation

Mohan and Shankar [13] first introduced Random Search Technique (RST) for global optimization based on Quadratic Approximation (QA). Later, it is modified and named as modified Quadratic Approximation (mQA). In QA, one *child* is produced from the selection of three parents. The concept behind this is simply to maintain the diversity for exploring the search space with the help of randomness mechanism. 'Non-redundant search (NRS)' [11] helps to replace redundant strings in a certain population. The replacement is made with the help of modified Quadratic Approximation (mQA) [12, 14]. Thus, by means of NRS it is understood that by keeping one copy of the redundant strings, rest will be replaced by mQA. This concept is being used in the proposed algorithm in the following section.

3 Concept of Troop Search Optimization Algorithm

3.1 Motivation and Methodology

It is observed that in ABC and its variants, no mechanism is there to care for multiple copies of similar food sources that exist in a population. Over the time, they can neither be destroyed nor be improved with more nectar amount (i.e. with more fitness). Therefore, the existences of food sources with equal nectar amount possibly cause unnecessary increase in the function evaluations and computational time as well.

Further, it is quite interesting to observe that the working principle of ABC resembles that of the mechanism of selecting efficient troop in defence sector before/after combat operation. Some efficient operators are also embedded in troop search mech-

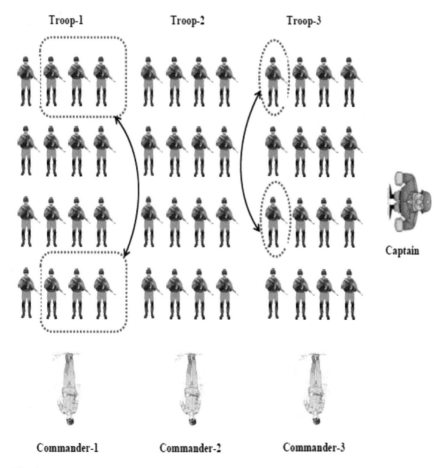

Fig. 2 Arrangement of combatants, commander/captain in troops in a battalion

anism adopted by the commander/captain before/during combat operation in defence sector.

Motivated by the above fact, the strategy mapping and concept development of an entirely new and more robust algorithm, namely 'Troop Search Optimization Algorithm', are proposed as follows.

Strategies of Troop Search:

A commander/captain attempts to prepare a bravery battalion with the help of efficient troops before/during each combat operation in the defence sector. A commander arranges the combatants in the troop with their ability in such a way that it is expected to impact maximum output of killing enemies while defending themselves. During the process of preparing the *best troop*, there could be a number of rearrangements, shuffling and swapping mechanisms applied over the combatants. These mechanisms or the course of actions taken by a commander/captain has been modelled to an optimization algorithm, proposed as 'Troop Search Optimization (TSO) algorithm'. TSO takes utmost care to balance both exploration and exploitation in the population during simulation.

The main idea of the proposed algorithm is based on the mechanism of selecting best troops before/during each combat operation. In defence sector, a battalion consists of many troops and a troop consists of many combatants. A troop is led by a commander and a battalion by a captain. The process is shown in Fig. 2. Efficient supervision of combatants in a troop and troops in a battalion plays a significant role in optimizing the fortitude of battalion, leading them to vanquish enemies and to achieve success triumphs over every contingent.

In TSO, the mechanism is conceptualized as follows. A **battalion** is treated as a **group of populations** of individuals, a **troop** is treated as a **population of strings**, a **combatant** is treated as a **string** in the population, and the **strength** of a particular combatant is being interpreted as the **fitness** of that string. Concept mapping of working mechanism between ABC and TSO is compared in Table 1.

3.2   TSO Algorithm

The TSO algorithm is the proposed algorithm in this paper, which consists of the following basic steps.

Step 1:   Initialization of population.
Step 2:   Repeat steps 3–8, while stopping criteria is not satisfied.
Step 3:   Apply swapping crossover.
Step 4:   Apply greedy selection.
Step 5:   Apply cut and fill method.
Step 6:   Apply simplex search method.
Step 7:   Apply elitism.
Step 8:   Apply NRS by mQA.
Step 9:   Return the optimal solution and stop.

Table 1 Conceptual mapping of ABC to TSO algorithm

ABC algorithm	Real troops in defence sector	TSO algorithm
Initialization of population: Creation of random food sources and determination of their nectar amount (i.e. calculation of fitness)	Initially, the combatants are arranged randomly in different troops	The **initial population** is being generated randomly
Employed bee phase: Creation of random food sources. Apply greedy selection process, and select the better solution between the new and old food source (i.e. sharing of information through waggle dance of bees and selecting better food source)	Later, the positions of the combatants are being interchanged/swapped by the commander within the troop to produce a balanced strength in it	**Swapping crossover** in between the strings is applied
	Better combatants have better chance of survival, depending upon their position in the troop and the defence ability	Better individuals are being selected to the next generation by applying **greedy selection**
Onlooker bee phase: Selection of better food sources based on their nectar amount in terms of the probability of individual fitness. By this process, the worst individuals having less nectar amount will die off. Apply greedy selection again between old and new food sources available	During the combat operation, a few of the combatants may die and the positions will be replaced by fresh combatants	The set of worst 20% individuals will die off and will be replaced by new ones, using **cut and fill** technique
Scout bee phase: A food source will be declared abandoned if no further improvement found in it over a certain time period. In such case, the employed bee acts as a scout bee to locate a new food source	A very few numbers of combatants getting injured in the battle. They may recollect their courage and participate back in the war with a modified strength. Alternatively, it seeks for a locally better combatant to replace the position	A few of the worst individuals will be replaced by the better ones obtained by **simplex search method**
	In case, commander feels that no much success is expected with retaining a particular arrangement of combatants in a troop, and a quick decision is to be sought to break up the existing arrangement of combatants in the troop. A few inefficient combatants can be replaced by some locally available efficient combatant, likely to be better fit for that place	**Non-redundant search (NRS)** is applied where the replacement of the equal performers is being broken by new strings generated by the local search, namely **modified QA**

Table 2 Assumption under different case studies

Case study	Problem domain	Compared paper with year	Algorithms compared with TSO
1	Unconstrained benchmark functions with dimensions 30, 50 and 100	Yu et al. [15] (2016)	DE, ABC, GABC, TLBO, ITLBO, HS, IBA
2	Unconstrained benchmark functions with dimensions from 2 to 60	Gao et al. [16] (2012)	GABC(C = 1.5), E-ABC, ABC/best/2, ABC/best/1

3.3 Comparative Study of TSO with Existing Algorithms

In order to realize the better strength of TSO, a set of 11 nonlinear unconstrained benchmark problems have been solved. For this, the following two case studies with different categories as described in Table 2 have been considered. The results are compared with ABC and recent state-of-the-art algorithms. The algorithm of TSO is coded in C++ and simulated in Linux platform.

Case Study-1:
In case study-1, a set of 11 unconstrained benchmark functions (listed in Table 3) are picked from [15]. For a fair comparison of the proposed TSO with the algorithms discussed in [15], the same sequence of experimental set-up is being retained here. The parameter setting remains the same as in [15]. The problems in Table 1 are picked under two categories as follows in same line of approach as discussed in [15].

Category-1: As in [15], a total of ten functions from the set in Table 3 are solved by TSO with $30D$ by fixing the population size (NS) to 40, except f10 for which it is 200. Stopping criteria also kept the same as to attain a maximum of 2000 generations. For 30 independent runs, the means and standard deviation (SD) are compared in Table 4a. The gist of Table 4a is cited in Table 4b. It is worth noting that the boldface letters in the entire study represent the best value achieved for that particular function over all compared algorithms under consideration.

Category-2: A total of 5 functions from the problem listed in Table 3 are solved by TSO for $50D$ and $100D$ by fixing the NS at 100 and 200, respectively. As in [15], the stopping criterion is fixed to either the maximum number of function evaluations 50,000 is achieved or the maximum number of generations 2500 is attained whichever is earlier. For 30 independent runs, the means and SDs are compared in Table 5a. The gist of Table 5a is reflected in Table 5b.

Discussion of Results (for Case Study-1):
This section contains the analysis of result (Tables 4 and 5) for both the categories of case study-1. From Table 4a, it can be observed that in the total of ten functions (f1–f10), TSO outperforms all rest of algorithms in solving all the functions (except f1, f7, f8 and f9), in terms of best achieved mean objective function value. TSO ties with TLBO and ITLBO for function f4. However, according to Table 4b, TSO

Table 3 Benchmark functions used in experiment

Test function	Formulation	Search range	Min. value
Sphere	$f1 = \sum_{i=1}^{D} x_i^2$	$[-100, 100]^D$	0
Rosenbrock	$f2 = \sum_{i=1}^{D-1}\left[100\left(x_i^2 - x_{i+1}\right)^2 + (x_i - 1)^2\right]$	$[-30, 30]^D$	0
Ackley	$f3 = -20\exp\left(-0.2\sqrt{\frac{1}{D}\sum_{i=1}^{D} x_i^2}\right)$ $- \exp\left(\frac{1}{D}\sum_{i=1}^{D}\cos(2\pi x_i)\right) + 20 + e$	$[-32, 32]^D$	0
Griewank	$f4 = \frac{1}{4000}\sum_{i=1}^{D} x_i^2 - \prod_{i=1}^{D}\cos\left(\frac{x_i}{\sqrt{i}}\right) + 1$	$[-600, 600]^D$	0
Rastrigin	$f5 = \sum_{i=1}^{D}\left[x_i^2 - 10\cos(2\pi x_i) + 10\right]$	$[-5.12, 5.12]^D$	0
Step	$f6 = \sum_{i=1}^{D} |x_i + 0.5|^2$	$[-100, 100]^D$	0
Schwefel 2.22	$f7 = \sum_{i=1}^{D} |x_i| + \prod_{i=1}^{D} |x_i|$	$[-10, 10]^D$	0
Schwefel 1.22	$f8 = \sum_{i=1}^{D}\left(\sum_{j=1}^{i} x_j\right)^2$	$[-100, 100]^D$	0
Schwefel 2.21	$f9 = max\{|x_i|, \ 1 \le i \le D\}$	$[-100, 100]^D$	0
Quartic	$f10 = \sum_{i=1}^{D} ix_i^4 + rand(0, 1)$	$[-1.28, 1.28]^D$	0
Schaffer	$f11 = 0.5 + \dfrac{\sin^2\left(\sqrt{\sum_{i=1}^{D} x_i^2}\right) - 0.5}{\left(1 + 0.001\left(\sqrt{\sum_{i=1}^{D} x_i^2}\right)\right)^2}$	$[-100, 100]^D$	0

performs well for rest of algorithms clearly for $30D$ benchmark functions, as it wins in most cases. TSO is equally stable with TLBO and ITLBO due to its equal wins in SDs.

Similarly in Table 5a, the result of TSO is being compared with five new algorithms over five problems and each with $50D$ and $100D$. Therefore, there arise a total of ten cases, out of which the best objective function value obtained by TSO is well comparable in eight cases. Only in two cases, TSO fails to achieve the best value. In addition, TSO impacts higher SD in four cases only out of ten. Hence as a whole, the performance of TSO is better than algorithms reflected in Table 5b as the rate of wining of TSO is mostly high.

Case Study-2:
In this case study, problems with varying dimensions (2–60) have been considered in order to check the stability of TSO. A set of six problems f1–f5 and f11 are taken from Table 3. This case study differs from the earlier one by adding an extra function f11 to the previous set of five functions. The various improved and hybrid versions of ABC algorithms are considered for comparison, namely GABC, E-ABC, ABC/best/2 and ABC/best/1 taken from [16] (which is different from those in [15]). For a fair

Table 4a Comparison of TSO with the algorithms in [15] for 30D

Function	Statistical features	DE	ABC	GABC	TLBO	ITLBO	TSO
Sphere (f1)	Mean	2.19E−27	3.62E−09	6.26E−16	0	0	1.22E−38
	SD	2.02E−27	5.85E−09	1.08E−16	0	0	1.67E−35
Rosenbrock (f2)	Mean	5.71E+01	4.55E+00	7.47E+00	2.65E+01	2.23E+01	**4.60E−07**
	SD	2.65E+01	4.88E+00	1.91E+01	3.90E−01	1.46E−01	**6.49E−07**
Ackley (f3)	Mean	1.93E−14	2.75E−05	7.78E−10	3.55E−15	3.55E−15	**4.22E−17**
	SD	4.81E−15	2.31E−05	2.98E−10			4.71E−17
Griewank (f4)	Mean	0	3.81E−03	6.96E−04	0	0	0
	SD	0	8.45E−03	2.26E−03	0	0	0
Rastrigin (f5)	Mean	1.98E−01	4.53E−01	3.31E−02	1.57E+01	2.24E+01	0
	SD	4.81E−01	5.15E−01	1.81E−01	1.31E+01	1.66E+01	0
Step (f6)	Mean	1.72E−27	2.49E−09	6.45E−16	2.29E−02	1.34E−07	0
	SD	1.39E−27	3.68E−09	1.11E−16	6.21E−02	4.34E−08	0
Schwefel2.22 (f7)	Mean	1.44E−16	5.11E−06	1.31E−10	2.12E−199	0	2.18E−42
	SD	6.38E−17	2.23E−06	4.69E−11	0	0	2.71E−21
Schwefel1.2 (f8)	Mean	5.66E+04	1.24E+04	1.09E+04	5.00E−113	**4.38e−322**	5.35E−19
	SD	1.82E+04	3.01E+03	2.57E+03	2.72E−112	0	3.79E−19
Schwefel2.21 (f9)	Mean	6.34E−56	2.45E+01	1.26E+01	0	0	3.02E−03
	SD	2.46E−55	5.66E+00	2.66E+00	0	0	2.30E−03
Quartic (f10)	Mean	1.61E−02	1.56E−01	8.48E−02	1.27E−03	2.05E−04	**1.71E−04**
	SD	3.45E−03	4.56E−02	2.79E−02	3.15E−03	1.96E−04	**1.62E−04**

Table 4b Gist of Table 4a (out of ten problems)

TSO	DE		ABC		GABC		TLBO		ITLBO	
	Mean	SD	Mean	SD	Mean	SD	Mean	SD	Mean	SD
Wins	9	9	10	10	10	10	5	4	5	4
Equals	–	–	–	–	–	–	1	1	1	1
Losses	1	1	–	–	–	–	4	5	4	5

Table 5a Comparison of TSO with algorithm in [15] for 50D and 100D

Function	Dim.	Statistical features	HS	IBA	ABC	TLBO	ITLBO	TSO
Sphere (f1)	50	Mean	5.46E+02	5.39E−16	1.19E−15	0	0	1.38E−147
		SD	9.27E+01	1.07E−16	4.68E−16	0	0	3.94E−147
	100	Mean	1.90E+04	1.45E−15	1.99E−06	0	0	1.70E−97
		SD	1.78E+03	1.63E−16	2.26E−06	0	0	2.54E−97
Rosenbrock (f2)	50	Mean	2.47E+04	6.30E+02	4.33E+00	4.44E+01	3.68E+01	**3.96E+00**
		SD	1.02E+04	1.20E+03	5.48E+00	6.61E−01	5.56E−01	**3.28E+00**
	100	Mean	1.45E+07	6.42E+02	1.12E+02	9.48E+01	9.09E+01	**6.16E+01**
		SD	2.16E+06	8.20E+02	6.92E+01	9.44E−01	**6.46E−01**	1.47E−01
Ackley (f3)	50	Mean	5.28E+00	8.43E+00	4.38E−08	3.55E−15	3.55E−15	**4.68E−17**
		SD	4.03E−01	7.70E+00	4.65E−08	0	0	8.69E−18
	100	Mean	1.32E+01	1.89E+01	1.32E−02	4.26E−15	3.78E−15	**1.48E−16**
		SD	4.90E−01	8.50E−01	1.30E−02	1.44E−15	9.01E−16	**2.35E−17**
Griewank (f4)	50	Mean	5.81E+00	1.34E+02	5.72e−01	0	0	0
		SD	9.13e−01	2.41E+01	9.22e−01	0	0	0
	100	Mean	1.78E+02	7.93E+02	1.31E+01	0	0	0
		SD	1.98E+01	7.96E+01	6.30E+00	0	0	0
Rastrigin (f5)	50	Mean	3.76E+01	2.72E+02	4.72e−01	1.63E+01	6.50E+00	0
		SD	4.87E+00	3.27E+01	4.92e−01	2.79E+01	1.99E+01	0
	100	Mean	3.15E+02	6.49E+02	1.46E+01	0	0	0
		SD	2.33E+01	4.52E+01	4.18E−00	0	0	0

Table 5b Gist of Table 5a (out of five problems)

TSO	HS		IBA		ABC		TLBO		ITLBO	
	Mean	SD	Mean	SD	Mean	SD	Mean	SD	Mean	SD
Wins	**10**	**10**	**10**	**10**	**10**	**10**	**5**	3	**5**	3
Equals	–	–	–	–	–	–	3	3	3	3
Losses	–	–	–	–	–	–	2	**4**	2	**4**

Table 6a Comparison of TSO with the algorithms in [16]

Functions	Dim.	Statistical features	GABC (C-1.5)	E-ABC	ABC/best/2	ABC/best/1	TSO
Sphere (f1)	30	Mean	4.17E−16	1.67E−16	1.70E−126	1.10E−150	**5.87312e−397**
		SD	7.36E−17	2.70E−16	2.70E−126	1.40E−150	**2.84687e−396**
	60	Mean	1.43E−15	1.41E−15	3.72E−58	4.40E−69	**1.48E−245**
		SD	1.37E−16	1.82E−15	2.67E−58	2.56E−69	**3.13E−245**
Rosenbrock (f2)	2	Mean	1.68E−04	4.63E−04	4.42E−04	4.99E−06	**0**
		SD	4.42E−04	4.57E−04	2.39E−04	8.22E−06	**0**
	3	Mean	2.65E−03	1.20E−02	9.90E−04	5.52E−06	**0**
		SD	2.22E−03	7.06E−03	6.92E−04	3.03E−06	**0**
Ackley (f3)	30	Mean	3.21E−14	1.22E−10	2.50E−14	1.72E−14	**3.16E−17**
		SD	3.25E−15	4.86E−11	3.48E−15	2.84E−15	**1.07E−17**
	60	Mean	1.66E−13	1.55E−07	7.12E−14	6.62E−14	**8.83E−17**
		SD	2.21E−14	2.84E−08	4.14E−15	1.74E−15	**3.33E−17**
Griewank (f4)	30	Mean	2.96E−17	4.90E−14	**0**	**0**	**0**
		SD	4.99E−17	7.31E−03	**0**	**0**	**0**
	60	Mean	7.54E−16	4.19E−14	**0**	**0**	**0**
		SD	4.12E−16	9.05E−03	**0**	**0**	**0**
Rastrigin (f5)	30	Mean	1.32E−14	9.97E−15	**0**	**0**	**0**
		SD	2.44E−14	3.87E−15	**0**	**0**	**0**
	60	Mean	3.52E−13	7.51E−13	**0**	**0**	**0**
		SD	1.24E−13	6.15E−13	**0**	**0**	**0**
Schaffer (f11)	2	Mean	**0**	**0**	**0**	**0**	**0**
		SD	**0**	**0**	**0**	**0**	**0**
	3	Mean	1.85E−18	2.79E−07	3.56E−06	**0**	**0**
		SD	1.01E−17	2.24E−07	1.27E−06	**0**	**0**

Table 6b Gist of Table 6a (out of six problems)

TSO	GABC (C-1.5)		E-ABC		ABC/best/2		ABC/best/1	
	Mean	SD	Mean	SD	Mean	SD	Mean	SD
Wins	**12**	**12**	**12**	**12**	**12**	**12**	**12**	**12**
Equals	–	–	–	–	–	–	–	–
Losses	–	–	–	–	–	–	–	–

comparison, all the parameters are kept the same as in [16], viz. the stopping criteria is either to get maximum number of function evaluation 400,000 or a maximum of 5000 generations is obtained. For all the problems, the NS is fixed at 80. Each of the six problems has been solved for two different dimensions as reported in Table 6a. A total of 30 runs are executed to solve the problems by TSO, and the performance of TSO is compared in Table 6a. The gist of Table 6a is presented in Table 6b.

Discussion of Results (for Case Study-2):
From Table 6a, it is clearly observed that TSO provides a reasonable improved result over the others in terms of both mean and SD of the objective function for different dimensions of each considered problem. TSO never goes worst in all the cases. Moreover, in order to observe the convergence rate of the proposed TSO over different variants of ABC are considered for comparison. Starting from the same seed, both the programmes are allowed to run. The reason behind the selection of such a set is that it is just the mixture of each type of functions like four functions from Set-1 and one function from each of the rest sets. Therefore, even under diversified dimensions, the better stability of TSO is concluded as it impacts either equal or very less SDs everywhere.

4 Conclusion

In this paper, an entirely new and robust optimization technique, namely 'Troop Search Optimization (TSO) algorithm', is proposed. Employment of efficient operators in TSO like 'swapping crossover', 'simplex search method' and 'cut and fill' possibly makes TSO faster and better. Incapable of eradicating/destroying multiple food sources of same nectar amounts in ABC has been taken care by TSO by using non-redundant search through modified Quadratic Approximation. From the result and discussion, it can be concluded that TSO is more robust than ABC. TSO is **efficient** as it provides better objective function values in most of the cases and it is also **reliable** because of high success rate. The impact of mostly less SDs of TSO implies better **stability** over others. Therefore as a whole, TSO is more efficient, more reliable and more stable than the recent variants of the state-of-the-art algorithms.

References

1. Rao, R.V., Savsani, V.J., Vakharia, D.P.: Teaching-learning-based optimization: a novel method for constrained mechanical design optimization problems. Comput. Aided Des. **43**(3) (2011)
2. Rao, R.V., Savsani, V.J., Vakharia, D.P.: Teaching-learning-based optimization: a novel optimization method for continuous non-linear large scale problems. Inf. Sci. **183**(1) (2012)
3. Rao, R.V., Patel, V.: An elitist teaching-learning-based optimization algorithm for solving complex constrained optimization problems. Int. J. Ind. Eng. Comput. **3**(4) (2012)
4. Rao, R.V., Patel, V.: Multi-objective optimization of heat exchangers using a modified teaching-learning-based optimization algorithm. Appl. Math. Model. **37**(3) (2013)
5. Rao, R.V., Patel, V.: Multi-objective optimization of two stage thermoelectric cooler using a modified teaching–learning-based optimization algorithm. Eng. Appl. Artif. Intell. **26**(1) (2013)
6. Baykasoˇglu, A., Hamzadayi, A., Köse, S.Y.: Testing the performance of teaching-learning based optimization (TLBO) algorithm on combinatorial problems: flowshop and job shop scheduling cases. Inf. Sci. **276** (2014)
7. Satapathy, S.C., Naik, A.: Modified Teaching-learning-based optimization algorithm for global numerical optimization–a comparative study. Swarm Evol. Comput. **16**, 2014 (2014)
8. Waghmare, G.: Comments on a note on teaching-learning based optimization algorithm. Inf. Sci. **229**(20), 2013 (2013)
9. Zou, F., Wang, L., Hei, X.L., Chen, D.B., Yang, D.D.: Teaching-learning-based optimization with dynamic group strategy for global optimization. Inf. Sci. **273** (2014)
10. Deb, K.: Optimization for Engineering Design. PHI (2010)
11. Zhang, H., Ishikawa, M.: An extended hybrid genetic algorithm for exploring a large search space. In: 2nd International Conference on Autonomous Robots and Agents, Palmerston, North New Zealand (2014)
12. Das, K.N., Singh, T.K.: Drosophila food-search optimization. Appl. Math. Comput. **231** (2014)
13. Mohan, C., Shankar, K.: A random search technique for global optimization based on quadratic approximation. Asia Pac. J. Oper. Res. **11** (1994)
14. Deep, K., Das, K.N.: Quadratic approximation based hybrid genetic algorithm for function optimization. Appl. Math. Comput. **203** (2008)
15. Yu, K., Wang, X., Wang, Z.: An improved teaching-learning based optimization algorithm for numerical and engineering optimization problems. J. Intell. Manuf. **27**(4) (2016)
16. Gao, W., Liu, S., Huang, L.: A global best artificial bee colony algorithm for global optimization. J. Comput. Appl. Math. **236** (2012)

Hybrid Grey Wolf Optimizer with Mutation Operator

Shubham Gupta and Kusum Deep

Abstract Grey Wolf Optimizer (GWO), developed by Mirjalili et al. (Adv Eng Softw 69:46–61, 2014 [1]), is a recently developed nature-inspired technique based on leadership hierarchy of grey wolves. In this paper, Grey Wolf Optimizer has been hybridized with differential evolution (DE) mutation, and two versions, namely DE-GWO and gDE-GWO, have been proposed to avoid the stagnation of the solution. To evaluate the performance of both the proposed versions, a set of 23 well-known benchmark problems has been taken. The comparison of obtained results between original GWO and proposed hybridized versions of GWO is done with the help of Wilcoxon signed-rank test. The results conclude that the proposed hybridized version gDE-GWO of GWO has better potential to solve these benchmark test problems compared to GWO and DE-GWO.

Keywords Grey Wolf Optimizer (GWO) · Differential evaluation (DE)
Numerical optimization

1 Introduction

In order to solve complex real-life optimization problems, many nature-inspired optimization algorithms have been used, and they show their great potential to solve optimization problems, for example genetic algorithm (GA) [2], differential evolution (DE) [3], particle swarm optimization (PSO) [4], Grey Wolf Optimizer (GWO) [1], artificial bee colony (ABC) algorithm [5], harmony search [6]. The advantage of these nature-inspired numerical approaches is that they are applicable to those problems

S. Gupta (✉) · K. Deep
Department of Mathematics, Indian Institute of Technology Roorkee, Roorkee 247667,
Uttarakhand, India
e-mail: g.shubh93@gmail.com

K. Deep
e-mail: kusumfma@iitr.ac.in; kusumdeep@gmail.com

© Springer Nature Singapore Pte Ltd. 2019
J. C. Bansal et al. (eds.), *Soft Computing for Problem Solving*, Advances in Intelligent
Systems and Computing 817, https://doi.org/10.1007/978-981-13-1595-4_75

also where the objective functions are non-differentiable even they are discontinuous as deterministic techniques fail in that cases.

Grey Wolf Optimizer (GWO) developed by Mirjalili et al. [1] is recently developed algorithm to solve real-world optimization problems in the field of nature-inspired techniques. This algorithm like other nature-inspired algorithms starts with a randomly generated population of individual solutions; here the individuals are termed as wolf. As described by Mirjalili et al. [1], grey wolf pack consists of four types of wolf: alpha, beta, delta and omega wolves, in which alpha, beta and delta are called leaders and omega wolf updates their positions with the help of these leaders. GWO has been applied to various optimization application problems [7–10], in which GWO shows their potential as it is very efficient algorithm. One of the recent improvements based on random walk has been done [11] to enhance the performance of GWO algorithm. There are many other modifications that have been done in the literature [12–15] to improve the performance of GWO algorithm.

Mirjalili et al. [1] mathematically model the hunting strategies and social behaviour of wolves as follows.

In order to mathematically model the social behaviour of wolves when GWO is designed, the best solutions are considered as alpha, second and third best solutions are called beta, and delta and rest solutions are assumed as omega solutions.

The hunting strategy of the wolves starts with encircling the prey that was mathematically modelled as

$$y(t + 1) = y(t) - A * d \tag{1}$$
$$d = |c * P(t) - y(t)| \tag{2}$$

where t is iteration count and $P(t)$ is a position of the prey at tth iteration. $y(t)$ and $y(t + 1)$ are the positions of wolf at tth and $t + 1$th iteration. The coefficients c and A are defined as

$$c = 2 * rand_1 \tag{3}$$
$$A = 2 * a * rand_2 - a \tag{4}$$

where a is linearly decreasing vector from 2 to 0 and $rand_1$ and $rand_2$ are the uniformly distributed random vectors between 0 and 1. These coefficient vectors c and A are responsible for exploration and exploitation which are the important factors in any nature-inspired technique.

In the second phase of hunting process of wolves, prey position can be approximated with the help of leaders, alpha, beta and delta, and the position of the wolf for the next iteration can be calculated as

$$y_1 = y_{alpha} - A_1 * d_1 \tag{5}$$
$$y_2 = y_{beta} - A_2 * d_2 \tag{6}$$
$$y_3 = y_{delta} - A_3 * d_3 \tag{7}$$

$$y(t + 1) = (y_1 + y_2 + y_3)/3 \tag{8}$$

In this way, each wolf updates their position with the help of above equations.

2 Hybridized GWO with DE Mutation

On analyzing the performance of GWO on test problem, it can be seen that in some cases GWO is trapped in local optima, and therefore, GWO can be hybridized with some genetic operators so that the problem of stagnation can be avoided. One attempt in the present work has been done in this direction by incorporating differential evolution mutation.

As in differential evolution (DE) [3], a mutant vector v for any target vector x is generated as

$$v = x_{r_1} + F * \left(x_{r_2} - x_{r_3} \right) \tag{9}$$

where x_{r_1}, x_{r_2} and x_{r_3} are the random vectors of the population, and $F \in [0\ 2]$ is a factor that controls the differential variation $\left(x_{r_2} - x_{r_3} \right)$. There are many mutation strategies available in the literature [16] to generate a mutant vector. To hybridize GWO with DE mutation, DE/best/1 strategy has been adopted to generate a mutant vector (wolf) as

$$v = x_{best} + F * \left(x_{r_2} - x_{r_3} \right) \tag{10}$$

where x_{best} is the best wolf having best fitness, and x_{r_1} and x_{r_2} are the random wolves of the pack.

In this paper, this hybridized version is named as DE-GWO. To maintain a balance between exploration and exploitation, greedy approach has been incorporated in DE-GWO. This hybridized version with greedy approach is named as gDE-GWO. The pseudo code of gDE-GWO has been presented in Fig. 1.

3 Analysis of the Results

A set of 23 well-known benchmark problems has been taken to evaluate the performance of GWO, DE-GWO and gDE-GWO algorithms. These benchmark problems also have been used in original paper to evaluate the performance of GWO. For the fair comparison, population size and maximum number of iterations have been taken same as in original paper [1].

Initialize the population of wolves
Evaluate the fitness of population of wolves
Pick the alpha, beta and delta solutions from the population
while *current itrration < max iteration*
    for each wolf of the population
        *find the new position with the help of equation (**8**)*
        apply greedy approach for the new position and previous
        position
        apply DE mutation for randomly selected target wolf y_{rand}
        If this mutant wolf is better than replace it with the wolf y_{rand}
    end *of for loop*
end *of while loop*

Fig. 1 Pseudo code of gDE-GWO

The obtained results (average and standard deviation of objective function values) on these benchmark problems by GWO, DE-GWO and gDE-GWO are reported in Table 1. Functions from F1 to F7 are unimodal problems, F8–F13 are multimodal functions, and problems from F14 to F23 are fixed-dimension multimodal functions in the benchmark set. In the unimidal problems F1–F4, gDE-GWO and DE-GWO outperform GWO, and in F7, gDE-GWO performs better compared to other algorithms. In F5 and F6, GWO is better compared to other algorithms.

gDE-GWO performs better compared to other algorithms in all the multimodal problems except F13. In F13, GWO is better than other algorithms. Also in most of the fixed-dimensional multimodal problems, gDE-GWO performs better compared to DE-GWO and GWO except F19 and F21. In these problems, GWO has better performance than the others.

In problems F1–F4, F9, F1, F16–F19, gDE-GWO gives optima. From analyzing the results reported in Table 1, it can be concluded that gDE-GWO outperforms GWO and DE-GWO.

4 Statistical Analysis

For the analysis of the results that are reported in Table 1, Wilcoxon signed-rank test has been used. The obtained results are presented in Table 2. '+' indicates the better performance, '−' indicates the worst performance, and '=' sign shows that both the algorithms are statistically same. Statistical analysis of the results obtained by algorithms is necessary to observe the significant performance.

Table 1 Average and standard values of the objective function on 23 benchmark problems by GWO, DE-GWO and gDE-GWO algorithms

Category	Function	Optima	GWO		DE-GWO		gDE-GWO	
			Average	STD	Average	STD	Average	STD
Unimodal functions	F1	0	7.17E−28	1.33E−27	**0**	0	**0**	0
	F2	0	7.56E−17	5.92E−17	**0**	0	**0**	0
	F3	0	5.77E−05	2.41E−04	**0**	0	**0**	0
	F4	0	9.50E−07	1.71E−06	**0**	0	**0**	0
	F5	0	**27.11930**	0.84	27.64229	0.59	27.6639	1.01
	F6	0	**0.77110**	0.44	1.04262	0.43	1.06009	0.49
	F7	0	0.0018395	9.67E−04	0.030967	0.03	**0.00083**	6.42E−04
Multimodal functions	F8	−12569.487	−6122.073	764.68	−1244.766	307.61	**−6187.133**	1601.37
	F9	0	2.98046	3.57	**0**	0	**0**	0
	F10	0	1.05E−13	1.84E−14	**8.88E−16**	4.01E−31	**8.88E−16**	4.01E−31
	F11	0	0.00540	0.01	**0**	0	**0**	0
	F12	0	0.06485	0.09	0.07677	0.03	**0.04289**	0.02
	F13	0	**0.72232**	0.19	1.35758	0.49	2.55692	0.28
Fixed-dimension multimodal functions	F14	1	3.61588	3.50	12.35073	1.22	**2.86225**	3.27
	F15	0.0003	0.0050515	8.59E−03	0.0027990	6.11E−03	**0.0024592**	6.07E−03
	F16	−1.0316	**−1.0316**	6.78E−16	**−1.0316**	6.78E−16	**−1.0316**	6.78E−16
	F17	0.398	**0.39789**	1.69E−16	**0.39789**	1.69E−16	**0.39789**	1.69E−16
	F18	3	3.00005	6.30E−05	3.0001	4.58E−04	**3**	0
	F19	−3.86	**−3.86178**	2.20E−03	−3.85598	2.22E−03	−3.86191	2.16E−03
	F20	−3.32	−3.26028	0.08	−3.19798	0.13	**−3.26250**	0.08
	F21	−10.1532	**−9.64456**	1.54	−7.60391	2.59	−8.97953	2.44
	F22	−10.4028	−10.11598	1.56	−9.64828	1.97	**−10.14787**	1.39
	F23	−10.5363	−10.29087	1.33	−10.53565	5.03E−04	**−10.53575**	3.92E−04

5 Convergence Graphs

We have plotted convergence graph for the functions F7, F10, F12, F14, F15 and F23 corresponding to average of the objective function values obtained in 30 runs for different iterations in Fig. 2. These graphs show that gDE-GWO has better convergence than GWO. As from the statistical analysis, it is clear that gDE-GWO performed better than GWO and DE-GWO. Therefore, these convergence graphs show that gDE-GWO speed up the convergence of GWO with the incorporation of mutation operator and greedy approach.

Table 2 Pairwise comparison of statistical results

Function	DE-GWO/GWO	gDE-GWO/GWO	gDE-GWO/GWO
F1	+	+	=
F2	+	+	=
F3	+	+	=
F4	+	+	=
F5	−	−	=
F6	−	−	=
F7	−	+	+
F8	−	=	+
F9	+	+	=
F10	+	+	=
F11	+	+	=
F12	−	=	+
F13	−	−	−
F14	−	=	+
F15	=	=	+
F16	=	=	=
F17	=	=	=
F18	=	+	=
F19	−	=	+
F20	−	=	+
F21	−	=	+
F22	+	+	=
F23	+	+	=
±	9/10	11/3	8/1

6 Conclusion

This paper presents hybridized versions of Grey Wolf Optimizer with DE mutation, namely DE-GWO and gDE-GWO. These hybridized versions have been evaluated on 23 well-known benchmark problems. The analysis of the obtained results has been done by Wilcoxon signed-rank test that concludes that gDE-GWO outperforms GWO and DE-GWO. Therefore, to solve these benchmark problems, gDE-GWO is recommended over GWO. For the problem F5, all the algorithms reported in this paper completely fail. Therefore, another genetic operator can be incorporated in GWO to improve its performance.

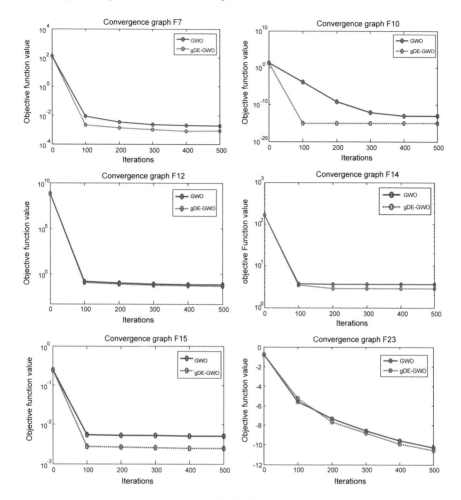

Fig. 2 Convergence graph of the functions F7, F10, F15, F14, F15 and F23

References

1. Mirjalili, S., Mirjalili, S.M., Lewis, A.: Grey wolf optimizer. Adv. Eng. Softw. **69**, 46–61 (2014)
2. Holland, J.H.: Adaptation in natural and artificial systems: an introductory analysis with applications to biology, control, and artificial intelligence. MIT press (1992)
3. Holland, J.H.: Adaptation in Natural and Artificial Systems: An Introductory Analysis with Application to Biology, Control, and Artificial Intelligence. Ann Arbor, MI: University of Michigan Press (1975)
4. Kennedy, J.: Particle swarm optimization. In: Encyclopedia of Machine Learning, pp. 760–766. Springer US (2011)
5. Karaboga, D.: Artificial bee colony algorithm. Scholarpedia **5**(3), 6915 (2010)
6. Geem, Z.W., Kim, J.H., Loganathan, G.V.: A new heuristic optimization algorithm: harmony search. Simulation **76**(2), 60–68 (2001)

7. Mirjalili, S.: How effective is the grey wolf optimizer in training multi-layer perceptrons. Appl. Intell. **43**(1), 150–161 (2015)
8. Song, X., Tang, L., Zhao, S., Zhang, X., Li, L., Huang, J., Cai, W.: Grey wolf optimizer for parameter estimation in surface waves. Soil Dyn. Earthq. Eng. **75**, 147–157 (2015)
9. Hong M.S., Mohd Herwan, S., Mohd Rusllim, M.: An application of grey wolf optimizer for solving combined economic emission dispatch problems. Int. Rev. Model. Simul. (IREMOS) **7**(5), 838–844 (2014)
10. Madadi, A., Motlagh, M.M.: Optimal control of DC motor using grey wolf optimizer algorithm. TJEAS J. 2014-4-04/373-379 4 (4), 373–379 (2014)
11. Gupta, S., Deep, K.: A novel random walk grey wolf optimizer. Swarm Evol. Comput. BASE DATA (2018). https://doi.org/10.1016/j.swevo.2018.01.001
12. Saremi, S., Mirjalili, S.Z., Mirjalili, S.M.: Evolutionary population dynamics and grey wolf optimizer. Neural Comput. Appl. **26**(5), 1257–1263 (2015)
13. Muangkote, N., Sunat, K., Chiewchanwattana, S.: An improved grey wolf optimizer for training q-Gaussian Radial Basis Functional-link nets. In: 2014 International Computer Science and Engineering Conference (ICSEC), pp. 209–214. IEEE (2014)
14. Heidari, A.A., Pahlavani, P.: An efficient modified grey wolf optimizer with Lévy flight for optimization tasks. Appl. Soft Comput. **60**, 115–134 (2017)
15. Zhu, A., Chuanpei, X., Li, Z., Jun, W., Liu, Z.: Hybridizing grey wolf optimization with differential evolution for global optimization and test scheduling for 3D stacked SoC. J. Syst. Eng. Electron. **26**(2), 317–328 (2015)
16. Qin, A.K., Huang, V.L., Suganthan, P.N.: Differential evolution algorithm with strategy adaptation for global numerical optimization. IEEE Trans. Evol. Comput. **13**(2), 398–417 (2009)

Design a New Protocol and Compare with BB84 Protocol for Quantum Key Distribution

Manish Kalra and Ramesh C. Poonia

Abstract Quantum key distribution is the latest advancement in quantum cryptography. There are several QKD protocols like BB84, B92, Ekert91, COW, SARG04, out of which BB84 is the first protocol developed in 1984. In this paper, we are discussing first about the working of BB84 protocol and then proposing a new protocol which is a variation over BB84 protocol, second the design of simulation setup is discussed, and then we compared the performance of BB84 with the proposed protocol and proved it much better in case of capacity and error estimation. Object-oriented approach is used in the simulation designing for new protocol and BB84 protocol.

Keywords BB84 simulation · QKD protocols · Quantum cryptography

1 Introduction

Quantum computing is the latest advancement in the field of computing. Besides, the computing quantum communication is also the latest theory in the communication field. Quantum communication is based on no-cloning [1] theory of quanta, and quanta are a packet of energy associated with photon. No-Cloning theorem states that when one sender send a polarized photon to the receiver and interrupted by the eavesdropper. Eavesdropper cannot regenerate the polarized photon as same as it was before detection. Quantum communication involves quantum key distribution, quantum teleportation, and quantum repeaters [2]. QKD is the part of quantum cryptography. There are several protocols for QKD [3] which includes BB84, BBM92, Ekert91, COW, SARG04. These protocols can be classified according to the principle used, i.e., based on Heisenberg's uncertainty principle, based on quantum entanglement and based on public–private cryptography.

M. Kalra (✉) · R. C. Poonia
Amity Institute of Information Technology, Amity University Rajasthan, Jaipur, India
e-mail: kalramanish83@gmail.com

R. C. Poonia
e-mail: rameshcpoonia@gmail.com

© Springer Nature Singapore Pte Ltd. 2019
J. C. Bansal et al. (eds.), *Soft Computing for Problem Solving*, Advances in Intelligent Systems and Computing 817, https://doi.org/10.1007/978-981-13-1595-4_76

Further, the paper is planned as follows: Firstly in Sect. 2, we have related work of research. Section 3 highlights the QKD protocol like BB84 protocol and proposed protocol, i.e., used in our research simulation and analysis to help simulation and different environments, performance metrics described. Section 4 represents simulation design. Section 5 shows the simulation result, and at the end of this paper conclusion and future work is included in Sect. 6.

2 Related Work

The revolution in quantum cryptography starts in 1983 when S. Wiesner gave the idea of conjugate coding [4]. It shows that in compensation for "quantum noise," quantum mechanics allows novel forms of coding without analog in communication channels adequately described by classical physics. In 1984, C. H. Bennett, G. Brassard used an uncertainty principle of quantum physics to encode the information in non-orthogonal quantum states and showed that quantum coding has been used in conjunction with public key cryptography by permitting secure distribution of random key information between parties who share no secret information initially [5]. Then, the "Quantum Cryptography Based on Bell's Theorem," was proven by Artur K. Ekert as Physics Review Letters in August 1991 [6]. "Experimental Quantum Cryptography," by Charles H. Bennett, John Smith, in Journal of Cryptology, 1992, shows the practical aspect of quantum cryptography. Again in August 1992, "Practical Quantum Cryptography Based on Two-Photon Interferometry" [7] by Artur K. Ekert, G. Massimo Palma in Physics Review Letters, of the American Physical Society gives an added proof. The article "Quantum Information and Computation" by Bennett and DiVincenzo, Macmillan [2] shows that quantum information theory extends the classical information theory. Quantum properties are uncertainty, interference, and entanglement. "Integration of Quantum Cryptography in 802.11 Networks" by Nguyen et al. [8] analyzes the interest of using quantum technique for the distribution of encryption keys in 802.11 wireless networks. They also proposed a scheme for the integration of quantum cryptography in 802.11i security mechanism for the establishment of the pairwise transient key (PTK). Quantum communication is the art of transferring a quantum state from one place to another. Examples are: quantum key distribution, quantum non-locality, quantum teleportation, communication complexity, and quantum bit string commitment. Entanglement is exploited to prepare the desired quantum state at a distance [9]. BB84 and B92 [10] protocols were discussed in context of Wi-Fi networks. Capacity measures the amount of information that can be stored in a quantum memory or transmitted through a quantum communication channel [11].

A new scheme for quantum key distribution in free space is proposed by Cui et al. [12] in Proceedings of the 15th Asia-Pacific Conference on Communications in year 2009. This scheme was combining MIMO system and wireless networks for QKD in free space networks, and with this approach a higher bit rate QKD in networks in free space could be realized theoretically. "Fuzzy Dynamic Switching in Quantum Key Distribution for Wi-Fi Networks" by Huang et al. [13], 2009 Sixth International Conference on Fuzzy Systems and Knowledge Discovery, covers the focus on two-dimensional (2D) fuzzy dynamic switching between the transmission and receiver for long distance. Xu Huang, Shirantha Wijesekera, and Dharmendra Sharma, in ICNSS, 2010, first discussed how QKD can be used in IEEE 802.11 wireless networks to securely distribute the keys, and second they used a new QKD protocol that takes the advantage of mutual authentication features offered by some ERP variants of 802.1X port-based network access control.

Finally, they present a new code called quantum message integrity code (Q-MIC) which provides mutual authentication between the two communication parties. Experimental results are presented with Simulink model [14].

A survey is done on quantum key distribution protocols by Elboukhari et al. [15] in 2010 which includes the protocols like BB84, B92, EPR protocol, differential phase shift quantum key distribution, COW protocol, SARG04 protocol, protocol with private–public key.

Lalu Naik et al. [16] discussed the first two stages of QKD for Wi-Fi are implemented for B92 protocol in C++ language on Linux platform. GNU has been used as the compiler. This setup has been successfully tested with multiple index file at the University of Canberra test lab.

"Cryptography using QKD in Wireless Networks" [17] by Katakam and Lakshmi Reddy introduced a novel security primitive that enables message authentication in wireless networks without the use of pre-established or pre-certified keys.

The title of research paper "Quantum Cryptography with Key Distribution in Wireless Network" [18] written by Premlata Sonawane, Leena Ragha replaced the four-way handshake protocol of the existing IEEE 802.11 with the QKD system-based four-phase handshake protocol. The main purpose of proposed QKD protocol is to make it much harder for an eavesdropper to conceal his presence.

3 QKD Protocol

QKD has some basic protocols and several modifications on these protocols out of which here is discussion on one basic and a proposed protocol for QKD.

3.1 BB84 Protocol

It is first protocol for QKD given by Brassard and Bennett [5, 6] in 1984. It uses Heisenberg's uncertainty principle. It is a four-state protocol and uses two polarization bases: First is rectilinear 0 and 1, and another is diagonal 0 and 1. The working of BB84 includes four states of quantum information transmission, error correction, key shifting, and privacy amplification. In quantum information transmission, first sender selects random basis for modulation and encodes on basis of random bits.

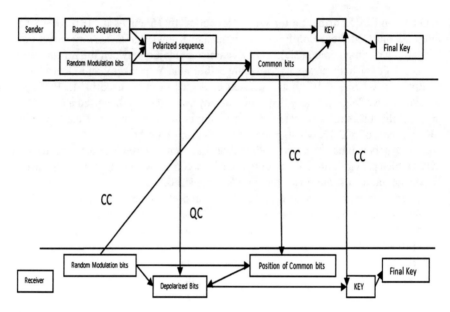

Fig. 1 Working of BB84 protocol. *Note QC* quantum channel; *CC* classical channel

After polarization, it sends quantum bits over quantum channel. On the other side, receiver selects its own random basis for depolarization and decodes on basis of random bits. This complete process happens on quantum channel. After this, classical channel is used in transmission. Receiver shares random basis with sender on classical channel, and sender corrects the positions of common bits and sends back to receiver. Finally, some bit positions are ignored to amplify the privacy. The complete working of BB84 protocol is illustrated in Fig. 1.

3.2 Proposed Protocol

In proposed protocol, both sender and receiver select random basis for modulation and encode on basis of random bits, both send qbits over quantum channel to each other, and then both decode on basis of their own random bits. Receiver and sender exchange their random basis and correct the positions of common bits. Later, some bits can be ignored to amplify the privacy. In this process, both sender and receiver get two keys and now the final key can be generated by combining them. The complete working of proposed protocol is illustrated in Fig. 2.

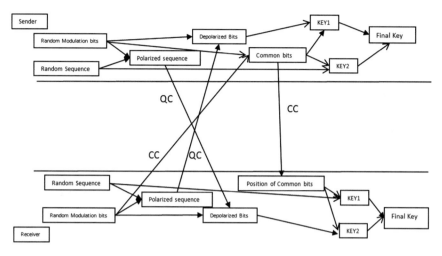

Fig. 2 Working of proposed protocol. *Note QC* quantum channel; *CC* classical channel

4 Simulation Design

Simulation setup is designed using DOS environment and C++ language as front end. Random bits are generated, and using them the protocols are implemented. Simulation is designed so that one can run the code for different lengths of bits, i.e., for 8, 16, 32, 64, 128, 256 bits, and for each bit length the code is executed for 50, 100, 150, 200, 250, 300 times.

4.1 Design of BB84 Protocol

Object-oriented approach is used in designing the code for BB84 protocol. Three classes were implemented SENDER, RECEIVER and EAVESDROPPER in C++ with following functionality. SENDER uses one random sequence bits and other random modulation bits, while RECEIVER uses only random modulation/demodulation bits. The functions, data members, and their description are shown in Tables 1, 2, and 3, respectively.

4.2 Design of Proposed Protocol

Object-oriented approach is used in designing the code for proposed protocol. Two classes were implemented ENTITY and EAVESDROPPER in C++ with following functionality. ENTITY plays the role of both sender and receiver in communication,

Table 1 SENDER class members and their description

SENDER class members	Description
int init_qbit[MAX_BIT]	Initial sequence at sender
int rand_modu[MAX_BIT]	Random modulation selector bits
int rr_demodu[MAX_BIT]	Random modulation bits send by receiver
int polarized[MAX_BIT]	Quantum bits to be sent by sender
int KEYQ[MAX_BIT], key_lenQ	Key as identified through QC process
void randgenr()	Function generating random modulation bits
SENDER()	Constructor
int* qbit_send()	Sending quantum bits
void modulate()	Modulating the sequence
void classic_rece(int rr[])	Receiving of random sequence of receiver through classical channel
int sharekeypos(int a[])	Sharing key positions on classical channel
int* sendkey()	Sending the original key on classical channel
int sendkeylen()	Sending the original key length on classical channel

Table 2 RECEIVER class members and their description

RECEIVER class members	Description
int rand_demodu[MAX_BIT]	Random bits for demodulation
int rece_qbit[MAX_BIT]	Qbits received
int depolarized[MAX_BIT]	Generated sequence after demodulation
int rec_keypos[MAX_BIT + 1]	Received key positions by sender
int KEYQ[MAX_BIT]	KEY as received by quantum channel
int KEYC[MAX_BIT]	KEY as received by classical channel
int key_lenQ, key_lenC	Length of keys
void randgenr()	Function generating random bits for demodulation
void qreceive(int r[])	Receiving the bits through quantum channel
void depolarize()	Extracting the encrypted data
int* classic_send()	Sending the random sequence to sender
void reckey(int* key, int len)	Receiving key on classical channel
void getkeypos(int* key, int len)	Receive key positions
int checkifeve()	Checking the presence of EVE

Table 3 EAVESDROPPER class members and description

EAVESDROPPER class members	Description
int eve_rand[MAX_BIT]	Eavesdropper random bits
int eve_qbits[MAX_BIT]	Eavesdropper sequence detected
int eve_polarize[MAX_BIT]	Quantum bits send by Eve
Eavesdropper()	Constructor
void eve_capture(int* sa)	Eve capturing in between
int* eve_resend()	Eve resending quantum bits after generating

and EAVESDROPPER is the interceptor between two ENTITIES. The functions, data members, and their description are shown in Table 4.

Table 4 ENTITY class members and description

ENTITY class members	Description
int init_qbit[MAX_BIT]	Initial sequence at sender
int rand_modu[MAX_BIT]	Random modulation selector bits
int rr_demodu[MAX_BIT]	Random modulation bits send by receiver
int polarized[MAX_BIT]	Quantum bits to be sent by sender
int KEYQ1[MAX_BIT], key_lenQ1, KEYQ2[MAX_BIT], key_lenQ2	Keys as identified through QC process
int rece_qbit[MAX_BIT]	Qbits received
int depolarized[MAX_BIT]	Generated sequence after demodulation
int rec_keypos[MAX_BIT + 1]	Received key positions by sender
int KEYC[MAX_BIT]	KEY as received by classical channel
int key_lenC	Length of keys
void randgenr()	Function generating random modulation bits
ENTITY()	Constructor
int* qbit_send()	Sending quantum bits
void modulate()	Modulating the sequence
void classic_rece(int rr[])	Receiving of random sequence of receiver through classical channel
int sharekeypos(int a[])	Sharing key positions on classical channel
int* sendkey()	Sending the original key on classical channel
int sendkeylen()	Sending the original key length on classical channel

(continued)

Table 4 (continued)

ENTITY class members	Description
void qreceive(int r[])	Receiving the bits through quantum channel
void depolarize()	Extracting the encrypted data
int* classic_send()	Sending the random sequence to sender
void reckey(int* key, int len)	Receiving key on classical channel
void getkeypos(int* key, int len)	Receive key positions
int checkifeve()	Checking the presence of EVE

5 Simulation Results

In the experiment, random bits are generated initially using different initial lengths starting from 8 bits to 256 bits in multiple of 2.

Experiment is repeated 300 times in slots of 50–300 with step value 50. Simulation is done for both BB84 protocol and proposed protocol in DOS environment in C++ language.

The results are shown in Tables 5 and 6. Table 5 has data corresponding to the length of final key received, and Table 6 has data corresponding to the error rate estimated. The first graph is between number of bits generated initially and the average key length received for BB84 protocol and proposed protocol (MKP16) in Fig. 3.

Graph shows that the proposed protocol is twice as capacitive in nature as compared to BB84 protocol. As the value of Average Key Length for 8 bit is 4.077381 for the BB84 protocol and 7.666667 for MKP16 which is almost double. The second graph is between number of bits generated initially and the error rate estimated in Fig. 4. The error rate is almost half in case of proposed protocol compared to BB84 protocol, as the error rate for 256 bits is 0.4964974 for BB84 and is 0.254688 (half) for proposed protocol.

Table 5 Average key length for BB84 and MKP16

Number of bits generated initially	Average key length (BB84)	Average key length (MKP16)
8	4.077381	7.666667
16	7.650685	15.903614
32	15.78378	32.953488
64	31.58824	63.472527
128	63.89	126.722892
256	127.1316	256.444444

Table 6 Error rate estimated for BB84 and MKP16

Number of bits generated initially	Error rate estimated (BB84)	Error rate estimated (MKP16)
8	0.538203	0.5625
16	0.487781	0.386029
32	0.493076	0.345982
64	0.493897	0.298611
128	0.515025	0.266085
256	0.496497	0.254688

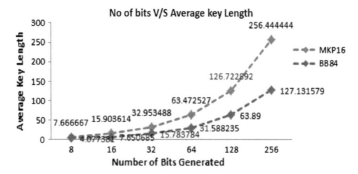

Fig. 3 Graph between numbers of bits generated and average key length

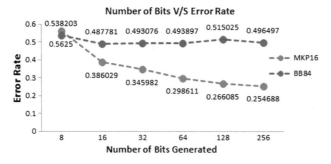

Fig. 4 Graph between number of bits generated and error rate

6 Conclusion and Future Work

After the complete process of simulation and finding the results here we can conclude that the proposed protocol is far better than the BB84 in terms of capacity and error estimation. In this simulation, the final key is generated using the concatenation of two keys, and there are possibilities for combining two keys in other ways also, i.e., multiplying two keys instead of adding them. The future work can be done with altering the B92 protocol with same proposed scheme and then comparing their performance.

References

1. Cobourne, S.: Quantum Key Distribution Protocols and Applications, Surrey TW20 0EX, England (2011)
2. Bennett, C.H., DiVincenzo, D.P.: Quantum information and computation. Nature **404**(6775), 247–255 (2000)
3. Elboukhari, M., Azizi, M., Azizi, A.: Quantum key distribution protocols: a survey. Int. J. Univers. Comput. Sci. **1**(2), 59–67 (2010)
4. Wiesner, S.: Conjugate coding. SIGACT News **15**(1), 78–88 (1983)
5. Bennett, C.H.: Quantum cryptography: public key distribution and coin tossing. In: International Conference on Computer System and Signal Processing, pp. 175–179. IEEE (1984)
6. Ekert, A.K.: Quantum cryptography based on Bell's theorem. Phys. Rev. Lett. **67**(6), 661 (1991)
7. Bennett, C.H., Bessette, F., Brassard, G., Salvail, L., Smolin, J.: Experimental quantum cryptography. J. Cryptol. **5**(1), 3–28 (1992)
8. Nguyen, T.M.T., Sfaxi, M.A., Ghernaouti-Hélie, S.: Integration of quantum cryptography in 802.11 networks. In: The First International Conference on Availability, Reliability and Security, ARES 2006, 8 pp. IEEE, Apr 2006
9. Gisin, N., Thew, R.: Quantum communication. Nat. Photonics **1**(3), 165–171 (2007)
10. Huang, X., Wijeseker, S., Sharma, D.: Implementation of quantum key distribution in Wi-Fi (IEEE 802.11) wireless networks. In: 10th International Conference on Advanced Communication Technology, ICACT 2008, vol. 2, pp. 865–870. IEEE, Feb 2008
11. Reinhard, F., Wenner, A.M.: Quantum Communication—Quantum Channel Capacities (2009)
12. Cui, G., Lu, Y., Zeng, G.: A new scheme for quantum key distribution in free-space. In: 15th Asia-Pacific Conference on Communications, APCC 2009, pp. 637–640. IEEE, Oct 2009
13. Huang, X., Wijesekera, S., Sharma, D.: Fuzzy dynamic switching in quantum key distribution for Wi-Fi networks. In: Sixth International Conference on Fuzzy Systems and Knowledge Discovery, FSKD'09, vol. 3, pp. 302–306. IEEE, Aug 2009
14. Huang, X., Wijesekera, S., Sharma, D.: Secure communication in 802.11 networks with a novel protocol using quantum cryptography. In: 2010 4th International Conference on Network and System Security (NSS), pp. 594–599. IEEE, Sept 2010
15. Elboukhari, M., Azizi, M., Azizi, A.: Quantum key distribution protocols: a survey. Int. J. Univers. Comput. Sci. **1**(2), 59–67 (2010)
16. Lalu Naik, R., Chenna Reddy, P., Sathish Kumar, U.: Implementation of QKD in Wi-Fi (IEEE 802.11) wireless networks. IJCSIT **2**(6) (2011)
17. Katakam, S.R., Lakshmi Reddy, M.S.R.: Cryptography using QKD in wireless networks. IJARCET **2**(1) (2013)
18. Naik, R.L., Reddy, D.P.C., Kumar, U.S., Narayana, D.Y.: Quantum cryptography with key distribution in wireless networks on privacy amplification. IRACST—Int. J. Comput. Netw. Wirel. Commun. (IJCNWC) **1**(1) (2011)
19. Bennett, C.H.: Phys. Rev. Lett. **68**, 3121 (1992)

MOBI-CLASS: A Fuzzy Knowledge-Based System for Mobile Handset Classification

Prabhash Chandra, Devendra Agarwal and Praveen Kumar Shukla

Abstract Fuzzy logic is a technique that provides a mathematical framework to deal with imprecise and uncertain information existing in the real-world decision-making systems. The objective is to integrate linguistic computation in decision making. Several rule-based systems are being developed using the subjective knowledge in the form of fuzzy if-then rules which are also known as 'Fuzzy Knowledge Base Systems (FKBS)'. In this paper, a new FKBS titled MOBI-CLASS is proposed and implemented using open-access software Guaje. The interpretability and accuracy parameters are studied.

Keywords Fuzzy logic · Fuzzy rule-based systems (FRBS) · Fuzzy knowledge-based systems (FKBS) · Interpretability · Accuracy

1 Introduction

Fuzzy logic [1, 2] establishes a mathematical framework to deal with information imprecision and uncertainty in the system to be designed. It is observed that human being has an excellent capability of reasoning and making decisions in the environment flooded with the imprecision, uncertainty, ambiguity partial truth, etc. [3]. This capability of human being is used to implement decision making in the machines also with the help of fuzzy logic. This leads to the development of fuzzy rule-based sys-

P. Chandra (✉) · D. Agarwal
Department of Computer Science & Engineering, Babu Banarasi Das University,
Lucknow, India
e-mail: pathakprabhash2@gmail.com

D. Agarwal
e-mail: dev_bbd@yahoo.com

P. K. Shukla
Department of Information Technology, Babu Banarasi Das Northern
India Institute of Technology, Lucknow, India
e-mail: drpraveenkumarshukla@gmail.com

© Springer Nature Singapore Pte Ltd. 2019   979
J. C. Bansal et al. (eds.), *Soft Computing for Problem Solving*, Advances in Intelligent
Systems and Computing 817, https://doi.org/10.1007/978-981-13-1595-4_77

tems (FRBS) [4–6], alternatively known as fuzzy knowledge-based systems (FKBS) [7, 8]. Two most commonly used FRBS are:

1. Mamdani-type FRBS [9]
2. Takagi–Sugeno FRBS [10].

The proposed model is based on Mamdani-type FKBS, and a specific concentration has been paid on the interpretability and accuracy of the model. Basically, the interpretability [11–13] and accuracy [14] are two very important features of the fuzzy system. Interpretability of the system explains the capacity of a model that permits a human being to understand its behavior by inspecting the functioning or its rule base, while accuracy shows the capacity of a system which makes it capable to faithfully represent the real system. Alternatively, accuracy is the measure of closeness between real system and its modeled fuzzy system. The relation between interpretability and accuracy is contradictory which means one can be improved at the cost of the other. This leads to the situation called 'Interpretability—Accuracy Trade-off' which is a challenging research issue [15].

The proposed model is implemented using Guaje [16, 17], an open-access software which is a Java-based framework.

The paper is divided into five sections. Section 2 is the background of the proposed model and work. Section 3 discusses the proposed model. The implementation and data analysis are explained in Sect. 4. Conclusion and future scope are illustrated in Sect. 5.

2 Related Work

FKBS as previously discussed plays a vital role in decision making in the imprecise and uncertain environment. Several applications are identified for the FKBS which are listed in Table 1.

Table 1 Applications of FKBS

S. no.	Application area	References
1	Stock price prediction	[5]
2	Sewage treatment	[6]
3	Medical (breast cancer diagnosis)	[8]
4	Medical (seizure detection in intracranial EEG)	[18]
5	Generation of facial expression from emotions	[19]
6	Modeling of reservoir operation	[20]
7	Land stability evaluation in agriculture	[21]
8	Decision support in ecosystem management	[22]

As discussed in [23], the Mamdani-type fuzzy system has following major components.

1. Fuzzification interface (converts crisp information into fuzzy)
2. Information engine (responsible to make decisions)
3. Knowledge base (repository of knowledge in problem domain)
4. Defuzzification interface (converts the fuzzy information into crisp information).

At present, the focus is shifted toward the development of type 2 fuzzy systems to improve the performance of the system. A type 2 FKBS has been developed in [24] for the financial data classification. Another huge application of FKBS is reported in the area of big data analytics and has become a trending research issue. A similar work has been reported in [25] that uses the MapReduce concept in analyzing big data using FKBS. The concept of granularity analysis is also used. The multi-objective optimization approaches inspired with evolutionary techniques are also used in developing interval type 2 fuzzy systems [26, 27].

3 Proposed FKBS MOBI-CLASS

A FRBS is proposed and implemented in this paper. The objective is to predict the quality class of the mobile phone based on five identified parameters which are as follows.

1. **Display (DI)**: This parameter is concerned with the quality and size of the display screen of the mobile phone.
2. **Camera Quality (CQ)**: This parameter is related to image quality captured by the camera and also concerns with the multiple functions in the software during capturing of the mage like color change, night vision mode.
3. **Battery Performance (BP)**: As multimedia phone requires an excellent backup, it is an important parameter. This depends on the backup time which is depending on the mAh (milli-Ampere-hours) of the battery.
4. **Biometric Security Features (BS)**: Biometric features of the security are the advanced facilities in the mobile devices. Several mechanisms, like fingerprints and face recognition, are implemented in the mobile device.
5. **Phone Look (PL)**: This parameter is concerned with the physical beauty of the phone like metallic body, cutting patterns of the edges, placement of camera.

On behalf of the above five parameters, all the mobile devices are divided into three classes, i.e., Class 1, Class 2, and Class 3 as per the description given in Table 2. The block diagram of the proposed system is shown in Fig. 1.

Table 2 Details of mobile handset classification

S. no.	Class number	Value
1	Class 1	Low quality
2	Class 2	Medium quality
3	Class 3	High quality

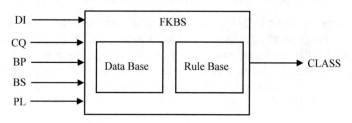

Fig. 1 Block diagram of the proposed system 'MOBI-CLASS'

4 Implementation of the Proposed System

The proposed system is implemented using Guaje open-access software. Three experiments have been carried out on different settings of the FIS operators which are explained in Table 3 (Figs. 2, 3, 4, 5, and 6).

Table 3 Settings of experiments

Experiment no.	Membership function	Conjunction value	Disjunction value	Defuzzification method
Experiment 1	Triangular and trapezoidal	Minimum	Maximum	Maxcrisp
Experiment 2	Triangular and trapezoidal	Minimum	Sum	Maxcrisp
Experiment 3	Triangular and trapezoidal	Product	Sum	Maxcrisp

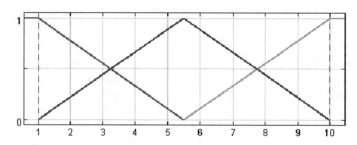

Fig. 2 Membership function of the input variable 'Display'

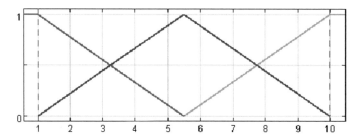

Fig. 3 Membership function of the input variable 'Camera Quality'

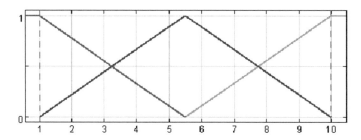

Fig. 4 Membership function of the input variable 'Battery Performance'

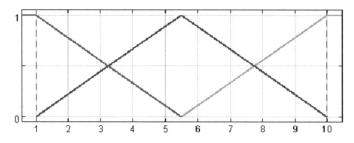

Fig. 5 Membership function of the input variable 'Biometric Security'

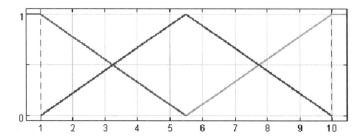

Fig. 6 Membership function of the input variable 'Phone Look'

Table 4 Accuracy parameters for the proposed MOBI-CLASS

Accuracy parameters	Experiment 1	Experiment 2	Experiment 3
Accuracy (AC) (%)	98.7	78.2	98.7
Consolidated accuracy (CA) (%)	98.7	78.2	98.7
Mean square classification error (ME)	0.065	0.604	0.178

Coverage: 100%

Table 5 Interpretability measures for the proposed MOBI-CLASS

Interpretability parameters	Experiment 1	Experiment 2	Experiment 3
Nauck's index (NI)	0.017	0.017	0.008
Number of rules (NR)	18	18	36
Total rule length (TL)	90	90	180
Average rule length (AL)	5	5	5
Accumulated rule complexity (AC)	18.124	18.124	36.248
Inferential fired rules (IR)	7.449	7.449	14.897

Fig. 7 Different accuracy parameters 'Accuracy' and 'Consolidated Accuracy (CA)'

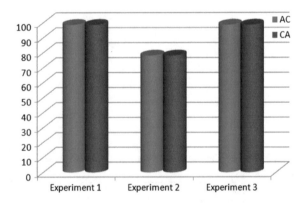

In all the experiments, the Wang–Mendel method has been used for the rule generation. The accuracy parameters of the proposed system are detailed in Table 4; however, Table 5 contains the values of the parameter interpretability.

The comparative study of the accuracy parameter is done in Figs. 7 and 8; however, the comparative study of the interpretability parameters is done in Figs. 9 and 10.

Fig. 8 Accuracy parameter 'Mean Square Classification Error (ME)'

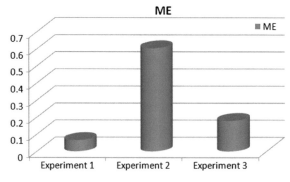

Fig. 9 Different interpretability parameters

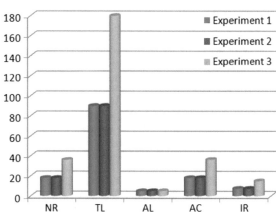

Fig. 10 Interpretability in terms of Nauck's index

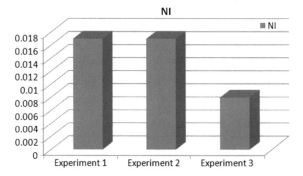

5 Conclusion and Future Scope

Fuzzy systems have proven capability to make decisions in the imprecise and uncertain environments. This paper also introduces a new FKBS for the classification of mobile devices into three classes based on certain parameters. The performance of the proposed system is analyzed using two parameters, i.e., interpretability and accuracy. The system is implemented using Guaje open-access software. Wang–Mendel method has been used to generate the rules which are further used to take decisions using inference and defuzzification process.

In future, the major focus will be on improving the performance of the system by replacing the implementation using interval type 2 fuzzy theory.

References

1. Zadeh, L.A.: Fuzzy Sets. Inf. Control **8**, 338–353 (1965)
2. Mendel, J.M.: Uncertain Rule Based Fuzzy Logic System: Introduction and New Directions. Prentice Hall (2001)
3. Klir, G.J., Yuan, B., Fuzzy Sets and Fuzzy Logic: Theory and Applications. Prentice Hall (1995)
4. Magdalena, L.: Fuzzy Rule Based Systems. Springer Handbook of Computational Intelligence, pp. 203–218 (2015)
5. Chang, P.-C., Liu, C.-H.: ATSK fuzzy rule based system for stock price prediction. Expert Syst. Appl. **34**(1), 135–144 (2008)
6. Dange, P.S., Lad, R.K.: A fuzzy rule based system for an environmental acceptability of sewage treatment plant. KSCE J. Civil Eng. **21**(7), 2590–2595 (2017)
7. Shukla, P.K., Tripathi, S.P.: New approach for tuning interval type-2 fuzzy knowledge based using genetic algorithm. J. Uncertain. Anal. Appl. **2**, 1–15 (2014)
8. Nilashi, M., Ibrahim, O., Ahmadi, H., Shahmoradi, L.: A knowledge based system for breast cancer classification using fuzzy logic method. Telem. Inf. **34**(4), 133–144 (2017)
9. Mamdani, E.H., Assilian, S.: An experiment in linguistic synthesis with fuzzy logic controllers. Int. J. Men-Mach. Stud. **7**(1), 1–13 (1975)
10. Takagi, T., Sugeno, M.: Fuzzy identification of systems and its application to modeling and control. IEEE Trans. Syst. Man Cybern. **15**(1), 116–132 (1985)
11. Shukla, P.K., Tripathi, S.P.: A survey on interpretability-accuracy (I-A) trade-off in evolutionary fuzzy systems. In: IEEE International Conference on Genetic and Evolutionary Computation (ICGEC 2011), Japan, 29 August–01 September 2011
12. Shukla, P.K., Tripathi, S.P.: On the design of interpretable evolutionary fuzzy system (1-EFS) with improved accuracy. In: International Conference on Computing Science, L. P. University, India (2012)
13. Shukla, P.K., Tripathi, S.P.: Interpretability issues evolutionary multi objective fuzzy knowledge based system. In: 7th International Conference on Bio-Inspired Computing: Theories and Applications (BIC-7A2012), ABV-IIIT, Gwalior, India, 14–16 December 2012
14. Cassils, J., Cordon, O., Herrera F., Magdalena, L.: Accuracy Improvement in Linguistic Fuzzy Modeling. Springer, Newyork, NY, USA (2013)
15. Shukla, P.K., Tripathi, S.P.: A review on the interpretability-accuracy trade-off in evolutionary multi-objective fuzzy systems (EMOFS). Information **3**, 256–277 (2012)
16. Alonso, J.M., Magdalena, L., Generating understandable and accurate fuzzy rule based system in a Java environment. In: 9th International Workshop on Fuzzy Logic and Applications, pp. 212–219, Trani, Italy 29–31 August 2011
17. Alonso, J.M., Magdalena, L.: HILK++: an interpretability guided fuzzy modeling methodology for learning readable and comprehensible fuzzy rule base classifiers. Soft. Comput. **15**(10), 1959–1980 (2011)
18. Aarabi, R., Rezai, F., Aghakhani, Y.: A fuzzy rule based system for epileptic seizure detection in intra-cranial EEG. Clin. Neurophysiol. **120**(9), 1648–1657 (2009)
19. Bui, T.D., Heylen, D., Poel, M., Nijhot, A.: Generation of facial expression from emotions using a fuzzy rule based system. In: Australian Joint Conference on Artificial Intelligence. Lecture Notes in Computer Science (LNCS), vol. 2256, pp. 83–94 (2002)

20. Shreshtha, B.P., Duckstein, L., Stakhiv, E.Z.: Fuzzy rule based modeling of reservoir operation. J. Water Resour. Plan. Manag. **122**(4), 212–218 (1996)
21. Rashmi Devi, T.V., Eldho, T.I., Jana, R.: A GIS integrated fuzzy rule based inference system for land suitability evaluation in agriculture watersheds. Agric. Syst. **101**(1–2), 101–109 (2009)
22. Adrinoenssens, V., De Baets, B., Goethals, P.L.M., De Pauw, N.: Fuzzy rule based model for decision support in eco system management. Sci. Total Environ. **319**(1–3), 1–12 (2004)
23. Cordon, O., Herrera, F., Hoffmann, F., Magdalena, L.: Genetic Fuzz Systems: Evolutionary Tuning and Learning of Fuzzy Knowledge Based. World Scientific (2001)
24. Antonelli, M., Bernardo, D., Hagras, M., Marcelloni, F.: Multiobjective optimization of type-2 fuzzy rule based systems for financial data classification. IEEE Trans. Fuzzy Syst. **25**(2), 249–264 (2017)
25. Fernandez, A., del Rio, S., Bawakid, A., Herrera, F.: Fuzzy rule based classification systems for big data with MapReduce: granularity analysis. Adv. Data Anal. Classif. **11**(4), 711–730 (2017)
26. Shukla, P.K., Tripathi, S.P.: Handling high dimensionality and interpretability accuracy trade-off issues in evolutionary multi-objective fuzzy classifies. Int. J. Sci. Eng. Res. **5**(6), 665–671 (2014)
27. Shukla, P.K., Tripathi, S.P.: Interpretability and accuracy issues in evolutionary multi-objective fuzzy classifies. Int. J. Soft Comput. Netw. **1**(1), 55–69 (2016)

Author Index

© Springer Nature Singapore Pte Ltd. 2019
J. C. Bansal et al. (eds.), *Soft Computing for Problem Solving*, Advances in Intelligent
Systems and Computing 817, https://doi.org/10.1007/978-981-13-1595-4

Printed in the United States
By Bookmasters